Metal Metabolism in Animals

Special Issue Editor
Reinhard Dallinger

MDPI • Basel • Beijing • Wuhan • Barcelona • Belgrade

MDPI

Special Issue Editor
Reinhard Dallinger
University of Innsbruck
Austria

Editorial Office
MDPI AG
St. Alban-Anlage 66
Basel, Switzerland

This edition is a reprint of the Special Issue published online in the open access journal *IJMS* (ISSN 1422-0067) from 2015–2017 (available at: http://www.mdpi.com/journal/ijms/special_issues/metal-metab-animal and http://www.mdpi.com/journal/ijms/special_issues/metal_metab_animal_II).

For citation purposes, cite each article independently as indicated on the article page online and as indicated below:

Lastname, F.M.; Lastname, F.M. Article title. *Journal Name* **Year**, *Article number*, page range.

First Edition 2018

ISBN 978-3-03842-843-5 (Pbk)
ISBN 978-3-03842-844-2 (PDF)

Table of Contents

About the Special Issue Editor

Reinhard Dallinger, Ph.D., served as a Professor of Ecotoxicology (now retired) at the University of Innsbruck (Austria). He received his Ph.D. biology degree from the University of Innsbruck with a focus on zoology (animal physiology) and minor in microbiology. In the past, Dr. Dallinger has repeatedly been appointed as a guest professor at Austrian Universities and higher Education colleges. He is a member of several Scientific Societies and served, for example, as a Board member and President of the new European Society for Comparative Physiology and Biochemistry. Dr. Dallinger is a member of several editorial boards. He is still leading the research group of Ecotoxicology and Molecular Physiology at the Department of Zoology of the University of Innsbruck. So far, Dr. Dallinger has conducted research on toxicology and ecotoyicology of metals, with applications in the field of metal-related biomonitoring. His scientific interests concentrate, among others, on mechanisms of metabolism and toxicology of metals in mainly invertebrate animals, with a special focus on cellular compartmentalization and biochemical, molecular and physiological functions of metallothioneins. Recently, he has also become involved in research on evolutionary diversification of mollusc metallothioneins.

Preface to "Metal Metabolism in Animals"

Metal metabolism in animals has been in the spotlight for more than 100 years. Never before, however, has the opportunity to shed light on metal interactions and their implications been as promising as it is today. Several important insights and achievements may have contributed to this development. One important step has been the realization that intracellular concentrations and activities of metal ions are strictly controlled by the cell, with the consequence that intracellular free metal ion levels are regulated at very low concentrations. On the other hand, metal ions are dynamically involved in cellular processes through continuous beneficial, or detrimental, interaction with intracellular ligands, co-factors and metabolites. The improvement of analytical techniques has greatly assisted in the discovery of these insights. However, our understanding of metal interactions with biological systems has been most notably expanded by the sophistication of molecular biology tools and bioinformatics. One of the most promising developments is the new integrative approach of metallomics, which spans the entirety of metal interactions with living systems, including their environmental interdependencies.

In this light, study of metal interactions in animals and animal cells offer the exciting possibility of resolving many molecular, biochemical, biomedical, physiological and environmental questions which remain open. The present issue aims at bringing together, under a metallomics perspective, important contributions to the field of metal metabolism in animals.

Reinhard Dallinger
Special Issue Editor

International Journal of
Molecular Sciences

MDPI

Review

The Metals in the Biological Periodic System of the Elements: Concepts and Conjectures

Wolfgang Maret

London Iron Metabolism Group, Division of Diabetes and Nutritional Sciences and Department of Biochemistry, Faculty of Life Sciences and Medicine, King's College London, 150 Stamford St., London SE1 9NH, UK; wolfgang.maret@kcl.ac.uk; Tel.: +44-20-7848-4264; Fax: +44-20-7848-4195

Academic Editor: Reinhard Dallinger
Received: 30 November 2015; Accepted: 22 December 2015; Published: 5 January 2016

Abstract: A significant number of chemical elements are either essential for life with known functions, or present in organisms with poorly defined functional outcomes. We do not know all the essential elements with certainty and we know even less about the functions of apparently non-essential elements. In this article, I discuss a basis for a biological periodic system of the elements and that biochemistry should include the elements that are traditionally part of inorganic chemistry and not only those that are in the purview of organic chemistry. A biological periodic system of the elements needs to specify what "essential" means and to which biological species it refers. It represents a snapshot of our present knowledge and is expected to undergo further modifications in the future. An integrated approach of biometal sciences called metallomics is required to understand the interactions of metal ions, the biological functions that their chemical structures acquire in the biological system, and how their usage is fine-tuned in biological species and in populations of species with genetic variations (the variome).

Keywords: essential metals; non-essential metals; periodic system of the elements; metallomics

---

1. Introduction

Many scientific disciplines address which chemical elements are present in organisms and which functions they have. Chemistry and biology are contributing to such investigations. With time, more specialized disciplines developed with emphasis on either organic or inorganic chemistry (Figure 1).

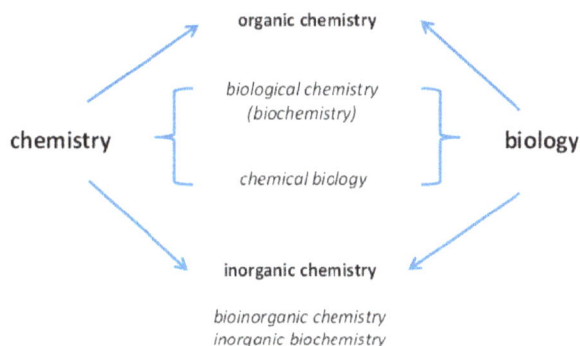

Figure 1. Both chemistry and biology come to bear on investigations of the roles of chemical elements in organisms.

The names of the sub-disciplines can be misleading when organic chemistry is associated with life and inorganic chemistry with the inanimate world only. Life is built on organic as well as inorganic compounds, and, as I shall discuss in this article, significant more chemical elements than the ones commonly treated in organic chemistry are important for life. Among them are many metal ions that have contributed critically to the evolution of life—life would not be possible without them. Metallomics, an integrated biometal science, aims to serve as a discipline that covers all aspects of how metals function in biological systems.

One can distinguish three developments that led to our present knowledge, all connected to a shift in the types of questions addressed and all on-going.

First, attempts were made to determine which elements occur in living organisms and which ones are essential for life. In a 70 kg human, the bulk elements, hydrogen (H), carbon (C), nitrogen (N), oxygen (O), and sulphur (S) are over 1 kg, except for sulphur, which is only about 100 g. Macrominerals were identified, for instance the metal cations of sodium (Na), potassium (K), magnesium (Mg), and calcium (Ca) and the non-metals phosphorous (P) and chlorine (Cl), which are in the gram range except for calcium, which is about 1.7 kg. Trace element research matured with advances in analytical chemistry and instrumentation and provided evidence for additional essential elements, the metal ions of iron (Fe), zinc (Zn), manganese (Mn), copper (Cu), cobalt (Co) in the form of vitamin B_{12}, molybdenum (Mo) and the non-metals selenium (Se) and iodine (I). They are in the mg range, except for iron and zinc which are 2–5 g and thus strictly speaking not trace elements. Yet other elements are present at even lower levels but noteworthy some are present above the levels of essential elements such as Mn and Mo, e.g., bromine (Br), rubidium (Rb), aluminium (Al), nickel (Ni), titanium (Ti) and barium (Ba).

Second, we witnessed the isolation of metalloproteins and the characterization of their coordination environments in which the metal ions function. Bioinformatics allowed for estimates that about 50% of all enzymes depend on a metal ion for catalysis, demonstrating the significance of metal ions in biochemistry. For zinc alone, about 3000 human proteins have the structural features to be zinc metalloproteins, meaning that about every tenth protein encoded in the human genome contains zinc [1]. Knowledge about interactions of metals with biological macromolecules other than proteins *in vivo* is scarce. An exception is magnesium, which has a role in RNA structure and function.

Third, proteins with specific roles in metal metabolism such as membrane metal transporters, metalloregulators, and metallochaperones were characterized. These discoveries brought insight into the complex biological mechanisms of metal regulation and selectivity. It turned out that each metal ion is controlled in a characteristic range of concentrations that is determined by the affinities of metal ions for their ligands. Thus, in the series Mg, Ca, Mn, Fe, Co, Ni, Zn, Cu, the binding strength (affinity) increases and accordingly, the free metal ion concentrations-metal ions not bound to proteins-decreases. In principle, this series conforms to the Irving-Williams series for divalent metal ions [2]. Although Mg complexes are usually stronger than Ca complexes, Ca is stabilized over Mg in biology by using additional ligands and thus coordination numbers higher than six. This stabilization is necessary because Ca^{2+} has been adopted as a cellular signalling ion controlling many processes. Zn is positioned after Cu in the periodic system of the elements (PSE), but complexes of Cu have higher stability. In the cell, Cu is the only metal ion in this series that is primarily monovalent (Cu(I)). A most remarkable consequence of these ranges of free metal ion concentrations is that they cover about 15 orders of magnitude, from millimolar (Mg) to attomolar (Cu). Thus metal ions can be used in biological processes over an extremely wide range, demonstrating in essence the enormous contribution that metal ions make to biological function. Total cellular metal ion concentrations in humans, however, follow a different series: Mg, Ca, Fe, Zn, Cu, Mn, Co. Hence, to maintain the trend expressed by the Irving-Williams series for characteristic free metal ion concentrations, biology has to adjust for the rather large differences in total metal concentrations. It also has to deal with the fluctuations/oscillations of signalling metal ions. In addition to Ca^{2+}, Zn^{2+} is also a cellular signalling ion. Whereas Ca^{2+} activates many processes, Zn^{2+} seems to be primarily an inhibitory ion and due to its preference for different ligands and different coordination environments it targets other sets of proteins [3].

2. The Essential Elements

Considering our vast knowledge about genes and proteins, it is quite remarkable that our knowledge about the role of chemical elements in life remains limited and open-ended. Only last year, bromine was added to the list of essential elements with its function in collagen metabolism [4]. Hence we should not assume that we know all the elements that are essential for animals and humans as uncertainties about the biochemical functions of some chemical elements persist. Striking as this statement may appear, the lack of knowledge provides ample opportunities for discoveries with significant implications for biochemistry and improving health.

We now know that 11 metals and 10 non-metals, *i.e.*, 21 elements, are essential for humans. The count includes chromium whose status as an essential element is controversial (see below). Additional elements are essential for other forms of life outside the kingdom animalia. Therefore when essential elements are listed in a "biological periodic system of the elements (PSE)" the meaning and implications usually are not clear. Conjectures arise mainly because of two issues:

"Essential" should include a reference to the biological species. Biological PSEs are usually presented with all the elements identified as essential in all species. However, some elements are essential to only some organisms. I propose to categorize in the following way: (i) elements that are essential to all species; (ii) elements that are used in a significant number of species (V, Ni); and (iii) elements that are used only in a few organisms in special ecological niches (W, Cd, Lanthanides). Vanadium is used in some nitrogenases and haloperoxidases in algae and fungi. Nickel has been discovered in only nine enzymes in some bacteria and plants. Tungsten is used instead of molybdenum in enzymes of some thermophilic bacteria. Ni, V, W are not established as being essential for animals and humans. Orthologous gene products requiring these elements have not been found in animals and humans. Carbonic anhydrase, which is usually a zinc enzyme, is a cadmium enzyme in the marine diatom *Thalassiosira weissflogii* [5]. The US environmental protection agency (EPA) has classified cadmium as a Group B1 probable human carcinogen. Lanthanides (rare earth elements, REE) instead of calcium were found as cofactors of methanol dehydrogenase of particular methanotrophic bacteria [6]. While this wider usage of metal ions in life is interesting from the standpoint of evolution and certainly revealed fascinating chemistries, the utilization of specific elements such as Cd and REE, is of limited significance for nutrition of animals and humans.

"Essential" should include a definition. Elements that are essential for survival should be distinguished from others that have limited functions and some health benefits only. An example is fluorine which as fluoride prevents dental caries and perhaps is beneficial for bone health but otherwise exhibits toxicity. A broader definition had been introduced: "An element is essential when a deficient intake consistently results in an impairment of a function from optimal to suboptimal and when supplementation with physiological levels of this element, but not others, prevents or cures this impairment" [7]. Following this definition, additional elements were classified as essential, but they are not necessarily essential for life [8]. Several of them we know to be beneficial for animals and humans, e.g., B, Cr, Ni, Si, and accordingly, there is an on-going discussion whether guidelines for their intake should be given [9–11].

Biological chromium research illustrates the difficulties in defining structure and function of biologically active metal ions. Chromium is an example of the importance of chemical speciation in biology: Chromium(III) complexes are the ones having beneficial functions and little toxicity, whereas Cr(VI), chromate, is a human carcinogen. Chromium(III) deserves further discussion as it is widely accepted as an essential trace metal and accordingly dietary reference intakes (DRIs) were issued and it is available as a dietary supplement [12]. However, its status as an essential element was questioned recently [13]. After sixty years of chromium research some crucial questions remain unanswered: (i) the structure of a biologically active chromium complex; (ii) ways to determine chromium status in humans; and (iii) elucidation of its exact mechanism of action in glucose and lipid metabolism [14]. One should concede that not having succeeded in isolating a biologically active chromium complex would seem not to be an issue: absence of evidence is not evidence for absence. The low amount of chromium per se should also not be a concern as molybdenum and cobalt occur at equally low concentrations.

A biological PSE that considers the two issues associated with the meaning of "essential" and separates general from specific cases (Figure 2) demonstrates that biochemistry, the chemistry of life, depends on a significant number of elements. The main features of such a PSE are that (i) most of the non-metals, forming a triangle in the upper right part of the PSE, are used (except the noble gases); (ii) from the metalloids (B, Si, Ge, As, Sb, Te) only B and Si are known to have beneficial effects for humans and are essential for some species; and (iii) the entirety of group 13 with the exception of boron remains unused. The individual elements in a biological periodic table have been discussed in detail [15].

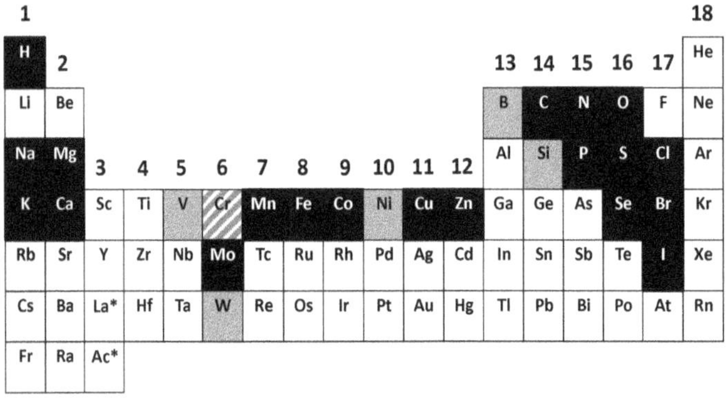

1	2	3	4	5	6	7	8	9	10	11	12	13	14	15	16	17	18
H																	He
Li	Be											B	C	N	O	F	Ne
Na	Mg											Al	Si	P	S	Cl	Ar
K	Ca	Sc	Ti	V	Cr	Mn	Fe	Co	Ni	Cu	Zn	Ga	Ge	As	Se	Br	Kr
Rb	Sr	Y	Zr	Nb	Mo	Tc	Ru	Rh	Pd	Ag	Cd	In	Sn	Sb	Te	I	Xe
Cs	Ba	La*	Hf	Ta	W	Re	Os	Ir	Pt	Au	Hg	Tl	Pb	Bi	Po	At	Rn
Fr	Ra	Ac*															

Figure 2. A biological periodic system of the elements (PSE) indicating the essential elements. The essential elements for most forms of life are shown in black with the exception of chromium (Cr), which is shown with an upward diagonal pattern (see text), and the essential elements that are more restricted for some forms of life are shown in grey. Not shown are the f-group elements: lanthanides and actinides (asterisk after lanthanum (La) and actinium (Ac). The groups are numbered 1–18.

Given the heightened interest in the fascinating and challenging chemistry of some metals in special microbes, it is instructive to recall the relative significance of the transition metals and zinc in mammalian biology: Fe ≈ Zn > Cu > Mn > Mo (in the pterin cofactor) and Co (in the corrin cofactor). A 70 kg human contains about 5 g Fe, 2 g Zn, 100 mg Cu, 12–20 mg Mn, 5 mg Mo, and 2 mg Cr and Co.

Mining genome database for signatures of metal binding sites in protein sequences allowed for estimates of the number of metalloproteins containing zinc, iron, and copper. With such predictions a true understanding of these three metals at the systems level, *i.e.*, their metalloproteomes, is emerging. Efforts are under way to annotate functions of these metalloproteins on the basis of similarities with structures of proteins with known functions. In humans, there are only four known molybdoenzymes and only two vitamin B_{12}-dependent enzymes. However, estimates of the number of enzymes that use manganese as a cofactor are lacking.

It seems that we do not even have a complete list of essential elements for one given species. The identification of roles of elements such as Cd and REE in metalloproteins (see above) indicates that our knowledge about essentiality of chemical elements in different forms of life is incomplete. An "omics" study also came to the conclusion that metalloproteomes of microorganisms are largely uncharacterized [16]. And then there is the issue what functions the other elements present have.

3. The Non-Essential Elements

Instrumental analytical chemistry for metal determination has advanced significantly. Inductively-coupled plasma mass spectrometry (ICP-MS) has sub-ppt detection limits, allowing for detection of virtually all natural occurring elements in biological samples. Even radioactive uranium can be detected. Thus, 74 out of 78 elements were found in salmon eggs [17]. The abundance of chemical elements in humans has been summarized in a periodic table (Figure 3).

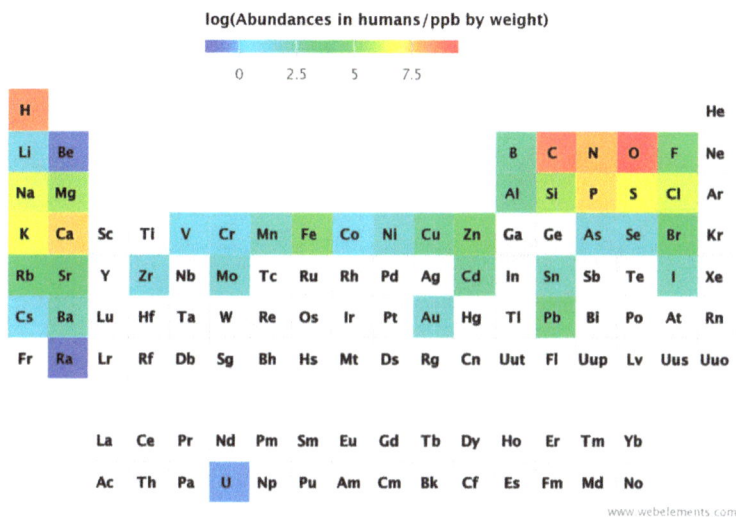

Figure 3. Abundance of the chemical elements in humans [18].

Some non-essential elements are present at significantly higher concentrations than the essential ones present at the lowest concentration. Sr is present at ppm levels, and Rb, Ba, Ni, As, Al, and Ti (not shown in Figure 3) at ppb levels, but also others that are considered to have mainly toxic actions, e.g., Cd, Hg (not shown in Figure 3), and Pb are present, but they are variable and depend on exposure. Some elements can accumulate with age, while others are rather tightly regulated. Distribution of concentrations has been suggested as a criterion for elements being essential, namely homeostatic control for essential elements leading to a normal distribution but the presence of non-essential elements following a skewed distribution depending on exposure [8]. While some elements are bioactive with positive effects on health, many non-essential elements nevertheless are bioreactive [16]. They are not chemically inert, and can serve as catalysts, be pharmacologically active, or toxic. Therefore, it is important to adopt a broader view of metal ions in biology, beyond a focus on essential metal ions, to include the many elements that are present with poorly understood functional consequences. We know very little how and to which extent animals and humans can cope with these apparently non-essential elements, whether the presence of some reflects a lack of selectivity in uptake mechanisms, whether there are any specific mechanisms of detoxification, and what the exposure limits and biological effects are.

The delicate balance among the essential elements extends to complex interactions with non-essential elements and other nutrients. Too much (overload) or too little (deficiency) of one has conditioning effects on others. For example, zinc supplementation can cause copper deficiency [19]. Deficiency of an essential element can lead to the uptake of a non-essential element with potential toxicity. For instance, under iron deficiency, uptake of cadmium increases and interferes with the metabolism of essential elements, such as zinc.

A focus on essential elements only is not sufficient. We need to monitor the presence of non-essential elements, how their concentrations vary and how they affect functions of the biological system. For the essential elements, we have some understanding of how deficiencies or overload perturb functions though cause and effect in chronic diseases is often not clear. For the non-essential elements effects at sub-toxic concentrations are more difficult to assess. For example, in 2012 the Centers of Disease Control and Prevention lowered blood lead levels of concern from 10 μg/dL to ≥5 μg/dL for children aged 1–5 to identify children and environments with lead-exposure hazards because neurobiological parameters are affected negatively even at previously accepted blood lead levels [20]. Average normal levels in the US population are 1.6 μg/dL for adults and 1.9 μg/dL for children. Clearly, we need additional solid data how each individual element affects animals and humans.

We also are beginning to realize that there are a significant number of inherited diseases of human metal metabolism. Comparatively fewer variations have been documented in animals. Our knowledge is very limited on how mutations in the many genes that encode metal regulatory processes affect the toxicity of metals, and metal dose-responses quantitatively. Polymorphisms in the about 12 human metallothionein genes suggest an effect on metal metabolism [21]. For example, an Asn27Thr substitution in metallothionein-1A affects its zinc-binding capacity and together with polymorphisms of other human metallothionein genes is associated with the risk of developing type 2 diabetes, coronary heart disease, and other complications of diabetes [22,23] and with altered metabolism of toxic metal ions, *i.e.*, cadmium, lead, and mercury [24–26]. Levels for toxic metals in blood are associated with polymorphisms in metal transporter genes and other genes, thus providing further evidence that subsets of individuals are more susceptible to the toxic effects of some metal ions [27].

New applications and manufacturing processes expose animals and humans increasingly to a number of metals to which they have not been exposed to in the past, in particular those from the bottom part of the PSE. Food chains and food webs amplify some exposures, which has largely unexplored effects for more recently employed metal ions. REE find applications in batteries, magnets, lighting, optical fibers; gallium, indium, and tellurium in solar cells; hafnium and tantalum in optical devices, and elements from the platinum group ruthenium, rhodium, palladium and their analogues osmium, iridium and platinum are used in fuel cells [28]. Another issue is the increasing applications for nanomaterials. They have different and often enhanced reactivities. Also, we are exposed to metallodrugs for therapeutic or diagnostic purposes, and use implants (Si), metal prostheses (Co), and dental fillings (Hg) with functional consequences for living systems. The increased use of additional elements adds a new dimension to exposure and pollution in air, soil, and water with fundamental implications for health and understanding etiology and pathology.

Most organisms live in a symbiotic relationship with others and need to defend themselves against parasites. The gut harbours a huge number of commensal bacteria (the microbiome) that are involved in processing food. An intriguing aspect is that microorganisms rely on a different complement of metal ions and utilize metal ions that animals and humans apparently do not use. For example, many bacteria need nickel, including the ulcer-causing pathogen *Helicobacter pyloris*. Hence depriving this organism of nickel provides a selective therapeutic means for its eradication in the upper digestive tract. Some of the trace element deficiencies observed could be due to an effect on symbiotic organisms that require particular trace elements, rather than on the host directly.

4. Conclusions

Chemistry and biology must come together in investigations of biometals. Neither establishing only structures nor finding only functions is sufficient. For example, sugars and lipids incorporate arsenic, but the characterization of such structures is not proof for arsenic being essential.

We seem to be preoccupied with controlling energy from dietary intake of protein, fat, and sugar but pay correspondingly very little attention to the intake of essential elements that make metabolism possible, or to the presence of apparently non-essential elements. With modern instrumentation, it has become relatively easy to measure metals with high sensitivity and accuracy. In contrast to the detection of millions of organic compounds to which we are exposed and which generally require quite sophisticated and hence expensive methods of analysis, the number of metals to be screened is less than a hundred and they are relatively inexpensive to monitor. Due to the high reactivity and catalytic potential of metal ions, their levels in our diet and in our environment are an important factor for health, for causing disease or affecting its progression, and for influencing healthy ageing. It is as valid in 2015 as it was discussed in 1951: The functions of trace elements in biology "may hold the answers to many unsolved biochemical and biological problems" [29].

Acknowledgments: Acknowledgments: Work in the author's laboratory is supported by a research grant from the Biotechnology and Biological Sciences Research Council (BB/K001442/1).

Conflicts of Interest: Conflicts of Interest: The author declares no conflict of interest.

References

1. Andreini, C.; Bertini, I.; Rosato, A. Metalloproteomes: A bioinformatic approach. *Acc. Chem. Res.* **2009**, *42*, 1471–1479. [CrossRef] [PubMed]
2. Irving, H.; Williams, R.J.P. Order of stability of metal complexes. *Nature* **1948**, *162*, 746–747. [CrossRef]
3. Maret, W. Molecular aspects of zinc signals. In *Zinc Signals in Cellular Functions and Disorders*; Fukuda, T., Kambe, T., Eds.; Springer Japan KK: Tokyo, Japan, 2014; pp. 7–26.
4. McCall, A.S.; Cummings, C.F.; Bhave, G.; Vanacore, R.; Page-McCaw, A.; Hudson, B.G. Bromine is an essential trace element for assembly of collagen IV scaffolds in tissue development and architecture. *Cell* **2014**, *157*, 1380–1392. [CrossRef] [PubMed]
5. Lane, T.W.; Morel, F.M.M. A biological function for cadmium in marine diatoms. *Proc. Natl. Acad. Sci. USA* **2000**, *97*, 4627–4631. [CrossRef] [PubMed]
6. Pol, A.; Barends, T.R.M.; Dietl, A.; Khadem, A.F.; Eygensteyn, J.; Jetten, M.S.M.; Op den Camp, H.J.M. Rare earth metals are essential for methanotrophic life in volcanic mudpots. *Environ. Microbiol.* **2014**, *16*, 255–264. [CrossRef] [PubMed]
7. Mertz, W. The essential trace elements. *Science* **1981**, *213*, 1332–1338. [CrossRef] [PubMed]
8. Underwood, E. *Trace Elements in Human and Animal Nutrition*; Academic Press: San Diego, CA, USA, 1977.
9. Nielsen, F.H. Should bioactive trace elements not recognized as essential, but with beneficial health effects, have intake recommendations. *J. Trace Elem. Med. Biol.* **2014**, *28*, 406–408. [CrossRef] [PubMed]
10. Nielsen, F.H. Update on the possible nutritional importance of silicon. *J. Trace Elem. Med. Biol.* **2014**, *28*, 379–382. [CrossRef] [PubMed]
11. Nielsen, F.H. Update on human health effects of boron. *J. Trace Elem. Med. Biol.* **2014**, *28*, 383–387. [CrossRef] [PubMed]
12. Food and Nutrition Board, Institute of Medicine. *Dietary Reference Intakes for Vitamin A, Vitamin K, Arsenic, Boron, Chromium, Copper, Iodine, Iron, Manganese, Molybdenum, Nickel, Silicon, Vanadium, and Zinc*; National Academy Press: Washington, DC, USA, 2001.
13. Mertz, W. Chromium in human nutrition. *J. Nutr.* **1993**, *123*, 626–633. [PubMed]
14. Vincent, J.B. Is chromium pharmacologically relevant? *J. Trace Elem. Med. Biol.* **2014**, *28*, 397–405. [CrossRef] [PubMed]
15. Chellan, P.; Sadler, P.J. The elements of life and medicines. *Philos. Trans. R. Soc.* **2015**, *A373*, 21040182. [CrossRef] [PubMed]
16. Cvetkovic, A.; Menon, A.L.; Thorgersen, M.P.; Scott, J.W.; Poole, F.L., II; Jenney, F.E., Jr.; Lancaster, W.A.; Praissman, J.L.; Shanmukh, S.; Vaccaro, B.J.; *et al.* Microbial metalloproteomes are largely uncharacterized. *Nature* **2010**, *466*, 779–784. [CrossRef] [PubMed]
17. Haraguchi, H.; Ishii, A.; Hasegawa, T.; Matsuura, H.; Umeura, T. Metallomics study on all-elements analysis of salmon egg cells and fractionation analysis of metal in cell cytoplasm. *Pure Appl. Chem.* **2008**, *80*, 2595–2608. [CrossRef]
18. Winter, M.F. Available online: http://www.webelements.com/hydrogen/biology.html (accessed on 4 January 2016).

19. Maret, W.; Sandstead, H. Zinc requirements and the risks and benefits of zinc supplementation. *J. Trace Elem. Med. Biol.* **2006**, *20*, 3–18. [CrossRef] [PubMed]
20. CDC response to advisory committee on childhood lead poisoning prevention recommendations in "low level lead exposure harms children: a renewed call of primary prevention". Available online: http://www.cdc.gov/nceh/lead/acclpp/cdc_response_lead_exposure_recs.pdf (accessed on 30 December 2015).
21. Raudenska, M.; Gumulec, J.; Podlaha, O.; Szatlmachova, M.; Babula, P.; Eckschlager, T.; Adam, V.; Krizek, R.; Masarik, M. Metallothionein polymorphisms in pathological processes. *Metallomics* **2014**, *6*, 55–68. [CrossRef] [PubMed]
22. Giacconi, R.; Bonfigli, A.R.; Testa, R.; Sirolla, C.; Cipriano, C.; Marra, M.; Muti, E.; Malavolta, M.; Costarelli, L.; Piacenza, F.; *et al.* +647A/C and +1245 MT1A polymorphisms in the susceptibility of diabetes mellitus and cardiovascular complications. *Mol. Genet. Metab.* **2008**, *94*, 98–104. [CrossRef] [PubMed]
23. Yang, L.; Li, H.; Yu, T.; Zhao, H.; Cherian, M.G.; Cai, L.; Lui, Y. Polymorphisms in metallothionein-1 and -2 genes associated with the risk of type 2 diabetes mellitus and its complications. *Am. J. Physiol. Endocrinol. Metab.* **2008**, *294*, E987–E992. [CrossRef] [PubMed]
24. Wang, Y.; Goodrich, J.M.; Gillespie, B.; Werner, R.; Basu, N.; Franzblau, A. An investigation of modifying effects of metallothionein single-nucleotide polymorphisms on the association between mercury exposure and biomarker levels. *Environ. Health Perspect.* **2012**, *120*, 530–534. [CrossRef] [PubMed]
25. Lei, L.; Chang, X.; Rentschler, G.; Tian, L.; Zhu, G.; Chen, X.; Jin, T.; Broberg, K. A polymorphism in metallothionein 1A (*MT1A*) is associated with cadmium-related excretion of urinary beta 2-microglobulin. *Toxicol. Appl. Pharmacol.* **2012**, *265*, 373–379. [CrossRef] [PubMed]
26. Fernandes, K.C.M.; Martins, A.C., Jr.; Soares de Oliveira, A.A.; Antunes, L.M.G.; de Syllos Colus, I.M.; Barbosa, F., Jr.; Barcelos, G.R.M. Polymorphism of metallothionein 2A modifies lead body burden in workers chronically exposed to the metal. *Public Health Genom.* **2015**, *19*, 47–52. [CrossRef] [PubMed]
27. Ng, E.; Lind, P.E.; Lindgren, C.; Ingelsson, E.; Mahajan, A.; Morris, A.; Lind, L. Genome-wide association study of toxic metals and trace elements reveals novel associations. *Hum. Mol. Genet.* **2015**, *24*, 4739–4754. [CrossRef] [PubMed]
28. Ridgway, A.; Webb, R. The materials bonanza. *New Sci.* **2015**, *225*, 35–41. [CrossRef]
29. Vallee, B.L. The functions of trace elements in biology. *Sci. Mon.* **1951**, *72*, 368–376.

International Journal of
Molecular Sciences

MDPI

Article

Deletion of Phytochelatin Synthase Modulates the Metal Accumulation Pattern of Cadmium Exposed *C. elegans*

Yona J. Essig [1,2], Samuel M. Webb [3] and Stephen R. Stürzenbaum [1,2,*]

[1]   Analytical and Environmental Sciences Division, Faculty of Life Sciences & Medicine,
      King's College London, London SE1 9NH, UK; julie.essig@kcl.ac.uk
[2]   Medical Research Council-Public Health England (MRC-PHE) Centre for Environment & Health,
      King's College London, London SE1 9NH, UK
[3]   Stanford Synchrotron Radiation Lightsource, SLAC National Accelerator Laboratory, 2575 Sand Hill Road,
      Menlo Park, CA 94025, USA; samwebb@slac.stanford.edu
*   Correspondence: stephen.sturzenbaum@kcl.ac.uk; Tel.: +44-207-848-4406

Academic Editor: Reinhard Dallinger
Received: 14 January 2016; Accepted: 14 February 2016; Published: 19 February 2016

Abstract: Environmental metal pollution is a growing health risk to flora and fauna. It is therefore important to fully elucidate metal detoxification pathways. Phytochelatin synthase (PCS), an enzyme involved in the biosynthesis of phytochelatins (PCs), plays an important role in cadmium detoxification. The PCS and PCs are however not restricted to plants, but are also present in some lower metazoans. The model nematode *Caenorhabditis elegans*, for example, contains a fully functional phytochelatin synthase and phytochelatin pathway. By means of a transgenic nematode strain expressing a *pcs-1* promoter-tagged GFP (*pcs-1*::GFP) and a *pcs-1* specific qPCR assay, further evidence is presented that the expression of the *C. elegans* phytochelatin synthase gene (*pcs-1*) is transcriptionally non-responsive to a chronic (48 h) insult of high levels of zinc (500 µM) or acute (3 h) exposures to high levels of cadmium (300 µM). However, the accumulation of cadmium, but not zinc, is dependent on the *pcs-1* status of the nematode. Synchrotron based X-ray fluorescence imaging uncovered that the cadmium body burden increased significantly in the *pcs-1*(tm1748) knockout allele. Taken together, this suggests that whilst the transcription of *pcs-1* may not be mediated by an exposure zinc or cadmium, it is nevertheless an integral part of the cadmium detoxification pathway in *C. elegans*.

Keywords: *C. elegans*; nematode; phytochelatin synthase; X-ray fluorescence microscopy

1. Introduction

Phytochelatin (PC), a non-ribosomal heavy-metal complexing peptide, was first identified in monocot and dicot plants [1]. They are characterized by a repetitive $(\gamma\text{-GluCys})_n\text{Gly}$ motif (where *n* typically ranges from 2 to 5) and their ability to bind to cadmium and other non-essential metals (e.g., arsenic and mercury) but also essential heavy metals (such as copper and zinc) via cysteine thiol residues [1]. PCs are biosynthesised by the enzyme glutathione gamma-glutamylcysteinyltransferase, which was isolated in 1989 from *Silene cucubalus*, a plant species belonging to the Caryophyllaceae family [2], and thus was named phytochelatin synthase (PCS). The phytochelatin nomenclature is however misleading, as PCs and PCS have since been identified in numerous taxonomic groups of the (lower) animal kingdom, including free-living nematodes, parasitic nematodes, flatworms, segmented worms and molluscs (Table 1) and therefore PCS is also frequently referred to as a glutathione gamma-glutamylcysteinyltransferase, to reflect its function rather than phylogenetic origin. Based on sequence homology, the N-terminal is more conserved than the C-terminal. It has been suggest that the N-terminal is may be linked to the activity of the enzyme and the C-terminal might be important for the regulation of the enzyme but not for the activity [3].

Table 1. Phytochelatin synthases/glutathione gamma-glutamylcysteinyltransferases identified within the Kingdom Animalia.

Phylum Nematoda (Nematodes)			
Class	Order	Species	Accession number
Chromadorea	Rhabditida	*Caenorhabditis elegans*	AF299333.1
Chromadorea	Rhabditida	*Pristionchus pacificus*	ABKE02006821
Chromadorea	Spirurida	*Loa loa*	XM_003138991.1
Chromadorea	Ascaridida	*Ascaris suum*	AEUI02000059.1
Secernentea	Rhabditida	*Caenorhabditis briggsae*	FR847113.2
Secernentea	Rhabditida	*Caenorhabditis remanei*	DS268408.1
Secernentea	Rhabditida	*Caenorhabditis brenneri*	XM_003117257.1
Secernentea	Rhabditida	*Haemonchus contortus*	HF966434.1
Secernentea	Strongylida	*Ancylostoma ceylanicum*	KE125293.1
Secernentea	Strongylida	*Necator americanus*	XM_013449818.1
Secernentea	Strongylida	*Ancylostoma duodenale*	KN734493.1
Secernentea	Strongylida	*Oesophagostomum dentatum*	KN567239.1
Secernentea	Strongylida	*Ancylostoma ceylanicum*	KC914882
Secernentea	Spirurida	*Brugia malayi*	XM_001902065.1
Secernentea	Ascaridida	*Toxocara canis*	JPKZ01001678.1

Phylum Annelida (Segmented Worms)			
Class	Order	Species	Accession number
Clitellata	Haplotaxida	*Lumbricus rubellus*	KC981075.1 / KC981074.1
Clitellata	Haplotaxida	*Eisenia fetida*	EF433776.1
Clitellata	Haplotaxida	*Eisenia andrei*	KP770990.1
Clitellata	Rhynchobdellida	*Helobdella robusta*	XM_009013566
Polychaeta	Capitellida	*Capitella teleta*	KB309928

Phylum Gastropoda (Molluscs)			
Class	Order	Species	Accession number
Gastropoda	Lottiida	*Lottia gigantea*	XM_009047767
Bivalvia	Ostreoida	*Crassostrea gigas*	XM_011447075

Phylum Platyhelminthes (flatworms)			
Class	Order	Species	Accession number
Trematoda	Strigeidida	*Schistosoma mansoni*	CABG01000042.1
Trematoda	Prosostomata	*Schistosoma haematobium*	XM_012938931.1
Trematoda	Opisthorchiida	*Opisthorchis viverrini*	XM_009168655
Trematoda	Opisthorchiida	*Clonorchis sinensis*	DF143054
Cestoda	Cyclophyllidea	*Echinococcus multilocularis*	LK031422.1
Cestoda	Cyclophyllidea	*Echinococcus granulosus*	LK028580.1
Cestoda	Cyclophyllidea	*Hymenolepis microstoma*	LK053266.1

Clemens *et al.* [4] and Vatamaniuk *et al.* [5] discovered the presence of a functional phytochelatin synthase in the model nematode *Caenorhabditis elegans*, now one of the most studied metazoan phytochelatin synthases. The fully sequenced *C. elegans* genome contains two *pcs-1* splice variants encoding a 426aa and 418aa protein, respectively, which are approximately 30% identical to the *Arabidopsis thaliana* PCS [3]. Knocking out or knocking down (by RNA interference) the *C. elegans* *pcs-1* gene results in a hypersensitive Cd^{2+} phenotype [5,6], leading to developmental retardation and early death. Intriguingly, the exposure to cadmium increases the concentration of PCs in the tissue of *C. elegans*, but *pcs-1* is notably transcriptionally non-responsive to metal exposures [6], a finding which was also observed in earthworms [7]. Although the PCS/PC pathway has been linked to cadmium detoxification [7,8], it has, to date, not been possible to provide evidence that the impairment of the phytochelatin synthase pathway influences directly the accumulation of metals in animals, a notion this study aimed to investigate.

The transcription of *C. elegans pcs-1* was quantified in animals challenged with high levels of cadmium or zinc by means of *in vivo* imaging using transgenic nematodes expressing a P*pcs-1*::GFP and by quantitative RT-PCR. In addition, synchrotron based X-ray fluorescence imaging was optimized to facilitate the measurement of the cadmium and zinc body burden in individual nematodes. The latter enabled the means to assess to what extent metal load changes in wild-type and *pcs-1* knockout nematodes raised in the presence or absence of high levels of cadmium or zinc, thereby providing first tantalizing insights into the *pcs-1* mediated alteration in metal accumulation.

2. Results and Discussion

Cadmium and zinc are chemically similar and are both positioned in group IIb of the periodic table. Two electrons are found on their most outer shell, hence the valencies of both elements is "2+". However, the affinity towards -SH groups is higher in the case of cadmium, which is an important fact given that cadmium is a toxic, non-essential metal, but zinc essential to a myriad of proteins and enzymes. Indeed, metal ions often interact and it is therefore important to define their precise biological function within biological systems, and in particular individuals with genetic variations within key metallo-pathways [9]. Defining the interaction between cadmium and zinc within an organism is therefore deemed to be crucial and has been nicely reviewed by Brzoska and Moniuszko-Jakoniuk [10].

The difference in chronic and acute exposure conditions were due to the different toxicities of the respective metals (based on the molar concentration zinc is less toxic to nematodes than cadmium, indeed exposure to 300 μM cadmium for extended time periods kills the nematode) and the need to apply the same exposure conditions to all experimental platforms (a relatively high concentration of metal is required to obtain a sufficiently strong signal in individual nematodes to allow the detection by X-ray fluorescence microscopy, see later). The concentrations of cadmium is high, but within the same order of magnitude as reported in Cui *et al.* [11], who exposed *C. elegans* to 100 μM for up to 36 h or 200 μM cadmium for 24 h. Cui *et al.* [11] were not able to observe a significantly change the expression of *pcs-1* at these concentrations and to confirm that this was not merely due to a threshold effect, we decided to determine whether an exposure to higher levels of cadmium (300 μM) or zinc (500 μM) may influence the expression of phytochelatin synthase. It should however be noted that although transcription did not change at the stated concentrations of zinc or cadmium, it is conceivable that an exposure to lower concentrations may induce the transcription of *pcs-1*.

2.1. Analysis of Ppcs-1::GFP Expression by Fluorescence Microscopy

The expression of *pcs-1* was quantified, *in vivo*, utilizing a transgenic *C. elegans* which bears extra-chromosomal copies of the *pcs-1* promoter fused to the coding sequence of the Green Fluorescent Protein, GFP (P*pcs-1*::GFP). The constitutive expression of P*pcs-1*::GFP was confirmed, as reported in [12,13] and shown to be limited to the cells of the pharyngeal intestinal valve and the rectal valve (Figure 1A) and whilst we are not able to hypothesize why the GFP signal was higher in the pharyngeal intestinal valve than in the rectal calve, quantitative assessment confirmed that the fluorescence intensity did not change within the two locations in nematodes exposed to 500 μM zinc or 300 μM cadmium (Figure 1B). The analysis of GFP signal intensity in individual nematodes suggests that the transcription of *pcs-1* is not dependent on the exposure to cadmium or zinc, at least at the concentrations indicated.

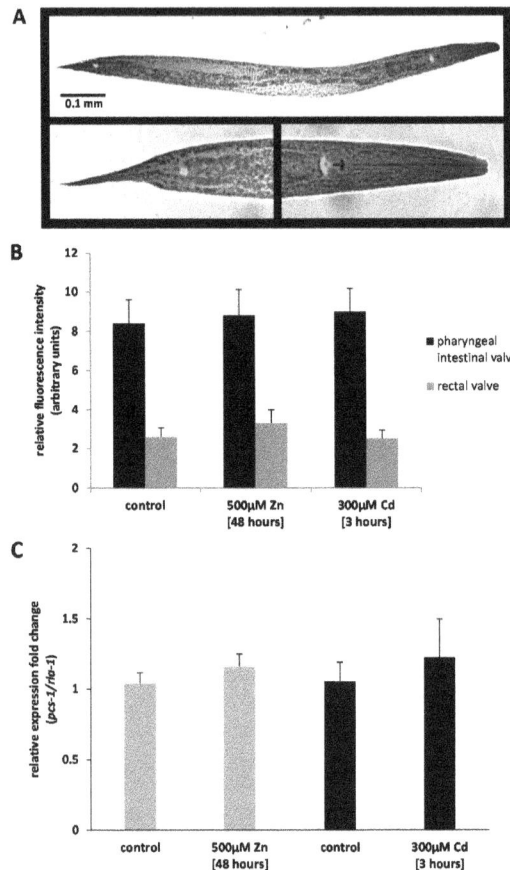

Figure 1. The transcriptional activity of the *pcs-1* promoter. The expression pattern of the *pcs-1* promoter was visualized in P*pcs-1*::GFP transgenic nematodes (synchronized at L4 stage) by means of confocal microscopy (Leica DMIRE2) (**A**); fluorescence in the head and tail region was quantified using ImageJ (**B**); Note, the fluorescence intensity, though universally higher in the pharyngeal intestinal valve compared to the intensity in the rectal valve, were statistically indistinguishable in nematodes raised under control conditions or nematodes exposed to 500 μM zinc for 48 h or 300 μM cadmium for 3 h. The error bars represent the standard error of the mean ($\pm$SEM, biological repeats $n = 9$). A quantitative RT-PCR confirmed the transcriptional invariance of *pcs-1* in wild-type nematodes challenged with metals (**C**). The error bars denote $\pm$SEM (technical repeats $n = 3$, biological repeats $n = 3$). Statistical analyses were performed using a one way ANOVA). Note, P*pcs-1*::GFP fluorescence signal intensity or *pcs-1* transcription did not differ ($p > 0.05$) in control and metal exposed nematodes.

2.2. Analysis of Pcs-1 Expression by Quantitative PCR

To independently validate the results obtained in the GFP assay, the expression fold-change of *pcs-1* was determined by qPCR in wild-type (N2) nematodes exposed to either 500 μM zinc for 48 h or 300 μM cadmium for 3 h, respectively. Exposures were conducted as true biological replicates, and synchronized (fourth larval stage, L4) nematodes harvested that had either not been subjected to metals, acutely dosed with cadmium or chronically exposed to zinc. Following normalization with the invariant ribosomal *rla-1*, analysis confirmed that the transcriptional response to metal exposure was, at best, negligible. In detail, nematodes challenged with 500 μM zinc for 48 h prior to sample collection resulted in a 1.2-fold increase in *pcs-1* and 300 μM cadmium for 3 h in a 1.8-fold increase, both values are statistically not different to the baseline levels observed in the control exposures (Figure 1C). Metallothionein (*mtl-2*) was utilized as a positive control to validate that the cDNA synthesized originated from successful exposures. The *mtl-2* transcript increased 5.2-fold and *mtl-1* transcript 3.5-fold in the respective cadmium and zinc exposed samples, which confirmed that the lack of *pcs-1* response was genuine and not due to a failed exposure. These results not only confirm the results obtained using the P*pcs-1*::GFP strain, but substantiate that even high (near toxic) levels of zinc or cadmium are not able to activate the transcription of *pcs-1* in nematode and extends previous findings by [14–16].

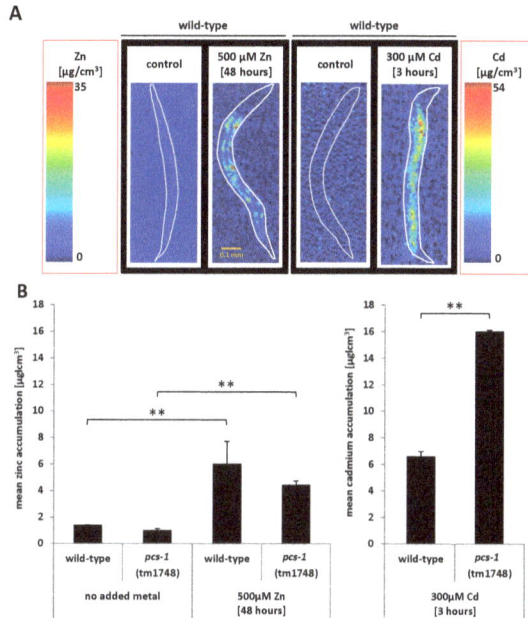

Figure 2. Accumulation of metals in the nematode body. X-ray fluorescence imaging (XFI) was utilized to assess the distribution and accumulation of zinc and cadmium in the body of *C. elegans* raised either on control plates or transferred to plates supplemented with 500 μM zinc for 48 h or 300 μM cadmium for 3 h (**A**); Note, base-line levels of zinc (but not cadmium) were observed in nematodes raised on control plates. All quantitative analyses of metal load within the body of nematodes were performed using the MicroAnalysis Toolkit (**A,B**). A highly significant difference ($p = 0.002$) was apparent in nematodes exposed to zinc (500 μM zinc for 48 h), however no significant difference was observed between the strains (wild-type and the *pcs-1*(tm1748) mutant). Likewise, exposure to cadmium resulted in a highly significant ($p = 0.002$) increase in metal load in wild-type and *pcs-1*(tm1748). However, strain specific differences were also observed, where the *pcs-1*(tm1748) accumulated significantly ($p = 0.002$) more cadmium than the respective wild-type (**B**). Note: whilst zinc could be measured in nematodes raised on control plates, the cadmium signal was below the detection limit. Statistical analyses were performed using a factorial ANOVA. Note, the pixel densities differ because Cd and Zn quantifications were performed at different beamlines (Cd: 14-3; Zn 2-3), different incident energies (Cd: 3.575 keV; Zn 10 keV), and different spot sizes (Cd: 5 × 5 μM; Zn 2 × 2 μM). ** denotes $p \leqslant 0.01$.

2.3. Analysis of Metal Accumulation by X-ray Fluorescence Imaging

To determine the accumulation and distribution pattern of zinc and cadmium within single *C. elegans*, high resolution X-ray fluorescence imaging (XFI) was applied using the facilities at the

Stanford Synchrotron Radiation Lightsource (SSRL). Initial trials confirmed that a strong base-line and exposure signals could be obtained from nematodes chronically exposed to zinc, however the chronic exposure to lower concentrations of cadmium (30 µM) did not return a signal above the detection limit. Therefore a higher concentration of cadmium, albeit via a more acute exposure route, was chosen. The detection limit requirements of XFI thus dictated the exposure conditions of the experiment conducted in this paper, namely 300 µM cadmium for 3 h and 500 µM zinc for 48 h. Wild-type (N2) and the *pcs-1* knock out mutant (*pcs-1*(tm1748)) were exposed to cadmium and zinc using the identical exposure conditions as in the previous experiments and the metal concentration was visualised in individual nematodes by means of a heat map scaled from 0 (blue) to 54 µg/cm^3 (red) for cadmium and 0 (blue) to 35 µg/cm^3 (red) for zinc (Figure 2A). Although no additional zinc was added to the control plates or the bacteria, nematodes were seemingly able to obtain sufficient levels of zinc from the media to maintain essential baseline levels of zinc, as evidenced by the detection of a zinc signal (compared to the absence of cadmium signal) in control conditions. Following the exposure to 500 µM zinc, a significant accumulation was observed in the intestinal cells of nematodes (Figure 2A). It should be noted that this signal is not free zinc located in the lumen of the gut (as this was cleared prior to analysis) but a genuine signal originating from an uptake of zinc. The presence or absence of the gene *pcs-1* did not affect the accumulation pattern or concentration of zinc in nematodes. This suggests that the phytochelatin synthase is not involved in zinc sensing or detoxification, at least at a concentration of 500 µM zinc, and contrasts studies in plants which have shown that zinc exposure can influence the phytochelatin synthase [1,17]. During the course of acute dosing, cadmium also increased throughout the nematodes body. However, unlike zinc, a considerable difference in cadmium accumulation was observed between wild-type and the *pcs-1*(tm 1748) knock out strain, namely a significant cadmium bio-concentration (wild-type: (6.6 ± 0.4) µg/cm^3; *pcs-1*(tm 1748): (16.0 ± 0.1) µg/cm^3) was observed upon deletion of the phytochelatin (Figure 2B).

3. Materials and Methods

3.1. Maintenance and Strains

C. elegans were maintained at 20 °C on Nematode Growth Media (NGM) [9,18] plates seeded with *Escherichia coli* OP50 [19]. Three different strains were used, namely the wild-type N2 Bristol strain and the knock out strain *pcs-1*(tm1748) originally generated by the Mitani laboratory, Tokyo Women's Medical University, Japan (both strains were obtained from the *Caenorhabditis* Genetics Center stock collection at the University of Minnesota, St. Paul, MN, USA), and UL1109 a transgenic strain carrying a *Ppcs-1*::GFP extrachromosomal construct (kindly donated by Prof. I. Hope, Leeds University, Leeds, UK). The *pcs-1*(tm1748) is deemed to be a genuine null-allele [13].

3.2. Metal Exposures

Metals (500 µM zinc or 300 µM cadmium) were added to the molten NGM agar as well as the OP50. The OP50 was spread onto NGM plates which were then incubated at room temperature overnight to allow the formation of a bacterial lawn. The chronic zinc exposure samples were generated by placing synchronized L1 nematodes onto NGM plates containing either no added zinc or 500 µM zinc, and then incubated at 20 °C for 48 h (until the nematodes had developed into pre-adult L4 stage). The acute cadmium exposures required staged L1 nematodes to be plated initially onto non-metal supplemented plates (control plates) for 45 h and then transferred to plates containing 300 µM for 3 h.

3.3. Confocal Fluorescence Microscopy

To visualize and quantify the expression of the *pcs-1* transcript, strain UL1109 was used which carries an extrachromosomal P*pcs-1*::GFP construct (1543 bp of the *pcs-1* promoter and 851 bp of the *pcs-1* coding sequence (spanning the first three exons and introns) fused to the coding sequence of GFP) made by Gateway recombination using pDEST-R4R2 and a GFP entry clone with pRF4 (*rol-6*) serving as co-transformant. The transgenic nematodes were maintained on control plates (with no added metals) or exposed to either 500 μM zinc for 48 h or 300 μM cadmium for 3 h. L4 nematodes were picked into a drop of M9 and sodium azide (to immobilize the nematode) on a glass slide and imaged using a confocal Leica DMIRE2 microscope (Leica Microsystems, Milton Keynes, UK); fluorescence was captured by using an argon laser (λ_{ex}: 488) and the quantitative analysis was performed by means of ImageJ ($n = 9$).

3.4. Quantitative PCR (qPCR)

To determine the expression of *pcs-1* by quantitative PCR, each condition/replicate consisted of 7000 synchronized L4 stage *C. elegans*, which were washed off NGM plates and collected in 15 mL centrifuge tubes. To eliminate the *Escherichia coli* OP50, nematodes were washed with M9 buffer four times, and then pelleted, shock frozen in liquid nitrogen and stored at $-80\ ^\circ$C for at least 24 h. RNA was extracted using the standard Tri Reagent® (Sigma-Aldrich, St. Louis, MO, USA) protocol, bar one modification, namely the frozen nematode pellet with Tri-reagent was vortexed with an equal quantity of acid-washed glass beads (particle size 425–600 μM, Sigma-Aldrich) for 4 min. The concentration and integrity of total RNA was determined with a NanoDrop 1000 Spectrophotometer (NanoDrop Technologies, Inc., Wilmington, DE, USA) and by agarose gel (2%) electrophoresis.

cDNA was synthesised from 1000 ng RNA using an oligo dT primer (5'-(T)20VN-3') and M-MLV reverse transcriptase (Promega, Southampton, UK) applying standard incubation condition. The quantity of *pcs-1* was measured on an ABI Prism 7500 Fast (Applied Biosystems®, Paisley, UK) utilizing the following cycling conditions: 2 min at 50 °C, 10 min at 95 °C, followed by 40 cycles of 15 s 95 °C and 1 min 60 °C. The housekeeping gene *rla-1* (acidic ribosomal subunit protein P1) was used for normalisation purposes, as it was previously shown to be invariant in nematodes subjected to a metal exposure [18,20]. All probes and primers were designed to be compatible with the Universal Probe Library (Roche Applied Sciences, Burgess Hill, UK). Each qPCR reaction contained 5 μL ROX Buffer (Roche Applied Sciences), 0.1 μL Probe (*pcs-1* probe #159 or *rla-1* probe #162), 0.4 μL forward and reverse primers (*pcs-1*: forward 5'-AAGCGCCGTGGAGATTCTA-3' and reverse 5'-TATTTTCCAAAGGCACACAACA-3' or *rla-1*: forward 5'-ACGTCGAGTTCGAGCCATA-3' and reverse 5'-GAAGTGATGAGGTTCTTCAC-3') and 2 μL diluted cDNA (100 ng/μL). The total volume was made up to 10.8 μL with nuclease free water. The CT values were determined using the 7500 Fast System SDS Software (Applied Biosystems®) and the fold changes in gene expression were calculated by applying the $2^{-\Delta\Delta Ct}$ method. Statistical analysis was performed on three independent biological replicates, each consisting of three technical repeat measurements.

3.5. X-ray Fluorescence Imaging (XFI)

The zinc and cadmium accumulated in the body of the nematode was visualized and quantified by X-ray fluorescence imaging (XFI) at the Stanford Synchrotron Radiation Laboratory (SSRL). *C. elegans* wild-type and *pcs-1* knockout (*pcs-1*(tm1748)) strains were placed, at room temperature, into a drop of sodium azide on a metal-free plastic microscope slide. For zinc measurements, beam line 2-3 was used, with an incident energy of 10 keV, selected with a Si(111) double crystal monochromator. The X-ray beam was focused to a 2×2 μm spot size using a Rh coated Kirkpatrick-Baez (KB) mirror pair (Xradia Inc (now Zeiss), Pleasanton, CA, USA). The nematodes were rastered in a continuous motion across the beam and the intensity of the fluorescence lines of the elements of interest from the sample were monitored at each pixel using a silicon drift Vortex detector (Hitachi, Northridge, CA, USA) equipped with Xspress3 electronics (Quantum Detectors). Dwell time on an individual pixel was approximately on the order of 75 ms. Cadmium measurements were performed at beam line 14-3, with the major differences being that the incident energy was selected to be 3.575 keV, the KB focussing mirrors were Ni coated, and the X-ray spot size was 5×5 μm. The resulting images were analysed with the Micro Analysis Toolkit (Webb 2011, Palo Alto, CA, USA). Elemental concentrations were determined by calibration with standard X-ray fluorescence concentration thin film standards

(MicroMatter, Vancouver, BC, Canada) and lines added to indicate the shape of the nematode. The latter was generated by taking the phosphorous and potassium signals as the template which is ubiquitous throughout the body of the nematode.

3.6. Statistical Analysis

Statistical analysis was performed using IBM SPSS. The qPCR was assessed by a one way ANOVA, the fluorescence microscopy was analysed using an independent *t*-test and the XFM was scrutinized via a factorial ANOVA. All error bars represent standard errors of the mean (SEM).

4. Conclusions

In conclusion, this paper substantiates the notion that *pcs-1* transcription is not activated by zinc or cadmium, even at high concentrations. The finding that the cadmium body burden increases in nematodes characterized by a defective *pcs-1*/PC pathway, suggests that the phytochelatin synthase and possibly phytochelatin may actively be involved in the binding, transport, excretion and/or detoxification of cadmium, rather than zinc.

Acknowledgments: Acknowledgments: We acknowledge funding through King's College London (to SRS) and thank the King's Genomics Centre for access to equipment. Use of the Stanford Synchrotron Radiation Lightsource, SLAC National Accelerator Laboratory, is supported by the U.S. Department of Energy, Office of Science, Office of Basic Energy Sciences under Contract No. DE-AC02-76SF00515. The SSRL Structural Molecular Biology Program is supported by the DOE Office of Biological and Environmental Research, and by the National Institutes of Health, National Institute of General Medical Sciences (including P41GM103393). The contents of this publication are solely the responsibility of the authors and do not necessarily represent the official views of NIGMS or NIH. Finally, we acknowledge Gladio Giannitelli and Garyfalia Vouloutsi, two MSc students who contributed towards the initial optimization of XFM technique with nematodes.

Author Contributions: Author Contributions: Stephen R. Stürzenbaum conceived the study and Yona J. Essig performed the experiments. Yona J. Essig, Samuel M. Webb and Stephen R. Stürzenbaum designed the experiments, analyzed the data and wrote the paper.

Conflicts of Interest: Conflicts of Interest: The authors declare no conflict of interest.

References

1. Grill, E.; Winnacker, E.L.; Zenk, M.H. Phytochelatins: The principal heavy-metal complexing peptides of higher plants. *Science* **1985**, *230*, 674–676. [CrossRef] [PubMed]
2. Grill, E.; Loffler, S.; Winnacker, E.L.; Zenk, M.H. Phytochelatins, the heavy-metal-binding peptides of plants, are synthesized from glutathione by a specific γ-glutamylcysteine dipeptidyl transpeptidase (phytochelatin synthase). *Proc. Natl. Acad. Sci. USA* **1989**, *86*, 6838–6842. [CrossRef] [PubMed]
3. Rea, P.A.; Vatamaniuk, O.K.; Rigden, D.J. Weeds, worms, and more. Papain's long-lost cousin, phytochelatin synthase. *Plant Physiol.* **2004**, *136*, 2463–2474. [CrossRef] [PubMed]
4. Clemens, S.; Schroeder, J.I.; Degenkolb, T. *Caenorhabditis elegans* expresses a functional phytochelatin synthase. *Eur. J. Biochem.* **2001**, *268*, 3640–3643. [CrossRef] [PubMed]
5. Vatamaniuk, O.K.; Bucher, E.A.; Ward, J.T.; Rea, P.A. A new pathway for heavy metal detoxification in animals. Phytochelatin synthase is required for cadmium tolerance in *Caenorhabditis elegans*. *J. Biol. Chem.* **2001**, *276*, 20817–20820. [CrossRef] [PubMed]
6. Hughes, S.L.; Bundy, J.G.; Want, E.J.; Kille, P.; Sturzenbaum, S.R. The metabolomic responses of *Caenorhabditis elegans* to cadmium are largely independent of metallothionein status, but dominated by changes in cystathionine and phytochelatins. *J. Proteom. Res.* **2009**, *8*, 3512–3519. [CrossRef] [PubMed]
7. Bundy, J.G.; Kille, P. Metabolites and metals in Metazoa—What role do phytochelatins play in animals? *Metallomics* **2014**, *9*, 1576–1582. [CrossRef] [PubMed]
8. Howden, R.; Andersen, C.R.; Goldsbrough, P.B.; Cobbett, C.S. A cadmium-sensitive, glutathione-deficient mutant of *Arabidopsis thaliana*. *Plant Physiol.* **1995**, *107*, 1067–1073. [CrossRef] [PubMed]
9. Maret, W. The metals in the biological periodic system of the elements: Concepts and conjectures. *Int. J. Mol. Sci.* **2016**. [CrossRef] [PubMed]

10. Brzoska, M.M.; Moniuszko-Jakoniuk, J. Interactions between cadmium and zinc in the organism. *Food Chem. Toxicol.* **2001**, *39*, 967–980. [CrossRef]
11. Cui, Y.; McBride, S.; Boyd, W.; Alper, S.; Freedman, J. Toxicogenomic analysis of *Caenorhabditis elegans* reveals novel genes and pathways involved in the resistance to cadmium toxicity. *Genome Biol.* **2007**, *8*, 90–105. [CrossRef] [PubMed]
12. Hope, I.A.; Stevens, J.; Garner, A.; Hayes, J.; Cheo, D.L.; Brasch, M.A.; Vidal, M. Feasibility of genome-scale construction of promoterreporter gene fusions for expression in *Caenorhabditis elegans* using a multisite gateway recombination system. *Genome Res.* **2004**, *14*, 2070–2075. [CrossRef] [PubMed]
13. Schwartz, M.S.; Benci, J.L.; Selote, D.S.; Sharma, A.K.; Chen, A.G.; Dang, H.; Fares, H.; Vatamaniuk, O.K. Detoxification of multiple heavy metals by a half-molecule ABC transporter, HMT-1, and coelomocytes of *Caenorhabditis elegans. PLoS ONE* **2010**, *5*. [CrossRef] [PubMed]
14. Vatamaniuk, O.K.; Mari, S.; Lu, Y.P.; Rea, P.A. AtPCS1, a phytochelatin synthase from *Arabidopsis:* Isolation and *in vitro* reconstitution. *Proc. Natl. Acad. Sci. USA* **1999**, *96*, 7110–7115. [CrossRef] [PubMed]
15. Clemens, S.; Kim, E.J.; Neumann, D.; Schroeder, J.I. Tolerance to toxic metals by a gene family of phytochelatin synthases from plants and yeast. *EMBO J.* **1999**, *18*, 3325–3333. [CrossRef] [PubMed]
16. Ha, S.B.; Smith, A.P.; Howden, R.; Dietrich, W.M.; Bugg, S.; O'Connell, M.J.; Goldsbrough, P.B.; Cobbett, C.S. Phytochelatin synthase genes from *Arabidopsis* and the yeast *Schizosaccharomyces pombe. Plant Cell* **1999**, *11*, 1153–1164. [CrossRef] [PubMed]
17. Grill, E.; Winnacker, E.L.; Zenk, M.H. Phytochelatins, a class of heavy-metal-binding peptides from plants, are functionally analogous to metallothioneins. *Proc. Natl. Acad. Sci. USA* **1987**, *84*, 439–443. [CrossRef] [PubMed]
18. Polak, N.; Read, D.S.; Jurkschat, K.; Matzke, M.; Kelly, F.J.; Spurgeon, D.J.; Sturzenbaum, S.R. Metalloproteins and phytochelatin synthase may confer protection against zinc oxide nanoparticle induced toxicity in *Caenorhabditis elegans. Comp. Biochem. Physiol. C* **2014**, *160*, 75–85. [CrossRef] [PubMed]
19. Brenner, S. The genetics of *Caenorhabditis elegans. Genetics* **1974**, *77*, 71–94. [PubMed]
20. Swain, S.C.; Keusekotten, K.; Baumeister, R.; Stürzenbaum, S.R. *C. elegans* metallothioneins: New insights into the phenotypic effects of cadmium toxicosis. *J. Mol. Biol.* **2004**, *341*, 951–959. [CrossRef] [PubMed]

International Journal of
Molecular Sciences

MDPI

Article

Earthworm *Lumbricus rubellus* MT-2: Metal Binding and Protein Folding of a True Cadmium-MT

Gregory R. Kowald [1], Stephen R. Stürzenbaum [2,3] and Claudia A. Blindauer [1,*]

1   Department of Chemistry, University of Warwick, Coventry CV4 7AL, UK; GKowald@acciuk.co.uk
2   MRC-PHE Centre for Environment & Health, King´s College London, London SE1 9NH, UK;
    stephen.sturzenbaum@kcl.ac.uk
3   Analytical and Environmental Sciences Division, Faculty of Life Science & Medicine, King's College London,
    London SE1 9NH, UK
*   Correspondence: c.blindauer@warwick.ac.uk; Tel.: +44-24-765-282-64; Fax: +44-24-765-241-12

Academic Editor: Reinhard Dallinger
Received: 30 November 2015; Accepted: 24 December 2015; Published: 5 January 2016

Abstract: Earthworms express, as most animals, metallothioneins (MTs)—small, cysteine-rich proteins that bind d^{10} metal ions (Zn(II), Cd(II), or Cu(I)) in clusters. Three MT homologues are known for *Lumbricus rubellus*, the common red earthworm, one of which, wMT-2, is strongly induced by exposure of worms to cadmium. This study concerns composition, metal binding affinity and metal-dependent protein folding of wMT-2 expressed recombinantly and purified in the presence of Cd(II) and Zn(II). Crucially, whilst a single Cd_7wMT-2 species was isolated from wMT-2-expressing *E. coli* cultures supplemented with Cd(II), expressions in the presence of Zn(II) yielded mixtures. The average affinities of wMT-2 determined for either Cd(II) or Zn(II) are both within normal ranges for MTs; hence, differential behaviour cannot be explained on the basis of overall affinity. Therefore, the protein folding properties of Cd- and Zn-wMT-2 were compared by ^{1}H NMR spectroscopy. This comparison revealed that the protein fold is better defined in the presence of cadmium than in the presence of zinc. These differences in folding and dynamics may be at the root of the differential behaviour of the cadmium- and zinc-bound protein *in vitro*, and may ultimately also help in distinguishing zinc and cadmium in the earthworm *in vivo*.

Keywords: metallothionein; cadmium; zinc; selectivity; protein folding; mass spectrometry; NMR spectroscopy

1. Introduction

One of the most intriguing questions in the field of metal homeostasis concerns how biological systems distinguish and discriminate between different metal ions. This is important for understanding not only the metabolic pathways of essential metal ions, but also mechanisms of tolerance and detoxification including those for xenobiotic ions. As far as essential metal ions are concerned, it has been recognised that the cytosolic concentrations of essential metal ions are, in healthy conditions, regulated according to their relative position within the Irving-Williams series, and the proteins involved in their homeostasis typically have metal affinities to match these concentrations [1]. This ensures, for example, that Cu(I), the most competitive essential metal ion, is kept out of the binding sites of all other metal ions—even when, and this is frequently the case, the affinity of the respective protein is higher for copper than for the "correct" metal ion. In the case of a non-essential, toxic metal ion such as cadmium, it may be inferred that at least one protein with sufficiently high affinity is required, with the added proviso that this protein, once expressed, should ideally not significantly interfere with the metabolism of other metal ions. This becomes particularly important when the toxic and essential metal ions have relatively similar coordination chemistry, such as in the case of essential

Zn(II) and toxic Cd(II). The current study will highlight that certain metallothioneins (MTs), small proteins with an extraordinarily high proportion of cysteine thiols that endows them with high affinity towards both Zn(II) and Cd(II), may display distinctly different behaviour towards these two closely related ions.

MTs were one of the first protein families to be associated with metal metabolism in animals [2–4]. Initially discovered in horse kidney cortex as major cadmium-binding proteins [5], their physiological functions in mammals are now thought by most workers in the field to be predominantly if not exclusively concerned with the metabolism of the essential zinc and copper metal ions. In addition, because of their high thiol content, they can also function as antioxidants, and may also link cellular redox status to zinc dynamics [6].

Unusually, and in contrast to most other protein families, MTs are a polyphyletic group of proteins held together not by significant similarities in protein sequence, but rather by a number of descriptors that refer to their overall composition (high proportion of sulfur and metals), and biophysical characteristics such as peculiar spectroscopic features indicative of metal–sulfur clusters. This fact should be reason enough to refrain from extrapolating the biological functions, structural features, or chemical reactivity of an MT from one phylum to another, but sadly, this is a frequent occurrence in the literature.

It has been pointed out that even though mammalian MTs undoubtedly bind cadmium *in vivo* and, hence, play a role in cadmium metabolism [7], it is unlikely that this action is a true, evolutionarily constrained function [8,9]. This may however be distinctly different for organisms in close contact with soils, *i.e.*, plants [10] and terrestrial invertebrates [11–13]. Typically, due to their relatively similar chemistries, toxic cadmium occurs in soils at concentrations that are only two to three orders of magnitude smaller than those of essential zinc. Moreover, in topsoils treated with rock phosphate fertilisers, cadmium levels can be significantly higher and reach up to 14 ppm [14]. With 2015 being the international year of soil [15], it seems appropriate to devote some attention to a group of very important soil organisms supremely adept at coping with cadmium: the earthworm.

Earthworms, master "soil engineers", directly ingest and process soil. This does not only mean that they play a major role in the dynamics of nutrients and essential elements present in soil [16], but also that they are exposed to any compound present, both through their digestive tract as well as through their skin. Earthworms are therefore also used for biomonitoring purposes in ecotoxicology, and have also been dubbed "soil sentinels" [17].

At least some species, for example the common red earthworm *Lumbricus rubellus*, can survive in the presence of 600 µg cadmium per g dry weight of soil [18]. Intriguingly, *L. rubellus* also specifically bio-accumulates cadmium to a staggering ratio of up to 1 mg per gram dry body weight, whilst lead, zinc and copper are not bio-accumulated [18]. This difference clearly indicates that there must be distinct metabolic pathways for the chemically closely related Zn(II) and Cd(II). A similar conclusion has been drawn for other soil organisms; for example, the nematode *Caenorhabditis elegans* uses a system that includes phytochelatins, cystathionine, and its two metallothioneins MTL-1 and MTL-2 to discriminate between zinc and cadmium [19,20], with subsequent distinct pathways for utilisation and detoxification [21]. Given the differences in zinc and cadmium accumulation, it is conceivable that similar mechanisms may exist in earthworms, with a potential function for MTs in the discrimination between essential zinc and toxic cadmium.

Many eukaryotic species, including invertebrates, express several different MT homologues (In MT literature, it is common to call these different forms "isoforms", although strictly speaking this is a term that should refer to different forms derived from a single gene. However, in most if not all cases, different forms of MTs in one particular species are derived from different genes; therefore we prefer to refer to these as "homologues" provided that an evolutionary relationship between them is likely.), sometimes, but not always in a tissue-specific manner. Evidence has been accumulating that different homologues in the same species can exhibit different metal selectivities; prime examples are the Cu- and Cd-MTs from snails [22–24] as well as the *C. elegans* MTs mentioned above [19,20]. MTs are being studied in several annelids, mainly in terms of gene expression [25–28]. Perhaps the best-studied system in terms of MTs is *L. rubellus*; for this species, at least three MT genes are known [29]. The corresponding wMT-1 and wMT-2 proteins have been isolated from adult earthworms [18], whilst the protein sequence of a third homologue, wMT-3, has been derived from an EST library generated from developing cocoons [29] (Figure 1).

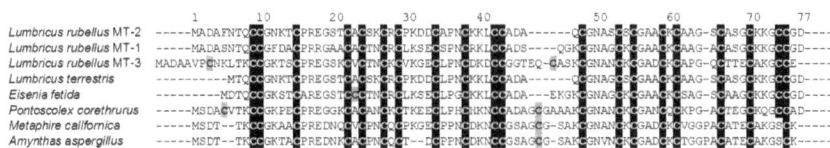

Figure 1. Sequence alignment of selected annelid MTs, showing conservation of two blocks of Cys-rich regions with 11–12 and 8–9 Cys residues (highlighted by black boxes). All three sequences from *L. rubellus* are shown; the numbering refers to wMT-2. Non-conserved Cys residues are highlighted in grey.

The protein sequences of wMT-1 and wMT-2 are 74.7% identical and 91.1% similar, whilst that of wMT-3 is only 56% identical and 67% similar to wMT-1 or wMT-2. The expression patterns of these three homologues also differ considerably; the *wMT-3* gene has been suggested to be highly expressed during embryonic development [29], whereas *wMT-1* and *wMT-2* are both responsive to metal exposure, but to different extents. *wMT-2* is the homologue with the most pronounced responsiveness to cadmium—its expression may be upregulated several hundred- to thousand-fold in response to high cadmium levels in soil [30]. The non-transience of this increased expression suggested that this constitutes the primary response to cadmium [31]. Accordingly, *wMT-2* was also one of the most upregulated genes in response to chronic cadmium exposure, as identified in an extensive transcriptomic study [32]. The mechanism for metallo-regulation of *wMT-2* genes—and indeed many other invertebrate MT genes—has puzzled researchers for some while, because even though three recognisable metal-response elements (MREs) are present in the pertinent upstream regions in all three identified *wMT-2* loci [29], a protein corresponding to MTF-1, the transcription factor that recognises MREs in vertebrates and *Drosophila melanogaster* [33], could not be identified. A recent study involving EMSA and DNAse I footprinting revealed that cytosolic, but not nuclear extracts from *L. rubellus* cells contain proteins capable of binding to the *wMT-2* promoter region in a zinc-dependent manner. The DNA footprinting experiments identified cAMP-responsive elements (CRE) as putative candidates for MT gene regulation in invertebrates [34].

The *wMT-2* gene is not only supremely responsive to cadmium exposure, its product was also the major Cd-binding protein isolated from earthworms from a contaminated site [18]. The highest levels of wMT-2 protein were found in Cd-exposed worms in the thyphlosole and gut epithelium (both alimentary surfaces), chloragogenous tissues (these have been likened to vertebrate livers), coelomocytes (a type of invertebrate immune cells) and nephridia (analogues of vertebrate kidneys) [35]. Thus, even though wMT-2, like any other MT, is also capable of binding other metal ions such as Zn(II) and Cu(I) [36], and even though gene expression may also be induced by metals other than cadmium [37], there are multiple lines of evidence, that as far as biological function is concerned, wMT-2 is a "cadmium-MT". We have used a combination of mass spectrometry and NMR spectroscopy to study whether and how biological function may be reflected in the biophysical properties, in particular metal affinities and protein folding, of this invertebrate MT.

2. Results and Discussion

2.1. Production of Untagged Recombinant Zinc- and Cadmium-Bound wMT-2

Earthworm MT proteins have previously been studied *in vitro*; data for proteins isolated from their native host are available for *L. rubellus* [18] and *Eisenia fetida* [38], and recombinant, S-tagged wMT-2 has been studied as well [31,39].

In the present study, wMT-2 was also expressed recombinantly in *E. coli* as an S-tagged fusion protein. Although the S-tag is relatively short and is not expected to impact on metal-binding abilities, in the context of protein structure and folding, it is still preferable that the studied protein is as similar to the native form as possible. Therefore, the tag was removed by cleavage with thrombin, after a first purification step. Contrary to its intended purpose, the S-tag was not used to aid purification; instead, a three-step procedure involving gel filtration, thrombin cleavage, and a second round of gel filtration chromatography was found to give higher protein yields compared to attempts that included S-tag affinity-based purification. During expression, cultures were supplemented with either Zn(II) or Cd(II). The resulting purified proteins were analysed by electrospray ionisation mass spectrometry (Figure 2).

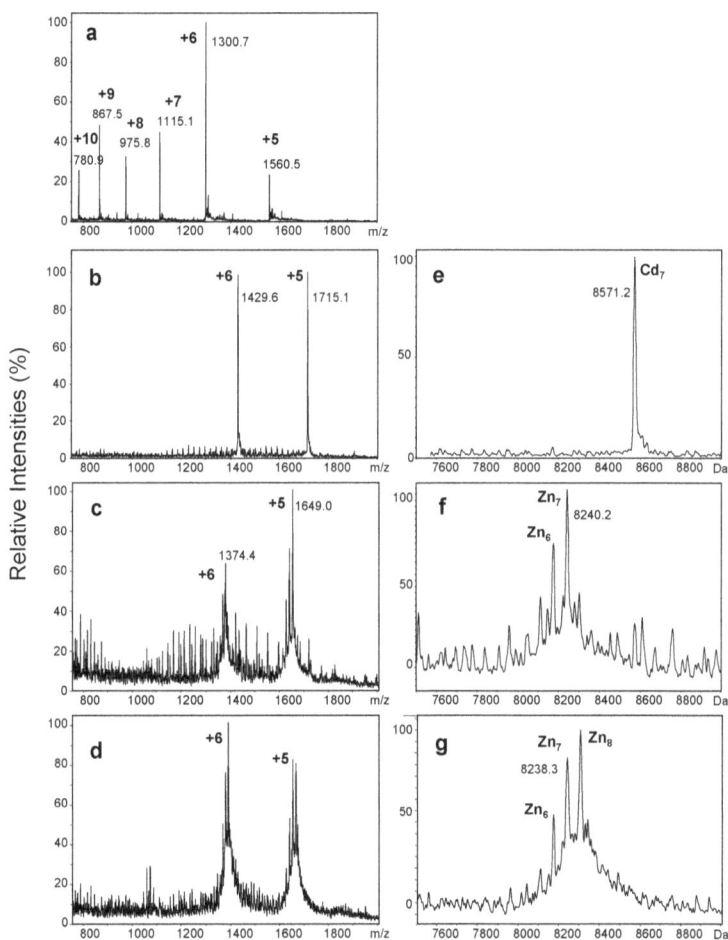

Figure 2. Raw (**a**–**d**) and deconvoluted (**e**–**g**) mass spectra of recombinant wMT-2 (10 mM NH_4HCO_3, 10% MeOH) (**a**) Raw mass spectrum showing charge states for demetallated wMT-2 at pH < 2.0. Deconvolution onto a true mass scale gives a neutral mass of 7798.4 Da; (**b,e**) MS data for wMT-2 expressed and purified in the presence of Cd(II); (**c**–**g**) MS data from two different expressions in the presence of Zn(II). Spectra (**b**–**f**) were recorded at pH 8.5, spectra (**d,g**) at pH 7.0. The low signal to noise ratio in spectra (**c**–**g**) is due to the low concentration of the protein, which in turn was due to rapid sample degradation. The series of additional peaks between *m/z* 1000 and 1700 seen in (**b,c**) correspond to polyethylene glycol, contamination with which can occur through use of ultracentrifugation filters.

A mass of 7798.4 Da for the charge-neutral species, obtained under denaturing and hence demetallation conditions (pH < 2), agrees well with the theoretical mass of 7797.9 Da for apo-wMT-2, calculated from the sequence (UniProt entry O76955_LUMRU; Figure 1) plus the mass of an additional glycine and serine, which remain at the N-terminus after cleavage with thrombin. The low baseline and absence of significant peaks not belonging to the charge state series for apo-wMT-2 is an indication for the purity of this preparation. The mass spectral data shown in Figure 2b–g were obtained under "native" ESI conditions; this means that the proteins are brought into the gas phase in their folded, metallated forms, which can be achieved by working at non-acidic pH, and using no or only small quantities of organic solvent [40]. This approach is in principle applicable to non-covalent protein complexes including metalloproteins in general, but has proven outstandingly useful for the study of MTs, as it is the only method that allows distinction between different metallospecies [41–46]. Furthermore, MTs are particularly amenable to this analysis, as they are small proteins that ionise well. Only two charge states, +5 and +6, are observed for metallated wMT-2 irrespective of whether Zn(II)- or Cd(II)-bound (Figure 2b–d). A small number of charge states is indicative of a folded state [47], as not only are there a smaller number of protonatable groups exposed in a folded protein, but the latter is also more compact and, hence, can accommodate fewer charges of the same sign. Figure 2e shows the result of deconvolution of the data shown in Figure 2b onto a true mass scale. One single metallospecies—Cd_7wMT-2 with a neutral mass of 8571.2 Da (theoretical: 8570.8 Da)—was observed for wMT-2 expressed in the presence of cadmium. A Cd_7 species was also the major species previously observed for S-tagged Cd-wMT-2 at neutral pH [39]. Figure 2c–g demonstrate that MS data obtained for protein expressed in the presence of Zn(II) did not only present a lower signal-to-noise ratio, but that invariably, more than one metallospecies was observed. This was indeed also the case before cleavage of the S-tag (data not shown). Zn_7 was a major species, but was accompanied by species with both higher and lower metal contents, depending on batch and sample pH. The significance of this observation will be discussed later on.

2.2. Proton-Driven Metal Loss

A pH titration monitored by mass spectrometry was undertaken for both the zinc and cadmium forms of wMT-2 (Figure 3). Both zinc- and cadmium-loaded wMT-2 eventually lost their bound metals due to competition from protons for the thiolate sulfurs, but like every MT studied so far by this method, the cadmium-bound form exhibited substantially higher pH-stability, with the Cd_7 species still dominant at pH 3.9, whilst the Zn(II)-containing preparation at similar pH was dominated by the apo-form. This lower stability of Zn-MTs *versus* Cd-MTs is expected, as Cd-S bonds are inherently stronger than Zn-S bonds [3]. From these data (note that the full dataset contained more datapoints), it was possible to estimate pH-of-half-displacement values (pH(1/2); also called apparent pK_a values) of 2.8 for Cd-wMT-2 and 4.2 for Zn-wMT-2. It should be pointed out that ESI-MS is not the method of choice for the determination of pH(1/2) values, as m/z peak intensities do not only depend on the concentration of the respective species in solution, but also their ionisation efficiency, which may or may not be comparable for different species [48], depending on their degree of (un-)folding. In the present case though, the value obtained for Cd-wMT-2 agreed within error limits with the pH(1/2) value of 2.9 previously determined via UV–Vis spectroscopy for S-tagged Cd-wMT-2 [31]. Both values for Cd- and Zn-wMT-2 are within ranges expected for MTs [3], with both being on the low side, which might be taken for an indication of higher stability—but only as far as competition with protons is concerned (as will be seen in Section 2.3).

As indicated before, ESI-MS is the only technique that is able to directly give detailed speciation information, and as such, can be used to explore cooperativity in the binding of multiple metals. The data shown in Figure 3 provide a classical example of cooperative (for Cd-wMT-2) *versus* largely non-cooperative (for Zn-wMT-2) behaviour. Whilst essentially all possible species (Zn_1-Zn_8) are observed for the zinc form during the pH titration, only two metallospecies (Cd_7 and Cd_4) are dominating the speciation at different pH values. The abrupt transition from Cd_7 to Cd_4 below pH 3.9

is highly reminiscent of previous observations made for the cadmium-bound forms of mammalian MTs, where also pH-dependent cooperative loss of first three then four metal ions occurs [41,49]. This behaviour has been interpreted—e.g., for mammalian MTs, and in the previous study on S-tagged Cd-wMT-2 [39]—as the cooperative loss of a 3-metal cluster with lower pH stability than the remaining 4-metal cluster. The two-domain structure of mammalian MTs with an N-terminal $M(II)_3Cys_9$ and a C-terminal $M(II)_4Cys_{11}$ cluster is well established [50], and several other animal MTs are known to be two-domain proteins [3]. The concerted loss of three Cd(II) ions hence may indicate that the 20 Cys residues in wMT-2 probably also form a Cd_4Cys_{11} and a Cd_3Cys_9 cluster in two separate domains. This hypothesis is also compatible with the sequence data shown in Figure 1; given the location of the gaps in the majority of sequences in the upper part of the alignment, and the high level of conservation of numbers and positions of Cys residues in the regions separated by these gaps, it seems highly probable that the region containing the first eleven Cys residues constitutes an N-terminal domain with a Cd_4Cys_{11} cluster, and that the remaining nine Cys residues belong to a separate domain with a Cd_3Cys_9 cluster, with the less well conserved region between them constituting a linker between two domains. Further support for a two-domain structure comes from work on *Eisenia fetida* MT [38]; efforts to purify Cd-MT directly from worms led to the isolation of a 41-residue protein (*i.e.*, all residues before the gap in the alignment) harbouring four Cd(II) ions.

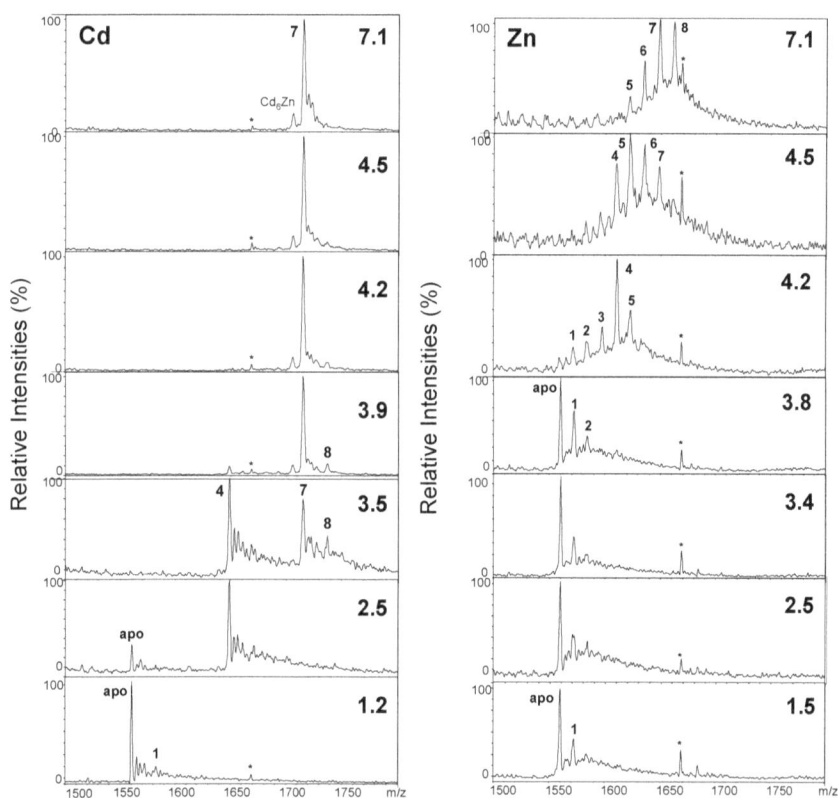

Figure 3. +5 charge state of raw, smoothed mass spectra showing speciation of Cd-wMT-2 and Zn-wMT-2 at different pH values (10 mM NH_4HCO_3, 10% MeOH). The peak labelled in both series with an asterisk is an unknown contaminant. The numbers indicate how many metal ions are bound in the respective species. All observed masses are compiled in Table 1.

Table 1. Theoretical and observed masses for species detected during pH titrations of Cd-and Zn-wMT-2.

Metals Bound (n)	Neutral Masses for Cd$_n$wMT-2		Neutral Masses for Zn$_n$wMT-2	
	Observed	Theoretical	Observed	Theoretical
8	8678.0	8681.1	8303.5	8305.1
7	8570.5	8570.8	8240.0	8241.7
6	–	–	8175.0	8178.3
5	–	–	8112.0	8114.9
4	8239.9	8239.6	8050.5	8051.5
3	–	–	7984.0	7988.1
2	–	–	7916.0	7924.7
1	7908.5	7908.3	7856.5	7861.3
0 (apo)	7798.0	7797.9	7796.0	7797.9

Whilst conclusions regarding the two-domain structure of wMT-2, which have been reached previously already [39], are probably valid in the present case, we would caution that seemingly cooperative loss of metal ions may not necessarily correlate with loss of a cluster. An example to the contrary concerns a type 4 MT from plants, wheat E$_C$, where the dominance of a Zn$_4$ species does neither correlate with the loss of a Zn$_2$ cluster nor the presence of a fully loaded Zn$_4$ cluster or domain [44].

In contrast to the cadmium-bound form, species from Zn$_8$ to Zn$_5$ were observed for the zinc-bound protein, even at neutral pH. Over the past decade, it has become clear that the formation of "correctly" metallated species depends on the nature of the metal ions that are to be bound to the MT [51]. In general, mis-metallation and/or a broad speciation even at neutral pH (rather than one or two species as observed for Cd-wMT-2), are typical observations for MTs bound to "non-cognate" metal ions [51]. The low degree of cooperativity also testifies to a comparably low thermodynamic stabilisation of intact zinc-thiolate clusters, a notion that is also compatible with this view. Having noted this, it should be acknowledged that even MTs that are thought to be zinc-specific might not show much cooperativity towards zinc [52]; therefore, the degree of cooperativity in metal binding does not appear to be a suitable criterion to classify specificity *per se*.

Interestingly, the Zn$_8$-species was not a major species at higher pH (see Figure 2f; pH 8.5), but became apparent only around neutral pH, together with larger quantities of under-metallated species. A closer look at the data for the Cd(II) form also reveals that once appreciable amounts of under-metallated (*i.e.*, Cd$_4$) species are being formed, small amounts of a Cd$_8$ species are also evident (Figure 3, pH 3.9 and 3.5). Considering the clean formation of Cd$_7$ (Figure 2), the M$_8$ species should be considered as an "over-metallated" species. Such over-metallated species have previously been observed, including for mammalian MTs [43,53]. In the present case, the existence of these species may indicate that these over-metallated species are more favourable than various possible under-metallated (and protonated) species, at least in the gas phase. The available data do not allow drawing conclusions regarding domain or cluster structures for these species, but in the case of mammalian MTs, it has been suggested that these "supermetallated" species may correspond to alternatively folded, single-domain structures [53].

Finally, the dominance of the Zn$_4$ species observed at pH 4.2, and its subsequent rapid disappearance, may suggest a small degree of cooperativity in the predicted Zn$_4$ cluster. This species predominated from pH 4.2–4.0, but at just 0.2 pH units lower, the likely Zn$_4$ cluster collapsed, and the apo form dominated, with minor amounts of Zn$_1$ and Zn$_2$ present.

In summary, the data in Figures 2 and 3 indicate considerable differences between the binding behaviour of wMT-2 towards Cd(II) and Zn(II). Whilst it needs to be acknowledged that in general, binding affinities of MTs for Zn(II) are several orders of magnitude lower than for Cd(II), this does not mean that the binding of Zn(II) is too weak for isolation and observation of fully metallated

Zn-MTs —indeed, there are many examples to the contrary [43,51,54,55]. We hold that the observation of significant quantities of under- and over-metallated species at neutral pH or above is an indication for the fact that the Zn_7 species is, overall, thermodynamically not much more favoured than the alternative species observed. To gain further insight into the origin of the observed differential behaviour, we next determined the average stability constants of the Zn(II) and Cd(II) complexes of wMT-2.

2.3. Affinities of wMT-2 for Cadmium and Zinc

Whilst competition with protons constitutes one possible measure of affinity and cluster stability that allows facile comparison between different MTs, it is desirable to also determine overall average stability constants. For both Zn- and Cd-wMT-2, this was achieved using a method based on competition with the chelator 5F-BAPTA (5-fluoro-1,2-bis(2-aminophenoxy)-ethane-N,N, N',N'-tetraacetic acid) [56]. Figure 4 compares the values obtained (log K_{Cd-MT} = 13.5 ± 0.5 and log K_{Zn-MT} = 10.9 ± 0.5) with those determined for other MTs by the same method and under the same experimental conditions. It should be noted that these conditions include a relatively low ionic strength (I = 4 mM) and high pH (8.1), both factors that strengthen coordinative bonds in general, as low ionic strength renders charge recombination between metals and ligands more favourable, and high pH reduces the competition from protons for lone pairs.

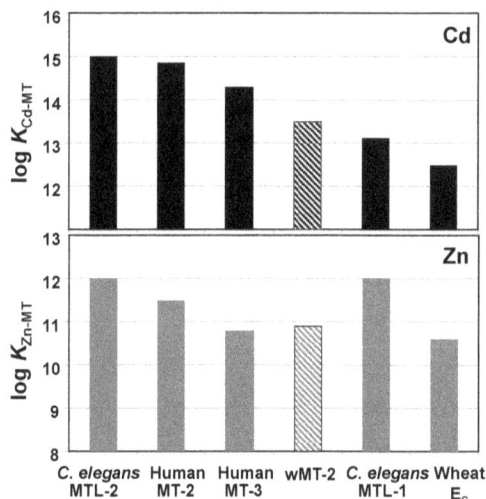

Figure 4. Comparison of stability constants determined by competition with 5F-BAPTA at pH 8.1 and I = 4 mM (10 mM Tris) [19,20,55,56]. The proteins are ordered by falling affinity for Cd(II). Hatched bars highlight wMT-2 data determined in the present study. Note that the scales do not start at 0.

The most salient point from the comparison shown in Figure 4 is that the affinity of wMT-2 for neither Zn(II) nor Cd(II) stands out. Its affinity for cadmium is not only lower than that of *C. elegans* MTL-2, another MT that is strongly induced by Cd(II) and with proven effects on Cd(II) homeostasis [19,20], but also lower than those measured for mammalian MTs, which are thought not to have evolved to serve in biological cadmium handling. Contrarily, its affinity for Zn(II) is even a little higher than that of the zinc-specific [57] wheat E_C.

This comparison highlights that it is not the absolute affinity of an MT for a particular metal ion that is the most important parameter that relates to biological function, but the relative stabilities of the complexes with different metal ions, in a context of an ensemble of other proteins [1,19]. As long as the MT's affinity for Cd(II) is higher than those of most other cellular proteins in *L. rubellus* (including those for handling zinc), and as long as the resulting complex has an appropriate lifetime that allows safe shuttling of the toxicant, wMT-2 will be able to effectively protect against toxic effects of cadmium. Furthermore, both pH-of-half-displacement values (Section 2.2) as well as competition reactions with a chelator (this section) indicate that overall affinities are not a major cause for the heterogeneity observed for the zinc-bound samples (Figure 2). Protein folds in the presence of Zn(II) and Cd(II) were studied next.

2.4. Folding Behaviour of Zn- and Cd-wMT-2

The folding behaviour of wMT-2 was studied by ^{1}H NMR spectroscopy. Figure 5 shows a [^{1}H,^{15}N] HSQC spectrum of ^{15}N,^{13}C-labelled Cd$_7$wMT-2. The dispersion and linewidths of the N–H crosspeaks indicate an overall well-folded protein. Figure 6a shows 1D ^{1}H NMR spectra for Zn- and Cd-wMT-2. It is immediately evident that the number of resolved backbone N–H resonances is higher for Cd-wMT-2 than for Zn-wMT-2. The backbone N–H peaks for Zn-wMT-2 are generally broader, and there is a large amount of unresolved intensity in the random-coil region (around 8.0–8.4 ppm) for the Zn-loaded form. From these data, it is possible to conclude qualitatively that the Zn-loaded protein is much less well folded than the Cd-loaded protein. This is also borne out in the 2D [^{1}H,^{1}H] TOCSY NMR spectra for Zn- and Cd-wMT-2 compared in Figure 6b. Nevertheless, at least some ^{1}H resonances for Zn-wMT-2 show considerable dispersion (see, e.g., the low-field-shifted peaks above 10.0 ppm), indicating that the Zn-loaded protein form is at least partially folded. Furthermore, and despite the heterogeneity in speciation as observed by mass spectrometry for Zn-wMT-2, the spectrum is dominated by just one species, as indicated by the almost complete absence of "split" peaks.

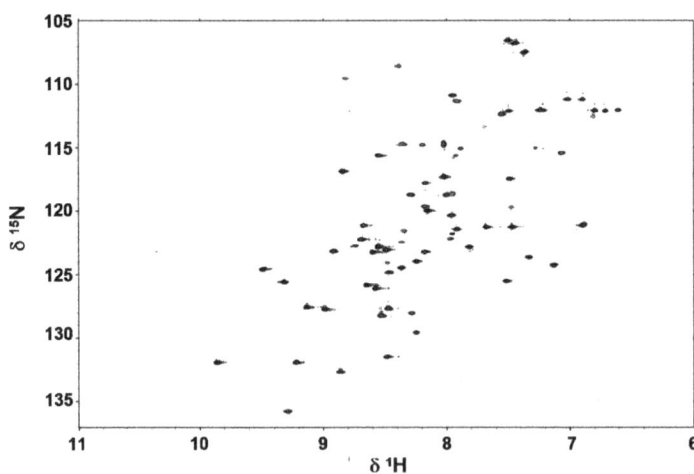

Figure 5. [^{1}H,^{15}N] HSQC spectrum (700 MHz) of ^{13}C,^{15}N-labelled Cd$_7$wMT-2 (25 °C, 0.5 mM protein, 20 mM NH$_4$HCO$_3$, pH 6.8, 10% D$_2$O). The broad dispersion of backbone N–H crosspeaks, and the narrow peak shapes indicate a well-folded protein.

We have obtained a near-complete sequential assignment for Cd-wMT-2, and a partial assignment for Zn-wMT-2. Surprisingly, residues that could be assigned for Zn-wMT-2 are distributed throughout the entire sequence (a few of these are indicated in Figure 6b); therefore, we can exclude the possibility of just one of the two domains being folded in Zn-wMT-2. Thus, it appears that the Zn-bound protein is much more dynamic than its Cd-bound form, leading to significant portions of backbone ^{1}H resonances not being resolvable or assignable.

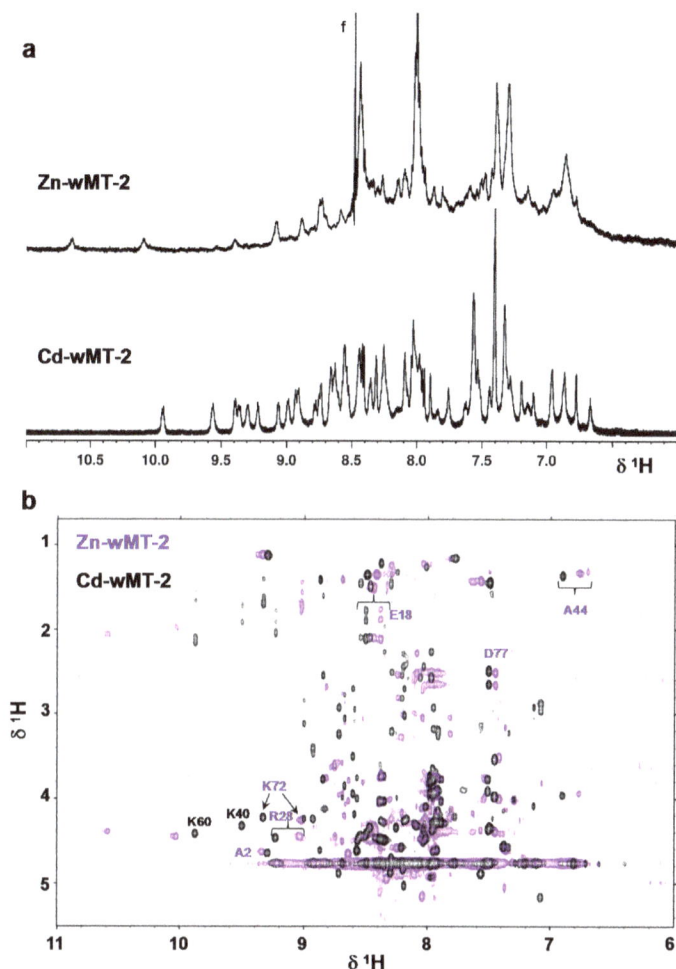

Figure 6. Comparison of 1D ^{1}H (**a**) and 2D [^{1}H,^{1}H] TOCSY (**b**) NMR data for Zn- and Cd-wMT-2. Only the fingerprint regions (consisting mainly of N–H backbone resonances) are shown. Although all four spectra show considerable dispersion of N–H resonances, especially the comparison in (**a**) demonstrates that Zn-wMT-2 is considerably less well folded than Cd-wMT-2. Selected assigned residues are labelled in (**b**) for both proteins; labels in purple refer to residues observed for both forms; black labels refer to residues that were assignable for Cd-wMT-2 only. The "f" in (**a**) refers to the signal for formate at 8.48 ppm.

Another conclusion that can be drawn from the 2D data in Figure 6b is that although some residues are in a similar environment in both forms (as indicated by the similarity in chemical shifts of their backbone N–H protons; see for example residues A2 and D77, both of which are close to the termini), there are also many residues with considerable (e.g., R28, K72), or even extremely large changes in chemical shift (e.g., the two unassigned, low-field shifted N-H resonances above 10 ppm for Zn-wMT-2). This is a strong indication for non-isostructural binding of Zn(II) and Cd(II). Typically, MTs, at least those for M(II) metal ions, are expected to bind these two metal ions in a structurally similar manner, although examples to the contrary are known, e.g., mammalian MT-3 [58], and plant type 4 MTs [57,59].

In conclusion, it is evident that the Cd-bound form is rather well and stably folded, whilst the Zn-bound form differs substantially from this form, both in terms of structure and dynamics.

3. Experimental Section

3.1. Materials

Unless stated otherwise, all chemicals and reagents were obtained from Fisher Scientific, Loughborough, UK. All solutions were prepared with purified water (MilliQ, Millipore, Nottingham, UK) and other reagents used were of analytical grade or better.

3.2. Protein Expression and Purification

Lumbricus rubellus MT-2 was recombinantly expressed in *E. coli* Rosetta 2(DE3)pLysS (Novagen, Nottingham, UK) cells as a fusion protein with an N-terminal S-peptide tag, using a pET29a-derived plasmid, with the *mt2* gene cloned between *Sal*I and *Nco*I restriction sites [31]. Cells were grown in standard Luria-Bertani medium at 37 °C (180 rpm), and Kanamycin (50 µg/mL) and Chloramphenicol (34 µg/mL) were used as selective antibiotics. Once an optical density at 600 nm of 0.6–0.8 was reached, protein over-expression was induced by 1 mM IPTG (Isopropyl-D-thiogalactopyranoside), and 0.5 mM ZnSO$_4$ or 0.2 mM CdSO$_4$ (both Sigma Aldrich, Dorset, UK) were added at the same time. Cells were harvested after 5–6 h of induction by centrifugation (5000× *g*, 10 min, 4 °C). Cell pellets were resuspended in 4–8 mL sonication buffer (50 mM Tris–HCl; 0.1 M KCl; 3 mM β-mercaptoethanol, 1% TWEEN-20, pH 8.5) per g wet cell weight. To prevent metal loss during sonication, 1 mM ZnSO$_4$ or CdCl$_2$ was added. The lysate was separated from cell debris by centrifugation for 30 min at 30,000× *g*, 4 °C. S-tagged wMT-2 was separated from the bulk of other contents of the cell lysate by gel filtration (GE Healthcare HiLoad 16/60, Superdex G75). After concentration to 1 mg/mL, the S-peptide tag was cleaved using bovine thrombin (50 units per mg protein; Sigma Aldrich, Dorset, UK) following manufacturer's protocols, and the mixture was subjected to a second round of gel filtration. Protein concentrations were determined via measuring sulfur concentration by Inductively-Coupled Plasma Optical Emission spectroscopy (ICP-OES, Perkin Elmer, Seer Green, UK), or by determining thiol content via Ellman's method [60], after removal of bound metal ions by incubation with EDTA (2,2′,2″,2‴-(Ethane-1,2-diyldinitrilo)tetraacetic acid, Sigma-Aldrich, Dorset, UK).

3.3. Determination of Protein Concentration and Metal-Protein Stoichiometries by ICP-OES

Samples of approximately 1 ppm [S] were prepared in 0.1 M ultrapure HNO$_3$ (prepared in-house by sub-boiling point distillation; DuoPUR, Milestone S.R.L, Sorisole, Italy). Each sample was simultaneously assessed for Zn, Cd, Cu and S against mixed-element standards of concentrations between 0 and 2 ppm, which were prepared gravimetrically from high grade commercial stocks (1000 ppm, Fluka, Buchs, Switzerland). ICP-OES measurements were performed on an Optima 5300 DV instrument (Perkin Elmer, Seer Green, UK).

3.4. Mass Spectrometry

Samples were concentrated to 30–50 µM protein concentration using either Amicon Ultra centrifugal filtration units, with 3 kDa molecular weight cut-off (Fisher Scientific, Loughborough, UK), or Vivaspin centrifugal concentrators, with 5 kDa molecular weight cut-off (Sigma-Aldrich, Dorset, UK), using 20 mM ammonium bicarbonate (pH 8.35) buffer. Prior to mass spectrometric analysis, 10% (*v/v*) HPLC grade methanol was added to the samples, to obtain mass spectra of metallated species. The pH was gradually deceased using either acetic acid or formic acid. All samples were analysed on an HCTultra ion-trap mass spectrometer (Bruker Daltonics, Coventry, UK) equipped with an ESI source. Samples were directly infused using a syringe pump at 240 µL/h. Typically, data were acquired for 0.5–1.0 min in the positive mode over an *m/z* range of 500–3000. Resulting mass spectra were averaged, smoothed using the Savitzky-Golay algorithm, and where required deconvoluted onto a true mass scale using the Bruker Data Analysis Suite (Bruker Daltonics, Coventry, UK).

3.5. Affinity Measurements by ^{19}F NMR Spectroscopy

Zn- and Cd-loaded forms of recombinant wMT-2 were buffer-exchanged into 10 mM Tris–HCl and 10% D_2O, pH 8.1, using Amicon ultrafiltration devices (3 kDa molecular weight cut-off). Samples, approximately 450 µM in metal ion concentration, were incubated with 5-fluoro-1,2-bis(2-aminophenoxy)-ethane-N,N,N',N'-tetraacetic acid (5F-BAPTA; 3 mM; Molecular Probes™, Invitrogen) overnight at 25 °C. Direct observe proton-decoupled 1D ^{19}F NMR spectroscopy was carried out on a DRX400 spectrometer (Bruker) fitted with a QNP probe-head operating at 375.91 MHz for ^{19}F. Datasets of 12,288 scans were acquired with a spectral width of 200 ppm, an acquisition time of 0.87 s and relaxation delay of 1.0 s. All spectra were acquired at 25 °C. FIDs were apodized with squared sine-bell functions, Fourier transformed with 64k data points and baseline corrected with TOPSPIN v. 2.1 software (Bruker Biospin, Coventry, UK). The apparent stability constants for metal-MT complexes were calculated using the method published by Hasler *et al.* [56]. The value for $K_{Cd(BAPTA)}$ at 30 °C and I = 138 mM was corrected for temperature (25 °C) and ionic strength (I = 4 mM) used in our experiments to give a log K value of 11.75. The value for $K_{Zn(BAPTA)}$ was also recalculated and a log K value of 9.91 was used. Metal and protein concentrations used in the calculations were determined accurately by ICP-OES.

3.6. ^1H NMR Spectroscopy

Samples for ^1H NMR spectroscopy were prepared in 20 mM NH_4HCO_3 buffer, pH 6.9, 10% D_2O. All data were acquired for 0.5–1.0 mM samples on an AV II 700 (Bruker Biospin, Coventry, UK) equipped with a TCI cryoprobe.

Spectral conditions for TOCSY and NOESY experiments were: 25 °C; 48 scans; 4096 data points in F2 and 512 increments in F1, 90° pulse width ≈ 8.00 µs; spectral width 16.0 ppm in both F1 and F2. A mixing time of 60 ms was used for the TOCSY experiments; for NOESY experiments, several experiments with different mixing times were acquired (60–120 ms). Data were processed with TOPSPIN v. 2.1 (Bruker Biospin, Coventry, UK). Raw data were apodized using squared sine-bell functions and Fourier transformed with 2048 × 2048 data points, and visualised using SPARKY v.3.114 software [61].

4. Conclusions

We have studied the composition, metal affinity, and folding behaviour of recombinant *L. rubellus* wMT-2 in its Zn(II)- and Cd(II)-bound forms. As previously found for the S-tagged protein [39], wMT-2 binds seven Cd(II) ions optimally, with a high likelihood for the formation of two separate domains—an N-terminal domain harbouring a $Cd(II)_4Cys_{11}$ cluster and a C-terminal domain with a $Cd(II)_3Cys_9$ cluster. Two-domain MTs are common in animals: vertebrate, crustacean, sea urchin, and mussel MTs all have two domains, despite the absence of significant sequence similarity between these MTs [3]. At present, there is no clear opinion as to why such a two-domain arrangement may be advantageous—especially not in the case of MTs with a predominant role in detoxification. In MTs dealing with essential Zn(II) or Cu(I), two domains might serve in supporting interactions with different partners, and in some cases, the two domains might even handle different metal ions—a prime example is mammalian brain-specific MT-3 [62,63].

We have found significant differences in protein yields, metal stoichiometry, and folding behaviour between Zn-wMT-2 and Cd-wMT-2, which are congruent with the major biological function of wMT-2, which entails protection against Cd(II) toxicity [29]. The concept that MTs have evolved to optimally bind their cognate metal ion is a still emerging and hence debated concept. It is clear that metal selectivity and discrimination cannot be based on overall affinity [1], as all true MTs studied thus far show a similar order of binding affinity as small-molecule thiolate chelators, *i.e.*,

Cu(I) > Cd(II) > Zn(II). This general trend, together with the knowledge that the protein backbones of MTs tend to be very flexible, may tempt one to regard MTs as not much more than larger versions of such small-molecule chelators, and to dismiss the relevance of protein folding. However, several recent studies on MTs from plants [44,55,59,64], snails [22–24,65–67], and nematodes [19,20,68] have demonstrated a correlation between the biologically "correct" metal ion and the "foldedness" of the respective MT. We consider that a correlation between cognate metal and protein folding also makes sense in the case of wMT-2, both in terms of fundamental (bio-)physical principles, as well as biological function: (i) even in cases where Zn(II) and Cd(II) are bound "isostructurally", the difference in ionic radii (0.74 *vs.* 0.92 Å), and hence bond lengths, will require small but potentially significant adaptations in side-chain conformation and backbone fold, and therefore, differences in the energetics for fold stability depending on which metal is bound can be expected; (ii) In recognition of the fact that the "foldedness" of proteins impacts on their persistence *in vivo* [69], we expect that MTs synthesised in the presence of the "wrong" metal ion (in the present case Zn(II)) will have shorter *in vivo* lifetimes. This may help to remobilise the mis-incorporated metal ion, and prevent it from going down the wrong pathway.

These ideas are principally compatible with the finding that cadmium, but not other metal ions, are bio-accumulated in sulfur-rich vesicles by *L. rubellus*. As a next step, it would be desirable to test their validity by carrying out further metalloproteomic experiments in earthworms.

Acknowledgments: Acknowledgments: We would like to thank the Engineering and Physical Sciences Research Council for a studentship to Gregory R. Kowald. Part of the equipment used in this research was obtained through Birmingham Science City with support from Advantage West Midlands and the European Regional Development Fund.

Author Contributions: Author Contributions: Claudia A. Blindauer, Gregory R. Kowald, and Stephen R. Stürzenbaum conceived the study, designed experiments, and wrote the paper. Gregory R. Kowald performed the experiments, Gregory R. Kowald and Claudia A. Blindauer analyzed the data.

Conflicts of Interest: Conflicts of Interest: The authors declare no conflict of interest.

References

1. Foster, A.W.; Robinson, N.J. Promiscuity and preferences of metallothioneins: The cell rules. *BMC Biol.* **2011**, *9*, 3. [CrossRef] [PubMed]
2. Blindauer, C.A.; Leszczyszyn, O.I. Metallothioneins: Unparalleled diversity in structures and functions for metal ion homeostasis and more. *Nat. Prod. Rep.* **2010**, *27*, 720–741. [CrossRef] [PubMed]
3. Blindauer, C.A. Metallothioneins. In *Binding, Transport and Storage of Metal Ions in Biological Cells*; Maret, W., Wedd, A.G., Eds.; The Royal Society of Chemistry: Cambridge, UK, 2014; pp. 606–665.
4. Capdevila, M.; Bofill, R.; Palacios, O.; Atrian, S. State-of-the-art of metallothioneins at the beginning of the 21st century. *Coord. Chem. Rev.* **2012**, *256*, 46–62. [CrossRef]
5. Margoshes, M.; Vallee, B.L. A cadmium protein from equine kidney cortex. *J. Am. Chem. Soc.* **1957**, *79*, 4813–4814. [CrossRef]
6. Maret, W. Redox biochemistry of mammalian metallothioneins. *J. Biol. Inorg. Chem.* **2011**, *16*, 1079–1086. [CrossRef] [PubMed]
7. Klaassen, C.D.; Liu, J.; Diwan, B.A. Metallothionein protection of cadmium toxicity. *Toxicol. Appl. Pharmacol.* **2009**, *238*, 215–220. [CrossRef] [PubMed]
8. Vallee, B.L. Implications and inferences of metallothionein structure. *Exp. Suppl.* **1987**, *52*, 5–16.
9. Palmiter, R.D. The elusive function of metallothioneins. *Proc. Natl. Acad. Sci. USA* **1998**, *95*, 8428–8430. [CrossRef] [PubMed]
10. Blindauer, C.A.; Schmid, R. Cytosolic metal handling in plants: Determinants for zinc specificity in metal transporters and metallothioneins. *Metallomics* **2010**, *2*, 510–529. [CrossRef] [PubMed]
11. Dallinger, R.; Berger, B.; Gruber, C.; Hunziker, P.; Stürzenbaum, S. Metallothioneins in terrestrial invertebrates: Structural aspects, biological significance and implications for their use as biomarkers. *Cell. Mol. Biol.* **2000**, *46*, 331–346. [PubMed]

12. Janssens, T.K.S.; Roelofs, D.; van Straalen, N.M. Molecular mechanisms of heavy metal tolerance and evolution in invertebrates. *Insect Sci.* **2009**, *16*, 3–18. [CrossRef]

13. Morgan, A.J.; Kille, P.; Stürzenbaum, S.R. Microevolution and ecotoxicology of metals in invertebrates. *Environ. Sci. Technol.* **2007**, *41*, 1085–1096. [CrossRef] [PubMed]

14. Pan, J.; Plant, J.A.; Voulvoulis, N.; Oates, C.J.; Ihlenfeld, C. Cadmium levels in Europe: Implications for human health. *Environ. Geochem. Health* **2010**, *32*, 1–12. [CrossRef] [PubMed]

15. 2015 International Year of Soils. Available online: http://www.fao.org/soils-2015/ (accessed on 30 November 2015).

16. Dominguez, J.; Gomez-Brandon, M. The influence of earthworms on nutrient dynamics during the process of vermicomposting. *Waste Manag. Res.* **2013**, *31*, 859–868. [CrossRef] [PubMed]

17. Stürzenbaum, S.R.; Andre, J.; Kille, P.; Morgan, A.J. Earthworm genomes, genes and proteins: The (re)discovery of Darwin's worms. *Proc. R. Soc. B* **2009**, *276*, 789–797. [CrossRef] [PubMed]

18. Stürzenbaum, S.R.; Kille, P.; Morgan, A.J. The identification, cloning and characterization of earthworm metallothionein. *FEBS Lett.* **1998**, *431*, 437–442. [CrossRef]

19. Zeitoun-Ghandour, S.; Charnock, J.M.; Hodson, M.E.; Leszczyszyn, O.I.; Blindauer, C.A.; Stürzenbaum, S.R. The two *Caenorhabditis elegans* metallothioneins (CeMT-1 and CeMT-2) discriminate between essential zinc and toxic cadmium. *FEBS J.* **2010**, *277*, 2531–2542. [CrossRef] [PubMed]

20. Leszczyszyn, O.I.; Zeitoun-Ghandour, S.; Stürzenbaum, S.R.; Blindauer, C.A. Tools for metal ion sorting: *In vitro* evidence for partitioning of zinc and cadmium in *C. elegans* metallothionein isoforms. *Chem. Commun.* **2011**, *47*, 448–450. [CrossRef] [PubMed]

21. Hughes, S.L.; Bundy, J.G.; Want, E.J.; Kille, P.; Stürzenbaum, S.R. The metabolomic responses of *Caenorhabditis elegans* to cadmium are largely independent of metallothionein status, but dominated by changes in cystathionine and phytochelatins. *J. Proteome Res.* **2009**, *8*, 3512–3519. [CrossRef] [PubMed]

22. Dallinger, R.; Berger, B.; Hunziker, P.; Kägi, J.H.R. Metallothionein in snail Cd and Cu Metabolism. *Nature* **1997**, *388*, 237–238. [CrossRef] [PubMed]

23. Palacios, O.; Pagani, A.; Perez-Rafael, S.; Egg, M.; Höckner, M.; Brandstätter, A.; Capdevila, M.; Atrian, S.; Dallinger, R. Shaping mechanisms of metal specificity in a family of metazoan metallothioneins: Evolutionary differentiation of mollusc metallothioneins. *BMC Biol.* **2011**, *9*, 20. [CrossRef] [PubMed]

24. Palacios, O.; Perez-Rafael, S.; Pagani, A.; Dallinger, R.; Atrian, S.; Capdevila, M. Cognate and noncognate metal ion coordination in metal-specific metallothioneins: The *Helix pomatia* system as a model. *J. Biol. Inorg. Chem.* **2014**, *19*, 923–935. [CrossRef] [PubMed]

25. Mustonen, M.; Haimi, J.; Vaisanen, A.; Knott, K.E. Metallothionein gene expression differs in earthworm populations with different exposure history. *Ecotoxicology* **2014**, *23*, 1732–1743. [CrossRef] [PubMed]

26. Fisker, K.V.; Holmstrup, M.; Sørensen, J.G. Variation in metallothionein gene expression is associated with adaptation to copper in the earthworm *Dendrobaena octaedra*. *Comp. Biochem. Phys. C* **2013**, *157*, 220–226. [CrossRef] [PubMed]

27. Homa, J.; Rorat, A.; Kruk, J.; Cocquerelle, C.; Plytycz, B.; Vandenbulcke, F. Dermal exposure of *Eisenia andrei* earthworms: Effects of heavy metals on metallothionein and phytochelatin synthase gene expressions in coelomocytes. *Environ. Toxicol. Chem.* **2015**, *34*, 1397–1404. [CrossRef] [PubMed]

28. Irizar, A.; Rodriguez, M.P.; Izquierdo, A.; Cancio, I.; Marigomez, I.; Soto, M. Effects of soil organic matter content on cadmium toxicity in *Eisenia fetida*: Implications for the use of biomarkers and standard toxicity tests. *Arch. Environ. Contam. Toxicol.* **2015**, *68*, 181–192. [CrossRef] [PubMed]

29. Stürzenbaum, S.R.; Georgiev, O.; Morgan, A.J.; Kille, P. Cadmium detoxification in earthworms: From genes to cells. *Environ. Sci. Technol.* **2004**, *38*, 6283–6289. [CrossRef] [PubMed]

30. Galay-Burgos, M.; Spurgeon, D.J.; Weeks, J.M.; Stürzenbaum, S.R.; Morgan, A.J.; Kille, P. Developing a new method for soil pollution monitoring using molecular genetic biomarkers. *Biomarkers* **2003**, *8*, 229–239. [CrossRef] [PubMed]

31. Stürzenbaum, S.R.; Winters, C.; Galay, M.; Morgan, A.J.; Kille, P. Metal ion trafficking in earthworms—Identification of a cadmium-specific metallothionein. *J. Biol. Chem.* **2001**, *276*, 34013–34018. [CrossRef] [PubMed]

32. Owen, J.; Hedley, B.A.; Svendsen, C.; Wren, J.; Jonker, M.J.; Hankard, P.K.; Lister, L.J.; Stürzenbaum, S.R.; Morgan, A.J.; Spurgeon, D.J.; *et al.* Transcriptome profiling of developmental and xenobiotic responses in a keystone soil animal, the oligochaete annelid *Lumbricus rubellus. BMC Genom.* **2008**, *9*, 21. [CrossRef] [PubMed]

33. Günther, V.; Lindert, U.; Schaffner, W. The taste of heavy metals: Gene regulation by MTF-1. *Biochim. Biophys. Acta* **2012**, *1823*, 1416–1425. [CrossRef] [PubMed]

34. Höckner, M.; Dallinger, R.; Stürzenbaum, S.R. Metallothionein gene activation in the earthworm (*Lumbricus rubellus*). *Biochem. Biophys. Res. Commun.* **2015**, *460*, 537–542. [CrossRef] [PubMed]

35. Morgan, A.J.; Stürzenbaum, S.R.; Winters, C.; Grime, G.W.; Abd Aziz, N.A.; Kille, P. Differential metallothionein expression in earthworm (*Lumbricus rubellus*) tissues. *Ecotoxicol. Environ. Saf.* **2004**, *57*, 11–19. [CrossRef] [PubMed]

36. Marino, F.; Stürzenbaum, S.R.; Kille, P.; Morgan, A.J. Cu–Cd interactions in earthworms maintained in laboratory microcosms: The examination of a putative copper paradox. *Comp. Biochem. Physiol. C* **1998**, *120*, 217–223. [CrossRef]

37. Bundy, J.G.; Sidhu, J.K.; Rana, F.; Spurgeon, D.J.; Svendsen, C.; Wren, J.F.; Stürzenbaum, S.R.; Morgan, A.J.; Kille, P. "Systems toxicology" approach identifies coordinated metabolic responses to copper in a terrestrial non-model invertebrate, the earthworm *Lumbricus rubellus. BMC Biol.* **2008**, *6*, 21. [CrossRef] [PubMed]

38. Gruber, C.; Stürzenbaum, S.; Gehrig, P.; Sack, R.; Hunziker, P.; Berger, B.; Dallinger, R. Isolation and characterization of a self-sufficient one-domain protein—(Cd)-metallothionein from *Eisenia foetida. Eur. J. Biochem.* **2000**, *267*, 573–582. [CrossRef] [PubMed]

39. Ngu, T.T.; Stürzenbaum, S.R.; Stillman, M.J. Cadmium binding studies to the earthworm *Lumbricus rubellus* metallothionein by electrospray mass spectrometry and circular dichroism spectroscopy. *Biochem. Biophys. Res. Commun.* **2006**, *351*, 229–233. [CrossRef] [PubMed]

40. Loo, J.A. Electrospray ionization mass spectrometry: A technology for studying noncovalent macromolecular complexes. *Int. J. Mass Spectrom.* **2000**, *200*, 175–186. [CrossRef]

41. Gehrig, P.M.; You, C.H.; Dallinger, R.; Gruber, C.; Brouwer, M.; Kägi, J.H.R.; Hunziker, P.E. Electrospray ionization mass spectrometry of zinc, cadmium, and copper metallothioneins: Evidence for metal-binding cooperativity. *Protein Sci.* **2000**, *9*, 395–402. [CrossRef] [PubMed]

42. Zaia, J.; Fabris, D.; Wei, D.; Karpel, R.L.; Fenselau, C. Monitoring metal ion flux in reactions of metallothionein and drug-modified metallothionein by electrospray mass spectrometry. *Protein Sci.* **1998**, *7*, 2398–2404. [CrossRef] [PubMed]

43. Palumaa, P.; Tammiste, I.; Kruusel, K.; Kangur, L.; Jornvall, H.; Sillard, R. Metal binding of metallothionein-3 *versus* metallothionein-2: Lower affinity and higher plasticity. *Biochim. Biophys. Acta Proteins Proteom.* **2005**, *1747*, 205–211. [CrossRef] [PubMed]

44. Leszczyszyn, O.I.; Blindauer, C.A. Zinc transfer from the embryo-specific metallothionein E_C from wheat: A case study. *Phys. Chem. Chem. Phys.* **2010**, *12*, 13408–13418. [CrossRef] [PubMed]

45. Sutherland, D.E. K.; Willans, M.J.; Stillman, M.J. Single domain metallothioneins: Supermetalation of human MT 1A. *J. Am. Chem. Soc.* **2012**, *134*, 3290–3299. [CrossRef] [PubMed]

46. Palacios, O.; Espart, A.; Espin, J.; Ding, C.; Thiele, D.J.; Atrian, S.; Capdevila, M. Full characterization of the Cu-, Zn-, and Cd-binding properties of CnMT1 and CnMT2, two metallothioneins of the pathogenic fungus *Cryptococcus neoformans* acting as virulence factors. *Metallomics* **2014**, *6*, 279–291. [CrossRef] [PubMed]

47. Kebarle, P.; Verkerk, U.H. Electrospray: From ions in solution to ions in the gas phase, what we know now. *Mass Spectrom. Rev.* **2009**, *28*, 898–917. [CrossRef] [PubMed]

48. Perez-Rafael, S.; Atrian, S.; Capdevila, M.; Palacios, O. Differential ESI-MS behaviour of highly similar metallothioneins. *Talanta* **2011**, *83*, 1057–1061. [CrossRef] [PubMed]

49. Yu, X.L.; Wojciechowski, M.; Fenselau, C. Assessment of metals in reconstituted metallothioneins by electrospray mass-spectrometry. *Anal. Chem.* **1993**, *65*, 1355–1359. [CrossRef] [PubMed]

50. Braun, W.; Wagner, G.; Wörgötter, E.; Vašák, M.; Kägi, J.H.R.; Wüthrich, K. Polypeptide fold in the 2 metal-clusters of metallothionein-2 by nuclear-magnetic-resonance in solution. *J. Mol. Biol.* **1986**, *187*, 125–129. [CrossRef]

51. Bofill, R.; Capdevila, M.; Atrian, S. Independent metal-binding features of recombinant metallothioneins convergently draw a step gradation between Zn- and Cu-thioneins. *Metallomics* **2009**, *1*, 229–234. [CrossRef] [PubMed]

52. Sutherland, D.E.K.; Summers, K.L.; Stillman, M.J. Noncooperative metalation of metallothionein 1A and its isolated domains with zinc. *Biochemistry* **2012**, *51*, 6690–6700. [CrossRef] [PubMed]
53. Sutherland, D.E.K.; Stillman, M.J. Challenging conventional wisdom: Single domain Metallothioneins. *Metallomics* **2014**, *6*, 702–728. [CrossRef] [PubMed]
54. Blindauer, C.A.; Harrison, M.D.; Robinson, A.K.; Parkinson, J.A.; Bowness, P.W.; Sadler, P.J.; Robinson, N.J. Multiple bacteria encode metallothioneins and SmtA-like zinc fingers. *Mol. Microbiol.* **2002**, *45*, 1421–1432. [CrossRef] [PubMed]
55. Leszczyszyn, O.I.; Schmid, R.; Blindauer, C.A. Toward a property/function relationship for metallothioneins: Histidine coordination and unusual cluster composition in a zinc-metallothionein from plants. *Proteins* **2007**, *68*, 922–935. [CrossRef] [PubMed]
56. Hasler, D.W.; Jensen, L.T.; Zerbe, O.; Winge, D.R.; Vašák, M. Effect of the two conserved prolines of human growth inhibitory factor (metallothionein-3) on its biological activity and structure fluctuation: Comparison with a mutant protein. *Biochemistry* **2000**, *39*, 14567–14575. [CrossRef] [PubMed]
57. Leszczyszyn, O.I.; White, C.R.J.; Blindauer, C.A. The isolated Cys(2)His(2) site in E_C metallothionein mediates metal-specific protein folding. *Mol. BioSyst.* **2010**, *6*, 1592–1603. [CrossRef] [PubMed]
58. Palumaa, P.; Njunkova, O.; Pokras, L.; Eriste, E.; Jornvall, H.; Sillard, R. Evidence for non-isostructural replacement of Zn^{2+} with Cd^{2+} in the beta-domain of brain-specific metallothionein-3. *FEBS Lett.* **2002**, *527*, 76–80. [CrossRef]
59. Leszczyszyn, O.I.; Imam, H.T.; Blindauer, C.A. Diversity and distribution of plant metallothioneins: A review of structure, properties and functions. *Metallomics* **2013**, *5*, 1146–1169. [CrossRef] [PubMed]
60. Ellman, G.L. Tissue sulfhydryl groups. *Arch. Biochem. Biophys.* **1959**, *82*, 70–77. [CrossRef]
61. Sparky 3. Available online: https://www.cgl.ucsf.edu/home/sparky/ (accessed on 30 November 2015).
62. Bogumil, R.; Faller, P.; Pountney, D.L.; Vašák, M. Evidence for Cu(I) clusters and Zn(II) clusters in neuronal growth-inhibitory factor isolated from bovine brain. *Eur. J. Biochem.* **1996**, *238*, 698–705. [CrossRef] [PubMed]
63. Artells, E.; Palacios, O.; Capdevila, M.; Atrian, S. *In vivo*-folded metal-metallothionein 3 complexes reveal the Cu-thionein rather than Zn-thionein character of this brain-specific mammalian metallothionein. *FEBS J.* **2014**, *281*, 1659–1678. [CrossRef] [PubMed]
64. Peroza, E.A.; Schmucki, R.; Güntert, P.; Freisinger, E.; Zerbe, O. The β(e)-domain of wheat E_C-1 metallothionein: A metal-binding domain with a distinctive structure. *J. Mol. Biol.* **2009**, *387*, 207–218. [CrossRef] [PubMed]
65. Perez-Rafael, S.; Monteiro, F.; Dallinger, R.; Atrian, S.; Palacios, O.; Capdevila, M. *Cantareus aspersus* metallothionein metal binding abilities: The unspecific CaCd/CuMT isoform provides hints about the metal preference determinants in metallothioneins. *Biochim. Biophys. Acta Proteins Proteom.* **2014**, *1844*, 1694–1707. [CrossRef] [PubMed]
66. Perez-Rafael, S.; Mezger, A.; Lieb, B.; Dallinger, R.; Capdevila, M.; Palacios, O.; Atrian, S. The metal binding abilities of *Megathura crenulata* metallothionein (McMT) in the frame of gastropoda MTs. *J. Inorg. Biochem.* **2012**, *108*, 84–90. [CrossRef] [PubMed]
67. Höckner, M.; Stefanon, K.; de Vaufleury, A.; Monteiro, F.; Perez-Rafael, S.; Palacios, O.; Capdevila, M.; Atrian, S.; Dallinger, R. Physiological relevance and contribution to metal balance of specific and non-specific metallothionein isoforms in the garden snail. *Biometals* **2011**, *24*, 1079–1092. [CrossRef] [PubMed]
68. Bofill, R.; Orihuela, R.; Romagosa, M.; Domenech, J.; Atrian, S.; Capdevila, M. *Caenorhabditis elegans* metallothionein isoform specificity—Metal binding abilities and the role of histidine in CeMT1 and CeMT2. *FEBS J.* **2009**, *276*, 7040–7069. [CrossRef] [PubMed]
69. Wickner, S.; Maurizi, M.R.; Gottesman, S. Posttranslational quality control: Folding, refolding, and degrading proteins. *Science* **1999**, *286*, 1888–1893. [CrossRef] [PubMed]

International Journal of
Molecular Sciences

MDPI

Article

Does Variation of the Inter-Domain Linker Sequence Modulate the Metal Binding Behaviour of *Helix pomatia* Cd-Metallothionein?

Selene Gil-Moreno [1,†], Elena Jiménez-Martí [2,†], Òscar Palacios [1], Oliver Zerbe [3], Reinhard Dallinger [4], Mercè Capdevila [1] and Sílvia Atrian [2,*]

[1] Departament de Química, Facultat de Ciències, Universitat Autònoma de Barcelona, E-08193 Cerdanyola del Vallès, Spain; selenebdn89@gmail.com (S.G.-M.); oscar.palacios@uab.cat (O.P.); merce.capdevila@uab.cat (M.C.)

[2] Departament de Genètica, Facultat de Biologia, Universitat de Barcelona, Av. Diagonal 643, E-08028 Barcelona, Spain; ejimenezmarti@gmail.com

[3] Institute of Organic Chemistry, University of Zurich, 8057 Zurich, Switzerland; zerbe@oci.uzh.ch

[4] Institute of Zoology, University of Innsbruck, Technikerstraße 25, A-6020 Innsbruck, Austria; reinhard.dallinger@uibk.ac.at

* Correspondence: satrian@ub.edu; Tel.: +34-93-4021501; Fax: +34-93-4034420

† These authors contributed equally to this work.

Academic Editor: Nick Hadjiliadis

Received: 17 November 2015; Accepted: 14 December 2015; Published: 22 December 2015

Abstract: Snail metallothioneins (MTs) constitute an ideal model to study structure/function relationships in these metal-binding polypeptides. *Helix pomatia* harbours three MT isoforms: the highly specific CdMT and CuMT, and an unspecific Cd/CuMT, which represent paralogous proteins with extremely different metal binding preferences while sharing high sequence similarity. Preceding work allowed assessing that, although, the Cys residues are responsible for metal ion coordination, metal specificity or preference is achieved by diversification of the amino acids interspersed between them. The metal-specific MT polypeptides fold into unique, energetically-optimized complexes of defined metal content, when binding their cognate metal ions, while they produce a mixture of complexes, none of them representing a clear energy minimum, with non-cognate metal ions. Another critical, and so far mostly unexplored, region is the stretch linking the individual MT domains, each of which represents an independent metal cluster. In this work, we have designed and analyzed two HpCdMT constructs with substituted linker segments, and determined their coordination behavior when exposed to both cognate and non-cognate metal ions. Results unequivocally show that neither length nor composition of the inter-domain linker alter the features of the Zn(II)- and Cd(II)-complexes, but surprisingly that they influence their ability to bind Cu(I), the non-cognate metal ion.

Keywords: Cd-isoform; domain linker sequence; *Helix pomatia*; metallothionein; metal binding

1. Introduction

Metallothioneins (MTs) are a super-family of mostly small, ubiquitous, but highly heterogeneous, proteins that coordinate heavy-metal ions owing to the metal-thiolate bonds contributed by their abundant cysteine residues (recent reviews in [1,2]). They have been traditionally associated with different biological roles mainly related to physiological metal (Zn and Cu) homeostasis and/or to toxic heavy metal chelation, but also to different stress responses, such as free radical scavenging. It has been hypothesized that this multitude of possible functions may be the basis of the high heterogeneity of the MT proteins reported up to now, so that MTs may have evolved differently in certain groups of organisms according to precise physiological requirements. Hence, the extraordinary diversity of MT isoforms -MTs are polymorphic in almost all organisms analyzed up to now—along all kinds of taxa seem to be related to their plasticity to perform a great multiplicity of functions [3]. At first, several classifications of this heterogeneous group of proteins had been first proposed on the basis of sequence similarity and taxonomic criteria [4,5], but most significantly, our group later proposed a functional classification of MTs founded on the analysis of their preference for divalent metal ion coordination, grouping them into the so-called Zn-thioneins (accounting for both Zn(II) or Cd(II)-binding MTs), and the so called Cu-thioneins (accounting for monovalent ion binding MTs). Each type of MT is characterized by yielding unique, well-folded, homometallic complexes when it coordinates its cognate metal ion [6]. Although originally this classification only proposed these two MT categories [7], it was later extended to a step-wise gradation between extreme, or genuine, Zn- (or divalent metal-ions)-thioneins and Cu-thioneins [8].

Gastropoda pulmonates is one of the Mollusca classes with a higher number of species, and they constitute an ideal model system to study the structure/function relationship and the evolutionary differentiation of polymorphic MTs. The different MT isoforms combine two valuable properties that allow to precisely recognize the features that confer the Zn- or Cu-thionein character to an MT polypeptide: the paralogous proteins are highly specialized for binding distinct metal ions while retaining high sequence similarities. Hence, the best characterized snail MT systems, those of the terrestrial snails *Helix pomatia* and *Cantareus aspersus* include three paralogous MT peptides with different metal binding preferences: the CdMT and CuMT isoforms which, respectively, bind cadmium and copper with high specificity, and an unspecific Cd/CuMT isoform that was isolated as a mixed Cd and Cu native complex. The CdMT and CuMT proteins were first extensively characterized in the species *Helix pomatia* [9], whereas the unspecific Cd/CuMT was initially isolated from cadmium-intoxicated garden snails (*Cantareus aspersus*) [10]. Nevertheless, its presence was later also corroborated in *H. pomatia* [11]. Since the synthesis of the *H. pomatia* Cd-specific isoform (HpCdMT) was shown to be inducible by cadmium food supplementation, and since it yielded homonuclear Cd_6-complexes, a metal detoxification role in the snail digestive tract was assigned to this peptide [12,13]. Contrarily, the Cu-specific isoform (HpCuMT), natively isolated as homonuclear Cu_{12}-complexes, is constitutively synthesized in the rhogocytes, which suggested a possible involvement in hemocyanin synthesis through storage and delivery of the required copper [14]. Further data from recombinantly-synthesized metal-complexes allowed to demonstrate that variations of the amino acid sequence interspersed between their fully conserved cysteines had led to the metal binding specificity these two *H. pomatia* MTs [11]. More recently, studies of the metal-binding behavior towards either cognate or non-cognate metal ions revealed that MT biosynthesis in the presence of the former renders unique, energetically optimized complexes, which is what outlines their metal specificity or preference. In contrast, the binding of non-cognate metal ions results in a mixture of complexes, with varied stoichiometries and folds [15]. Thus, the thermodynamic stability of the metal-MT complexes appears not exclusively related to their metal-thiolate bonds, as could have been theorized from strict chemical considerations, but is rather determined by the nature of the non-coordinating residues of each MT sequence [15].

In addition to Cys patterns and the nature of the intercalated residues, a third element that may modulate the binding behavior of an MT polypeptides are the linker stretches between the domains (metal-clusters). Their length and composition may influence the stability and independency of the metal-MT structural domains. This is a rather unexplored aspect of MT structures, mainly because the number of 3D structures available is still limited and, therefore, does not provide a sound statistical basis to study this aspect. In fact, among the 16 MT structures available in PDB [2], only one—the rat liver Zn_2, Cd_5–MT2 complex—reveals the relative orientation of the two domains. This structure displays the paradigmatic dumbbell shape that vertebrate MTs yield when they coordinate divalent metal ions: the N-terminal segment (β domain), with 9 cysteines in Cys–X–Cys arrays, which binds three M(II) ions, and the C-terminal segment (α domain), with 11 cysteines, which binds four M(II) ions [16]. In all the other proteins, the putative independent domains have been solved independently by NMR, and were assumed to be connected more or less flexibly by the linker residues. However, for some MTs despite the absence of resolved 3D structures, some clear data have been reported, about the influence of both the linker composition and the N-term and C-term MT flanking regions for their metal binding capabilities, such as in the case of arsenic chelation by *Fucus vesiculosus* MT [17], cadmium scavenging by the type 2 *Quercus suber* QsMT [18] or, most recently, copper coordination by the two fungal *Cryptococcus neoformans* CnMT1 and CnMT2 isoforms [19].

In this report, we aimed at investigating the influence of the amino acid sequence in the linker connecting the two nine-Cys moieties of HpCdMT for the stoichiometry and folding of the corresponding Zn(II), Cd(II), or Cu(I) complexes. Although no crystal or solution structure of the Cd-HpCdMT complex is yet available (work is in progress), the results from its spectroscopic characterization fully support the existence of two domains constituting separate Cd_3Cys_9 clusters [20], as observed for the marine crustaceans [21,22] and the *C. elegans* MTs [23]. To this end, two mutant HpCdMT proteins with prolonged linkers (eight instead of the two native residues) were designed and expressed in *E. coli*. Thereafter, the metal binding behavior of these two mutants (called HpCdMcMT and HpCdPlMT from now on) was assessed for recognizing both their cognate metal ions (*i.e.*, Zn(II) and Cd(II)), but also the non-cognate monovalent Cu(I) ions, and all the data were compared with those from the wild-type HpCdMT isoform.

2. Results and Discussion

2.1. The HpCdMcMT and HpCdPlMT Recombinant Peptides

The two HpCdMT mutants designed contained longer linkers than the -KT- dipeptide of the wild type protein: one of them—that of HpCdMcMT—reproduces the one of another gastropod MT, *M. crenulata*, and exhibits a clear polar character (-VKTEAKTT-) [24]. The other linker—that of HpCdPlMT—derives from a plant MT (the wheat Ec-1 protein), and is of clear apolar composition (-SARSGAAA-) [25]. DNA sequencing of the HpCdMcMT- and HpCdPlMT-coding pGEX-4T-1 constructs ruled out any nucleotide mutation, and confirmed that the cDNAs were cloned in correct frame. After expression in *E. coli* and purification, acidification of the recombinant Zn-HpCdMcMT and Zn-HpCdPlMT samples yielded the corresponding apo-forms, with respective molecular masses of 7254.7 and 7066.6 Da, in accordance with the respective theoretical values of 7255.7 and 7067.9 Da (Figures 1 and 2). This confirmed the correctness of both synthesized proteins.

```
HpCdMT     GSGKGKGEKCTSACRSEPCQCGSKCQCGEGCTCAAC---KT---CNCTSDGCKCGKECTGPDSCKCGSSCSCK
HpCdMcMT   GSGKGKGEKCTSACRSEPCQCGSKCQCGEGCTCAACVKTEAKTTCNCTSDGCKCGKECTGPDSCKCGSSCSCK
HpCdPlMT   GSGKGKGEKCTSACRSEPCQCGSKCQCGEGCTCAACSARSGAAACNCTSDGCKCGKECTGPDSCKCGSSCSCK
```

Figure 1. Sequence alignment of the recombinant proteins studied in this work: the constructs HpCdMcMT and HpCdPlMT are aligned with the HpCdMT wild-type form. The Cys residues are written in red, and the linker residues are shaded in grey. The initial Gly, which is a remainder from the thrombin cleavage site, is printed in italics.

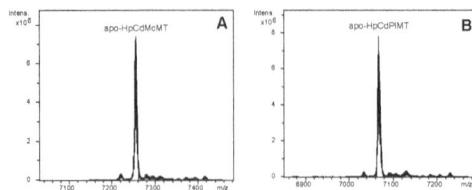

Figure 2. Deconvoluted ESI-MS spectra of the recombinant preparations of (**A**) HpCdMcMT; and (**B**) HpCdPlMT purified from bacterial cultures grown under Zn-supplementation, analyzed at acid pH (2.4).

2.2. Zn and Cd Binding Abilities of HpCdMcMT and HpCdPlMT

Both HpCdMcMT and HpCdPlMT polypeptides synthesized in Zn-supplemented *Escherichia coli* cultures folded into unique Zn_6-complexes, (*cf.* the ESI-MS spectra shown in Figure 3A, where only very minor, negligible accompanying peaks are detected, attributable to frequently observed NH_4^+ adducts, and Table 1). The syntheses in Cd-supplemented cultures equally yielded almost unique peaks corresponding to the Cd_6-complexes of both peptides, as identified in the respective ESI-MS analyses at neutral pH (Figure 3A and Table 1). These results fully coincide with the Zn- and Cd-species afforded by the wild-type HpCdMT synthesized under equivalent conditions, that affords unique $M(II)_6$ complexes, as we demonstrated in [11]. Analysis of the CD spectra of the Zn- and Cd-preparations of HpCdMcMT and HpCdPlMT totally confirmed that these two mutants exhibit equivalent folds when coordinating Zn(II) ions, and also when coordinating Cd(II) ions, which are also practically indistinguishable from those of the respective wild-type HpCdMT complexes (Figure 3B). For example, the Zn-MT complexes show the typical Gaussian band centred at *ca.* 240 nm, while the Cd-MT species display the exciton coupling envelop at *ca.* 250 nm characteristic of the Zn- and Cd-thiolate chromophores.

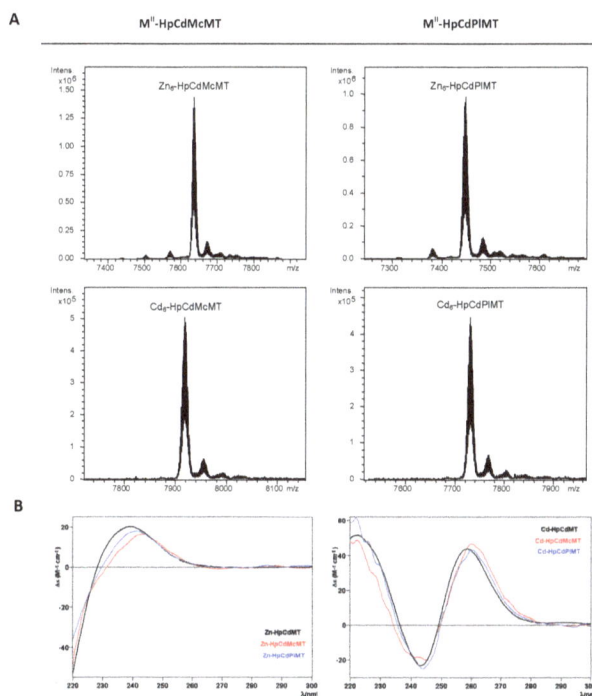

Figure 3. Analysis of the Zn- and Cd-HpCdMcMT and -HpCdPlMT complexes. (**A**) Deconvoluted ESI-MS spectra of the recombinant preparations of HpCdMcMT and HpCdPlMT purified from Zn- and Cd-supplemented cultures, analyzed at neutral pH (7.0); (**B**) CD spectra of the corresponding recombinant preparations. For comparative purposes, the CD spectra of the recombinant preparations yielded by the wild-type HpCdMT protein [11] have been included.

Table 1. Analytical characterization of the recombinant Zn(II)- and Cd(II)-complexes of the HpCdMT mutants studied in this work. For comparative purposes, data for the recombinantly-synthesized wild-type HpCdMT are included [11].

MT	ICP-AES [a]	Neutral ESI-MS [b]	Experimental MM [c]	Calculated MM [d]
HpCdMT [11]	5.8	Zn_6-MT	7005	7005.6
HpCdMcMT	5.9	Zn_6-MT	7635	7635.6
HpCdPlMT	6.0	Zn_6-MT	7448	7448.3
HpCdMT [11]	6.2	Cd_6-MT	7287	7287.8
HpCdMcMT	6.1	Cd_6-MT	7917	7917.7
HpCdPlMT	6.6	Cd_6-MT	7730	7730.4

[a] Zn(II)-to-peptide ratio calculated from S and Zn content (ICP-AES data); [b] The deduced Zn(II)-species were calculated from the mass difference between the holo- and apo-peptides; [c] experimental molecular masses corresponding to the detected M(II)-complexes. The corresponding ESI-MS spectra are shown in Figure 3; [d] theoretical molecular masses corresponding to the M(II)-complexes.

Furthermore, the Zn^{2+}/Cd^{2+} displacement process in Zn_6-HpCdMcMT and Zn_6-HpCdPlMT was a straight reaction that exclusively yielded Cd_6-complexes after the addition of 6 Cd^{2+} eq (Figure 4A,B, respectively), in agreement with the behavior in wild-type Zn_6-HpCdMT [15]. Most significantly, not only the final step of this reaction was the same, and also equivalent to the respective recombinant Cd-complexes, but CD spectra recorded at progressive stages of the reaction revealed identical profiles (*cf.* Figure 4). These basically consist in the evolution of the initial Gaussian band at *ca.* 240 nm characteristic of the Zn-complexes to the exciton-coupling signal centered at *ca.* 250 nm, typical of Cd-complexes. This suggests that the Zn(II)/Cd(II) substitution proceeds in an almost parallel way in all three cases, *i.e.*, for the wild-type HpCdMT and for the two mutant forms.

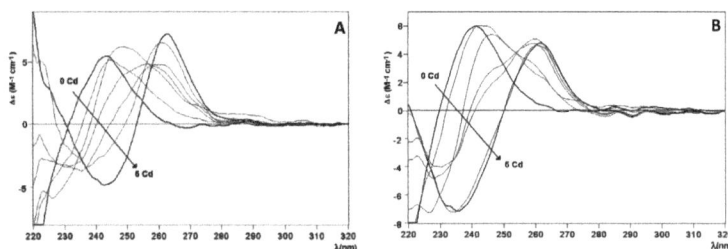

Figure 4. Zn(II)/Cd(II) replacement reaction of the Zn-HpCdMcMT and Zn-HpCdPlMT complexes. (**A**) CD spectra of a 10 µM solution of the Zn-HpCdMcMT sample titrated with CdCl$_2$ at neutral pH up to six Cd(II) equivalent; (**B**) CD spectra of a 10 µM solution of the Zn-HpCdPlMT sample titrated with CdCl$_2$ at neutral pH, up to six Cd(II) equivalent.

Therefore, it can be concluded that the different composition of the amino acid sequence linking the two putative domains of the HpCdMT proteins does affect neither the stoichiometry nor the folding of the complexes formed when coordinating divalent metal ions. To conclude, the linker length and compositions do not seem to alter the binding behavior towards a cognate metal ion.

2.3. Cu Binding Abilities of HpCdMcMT and HpCdPlMT

Before going into the details of the Cu(I) binding analysis of the two mutant constructs HpCdMcMT and HpCdPlMT, it is worth remembering that the data of the previous Cu(I) binding study performed with the wild-type HpCdMT form already exhibited a high degree of complexity, typical of the recombinant samples obtained when synthesizing a MT protein (here HpCdMT) in the presence of its non-cognate metal ion (here Cu(I)) [15]. As previously described [26], we perform two types of Cu-supplemented productions: at standard and at low aeration conditions. This responds to the known influence of culture oxygenation on the amount of internal copper in bacteria, which determines the composition of the final Cu-species. But unfortunately, several efforts to purify HpCdMcMT and HpCdPlMT from *E. coli* cultures grown at low oxygen conditions were not successful. Contrarily, the synthesis of both polypeptides performed at regular oxygen conditions yielded preparations that allowed their analysis by ESI-MS and CD, and facilitated the comparison of all their features with those of the complexes obtained from HpCdMT.

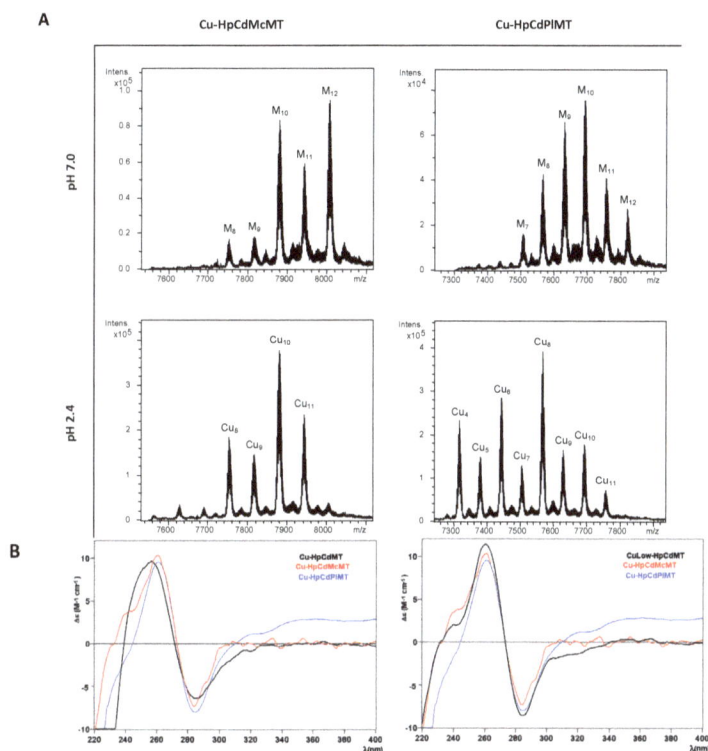

Figure 5. Analysis of the Cu-HpCdMcMT and Cu-HpCdPlMT complexes. (**A**) Deconvoluted ESI-MS spectra of the recombinant preparations of HpCdMcMT and HpCdPlMT purified from Cu-supplemented cultures, analyzed at neutral pH (7.0); (**B**) CD spectra of the same recombinant preparations. For comparative purposes, the CD spectra of the corresponding recombinant preparations yielded by the wild-type HpCdMT protein [15] have been included.

The first noticeable observation was that the composition of the Cu-HpCdMcMT and Cu-HpCdPlMT purified samples was significantly different, and in turn different from that of the wild-type Cu-HpCdMT preparations [15]. For example, the ESI-MS spectra of the Cu-HpCdMcMT sample recorded at neutral pH exhibited two clear major peaks corresponding to M_{12}- and M_{10}-HpCdMcMT complexes, an intermediate M_{11}-HpCdMT and very minor M_9- and M_8-HpCdMcMT species (Figure 5A, Table 2), where M can be either Zn(II) or Cu(I) because the similarity of their atomic masses. When this same preparation was analyzed at pH 2.4—which allows the release of all bound Zn(II) but not of Cu(I) [27,28]—one major peak corresponding to homonuclear Cu_{10}-HpCdMT together with minor peaks of Cu_{11}-, Cu_8- and Cu_9-complexes were observed (Figure 5A, Table 2). Taking into consideration that the ICP-AES analyses of the purified Cu-HpCdMcMT sample yielded an average metal content of 1.32 Zn and 12.25 Cu per MT, the most straightforward explanation for these results is to assume the presence of a mixture of major homonuclear complexes (Cu_{11}-, Cu_{10}-, Cu_9- and Cu_8-HpCdMcMT) together with some minor amounts of heteronuclear CuZn-HpCdMcMT species, mainly accounting for the M_{12}-HpCdMcMT species. On the other hand, a slightly poorer Cu-binding capacity can be attributed to HpCdPlMT for the following reasons: First, the ICP-AES analysis indicated a Zn:Cu mean content in the purified preparation of 2.17 Zn:7.87 Cu, corresponding to higher Zn:Cu ratio than in the HpCdMcMT sample. Second, the Cu-HpCdPlMT sample presents a more complex mixture, in terms of the number of ESI-MS detected species, both at neutral and acid pH. Thus, the ESI-MS spectrum at pH 7.0 allows the detection of a major M_{10}-HpCdPlMT peak, followed by M_9-, M_{11}-, M_8-, M_{12}-, and M_7-HpCdPlMT species, in decreasing order of intensity. This sample exhibited a mixture of major Cu_8-, and then minor Cu_6-, Cu_4-, and all the rest of peaks between Cu_4- and Cu_{11}-HpCdPlMT when analyzed at acidic ESI-MS, which would point to a continuum of heterometallic ZnCu-HpCdPlMT species (Figure 5A, Table 2).

Table 2. Analytical characterization of the recombinant complexes obtained from Cu supplemented cultures of the HpCdMT mutants studied in this work. All data for the two mutant proteins correspond to normal aerated cultures, since no complexes could be recovered from low aeration conditions. For comparative purposes, data for the wild-type HpCdMT, recombinantly-synthesized in Cu-supplemented cultures grown at both aeration conditions, are included [15].

MT	ICP-AES [a]	Neutral ESI-MS [b]	MM_{Exp} [c]	MM_{Theor} [d]	Acidic ESI-MS [b]	MM_{Exp} [c]	MM_{Theor} [d]
HpCdMT Normal aeration [15]	2.6 Zn 1.9 Cu	**M_5-MT**	6932	6938.1	**apo-MT**	6623	6625.5
		M_4-MT	6867	6875.5	Cu_4-MT	6873	6875.5
		M_6-MT	6996	7000.6	Cu_5-MT	6930	6938.1
		M_7-MT	7062	7063.2			
		M_8-MT	7124	7125.7			
HpCdMT Low aeration [15]	0.8 Zn 8.3 Cu	**M_{10}-MT**	7251	7250.8	**Cu_8-MT**	7120	7,125.7
		M_{11}-MT	7313	7313.4	Cu_{10}-MT	7248	7,250.8
		M_{12}-MT	7379	7375.9	Cu_{11}-MT	7312	7313.4
		M_8-MT	7122	7125.9	Cu_9-MT	7184	7188.3
		M_9-MT	7182	7188.3	Cu_5-MT	6929	6938.1
HpCdMcMT	1.3 Zn 12.3 Cu	M_{12}-MT	8006	8005.8	**Cu_{10}-MT**	7879	7880.7
		M_{10}-MT	7878	7880.7	Cu_{11}-MT	7942	7943.3
		M_{11}-MT	7942	7943.3	Cu_8-MT	7753	7755.6
		M_9-MT	7816	7818.2	Cu_9-MT	7816	7818.2
		M_8-MT	7753	7755.6			
HpCdPlMT	2.2 Zn 7.9 Cu				**Cu_8-MT**	7566	7568.3
		M_{10}-MT	7692	7693.4	Cu_6-MT	7442	7443.2
		M_9-MT	7631	7630.9	Cu_4-MT	7316	7318.1
		M_8-MT	7566	7568.3	Cu_{10}-MT	7692	7693.4
		M_{11}-MT	7755	7755.9	Cu_9-MT	7629	7630.9
		M_{12}-MT	7818	7818.5	Cu_5-MT	7377	7380.7
		M_7-MT	7505	7505.8	Cu_7-MT	7503	7505.8
					Cu_{11}-MT	7755	7755.9

[a] Zn(II) and Cu(I)-to-peptide ratio calculated from S, Zn, and Cu content (ICP-AES data); [b] the deduced species (M = Zn or Cu) were calculated from the mass difference between the holo- and the respective apo-peptides. The major species are indicated in bold, and the rest are in decreasing order of ESI-MS peak intensity; [c] experimental molecular masses corresponding to the detected complexes. The corresponding ESI-MS spectra are shown in Figure 5; [d] theoretical molecular masses corresponding to the metal-complexes.

Finally, comparison of these results with those from the wild-type HpCdMT protein also differed considerably (included in Table 2 for comparative purposes). Synthesis of Cu-HpCdMT yielded very poor results when performed in normally aerated cultures [15], only allowing recognizing major M_5-HpCdMT and minor M_4- to M_8-complexes that contained Cu_4- or Cu_5-cores. In fact, the results for the two mutant constructs analyzed in this work resembled far more those of Cu-HpCdMT produced in low aerated cultures, this means in the presence of higher intracellular copper levels, because in that case also major M_{10}-HpCdMT, together with minor M_8- to M_{12}-complexes were identified by neutral

ESI-MS, which were composed of major Cu$_8$-cores [15]. However, despite the diversity of results about the species composition yielded from the three HpCdMT proteins when coordinating Cu(I), they all gave rise to comparable CD signals, both in shape and intensity, which display the typical envelope of the Cu-MT complexes with absorption maxima at *ca.* 260 nm and minima at *ca.* 280 nm. The CD fingerprint of the Cu-HpCdMcMT and Cu-HpCdPlMT samples was more similar to that of the Cu-HpCdMT synthesized at low aeration [15] (Figure 5B), in concordance with the above-commented species composition.

Therefore, it can be unambiguously concluded that the inter-domain linker features (length and composition) greatly influences the Cu(I) binding behavior of the HpCdMT protein, so that the capacity to bind Cu(I) for the mutant HpCdMcMT and HpCdPlMT constructs, with eight amino acid-long linkers between the ninth and tenth Cys in their sequence, appears considerably enhanced in comparison with HpCdMT whose linker contains only two residues between the homologous Cys residues (*cf.* Figure 1). However, the differences in Cu-binding capabilities between HpCdMcMT and HpCdPlMT reveals that not only the length, but also the composition of the linker is important. Hence, the linker of the HpCdMcMT construct, containing two lysine, one glutamic acid and three threonine residues, seems to favor the formation of higher nuclearity homonuclear Cu(I) complexes, over the highly apolar HpCdPlMT linker, that is composed of only one arginine and two serine residues and that also contains five apolar amino acids (four alanine and a glycine).

3. Experimental Section

3.1. Construction and Cloning of the cDNAs Encoding the HpCdMT Linker Mutants

Two HpCdMT mutants were designed to replace the only two residues (-KT-) that act as a linker between the two nine-Cys domains in the HpCdMT sequence with the corresponding linkers from the (giant keyhole limpet) *M. crenulata* MT [24] (-VKTEAKTT-; HpCdMcMT protein) or from wheat (*T. aestivum*) Ec-1 MT [25] (-SARSGAAA-; HpCdPlMT protein) (Figure 1). The cDNAs encoding for these sequences were designed on the basis of the flanking regions in wild-type HpCdMT cDNA, and to encode the new linkers in the corresponding limpet or wheat cDNAs. Additionally, the restriction sites for *Bam*HI and *Xho*I were added to the 5′ and 3′ ends respectively for cloning purposes (Supplementary Materials, Figure S1). The HpCdMcMT and HpCdPlMT coding sequences designed this way were purchased as synthetic DNAs from ID & Conda Labs (Spain). After PCR amplification (35 cycles: 95 °C 30 s, 50 °C 30 s, and 72 °C 30 s, using Expand High Fidelity (Roche) thermostable DNA polymerase) of the synthetic cDNAs using the flanking primers: 5′-TTTATTGGATCCGGTAAGGG-3′ (HpCdMcMT forward); 5′-TTTCTCGAGTTACTTACAGG-3′ (HpCdMcMT reverse); 5′-TTTATTGGATCCGGCAAAGGG-3′ (HpCdPlMT forward); and 5′-TTTCTCGAGTTATTTGCAAG-3′ (HpCdPlMT reverse). They were subsequently digested with *Bam*HI and *Xho*I restriction enzymes, and the resulting products were ligated in-frame (DNA ligation kit, Takara Bio, Kusatsu, Shiga, Japan) in the pGEX-4T-1 (Amersham-GE Healthcare Europe, Cerdanyola del Valles, Spain) *E. coli* expression vector, which yields GST-fusion proteins. DNA sequencing allow confirming all the DNA constructs (ABIPRISM 310, Applied Biosystems, Foster City, CA, USA), using BigDye Terminator. The *E. coli* MachI strain was used for cloning and sequencing. The expression plasmids were then transformed into the protease-deficient strain BL21 for protein synthesis. The construction of the pGEX plasmid encoding for the wild type HpCdMT isoform has been previously described [11].

3.2. Synthesis and Purification of the Recombinant Zn-, Cd-, and Cu-Complexes of the HpCdMT Linker Mutants

All the purifications of the metal-MT complexes were carried out as reported in [29] for the wild-type HpCdMT isoform, which ensured full comparative results. Hence, the GST-HpCdMT fusions were produced in 5-L cultures (Luria Bertani medium) of transformed *E. coli* BL21 bacteria. Induction of gene expression was achieved with 100 μM (final concentration) of isopropyl β-D-thiogalactopyranoside (IPTG). After 30 min of induction, 300 μM $ZnCl_2$ or 500 μM $CuSO_4$ (final concentration) were supplemented to the cultures, which grew for a further 2.5 h, for the synthesis of the respective metal complexes. The Cu-cultures were grown both under normal (1-L medium in a 2-L Erlenmeyer flask, at 250 rpm) and low oxygen conditions (1.5-L medium in a 2-L Erlenmeyer flask at 150 rpm). It is well known that the culture aeration determines the amount of intracellular copper in the host cells available for the newly synthesized MTs [26].

Harvesting and centrifugation of the grown cells renders a cell mass that, resuspended in ice-cold PBS (1.4 M NaCl, 27 mM KCl, 101 mM Na_2HPO_4, 18 mM KH_2PO_4) with 0.5% v/v β-mercaptoethanol, is disrupted by sonication (20 s pulses for 5 min). All solutions used were oxygen-purged by saturating them with pure-grade argon to prevent metal-MT oxidation. The suspension was centrifuged at $12,000 \times g$ for 30 min, and the incubation of the resulting supernatant (gentle agitation for 60 min at room temperature) with Glutathione-Sepharone 4B (GE Healthcare) allowed batch affinity purification of the GST-HpCdMT species. The MT portion was recovered after thrombin cleavage (10 μ per mg of fusion protein at 17 °C over-night). The solution containing the cleaved metal-MT complexes was concentrated by Centriprep Microcon 3 (Amicon, cut-off of 3 kDa, Merck-Millipore, Darmstadt, Germany) centrifugation. The final metal complexes were purified through FPLC size-exclusion chromatography in a Superdex75 column (GE Healthcare) equilibrated with 50 mM Tris-HCl, pH 7.0 and run at 0.8 mL·min^{-1}. Absorbances at 254 and 280 nm signaled the fractions to be collected and analyzed for protein content.

3.3. Cd(II) Replacement Reactions with the Zn(II)-HpCdMT Mutants

The so-called "*in vitro* complexes" were obtained by metal displacement reactions using the recombinant Zn-HpCdMT preparations, by adding several molar equivalents of Cd^{2+} ions from a standard solution. As described elsewhere [29], the titrations were performed at pH 7.0, and all assays were performed under an Ar atmosphere. The pH remained constant throughout all experiments without the addition of any extra buffers.

3.4. Spectroscopic Analyses (ICP-AES and CD) of the Metal Complexes Formed by the HpCdMT Linker Mutants

The sulfur and metal content of all the metal-MT samples was determined by Inductively Coupled Plasma Atomic Emission Spectroscopy (ICP-AES) in a Polyscan 61E (Thermo Jarrel Ash, Franklin, MA, USA) spectrometer, measuring S at 182.040 nm, Zn at 213.856 nm, and Cu at 324.803 nm. Conventional treatment [30] and incubation in 1 M HNO_3 at 65 °C for 10 min before measurements to avoid possible traces of labile sulfide anions [31] were used to obtain the protein concentration, by considering that all S atoms were provided by the MT peptides.

CD measurements were performed at 25 °C in a Jasco spectropolarimeter (Model J-715, JASCO, Groß-Umstadt, Germany) interfaced to a computer (J700 software, JASCO, Groß-Umstadt, Germany) by using Peltier PTC-351S equipment (TE Technology, Traverse City, MI, USA). An HP-8453 Diode array UV-VIS spectrophotometer (GIM, Ramsey, MN, USA) was used for the electronic absorption measurements. 1-cm capped quartz cuvettes were employed for spectra recording, and the dilution effects were corrected and processed using the GRAMS 32 software (Thermo Fisher Scientific, Waltham, MA, USA).

3.5. Electrospray Ionization Time-of-Flight Mass Spectrometry (ESI-TOF MS) of the Metal Complexes Obtained from the HpCdMT Linker Mutants

A Micro TOF-Q instrument (Bruker Daltonics, Bremen, Germany) interfaced with a Series 1200 HPLC Agilent pump and equipped with an autosampler, all of which controlled by the Compass Software, allowed MW determinations by Electrospray ionization time-of-flight mass spectrometry (ESI-TOF MS). ESI-L Low Concentration Tuning Mix (Agilent Technologies, Santa Clara, CA, USA) was used for calibration. The conditions for the Zn-MT complex analyses were the following: 20 µL of sample solution injected through a PEEK (polyether heteroketone) tubing (1.5 m–0.18 mm i.d.) at 40 µL·min^{-1}; capillary counter-electrode voltage 5 kV; desolvation temperature 90–110 °C; dry gas 6 L·min^{-1}; spectra collection range 800–2500 *m/z*. A 5:95 mixture of acetonitrile:ammonium acetate (15 mM, pH 7.0) was the carrier buffer.

The conditions for the Cu-MT complex analyses were: 20 µL of sample solution injected at 40 µL·min^{-1}; capillary counter-electrode voltage 3.5 kV; lens counter-electrode voltage 4 kV; dry temperature 80 °C; dry gas 6 L·min^{-1}; and a 10:90 acetonitrile:ammonium acetate 15 mM, pH 7.0 mixture as carrier. The apo-proteins and the Cu-complexes at acid pH were analyzed following the same conditions previously described, but using a liquid carrier consisting of a 5:95 acetonitrile:formic acid mixture at pH 2.4. This causes the release of Zn(II), but keeps Cu(I) bound to the peptides. Experimental mass values were calculated as described in [32], and the error associated with the measurements resulted to be always smaller than 0.1%.

4. Conclusions

In summary, in this study we investigated the influence of the polypeptide region linking the two nine-Cys moieties of the CdHpMT isoform, hypothesized to give rise to two independent domains when coordinating divalent metal ions, into the stoichiometry and folding of the corresponding complexes. This was performed by constructing two mutant CdHpMT proteins with longer linkers than the wild-type (eight instead of the two residues), one of them derived from another snail MT sequence: *M. crenulata*, with a clear polar character (HpCdMcMT); and the other derived from a plant MT (the wheat Ec-1 protein), of clear apolar composition (HpCdPlMT). Thereafter, the metal binding behavior of the two mutants was assessed not only for the CdHpMT cognate metal ions (*i.e.*, the divalent Zn(II) and Cd(II)), but also for the non-cognate monovalent Cu(I) ions. The results clearly show that both HpCdMcMT and HpCdPlMT form unique Zn$_6$- and Cd$_6$-complexes, of the same stoichiometry than HpCdMT, and with indistinguishable CD fingerprints. On the contrary, when synthesized in normally-aerated Cu-enriched cultures, the two mutants behave like the wild-type form only in the sense that they yielded a mixture of heterometallic species—expected from their character as non-Cu-thioneins—but otherwise they differed a lot, in comparison to each other as well as in comparison to the wild-type protein. Hence, HpCdMcMT (with a polar linker) yields major M$_{12}$-species (M = Zn or Cu), with Cu$_{10}$ and Cu$_{11}$ cores, and HpCdPlMT (with an apolar linker) yields major M$_{10}$-species (M = Zn or Cu), with a major Cu$_8$ core (M = Zn or Cu). Under the same culture conditions, HpCdMT was only capable of folding into major M$_5$-species (M = Zn or Cu), with a Cu$_4$ core, and only when synthesized at high intracellular Cu concentrations (low culture aeration), similarly to HpCdPlMT major M$_{10}$-species (M = Zn or Cu) with Cu$_8$ cores were observed. No Cu(I)-MT complexes were obtained for the mutants at low oxygenation conditions, a fact that is in agreement with the observation that Cu(I)-complexes formed by Cu-thioneins unfold at high Cu concentrations, as has been described for *Drosophila* MtnE [33] or even *M. crenulata* MT [24].

To summarize, variation of the linker (length and amino acid features) does not alter the divalent metal ion (Zn(II) or Cd(II)) binding behavior of HpCdMT, probably because of the independence of the two separate domains. In contrast, variation of the linker (length and sequence) has an influence on the Cu(I) binding behavior of this MT, so that the mutants with the elongated linker better bind Cu(I) ions compared to the wild-type form. Moreover, the more polar the linker, the higher the Cu-thionein character of the protein, as shown by the higher nuclearity and the higher Cu *vs.* Zn content of the heterometallic species resulting from biosynthesis in Cu-enriched cultures. Interestingly, the *C. aspersus* Cu-specific MT (CaCuMT), which has a four-amino acid linker, exhibits a better Cu-binding behavior than the orthologous HpCuMT isoform, which possesses a two-residue linker [34]. This fact is worth considering even if Cu(I) is the non-cognate metal ion for this isoform, because it can provide valuable information about the determinants of the Cu(I) binding capabilities of a great number of MTs with intermediate Zn- *vs.* Cu-thionein character [8].

Supplementary Materials: **Supplementary Materials:** Supplementary materials can be found at http://www.mdpi.com/1422-0067/17/1/6/s1.

Acknowledgments: Acknowledgments: This work was supported by the DACH (International Cooperation) Project ref. I-1482-N28 of the Austrian Science Fund (FWF) to RD and the Swiss National Science Foundation (SNF) to OZ. Authors from both Barcelona universities are members of the 2014SGR-423 Grup de Recerca de la Generalitat de Catalunya, and they are recipients of MINECO-FEDER grants (BIO2012-39682-C02-01 to SA), and (BIO2012-39682-C02-02 to MC). We thank the Centres Científics i Tecnològics (CCiT) de la Universitat de Barcelona (ICP-AES, DNA sequencing) and the Servei d'Anàlisi Química (SAQ) de la Universitat Autònoma de Barcelona (CD, UV-vis, ESI-MS) for allocating instrument time.

Author Contributions: Author Contributions: Oliver Zerbe, Reinhard Dallinger, Mercè Capdevila and Sílvia Atrian designed experiments and discussed the results. Elena Jiménez-Martí performed the cloning and recombinant synthesis of the analyzed proteins and Selene Gil-Moreno and Òscar Palacios performed their ESI-MS and CD characterization. Òscar Palacios, Mercè Capdevila and Sílvia Atrian prepared the manuscript.

Conflicts of Interest: Conflicts of Interest: The authors declare no conflict of interest.

References

1. Capdevila, M.; Bofill, R.; Palacios, O.; Atrian, S. State-of-the-art of metallothioneins at the beginning of the 21st century. *Coord. Chem. Rev.* **2012**, *256*, 46–62. [CrossRef]

2. Blindauer, C. Metallothioneins. In *RSC Metallobiology: Binding, Transport and Storage of Metal Ions in Biological Cells*; Maret, W., Wedd, A., Eds.; The Royal Society of Chemistry: Cambridge, UK, 2014; Volume 2, pp. 594–653.

3. Capdevila, M.; Atrian, S. Metallothionein protein evolution: A miniassay. *J. Biol. Inorg. Chem.* **2011**, *16*, 977–989. [CrossRef] [PubMed]

4. Kägi, J.H.R.; Kojima, Y. Chemistry and biochemistry of metallothionein. In *Metallothionein II*; Kägi, J.H.R., Kojima, Y., Eds.; Birkhäuser Verlag: Basel, Switzerland, 1987; pp. 25–61.

5. Binz, P.A.; Kägi, J.H.R. Metallothionein: Molecular evolution and classification. In *Metallothionein IV*; Klassen, C.D., Ed.; Birkhäuser Verlag: Basel, Switzerland, 1999; pp. 7–13.

6. Palacios, O.; Atrian, S.; Capdevila, M. Zn- and Cu-thioneins: A functional classification for metallothioneins? *J. Biol. Inorg. Chem.* **2011**, *16*, 991–1009. [CrossRef] [PubMed]

7. Valls, M.; Bofill, R.; González-Duarte, R.; González-Duarte, P.; Capdevila, M.; Atrian, S. A new insight into metallothionein classification and evolution. The *in vivo* and *in vitro* metal binding features of *Homarus americanus* recombinant MT. *J. Biol. Chem.* **2001**, *276*, 32835–32843. [CrossRef] [PubMed]

8. Bofill, R.; Capdevila, M.; Atrian, S. Independent metal-binding features of recombinant metallothioneins convergently draw a step gradation between Zn- and Cu-thioneins. *Metallomics* **2009**, *1*, 229–234. [CrossRef] [PubMed]

9. Dallinger, R.; Berger, B.; Hunziker, P.E.; Kägi, J.H.R. Metallothionein in snail Cd and Cu metabolism. *Nature* **1997**, *388*, 237–238. [CrossRef] [PubMed]

10. Hispard, F.; Schuler, D.; de Vaufleury, A.; Scheifler, R.; Badot, P.M.; Dallinger, R. Metal distribution and metallothionein induction after cadmium exposure in the terrestrial snail *Helix aspersa* (Gastropoda, Pulmonata). *Environ. Toxicol. Chem.* **2008**, *27*, 1533–1542. [CrossRef] [PubMed]

11. Palacios, O.; Pagani, A.; Pérez-Rafael, S.; Egg, M.; Höckner, M.; Brandstätter, A.; Capdevila, M.; Atrian, S.; Dallinger, R. Shaping mechanisms of metal specificity in a family of metazoan metallothioneins: Evolutionary differentiation of mollusc metallothioneins. *BMC Biol.* **2011**, *9*, 4. [CrossRef] [PubMed]

12. Chabicovsky, M.; Klepal, W.; Dallinger, R. Mechanisms of cadmium toxicity in terrestrial pulmonates: Programmed cell death and metallothionein overload. *Environ. Toxicol. Chem.* **2004**, *23*, 648–655. [CrossRef] [PubMed]

13. Chabicovsky, M.; Niederstaetter, H.; Thaler, R.; Hödl, E.; Parson, W.; Rossmanith, W.; Dallinger, R. Localisation and quantification of Cd- and Cu-specific metallothionein isoform mRNA in cells and organs of the terrestrial gastropod *Helix pomatia*. *Toxicol. Appl. Pharmacol.* **2003**, *190*, 25–36. [CrossRef]

14. Dallinger, R.; Chabicovsky, M.; Hödl, E.; Prem, C.; Hünziker, P.; Manzl, C. Copper in *Helix pomatia* (Gastropoda) is regulated by one single cell type: Differently responsive metal pools in rhogocytes. *Am. J. Physiol.* **2005**, *189*, R1185–R1195. [CrossRef] [PubMed]

15. Palacios, O.; Pérez-Rafael, S.; Pagani, A.; Dallinger, R.; Atrian, S.; Capdevila, M. Cognate and noncognate metal ion coordination in metal-specific metallothioneins: The *Helix pomatia* system as a model. *J. Biol. Inorg. Chem.* **2014**, *19*, 923–935. [CrossRef] [PubMed]

16. Braun, W.; Vasak, M.; Robbins, A.H.; Stout, C.D.; Wagner, G.; Kagi, J.H.; Wuthrich, K. Comparison of the NMR solution structure and the x-ray crystal structure of rat metallothionein-2. *Proc. Natl. Acad. Sci. USA* **1992**, *89*, 10124–10128. [CrossRef] [PubMed]

17. Ngu, T.T.; Lee, J.A.; Rushton, M.K.; Stillman, M.J. Arsenic metalation of seaweed *Fucus vesiculosus* metallothionein: The importance of the interdomain linker in metallothionein. *Biochemistry* **2009**, *48*, 8806–8816. [CrossRef] [PubMed]

18. Domenech, J.; Orihuela, R.; Mir, G.; Molinas, M.; Atrian, S.; Capdevila, M. The Cd^{2+}-binding abilities of recombinant *Quercus suber* metallothionein, QsMT: Bridging the gap between phytochelatins and metallothioneins. *J. Biol. Inorg. Chem.* **2007**, *12*, 867–882. [CrossRef] [PubMed]

19. Espart, A.; Gil-Moreno, S.; Palacios, P.; Capdevila, M.; Atrian, S. Understanding the internal architecture of long metallothioneins: 7-Cys building blocks in fungal (*C. neoformans*) MTs. *Mol. Microbiol.* **2015**. [CrossRef]

20. Dallinger, R.; Wang, Y.; Berger, B.; Mackay, E.A.; Kägi, J.H.R. Spectroscopic characterization of metallothionein from the terrestrial snail, *Helix pomatia*. *Eur. J. Biochem.* **2001**, *268*, 4126–4133. [CrossRef] [PubMed]

21. Zhu, Z.; de Rose, E.F.; Mullen, G.P.; Petering, D.H.; Shaw, C.F., III. Sequential proton resonance assignments and metal cluster topology of lobster metallothionein-1. *Biochemistry* **1994**, *33*, 8858–8865. [CrossRef] [PubMed]

22. Narula, S.S.; Brouwer, M.; Hua, Y.; Armitage, I.M. Three-dimensional structure of *Callinectes sapidus* metallothionein-1 determined by homonuclear and heteronuclear magnetic resonance spectroscopy. *Biochemistry* **1995**, *34*, 620–631. [CrossRef] [PubMed]

23. Bofill, R.; Orihuela, R.; Romagosa, M.; Domenech, J.; Atrian, S.; Capdevila, M. *C. elegans* metallothionein isoform specificity: Metal binding abilities and histidine role in CeMT1 and CeMT2. *FEBS J.* **2009**, *276*, 7040–7056. [CrossRef] [PubMed]

24. Perez-Rafael, S.; Mezger, A.; Lieb, B.; Dallinger, R.; Capdevila, M.; Palacios, O.; Atrian, S. The metal binding abilities of *Megathura crenulata* metallothionein (McMT) in the frame of gastropoda MTs. *J. Inorg. Biochem.* **2012**, *108*, 84–90. [CrossRef] [PubMed]

25. Peroza, E.A.; Schmucki, R.; Güntert, P.; Freisinger, E.; Zerbe, O. The β(E)-domain of wheat E(c)-1 metallothionein: A metal-binding domain with a distinctive structure. *J. Mol. Biol.* **2009**, *387*, 207–218. [CrossRef] [PubMed]

26. Pagani, A.; Villarreal, L.; Capdevila, M.; Atrian, S. The *Saccharomyces cerevisiae* Crs5 metallothionein metal-binding abilities and its role in the response to zinc overload. *Mol. Microbiol.* **2007**, *63*, 256–269. [CrossRef] [PubMed]

27. Orihuela, R.; Domenech, J.; Bofill, R.; You, C.; Mackay, E.A.; Kägi, J.H.R.; Capdevila, M.; Atrian, S. The metal-binding features of the recombinant mussel *Mytilus edulis* MT-10-IV metallothionein. *J. Biol. Inorg. Chem.* **2008**, *13*, 801–812. [CrossRef] [PubMed]

28. Palacios, O.; Espart, A.; Espín, J.; Ding, C.; Thiele, D.J.; Atrian, S.; Capdevila, M. Full characterization of the Cu-, Zn- and Cd-binding properties of CnMT1 and CnMT2, two metallothioneins of the pathogenic fungus *Cryptococcus neoformans* acting as virulence factors. *Metallomics* **2014**, *6*, 279–291. [CrossRef] [PubMed]

29. Capdevila, M.; Cols, N.; Romero-Isart, N.; Gonzalez-Duarte, R.; Atrian, S.; Gonzalez-Duarte, P. Recombinant synthesis of mouse Zn3-β and Zn4-α metallothionein 1 domains and characterization of their cadmium(II) binding capacity. *Cell. Mol. Life Sci.* **1997**, *53*, 681–688. [CrossRef] [PubMed]

30. Bongers, J.; Walton, C.D.; Richardson, D.E.; Bell, J.U. Micromolar protein concentrations and metalloprotein stoichiometries obtained by inductively coupled plasma. Atomic emission spectrometric determination of sulfur. *Anal. Chem.* **1988**, *60*, 2683–2686. [CrossRef] [PubMed]

31. Capdevila, M.; Domenech, J.; Pagani, A.; Tio, L.; Villarreal, L.; Atrian, S. Zn- and Cd-metallothionein recombinant species from the most diverse phyla may contain sulfide (S^{2-}) ligands. *Angew. Chem. Int. Ed. Engl.* **2005**, *44*, 4618–4622. [CrossRef] [PubMed]

32. Fabris, D.; Zaia, J.; Hathout, Y.; Fenselau, C. Retention of thiol protons in two classes of protein zinc ion coordination centers. *J. Am. Chem. Soc.* **1996**, *118*, 12242–12243. [CrossRef]

33. Perez-Rafael, S.; Kurz, A.; Guirola, M.; Capdevila, M.; Palacios, O.; Atrian, S. Is MtnE, the fifth *Drosophila metallothionein*, functionally distinct from the other members of this polymorphic protein family? *Metallomics* **2012**, *4*, 342–349. [CrossRef] [PubMed]

34. Perez-Rafael, S.; Monteiro, F.; Dallinger, R.; Atrian, S.; Palacios, O.; Capdevila, M. *Cantareus aspersus* metallothionein metal binding abilities: The unspecific CaCd/CuMT isoform provides hints about the metal preference determinants in metallothioneins. *Biochim. Biophys. Acta* **2014**, *1884*, 1694–1707. [CrossRef] [PubMed]

International Journal of
Molecular Sciences

MDPI

Article

The Construction and Characterization of Mitochondrial Ferritin Overexpressing Mice

Xin Li [1],[†], Peina Wang [1],[†], Qiong Wu [1], Lide Xie [2], Yanmei Cui [1], Haiyan Li [1], Peng Yu [1] and Yan-Zhong Chang [1],*

[1] Laboratory of Molecular Iron Metabolism, The Key Laboratory of Animal Physiology, Biochemistry and Molecular Biology of Hebei Province, College of Life Science, Hebei Normal University, Shijiazhuang 050024, China; 18333159079@163.com (X.L.); hbsdwpn@163.com (P.W.); 15933623497@163.com (Q.W.); 18330117356@163.com (Y.C.); lihaiyan606@163.com (H.L.); yupeng0311@hebtu.edu.cn (P.Y.)

[2] Department of Biomedical Engineering, Chengde Medical University, Chengde 067000, China; xielide65@163.com

* Correspondence: yzchang@hebtu.edu.cn; Tel.: +86-311-8078-6311

† These authors contributed equally to this work.

Received: 31 May 2017; Accepted: 10 July 2017; Published: 13 July 2017

Abstract: Mitochondrial ferritin (FtMt) is a H-ferritin-like protein which localizes to mitochondria. Previous studies have shown that this protein can protect mitochondria from iron-induced oxidative damage, while FtMt overexpression in cultured cells decreases cytosolic iron availability and protects against oxidative damage. To investigate the in vivo role of FtMt, we established FtMt overexpressing mice by pro-nucleus microinjection and examined the characteristics of the animals. We first confirmed that the protein levels of FtMt in the transgenic mice were increased compared to wild-type mice. Interestingly, we found no significant differences in the body weights or organ to body weight ratios between wild type and transgenic mice. To determine the effects of FtMt overexpression on baseline murine iron metabolism and hematological indices, we measured serum, heart, liver, spleen, kidney, testis, and brain iron concentrations, liver hepcidin expression and red blood cell parameters. There were no significant differences between wild type and transgenic mice. In conclusion, our results suggest that FtMt overexpressing mice have no significant defects and the overexpression of FtMt does not affect the regulation of iron metabolism significantly in transgenic mice.

Keywords: iron; mitochondrial ferritin; overexpression

---

1. Introduction

Iron is an essential trace element for cell metabolism. Many essential biochemical processes require iron, such as oxygen transport, DNA synthesis, and electron transport [1,2]. In the brain, iron homeostasis is tightly regulated. Dysregulation of brain iron homeostasis can lead to severe pathological changes in the nervous system [3,4]. Iron deficiency can (1) affect neurotransmitter synthesis and cause language and motion deficiencies during brain development, (2) trigger iron deficiency anemia [5], and (3) be the underlying disorder in other diseases [6,7]. In addition, iron overload can lead to the death of neurons and induce neurodegenerative diseases such as Parkinson's disease (PD) [8] and Alzheimer disease (AD) [9]. Excess iron, left unchecked, can catalyze the production of hydrogen peroxide (H_2O_2) and other damaging reactive oxygen species (ROS), including the highly reactive hydroxyl radical, through a Fenton reaction [10]. The formation of an unmanageable level of hydroxyl radicals generated as the result of ferrous iron accumulation can lead to a cascade of reactions, culminating in the destruction of cell structure, and leading to cell damage.

Mitochondrial ferritin (FtMt) is a mitochondria-localized iron storage protein encoded by an intron-less gene on chromosome 5q23.1 [11]. Levi et al. [11] have found that FtMt possesses high homology to H-ferritin with ferroxidase activity [12,13], which can help to store iron in the shell structure of FtMt. In contrast to H-ferritin, FtMt lacks an iron responsive element (IRE) consensus sequence for iron-dependent translational control. Thus, the translation of mitochondrial ferritin is not regulated by the IRE–IRP machinery, which generally controls cellular iron homeostasis [11,14,15]. FtMt is mainly expressed in cells with high oxygen consumption and high metabolic activity, such as spermatocytes, neurons, and cardiomyocytes [16]. Additionally, the mitochondrial demand for iron shows a significantly increasing trend as cellular respiration increases in cells. However, there is relatively low expression of FtMt in the liver and spleen, which store considerable amounts of iron in the form of cytosolic ferritin [14]. Studies in HeLa and H1299 cells overexpressing FtMt have shown that the potency of FtMt as a sink for iron is greater than that of cytoplasmic ferritin and that the sequestration of iron in overexpressed FtMt elicits a cellular iron-deficient phenotype [14,17,18].

FtMt may exert a protective role in mitochondria against iron-dependent oxidative damage. It has been found that in some neurodegenerative diseases, such as AD, PD, and Friedreich's ataxia, the expression of FtMt is induced [19,20]. Furthermore, our previous studies have shown that FtMt can decrease the cellular damage induced by 6-hydroxydopamine- (6-OHDA-) [21], and attenuate Aβ-induced neurotoxicity in PD pathogenesis [22,23].

At present, the specific mechanisms regulating FtMt in iron metabolism are not clear. To investigate the in vivo role of FtMt, we established and characterized FtMt-overexpressing mice. We examined baseline murine iron metabolism and hematological phenotype. Surprisingly, our data show no overt phenotypic differences between wild type and transgenic mice.

2. Results

2.1. Generation of Transgenic Mice

C57BL/6 FtMt overexpressing mice were established by pro-nucleus microinjection with the PiggyBac (PB) System by Cyagen Biosciences Inc. (Suzhou, China). A mouse *Ftmt*-pcDNA3.1(−) construct was established as previously described [21]. Briefly, mouse *Ftmt* cDNA encoding the full *Ftmt* open reading frame was amplified by polymerase chain reaction using 5′-AGGGAATTC ACCATGGGCCTGTCCTGCTTTTGGTTCTTCTC-3′, and 5′-GGCGGATCCTATTTAAGCGTAATCTGG AACATCGTATGGGTAGTGCTTGCTCTCGCTTCCAA-3′ primers. The PCR product was inserted into the pGEM-T vector to obtain the plasmid containing a 768-bp fragment including the entire *Ftmt* sequence with a C-terminal hemagglutinin (HA) epitope and an *Eco*RI site at the 5′ end, and a *Bam*HI site downstream of the stop codon at the 3′ end. The 768-bp fragment was excised and sub-cloned into the *Eco*RI/*Bam*HI sites of pcDNA3.1(−) to obtain pcDNA3.1(−)-*Ftmt* [18] (Figure 1). The linearized and purified plasmid was diluted to 1 to 5 ng/μL before it was loaded into the microinjection needle. Morphologically normal zygotes were selected to perform the microinjection and then cultured in an incubator for 1 h. Next, the embryos were transplanted into pseudo-pregnant female mice where the development of embryo proceeded to term, twenty days following the procedure [24]. The pups were genotyped by PCR and positive male mice were viable (F$_0$). The foreign gene is usually incorporated into only one chromosome, so the founder mouse is heterozygous. Heterozygous offspring (F$_1$) of both sexes were obtained through founder male mouse crosses with wild-type female mice (Figure 1). We were able to obtain the homozygous mice by sib mating of the heterozygous ones. Litters were genotyped to identify male Tg mice to be used in experiments.

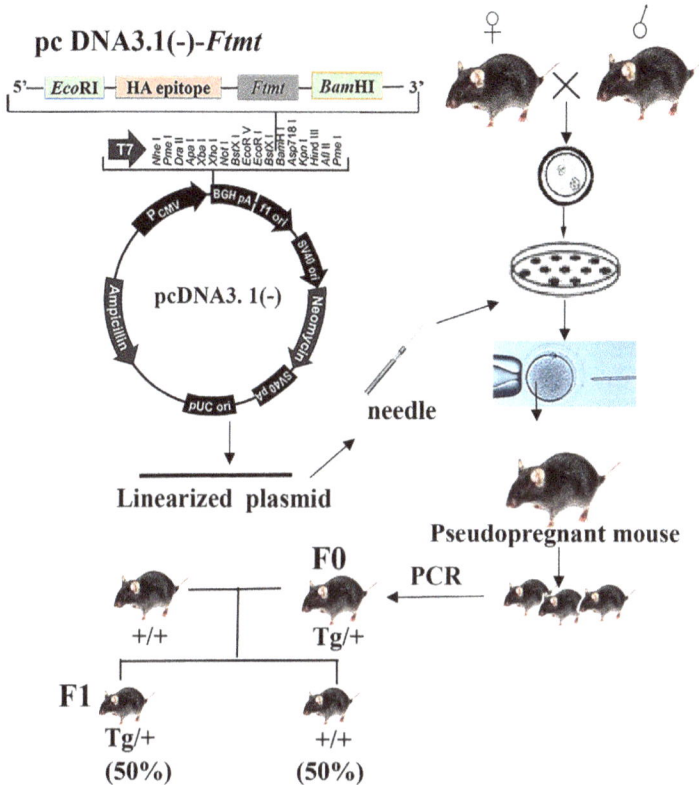

Figure 1. The construction of the transgenic mice. Tg/+, heterozygous; +/+, wild type.

2.2. Expression of FtMt is Increased in the Transgenic Mice

After PCR genotyping FtMt overexpressing mice (Figure 2a), we used western blot analysis to measure FtMt expression. As FtMt has high homology with H-Ferritin, the specificity of the FtMt antibody was confirmed by the detection of FtMt and H-Ferritin from testis with anti-FtMt and anti-H-Ferritin antibodies, respectively [25] (Figure 2b). Using the anti-FtMt antibody to decorate western blots, we observed an increased expression of FtMt in transgenic mice compared to wild-type mice in different regions of brain including cerebellum, cortex, and striatum. Although there was no significant difference in the hippocampus, the expression of FtMt showed a rising trend in this region. The expression of FtMt in the testis and heart also increased in the transgenic mice (Figure 2b–g). Protein extracted from FtMt-overexpressing SH-SY5Y cells was used as positive controls, while extracts from FtMt-deficient mice served as negative controls. Taken together, our results demonstrate the successful generation of FtMt-overexpressing mice.

2.3. FtMt Overexpressing Mice Have Normal Body and Organ Weights

We determined the body weights of sixteen wild type and seventeen transgenic mice and found that there were no significant differences between the two groups (Figure 3a). We also analyzed the organ to body weight ratios of the mice. The results showed that there were no significant differences in heart, liver, spleen, thymus, kidney, and testis (Figure 3b–g).

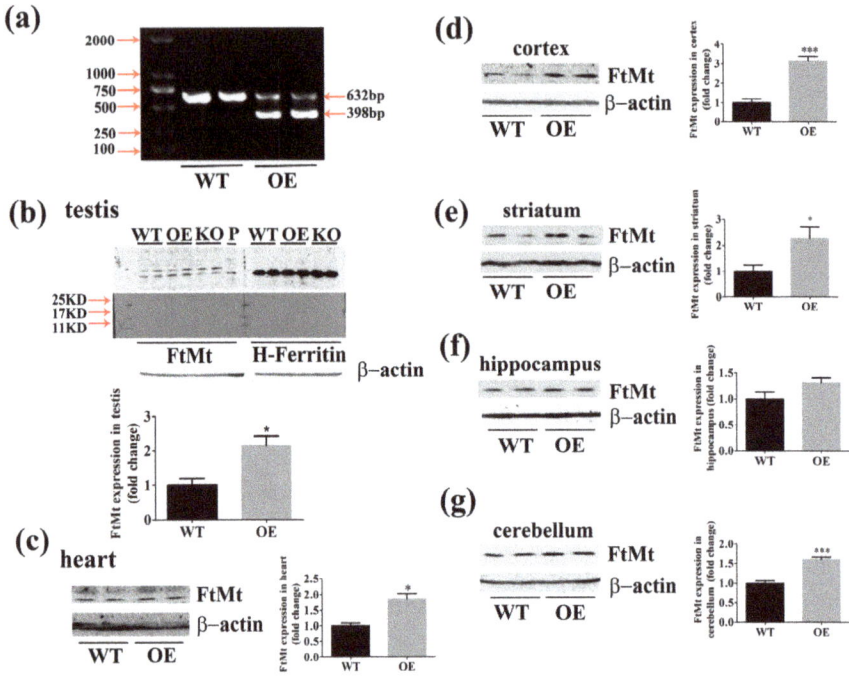

Figure 2. The expression of Mitochondrial ferritin (FtMt) in transgenic mice. (**a**) Genotyping agarose gel electrophoresis analysis of wild-type (WT) and overexpression (OE) mice. Transgenic mice were identified by the presence of 398 and 632 bp PCR products. (**b–g**) Western blot and subsequent densitometric analysis of FtMt in (**b**) testis, (**c**) heart, (**d**) cortex, (**e**) striatum, (**f**) hippocampus, (**g**) cerebellum. $n = 3$, * $p < 0.05$, *** $p < 0.001$ versus WT group.

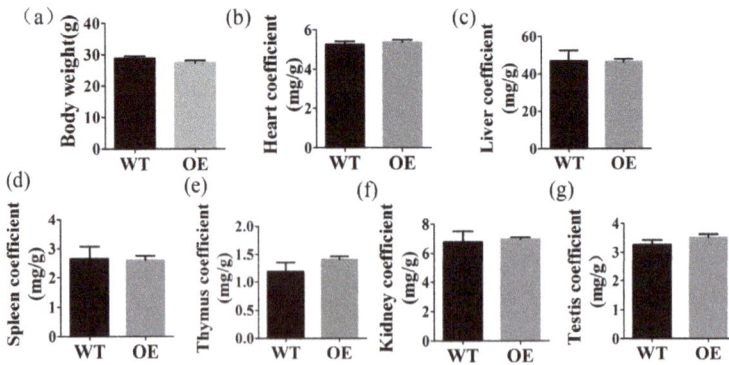

Figure 3. Body weights and organ to body weight ratios in transgenic mice. (**a**) The weights of age-matched (5-month-old) FtMt overexpressing ($n = 17$) and wild-type ($n = 16$) mice were examined. (**b–g**) The ratios of heart, liver, spleen, thymus, kidney, and testis weight to body weight ratios were determined in FtMt overexpressing ($n = 11$) and wild-type ($n = 10$) mice. The data are presented as the mean ± SEM.

2.4. FtMt Overexpressing Mice Exhibit Normal Hematological Parameters and Blood Pressure

To determine the effect of FtMt overexpression on murine hematology, we examined the red blood cell parameters in wild-type and transgenic mice. The parameters we analyzed were the following: red blood cell count (RBC, $\times 10^{12}$/L), hemoglobin (HGB, g/L), hematocrit (HCT, %), mean corpuscular hemoglobin (MCH, pg), mean corpuscular hemoglobin concentration (MCHC, g/L), mean corpuscular volume (MCV, fL) and coefficient variation of red blood cell volume distribution width (RDW-CV, %). We found no significant differences (Figure 4). We also measured the blood pressures of the two groups of mice. No significant differences were found in systolic or diastolic blood pressures and heart rates between the two groups.

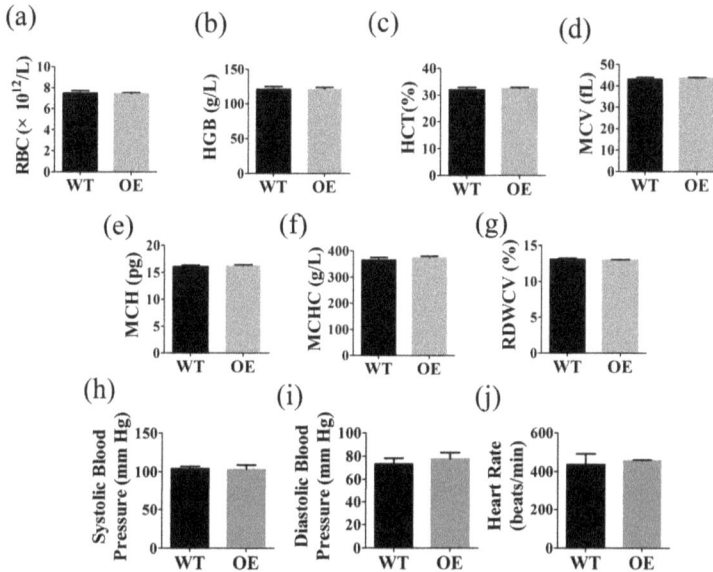

Figure 4. Hematological parameters and blood pressure in the wild type (WT) and transgenic mice (OE). 50 µL fresh blood was obtained from each mouse for hematological parameters measurement. Parameters measured were (**a**) Red blood cell (RBC) count, (**b**) Hemoglobin (HGB), (**c**) Hematocrit (HCT), (**d**) Mean corpuscular volume (MCV), (**e**) Mean corpuscular hemoglobin (MCH), (**f**) Mean corpuscular hemoglobin concentration (MCHC), and (**g**) Coefficient variation of red blood cell volume distribution width (RDWCV). Wild type mice, $n = 16$; transgenic mice, $n = 17$. Data of blood pressure in WT and OE mice: (**h**) Systolic blood pressure, (**i**) Diastolic blood pressure, (**j**) Heart rate, $n = 3$. The data are presented as the mean $\pm$ SEM.

2.5. Transgenic Mice Have Normal Iron Distribution

To determine the effect of FtMt overexpression on baseline murine iron metabolism, we measured the iron concentrations in the serum, heart, liver, spleen, kidney, testis, cortex, hippocampus, striatum and cerebellum (Figure 5). We found no significant changes between wild-type and FtMt overexpressing mice. Additionally, the expression of hepcidin in the liver was unchanged between the two groups.

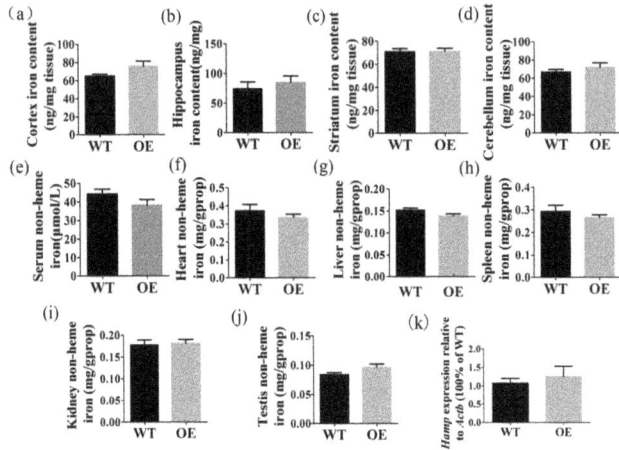

Figure 5. Iron distribution in transgenic mice. The total iron content in the (**a**) cortex, (**b**) hippocampus, (**c**) striatum, (**d**) cerebellum was measured by inductively coupled plasma mass spectrometry (ICP-MS). The total iron concentration in (**e**) serum, (**f**) heart, (**g**) liver, (**h**) spleen, (**i**) kidney, and (**j**) testis. (**k**) *Hamp* expression in liver. Six wild-type and six FtMt overexpression mice were analyzed. The data are presented as the mean ± SEM.

2.6. The Effects of FtMt Overexpression on the Levels of Ferritin, TfR1, FPN1, and DMT1

We next examined the levels of Ferritin, TfR1, FPN1, and DMT1 in testis of wild type and transgenic mice. The levels of L-Ferritin and H-Ferritin in FtMt overexpressing mice were decreased significantly (Figure 6a,c), and those of TfR1 increased (Figure 6d), compared to wild-type mice. There were no significant differences in the expression of FPN1, DMTI(+IRE), and DMTI(−IRE) between the two groups (Figure 6a–c).

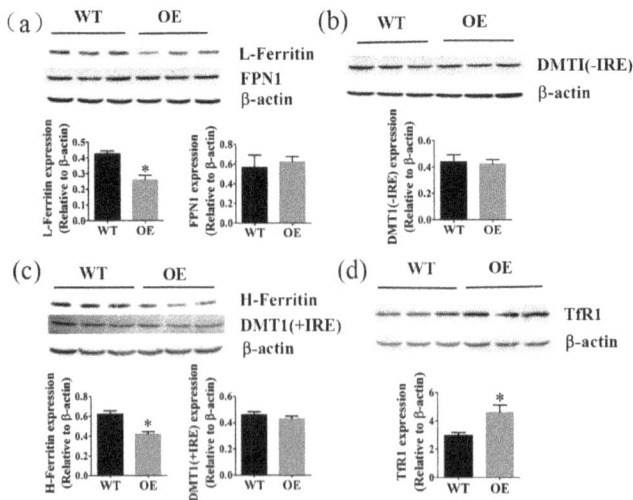

Figure 6. The effect of FtMt overexpression on Ferritin, TfR1, FPN1 and DMT1. Western blot analysis, and subsequent densitometry of (**a**) L-Ferritin, FPN1, (**b**) DMT1(−IRE), (**c**) H-Ferritin, DMT1(+IRE), (**d**) TfR1 in testis. The data are presented as the mean ± SEM, $n = 3$. * $p < 0.05$ versus WT group.

2.7. The Locomotor Activity Was Not Changed in FtMt Overexpressing Mice

Our data showed that FtMt overexpressing mice exhibited normal body weights, blood pressure and hematological parameters. The iron contents of brain, liver, testis, heart, kidney and spleen were also not significantly changed in transgenic mice. Was there any behavioral differences between the two types of mice? We tested the activities of the two groups of mice using a mouse autonomic activity tester. The locomotor activity counts and stand-up times of the mice in ten minutes were measured. The data showed that there were no significant differences between wild type and transgenic mice. (Figure 7).

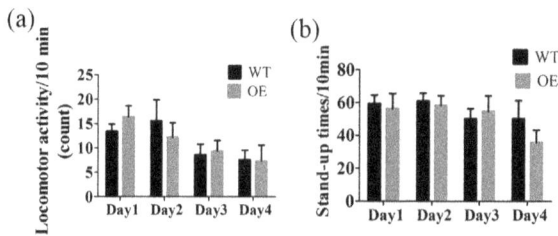

Figure 7. Locomotor activity of wild type (WT) and transgenic mice (OE). (**a**) Locomotor activity counts in ten minutes, (**b**) Stand-up times in ten minutes. The data are presented as the mean ± SEM, $n = 5$.

3. Discussion

As an essential trace element in biology, iron participates in numerous metabolic processes in humans. The metal plays an important role in oxygen transport, redox reactions, neurotransmitter synthesis, and many other fundamental metabolic processes in cells and organisms [26–28]. Though it is essential to mitochondrial and cellular function, iron can also be toxic when in excess through the Fenton reaction, which produces ROS. The labile plasma iron catalyzing this reaction is able to target cell membranes, a process that is considered the main culprit for iron-related cellular and organ damage [29–31]. Furthermore, excess iron is considered a contributing neurotoxic factor in several neurodegenerative disorders, including AD [32]. Thus, the levels of iron must be tightly controlled to prevent cellular damage in the brain and maintain normal functions throughout the body.

Mitochondria is an essential and dynamic component of cellular biochemistry. Numerous studies have shown these organelles to be vital to iron metabolism, not only for heme synthesis but also for the biogenesis of [Fe-S] clusters [33]. FtMt, an H-ferritin-like protein targeted to mitochondria, has been shown to protect mitochondria from iron-induced oxidative damage in cells with high metabolic activity and oxygen consumption [23]. Overexpression of FtMt can modulate intracellular iron distribution to decrease cytosolic iron and protect cultured cells against oxidative damage [17,18,21]. Thus, a better understanding of the role of FtMt in mitochondrial iron homeostasis may provide new insights into the treatment of diseases associated with abnormal iron homeostasis.

In order to investigate the in vivo role of FtMt, we established a FtMt overexpressing mouse by pro-nucleus microinjection of pcDNA3.1(−)-*Ftmt* and characterized the animal model. Previous studies have shown that FtMt is highly expressed in the testis in mammals. It is also expressed in the kidney, heart, brain and thymus in mice [34]. There is a relationship between the high metabolic rate of tissues and FtMt expression [14]. Therefore, we determined the expression of FtMt in the heart, testis and four areas of the brain including the cerebellum, cortex, striatum and hippocampus. We found that the FtMt was significantly overexpressed in the heart, testis and the areas of the cortex, striatum, and cerebellum of brain. Interestingly, the increased level of FtMt in different regions of the brain is different, and there is especially no significant change in the hippocampus. It is probably due to the fact that the cell types from different regions are various, which induces different degrees of effect on FtMt overexpression. There may have been some mechanisms involved in the regulation of FtMt overexpression in disparate regions and tissue. Further study needs to be explored. In our study, FtMt overexpression did not affect the hematological indices, blood pressure and heart rates of the mice. There was an increased tendency of iron content in the testis, and a decreased tendency in serum iron, though there were no significant differences between these under baseline conditions. We tested the locomotor activities of the two groups, and we found no behavioral differences in our transgenic mice. In general, FtMt overexpressing mice are healthy and do not show any overt phenotypic differences from wild types under baseline feeding conditions at the age of five months.

We next considered whether overexpression of FtMt altered the iron metabolism in the transgenic mice by measuring the levels of Ferritin, TfR1, FPN1, and DMT1 in the testis, where FtMt is normally highly expressed. Ferritin is the major intracellular iron storage protein [35,36]. It plays an important role in iron metabolism, sequestering excess cytosolic iron and serving as a systemic iron storage depot. Ferritin is composed of two subunits, H and L chains. In ferritin shells, the H subunit mainly exerts an important protective function against oxidative damage because of the ferroxidase activity that converts soluble, catalytic ferrous ions into inert ferric hydroxides [37–39]. The L chain lacks the ferroxidase center, but it provides sites for iron nucleation and mineralization, facilitating efficient iron storage [18,40,41]. The Fe-binding protein, Transferrin (Tf) is the major vehicle for iron transport in the body with the help of transferrin receptor 1 (TfR1) [28,42]. And the iron enters cells in the form of receptor-mediated endocytosis of the Tf-TfR1-Fe complex. A reduction in endosomal PH mediates the release of Fe from Tf.

Here we showed that FtMt overexpressing mice, as compared to the wild type, had decreased L-Ferritin and H-Ferritin expression, and increased TfR1 expression. This phenotype was consistent with a deficiency of iron in the cytoplasm, which stimulated a decrease in the levels of L- and H-Ferritin translation in HeLa cells overexpressing human FtMt [17,18]. We suspected that the overexpression of FtMt might cause a redistribution of cellular iron from the cytoplasm to the mitochondria. The iron from the blood was then transported to the cytoplasm with the help of Tf-TfR1 to maintain the balance of iron in the cytoplasm. And there was an increased tendency of iron content in testis. There were no significant differences in the expression of FPN1, DMT1(+IRE) and DMT1(−IRE). In total, the overexpression of FtMt did not affect the regulation of iron metabolism significantly in the transgenic mice.

Our results showed that there were no significant differences in body weight or the ratio of organs to body weight in FtMt overexpressing mice. The changes to the hematological parameters, blood pressure and iron contents were not noticeable. Thus, in normal conditions, the overexpression of FtMt in transgenic mice has no obvious effects compared to wild type mice. Bartnikas et al. [43] found that there were no significant defects in mice lacking mitochondrial ferritin. Our previous studies [44] showed that FtMt disruption decreased the exhaustion exercise time and altered heart morphology with severe cardiac mitochondrial injury. The absence of FtMt increased the sensitivity of mitochondria to cardiac injury via oxidative stress. It suggests that FtMt has a protective role regarding cell damage when responding to oxidative stress. Therefore, we speculate that FtMt overexpression may play a very important role when one suffers from the risk.

In conclusion, there were no pronounced differences between the wild type and FtMt overexpressing mice. The transgenic mice did not exhibit abnormalities, and iron metabolism was also not significantly influenced Generating FtMt overexpressing mice can supply a new model for us to explore the function of mitochondrial ferritin.

4. Materials and Methods

4.1. Animals

Mice were kept in accordance with the National Institutes of Health Guide for the Care and Use of Laboratory Animals, and by approval of the Animal Care and Use Committee of the Hebei Science and Technical Bureau, and the Laboratory Animal Ethical and Welfare Committee (AEWC) of Hebei

Normal University (8 March 2015, Number: 2015-003). Mice were housed in stainless steel cages at $21 \pm 2\,°C$ and provided free access to food and water. The rooms maintained a 12 h light and 12 h dark cycle [45]. Age-matched wild-type male mice and FtMt overexpressing male mice (5 months old) were used in this study.

The primers that we used to identify the genotype of mice were as follows:

Transgene PCR primer forward 5'-CCCACTGCTTACTGGCTTATCGAA-3', and reverse 5'-TAC ACGTAGGATGCGTAAAGCTC-3'.

Internal control PCR primer forward 5'-CAACCACTTACAAGAGACCCGTA-3', and reverse 5'-GAGCCCTTAGAAATAACGTTCACC-3'. The internal control PCR targets the endogenous mouse Rgs7 (G protein signaling 7) locus, which exerts the role as internal primer to exclude the influence of the mouse itself.

The mice were anesthetized with pentobarbital sodium (40 mg/kg) and then were perfused with 0.9% saline. The body weights of the mice were measured before they were killed. The peripheral tissues (heart, liver, spleen, kidney, testis, thymus) were separated and washed in normal saline solution, dried with sterilized filter paper, and then weighed to calculate the organ coefficient. The tissues were later used to measure iron concentration. The brains of the mice were dissected into cerebral cortex, hippocampus, and striatum, and used for western blot and iron concentration analyses.

4.2. RNA Isolation and Real-Time PCR

Liver hepcidin expression was measured by real-time PCR as previously described [44]. Total RNA was extracted from the liver with a TRIzol reagent (Ambion, Carlsbad, CA, USA) and 2 µg of the total RNA was reverse transcribed in a 20 µL reaction using TransScript One-Step gDNA Removal and cDNA Synthesis SuperMix (Transgen Biotech, Beijing, China) according to the manufacturer's instructions. Control reactions without reverse transcription were performed to ensure that PCR products did not represent an amplification of genomic DNA. The solution was diluted to 100 µL, and 4 µL cDNA was then used as the template for real-time PCR with SYBRGreen (KangWei, Beijing, China). PCR amplification was performed with the BIO-RAD CFX Connect Real-Time System (Hercules, CA, USA) with the following parameters: $95\,°C$ for 10 min, followed by $95\,°C$ for 15 s, $60\,°C$ for 1 min; 40 cycles. Each sample was repeated three times and the data were averaged. The primer sequences used for the PCR reaction were as follows:

Mouse hepcidin forward: 5'-AGACATTGCGATACCAATGCA-3', and reverse: 5'-GCAACA GATACCACACTGGGAA-3', β-actin forward: 5'-AGGCCCAGAGCAAGAGAGGTA-3', and reverse: 5'-TCTCCATGTCGTCCCAGTTG-3'.

4.3. Hematological Analyses

Blood samples (50 µL) were collected from each mouse after anaesthetizing with 8% pentobarbital sodium. An anticoagulant was used to prevent blood clotting. Quantitative determinations of blood indices were performed using a BM830 automatic blood cell analyzer (Bao Ling Man Technology Company, Beijing, China).

4.4. Blood Pressure and Heart Rate Measurement

Blood pressure and heart rate were measured in conscious animals using the tail-cuff method (CODA-2, Kent Scientific, Torrington, CT, USA). Briefly, mice were acclimatized for 6 consecutive days by measuring blood pressure and heart rate before the final results were recorded and in each session 15 consecutive readings were recorded. On the data collection day, the average of readings were used for sytolic pressure, diastolic pressure, and also heart rate.

4.5. Measurement of Serum and Tissue Iron

The total iron concentration of serum and tissues (heart, liver, spleen, kidney, testis) was measured using Serum Iron Assay Kit and Tissue Iron Assay Kit from the Nanjing Jiancheng Bioengineering

Institute (Nanjing, China) according to the manufacturer's instructions. The serum was obtained by retro orbital bleeding, and the supernatant was collected after centrifugation at $2000 \times g$ for 20 min at 4 °C.

4.6. Measurement of Brain Iron

The total iron content of the brain regions (cerebellum, cortex, hippocampus, and striatum) was measured by inductively coupled plasma mass spectrometry (ICP-MS) as previously described [46]. The tissues used for the experiment were incubated for at least 12 h at 106 °C to dry tissues [47]. Before the experiment, the Teflon digestion tubes were washed with tap water, then double distilled water, and finally deionized water. The tubes were then soaked in 15% nitric acid for 24 h, washed with deionized water and then rinsed with ultrapure water. Approximately 6 mg sample was added to 1.5 mL ultrapure nitric acid (69.9–70.0%; J.T. Baker, Phillipsburg, NJ, USA), and then digested using a microwave digestion system for 2 h at 100°C and then 4 h at 200 °C. The completely digested samples were diluted to 2.5 mL with deionized water. Standard curves ranging from 0 to 100 ppb were prepared by diluting an iron standard (1 mg iron/mL) with blanks prepared from homogenization reagents in 0.2% nitric acid [48].

4.7. Western Blot Analysis

Protein expression was assessed by western blot as previously described [49]. The tissues were homogenized in a RIPA buffer containing 1% NP40 and protease inhibitor cocktail tablets (Roche Diagnostics GmbH, Roche Applied Science, 68298 Mannheim, Germany). The supernatant was collected after centrifugation at $12,000 \times g$ for 20 min at 4 °C, and the protein content was measured using a Bicinchoninic acid (BCA) Protein Quantification Kit (Yeasen Biotechnology, Shanghai, China). The extract, containing 40 µg of protein, was diluted in a 2× sample buffer, and heated for 5 min to denature the protein. The samples were resolved by SDS-PAGE, and then transferred to nitrocellulose membranes. The blots were blocked in 5% nonfat milk containing 20 mM Tris-buffered saline solution (pH 7.6, 137 mM NaCl, and 0.05% Tween-20; TBS-T) for 1.5 h at room temperature, followed by incubation with FtMt (Abcam Inc., San Francisco, CA, USA), TfR1 (ThermoFisher, Waltham, MA, USA), L-Ferritin, H-Ferritin (Abcam Inc., USA), DMT1(+IRE), DMT1(−IRE) (Alpha Diagnostic Intl Inc., San Antonio, TA, USA) or β-actin (Alpha Diagnostic Intl Inc.) primary antibody overnight at 4 °C. After four washes with TBS-T, the membrane was incubated with an anti-rabbit (Ruiying Bio, Suzhou, China) or anti-mouse (Ruiying Bio, Suzhou, China) horseradish peroxidase-conjugated secondary antibody for 90 min at room temperature. The specific proteins were detected by enhanced chemiluminescence (ECL), and analyzed with ImageQuant (Fujifilm LAS4000, Tokyo, Japan). The relative band intensities of the proteins are presented in comparison to that of β-actin [46].

4.8. Locomotor Activity Test

The locomotor activity was tested by a zz-6 mouse autonomic activity tester (Chengdu Technology & Marker CO. LTD, Sichuan, China). Locomotor activity counts and the stand-up times of mice were tested by infrared array beam. The mice were placed in the testing chamber (330 mm × 100 mm × 110 mm) to adapt to the environment before the experiment. Five mice from each group were tested for ten minutes per day, for four days. A quiet environment was required during the experiment.

4.9. Statistical Analysis

All data are presented as the mean ± SEM. The differences between means was determined by one-way ANOVA, followed by post hoc Tukey tests corrected for multiple comparisons. Differences were considered significant if $p < 0.05$. All tests were performed using SPSS 21.0 (IBM SPSS21.0, Armonk, New York, NY, USA).

Int. J. Mol. Sci. **2017**, *18*, 1518

5. Conclusions

Our current study shows that FtMt overexpressing mice do not have an overt phenotype and that iron metabolism is not significantly influenced.

Acknowledgments: This work was supported by the National Science Foundation of China (31520103908, 31471035 and 31271473).

Author Contributions: Xin Li and Peina Wang performed the experiments, and contributed equally to this work; Yan-Zhong Chang and Qiong Wu participated in the design of the experiments; Xin Li and Peina Wang analyzed the data; Lide Xie, Yanmei Cui, Haiyan Li, and Peng Yu contributed materials/analysis tools; Yan-Zhong Chang revised the manuscript; Xin Li wrote the paper.

Conflicts of Interest: The authors declare no conflict of interest.

Abbreviations

FtMt	Mitochondrial ferritin
PD	Parkinson's disease
AD	Alzheimer disease
ROS	Reactive oxygen species
IRE	Iron responsive element
IRP	Iron regulation protein
SEM	Standard error of the mean
RBC	Red blood cell count
HGB	Hemoglobin
HCT	Hematocrit
MCH	Mean corpuscular hemoglobin
MCHC	Mean corpuscular hemoglobin concentration
MCV	Mean corpuscular volume
RDW-CV	Coefficient variation of red blood cell volume distribution width
ICP-MS	Inductively coupled plasma mass spectrometry
FPN	Ferroportin
TfR1	Transferrin receptor protein 1
DMT1	Divalent metal transporter 1

References

1. Aisen, P.; Enns, C.; Wessling-Resnick, M. Chemistry and biology of eukaryotic iron metabolism. *Int. J. Biochem. Cell Biol.* **2001**, *33*, 940–959. [CrossRef]
2. Hansen, J.B.; Moen, I.W.; Mandrup-Poulsen, T. Iron: The hard player in diabetes pathophysiology. *Acta Physiol.* **2014**, *210*, 717–732. [CrossRef] [PubMed]
3. Gao, G.; Chang, Y.Z. Mitochondrial ferritin in the regulation of brain iron homeostasis and neurodegenerative diseases. *Front. Pharmacol.* **2014**, *5*, 19. [CrossRef] [PubMed]
4. Zhao, N.; Enns, C.A. Iron transport machinery of human cells: Players and their interactions. *Curr. Top. Membr.* **2012**, *69*, 67–93. [PubMed]
5. Weiss, G.; Goodnough, L.T. Anemia of chronic disease. *N. Engl. J. Med.* **2005**, *352*, 1011–1023. [CrossRef] [PubMed]
6. Fraenkel, P.G. Understanding anemia of chronic disease. *Hematol. Am. Soc. Hematol. Educ. Program Book* **2015**, *2015*, 14–18. [CrossRef] [PubMed]
7. Cullis, J. Anaemia of chronic disease. *Clin. Med.* **2013**, *13*, 193–196. [CrossRef] [PubMed]
8. Lingor, P.; Carboni, E.; Koch, J.C. Alpha-synuclein and iron: Two keys unlocking Parkinson's disease. *J. Neural Transm.* **2017**. [CrossRef] [PubMed]
9. Wang, P.; Wang, Z.Y. Metal ions influx is a double edged sword for the pathogenesis of Alzheimer's disease. *Ageing Res. Rev.* **2017**, *35*, 265–290. [CrossRef] [PubMed]
10. Gil-Lozano, C.; Davila, A.F.; Losa-Adams, E.; Fairen, A.G.; Gago-Duport, L. Quantifying fenton reaction pathways driven by self-generated H_2O_2 on pyrite surfaces. *Sci. Rep.* **2017**, *7*, 43703. [CrossRef] [PubMed]

11. Levi, S.; Corsi, B.; Bosisio, M.; Invernizzi, R.; Volz, A.; Sanford, D.; Arosio, P.; Drysdale, J. A human mitochondrial ferritin encoded by an intronless gene. *J. Biol. Chem.* **2001**, *276*, 24437–24440. [CrossRef] [PubMed]
12. Honarmand Ebrahimi, K.; Hagedoorn, P.L.; Hagen, W.R. Unity in the biochemistry of the iron-storage proteins ferritin and bacterioferritin. *Chem. Rev.* **2015**, *115*, 295–326. [CrossRef] [PubMed]
13. Hagen, W.R.; Hagedoorn, P.L.; Honarmand Ebrahimi, K. The workings of ferritin: A crossroad of opinions. *Metallomics* **2017**, *9*, 595–605. [CrossRef] [PubMed]
14. Santambrogio, P.; Biasiotto, G.; Sanvito, F.; Olivieri, S.; Arosio, P.; Levi, S. Mitochondrial ferritin expression in adult mouse tissues. *J. Histochem. Cytochem.* **2007**, *55*, 1129–1137. [CrossRef] [PubMed]
15. Levenson, C.W.; Tassabehji, N.M. Iron and ageing: An introduction to iron regulatory mechanisms. *Ageing Res. Rev.* **2004**, *3*, 251–263. [CrossRef] [PubMed]
16. Yang, M.; Yang, H.; Guan, H.; Bellier, J.P.; Zhao, S.; Tooyama, I. Mapping of mitochondrial ferritin in the brainstem of *Macaca fascicularis*. *Neuroscience* **2016**, *328*, 92–106. [CrossRef] [PubMed]
17. Corsi, B.; Cozzi, A.; Arosio, P.; Drysdale, J.; Santambrogio, P.; Campanella, A.; Biasiotto, G.; Albertini, A.; Levi, S. Human mitochondrial ferritin expressed in hela cells incorporates iron and affects cellular iron metabolism. *J. Biol. Chem.* **2002**, *277*, 22430–22437. [CrossRef] [PubMed]
18. Nie, G.; Sheftel, A.D.; Kim, S.F.; Ponka, P. Overexpression of mitochondrial ferritin causes cytosolic iron depletion and changes cellular iron homeostasis. *Blood* **2005**, *105*, 2161–2167. [CrossRef] [PubMed]
19. Yang, H.; Yang, M.; Guan, H.; Liu, Z.; Zhao, S.; Takeuchi, S.; Yanagisawa, D.; Tooyama, I. Mitochondrial ferritin in neurodegenerative diseases. *Neurosci. Res.* **2013**, *77*, 1–7. [CrossRef] [PubMed]
20. Wang, L.; Yang, H.; Zhao, S.; Sato, H.; Konishi, Y.; Beach, T.G.; Abdelalim, E.M.; Bisem, N.J.; Tooyama, I. Expression and localization of mitochondrial ferritin mRNA in Alzheimer's disease cerebral cortex. *PLoS ONE* **2011**, *6*, e22325. [CrossRef] [PubMed]
21. Shi, Z.H.; Nie, G.; Duan, X.L.; Rouault, T.; Wu, W.S.; Ning, B.; Zhang, N.; Chang, Y.Z.; Zhao, B.L. Neuroprotective mechanism of mitochondrial ferritin on 6-hydroxydopamine-induced dopaminergic cell damage: Implication for neuroprotection in Parkinson's disease. *Antioxid. Redox Signal.* **2010**, *13*, 783–796. [CrossRef] [PubMed]
22. Wu, W.S.; Zhao, Y.S.; Shi, Z.H.; Chang, S.Y.; Nie, G.J.; Duan, X.L.; Zhao, S.M.; Wu, Q.; Yang, Z.L.; Zhao, B.L.; et al. Mitochondrial ferritin attenuates β-amyloid-induced neurotoxicity: Reduction in oxidative damage through the Erk/P38 mitogen-activated protein kinase pathways. *Antioxid. Redox Signal.* **2013**, *18*, 158–169. [CrossRef] [PubMed]
23. Wang, P.; Wu, Q.; Wu, W.; Li, H.; Guo, Y.; Yu, P.; Gao, G.; Shi, Z.; Zhao, B.; Chang, Y.Z. Mitochondrial ferritin deletion exacerbates β-amyloid-induced neurotoxicity in mice. *Oxid. Med. Cell. Longev.* **2017**, *2017*, 1020357. [CrossRef] [PubMed]
24. Doyle, A.; McGarry, M.P.; Lee, N.A.; Lee, J.J. The construction of transgenic and gene knockout/knockin mouse models of human disease. *Transgenic Res.* **2012**, *21*, 327–349. [CrossRef] [PubMed]
25. Yang, H.; Guan, H.; Yang, M.; Liu, Z.; Takeuchi, S.; Yanagisawa, D.; Vincent, S.R.; Zhao, S.; Tooyama, I. Upregulation of mitochondrial ferritin by proinflammatory cytokines: Implications for a role in Alzheimer's disease. *J. Alzheimer's Dis.* **2015**, *45*, 797–811.
26. Camaschella, C. Understanding iron homeostasis through genetic analysis of hemochromatosis and related disorders. *Blood* **2005**, *106*, 3710–3717. [CrossRef] [PubMed]
27. Hentze, M.W.; Muckenthaler, M.U.; Galy, B.; Camaschella, C. Two to tango: Regulation of mammalian iron metabolism. *Cell* **2010**, *142*, 24–38. [CrossRef] [PubMed]
28. Hentze, M.W.; Muckenthaler, M.U.; Andrews, N.C. Balancing acts: Molecular control of mammalian iron metabolism. *Cell* **2004**, *117*, 285–297. [CrossRef]
29. Brissot, P. Optimizing the diagnosis and the treatment of iron overload diseases. *Expert Rev. Gastroenterol. Hepatol.* **2016**, *10*, 359–370. [CrossRef] [PubMed]
30. Esposito, B.P.; Breuer, W.; Sirankapracha, P.; Pootrakul, P.; Hershko, C.; Cabantchik, Z.I. Labile plasma iron in iron overload: Redox activity and susceptibility to chelation. *Blood* **2003**, *102*, 2670–2677. [CrossRef] [PubMed]
31. Cabantchik, Z.I.; Breuer, W.; Zanninelli, G.; Cianciulli, P. LPI-labile plasma iron in iron overload. *Best Pract. Res. Clin. Haematol.* **2005**, *18*, 277–287. [CrossRef] [PubMed]

32. Mancuso, C.; Scapagini, G.; Curro, D.; Giuffrida Stella, A.M.; de Marco, C.; Butterfield, D.A.; Calabrese, V. Mitochondrial dysfunction, free radical generation and cellular stress response in neurodegenerative disorders. *Front. Biosci.* **2007**, *12*, 1107–1123. [CrossRef] [PubMed]
33. Napier, I.; Ponka, P.; Richardson, D.R. Iron trafficking in the mitochondrion: Novel pathways revealed by disease. *Blood* **2005**, *105*, 1867–1874. [CrossRef] [PubMed]
34. Della Porta, M.G.; Malcovati, L.; Invernizzi, R.; Travaglino, E.; Pascutto, C.; Maffioli, M.; Galli, A.; Boggi, S.; Pietra, D.; Vanelli, L.; et al. Flow cytometry evaluation of erythroid dysplasia in patients with myelodysplastic syndrome. *Leukemia* **2006**, *20*, 549–555. [CrossRef] [PubMed]
35. Peng, Y.Y.; Uprichard, J. Ferritin and iron studies in anaemia and chronic disease. *Ann. Clin. Biochem.* **2017**, *54*, 43–48. [CrossRef] [PubMed]
36. Waldvogel-Abramowski, S.; Waeber, G.; Gassner, C.; Buser, A.; Frey, B.M.; Favrat, B.; Tissot, J.D. Physiology of iron metabolism. *Transfus. Med. Hemother.* **2014**, *41*, 213–221. [CrossRef] [PubMed]
37. Arosio, P.; Carmona, F.; Gozzelino, R.; Maccarinelli, F.; Poli, M. The importance of eukaryotic ferritins in iron handling and cytoprotection. *Biochem. J.* **2015**, *472*, 1–15. [CrossRef] [PubMed]
38. Santambrogio, P.; Levi, S.; Cozzi, A.; Rovida, E.; Albertini, A.; Arosio, P. Production and characterization of recombinant heteropolymers of human ferritin H and L chains. *J. Biol. Chem.* **1993**, *268*, 12744–12748. [PubMed]
39. Honarmand Ebrahimi, K.; Bill, E.; Hagedoorn, P.L.; Hagen, W.R. The catalytic center of ferritin regulates iron storage via Fe(II)-Fe(III) displacement. *Nat. Chem. Biol.* **2012**, *8*, 941–948. [CrossRef] [PubMed]
40. Arosio, P.; Levi, S. Cytosolic and mitochondrial ferritins in the regulation of cellular iron homeostasis and oxidative damage. *Biochim. Biophys. Acta* **2010**, *8*, 783–792. [CrossRef] [PubMed]
41. de la Pena, T.C.; Carcamo, C.B.; Diaz, M.I.; Brokordt, K.B.; Winkler, F.M. Molecular characterization of two ferritins of the scallop argopecten purpuratus and gene expressions in association with early development, immune response and growth rate. *Comp. Biochem. Physiol. B Biochem. Mol. Biol.* **2016**, *198*, 46–56. [CrossRef] [PubMed]
42. Cheng, Y.; Zak, O.; Aisen, P.; Harrison, S.C.; Walz, T. Structure of the human transferrin receptor-transferrin complex. *Cell* **2004**, *116*, 565–576. [CrossRef]
43. Bartnikas, T.B.; Campagna, D.R.; Antiochos, B.; Mulhern, H.; Pondarre, C.; Fleming, M.D. Characterization of mitochondrial ferritin-deficient mice. *Am. J. Hematol.* **2010**, *85*, 958–960. [CrossRef] [PubMed]
44. Wu, W.; Chang, S.; Wu, Q.; Xu, Z.; Wang, P.; Li, Y.; Yu, P.; Gao, G.; Shi, Z.; Duan, X.; et al. Mitochondrial ferritin protects the murine myocardium from acute exhaustive exercise injury. *Cell Death Dis.* **2016**, *7*, e2475. [CrossRef] [PubMed]
45. You, L.H.; Li, F.; Wang, L.; Zhao, S.E.; Wang, S.M.; Zhang, L.L.; Zhang, L.H.; Duan, X.L.; Yu, P.; Chang, Y.Z. Brain iron accumulation exacerbates the pathogenesis of MPTP-induced Parkinson's disease. *Neuroscience* **2015**, *284*, 234–246. [CrossRef] [PubMed]
46. You, L.H.; Yan, C.Z.; Zheng, B.J.; Ci, Y.Z.; Chang, S.Y.; Yu, P.; Gao, G.F.; Li, H.Y.; Dong, T.Y.; Chang, Y.Z. Astrocyte hepcidin is a key factor in lps-induced neuronal apoptosis. *Cell Death Dis.* **2017**, *8*, e2676. [CrossRef] [PubMed]
47. Li, Y.; Yu, P.; Chang, S.Y.; Wu, Q.; Xie, C.; Wu, W.; Zhao, B.; Gao, G.; Chang, Y.Z. Hypobaric hypoxia regulates brain iron homeostasis in rats. *J. Cell. Biochem.* **2017**, *118*, 1596–1605. [CrossRef] [PubMed]
48. Chang, Y.Z.; Qian, Z.M.; Wang, K.; Zhu, L.; Yang, X.D.; Du, J.R.; Jiang, L.; Ho, K.P.; Wang, Q.; Ke, Y. Effects of development and iron status on ceruloplasmin expression in rat brain. *J. Cell. Physiol.* **2005**, *204*, 623–631. [CrossRef] [PubMed]
49. You, L.H.; Li, Z.; Duan, X.L.; Zhao, B.L.; Chang, Y.Z.; Shi, Z.H. Mitochondrial ferritin suppresses MPTP-induced cell damage by regulating iron metabolism and attenuating oxidative stress. *Brain Res.* **2016**, *1*, 33–42. [CrossRef] [PubMed]

International Journal of
Molecular Sciences

MDPI

Review

Anemia in Kawasaki Disease: Hepcidin as a Potential Biomarker

Ying-Hsien Huang [1,2] and Ho-Chang Kuo [1,2,*]

1   Department of Pediatrics, Kaohsiung Chang Gung Memorial Hospital and Chang Gung University College
    of Medicine, Kaohsiung 833, Taiwan; yhhuang123@yahoo.com.tw
2   Kawasaki Disease Center, Kaohsiung Chang Gung Memorial Hospital, Kaohsiung 833, Taiwan
*   Correspondence: erickuo48@yahoo.com.tw; Tel.: +886-7-7317-123 (ext. 8795); Fax: +886-7-733-8009

Academic Editor: Reinhard Dallinger
Received: 9 March 2017; Accepted: 11 April 2017; Published: 12 April 2017

Abstract: Kawasaki disease (KD) is an autoimmune-like disease and acute childhood vasculitis syndrome that affects various systems but has unknown etiology. In addition to the standard diagnostic criteria, anemia is among the most common clinical features of KD patients and is thought to have a more prolonged duration of active inflammation. In 2001, the discovery of a liver-derived peptide hormone known as hepcidin began revolutionizing our understanding of anemia's relation to a number of inflammatory diseases, including KD. This review focuses on hepcidin-induced iron deficiency's relation to transient hyposideremia, anemia, and disease outcomes in KD patients, and goes on to suggest possible routes of further study.

Keywords: anemia; hepcidin; Kawasaki disease; iron deficiency

1. Kawasaki Disease: The Most Common Acute Coronary Vasculitis Disease in Children

Although its etiology is yet unknown, Kawasaki disease (KD) is an acute childhood vasculitis syndrome that affects various systems [1]. The prevalence of KD in children under the age of 5 years is the highest in Japan with $218/10^5$, followed by Taiwan with $66/10^5$, and the lowest $(4.7/10^5)$ in Europe [2]. The prevalence of KD is more than 10 times higher in Asian children than in European and American children. KD presents as prolonged fever over five days, bulbar conjunctivitis, diffuse mucosal inflammation, unilateral neck nonsuppurative lymphadenopathy, polymorphous skin rashes, and indurative edema of the hands and feet associated with peeling of the fingertips [2]. Vascular involvement of KD occurs in small and medium-sized blood vessels, particularly the coronary arteries [2]. The most severe complications that KD patients experience are coronary artery lesions (CAL), including myocardial infarction and coronary artery aneurysm (CAA); sequelae of the vasculitis with CAA develop in 20% of untreated children [3]. A U.S. multicenter study group found that a single high dose of 2 g/kg intravenous immunoglobulin (IVIG) combined with aspirin can reduce the incidence of aneurysm from 20%–25% to 3%–5% [4,5]. However, previous studies have failed to determine a pathogen responsible for KD, or the identified pathogen did not agree among studies [6,7]. While the exact etiology of KD remains uncertain, we have reported that KD stimulates the extraordinary upregulation of TLR1, 2, 4, 5, 6, and 9, which correlates with bacteria-related pathogen-associated molecular patterns, except for the activation of TLR3 and 7, which relates to double-stranded RNA and single-stranded viral RNA in the acute stage of KD. [8]. This study's results support the idea that KD induces a bacterium-like inflammatory disease. Guo et al., reported that KD's trademark characteristic is an autoimmune-like disease rather than an infectious disease [9]. Furthermore, children with certain single nucleotide polymorphisms of immune genes (ex. *BLK, CD40, FCGR2A, ITPKC* and *IFNG*) are susceptible to triggering over activated inflammatory reactions through certain pathogens with a unique pathogen-associated molecular pattern, which may be KD's immunopathogenesis [10–14]. Therefore, KD may be attributed to not only genetic susceptibility, but also to environmental factors and host immune response [15,16]. Currently, no biological markers are available to differentially diagnose KD from other febrile diseases.

2. Anemia in Patients with Kawasaki Disease

In addition to standard diagnostic criteria, KD patients may experience a variety of nonspecific clinical features, including uveitis, aseptic meningitis, abdominal pain, gallbladder hydrops, rash at the bacillus Calmette-Guérin inoculation site, impaired liver function, hypoalbuminemia, and anemia [15,17–19]. Of these, anemia is the most common clinical feature in KD patients and is thought to have a more prolonged duration of active inflammation [20–23]. A dataset of 783 people, including 441 patients with KD and 342 febrile controls, demonstrated that hemoglobin level was among seven variables to have the largest differential diagnostic absolute values of coefficients [24]. Furthermore, Lin et al., observed that hemoglobin is a useful marker for differentiating KD shock syndrome from toxic shock syndrome in a pediatric intensive care unit [25]. Although severe hemolytic anemia requiring transfusion is rare, it may be related to IVIG infusion [21,22,26]. The major causes of hemolysis are generally associated with anti-A and anti-B IgM antibodies, as well as the anti-Rh IgG antibody [27]. In fact, the IVIG products that are used today are usually safe and effective; they are composed of at least 98% of IgG and very low titers of anti-A (1:8) and anti-B (1:4) IgM, but no anti-D IgG antibodies [19,27]. Furthermore, Rh negative blood types are much less common in Asian populations (0.3%) than in Caucasian populations (15%). The phenomena and literature regarding hemolysis after IVIG in KD patients may thus be more commonly reported in European ancestry than Asian ancestry. We also found no significant difference in total bilirubin and haptoglobin levels between KD patients before and after being treated with IVIG [19]. Therefore, we assume that a key pathogenic connection can explain the relationship between KD and anemia.

Inflammation-associated anemia represents a significant, highly prevalent clinical problem [28]. Chronic disease anemia is often observed in various inflammatory states, such as infections, inflammatory disorders, and certain cancers [29–32]. In 2000, Krause et al., described a peptide that was first referred to as liver-expressed antimicrobial peptide-1, or LEAP-1 and was later called 'hepcidin' due to its hepatic expression and antimicrobial activity [33]. Hepcidin is understood to have a crucial function in blocking the following iron flows into plasma: duodenal absorption, release from macrophages, mobilization of stored iron from hepatocytes, and all that being related to the anemia of inflammation [28,34]. Moreover, high fat diet-induced hepcidin expression is associated with steatosis development and hepatocellular iron accumulation [35,36]. Abnormally elevated hepcidin levels have also been observed in anemia associated with such inflammatory disorders as infections [37,38], autoimmune diseases [39,40], critical illnesses [41,42], obesity [43], and acute myocardial infarction [44].

3. Hepcidin Expression Is Correlated with Kawasaki Disease Outcomes

We have previously reported that both plasma hepcidin and IL-6 levels were elevated in KD patients before undergoing IVIG therapy than in febrile controls [45]. Following IVIG treatment, both hepcidin and IL-6 levels decreased significantly. Notably, the changes of hepcidin levels after IVIG administration were related to IVIG treatment resistance and CAL formation, which supports the theory that elevated inflammatory markers and IVIG non-responsiveness may be related to CAL development in KD patients [15,45].

Previous studies have proven that IVIG can effectively reduce the incidence of CAL [4], but the role and effective dose of aspirin for KD patients remains unclear. Aspirin-related practices have been administered in KD treatment for the past couple of decades, even prior to the administration of IVIG [3]. Furthermore, anemia and overt bleeding are correlated with aspirin use [46]. We reported in one study of a total of 851 KD patients that high-dose aspirin in acute-phase KD does not confer any benefits on disease outcomes and may even be harmful with regard to reducing disease inflammation [47]. Furthermore, this is the first study to show that high-dose aspirin actually results in lower hemoglobin levels and hinders the ability to decrease hepcidin levels after IVIG treatment. Therefore, high-dose aspirin may not be a necessary part of treatment in acute-phase KD. However, additional randomized placebo control studies are required to clarify the function of high-dose aspirin in KD.

4. Hepcidin-Induced Iron Deficiency Is Correlated with Transient Hyposideremia and Anemia in KD Patients

Hepcidin is vital in orchestrating both iron metabolism and the pathogenesis of the anemia of inflammation [48,49]. After hepcidin interacts with ferroportin, ferroportin becomes internalized and degraded, ultimately leading to intracellular iron sequestration and decreased iron absorption [50]. Currently, ferroportin is the only known mammalian iron exporter and is vital for transporting iron from one cell type to another [50]. Hepcidin not only controls iron absorption, but also has an effect on iron-restricted erythropoiesis [51]. Furthermore, hepcidin has also been demonstrated to directly influence erythroid precursor proliferation and survival as erythroid colony formation [52], which agrees with the observation of a transient erythroblastopenia in bone marrow aspiration in KD patients [53]. In our previous study, hemoglobin levels continued to decrease significantly following IVIG treatment, indicating that bone marrow suppression in KD patients is not rapidly reversed following IVIG treatment. Compared to 27 age-matched healthy controls, hemoglobin levels increased at three weeks after IVIG treatment, and hemoglobin levels fully recovered at the six-month follow-up in 117 KD patients. Therefore, we suggest that iron supplementation is not necessary for KD patients.

5. Additional Studies Regarding Hepcidin in Kawasaki Disease

Macrophages play a vital role in regulating iron homeostasis, which is closely connected to their polarization during innate immunity. Macrophage iron homeostasis is correlated with the functional polarization and plasticity of these cells, with extreme roles during inflammation, immune modulation, and inflammation resolution [54]. According to the Mosser and Edwards model, macrophage phenotypes are categorized based on their functional characteristics with host defense (M1), wound healing (M2a), and immune regulation (M2b/c), with the concept that the "hybrid-type" macrophages phenotype occurs [55]. Polarization characteristics often refer to cytokine profiles that have been extensively studied in KD patients. However, no studies have yet addressed exact macrophage polarization in KD. Since iron is an essential growth factor for most bacteria and parasites, they have developed various mechanisms to separate iron from the host. Doing so makes M1-macrophages a major iron storage site under inflammatory conditions [54]. In contrast, M2-macrophages increase ferroportin to promote iron release [56]. However, little is known about whether iron homeostasis influences the ability of the macrophage polarization program and molecular machinery involved in KD processes.

6. Conclusions

Inflammation-induced hepcidin can induce transient hyposideremia, anemia, and disease outcomes in acute-phase KD (Figure 1), and further insightful research is required to better clarify the role of hepcidin in the pathogenesis of KD.

Int. J. Mol. Sci. **2017**, *18*, 820

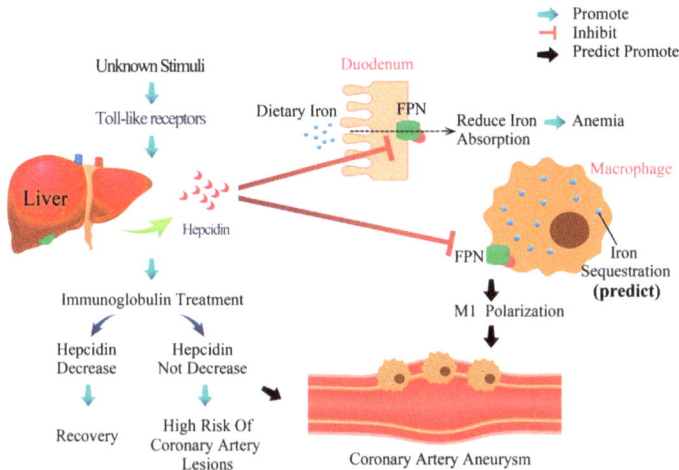

Figure 1. The proposed mechanism of hepcidin-induced transient anemia and coronary artery lesions in patients with Kawasaki disease. While the exact etiology of Kawasaki disease (KD) remains uncertain, we have reported that KD stimulates the extraordinary upregulation of most TLRs that upregulate hepcidin expression. After hepcidin interacts with ferroportin, ferroportin becomes internalized and degraded, ultimately leading to intracellular iron sequestration and decreased iron absorption from duodenum. Hepcidin not only controls iron absorption, but also has a direct inhibitory effect on erythropoiesis, which leads to transient hyposideremia and anemia in KD patients. Following intravenous immunoglobulin (IVIG) treatment, hepcidin levels decrease significantly. Notably, the changes of hepcidin levels after IVIG administration are related to IVIG treatment resistance and coronary artery lesions formation. Macrophage iron homeostasis is correlated with the functional polarization and plasticity of these cells. Doing so makes M1-macrophages a major iron storage site under inflammatory conditions. However, little is known about whether iron homeostasis influences the ability of the macrophage polarization program and molecular machinery involved in KD processes.

Acknowledgments: This study was supported by grants from Taiwan's Ministry of Science and Technology (MOST: 105-2314-B-182-050-MY3), Ministry of Health and Welfare (PMRPG8E0011), and Chang Gung Memorial Hospital (CMRPG8C1082, CMRPG8F1561, CMRPG8F1911, CMRPG8F1931and CORPG8F0011). The aforementioned organizations had no role in the study design, data collection and analysis, decision to publish, or preparation of this manuscript.

Conflicts of Interest: The authors declare no conflicts of interest.

Abbreviations

CAL	Coronary artery lesions
CAA	Coronary artery aneurysm
IVIG	Intravenous immunoglobulin
KD	Kawasaki disease

References

1. Kawasaki, T.; Kosaki, F.; Okawa, S.; Shigematsu, I.; Yanagawa, H. A new infantile acute febrile mucocutaneous lymph node syndrome (MLNS) prevailing in Japan. *Pediatrics* **1974**, *54*, 271–276. [PubMed]
2. Wang, C.L.; Wu, Y.T.; Liu, C.A.; Kuo, H.C.; Yang, K.D. Kawasaki disease: Infection, immunity and genetics. *Pediatr. Infect. Dis. J.* **2005**, *24*, 998–1004. [CrossRef] [PubMed]
3. Newburger, J.W.; Takahashi, M.; Burns, J.C.; Beiser, A.S.; Chung, K.J.; Duffy, C.E.; Glode, M.P.; Mason, W.H.; Reddy, V.; Sanders, S.P.; et al. The treatment of Kawasaki syndrome with intravenous gamma globulin. *N. Engl. J. Med.* **1986**, *315*, 341–347. [CrossRef] [PubMed]

4. Newburger, J.W.; Takahashi, M.; Beiser, A.S.; Burns, J.C.; Bastian, J.; Chung, K.J.; Colan, S.D.; Duffy, C.E.; Fulton, D.R.; Glode, M.P.; et al. A single intravenous infusion of gamma globulin as compared with four infusions in the treatment of acute Kawasaki syndrome. *N. Engl. J. Med.* **1991**, *324*, 1633–1639. [CrossRef] [PubMed]

5. Kuo, H.C.; Liang, C.D.; Wang, C.L.; Yu, H.R.; Hwang, K.P.; Yang, K.D. Serum albumin level predicts initial intravenous immunoglobulin treatment failure in Kawasaki disease. *Acta Paediatr.* **2010**, *99*, 1578–1583. [CrossRef] [PubMed]

6. Principi, N.; Rigante, D.; Esposito, S. The role of infection in Kawasaki syndrome. *J. Infect.* **2013**, *67*, 1–10. [CrossRef] [PubMed]

7. Rigante, D.; Tarantino, G.; Valentini, P. Non-infectious makers of Kawasaki syndrome: Tangible or elusive triggers? *Immunol. Res.* **2016**, *64*, 51–54. [CrossRef] [PubMed]

8. Huang, Y.H.; Li, S.C.; Huang, L.H.; Chen, P.C.; Lin, Y.Y.; Lin, C.C.; Kuo, H.C. Identifying genetic hypomethylation and upregulation of toll-like receptors in Kawasaki disease. *Oncotarget* **2017**. [CrossRef] [PubMed]

9. Guo, M.M.; Tseng, W.N.; Ko, C.H.; Pan, H.M.; Hsieh, K.S.; Kuo, H.C. Th17- and Treg-related cytokine and mRNA expression are associated with acute and resolving Kawasaki disease. *Allergy* **2015**, *70*, 310–318. [CrossRef] [PubMed]

10. Lee, Y.-C.; Kuo, H.-C.; Chang, J.-S.; Chang, L.-Y.; Huang, L.-M.; Chen, M.-R.; Liang, C.-D.; Chi, H.; Huang, F.-Y.; Lee, M.-L.; et al. Two new susceptibility loci for Kawasaki disease identified through genome-wide association analysis. *Nat. Genet.* **2012**, *44*, 522–525. [CrossRef] [PubMed]

11. Kuo, H.-C.; Yang, K.D.; Juo, S.-H.H.; Liang, C.-D.; Chen, W.-C.; Wang, Y.-S.; Lee, C.-H.; Hsi, E.; Yu, H.-R.; Woon, P.-Y.; et al. ITPKC single nucleotide polymorphism associated with the Kawasaki disease in a Taiwanese population. *PLoS ONE* **2011**, *6*. [CrossRef] [PubMed]

12. Kuo, H.C.; Yang, K.D.; Chang, W.C.; Ger, L.P.; Hsieh, K.S. Kawasaki disease: An update on diagnosis and treatment. *Pediatr. Neonatol.* **2012**, *53*, 4–11. [CrossRef] [PubMed]

13. Huang, Y.H.; Hsu, Y.W.; Lu, H.F.; Wong, H.S.; Yu, H.R.; Kuo, H.C.; Huang, F.C.; Chang, W.C.; Kuo, H.C. Interferon-gamma Genetic Polymorphism and Expression in Kawasaki Disease. *Medicine* **2016**, *95*, e3501. [CrossRef] [PubMed]

14. Kuo, H.C.; Chang, J.C.; Kuo, H.C.; Yu, H.R.; Wang, C.L.; Lee, C.P.; Huang, L.T.; Yang, K.D. Identification of an association between genomic hypomethylation of FCGR2A and susceptibility to Kawasaki disease and intravenous immunoglobulin resistance by DNA methylation array. *Arthritis Rheumatol.* **2015**, *67*, 828–836. [CrossRef] [PubMed]

15. Kuo, H.C.; Hsu, Y.W.; Wu, M.S.; Chien, S.C.; Liu, S.F.; Chang, W.C. Intravenous immunoglobulin, pharmacogenomics, and Kawasaki disease. *J. Microbiol. Immunol. Infect.* **2016**, *49*, 1–7. [CrossRef] [PubMed]

16. Kuo, H.C.; Wang, C.L.; Yang, K.D.; Lo, M.H.; Hsieh, K.S.; Li, S.C.; Huang, Y.H. Plasma Prostaglandin E2 Levels Correlated with the Prevention of Intravenous Immunoglobulin Resistance and Coronary Artery Lesions Formation via CD40L in Kawasaki Disease. *PLoS ONE* **2016**, *11*, e0161265. [CrossRef] [PubMed]

17. Newburger, J.W.; Takahashi, M.; Gerber, M.A.; Gewitz, M.H.; Tani, L.Y.; Burns, J.C.; Shulman, S.T.; Bolger, A.F.; Ferrieri, P.; Baltimore, R.S.; et al. Diagnosis, treatment, and long-term management of Kawasaki disease: A statement for health professionals from the Committee on Rheumatic Fever, Endocarditis and Kawasaki Disease, Council on Cardiovascular Disease in the Young, American Heart Association. *Circulation* **2004**, *110*, 2747–2771. [CrossRef] [PubMed]

18. Tseng, H.C.; Ho, J.C.; Guo, M.M.; Lo, M.H.; Hsieh, K.S.; Tsai, W.C.; Kuo, H.C.; Lee, C.H. Bull's eye dermatoscopy pattern at bacillus Calmette-Guerin inoculation site correlates with systemic involvements in patients with Kawasaki disease. *J. Dermatol.* **2016**, *43*, 1044–1050. [CrossRef] [PubMed]

19. Huang, Y.H.; Kuo, H.C.; Huang, F.C.; Yu, H.R.; Hsieh, K.S.; Yang, Y.L.; Sheen, J.M.; Li, S.C.; Kuo, H.C. Hepcidin-Induced Iron Deficiency Is Related to Transient Anemia and Hypoferremia in Kawasaki Disease Patients. *Int. J. Mol. Sci.* **2016**, *17*, 715. [CrossRef] [PubMed]

20. Alves, N.R.; Magalhaes, C.M.; Almeida Rde, F.; Santos, R.C.; Gandolfi, L.; Pratesi, R. Prospective study of Kawasaki disease complications: Review of 115 cases. *Rev. Assoc. Med. Bras.* **2011**, *57*, 295–300. [CrossRef]

21. Fukushige, J.; Takahashi, N.; Ueda, Y.; Ueda, K. Incidence and clinical features of incomplete Kawasaki disease. *Acta Paediatr.* **1994**, *83*, 1057–1060. [CrossRef] [PubMed]

22. Kuo, H.C.; Wang, C.L.; Liang, C.D.; Yu, H.R.; Chen, H.H.; Wang, L.; Yang, K.D. Persistent monocytosis after intravenous immunoglobulin therapy correlated with the development of coronary artery lesions in patients with Kawasaki disease. *J. Microbiol. Immunol. Infect.* **2007**, *40*, 395–400. [PubMed]

23. Kuo, H.C.; Yang, K.D.; Liang, C.D.; Bong, C.N.; Yu, H.R.; Wang, L.; Wang, C.L. The relationship of eosinophilia to intravenous immunoglobulin treatment failure in Kawasaki disease. *Pediatr. Allergy Immunol.* **2007**, *18*, 354–359. [CrossRef] [PubMed]

24. Ling, X.B.; Lau, K.; Kanegaye, J.T.; Pan, Z.; Peng, S.; Ji, J.; Liu, G.; Sato, Y.; Yu, T.T.; Whitin, J.C.; et al. A diagnostic algorithm combining clinical and molecular data distinguishes Kawasaki disease from other febrile illnesses. *BMC Med.* **2011**, *9*, 130. [CrossRef] [PubMed]

25. Lin, Y.J.; Cheng, M.C.; Lo, M.H.; Chien, S.J. Early Differentiation of Kawasaki Disease Shock Syndrome and Toxic Shock Syndrome in a Pediatric Intensive Care Unit. *Pediatr. Infect. Dis. J.* **2015**, *34*, 1163–1167. [CrossRef] [PubMed]

26. Nakagawa, M.; Watanabe, N.; Okuno, M.; Kondo, M.; Okagawa, H.; Taga, T. Severe hemolytic anemia following high-dose intravenous immunoglobulin administration in a patient with Kawasaki disease. *Am. J. Hematol.* **2000**, *63*, 160–161. [CrossRef]

27. Thorpe, S.J. Specifications for anti-A and anti-B in intravenous immunoglobulin: History and rationale. *Transfusion* **2015**, *55* (Suppl. S2), S80–S85. [CrossRef] [PubMed]

28. Nemeth, E.; Ganz, T. Anemia of inflammation. *Hematol. Oncol. Clin. N. Am.* **2014**, *28*, 671–681. [CrossRef] [PubMed]

29. Keel, S.B.; Abkowitz, J.L. The microcytic red cell and the anemia of inflammation. *N. Engl. J. Med.* **2009**, *361*, 1904–1906. [CrossRef] [PubMed]

30. Weiss, G.; Goodnough, L.T. Anemia of chronic disease. *N. Engl. J. Med.* **2005**, *352*, 1011–1023. [CrossRef] [PubMed]

31. Hohaus, S.; Massini, G.; Giachelia, M.; Vannata, B.; Bozzoli, V.; Cuccaro, A.; D'Alo, F.; Larocca, L.M.; Raymakers, R.A.; Swinkels, D.W.; et al. Anemia in Hodgkin's lymphoma: The role of interleukin-6 and hepcidin. *J. Clin. Oncol.* **2010**, *28*, 2538–2543. [CrossRef] [PubMed]

32. Lee, S.H.; Jeong, S.H.; Park, Y.S.; Hwang, J.H.; Kim, J.W.; Kim, N.; Lee, D.H. Serum prohepcidin levels in chronic hepatitis C, alcoholic liver disease, and nonalcoholic fatty liver disease. *Korean J. Hepatol.* **2010**, *16*, 288–294. [CrossRef] [PubMed]

33. Krause, A.; Neitz, S.; Magert, H.J.; Schulz, A.; Forssmann, W.G.; Schulz-Knappe, P.; Adermann, K. LEAP-1, a novel highly disulfide-bonded human peptide, exhibits antimicrobial activity. *FEBS Lett.* **2000**, *480*, 147–150. [CrossRef]

34. Girelli, D.; Nemeth, E.; Swinkels, D.W. Hepcidin in the diagnosis of iron disorders. *Blood* **2016**, *127*, 2809–2813. [CrossRef] [PubMed]

35. Meli, R.; Mattace Raso, G.; Irace, C.; Simeoli, R.; di Pascale, A.; Paciello, O.; Pagano, T.B.; Calignano, A.; Colonna, A.; Santamaria, R. High Fat Diet Induces Liver Steatosis and Early Dysregulation of Iron Metabolism in Rats. *PLoS ONE* **2013**, *8*, e66570. [CrossRef] [PubMed]

36. Dongiovanni, P.; Lanti, C.; Gatti, S.; Rametta, R.; Recalcati, S.; Maggioni, M.; Fracanzani, A.L.; Riso, P.; Cairo, G.; Fargion, S.; et al. High fat diet subverts hepatocellular iron uptake determining dysmetabolic iron overload. *PLoS ONE* **2015**, *10*, e0116855. [CrossRef] [PubMed]

37. Armitage, A.E.; Eddowes, L.A.; Gileadi, U.; Cole, S.; Spottiswoode, N.; Selvakumar, T.A.; Ho, L.P.; Townsend, A.R.; Drakesmith, H. Hepcidin regulation by innate immune and infectious stimuli. *Blood* **2011**, *118*, 4129–4139. [CrossRef] [PubMed]

38. De Mast, Q.; Nadjm, B.; Reyburn, H.; Kemna, E.H.; Amos, B.; Laarakkers, C.M.; Silalye, S.; Verhoef, H.; Sauerwein, R.W.; Swinkels, D.W.; et al. Assessment of urinary concentrations of hepcidin provides novel insight into disturbances in iron homeostasis during malarial infection. *J. Infect. Dis.* **2009**, *199*, 253–262. [CrossRef] [PubMed]

39. Abdel-Khalek, M.A.; El-Barbary, A.M.; Essa, S.A.; Ghobashi, A.S. Serum hepcidin: A direct link between anemia of inflammation and coronary artery atherosclerosis in patients with rheumatoid arthritis. *J. Rheumatol.* **2011**, *38*, 2153–2159. [CrossRef] [PubMed]

40. Demirag, M.D.; Haznedaroglu, S.; Sancak, B.; Konca, C.; Gulbahar, O.; Ozturk, M.A.; Goker, B. Circulating hepcidin in the crossroads of anemia and inflammation associated with rheumatoid arthritis. *Intern. Med.* **2009**, *48*, 421–426. [CrossRef] [PubMed]

41. Isoda, M.; Hanawa, H.; Watanabe, R.; Yoshida, T.; Toba, K.; Yoshida, K.; Kojima, M.; Otaki, K.; Hao, K.; Ding, L.; et al. Expression of the peptide hormone hepcidin increases in cardiomyocytes under myocarditis and myocardial infarction. *J. Nutr. Biochem.* **2010**, *21*, 749–756. [CrossRef] [PubMed]

42. Sihler, K.C.; Raghavendran, K.; Westerman, M.; Ye, W.; Napolitano, L.M. Hepcidin in trauma: Linking injury, inflammation, and anemia. *J. Trauma* **2010**, *69*, 831–837. [CrossRef] [PubMed]

43. Del Giudice, E.M.; Santoro, N.; Amato, A.; Brienza, C.; Calabro, P.; Wiegerinck, E.T.; Cirillo, G.; Tartaglione, N.; Grandone, A.; Swinkels, D.W.; et al. Hepcidin in obese children as a potential mediator of the association between obesity and iron deficiency. *J. Clin. Endocrinol. Metab.* **2009**, *94*, 5102–5107. [CrossRef] [PubMed]

44. Sasai, M.; Iso, Y.; Mizukami, T.; Tomosugi, N.; Sambe, T.; Miyazaki, A.; Suzuki, H. Potential contribution of the hepcidin-macrophage axis to plaque vulnerability in acute myocardial infarction in human. *Int. J. Cardiol.* **2017**, *227*, 114–121. [CrossRef] [PubMed]

45. Kuo, H.C.; Yang, Y.L.; Chuang, J.H.; Tiao, M.M.; Yu, H.R.; Huang, L.T.; Yang, K.D.; Chang, W.C.; Lee, C.P.; Huang, Y.H. Inflammation-induced hepcidin is associated with the development of anemia and coronary artery lesions in Kawasaki disease. *J. Clin. Immunol.* **2012**, *32*, 746–752. [CrossRef] [PubMed]

46. Gaskell, H.; Derry, S.; Moore, R.A. Is there an association between low dose aspirin and anemia (without overt bleeding)? Narrative review. *BMC Geriatr.* **2010**, *10*, 71. [CrossRef] [PubMed]

47. Kuo, H.C.; Lo, M.H.; Hsieh, K.S.; Guo, M.M.; Huang, Y.H. High-Dose Aspirin Is Associated with Anemia and Does Not Confer Benefit to Disease Outcomes in Kawasaki Disease. *PLoS ONE* **2015**, *10*, e0144603. [CrossRef] [PubMed]

48. Le, N.T.; Richardson, D.R. Ferroportin1: A new iron export molecule? *Int. J. Biochem. Cell. Biol.* **2002**, *34*, 103–108. [CrossRef]

49. Ward, D.M.; Kaplan, J. Ferroportin-mediated iron transport: Expression and regulation. *Biochim. Biophys. Acta* **2012**, *1823*, 1426–1433. [CrossRef] [PubMed]

50. Nemeth, E.; Tuttle, M.S.; Powelson, J.; Vaughn, M.B.; Donovan, A.; Ward, D.M.; Ganz, T.; Kaplan, J. Hepcidin regulates cellular iron efflux by binding to ferroportin and inducing its internalization. *Science* **2004**, *306*, 2090–2093. [CrossRef] [PubMed]

51. Kim, A.; Nemeth, E. New insights into iron regulation and erythropoiesis. *Curr. Opin. Hematol.* **2015**, *22*, 199–205. [CrossRef] [PubMed]

52. Dallalio, G.; Law, E.; Means, R.T., Jr. Hepcidin inhibits in vitro erythroid colony formation at reduced erythropoietin concentrations. *Blood* **2006**, *107*, 2702–2704. [CrossRef] [PubMed]

53. Frank, G.R.; Cherrick, I.; Karayalcin, G.; Valderrama, E.; Lanzkowsky, P. Transient erythroblastopenia in a child with Kawasaki syndrome: A case report. *Am. J. Pediatr. Hematol. Oncol.* **1994**, *16*, 271–274. [CrossRef] [PubMed]

54. Jung, M.; Mertens, C.; Brune, B. Macrophage iron homeostasis and polarization in the context of cancer. *Immunobiology* **2015**, *220*, 295–304. [CrossRef] [PubMed]

55. Mosser, D.M.; Edwards, J.P. Exploring the full spectrum of macrophage activation. *Nat. Rev. Immunol.* **2008**, *8*, 958–969. [CrossRef] [PubMed]

56. Corna, G.; Campana, L.; Pignatti, E.; Castiglioni, A.; Tagliafico, E.; Bosurgi, L.; Campanella, A.; Brunelli, S.; Manfredi, A.A.; Apostoli, P.; et al. Polarization dictates iron handling by inflammatory and alternatively activated macrophages. *Haematologica* **2010**, *95*, 1814–1822. [CrossRef] [PubMed]

International Journal of
Molecular Sciences

MDPI

Review

Iron Homeostasis in Health and Disease

Raffaella Gozzelino [1,*,†] **and Paolo Arosio** [2,*,†]

1   Inflammation and Neurodegeneration Laboratory, Chronic Diseases Research Center (CEDOC),
    Nova Medical School (NMS)/Faculdade de Ciências Médicas, University of Lisbon,
    Lisbon 1150-082, Portugal
2   Department of Molecular and Translational Medicine (DMMT), University of Brescia, Brescia 25123, Italy
*   Correspondence: raffaella.gozzelino@nms.unl.pt (R.G.); paolo.arosio@unibs.it (P.A.);
    Tel.: +351-218-803-038 (R.G.); +39-030-371-73-03 (P.A.)
†   These authors contributed equally to this work.

Academic Editor: Reinhard Dallinger
Received: 13 December 2015; Accepted: 12 January 2016; Published: 20 January 2016

Abstract: Iron is required for the survival of most organisms, including bacteria, plants, and humans. Its homeostasis in mammals must be fine-tuned to avoid iron deficiency with a reduced oxygen transport and diminished activity of Fe-dependent enzymes, and also iron excess that may catalyze the formation of highly reactive hydroxyl radicals, oxidative stress, and programmed cell death. The advance in understanding the main players and mechanisms involved in iron regulation significantly improved since the discovery of genes responsible for hemochromatosis, the IRE/IRPs machinery, and the hepcidin-ferroportin axis. This review provides an update on the molecular mechanisms regulating cellular and systemic Fe homeostasis and their roles in pathophysiologic conditions that involve alterations of iron metabolism, and provides novel therapeutic strategies to prevent the deleterious effect of its deficiency/overload.

Keywords: iron; iron metabolism; iron toxicity

1. Introduction

Iron (Fe) is a trace metal essential to ensure the survival of almost all organisms. Its participation in heme- and iron-sulfur cluster (ISC)-containing proteins allows the involvement of Fe in a variety of vital functions, including oxygen transport, DNA synthesis, metabolic energy, and cellular respiration [1,2]. However, the ability of Fe to exchange single electrons with a number of substrates can lead to the generation of reactive oxygen species (ROS), as a result of Fe participation in the Fenton chemistry [3]. This triggers oxidative stress, lipid peroxidation, and DNA damage, which can lead to genomic instability and DNA repair defects [4,5] that ultimately compromise cell viability and promote programmed cell death (PCD) [6]. Under physiologic conditions these deleterious effects are prevented by fine-tuned regulatory mechanisms, which maintain systemic and cellular Fe homeostasis [7] through the cooperation of functional compartments (erythroid and proliferating cells), uptake and recycling systems (enterocytes and splenic macrophages), storage elements (hepatocytes), and mobilization processes that allow Fe trafficking through polarized cells and their corresponding organs, presumably assisted by the poly-r(C)-binding protein-1 (PCBP-1)-mediated transport [8,9]. The treatments of pathological conditions associated with Fe overload strongly improved with the introduction of Fe chelators and the advance of chemical, immunohistochemical, scanning transmission X-ray microscopy, and magnetic resonance imaging (MRI), important tools to monitor these conditions. Synchrotron X-ray fluorescence (SXRF) and/or absorption (XAS) are among the best available techniques to determine Fe forms and tissues distribution [10]. In addition, MRI is a powerful tool to detect and assess the size of insoluble Fe deposits (hemosiderin) [11], although its sensitivity is not sufficient to visualize other potentially toxic forms of Fe, such as labile Fe or excess of cytosolic or soluble ferritin [12]. Growing efforts are being made to correlate the results obtained from spatial analyses with SXRF and XAS, and the distribution of Fe observed by MRI. These techniques mainly used in the diagnosis of hepatic disorders, neurodegenerative [10] and cardiovascular diseases [11], will probably be extended to other organs after further investigations. The growing number of data showing a role of Fe in pathologic conditions was a stimulus to this review (Figure 1).

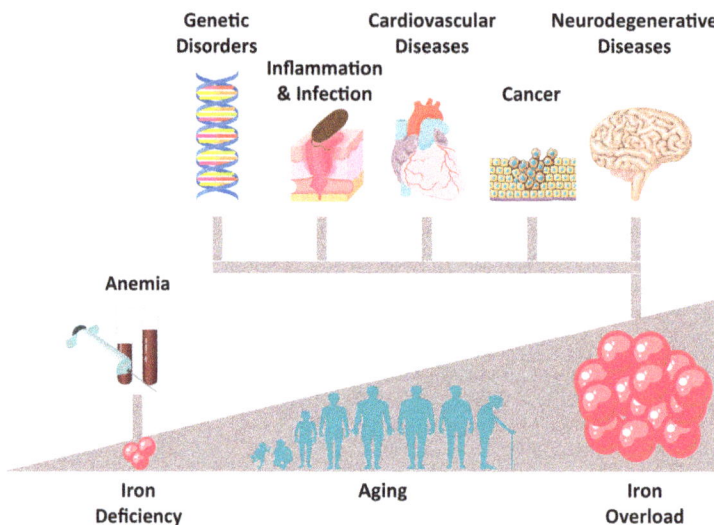

Figure 1. The importance of Iron in pathophysiologic conditions. Essential to ensure survival, disruption of iron homeostasis has been shown to be involved in a variety of pathophysiological conditions, which include anemia and iron-overload related disorders. In particular, the importance of tissue iron accumulation in inflammation and infection, cancer, genetic, cardiovascular and neurodegenerative diseases continuously increases.

2. Iron Metabolism

In healthy individuals, the amount of body Fe is maintained within a range of 4–5 g [13] by a strict control of its absorption, mobilization, storage, and recycling. Fe excretion is not actively controlled and skin desquamation is the major mechanism described so far, accounting for about 1–2 mg per day [14]. Contrarily, well studied are the processes of Fe uptake and recycling that supply the daily need for hemoglobin (Hb) synthesis (25–30 mg). Degradation of senescent red blood cells (RBCs) by splenic macrophages accounts for 90% of total Fe recycling, the remaining 10% comes from the diet [13]. This process occurs in the duodenum where the duodenal cytochrome b ferrireductase (Dcytb) reduces it to Fe(II), which is then offered to the divalent metal transporter-1 (DMT-1) for cellular uptake [15]. DMT-1 (SMF-3 in *C. elegans*) is located on the apical membrane of duodenal epithelial cells and is involved also in Fe reabsorption from glomeruli filtration, recycling from RBCs and transfered from endosomal to cytosolic compartments [16]. The important role of DMT-1 is revealed by the pathologic conditions caused by its mutations, including a severe form of microcytic hypochromic anemia extensively studied in humans and in the Belgrade rat [17].

Inside the cell, the Fe levels are controlled by an elegant machinery involving the Iron Regulatory Proteins (IRPs) and the Iron Responsive Elements (IREs). In low Fe conditions, the high affinity binding of IRPs to IREs inhibits mRNA translation when located in 5′UTR (e.g., Ferritins) and stabilizes the mRNA when in the 3′UTR (e.g., transferrin receptor 1, TfR1). In conditions of Fe excess, ferritins are derepressed and TfR1 is downregulated [18]. This Fe-mediated post-transcriptional regulation modulates the expression of a series of proteins responsible for internalization and storage or utilization of this metal. This ensures that when in excess, Fe is readily stored within the multimeric ferritin made of H- (FtH) and L-chains (FtL), from which is released to satisfy possible intracellular demands [19]. This system, partially conserved in animals, in *C. elegans* involves a cytosolic aconitase (ACO-1) activated in response to hypoxia, while ferritin expression is regulated via activation of the insulin/insulin-like (IIS) growth factor signaling pathway and is favored by increased intracellular stress [16].

Cellular Fe export is mediated by Ferroportin-1 (Fpn1), the only Fe exporter described so far [20]. Its function to release Fe into circulation is supported by the ferroxidase activity of multicopper proteins, namely hephaestin or ceruloplasmin (CP), that favor the formation of Fe-transferrin (Tf) complexes [12]. These bind to cellular surface TfRs for endocytosis and the Tf-Fe^{3+} complex is subsequently reduced by STEAP3 ferrireductase, a step necessary to deliver Fe to mitochondria for heme synthesis and ISC formation [21,22].

It was more recently shown that systemic Fe homeostasis is ensured by a body sensor, hepcidin, a peptide hormone produced by the liver in response to Fe and inflammation. It is controlled also by erythroferrone (ERFE), an erythroid protein produced upon erythropoietic stimulation that suppresses the synthesis of hepcidin [23]. Hepcidin acts as negative regulator of Fe uptake by binding the Fe exporter Fpn1, promoting its internalization and subsequent degradation [24], consequently, increased levels of ERFE enhance Fe availability and release the inhibitory effect of hepcidin on Fe absorption [25]. Indeed, upregulation of hepcidin prevents duodenal Fe uptake and entry into circulation, recycling from RBC heme-Fe and release from Fe stores. The important role of hepcidin in Fe metabolism leads to considering it one, if not the main, regulator of Fe homeostasis [26].

3. Heme-Iron Regulation

Most of the body Fe is contained within the protoporphyrin ring of heme, an insertion catalyzed by the mitochondrial enzyme ferrochelatase. Heme acts as prosthetic group in a variety of proteins essential for cellular proliferation, differentiation, and proper functioning, named hemoproteins, which include globins, cytochromes, myeloperoxidases, catalase, and guanylyl cyclase among others [27].

Similarly to Fe, accumulation of non-hemoprotein bound heme (referred as free heme) is cytotoxic, as sensitizes tissue parenchyma cells to undergo apoptosis in response to pro-inflammatory agonists [28,29]. Disruption of heme homeostasis in porphyrias, anemia, hemolytic diseases, hyperuricaemia, intracerebral hemorrhage and neurological disorders, atherogenesis, were demonstrated by its deficiency/accumulation. In addition, the ability of heme to elicit cytokine production, vascular permeability, and recruitment of immune cells to the inflamed tissue confirm its participation in the first stages of inflammatory conditions [30].

The deleterious effect of free heme relies on the capacity of Fe to participate in the Fenton chemistry. Thus, the maintenance of heme homeostasis is essential to prevent its cytotoxicity and, as such, the existence of a systemic and intracellular regulation is fundamental [31]. Two main circulating proteins are upregulated in response to stressful conditions and afford protection against the pro-oxidant effect of heme. Haptoglobin (Hp) immediately captures the cell-free Hb released in circulation from RBCs disruption, thus preventing its further oxidations, while hemopexin (Hx) is a heme scavenger that inhibits the cellular entry of free heme, delivering it safely to macrophages and hepatocytes for Fe recycling [32,33]. A number of membrane heme transporters were discovered so far to play an important role in heme homeostasis. Two Feline Leukemia Virus subgroup C cellular Receptor (FLVCR) isoforms control heme extracellular trafficking, FLVCR1a is essential for skeletal formation and vascular integrity, and FLVCR1b avoids heme accumulation during fetal erythropoiesis [34,35]. The ubiquitous location of ATP-Binding Cassette, subfamily G, member 2 (ABCG2) at the apical membrane of the cells also facilitates heme extracellular export and prevents the deleterious effects of its accumulation [36,37]. Intracellular heme trafficking is mainly mediated by the heme-importer heme-responsive gene 1 (HRG-1) [38], fundamental to ensure the recycling of this molecule by erythrophagocytic macrophages [39]. Roles of heme importers have also been proposed for the heme carrier protein 1 (HCP1), ABCB6, and FLVCR2 [40].

4. Genetic Disorders—Hemochromatosis

The importance of Fe in pathophysiologic conditions continuously increases [6] and the first type of inherited disorders in which the contribution of this metal was well-established is hereditary hemochromatosis (HH), caused by mutations in genes maintaining Fe homeostasis. Different types of HH have been discovered so far and the severity of the phenotypes observed varies with the gene(s) involved [41]. The most common type of HH (type I) is adult onset and accounts for >80% of all hemochromatosis patients, mostly Caucasian. It is caused by mutations in the hereditary hemochromatosis (HFE) protein and the most prevalent substitution is cysteine 282 to tyrosine. This leads to the inability of HFE to sense increased levels of Fe and interact with TfR1 [42], which causes decreased hepcidin expression, Fe overload, and possible liver dysfunction. The rare type III HH has a similar phenotype, but caused by mutations in TfR2 gene that abolish its capacity to sense Fe levels and interact with HJV and HFE in the liver. This results in a decrease of hepcidin expression and increase of Fe accumulation in the liver and heart. Type II (juvenile HH) is a more severe disorder that affects younger individuals and causes a fast and heavy Fe overload in the liver and parenchyma. If left untreated, it leads to Fe-mediated multi-organ dysfunction. It is associated with mutations in hemojuvelin (HJV) or hepcidin (type II a and type II b, respectively) and with extremely low levels of serum hepcidin. Type IV HH (ferroportin disease) differs from the other ones for having an autosomal dominant transmission and for not affecting hepcidin expression. It is caused by mutations in the SLC40A gene, which encodes the Fe exporter Fpn, namely the hepcidin target [43]. Known also as ferroportin disease, this is characterized by hyperferritinaemia, normal Tf saturation, and Fe-loaded macrophages [44]. Several mutations have been described, most cause a loss of function, while in few cases they cause a gain of function that results in high serum Ft concentration, elevated Tf saturation, and accumulation of Fe in parenchymal cells [15,41,44,45]. Among the various mutations, a characterized one is the Cys326 to Ser substitution [46], which abolishes a ferroportin crucial binding site to hepcidin [47]. The introduction of C326S in the endogenous ferroportin locus in a knockin mouse confirmed the phenotypic pattern of hepatic damage, increased risk of cirrhosis, and hepatocellular carcinoma [48] observed in human type IV HH, which is caused by an unrestrained Fe export from macrophages and enterocytes, and a lethal exocrine pancreatic function [47,49].

5. Iron Deficiency and Anemia

When reviewing the importance of Fe in pathophysiologic conditions it is common to refer mainly to Fe overload and its deleterious effects. However, we should remind that disruption of Fe homeostasis also implies a decrease in its level and subsequent biological consequences. Fe deficiency is the most common disease worldwide and affects mainly children, women, and the elderly [50]. Decreased Fe levels may be associated with mild or severe anemia, conditions that develops when body Fe stores are depleted and the supply of Fe to organs becomes compromised [51,52]. Various causes may contribute to this condition, including genetic defects in proteins responsible for DNA repair, as in Fanconi anemia [4,5,53], and rare genetic mutations occurring in *TMPRSS6* (alias matriptase-2) that cause iron refractory iron deficiency anemia (IRIDA). *TMPRSS6* is a negative regulator of hepcidin expression [45] and its single nucleotide polymorphisms positively correlate with low serum Fe and serum ferritin [54]. Anemic conditions are often caused when increased Fe demand do not meet the adequate supply and occur mainly in children and pregnant women. In addition, enhanced Fe loss affects mostly elderly and results from the use of drugs that cause gastrointestinal bleeding [55]. A contributor to Fe deficiency in elderly comes also from the continuous exposure to subclinical inflammation, which, as described below, modulates hepcidin expression, impairs the maintenance of Fe homeostasis, and leads to the development of anemia of chronic diseases (ACD) [56]. Novel therapeutic approaches to correct the anemia due to hepcidin excess are still under investigation. Among them, the ablation of ERFE in mice was shown not to improve inherited anemias with high hepcidin and ineffective erythropoiesis [57], while heparin administration effectively reduced hepcidin expression. In mice, the therapeutic use of heparins without anticoagulant activity increased the levels of serum Fe, ameliorating significantly the clinical symptoms associated with anemia of inflammation and IRIDA. These results strongly suggest that endogenous heparan sulfate proteoglycans contribute to regulate hepcidin expression. Humanized hepcidin monoclonal antibodies as well as structured L-oligoribonucleotide are currently tested in clinical trials for the treatment of cytokines-induced hypoferremia, aimed to prevent ferroportin–hepcidin interaction and to reduce hepcidin expression [58].

6. Iron and Cancer

Fe homeostasis is altered in most cancer patients, who are affected by anemia in more than 40% of cases [59] and up to 90% when undergoing chemotherapy [60]. The low availability of systemic Fe is partially promoted by the tumor itself that sequesters the metal to ensure its proper growth [61]. In fact, Fe availability is fundamental for cell proliferation and thus highly up-taken by neoplastic cells [62]. Changes in Fe metabolism characterize all phases of tumor development, from proliferation to metastasis [4,62]. Abnormal expression of TfR1 and FtH [63–65] are observed in a variety of cancers, as well as increased levels of circulating hepcidin [64], and these proteins are potential clinical predictors for the prognosis of lung, breast, prostate, liver, and pancreatic cancer [66–69]. The importance of Fe in tumor development has been confirmed in solid and blood tumors, including lymphoma, and multiple myeloma [70,71], where adverse prognosis correlates to Fe load in the affected tissue. The Fe deregulation in tumor cells is associated with ROS production, which promotes DNA modification, strand breaks and is potentially mutagenic [72–75]. Among the effects of the oxidative stress generated by Fe accumulation, which are relevant to malignant transformation, there is the activation of signaling transduction pathways that are essential for tumor growth [67]. These include p53, Wnt, NF-κB, Hypoxia-inducible factor (HIF), DNA replication and repair, cyclins and cell cycle regulation, AKT, and epidermal (EGF) and vascular endothelial growth factor (VEGF) [61,67]. Recent studies also indicate the capacity of Fe-driven ROS to induce epigenetic changes that favor tumor metastasis, as they trigger mutations in hot spots and the suppression and/or activation of tumor suppression genes and/or proto-oncogenes, respectively [4,5,61,67,75]. The inflammation associated with cancer also contributes to Fe deregulation and promotion of DNA damage. Tumor immunology is one of the most investigated fields known to play a dominant role in tumor growth [76] and also aimed to the development of therapeutic strategies eliciting anti-tumor responses [77,78]. Further investigations may be required to assess whether the potential combination of Fe chelators and immune therapy could prevent tumor growth and/or relapse.

7. The Iron-Inflammation Connection

Inflammation plays a critical role in controlling Fe metabolism, as the pro-inflammatory cytokines released upon immune cell activation alter the levels of proteins regulating Fe homeostasis [79]. Taking into account that Fe is essential for proliferation of both prokaryotes and eukaryotes, disruption of its homeostasis may either favor the establishment of the infection or act as a host defense mechanism to defeat pathogen invasion [80,81]. The ability of Fe to impair cytokine secretion renders individuals with Fe overload more susceptible to systemic infections than those with Fe deficiency, therefore excess Fe favors pathogen competition for Fe and increases the morbidity and mortality of infectious diseases [82]. An interesting hypothesis that was recently proposed is that macrophages and T-lymphocytes take up and accumulate non-transferrin bound Fe (NTBI), thus acting as circulating Fe storage compartments to protect different organs from Fe-dependent cytotoxicity [12,83]. The engagement of pattern recognition receptors (PPRs) and the release of pro-inflammatory cytokines (IL-6, IL-22, Oncostatin-M or Activin B) from these immune cells stimulate hepcidin expression, which is considered one of major host defense strategies against infection, an effect triggered by restricting Fe availability to pathogens [84,85]. Macrophage Fe retention is protective in most of the cases, but it may also inhibit protective anti-microbial strategies, which become no longer effective to prevent pathogen growth. This is the case for intracellular invading organisms, as *Salmonella typhimurium*, able to induce hepcidin expression and cellular Fe to ensure its own growth [84]. Much studied is the role of hepcidin in malaria, in which high levels are associated with diminished hepatic growth and differentiation of *Plasmodium* parasite [86]. Fe deficiency was shown to correlate with decreased malaria susceptibility in mice and humans, and consistently excess Fe positively correlates with increased lethality. This notion is also supported by the observation that, in a randomized scale trail, most children succumbing to the infection were the ones receiving Fe supplementation [51,85]. Mouse studies indicate that the disruption of Fe homeostasis could also underlie the poor outcome observed in pregnancy malaria, which results in abortions, stillbirths, underweight babies, and fetal and maternal mortality. The characteristic microvasculature of the placenta, composed of high and low blood flow regions, seems to favor the sequestration of infected RBCs, the lysis of which is a cause of heme/Fe-mediated cytotoxicity to trophoblasts and the fetus. While this would contribute to explain the correlation between Fe overload and the severity of placental malaria (PM), further investigations are required to assess whether this is due to a dysfunctional intracellular and extracellular heme/Fe trafficking [87]. An evolutionary defense strategy against malaria infection is provided by the inheritance of one copy of defective beta globin gene, which confers survival advantage against this disease. However, individuals carrying the two copies of the mutations and suffering from sickle cell disease (SCD) develop vaso-occlusion, endothelial cell dysfunction and chronic vasculopathy, all symptoms mediated by heme/Fe driven cytotoxicity that cannot be prevented by the low level of Hp and Hx plasma proteins observed in these conditions. Therapeutic approaches based on the re-establishment of Fe homeostasis, achieved by overexpression of FtH, are currently under investigations in mice, in which increased levels FtH were shown to prevent Hb-mediated microvascular stasis and ameliorate the symptoms associated [88].

8. Role of Iron in Cardiotoxicity

The involvement of inflammation in cardiotoxicity was extensively demonstrated in cancer patients, in which severe cardiomyopathy remains a major concern of chemotherapeutic drugs administration. The involvement of Fe metabolism in the adverse effects induced by anthracycline and doxorubicin (DOX) treatment, as cardiomyopathies and congestive heart failure, was confirmed by the protective role of Fe chelators. Preventing disruption of Fe homeostasis and restoring the normal expression of Fe transporters, Fe chelator therapy in anthracycline-induced cardiotoxicity suppresses the deleterious effects of Fe overload, of which its involvement was also demonstrated by the increased susceptibility to this treatment of HFE-deficient mice [89]. Whether the use of Fe chelators may also be applied to prevent common cardiac failure remains to be established [11]. Nevertheless, the role of Fe in cardiovascular diseases (CVD) was extensively demonstrated in epidemiological studies, reporting the existence of a positive correlation between Fe accumulation and CVD [90]. This notion was further supported by the lower incidence of ischemic heart disease in Fe deficient patients. The association between disruption of Fe homeostasis and atherosclerosis-driven ischemic CVD was first postulated by Sullivan in 1980s, who indicated that increased Fe levels may enhance the risk of CVD. This may explain the enhanced susceptibility to CVD of post-menopausal women when compared to pre-menopausal [91,92], for whom a reduced level of hepcidin was assumed to diminish more than 50% the risks of CVD when compared to men of the same age. In agreement with this observation, a number of investigations confirmed the protective effect of blood donations against CVD, as this reduces body Fe stores and improves vascular function [93]. The importance of Fe in cardiac function has been demonstrated also in studies using mice that were genetically deleted for TfR-1 or FPN in cardiomyocytes and shown to develop severe cardiomyopathy and heart failure [94,95]. While this is due to a phenotype associated with a heart Fe deficiency or Fe overload, respectively, the restoration of Fe homeostasis via supplementation or chelation of this metal further confirms its dominant role in the development of CVD. Although little is known about cardiac Fe utilization, the ratio between soluble TfR and serum ferritin has been shown indicative of acute myocardial infarction. In particular, the levels of serum ferritin were demonstrated to directly correlate with the incidence and progression of atherosclerosis, which leads ultimately to the occurrence of CVD [93,96–100]. This notion is further confirmed by the increased expression of FtH and FtL in atherosclerotic lesions. The involvement of Fe in modulating the activity of enzymes fundamental in the regulation of cholesterol and triglycerides as well as in the induction of low-density lipoprotein (LDL) oxidative modifications leads to formation of Fe-rich macrophages, *i.e.*, foam cells, responsible for atherosclerotic plaque development, progression and subsequent vulnerability for rupture [90,93,101]. Degradation of the cell-free hemoglobin subsequently to intraplaque hemorrhages results in the release into circulation of its heme prosthetic groups promotes unfettered production of free radicals and oxidative stress, strongly involved in the development of atherosclerotic lesions. The cytotoxicity induced by continuous release of heme in circulation, as in thalassemic and SCD suffering individuals, is apparently the leading cause of CVD, a belief also confirmed by the protective effect of the Hb/heme scavenger Hp and the heme degrading enzyme heme oxygenase 1 (HO-1) in the development of these pathologies. Interestingly, a strong correlation was found between Hp polymorphic genotype and the level of Fe within atherosclerotic plaques, macrophage infiltration, and plaque instability, all symptoms associated with enhanced CVD risks. Moreover, the anti-oxidant and vasodilator properties of HO-1 prevented heme/Fe-mediated endothelial dysfunction and atherogenic plaque formation [93,101]. The higher occurrence of cardiovascular death and non-ischemic cardiovascular abnormalities in hemochromatosis patients as well as the role of hepcidin in Fe retention by macrophages and their consequent transformation into foam cells is currently under investigation. Contrarily to the expectations, hepatic hepcidin expression does not seem to correlate with atherosclerosis progression, as increased accumulation of Fe in macrophages of atherosclerotic mice does not promote atherosclerotic lesions or calcification. However, despite the debate about whether Fe loaded macrophages would aggravate atherosclerosis-mediated CVD, the beneficial effect of the Fe chelator deferoxamine on the recovery of ischemia/reperfusion-induced animals and the outcome of patients submitted to coronary artery bypass surgery [90] reveals the crucial role of Fe in the pathogenesis of CVD.

9. Iron, Aging, and Neurodegeneration

The pathologies mentioned above imply systemic and local dysregulations of Fe homeostasis, but it is still not clear whether this is also the case for neurodegenerative diseases, in which Fe overload is postulated as one of the main contributors to neuronal death. Fe is required for normal brain functions, e.g., neural respiration and metabolic activities, myelin synthesis, production of neurotransmitters and synaptic plasticity [102]. The importance of brain Fe homeostasis is also indicated by the retardation and impaired cognitive abilities caused by its deficiency during early development as well as the axonal degeneration and neuronal death triggered by its overload [13]. Fe physiologically accumulates in aging brains via a mechanism that remains to be fully elucidated and that is accompanied by a reduced expression of anti-oxidant proteins and repair mechanisms, which contribute to Fe-mediated oxidative stress [102,103]. During aging, the disruption of Fe homeostasis affects all organs and contributes to the senescence process. Nutritional habits could contribute to increase Fe content, and meta-analysis studies indicated a positive correlation between elevated Fe consumption and increased incidence of age-related pathologies [71]. This notion is also supported by mouse experiments demonstrating that animals fed with an enriched Fe-diet senesce faster than those maintained under a restricted Fe diet. The high level of inflammation and circulating hepcidin, observed under these conditions, is accompanied by disrupted Fe homeostasis, altered levels of Fe, and organs senescence, as particularly evident in liver, spleen, and gut [103,104]. However, it remains poorly understood how increased dietary Fe uptake affects the brain. Nevertheless, it is well known that the levels of Fe increase disproportionally in neurodegenerative diseases, e.g., Parkinson's,

Alzheimer's, Huntington's, Prions, and neurodegeneration with brain Fe accumulation (NBIA), when compared to brains of elderly individuals [6,103,105]. It is established the involvement of this metal in Friedrich's ataxia, a disease characterized by progressive degeneration of sensory neurons in the dorsal root ganglia caused by a mutation in the frataxin gene. Frataxin plays an essential role in delivering Fe to mitochondrial pathways involved in ISC biogenesis [5] and when its functionality is reduced below a critical threshold, Fe accumulates in mitochondria [106]. Reduced mitochondrial functioning and impaired electron transport chain results in Fe-mediated ROS, which was shown to positively correlate with the severity of Friedrich's ataxia. Mitochondrial defects are also associated with the pathogenesis of Parkinson's disease (PD) [107], in which a compromised Fe transport to this organelle leads to its accumulation and contributes for the formation of alpha-synuclein aggregates, a hallmark of PD. Disruption of Fe homeostasis is also observed in patients suffering from early onset Alzheimer's disease (AD), where brain and cerebrospinal fluid (CSF) accumulation [108] promotes β-amyloid aggregation and hyperphosphorylated tau, two typical features of AD [105]. While different expression or mutations in genes regulating Fe metabolism underlie a number of neurodegenerative diseases, including restless syndrome, neuroferritinopathies, and aceruloplasminemia [109,110], the requirement of Fe also in maintaining lipids homeostasis may suggest a potential role for this metal in demyelinating conditions. However, the observation that most of the clinical symptoms associated with these diseases occur in adulthood implies the existence of protecting mechanisms that retard/prevent brain Fe toxicity. Whether the deleterious effect of brain Fe accumulation is associated with exposure to sub-lethal chronic or acute inflammation is currently the object of investigations [102]. A high level of pro-inflammatory cytokines has been detected in the affected brain regions and they may regulate local Fe homeostasis [111]. In addition, systemic inflammation significantly increases the expression of brain hepcidin, suggesting the existence of a hepcidin/ferroportin axis that actively regulate Fe in the brain. Whether the FPN1/hepcidin axis may be involved in a potential redistribution of Fe between the peripheral and central compartment remains to be elucidated [58,102]. Detoxification of brain Fe overload is a task pursued by Fe chelators, most of which have poor access to the brain. Promising results were obtained in clinical trials with deferiprone, a compound able to translocate across the blood brain barrier and scavenge excess Fe from regional foci of siderosis. By efficiently relocating this metal, deferiprone replenishes Fe-deprived regions and prevents the deleterious effects of Fe maldistribution [12,112].

10. Conclusions

Major advances in the understanding of the mechanisms regulating intracellular and systemic levels of Fe proved that the alteration of Fe homeostasis in mammals underlies a variety of pathological conditions. The oxidative damage generated by the participation of Fe in redox reactions was shown to be the leading cause of programmed cell death and tissue damage, which might also be enhanced by the improper compartmentalization of this metal rather than its total accumulation. Now, the challenge is in developing new approaches capable to restore the deregulation of Fe that accompanies and exacerbates these disorders, which are capable of specifically targeting the tissues involved.

Acknowledgments: Acknowledgments: The authors would like to thank Dr. Bahtiyar Yilmaz for his invaluable help with the figure. This work was partially supported by Fondazione Cariplo grant No. 2012-0570, MIUR-PRIN-11 and Telethon grant GGP10099 to Paolo Arosio, and by iNOVA4Health-UID Multi/04462, a program financially supported by Fundação para a Ciência e Tecnologia/Ministério da Educação e Ciência, through national funds and co-funded by FEDER under the PT2020 Partnership Agreement to Raffaella Gozzelino.

Author Contributions: Author Contributions: Raffaella Gozzelino and Paolo Arosio wrote the paper.

Conflicts of Interest: Conflicts of Interest: The authors declare no conflict of interest.

Abbreviations

Fe	Iron
ISC	Iron-Sulfur Cluster

ROS	Radical Oxygen Species
PCD	Programmed Cell Death
PCBP	Poly-r(C)-Binding Protein 1
MRI	Magnetic Resonance Imaging
SXRF	Synchrotron X-ray Fluorescence
XAS	X-ray Absorption Spectroscopy
RBCs	Red Blood Cells
Dcytb	duodenal cytochrome b ferrireductase
DMT-1	Divalent Metal Transported 1
IRP	Iron Regulatory Protein
IRE	Iron Responsive Element
TfR-1	Transferrin Receptor 1
FtH	Ferritin H Chain
FtL	Ferritin L Chain
ACO-1	Aconitase 1
Fpn	Ferroportin
CP	Ceruloplasmin
Tf	Transferrin
ERFE	Erythroferron
Hp	Haptoglobin
Hx	Hemopexin
FLVCR-1	Feline leukemia virus subgroup C cellular receptor 1
ABCG-2	ATP-Binding Cassette, subfamily G, member 2
HRG-1	Heme Responsive Gene 1
HH	Hereditary Hemochromatosis
HJV	Hemojuvelin
IRIDA	Iron refractory iron deficiency anemia
ACD	Anemia of Chronic Diseases
NTBI	Non-Transferrin Bound Iron
PPRs	Pattern Recognition Receptors
SCD	Sickle Cell Disease
DOX	Doxorubicin
CVD	Cardiovascular Diseases
LDL	Low-Density Lipoprotein
HO-1	Heme Oxygenase 1
NBIA	Neurodegeneration with Brain Iron Accumulation
PD	Parkinson's disease
AD	Alzheimer's disease
CSF	Cerebrospinal Fluid

References

1. Zhao, L.; Xia, Z.; Wang, F. Zebrafish in the sea of mineral (iron, zinc, and copper) metabolism. *Front. Pharmacol.* **2014**, *5*, 33. [CrossRef] [PubMed]
2. Loreal, O.; Cavey, T.; Bardou-Jacquet, E.; Guggenbuhl, P.; Ropert, M.; Brissot, P. Iron, hepcidin, and the metal connection. *Front. Pharmacol.* **2014**, *5*, 128. [PubMed]
3. Fenton, H.J.H. Oxidation of tartaric acid in presence of iron. *J. Chem. Soc.* **1894**, *65*, 899–910. [CrossRef]
4. Zhang, C. Essential functions of iron-requiring proteins in DNA replication, repair and cell cycle control. *Protein Cell* **2014**, *5*, 750–760. [CrossRef] [PubMed]
5. Paul, V.D.; Lill, R. Biogenesis of cytosolic and nuclear iron–sulfur proteins and their role in genome stability. *Biochim. Biophys. Acta* **2015**, *1853*, 1528–1539. [CrossRef] [PubMed]

6. Gozzelino, R.; Arosio, P. The importance of iron in pathophysiologic conditions. *Front. Pharmacol.* **2015**, *6*, 6. [CrossRef] [PubMed]
7. Andrews, N.C.; Schmidt, P.J. Iron homeostasis. *Annu. Rev. Physiol.* **2007**, *69*, 69–85. [CrossRef] [PubMed]
8. Meyron-Holtz, E.G.; Cohen, L.A.; Fahoum, L.; Haimovich, Y.; Lifshitz, L.; Magid-Gold, I.; Stuemler, T.; Truman-Rosentsvit, M. Ferritin polarization and iron transport across monolayer epithelial barriers in mammals. *Front. Pharmacol.* **2014**, *5*, 194. [CrossRef] [PubMed]
9. Philpott, C.C.; Ryu, M.S. Special delivery: Distributing iron in the cytosol of mammalian cells. *Front. Pharmacol.* **2014**, *5*, 173. [CrossRef] [PubMed]
10. Collingwood, J.F.; Davidson, M.R. The role of iron in neurodegenerative disorders: Insights and opportunities with synchrotron light. *Front. Pharmacol.* **2014**, *5*, 191. [CrossRef] [PubMed]
11. Baksi, A.J.; Pennell, D.J. Randomized controlled trials of iron chelators for the treatment of cardiac siderosis in thalassaemia major. *Front. Pharmacol.* **2014**, *5*, 217. [CrossRef] [PubMed]
12. Cabantchik, Z.I. Labile iron in cells and body fluids: Physiology, pathology, and pharmacology. *Front. Pharmacol.* **2014**, *5*, 45. [CrossRef] [PubMed]
13. Zhang, D.L.; Ghosh, M.C.; Rouault, T.A. The physiological functions of iron regulatory proteins in iron homeostasis—An update. *Front. Pharmacol.* **2014**, *5*, 124. [CrossRef] [PubMed]
14. Wright, J.A.; Richards, T.; Srai, S.K. The role of iron in the skin and cutaneous wound healing. *Front. Pharmacol.* **2014**, *5*, 156. [CrossRef] [PubMed]
15. Worthen, C.A.; Enns, C.A. The role of hepatic transferrin receptor 2 in the regulation of iron homeostasis in the body. *Front. Pharmacol.* **2014**, *5*, 34. [CrossRef] [PubMed]
16. Anderson, C.P.; Leibold, E.A. Mechanisms of iron metabolism in caenorhabditis elegans. *Front. Pharmacol.* **2014**, *5*, 113. [CrossRef] [PubMed]
17. Veuthey, T.; Wessling-Resnick, M. Pathophysiology of the belgrade rat. *Front. Pharmacol.* **2014**, *5*, 82. [CrossRef] [PubMed]
18. Torti, F.M.; Torti, S.V. Regulation of ferritin genes and protein. *Blood* **2002**, *99*, 3505–3516. [CrossRef] [PubMed]
19. Arosio, P.; Carmona, F.; Gozzelino, R.; Maccarinelli, F.; Poli, M. The importance of eukaryotic ferritins in iron handling and cytoprotection. *Biochem. J.* **2015**, *472*, 1–15. [CrossRef] [PubMed]
20. Drakesmith, H.; Nemeth, E.; Ganz, T. Ironing out ferroportin. *Cell Metab.* **2015**, *22*, 777–787. [CrossRef] [PubMed]
21. Ohgami, R.S.; Campagna, D.R.; Greer, E.L.; Antiochos, B.; McDonald, A.; Chen, J.; Sharp, J.J.; Fujiwara, Y.; Barker, J.E.; Fleming, M.D. Identification of a ferrireductase required for efficient transferrin-dependent iron uptake in erythroid cells. *Nat. Genet.* **2005**, *37*, 1264–1269. [CrossRef] [PubMed]
22. Sendamarai, A.K.; Ohgami, R.S.; Fleming, M.D.; Lawrence, C.M. Structure of the membrane proximal oxidoreductase domain of human steap3, the dominant ferrireductase of the erythroid transferrin cycle. *Proc. Natl. Acad. Sci. USA* **2008**, *105*, 7410–7415. [CrossRef] [PubMed]
23. Kautz, L.; Jung, G.; Valore, E.V.; Rivella, S.; Nemeth, E.; Ganz, T. Identification of erythroferrone as an erythroid regulator of iron metabolism. *Nat. Genet.* **2014**, *46*, 678–684. [CrossRef] [PubMed]
24. Nemeth, E.; Tuttle, M.S.; Powelson, J.; Vaughn, M.B.; Donovan, A.; Ward, D.M.; Ganz, T.; Kaplan, J. Hepcidin regulates cellular iron efflux by binding to ferroportin and inducing its internalization. *Science* **2004**, *306*, 2090–2093. [CrossRef] [PubMed]
25. Kim, A.; Nemeth, E. New insights into iron regulation and erythropoiesis. *Curr. Opin. Hematol.* **2015**, *22*, 199–205. [CrossRef] [PubMed]
26. Ganz, T.; Nemeth, E. Hepcidin and iron homeostasis. *Biochim. Biophys. Acta* **2012**, *1823*, 1434–1443. [CrossRef] [PubMed]
27. Ponka, P. Cell biology of heme. *Am. J. Med. Sci.* **1999**, *318*, 241–256. [CrossRef]
28. Larsen, R.; Gouveia, Z.; Soares, M.P.; Gozzelino, R. Heme cytotoxicity and the pathogenesis of immune-mediated inflammatory diseases. *Front. Pharmacol.* **2012**, *3*, 77. [CrossRef] [PubMed]
29. Gozzelino, R.; Soares, M.P. Heme sensitization to TNF-mediated programmed cell death. *Adv. Exp. Med. Biol.* **2011**, *691*, 211–219.
30. Dutra, F.F.; Bozza, M.T. Heme on innate immunity and inflammation. *Front. Pharmacol.* **2014**, *5*, 115. [CrossRef] [PubMed]
31. Gozzelino, R.; Soares, M.P. Coupling heme and iron metabolism via ferritin H chain. *Antioxid. Redox Signal.* **2014**, *20*, 1754–1769. [CrossRef] [PubMed]

32. Schaer, D.J.; Vinchi, F.; Ingoglia, G.; Tolosano, E.; Buehler, P.W. Haptoglobin, hemopexin, and related defense pathways-basic science, clinical perspectives, and drug development. *Front. Physiol.* **2014**, *5*, 415. [CrossRef] [PubMed]

33. Smith, A.; McCulloh, R.J. Hemopexin and haptoglobin: Allies against heme toxicity from hemoglobin not contenders. *Front. Physiol.* **2015**, *6*, 187. [CrossRef] [PubMed]

34. Chiabrando, D.; Marro, S.; Mercurio, S.; Giorgi, C.; Petrillo, S.; Vinchi, F.; Fiorito, V.; Fagoonee, S.; Camporeale, A.; Turco, E.; *et al.* The mitochondrial heme exporter flvcr1b mediates erythroid differentiation. *J. Clin. Investig.* **2012**, *122*, 4569–4579. [CrossRef] [PubMed]

35. Keel, S.B.; Doty, R.T.; Yang, Z.; Quigley, J.G.; Chen, J.; Knoblaugh, S.; Kingsley, P.D.; de Domenico, I.; Vaughn, M.B.; Kaplan, J.; *et al.* A heme export protein is required for red blood cell differentiation and iron homeostasis. *Science* **2008**, *319*, 825–828. [CrossRef] [PubMed]

36. Krishnamurthy, P.; Ross, D.D.; Nakanishi, T.; Bailey-Dell, K.; Zhou, S.; Mercer, K.E.; Sarkadi, B.; Sorrentino, B.P.; Schuetz, J.D. The stem cell marker BCRP/ABCG2 enhances hypoxic cell survival through interactions with heme. *J. Biol. Chem.* **2004**, *279*, 24218–24225. [CrossRef] [PubMed]

37. Latunde-Dada, G.O.; Laftah, A.H.; Masaratana, P.; McKie, A.T.; Simpson, R.J. Expression of ABCG2 (BCRP) in mouse models with enhanced erythropoiesis. *Front. Pharmacol.* **2014**, *5*, 135. [CrossRef] [PubMed]

38. Rajagopal, A.; Rao, A.U.; Amigo, J.; Tian, M.; Upadhyay, S.K.; Hall, C.; Uhm, S.; Mathew, M.K.; Fleming, M.D.; Paw, B.H.; *et al.* Haem homeostasis is regulated by the conserved and concerted functions of HRG-1 proteins. *Nature* **2008**, *453*, 1127–1131. [CrossRef] [PubMed]

39. Korolnek, T.; Hamza, I. Like iron in the blood of the people: The requirement for heme trafficking in iron metabolism. *Front. Pharmacol.* **2014**, *5*, 126. [CrossRef] [PubMed]

40. Chiabrando, D.; Vinchi, F.; Fiorito, V.; Mercurio, S.; Tolosano, E. Heme in pathophysiology: A matter of scavenging, metabolism and trafficking across cell membranes. *Front. Pharmacol.* **2014**, *5*, 61. [PubMed]

41. Silvestri, L.; Nai, A.; Pagani, A.; Camaschella, C. The extrahepatic role of TFR2 in iron homeostasis. *Front. Pharmacol.* **2014**, *5*, 93. [CrossRef] [PubMed]

42. Vujic, M. Molecular basis of HFE-hemochromatosis. *Front. Pharmacol.* **2014**, *5*, 42. [PubMed]

43. Donovan, A.; Lima, C.A.; Pinkus, J.L.; Pinkus, G.S.; Zon, L.I.; Robine, S.; Andrews, N.C. The iron exporter ferroportin/Slc40a1 is essential for iron homeostasis. *Cell Metab.* **2005**, *1*, 191–200. [CrossRef] [PubMed]

44. Pietrangelo, A. The ferroportin disease. *Blood Cells Mol. Dis.* **2004**, *32*, 131–138. [CrossRef] [PubMed]

45. Core, A.B.; Canali, S.; Babitt, J.L. Hemojuvelin and bone morphogenetic protein (BMP) signaling in iron homeostasis. *Front. Pharmacol.* **2014**, *5*, 104. [CrossRef] [PubMed]

46. Sham, R.L.; Phatak, P.D.; West, C.; Lee, P.; Andrews, C.; Beutler, E. Autosomal dominant hereditary hemochromatosis associated with a novel ferroportin mutation and unique clinical features. *Blood Cells Mol. Dis.* **2005**, *34*, 157–161. [CrossRef] [PubMed]

47. Fernandes, A.; Preza, G.C.; Phung, Y.; de Domenico, I.; Kaplan, J.; Ganz, T.; Nemeth, E. The molecular basis of hepcidin-resistant hereditary hemochromatosis. *Blood* **2009**, *114*, 437–443. [CrossRef] [PubMed]

48. Pietrangelo, A. Hereditary hemochromatosis: Pathogenesis, diagnosis, and treatment. *Gastroenterology* **2010**, *139*, 393–408, 408 e1–408 e2. [CrossRef] [PubMed]

49. Altamura, S.; Kessler, R.; Grone, H.J.; Gretz, N.; Hentze, M.W.; Galy, B.; Muckenthaler, M.U. Resistance of ferroportin to hepcidin binding causes exocrine pancreatic failure and fatal iron overload. *Cell Metab.* **2014**, *20*, 359–367. [CrossRef] [PubMed]

50. Lopez, A.; Cacoub, P.; Macdougall, I.C.; Peyrin-Biroulet, L. Iron deficiency anaemia. *Lancet* **2015**. [CrossRef]

51. Clark, M.A.; Goheen, M.M.; Cerami, C. Influence of host iron status on plasmodium falciparum infection. *Front. Pharmacol.* **2014**, *5*, 84. [CrossRef] [PubMed]

52. Sankaran, V.G.; Weiss, M.J. Anemia: Progress in molecular mechanisms and therapies. *Nat. Med.* **2015**, *21*, 221–230. [CrossRef] [PubMed]

53. Deans, A.J.; West, S.C. Fancm connects the genome instability disorders bloom's syndrome and fanconi anemia. *Mol. Cell* **2009**, *36*, 943–953. [CrossRef] [PubMed]

54. Wang, C.Y.; Meynard, D.; Lin, H.Y. The role of TMPRSS6/matriptase-2 in iron regulation and anemia. *Front. Pharmacol.* **2014**, *5*, 114. [CrossRef] [PubMed]

55. Busti, F.; Campostrini, N.; Martinelli, N.; Girelli, D. Iron deficiency in the elderly population, revisited in the hepcidin era. *Front. Pharmacol.* **2014**, *5*, 83. [CrossRef] [PubMed]

56. Nemeth, E.; Ganz, T. Anemia of inflammation. *Hematol. Oncol. Clin. N. Am.* **2014**, *28*, 671–681. [CrossRef] [PubMed]
57. Kautz, L.; Jung, G.; Du, X.; Gabayan, V.; Chapman, J.; Nasoff, M.; Nemeth, E.; Ganz, T. Erythroferrone contributes to hepcidin suppression and iron overload in a mouse model of β-thalassemia. *Blood* **2015**, *126*, 2031–2037. [CrossRef] [PubMed]
58. Poli, M.; Asperti, M.; Ruzzenenti, P.; Regoni, M.; Arosio, P. Hepcidin antagonists for potential treatments of disorders with hepcidin excess. *Front. Pharmacol.* **2014**, *5*, 86. [CrossRef] [PubMed]
59. Knight, K.; Wade, S.; Balducci, L. Prevalence and outcomes of anemia in cancer: A systematic review of the literature. *Am. J. Med.* **2004**, *116*, 11S–26S. [CrossRef] [PubMed]
60. Dicato, M.; Plawny, L.; Diederich, M. Anemia in cancer. *Ann. Oncol.* **2010**, *21*, vii167–vii172. [CrossRef] [PubMed]
61. Torti, S.V.; Torti, F.M. Ironing out cancer. *Cancer Res.* **2011**, *71*, 1511–1514. [CrossRef] [PubMed]
62. Munoz, M.; Gomez-Ramirez, S.; Martin-Montanez, E.; Auerbach, M. Perioperative anemia management in colorectal cancer patients: A pragmatic approach. *World J. Gastroenterol. WJG* **2014**, *20*, 1972–1985. [CrossRef] [PubMed]
63. Kukulj, S.; Jaganjac, M.; Boranic, M.; Krizanac, S.; Santic, Z.; Poljak-Blazi, M. Altered iron metabolism, inflammation, transferrin receptors, and ferritin expression in non-small-cell lung cancer. *Med. Oncol.* **2010**, *27*, 268–277. [CrossRef] [PubMed]
64. Xiong, W.; Wang, L.; Yu, F. Regulation of cellular iron metabolism and its implications in lung cancer progression. *Med. Oncol.* **2014**, *31*, 28. [CrossRef] [PubMed]
65. Aleman, M.R.; Santolaria, F.; Batista, N.; de La Vega, M.; Gonzalez-Reimers, E.; Milena, A.; Llanos, M.; Gomez-Sirvent, J.L. Leptin role in advanced lung cancer. A mediator of the acute phase response or a marker of the status of nutrition? *Cytokine* **2002**, *19*, 21–26. [CrossRef] [PubMed]
66. Orlandi, R.; de Bortoli, M.; Ciniselli, C.M.; Vaghi, E.; Caccia, D.; Garrisi, V.; Pizzamiglio, S.; Veneroni, S.; Bonini, C.; Agresti, R.; *et al.* Hepcidin and ferritin blood level as noninvasive tools for predicting breast cancer. *Ann. Oncol.* **2014**, *25*, 352–357. [CrossRef] [PubMed]
67. Zhang, C.; Zhang, F. Iron homeostasis and tumorigenesis: Molecular mechanisms and therapeutic opportunities. *Protein Cell* **2015**, *6*, 88–100. [CrossRef] [PubMed]
68. Wu, T.; Sempos, C.T.; Freudenheim, J.L.; Muti, P.; Smit, E. Serum iron, copper and zinc concentrations and risk of cancer mortality in us adults. *Ann. Epidemiol.* **2004**, *14*, 195–201. [CrossRef]
69. Van Asperen, I.A.; Feskens, E.J.; Bowles, C.H.; Kromhout, D. Body iron stores and mortality due to cancer and ischaemic heart disease: A 17-year follow-up study of elderly men and women. *Int. J. Epidemiol.* **1995**, *24*, 665–670. [CrossRef] [PubMed]
70. Caro, J.J.; Salas, M.; Ward, A.; Goss, G. Anemia as an independent prognostic factor for survival in patients with cancer: A systemic, quantitative review. *Cancer* **2001**, *91*, 2214–2221. [CrossRef]
71. Torti, S.V.; Torti, F.M. Iron and cancer: More ore to be mined. *Nat. Rev. Cancer* **2013**, *13*, 342–355. [CrossRef] [PubMed]
72. Inoue, S.; Kawanishi, S. Hydroxyl radical production and human DNA damage induced by ferric nitrilotriacetate and hydrogen peroxide. *Cancer Res.* **1987**, *47*, 6522–6527. [PubMed]
73. Dizdaroglu, M.; Rao, G.; Halliwell, B.; Gajewski, E. Damage to the DNA bases in mammalian chromatin by hydrogen peroxide in the presence of ferric and cupric ions. *Arch. Biochem. Biophys.* **1991**, *285*, 317–324. [CrossRef]
74. Dizdaroglu, M.; Jaruga, P. Mechanisms of free radical-induced damage to DNA. *Free Radic. Res.* **2012**, *46*, 382–419. [CrossRef] [PubMed]
75. Toyokuni, S. Molecular mechanisms of oxidative stress-induced carcinogenesis: From epidemiology to oxygenomics. *IUBMB Life* **2008**, *60*, 441–447. [CrossRef] [PubMed]
76. Landskron, G.; de la Fuente, M.; Thuwajit, P.; Thuwajit, C.; Hermoso, M.A. Chronic inflammation and cytokines in the tumor microenvironment. *J. Immunol. Res.* **2014**, *2014*, 149185. [CrossRef] [PubMed]
77. Silva-Santos, B.; Serre, K.; Norell, H. Gammadelta T cells in cancer. *Nat. Rev. Immunol.* **2015**, *15*, 683–691. [CrossRef] [PubMed]
78. Kitamura, T.; Qian, B.Z.; Pollard, J.W. Immune cell promotion of metastasis. *Nat. Rev. Immunol.* **2015**, *15*, 73–86. [CrossRef] [PubMed]

79. Ganz, T.; Nemeth, E. Iron homeostasis in host defence and inflammation. *Nat. Rev. Immunol.* **2015**, *15*, 500–510. [CrossRef] [PubMed]
80. Soares, M.P.; Weiss, G. The iron age of host-microbe interactions. *EMBO Rep.* **2015**, *16*, 1482–1500. [CrossRef] [PubMed]
81. Kosmidis, S.; Missirlis, F.; Botella, J.A.; Schneuwly, S.; Rouault, T.A.; Skoulakis, E.M. Behavioral decline and premature lethality upon pan-neuronal ferritin overexpression in Drosophila infected with a virulent form of wolbachia. *Front. Pharmacol.* **2014**, *5*, 66. [CrossRef] [PubMed]
82. Drakesmith, H.; Prentice, A.M. Hepcidin and the iron-infection axis. *Science* **2012**, *338*, 768–772. [CrossRef] [PubMed]
83. Pinto, J.P.; Arezes, J.; Dias, V.; Oliveira, S.; Vieira, I.; Costa, M.; Vos, M.; Carlsson, A.; Rikers, Y.; Rangel, M.; *et al.* Physiological implications of NTBI uptake by t lymphocytes. *Front. Pharmacol.* **2014**, *5*, 24. [CrossRef] [PubMed]
84. Nairz, M.; Haschka, D.; Demetz, E.; Weiss, G. Iron at the interface of immunity and infection. *Front. Pharmacol.* **2014**, *5*, 152. [CrossRef] [PubMed]
85. Spottiswoode, N.; Duffy, P.E.; Drakesmith, H. Iron, anemia and hepcidin in malaria. *Front. Pharmacol.* **2014**, *5*, 125. [CrossRef] [PubMed]
86. Portugal, S.; Carret, C.; Recker, M.; Armitage, A.E.; Goncalves, L.A.; Epiphanio, S.; Sullivan, D.; Roy, C.; Newbold, C.I.; Drakesmith, H.; *et al.* Host-mediated regulation of superinfection in malaria. *Nat. Med.* **2011**, *17*, 732–737. [CrossRef] [PubMed]
87. Penha-Goncalves, C.; Gozzelino, R.; de Moraes, L.V. Iron overload in Plasmodium berghei-infected placenta as a pathogenesis mechanism of fetal death. *Front. Pharmacol.* **2014**, *5*, 155. [PubMed]
88. Vercellotti, G.M.; Khan, F.B.; Nguyen, J.; Chen, C.; Bruzzone, C.M.; Bechtel, H.; Brown, G.; Nath, K.A.; Steer, C.J.; Hebbel, R.P.; *et al.* H-ferritin ferroxidase induces cytoprotective pathways and inhibits microvascular stasis in transgenic sickle mice. *Front. Pharmacol.* **2014**, *5*, 79. [CrossRef] [PubMed]
89. Gammella, E.; Maccarinelli, F.; Buratti, P.; Recalcati, S.; Cairo, G. The role of iron in anthracycline cardiotoxicity. *Front. Pharmacol.* **2014**, *5*, 25. [CrossRef] [PubMed]
90. Basuli, D.; Stevens, R.G.; Torti, F.M.; Torti, S.V. Epidemiological associations between iron and cardiovascular disease and diabetes. *Front. Pharmacol.* **2014**, *5*, 117. [PubMed]
91. Sullivan, J.L. Iron and the sex difference in heart disease risk. *Lancet* **1981**, *1*, 1293–1294. [CrossRef]
92. Sullivan, J.L. The iron paradigm of ischemic heart disease. *Am. Heart J.* **1989**, *117*, 1177–1188. [CrossRef]
93. Vinchi, F.; Muckenthaler, M.U.; da Silva, M.C.; Balla, G.; Balla, J.; Jeney, V. Atherogenesis and iron: From epidemiology to cellular level. *Front. Pharmacol.* **2014**, *5*, 94. [CrossRef] [PubMed]
94. Xu, W.; Barrientos, T.; Mao, L.; Rockman, H.A.; Sauve, A.A.; Andrews, N.C. Lethal cardiomyopathy in mice lacking transferrin receptor in the heart. *Cell Rep.* **2015**, *13*, 533–545. [CrossRef] [PubMed]
95. Lakhal-Littleton, S.; Wolna, M.; Carr, C.A.; Miller, J.J.; Christian, H.C.; Ball, V.; Santos, A.; Diaz, R.; Biggs, D.; Stillion, R.; *et al.* Cardiac ferroportin regulates cellular iron homeostasis and is important for cardiac function. *Proc. Natl. Acad. Sci. USA* **2015**, *112*, 3164–3169. [CrossRef] [PubMed]
96. Morrison, H.I.; Semenciw, R.M.; Mao, Y.; Wigle, D.T. Serum iron and risk of fatal acute myocardial infarction. *Epidemiology* **1994**, *5*, 243–246. [CrossRef] [PubMed]
97. Holay, M.P.; Choudhary, A.A.; Suryawanshi, S.D. Serum ferritin-a novel risk factor in acute myocardial infarction. *Indian Heart J.* **2012**, *64*, 173–177. [CrossRef]
98. Menke, A.; Fernandez-Real, J.M.; Muntner, P.; Guallar, E. The association of biomarkers of iron status with peripheral arterial disease in us adults. *BMC Cardiovasc. Disord.* **2009**, *9*, 34. [CrossRef] [PubMed]
99. Lauffer, R.B. Iron depletion and coronary disease. *Am. Heart J.* **1990**, *119*, 1448–1449. [CrossRef]
100. Kiechl, S.; Willeit, J.; Egger, G.; Poewe, W.; Oberhollenzer, F. Body iron stores and the risk of carotid atherosclerosis: Prospective results from the bruneck study. *Circulation* **1997**, *96*, 3300–3307. [CrossRef] [PubMed]
101. Habib, A.; Finn, A.V. The role of iron metabolism as a mediator of macrophage inflammation and lipid handling in atherosclerosis. *Front. Pharmacol.* **2014**, *5*, 195. [CrossRef] [PubMed]
102. Gozzelino, R. The pathophysiology of heme in the brain. *Curr. Alzheimer Res.* **2015**, *13*, 174–184. [CrossRef]
103. Arruda, L.F.; Arruda, S.F.; Campos, N.A.; de Valencia, F.F.; Siqueira, E.M. Dietary iron concentration may influence aging process by altering oxidative stress in tissues of adult rats. *PLoS ONE* **2013**, *8*, e61058. [CrossRef] [PubMed]

104. Nelson, R.L. Iron and colorectal cancer risk: Human studies. *Nutr. Rev.* **2001**, *59*, 140–148. [CrossRef] [PubMed]

105. Wong, B.X.; Duce, J.A. The iron regulatory capability of the major protein participants in prevalent neurodegenerative disorders. *Front. Pharmacol.* **2014**, *5*, 81. [CrossRef] [PubMed]

106. Isaya, G. Mitochondrial iron–sulfur cluster dysfunction in neurodegenerative disease. *Front. Pharmacol.* **2014**, *5*, 29. [CrossRef] [PubMed]

107. Martelli, A.; Puccio, H. Dysregulation of cellular iron metabolism in friedreich ataxia: From primary iron–sulfur cluster deficit to mitochondrial iron accumulation. *Front. Pharmacol.* **2014**, *5*, 130. [CrossRef] [PubMed]

108. Ali-Rahmani, F.; Schengrund, C.L.; Connor, J.R. Hfe gene variants, iron, and lipids: A novel connection in alzheimer's disease. *Front. Pharmacol.* **2014**, *5*, 165. [CrossRef] [PubMed]

109. Levi, S.; Finazzi, D. Neurodegeneration with brain iron accumulation: Update on pathogenic mechanisms. *Front. Pharmacol.* **2014**, *5*, 99. [CrossRef] [PubMed]

110. Gao, G.; Chang, Y.Z. Mitochondrial ferritin in the regulation of brain iron homeostasis and neurodegenerative diseases. *Front. Pharmacol.* **2014**, *5*, 19. [CrossRef] [PubMed]

111. Urrutia, P.J.; Mena, N.P.; Nunez, M.T. The interplay between iron accumulation, mitochondrial dysfunction, and inflammation during the execution step of neurodegenerative disorders. *Front. Pharmacol.* **2014**, *5*, 38. [CrossRef] [PubMed]

112. Cabantchik, Z.I.; Munnich, A.; Youdim, M.B.; Devos, D. Regional siderosis: A new challenge for iron chelation therapy. *Front. Pharmacol.* **2013**, *4*, 167. [CrossRef] [PubMed]

International Journal of
Molecular Sciences

MDPI

Review

Canine Models for Copper Homeostasis Disorders

Xiaoyan Wu, Peter A. J. Leegwater and Hille Fieten *

Department of Clinical Sciences of Companion animals, Faculty of Veterinary Medicine, Utrecht University,
Yalelaan 108, 3584 CM Utrecht, The Netherlands; x.wu@uu.nl (X.W.); P.A.J.Leegwater@uu.nl (P.A.J.L.)
* Correspondence: H.Fieten@uu.nl; Tel.: +31-30-253-9411; Fax: +31-30-253-9393

Academic Editor: Reinhard Dallinger
Received: 28 November 2015; Accepted: 25 January 2016; Published: 4 February 2016

Abstract: Copper is an essential trace nutrient metal involved in a multitude of cellular processes. Hereditary defects in copper metabolism result in disorders with a severe clinical course such as Wilson disease and Menkes disease. In Wilson disease, copper accumulation leads to liver cirrhosis and neurological impairments. A lack in genotype-phenotype correlation in Wilson disease points toward the influence of environmental factors or modifying genes. In a number of Non-Wilsonian forms of copper metabolism, the underlying genetic defects remain elusive. Several pure bred dog populations are affected with copper-associated hepatitis showing similarities to human copper metabolism disorders. Gene-mapping studies in these populations offer the opportunity to discover new genes involved in copper metabolism. Furthermore, due to the relatively large body size and long life-span of dogs they are excellent models for development of new treatment strategies. One example is the recent use of canine organoids for disease modeling and gene therapy of copper storage disease. This review addresses the opportunities offered by canine genetics for discovery of genes involved in copper metabolism disorders. Further, possibilities for the use of dogs in development of new treatment modalities for copper storage disorders, including gene repair in patient-derived hepatic organoids, are highlighted.

Keywords: copper toxicosis; nutrition; genetics; Wilson disease; Menkes disease; ATP7A; ATP7B; COMMD1; Bedlington terrier; Labrador retriever

---

1. Introduction

The essential micronutrient copper plays a key role in several vital biological processes including neurotransmitter synthesis, antioxidant defense, mitochondrial respiration, iron metabolism, pigmentation, and connective tissue formation [1–7]. As a transition metal, copper can be highly reactive, therefore copper levels need to be strictly regulated. A disruption in the function of proteins involved in the regulation of copper metabolism can lead to severe clinical phenotypes illustrated by the diseases caused by mutations in genes encoding the P-type ATPase copper transporters ATP7A and ATP7B.

Mutations in ATP7A give rise to copper deficiency disorders [8], of which Menkes disease (MD) is best described [9,10]. Menkes disease patients suffer from severe neurological impairment and failure to thrive. Treatment consists of parenteral copper supplementation, however, the disease is usually lethal in early childhood [9].

Mutations in the copper transporter ATP7B result in the copper overload disorder Wilson disease (WD) [11,12]. Wilson disease patients present with hepatitis resulting from hepatic copper accumulation and/or neurological or psychiatric symptoms [13]. Many different mutations in ATP7B may result in Wilson disease and there is a lack of a clear genotype-phenotype association [14]. Probably environmental factors or, yet unidentified, modifier genes contribute to the diverse manifestations of WD. Treatment consists of life long copper chelation therapy or liver transplantation [15,16].

Besides Wilson disease, non-Wilsonian forms or copper toxicosis leading to liver cirrhosis at young age include Indian childhood cirrhosis [17], endemic Tyrolean infantile cirrhosis [18], and idiopathic copper toxicosis [19]. In addition to genetic predisposition, increased uptake of copper via diet and drinking water predispose for development of disease. The underlying genetic defects have not been elucidated yet.

Naturally occurring copper-associated hepatitis occurs in high frequency in a number of pure bred dog populations [20]. Copper storage diseases in dogs mimic Wilson disease and ecogenetic forms of copper toxicosis with regard to hepatic copper accumulation resulting in liver cirrhosis.

One of the best described diseases is autosomal recessive copper toxicosis in the Bedlington terrier which is caused by a mutation in *COMMD1* [21]. More recently, both *ATP7A* and *ATP7B* were identified to be associated with copper toxicosis in Labrador retrievers [22]. Many more dog breeds are affected with copper toxicosis, offering the opportunity for genetic studies for identification of new genes involved in copper metabolism.

Pure bred dogs have a unique genetic structure, with large linkage disequilibrium blocks which makes them ideal for gene mapping studies [23,24]. In contrast, gene mapping studies in humans with Wilson disease or non-Wilsonian forms of copper toxicosis are difficult to perform due to low frequency of the diseases and phenotypic variability of patients. As the proteins involved in copper metabolism are highly conserved, newly identified genes in canines with copper toxicosis may be valuable for evaluation in human patients.

For the continuous development of new treatment strategies for people with copper metabolism disorders, animal models are needed. Several rodent models including naturally occurring models [25,26] as well as genetically engineered models [27,28] have proven to be of great value in the past years. Rodent models, however, have obvious limitations including short lifespan and an inappropriate body size for true longitudinal studies. As opposed to canines, a time course of liver biopsies cannot be collected from individual rodents, as they need to be sacrificed in order to obtain liver samples. Their lifespan of at most two years hampers evaluation of treatment side-effects on the long term. Canine models of copper toxicosis can be a valuable addition for development of new treatment strategies. The body size of dogs is more in the range of that of humans and facilitates translation of procedures (including those needed for application of stem-cell treatments such as catheterization of the vena porta for intraportal delivery of cell-transplants to the liver) and collection of multiple liver biopsies over time from the same animal. Furthermore, their long lifespan (up to 16 years) facilitates the evaluation of long-term treatment effects. To explore their full potential, a thorough genetic characterization of the canine models is a prerequisite, as it is in rodents.

In this review, we will discuss the present canine models of copper metabolism, and the role of the canine models in identification of new genes, the opportunities for development of new medical and dietary treatments, and the possibility of auto-transplantation of gene-corrected organoids in dogs as a large animal model for copper storage diseases.

2. Copper Homeostasis

Copper is an essential trace mineral which must be ingested from dietary sources and drinking water. Dietary copper absorption takes place in the stomach and small intestine [29]. After being released from intestinal cells, copper is transported to the liver via the portal circulation. Copper from the liver is redistributed to extra-hepatic tissues bound to ceruloplasmin, albumin, and transcuprein [30]. A minimal amount of copper is excreted via the kidney, but the main route of removal of excessive copper is via biliary excretion into the feces (Figure 1) [31]. By regulating copper storage, redistribution, and excretion, the liver is one of the main organs in homeostatic control of copper metabolism (Figure 1) [31]. Cytosolic free copper is toxic, because unbound copper generates highly reactive hydroxyl radicals that cause cell damage and lead to inflammatory reactions. Therefore, copper metabolism at the cellular level is tightly regulated to prevent the presence of free copper ions. Import of copper takes place via the plasma membrane transport protein Copper Transporter 1

(CTR1) [32]. Intracellularly, copper is bound to small copper scavengers like metallothionein (MT) [33] and glutathione (GSH) [34]. Specialized proteins, the copper chaperones, facilitate delivery of copper to various destination proteins [35]. Copper chaperone for SOD (CCS) is the copper chaperone for the detoxifying enzyme copper, zinc dependent superoxide dismutase (SOD1) [35]. SOD1 protects cells from reactive oxygen species and resides both in the cytoplasm and the intermembrane space of mitochondria [36]. Cox17 is the copper chaperone for cytochrome C oxidase (CCO), which is located in the mitochondrial inner membrane and is involved in cellular respiration [37]. Atox1 chaperones copper to the copper-transporting ATPases, ATP7A or ATP7B (Figure 1) [38]. Both ATP7A and ATP7B reside in the trans-Golgi network (TGN) and their basic functions include copper transport to newly synthesized cuproenzymes through the secretory pathway and copper export from cells [39]. In low copper circumstances, ATP7A and ATP7B reside in the trans-Golgi network, ATP7A is ubiquitously expressed in a number of cell-types, whereas ATP7B expression is detected in a selected number of cell-types, including hepatocytes. In enterocytes, copper uptake from the intestinal lumen is mediated by CTR1, which plays an important role in copper uptake by facilitating copper transport over the basolateral membrane towards the portal circulation (Figure 1). Recently, ATP7A was identified to be an important factor in mobilizing hepatic copper in response to peripheral tissue copper demand as well [40].

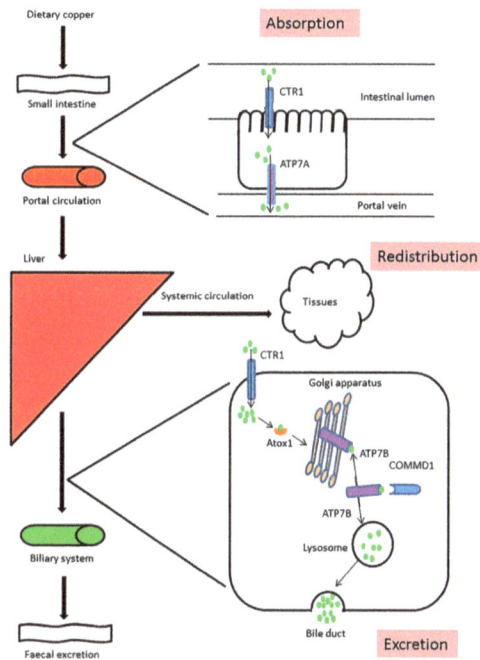

Figure 1. Dietary copper is absorbed in the small intestine via Ctr1. ATP7A facilitates copper transport from the enterocyte into the portal circulation for transportation to the liver. In the liver, copper is imported in the enterocytes by Ctr1. Here, copper is stored or redistributed via the systemic circulation for use in body tissues. Excretion of excess copper from hepatocytes takes place via copper transporter ATP7B. ATP7B resides in the trans-Golgi network under low copper conditions. It receives copper molecules from the chaperone ATOX1 and under high copper conditions it moves to a late endosome/lysosome compartment, from which copper is eventually excreted in the bile and expelled from the body with the feces. COMMD1 is believed to interact with ATP7B and to facilitate retrograde trafficking of ATP7B back to the trans-Golgi network.

The main functions of the copper transporter ATP7B are incorporation of copper into the ferroxidase ceruloplasmin [41] and biliary excretion of excess copper [42].

In order to fulfill these functions, ATP7A and ATP7B exhibit copper responsive intracellular trafficking. Upon a rise of intracellular copper, both proteins move away from the TGN. ATP7A is mainly delivered to the cell-surface or to a post-Golgi vesicular compartment in polarized intestinal cells [43]. In hepatocytes, ATP7B is moved to an endosome/lysosome compartment in response to copper, and subsequently removes excess copper into the bile through lysosomal exocytosis towards the apical membrane of hepatocytes [44].

COMMD1, the protein that is absent in Bedlington terriers with copper toxicosis, interacts with ATP7A and ATP7B and affects trafficking of both proteins [45,46]. The role of COMMD1 in the development of copper toxicosis is not completely elucidated but evidence exists that it is associated with the regulation of the retrograde transport of ATP7B from peripheral endosomes back to the TGN in low copper circumstances (Figure 1) [47,48].

3. Copper Metabolism Disorders in Humans

As illustrated in the previous paragraph, copper metabolism is a complex process involving many proteins. A disruption of this tightly controlled system will lead to severe clinical phenotypes in patients. Two well documented copper metabolism disorders in humans are Menkes disease and Wilson disease which are induced by mutations of genes coding for the copper transport proteins, ATP7A and ATP7B, respectively.

MD is an X-linked recessive and lethal neurodegenerative disorder resulting from a wide spectrum of mutations in *ATP7A* [8]. Clinical symptoms result from low copper levels in the liver and brain [8] and affected MD infants suffer from brittle hair, growth failure, neurodegeneration, arterial tortuosity, and hypopigmentation. The diagnosis can be confirmed by measurement of plasma copper and neurochemical levels [49]. Treatment currently available for MD is parenteral administration of copper, which may increase lifespan, however many patients die in early childhood [50].

In contrast to MD, WD is a result of copper overload [12]. WD is a rare, autosomal recessive disease. The onset of this disease occurs at a wide range of ages from childhood [51] to the elderly [52]. Patients can present with a variety of disease phenotypes including progressive hepatic disease, neurological diseases, and psychiatric illness [53]. Diagnosis of WD is based on the combination of clinical symptoms, biochemical features, histological findings, and mutation analysis of the *ATP7B* gene. Diagnostic criteria for WD include recognition of corneal Kayser-Fleischer rings, a decreased serum ceruloplasmin level, increased excretion of urinary copper and increased hepatic copper levels (>250 mg/kg) [54]. Further, mutation analysis of the *ATP7B* gene is usually performed and over 500 mutations in the *ATP7B* gene associated with WD have currently been identified [55] and many patients are compound heterozygotes. A clear discordance between genotype and phenotypic presentation exists for WD [14]. This is illustrated by the fact that even monozygotic twins may be discordant for the WD phenotype [56]. Environmental, epigenetic, and genetic modulations probably play a role in the phenotypic expression of WD [14]. Several attempts to identify modifier genes have been made [57–62], however overall contribution of these genes to the disease heterogeneity is uncertain.

Currently, the available treatments for WD are aimed at creating a negative copper balance by copper chelation therapy or blocking of copper uptake [63]. Drugs used for copper chelation include D-Penicillamine [64] and trientine [65,66] to promote urinary copper excretion. Side effects may occur after D-Penicillamine administration including hypersensitivity, nausea, proteinuria, and development of auto-immune disorders [15]. Also, worsening of neurological signs may occur after treatment with D-Penicillamine [67,68]. Zinc salts can be used as a maintenance therapy or in asymptomatic family members of affected individuals [54]. Zinc salts can be used to decrease the absorption of dietary copper [69]. In general, Wilson disease patients are instructed to avoid foods high in copper such as chocolate, liver, nuts, mushrooms, and shellfish. However, clinical effects of an adjusted diet in the long term have not been studied. Patients are not always responsive to medical therapy, or may even show worsening of clinical symptoms [15,67]. Liver transplantation is indicated for these patients and for patients that suffer from severe liver failure due to copper-induced cirrhosis [70–74]. Although good outcomes in most transplantation cases have been reported [71–74], the shortage of donors, complications associated with transplantations (*i.e.*, peri-operative complications) and the risk of graft rejection, requiring long term immunosuppressive therapy, obviously complicate liver transplantation as a general cure for Wilson disease.

Non-Wilsonian disorders of copper toxicosis, often occurring in early childhood, are Indian childhood cirrhosis [17], endemic Tyrolean infantile cirrhosis [18], and Idiopathic copper toxicosis [19]. These disorders are characterized by severe hepatic copper overload and a quick progression to liver cirrhosis. Neurological phenotypes, as observed in WD, are not recognized in these infant patients. The inherited origin of these diseases was indicated by pedigree studies [18]. However, the causal genes contributing to the non-Wilson copper overload diseases have not been identified yet. Increased dietary copper uptake was assumed to be associated with development of disease, although this subject was recently under debate [75].

4. Copper Metabolism Disorders in Dogs

In addition to the copper disorders in humans, hereditary copper metabolism disorders are recognized in other mammals including rodents [76], sheep [77], and dogs [20]. Copper toxicosis in a variety of dog breeds mimics copper overload disorders in humans. Like in humans, copper in dogs is absorbed from diet and drinking water and subsequently accumulates in the liver. In dogs, the diagnosis is made by histological evaluation and copper measurement in liver biopsies. Hepatic copper concentrations in dogs are normally higher than in humans, with a concentration of <400 mg/kg dry weight liver (dwl) being considered as normal [78]. Copper accumulation in affected dogs can range from 800 to 10,000 mg/kg dwl. Hepatic copper accumulation induces cell death, bridging fibrosis and will progress to liver cirrhosis and hepatic failure. The disease can also manifest more acutely as fulminant liver failure. In Bedlington terriers, massive release of copper into the circulation may lead to hemolysis and anemia. In other breeds, usually a more gradual hepatic copper accumulation occurs initially without overt clinical signs. Usually during middle age (median age 7 years, range 2–12 years) dogs will display clinical symptoms of liver failure including icterus, ascites, anorexia, vomiting, and development of hepato-encephalopathy. Besides neurological signs related to hepato-encephalopathy, obvious neurological signs, or behavioral changes were not recognized. Like in humans, copper toxicosis in dogs can be treated with copper chelators D-Penicillamine [79,80] and 2,3,2-tetramine [81], or with zinc-acetate [82].

One of the best described forms of copper-associated hepatitis in dogs is copper toxicosis in Bedlington terriers [83,84]. Bedlington terrier copper toxicosis is an autosomal recessive disease resulting from a deletion of exon 2 of *COMMD1*, encoding COMM domain-containing protein 1 [21,85]. The total lack of COMMD1 protein [86] results in massive hepatic copper accumulation, which can be more than 25 times higher than normal [87]. The *COMMD1*-deficient dog is now an established canine model for copper-induced human chronic hepatitis [88].

Labrador retrievers have been reported to suffer from copper-associated chronic hepatitis with high frequency [89–91]. In this breed a female predisposition and a polygenic inheritance pattern with high heritability for copper accumulating traits was observed [89–92]. Labrador retrievers do not have the exon 2 deletion in the *COMMD1* gene. Illustrating that copper-associated hepatitis has a different genetic origin in different dog breeds.

Besides a genetic predisposition to disease, a large influence of dietary intake of copper and zinc on hepatic copper levels was observed in Labrador retrievers. Although the etiology of copper toxicosis in Labrador retrievers is multifactorial, a genome wide association study showed a clear association of hepatic copper levels with variations in the Wilson disease gene *ATP7B*. Interestingly, not only the Wilson disease gene but also the Menkes disease gene was associated to variation in hepatic copper levels in this breed. The presence of the mutation in *ATP7A* seemed to attenuate hepatic copper levels in

Labrador retrievers, but did not induce obvious copper deficiency symptoms in this cohort. Functional assays in cell lines showed that the *ATP7B* mutation in the conserved arginine (ATP7B:p.Arg1453Gln in canine protein, corresponding to ATP7B:p.Arg1415Gln in human protein) resulted in an aberrant retention of the protein in the endoplasmic reticulum in high copper circumstances. The *ATP7A* mutation (ATP7A:p.Thr327Ile) did not induce aberrant trafficking of the protein, yet lead to abrogation of copper efflux in dermal fibroblasts, indicating a functional impairment of the protein [22].

The elucidation of part of the genetic background of copper-associated hepatitis in the Labrador retrievers, poses this breed as a new large animal model for Wilson disease. Furthermore, based on the observations in the Labrador retrievers, mutations in *ATP7A*, with a subtle effect, may be involved as modifiers in Wilson disease and may explain part of the observed difference in *ATP7B* mutation frequency and presence of clinical disease [22].

Copper storage disorders are present in a number of other pure bred dog populations including Skye terriers [93], West highland white terriers [94], Dobermanns [80], and Dalmatians [95]. The clinical phenotypes present in these dog breeds are slightly different compared to Bedlington terriers and Labrador retrievers. Future gene mapping studies in these dog breeds offer the opportunity to identify new genes and to unveil more complex copper associated phenotypes.

5. The Power of Gene-Mapping Studies in Dog for Identification of Genes Involved in Copper Metabolism

Illustrated by the examples of the identification of *COMMD1* in the Bedlington terriers and *ATP7A* and *ATP7B* in the Labrador retrievers, gene mapping studies in pure bred dog populations affected with copper toxicosis are a powerful tool to identify genes involved in copper metabolism. Genes involved in copper metabolism are highly conserved among species, implicating that identified genes and mutations in dogs are potentially valuable for evaluation in human patients.

One of the tools that can be used in dogs is the Genome Wide Association Study (GWAS) [96]. Typically, GWAS focuses on association between phenotype of diseases and single nucleotide polymorphisms (SNPs). SNPs are one base-pair variations in the DNA with variability in the population and a known chromosomal location. Phenotypic traits can be a binary (affected *vs.* unaffected) as well as quantitative (*i.e.*, hepatic copper levels). By using statistical methods, an association between a SNP variation and a phenotypic trait can be identified. The associated SNPs are considered to mark a region of the genome which influences the risk for disease or variation in a quantitative trait. Consecutive in-depth genetic analysis of the identified region by sequencing is needed for identification of the disease-causing variations [96].

Pure bred dogs possess some preferable characteristics for gene mapping studies [23,24,97,98]. Due to severe selection on external characteristics pure bred dogs have a simple genome build which is advantageous for unveiling the molecular genetics of complex diseases [97]. The linkage disequilibrium in dogs can extend over as much as 100 times longer distances than in humans, which reduces the number of SNPs that is needed for successful mapping of a phenotypic trait. Furthermore, the phenotypes of copper disease in a particular dog breed are much less diverse compared to those of human copper metabolism disorders, which makes accurate diagnosis much simpler.

The success of gene mapping studies in the Bedlington terrier (monogenetic disease) and the Labrador retriever (complex genetic disease) illustrate the power of the use of dogs in genetic studies for identification of new genes or modifier genes that may be involved in copper metabolism disorders in humans. GWAS in other breeds like Dobermanns, West Highland White terriers, and Dalmatians may help unveil other genes.

6. Canine Models for Development of New Chelation and Dietary Treatments

6.1. Chelation Therapy

Dogs with copper toxicosis are similar to humans with WD and ecogenetic forms of copper toxicosis with regard to hepatic copper accumulation [20]. Both in humans and in dogs, chelation therapy is the most important treatment for decreasing hepatic copper levels [79]. D-Penicillamine has been successfully used in Bedlington terriers [83], Dobermans [99], and Labrador retrievers [89] to reduce copper hepatic copper level and inflammatory lesions. Side effects from life-long systemic chelation therapy in humans may result in a decreased therapy compliance, and progression of Wilson disease [15]. Furthermore, current available chelation therapies may be ineffective for controlling clinical symptoms, or may even lead to deterioration of the disease [54]. In order to overcome the disadvantage of the current available chelators, the need exist to develop new chelators and test them for effectiveness and safety before application in human patients.

An example of a new promising drug is the small copper-binding peptide Methanobactin (MB) which was derived from methane-oxidizing bacteria [100].This unique protein has a very strong affinity for copper, and promotes copper excretion via the bile rather than via urine, which is the natural route for copper excretion. MB has shown its potential in a rat model for WD, where intraperitoneal infection resulted in a prompt, significant release of copper associated with MB into bile [101]. Intraperitoneal injections for long-term treatment in humans is not feasible. To test other routes of administration and to evaluate long-term effects of treatment, the canine models with longer life spans and larger body sizes can fulfill an important role in pre-clinical studies.

6.2. Dietary Strategy

Levels of dietary copper and zinc were identified to be of large influence on hepatic copper levels in Labrador retrievers [102].

This observation was explored further by the use of an adjusted diet in therapeutic protocols for dogs in a pre-clinical phase [103] and as a maintenance therapy in dogs successfully treated with D-Penicillamine [104,105]. Conclusions from these studies were that diet adaptation alone may be enough to normalize hepatic copper in a subpopulation of affected Labrador retrievers [103] and that dietary treatment may be a valuable alternative to lifelong, continuous D-Penicillamine therapy in this dog breed [104]. At this moment, diet trials were only performed in the Labrador retriever dog breed. Dietary effects on hepatic copper levels in other dog breeds affected with copper toxicosis require further study. Wilson disease patients are generally advised to avoid food with a high copper content, however the effects of specific dietary interventions have not yet been investigated.

Diet trials in canine patients illustrate the value of a large animal model for long-term dietary studies in which multiple liver biopsies over several years can be collected and evaluated. Such long-term trials are impossible in rodent models, due to their short lifespan. The small body size of rodents precludes the possibility for collection of follow-up liver biopsy specimens. Dogs can fulfill an important role for pre-clinical investigation of dietary components that may be beneficial for human patients. Dog food can be easily standardized and dietary intake can be controlled. In the near future, other dietary components, including soy protein isolates [106,107], can be evaluated in dogs prior to possible application in human patients.

7. Organoids and Transplantation Studies

Liver transplantation is indicated for Wilson disease patients refractory to chelation therapy or with fulminant liver failure [108]. The drawbacks of this procedure are its invasiveness, the limited availability of suitable liver transplants and the risk of graft rejection [109]. An alternative way for whole organ transplants is cell-based therapy, including transplantation of mature hepatocytes or adult stem-cells. Transplantation of primary hepatocytes is hampered by the lack of a sufficient supply of viable cells, due to the quick loss of viability and dedifferentiation in cell-cultures [110].

Recently, a long-term stable hepatic stem cell culture in 3D (hepatic organoids) was established [111,112]. Hepatic organoids are derived from the adult stem cell niche from the liver, the so called hepatic progenitor cells. Hepatic organoids are easily expandable and are stable in culture for several months which facilitates bulking of cells for re-transplantation [111,112]. To develop and evaluate new stem-cell based treatment strategies, relevant animal models are needed for testing the safety and efficacy of stem-cell treatment. In this way, canine models are invaluable to build the bridge between rodent models and human patients. Recently, a canine hepatic organoid culture was established for this purpose [113].

In order to avoid rejection of transplanted cells, autologous cell transplantation is preferable. Hepatic progenitor cells derived from a patient can be expanded in vitro before re-transplantation. Obviously, for patients with hereditary defects, gene-correction of cells before re-transplantation is then a necessary intermediate step.

Recently, we successfully demonstrated that gene supplementation, using lentiviral transduction, in hepatic organoids derived from *COMMD1* deficient dogs could rescue the copper accumulation phenotype [113]. Currently, the first experiments involving autologous organoid transplantations with *COMMD1* gene correction are being conducted in *COMMD1* deficient dogs with copper toxicosis. Herewith, we will test the safety and effectivity of organoid auto-transplantation, after gene correction as a new treatment modality for dogs and eventually humans with hepatic copper accumulation disorders [113]. At present, there is a continuing development in genome editing technologies, including TALENs [114] and CRISPR/Cas [115] that hold exciting promises for the future.

8. Conclusions

Recently established canine models for copper metabolism include the *COMMD1*-deficient dogs and Labrador retrievers with copper-associated hepatitis harboring both *ATP7A* and *ATP7B* mutations. Both canine models are invaluable for development and evaluation of new treatment strategies, including dietary treatment, chelation therapy, and autologous transplantation of hepatic organoids after gene-correction. Further, other canine populations affected with copper toxicosis may be explored by gene mapping studies for identification of new genes and mutations involved in copper metabolism.

Acknowledgments: Acknowledgments: The Department of Clinical Sciences of Companion Animals, Faculty of Veterinary Medicine, Utrecht University covered the costs for open access publication.

Conflicts of Interest: Conflicts of Interest: The authors declare no conflict of interest.

References

1. Osredkar, J.; Sustar, N. Copper and zinc, biological role and significance of Copper/Zinc imbalance. *J. Clin. Toxicol.* **2011**. [CrossRef]
2. Opazo, C.M.; Greenough, M.A.; Bush, A.I. Copper: From neurotransmission to neuroproteostasis. *Front. Aging Neurosci.* **2014**, *6*, 143. [CrossRef] [PubMed]
3. Chan, W.Y.; Rennert, O.M. The role of copper in iron metabolism. *Ann. Clin. Lab. Sci.* **1980**, *10*, 338–344. [PubMed]
4. Horn, D.; Barrientos, A. Mitochondrial copper metabolism and delivery to cytochrome C oxidase. *IUBMB Life* **2008**, *60*, 421–429. [CrossRef] [PubMed]
5. Miranda, M.; Bartoli, G.; Ragnelli, A.M.; Cittadini, A.; Palozza, P.; Aimola, P.; Zarivi, O.; Bonfigli, A. Copper deficiency and pigmentation in the Rat: Morphofunctional aspects. *J. Submicrosc. Cytol. Pathol.* **1992**, *24*, 273–279. [PubMed]
6. Bousquet-Moore, D.; Prohaska, J.R.; Nillni, E.A.; Czyzyk, T.; Wetsel, W.C.; Mains, R.E.; Eipper, B.A. Interactions of peptide amidation and copper: Novel biomarkers and mechanisms of neural dysfunction. *Neurobiol. Dis.* **2010**, *37*, 130–140. [CrossRef] [PubMed]
7. O'Dell, B.L. Roles for iron and copper in connective tissue biosynthesis. *Philos. Trans. R. Soc. Lond. B. Biol. Sci.* **1981**, *294*, 91–104. [CrossRef] [PubMed]
8. Kaler, S.G. ATP7A-related copper transport diseases-emerging concepts and future trends. *Nat. Rev. Neurol.* **2011**, *7*, 15–29. [CrossRef] [PubMed]
9. Menkes, J.H.; Alter, M.; Steigleder, G.K.; Weakley, D.R.; Sung, J.H. A Sex-linked recessive disorder with retardation of growth, peculiar hair, and focal cerebral and cerebellar degeneration. *Pediatrics* **1962**, *29*, 764–779. [PubMed]

10. Vulpe, C.; Levinson, B.; Whitney, S.; Packman, S.; Gitschier, J. Isolation of a Candidate gene for menkes disease and evidence that it encodes a copper-transporting ATPase. *Nat. Genet.* **1993**, *3*, 7–13. [CrossRef] [PubMed]
11. Bull, P.C.; Thomas, G.R.; Rommens, J.M.; Forbes, J.R.; Cox, D.W. The Wilson disease gene is a putative copper transporting P-type ATPase similar to the menkes gene. *Nat. Genet.* **1993**, *5*, 327–337. [CrossRef] [PubMed]
12. Tanzi, R.E.; Petrukhin, K.; Chernov, I.; Pellequer, J.L.; Wasco, W.; Ross, B.; Romano, D.M.; Parano, E.; Pavone, L.; Brzustowicz, L.M. The Wilson disease gene is a copper transporting ATPase with homology to the menkes disease gene. *Nat. Genet.* **1993**, *5*, 344–350. [CrossRef]
13. Roberts, E.A.; Schilsky, M.L. American association for study of liver diseases (AASLD). Diagnosis and treatment of Wilson disease: An update. *Hepatology* **2008**, *47*, 2089–2111. [CrossRef] [PubMed]
14. Ferenci, P. Phenotype-genotype correlations in patients with Wilson's disease. *Ann. N. Y. Acad. Sci.* **2014**, *1315*, 1–5. [CrossRef] [PubMed]
15. Shimizu, N.; Yamaguchi, Y.; Aoki, T. Treatment and management of Wilson's disease. *Pediatr. Int.* **1999**, *41*, 419–422. [CrossRef] [PubMed]
16. Sternlieb, I. Wilson's disease: Indications for liver transplants. *Hepatology* **1984**, *4*, 15S–17S. [CrossRef] [PubMed]
17. Tanner, M.S. Role of Copper in Indian childhood cirrhosis. *Am. J. Clin. Nutr.* **1998**, *67*, 1074S–1081S. [PubMed]
18. Muller, T.; Feichtinger, H.; Berger, H.; Muller, W. Endemic tyrolean infantile cirrhosis: An ecogenetic disorder. *Lancet* **1996**, *347*, 877–880. [CrossRef]
19. Scheinberg, I.H.; Sternlieb, I. Wilson disease and idiopathic copper toxicosis. *Am. J. Clin. Nutr.* **1996**, *63*, 842S–845S. [PubMed]
20. Fieten, H.; Leegwater, P.A.; Watson, A.L.; Rothuizen, J. Canine models of copper toxicosis for understanding mammalian copper metabolism. *Mamm. Genome* **2012**, *23*, 62–75. [CrossRef] [PubMed]
21. van De Sluis, B.; Rothuizen, J.; Pearson, P.L.; van Oost, B.A.; Wijmenga, C. Identification of a new copper metabolism gene by positional cloning in a purebred dog population. *Hum. Mol. Genet.* **2002**, *11*, 165–173. [CrossRef] [PubMed]
22. Fieten, H.; Gill, Y.; Martin, A.J.; Concilli, M.; Dirksen, K.; van Steenbeek, F.G.; Spee, B.; van den Ingh, T.S.; Martens, E.C.; Festa, P.; *et al.* The Menkes and Wilson disease genes counteract in copper toxicosis in labrador retrievers: A new canine model for copper-metabolism disorders. *Dis. Model. Mech.* **2016**, *9*, 25–38. [CrossRef] [PubMed]
23. Shearin, A.L.; Ostrander, E.A. Leading the way: Canine models of genomics and disease. *Dis. Model. Mech.* **2010**, *3*, 27–34. [CrossRef]
24. Lindblad-Toh, K.; Wade, C.M.; Mikkelsen, T.S.; Karlsson, E.K.; Jaffe, D.B.; Kamal, M.; Clamp, M.; Chang, J.L.; Kulbokas, E.J., 3rd; Zody, M.C.; *et al.* Genome Sequence, comparative analysis and haplotype structure of the domestic dog. *Nature* **2005**, *438*, 803–819. [CrossRef] [PubMed]
25. Kasai, N.; Osanai, T.; Miyoshi, I.; Kamimura, E.; Yoshida, M.C.; Dempo, K. Clinico-pathological studies of LEC rats with hereditary hepatitis and hepatoma in the acute phase of hepatitis. *Lab. Anim. Sci.* **1990**, *40*, 502–505. [PubMed]
26. Theophilos, M.B.; Cox, D.W.; Mercer, J.F. The toxic milk mouse is a murine model of wilson disease. *Hum. Mol. Genet.* **1996**, *5*, 1619–1624. [CrossRef] [PubMed]
27. Huster, D.; Finegold, M.J.; Morgan, C.T.; Burkhead, J.L.; Nixon, R.; Vanderwerf, S.M.; Gilliam, C.T.; Lutsenko, S. Consequences of copper accumulation in the livers of the ATP7B−/− (Wilson disease gene) knockout mice. *Am. J. Pathol.* **2006**, *168*, 423–434. [CrossRef] [PubMed]
28. Wang, Y.; Zhu, S.; Weisman, G.A.; Gitlin, J.D.; Petris, M.J. Conditional knockout of the Menkes disease copper transporter demonstrates its critical role in embryogenesis. *PLoS ONE* **2012**, *7*, e43039. [CrossRef] [PubMed]
29. Mason, K.E. A conspectus of research on copper metabolism and requirements of man. *J. Nutr.* **1979**, *109*, 1979–2066. [PubMed]
30. Moriya, M.; Ho, Y.H.; Grana, A.; Nguyen, L.; Alvarez, A.; Jamil, R.; Ackland, M.L.; Michalczyk, A.; Hamer, P.; Ramos, D.; *et al.* Copper is taken up efficiently from albumin and α2-macroglobulin by cultured human cells by more than one mechanism. *Am. J. Physiol. Cell Physiol.* **2008**, *295*, C708–C721. [CrossRef] [PubMed]
31. Van den Berghe, P.V.; Klomp, L.W. New Developments in the regulation of intestinal copper absorption. *Nutr. Rev.* **2009**, *67*, 658–672. [CrossRef] [PubMed]

32. Zhou, B.; Gitschier, J. hCTR1: A human gene for copper uptake identified by complementation in yeast. *Proc. Natl. Acad. Sci. USA.* **1997**, *94*, 7481–7486. [CrossRef] [PubMed]

33. Coyle, P.; Philcox, J.C.; Carey, L.C.; Rofe, A.M. Metallothionein: The multipurpose protein. *Cell. Mol. Life Sci.* **2002**, *59*, 627–647. [CrossRef] [PubMed]

34. Freedman, J.H.; Ciriolo, M.R.; Peisach, J. The role of glutathione in copper metabolism and toxicity. *J. Biol. Chem.* **1989**, *264*, 5598–5605. [PubMed]

35. Rosenzweig, A.C. Copper delivery by metallochaperone proteins. *Acc. Chem. Res.* **2001**, *34*, 119–128. [CrossRef] [PubMed]

36. Schmidt, P.J.; Kunst, C.; Culotta, V.C. Copper activation of superoxide dismutase 1 (SOD1) *in vivo*. Role for protein-protein interactions with the copper chaperone for SOD1. *J. Biol. Chem.* **2000**, *275*, 33771–33776. [CrossRef] [PubMed]

37. Amaravadi, R.; Glerum, D.M.; Tzagoloff, A. Isolation of a cDNA encoding the human homolog of COX17, a yeast gene essential for mitochondrial copper recruitment. *Hum. Genet.* **1997**, *99*, 329–333. [CrossRef] [PubMed]

38. Klomp, L.W.; Lin, S.J.; Yuan, D.S.; Klausner, R.D.; Culotta, V.C.; Gitlin, J.D. Identification and functional expression of HAH1, a novel human gene involved in copper homeostasis. *J. Biol. Chem.* **1997**, *272*, 9221–9226. [PubMed]

39. Wang, Y.; Hodgkinson, V.; Zhu, S.; Weisman, G.A.; Petris, M.J. Advances in the Understanding of Mammalian Copper Transporters. *Adv. Nutr.* **2011**, *2*, 129–137. [CrossRef] [PubMed]

40. Kim, B.E.; Turski, M.L.; Nose, Y.; Casad, M.; Rockman, H.A.; Thiele, D.J. Cardiac copper deficiency activates a systemic signaling mechanism that communicates with the copper acquisition and storage organs. *Cell Metab.* **2010**, *11*, 353–363. [CrossRef] [PubMed]

41. Yanagimoto, C.; Harada, M.; Kumemura, H.; Abe, M.; Koga, H.; Sakata, M.; Kawaguchi, T.; Terada, K.; Hanada, S.; Taniguchi, E.; *et al.* Copper incorporation into ceruloplasmin is regulated by Niemann-Pick C1 protein. *Hepatol. Res.* **2011**, *41*, 484–491. [CrossRef] [PubMed]

42. Cater, M.A.; La Fontaine, S.; Shield, K.; Deal, Y.; Mercer, J.F. ATP7B mediates vesicular sequestration of copper: Insight into biliary copper excretion. *Gastroenterology* **2006**, *130*, 493–506. [CrossRef] [PubMed]

43. Nyasae, L.; Bustos, R.; Braiterman, L.; Eipper, B.; Hubbard, A. Dynamics of endogenous ATP7A (Menkes Protein) in intestinal epithelial cells: Copper-dependent redistribution between two intracellular sites. *Am. J. Physiol. Gastrointest. Liver Physiol.* **2007**, *292*, G1181–G1194. [CrossRef] [PubMed]

44. Polishchuk, E.V.; Concilli, M.; Iacobacci, S.; Chesi, G.; Pastore, N.; Piccolo, P.; Paladino, S.; Baldantoni, D.; van IJzendoorn, S.C.; Chan, J.; *et al.* Wilson disease protein ATP7B utilizes lysosomal exocytosis to maintain copper homeostasis. *Dev. Cell* **2014**, *29*, 686–700. [CrossRef] [PubMed]

45. Phillips-Krawczak, C.A.; Singla, A.; Starokadomskyy, P.; Deng, Z.; Osborne, D.G.; Li, H.; Dick, C.J.; Gomez, T.S.; Koenecke, M.; Zhang, J.S.; *et al.* COMMD1 is linked to the WASH complex and regulates endosomal trafficking of the copper transporter ATP7A. *Mol. Biol. Cell* **2015**, *26*, 91–103. [CrossRef] [PubMed]

46. Vonk, W.I.; de Bie, P.; Wichers, C.G.; van den Berghe, P.V.; van der Plaats, R.; Berger, R.; Wijmenga, C.; Klomp, L.W.; van de Sluis, B. The copper-transporting capacity of ATP7A mutants associated with menkes disease is ameliorated by COMMD1 as a result of improved protein expression. *Cell. Mol. Life Sci.* **2012**, *69*, 149–163. [CrossRef] [PubMed]

47. De Bie, P.; van de Sluis, B.; Burstein, E.; van de Berghe, P.V.; Muller, P.; Berger, R.; Gitlin, J.D.; Wijmenga, C.; Klomp, L.W. Distinct Wilson's disease mutations in ATP7B are associated with enhanced binding to COMMD1 and reduced stability of ATP7B. *Gastroenterology* **2007**, *133*, 1316–1326. [CrossRef] [PubMed]

48. Miyayama, T.; Hiraoka, D.; Kawaji, F.; Nakamura, E.; Suzuki, N.; Ogra, Y. Roles of COMM-domain-containing 1 in stability and recruitment of the copper-transporting ATPase in a mouse hepatoma cell line. *Biochem. J.* **2010**, *429*, 53–61. [CrossRef] [PubMed]

49. Goldstein, D.S.; Holmes, C.S.; Kaler, S.G. Relative efficiencies of plasma catechol levels and ratios for neonatal diagnosis of menkes disease. *Neurochem. Res.* **2009**, *34*, 1464–1468. [CrossRef]

50. Kaler, S.G.; Holmes, C.S.; Goldstein, D.S.; Tang, J.; Godwin, S.C.; Donsante, A.; Liew, C.J.; Sato, S.; Patronas, N. Neonatal diagnosis and treatment of menkes disease. *N. Engl. J. Med.* **2008**, *358*, 605–614. [CrossRef] [PubMed]

51. Abdel Ghaffar, T.Y.; Elsayed, S.M.; Elnaghy, S.; Shadeed, A.; Elsobky, E.S.; Schmidt, H. Phenotypic and genetic characterization of a cohort of pediatric Wilson disease patients. *BMC Pediatr.* **2011**, *11*. [CrossRef] [PubMed]

52. Ferenci, P.; Czlonkowska, A.; Merle, U.; Ferenc, S.; Gromadzka, G.; Yurdaydin, C.; Vogel, W.; Bruha, R.; Schmidt, H.T.; Stremmel, W. Late-onset Wilson's disease. *Gastroenterology* **2007**, *132*, 1294–1298. [CrossRef] [PubMed]

53. Das, S.K.; Ray, K. Wilson's disease: An Update. *Nat. Clin. Pract. Neurol.* **2006**, *2*, 482–493. [CrossRef] [PubMed]

54. Ala, A.; Walker, A.P.; Ashkan, K.; Dooley, J.S.; Schilsky, M.L. Wilson's disease. *Lancet* **2007**, *369*, 397–408. [CrossRef]

55. Wilson Disease Mutation Database. Available online: http://www.Wilsondisease.Med.Ualberta.Ca/Database.Asp (accessed on 7 October 2009).

56. Czlonkowska, A.; Gromadzka, G.; Chabik, G. Monozygotic female twins discordant for phenotype of Wilson's disease. *Mov. Disord.* **2009**, *24*, 1066–1069. [CrossRef] [PubMed]

57. Gromadzka, G.; Rudnicka, M.; Chabik, G.; Przybylkowski, A.; Czlonkowska, A. Genetic variability in the methylenetetrahydrofolate reductase gene (MTHFR) affects clinical expression of Wilson's disease. *J. Hepatol.* **2011**, *55*, 913–919. [CrossRef]

58. Schiefermeier, M.; Kollegger, H.; Madl, C.; Polli, C.; Oder, W.; Kuhn, H.; Berr, F.; Ferenci, P. The impact of apolipoprotein E genotypes on age at onset of symptoms and phenotypic expression in Wilson's disease. *Brain* **2000**, *123 Pt 3*, 585–590. [CrossRef] [PubMed]

59. Merle, U.; Stremmel, W.; Gessner, R. Influence of homozygosity for methionine at codon 129 of the human prion gene on the onset of neurological and hepatic symptoms in Wilson disease. *Arch. Neurol.* **2006**, *63*, 982–985. [CrossRef]

60. Litwin, T.; Gromadzka, G.; Czlonkowska, A. Apolipoprotein E gene (APOE) genotype in Wilson's disease: impact on clinical presentation. *Parkinsonism Relat. Disord.* **2012**, *18*, 367–369. [CrossRef] [PubMed]

61. Gromadzka, G.; Czlonkowska, A. Influence of IL-1RN intron 2 variable number of tandem repeats (VNTR) polymorphism on the age at onset of neuropsychiatric symptoms in Wilson's disease. *Int. J. Neurosci.* **2011**, *121*, 8–15. [CrossRef]

62. Weiss, K.H.; Runz, H.; Noe, B.; Gotthardt, D.N.; Merle, U.; Ferenci, P.; Stremmel, W.; Fullekrug, J. Genetic analysis of BIRC4/XIAP as a Putative modifier gene of wilson disease. *J. Inherit. Metab. Dis.* **2010**, *33* (Suppl. 3), S233–S240. [CrossRef] [PubMed]

63. Schilsky, M.L. Treatment of Wilson's disease: What are the relative roles of penicillamine, trientine, and zinc supplementation? *Curr. Gastroenterol. Rep.* **2001**, *3*, 54–59. [CrossRef] [PubMed]

64. Czlonkowska, A.; Litwin, T.; Karlinski, M.; Dziezyc, K.; Chabik, G.; Czerska, M. D-Penicillamine *vs.* zinc sulfate as first-line therapy for Wilson's disease. *Eur. J. Neurol.* **2014**, *21*, 599–606. [CrossRef] [PubMed]

65. Boga, S.; Jain, D.; Schilsky, M.L. Trientine induced colitis during therapy for Wilson disease: A case report and review of the literature. *BMC Pharmacol. Toxicol.* **2015**, *16*. 30–015–0031-z. [CrossRef] [PubMed]

66. Dahlman, T.; Hartvig, P.; Lofholm, M.; Nordlinder, H.; Loof, L.; Westermark, K. Long-term treatment of Wilson's disease with Triethylene tetramine dihydrochloride (Trientine). *QJM* **1995**, *88*, 609–616. [PubMed]

67. Brewer, G.J.; Terry, C.A.; Aisen, A.M.; Hill, G.M. Worsening of neurologic syndrome in patients with Wilson's disease with initial Penicillamine therapy. *Arch. Neurol.* **1987**, *44*, 490–493. [CrossRef] [PubMed]

68. Brewer, G.J.; Turkay, A.; Yuzbaziyan-Gurkan, V. Development of neurologic symptoms in a patient with asymptomatic Wilson's disease treated with penicillamine. *Arch. Neurol.* **1994**, *51*, 304–305. [CrossRef] [PubMed]

69. Ranucci, G.; di Dato, F.; Spagnuolo, M.I.; Vajro, P.; Iorio, R. Zinc Monotherapy is effective in Wilson's disease patients with mild liver disease diagnosed in childhood: a retrospective study. *Orphanet J. Rare Dis.* **2014**, *9*. [CrossRef]

70. Marin, C.; Robles, R.; Parrilla, G.; Ramirez, P.; Bueno, F.S.; Parrilla, P. Liver Transplantation in Wilson's disease: Are its indications established? *Transplant. Proc.* **2007**, *39*, 2300–2301. [CrossRef] [PubMed]

71. Tamura, S.; Sugawara, Y.; Kishi, Y.; Akamatsu, N.; Kaneko, J.; Makuuchi, M. Living-related liver transplantation for Wilson's disease. *Clin. Transplant.* **2005**, *19*, 483–486. [CrossRef] [PubMed]

72. Emre, S.; Atillasoy, E.O.; Ozdemir, S.; Schilsky, M.; Rathna Varma, C.V.; Thung, S.N.; Sternlieb, I.; Guy, S.R.; Sheiner, P.A.; Schwartz, M.E.; *et al.* Orthotopic liver transplantation for Wilson's disease: A single-center experience. *Transplantation* **2001**, *72*, 1232–1236. [CrossRef] [PubMed]

73. Bax, R.T.; Hassler, A.; Luck, W.; Hefter, H.; Krageloh-Mann, I.; Neuhaus, P.; Emmrich, P. Cerebral manifestation of Wilson's disease successfully treated with liver transplantation. *Neurology* **1998**, *51*, 863–865. [CrossRef] [PubMed]

74. Podgaetz, E.; Chan, C. Liver transplant team. Liver transplantation for Wilson s disease: Our experience with review of the literature. *Ann. Hepatol.* **2003**, *2*, 131–134. [PubMed]

75. Nayak, N.C.; Chitale, A.R. Indian Childhood Cirrhosis (ICC) & ICC-like diseases: The changing scenario of facts *vs.* notions. *Indian J. Med. Res.* **2013**, *137*, 1029–1042. [PubMed]

76. Li, Y.; Togashi, Y.; Sato, S.; Emoto, T.; Kang, J.H.; Takeichi, N.; Kobayashi, H.; Kojima, Y.; Une, Y.; Uchino, J. Spontaneous hepatic copper accumulation in long-evans cinnamon rats with hereditary hepatitis. A model of Wilson's disease. *J. Clin. Investig.* **1991**, *87*, 1858–1861. [CrossRef] [PubMed]

77. Haywood, S.; Muller, T.; Muller, W.; Heinz-Erian, P.; Tanner, M.S.; Ross, G. Copper-Associated Liver Disease in north ronaldsay sheep: A possible animal model for non-wilsonian hepatic copper toxicosis of infancy and childhood. *J. Pathol.* **2001**, *195*, 264–269. [CrossRef] [PubMed]

78. Puls, R. *Mineral Levels in Animal Health: Diagnostic Data*, 2nd ed.; Clearbrook, B.C., Ed.; Sherpa international, cop.: Dalhousie, QC, Canada, 1994.

79. Fieten, H.; Dirksen, K.; van den Ingh, T.S.; Winter, E.A.; Watson, A.L.; Leegwater, P.A.; Rothuizen, J. D-Penicillamine treatment of copper-associated hepatitis in labrador retrievers. *Vet. J.* **2013**, *196*, 522–527. [CrossRef] [PubMed]

80. Mandigers, P.J.; van den Ingh, T.S.; Bode, P.; Teske, E.; Rothuizen, J. Association between Liver Copper Concentration and subclinical hepatitis in doberman pinschers. *J. Vet. Intern. Med.* **2004**, *18*, 647–650. [CrossRef] [PubMed]

81. Twedt, D.C.; Hunsaker, H.A.; Allen, K.G. Use of 2,3,2-Tetramine as a hepatic copper chelating agent for treatment of copper hepatotoxicosis in bedlington terriers. *J. Am. Vet. Med. Assoc.* **1988**, *192*, 52–56. [PubMed]

82. Brewer, G.J.; Dick, R.D.; Schall, W.; Yuzbasiyan-Gurkan, V.; Mullaney, T.P.; Pace, C.; Lindgren, J.; Thomas, M.; Padgett, G. Use of zinc acetate to treat copper toxicosis in dogs. *J. Am. Vet. Med. Assoc.* **1992**, *201*, 564–568. [PubMed]

83. Twedt, D.C.; Sternlieb, I.; Gilbertson, S.R. Clinical, morphologic, and chemical studies on copper toxicosis of bedlington terriers. *J. Am. Vet. Med. Assoc.* **1979**, *175*, 269–275. [PubMed]

84. Hyun, C.; Filippich, L.J. Inherited canine copper toxicosis in australian bedlington terriers. *J. Vet. Sci.* **2004**, *5*, 19–28. [PubMed]

85. Vonk, W.I.; Bartuzi, P.; de Bie, P.; Kloosterhuis, N.; Wichers, C.G.; Berger, R.; Haywood, S.; Klomp, L.W.; Wijmenga, C.; van de Sluis, B. Liver-specific commd1 knockout mice are susceptible to hepatic copper accumulation. *PLoS ONE* **2011**, *6*, e29183. [CrossRef] [PubMed]

86. Klomp, A.E.; van de Sluis, B.; Klomp, L.W.; Wijmenga, C. The ubiquitously expressed MURR1 protein is absent in canine copper toxicosis. *J. Hepatol.* **2003**, *39*, 703–709. [CrossRef]

87. Su, L.C.; Ravanshad, S.; Owen, C.A., Jr.; McCall, J.T.; Zollman, P.E.; Hardy, R.M. A Comparison of copper-loading disease in bedlington terriers and Wilson's disease in humans. *Am. J. Physiol.* **1982**, *243*, G226–G230. [PubMed]

88. Favier, R.P.; Spee, B.; Penning, L.C.; Rothuizen, J. Copper-induced hepatitis: The COMMD1 deficient dog as a translational animal model for human chronic hepatitis. *Vet. Q.* **2011**, *31*, 49–60. [CrossRef] [PubMed]

89. Hoffmann, G.; van den Ingh, T.S.; Bode, P.; Rothuizen, J. Copper-associated chronic hepatitis in labrador retrievers. *J. Vet. Intern. Med.* **2006**, *20*, 856–861. [CrossRef] [PubMed]

90. Shih, J.L.; Keating, J.H.; Freeman, L.M.; Webster, C.R. Chronic hepatitis in labrador retrievers: Clinical presentation and prognostic factors. *J. Vet. Intern. Med.* **2007**, *21*, 33–39. [CrossRef] [PubMed]

91. Smedley, R.; Mullaney, T.; Rumbeiha, W. Copper-Associated hepatitis in labrador retrievers. *Vet. Pathol.* **2009**, *46*, 484–490. [CrossRef] [PubMed]

92. Hoffmann, G.; Heuven, H.C.; Leegwater, P.A.; Jones, P.G.; van den Ingh, T.S.; Bode, P.; Rothuizen, J. Heritabilities of copper-accumulating traits in labrador retrievers. *Anim. Genet.* **2008**, *39*, 454. [CrossRef] [PubMed]

93. Haywood, S.; Rutgers, H.C.; Christian, M.K. Hepatitis and copper accumulation in skye terriers. *Vet. Pathol.* **1988**, *25*, 408–414. [CrossRef] [PubMed]
94. Thornburg, L.P.; Shaw, D.; Dolan, M.; Raisbeck, M.; Crawford, S.; Dennis, G.L.; Olwin, D.B. Hereditary copper toxicosis in West Highland white terriers. *Vet. Pathol.* **1986**, *23*, 148–154. [CrossRef] [PubMed]
95. Webb, C.B.; Twedt, D.C.; Meyer, D.J. Copper-associated liver disease in dalmatians: A review of 10 dogs (1998–2001). *J. Vet. Intern. Med.* **2002**, *16*, 665–668.
96. Karlsson, E.K.; Baranowska, I.; Wade, C.M.; Salmon Hillbertz, N.H.; Zody, M.C.; Anderson, N.; Biagi, T.M.; Patterson, N.; Pielberg, G.R.; Kulbokas, E.J., 3rd; *et al.* Efficient mapping of mendelian traits in dogs through genome-wide association. *Nat. Genet.* **2007**, *39*, 1321–1328. [CrossRef] [PubMed]
97. Karlsson, E.K.; Lindblad-Toh, K. Leader of the pack: Gene mapping in dogs and other model organisms. *Nat. Rev. Genet.* **2008**, *9*, 713–725. [CrossRef] [PubMed]
98. Parker, H.G.; Kim, L.V.; Sutter, N.B.; Carlson, S.; Lorentzen, T.D.; Malek, T.B.; Johnson, G.S.; DeFrance, H.B.; Ostrander, E.A.; Kruglyak, L. Genetic structure of the purebred domestic dog. *Science* **2004**, *304*, 1160–1164. [CrossRef] [PubMed]
99. Mandigers, P.J.; van den Ingh, T.S.; Bode, P.; Rothuizen, J. Improvement in Liver Pathology After 4 Months of D-penicillamine in 5 doberman pinschers with subclinical hepatitis. *J. Vet. Intern. Med.* **2005**, *19*, 40–43. [CrossRef] [PubMed]
100. Kim, H.J.; Graham, D.W.; DiSpirito, A.A.; Alterman, M.A.; Galeva, N.; Larive, C.K.; Asunskis, D.; Sherwood, P.M. Methanobactin, a copper-acquisition compound from methane-oxidizing bacteria. *Science* **2004**, *305*, 1612–1615. [CrossRef] [PubMed]
101. Summer, K.H.; Lichtmannegger, J.; Bandow, N.; Choi, D.W.; DiSpirito, A.A.; Michalke, B. The biogenic methanobactin is an effective chelator for copper in a rat model for wilson disease. *J. Trace Elem. Med. Biol.* **2011**, *25*, 36–41. [CrossRef] [PubMed]
102. Fieten, H.; Hooijer-Nouwens, B.D.; Biourge, V.C.; Leegwater, P.A.; Watson, A.L.; van den Ingh, T.S.; Rothuizen, J. Association of Dietary Copper and Zinc Levels with Hepatic Copper and Zinc Concentration in Labrador Retrievers. *J. Vet. Intern. Med.* **2012**, *26*, 1274–1280. [CrossRef] [PubMed]
103. Fieten, H.; Biourge, V.C.; Watson, A.L.; Leegwater, P.A.; van den Ingh, T.S.; Rothuizen, J. Dietary management of labrador retrievers with subclinical hepatic copper accumulation. *J. Vet. Intern. Med.* **2015**, *29*, 822–827. [CrossRef] [PubMed]
104. Fieten, H.; Biourge, V.C.; Watson, A.L.; Leegwater, P.A.; van den Ingh, T.S.; Rothuizen, J. Nutritional management of inherited copper-associated hepatitis in the labrador retriever. *Vet. J.* **2014**, *199*, 429–433. [CrossRef] [PubMed]
105. Hoffmann, G.; Jones, P.G.; Biourge, V.; van den Ingh, T.S.; Mesu, S.J.; Bode, P.; Rothuizen, J. Dietary management of hepatic copper accumulation in labrador retrievers. *J. Vet. Intern. Med.* **2009**, *23*, 957–963. [CrossRef] [PubMed]
106. Yonezawa, K.; Nakagama, H.; Tajima, R.; Ushigome, M.; Ogra, Y.; Suzuki, K.T.; Yoshikawa, K.; Nagao, M. Effects of soy protein isolate on LEC rats, a model of Wilson disease: Mechanisms underlying enhancement of liver cell damage. *Biochem. Biophys. Res. Commun.* **2003**, *302*, 271–274. [CrossRef]
107. Yonezawa, K.; Nunomiya, S.; Daigo, M.; Ogra, Y.; Suzuki, K.T.; Enomoto, K.; Nakagama, H.; Yoshikawa, K.; Nagao, M. Soy protein isolate enhances hepatic copper accumulation and cell damage in LEC rats. *J. Nutr.* **2003**, *133*, 1250–1254. [PubMed]
108. Schilsky, M.L.; Scheinberg, I.H.; Sternlieb, I. Liver transplantation for Wilson's disease: Indications and outcome. *Hepatology* **1994**, *19*, 583–587. [CrossRef] [PubMed]
109. Dalgetty, D.M.; Medine, C.N.; Iredale, J.P.; Hay, D.C. Progress and future challenges in stem cell-derived liver technologies. *Am. J. Physiol. Gastrointest. Liver Physiol.* **2009**, *297*, G241–G248. [CrossRef] [PubMed]
110. Van Laecke, S.; Desideri, F.; Geerts, A.; van Vlierberghe, H.; Berrevoet, F.; Rogiers, X.; Troisi, R.; de Hemptinne, B.; Vanholder, R.; Colle, I. Hypomagnesemia and the risk of new-onset diabetes after liver transplantation. *Liver Transpl.* **2010**, *16*, 1278–1287. [CrossRef] [PubMed]
111. Huch, M.; Dorrell, C.; Boj, S.F.; van Es, J.H.; Li, V.S.; van de Wetering, M.; Sato, T.; Hamer, K.; Sasaki, N.; Finegold, M.J.; *et al.* In Vitro Expansion of single Lgr5+ liver stem cells induced by Wnt-driven regeneration. *Nature* **2013**, *494*, 247–250. [CrossRef] [PubMed]

112. Huch, M.; Gehart, H.; van Boxtel, R.; Hamer, K.; Blokzijl, F.; Verstegen, M.M.; Ellis, E.; van Wenum, M.; Fuchs, S.A.; de Ligt, J.; *et al.* Long-term culture of genome-stable bipotent stem cells from adult human liver. *Cell* **2015**, *160*, 299–312. [CrossRef] [PubMed]
113. Nantasanti, S.; Spee, B.; Kruitwagen, H.S.; Chen, C.; Geijsen, N.; Oosterhoff, L.A.; van Wolferen, M.E.; Pelaez, N.; Fieten, H.; Wubbolts, R.W.; *et al.* Disease modeling and gene therapy of copper storage disease in canine hepatic organoids. *Stem Cell. Rep.* **2015**, *5*, 895–907. [CrossRef]
114. Boch, J. TALEs of Genome Targeting. *Nat. Biotechnol.* **2011**, *29*, 135–136. [CrossRef] [PubMed]
115. Horvath, P.; Barrangou, R. CRISPR/Cas, the immune system of bacteria and archaea. *Science* **2010**, *327*, 167–170. [CrossRef] [PubMed]

International Journal of
Molecular Sciences

MDPI

Review

The Complex Relationship between Metals and Carbonic Anhydrase: New Insights and Perspectives

Maria Giulia Lionetto *, Roberto Caricato, Maria Elena Giordano and Trifone Schettino

Department of Biological and Environmental Science and Technologies (DiSTeBA), University of Salento, Via Prov.le Lecce-Monteroni, 73100 Lecce, Italy; roberto.caricato@unisalento.it (R.C.); elena.giordano@unisalento.it (M.E.G.); trifone.schettino@unisalento.it (T.S.)

* Correspondence: giulia.lionetto@unisalento.it; Tel.: +39-0832-298694; Fax: +39-0832-298626

Academic Editor: Reinhard Dallinger
Received: 2 December 2015; Accepted: 11 January 2016; Published: 19 January 2016

Abstract: Carbonic anhydrase is a ubiquitous metalloenzyme, which catalyzes the reversible hydration of CO_2 to HCO_3^- and H^+. Metals play a key role in the bioactivity of this metalloenzyme, although their relationships with CA have not been completely clarified to date. The aim of this review is to explore the complexity and multi-aspect nature of these relationships, since metals can be cofactors of CA, but also inhibitors of CA activity and modulators of CA expression. Moreover, this work analyzes new insights and perspectives that allow translating new advances in basic science on the interaction between CA and metals to applications in several fields of research, ranging from biotechnology to environmental sciences.

Keywords: carbonic anhydrase; metals; inhibition; expression; biomarker; bioassay

---

1. Introduction

Carbonic anhydrase (CA) is a widely-distributed metalloenzyme, which catalyzes the reversible hydration of CO_2 to HCO_3^- and H^+. This biochemical reaction plays a key physiological role in diverse biological systems. Six distinct and unrelated CA families (α-, β-, γ-CA, δ, ζ and η-CAs) have been identified in animals, plants, algae and bacteria [1,2]. They all catalyze the same reaction of CO_2 hydration, but each family shows proper specific characteristics in primary amino acid sequence and 3D tertiary structure.

In animals, CA isoforms play a fundamental role in a number of physiological processes involving carbon dioxide and bicarbonate, such as transport of CO_2 and HCO_3^- between body tissues and respiratory surfaces, pH homeostasis, electrolyte transport in various epithelia, biosynthetic reactions (gluconeogenesis, lipogenesis and ureagenesis), bone resorption and calcification. In algae, plants and some bacteria, CA isoforms are fundamental for photosynthesis [3,4].

The α-carbonic anhydrases are monomeric or dimeric and are found in animals, some fungi, bacteria, algae and green plants [5]. In mammals, at least 16 different α CA isoforms were isolated. Mammalian CA isoforms CAI, II, III, IV, VA, VB, VI, VII, IX, XII, XIII, XIV and CA XV (not expressed in humans) have catalytic activity, while the remaining three CAs (CARP VIII, CARP X and CARP XI) have lost the catalytic activity and are known as CA-related proteins [6].

The β-carbonic anhydrases are dimers, tetramers or octamers and are expressed mainly in fungi, bacteria, archaea, algae and chloroplasts of monocotyledons and dicotyledons [7] and some prokaryotes [8].

The γ-anhydrase class is a homotrimer that has been described in bacteria, Archaea and plants [9]. It also includes a number of non-catalytically-active homologs present in diverse species. δ- and ζ-CAs are present in several classes of marine phytoplankton. The δ class has been described in diatoms, and its prototype is the CA TWCA1 from the marine diatom *Thalassiosira weissflogii* [10,11]. The ζ-CAs are probably monomers and have three slightly different active sites on the same protein molecule [12].

The η-CA was recently found in a number of species of the *Plasmodium* genus. These are a group of enzymes previously ascribed to the α family, but recently demonstrated to have a number of unique features, including their metal ion coordination pattern [2].

This review focuses on an interesting aspect of the research on CA, the relationships between carbonic anhydrase and metals, which play a fundamental role in the bioactivity of this metalloenzyme. The review points out the complexity and multi-aspect nature of these relationships, since metals can be cofactors of CA, but also inhibitors of CA activity and modulators of CA expression. New insights and perspectives are discussed encompassing several fields of research from biotechnological applications to environmental sciences.

2. Metals and CA Catalytic Site

All CA isoenzymes catalyze the reversible hydration of CO_2 to HCO_3 and H^+ through a metal-hydroxide [$Lig^3 M^{2+}(OH)^-$] mechanism [13–15] (Figure 1). The central catalytic step involves the reaction between CO_2 and the OH^- bound to the zinc ion, yielding a coordinated HCO_3^- ion, which is subsequently displaced from the metal by H_2O. In the α-, γ- and δ-CA classes, Lig^3 is represented by three key amino acid residues, which are three histidines in α-CA, γ-CA and δ-CA, one histidine and two cysteines in β-CA and ζ-CA and two His and one Gly residues in η-CA [16]. A fourth histidine, that is His 64 in human CAII (the most investigated CA isoform), not directly part of the active site, contributes to the catalytic process representing the so-called "proton shuttle". This allows the H^+ transfer from the metal-bound water molecule to buffer molecules located outside the active site and ensures the reaction of the metal-bound OH^- with CO_2 to produce HCO_3^-.

The metal (M) in the carbonic anhydrase metal-hydroxide [$Lig^3 M^{2+}(OH)^-$] mechanism is Zn^{2+} for all classes, but other transition metals have been demonstrated to bind to the catalytic site as physiologically-relevant metal cofactors or displacers of the native cofactor, producing in this case new CA metallovariants (Table 1).

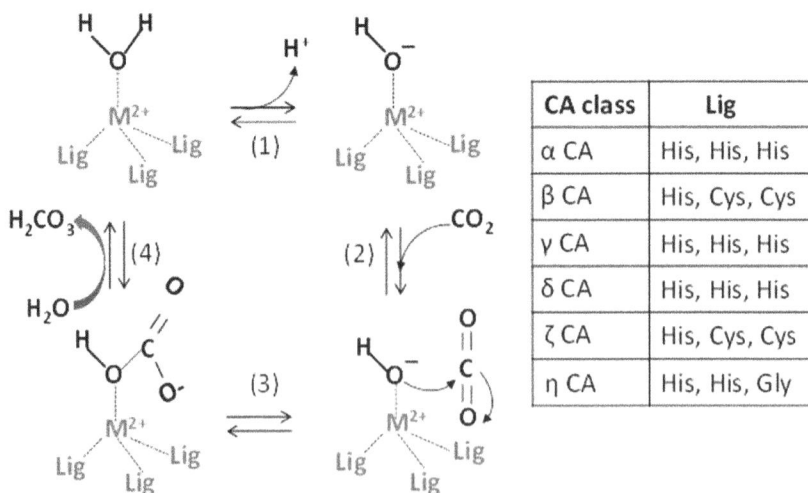

CA class	Lig
α CA	His, His, His
β CA	His, Cys, Cys
γ CA	His, His, His
δ CA	His, His, His
ζ CA	His, Cys, Cys
η CA	His, His, Gly

Figure 1. The reversible hydration of carbon dioxide to bicarbonate catalyzed by CAs by means of a metal (M)-hydroxide mechanism. Modified from Berg [17]. (**1**) The release of a proton from the zinc-bound water generates the zinc-bound OH^-; (**2**) A CO_2 molecule binds to the active site and is positioned for optimal interaction with the zinc-bound OH^-; (**3**) The hydroxide ion attacks the carbonyl of CO_2, producing HCO_3^-; (**4**) The release of HCO_3^- regenerates the enzyme.

Table 1. Metals as physiologically-relevant cofactors of CA.

CA Families	Metals as Physiologically-Relevant CA Cofactors	Ref.
α-CA	Zn^{2+}	[21]
β-CA	Zn^{2+}	[7]
γ-CA	Fe^{2+}; Zn^{2+}	[9,22]
δ-CA	Zn^{2+}	[10]
ζ-CA	Cd^{2+}; Zn^{2+}	[12]
η-CA	Zn^{2+}	[2]

Apart from zinc, other metals have been found to be physiologically-relevant cofactors of some CAs.

2.1. Metals as Physiologically-Relevant Cofactors of CA

Zn^{2+} is one of the most widely-used metallic elements as enzyme cofactor in nature, and its presence in all of the CA families is a successful confirmation of its peculiar properties. The reason for its success lies in the filled *d* orbital (d10). Unlike other first-row transition elements (e.g., Sc^{2+}, Ti^{2+}, V^{2+}, Cr^{2+}, Mn^{2+}, Fe^{2+}, Co^{2+}, Ni^{2+} and Cu^{2+}), Zn^{2+} is not involved in redox reactions, but rather, it acts as a Lewis acid accepting a pair of electrons [18]. This makes zinc a good metal cofactor for biochemical reactions requiring a redox-stable ion to function as a Lewis acid-type catalyst [19], such as proteolysis and carbon dioxide hydration. Zinc complexes have low thermodynamic stabilities, as well as variable geometries, which in turn account for low activation barriers. This makes zinc a versatile and suitable as an active site metal [20]. Zinc is in the +2 state, and it is positioned in a cleft in the center of the CA molecule (Figure 2). It is coordinated by the three key amino acid residues (see above). The fourth coordination site is occupied by a water molecule. The role of Zn^{2+} in the CA catalytic mechanism is to promote the deprotonation of H_2O with the production of the nucleophilic OH^-, which in turn can attack the carbonyl group of CO_2 to convert it into HCO_3^-. A water molecule subsequently displaces the bicarbonate at the metal (Figure 1).

Fe^{2+} has been demonstrated to be a physiological metal cofactor of γ-CAs [22,23]. The γ-CA class is among the most ancient, with homologs widespread in Archaea and Bacteria [24]. CAM, the prototypic γ-class CA from the anaerobic Archaea species *Methanosarcina thermophila*, has been demonstrated to contain zinc in the active site when overproduced in *Escherichia coli* and purified in aerobic experimental conditions [22]. On the other hand, when the enzyme is purified anaerobically, it has been demonstrated to contain Fe^{2+} in the active site (Fe-CAM) and to have a three-fold increased activity. These contrasting results are explained by the fact that in aerobic conditions, Fe^{3+} is oxidized and is rapidly substituted by Zn^{2+} present in the reaction buffers not treated with chelating agents. The stability of complexed Zn^{2+} is much greater than that for Fe^{2+} ligated with the nitrogen atoms of histidine residues coordinating the active-site metal in CAM. Thus, aerobic purification results in the loss of Fe^{3+} and substitution with Zn^{2+}. These findings clearly suggest Fe^{2+} as the physiologically-relevant metal [22–24] in the active site of the CAM enzyme. Considering that Fe^{2+} is available in oxygen-free environments, the finding of iron as a physiological metal cofactor in CAM from an anaerobic species suggests the possibility that iron could act as a metal cofactor in γ-class CA enzymes and possibly in other classes in anaerobic organisms [23]. Moreover, evidence for the role of Fe^{2+} in CA catalytic activity has been found also in the α-class. In fact, in duck erythrocytes, the addition of Fe^{2+} in the reaction medium increases CA activity [25].

Cd^{2+} has been demonstrated to be naturally used as a catalytic metal in the cadmium-carbonic anhydrase (CDCA1), a ζ-CA isolated from the marine diatom *Thalassiosira weissflogii* [26–28]. Its homolog gene has been found in a number of diatom species, as well as natural assemblages [12]. CDCA1 provides the first evidence of the biological role of cadmium, which is usually considered a toxic element associated with environmental pollution [29]. CDCA1 is considered a "cambialistic" enzyme because it can use either Zn or Cd for the catalytic reaction [12]. The ability to use Cd in CDCA allows the diatom *T. weissflogii* to support the needs of fast growth also in Zn^{2+} limiting conditions, and this has ecologically relevant implications. In fact, in the oceans, which are known to be poor in metals, this ability might have given a competitive advantage to diatoms with respect to other species, contributing to the radiation of diatoms during the Cenozoic Era and to the parallel decrease in atmospheric CO_2 [12]. In the ocean, Cd shows a nutrient-like concentration profile, with a low concentration at the surface, because of phytoplankton uptake, and an increased concentration at depth due to remineralization of sinking organic matter. This Cd profile is thought to be explained by the use of this metal by CDCA.

Co^{2+} has been demonstrated to *in vivo* substitute Zn^{2+} in the other genetically-distinct CA form (TWCA1) (δ-CA) isolated from the diatom *T. weissflogii* [30]. Although the affinity of TWCA1 for Co^{2+} is lower than for Zn^{2+} and the Co-substituted enzyme is less active than the Zn-form [30], it has been suggested that the Co^{2+} substitution can alleviate the Zn^{2+} limitation in diatoms in open oceans, as also suggested for Cd^{2+} [26,30].

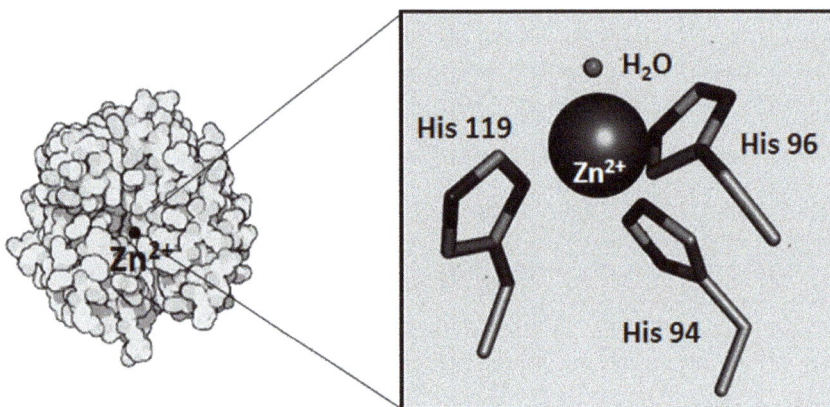

Figure 2. Human CAII: in detail, the metal binding site with the zinc ion as a sphere, the direct ligand histidines, H94, H96, H119, and the water molecule. Modified from Mahon *et al* [31] and from Dutta and Goodsell [32].

2.2. Metals as Displacers of the Native Metal Cofactor

In addition to the role of some metals as CA physiologically-relevant metal cofactors, many divalent metal ions, such as Co^{2+}, Ni^{2+}, Mn^{2+}, Cu^{2+}, Cd^{2+} and Hg^{2+}, have been demonstrated to easily bind to the three-histidine moiety within the CA active site in *in vitro* experimental conditions [33–35]. Most of this information comes from transmetallation experiments on the α-CAII. The CO_2 hydration catalytic activity of metal-substituted human CAIIs, relative to wild-type CAII activity, is about 7% for the Mn^{2+}-substituted CA, about 2% for the Cd^{2+}-substituted and about 0% for the Cu^{2+}-substituted and Hg^{2+}-substituted ones, respectively [36,37]. Only Co^{2+} is able to produce a catalytically-functioning enzyme with a wild-type catalytic efficiency ($K_{cat}/K_m = 8.7 \times 10^7$ $M^{-1} \cdot s^{-1}$ for Zn^{2+} *vs.* 8.8×10^7 $M^{-1} \cdot s^{-1}$ for Co^{2+}) [34] because only cobalt besides zinc has a tetrahedral coordination at a pH around eight. In particular, Co-CA represents an ideal metallovariant for studying the CA catalytic site. In contrast to Zn^{2+}, lacking a spectroscopic signature, the paramagnetic Co^{2+} is accessible to spectroscopic analysis, providing information about the environment of the metal ions in the CA active site [38].

2.3. Metal Binding to Other Sites in the CA Protein

In addition to the ability of some metals to bind the three key amino acid residues of the CA active site, metals are also able to bind CA protein in other sites elsewhere in the molecule thanks to their affinity for thiol and histidyl groups. Mercuric ions have been demonstrated by X-ray crystallography [39] to bind to the His-64 ring (the so-called proton shuttle) and also to Cys-206 of CAII [40]. Copper ion is known to bind to the imidazole of His-64 in CAII, preventing its function in the H^+ transfer from the zinc-bound H_2O to buffer molecules outside the active site [41]. In addition, Cu^{2+} is also able to bind CA in another binding site [42] nearby the N-terminus of the proteins, presumably in a four-coordinate tetragonal Cu^{2+} site with two histidine ligands and an N-terminal amine [43]. Because of the presence of this high affinity site, CA seems to be a member of a growing number of proteins and peptides that have been found to have an N-terminal Cu^{2+}-binding site, including serum albumin and amyloid β-peptide complexes [43]. It is unknown why CA has this N-terminal Cu^{2+} site that does not show any catalytic activity. It is possible that this site could function as a sequestering site of adventitious Cu^{2+}. In this respect, an intriguing hypothesis suggests that this N-terminal site can be effective in defending the native zinc(II)-(His)3 active site of CA, as well as other metalloproteins, from Cu^{2+} displacement, and in turn, inactivation [43].

3. Metals and CA Activity Inhibition

The ability of some transition metals either to displace Zn^{2+} in the active site or to bind histidine and cysteine residues in sites other than the active site accounts for the inhibitory effect of some metals on CA activity, reported by several authors in a number of tissues in a variety of animal species. Table 2 summarizes the results obtained in *in vitro* studies.

In the teleost fish *Ictalurus punctatus*, Cd^{2+}, Cu^{2+}, Ag^+ and Zn^{2+} were demonstrated to inhibit erythrocyte CA activity [44]. In the euryhaline teleost *Anguilla anguilla*, a significant *in vitro* tissue-specific inhibition of branchial and intestinal CA activity by Cd^{2+} was found [45,46]. In particular, in the intestine, a cytosolic and a membrane-bound CA were detected, with the cytosolic isoform more sensitive to the metal than the membrane-bound one. In the European seabass (*Dicentrarchus labrax*), Al^{+3}, Cu^{+2}, Pb^{+2}, Co^{+3}, Ag^{+1}, Zn^{+2} and Hg^{+2} were found to be competitive inhibitors of liver CA activity [47]. In the liver of sea bream (*Sparus aurata*) Ag^+, Ni^{2+}, Cd^{2+} and Cu^{2+} were shown to be noncompetitive inhibitors of CA activity [48].

In rainbow trout (*Oncorhynchus mykiss*), Co^{2+}, Cu^{2+}, Zn^{2+}, Ag^+ and Cd^{2+} inhibited brain CA activity. Ag^+ and Cd^{2+} were competitive inhibitors, Cu^{2+} was noncompetitive inhibitor, while Zn^{2+} was uncompetitive inhibitor [49].

In the estuarine crab *Chasmagnathus granulate*, Cd^{2+}, Cu^{2+} and Zn^{2+} inhibited branchial CA activity both *in vitro* and *in vivo* [50]. In other euryhaline crabs, *Callinectes sapidus* and *Carcinus maenas*, a significant *in vitro* inhibition of gill CA by Ag^+, Cd^{2+}, Cu^{2+} and Zn^{2+} was demonstrated, with strong differences between the gill CA of the two species, *Callinectes sapidus* CA being more sensitive than *Carcinus maenas* CA [51]. In humans, the cytosolic HCAI and II were *in vitro* inhibited by Pb^{2+}, Co^{2+} and Hg^{2+}. Pb^{2+} was noncompetitive inhibitor for HCAI and competitive for HCAII, Co^{2+} was competitive for HCAI and noncompetitive for HCAII, and Hg^{2+} was uncompetitive for both isoforms [52].

In the sturgeon *Acipenser gueldenstaedti*, Ag^+, Zn^{2+}, Cu^{2+} and Co^{2+} were weak inhibitors of the erythrocyte CA activity [53]. In sheep kidney, Pb^{2+}, Co^{2+}, Hg^{2+}, Cd^{2+}, Zn^{2+}, Se^{2+}, Cu^{2+} and Al^{3+} competitively inhibited CA activity in the low molar/millimolar range [54]. In the Turkish native chicken, Al^{3+}, Hg^{2+}, Cu^{2+}, Pb^{2+} and Cd^{2+} were competitive inhibitors of erythrocyte CA activity. Pb^{2+} exhibited the strongest inhibitory action; Cd^{2+} and Hg^{2+} were moderate inhibitors, while Al^{3+} and Cu^{2+} were weaker inhibitors [55].

Ki or IC_{50} values for *in vitro* metal inhibition of CA activity range from micromolar to millimolar depending on species, tissue and isoform. Although in some cases the Ki or IC_{50} values are in the submillimolar or millimolar range, suggesting a weak inhibition, they may be significant for infield inhibition due to the bioaccumulation process occurring for most metals in many exposed organisms under prolonged exposure [56].

Table 2. *In vitro* inhibition of CA activity by trace metals. IC_{50} and K_i values are expressed as mM concentrations.

Species	Tissue	Cd^{2+} (mM)	Cu^{2+}	Hg^{2+}	Zn^{2+}	Co^{2+}	Pb^{2+}	Ag^+	Ref.
Callinectes sapidus	gills	Ki 0.1×10^{-3}	Ki 3.6×10^{-3}		Ki 2–6×10^{-3}			Ki 0.05×10^{-3}	[51]
Carcinus maenas	gills	Ki 0.6–2.5	Ki 0.6–2.5		Ki 0.6–2.5				[51]
Anguilla anguilla	gills	IC_{50} 9.9×10^{-3}							[46]
Anguilla anguilla	intestine	IC_{50} 36.4×10^{-3}							[46]
Sparus aurata	liver	Ki 17.7 (non-competitive)	Ki 36.2 (non-competitive)					Ki 0.02 (non-competitive)	[48]
Ictalurus punctatus	erythrocyte	IC_{50} 0.9	IC_{50} 0.065		IC_{50} 0.7			IC_{50} 0.035	[44]
Oncorhynchus mykiss	brain	Ki 94.2×10^{-3}	Ki 27.6×10^{-3} (non-competitive)		Ki 1.20	Ki 0.035			[49]
Acipenser gueldenstaedti	erythrocyte		IC_{50} 5.2		IC_{50} 2.8			IC_{50} 1.7	[53]
Dicentrarchus labrax	liver			Ki 0.76	Ki 0.72 (competitive)	Ki 0.53 (competitive)	Ki 0.24 (competitive)		[47]
Gallus	erythrocyte		Ki 2.78 (competitive)	Ki 1.26 (competitive)			Ki 0.97 (competitive)		[55]
Ovis aries	kidney	Ki 1.04 (competitive)	Ki 4.70 (competitive)		Ki 0.96 (competitive)				[54]
Homo sapiens	Erythrocyte, CA I		Ki 3.22	Ki 3.22 (uncompetitive)		Ki 1.45 (competitive)	Ki 1 (non-competitive)		[52]
Homo sapiens	Erythrocyte, CA II		Ki 0.312	Ki 0.312 (uncompetitive)		Ki 1.7 (non-competitive)	Ki 0.056 (competitive)		[52]

The inhibitory effect of metals on CA activity demonstrated in *in vitro* studies has been confirmed by works carried out in *in vivo* exposure conditions. In the mantle of the filter-feeding mussel *Mytilus galloprovincialis*, CA was significantly inhibited by exposure to 1.78 µM Cd^{2+} for 15 days [57]. Considering the role of CA in carbonate salt deposition, the inhibitory effect of Cd^{2+} on mantle CA activity could explain the significant decrease in shell growth previously observed in *M. galloprovincialis* exposed to heavy metals. [58]. In the marine anemone *Aiptasia pallida*, a decrease of CA activity was observed after waterborne exposure to metal mixtures (Cu^{2+}, Zn^{2+}, Ni^{2+} and Cd^{2+}) with a significant CA activity reduction following treatment with 100 µg/L for three days [59]. CA activity in the gills of the freshwater bivalve *Anodonta anatine* was significantly inhibited by the exposure to 0.35 µM Cu^{2+} for 15 days [60]. A CA activity decrease with increasing exposure concentrations of Cu^{2+} was observed in two species of scleractinian corals, such as *Acropora cervicornis* and *Montastraea faveolata*. Significant effects were detected in *A. cervicornis* exposed to 10 and 20 µg/L Cu^{2+} and in M. faveolata exposed to 20 µg/L Cu^{2+} for five weeks [61], respectively. A significant decrease in CA activity was recorded also in the anemones *Condylactis gigantea* and *Stichodactyla helianthus* exposed to Cu^{2+}, Ni^{2+}, Pb^{2+} and V^{2+} for 48 h [62].

From all of the available studies, a great variability in the sensitivity of CA activity to *in vitro* and *in vivo* metal exposure was observed among species, tissues and metals. This great variability can be explained by the complexity and multifaceted aspects of the binding of metals with CA isoforms, producing a number of different inhibitory responses to different metals in different tissues and species. It is possible to hypothesize that structural differences in CA isoforms could generate different metal-binding affinities and in turn different inhibitory responses.

4. Metals and CA Protein Expression

The inhibition exerted by some metals on CA-specific isoforms, as a result of the displacement of the native metal cofactor or the binding to other sites in the CA molecule, did not explain all of the biological effects of metals on CA.

It is known in plants and phytoplankton that biosynthesis of CA may be regulated by some environmental factors, including environmental trace metal concentrations [26,63]. Concerning animals, less information is available on this aspect. Some works indicate CA protein expression to be influenced by Zn^{2+} availability. For example, in humans, patients with CAVI deficiency show stimulation of CAVI synthesis/secretion by Zn^{2+} treatment [64], presumably through metal-induced upregulation of the CAIV gene. In rats, zinc deficiency induced a decrease in CAII protein expression in the submandibular gland [65].

In the fish *Cyprinodon variegatus*, CAII gene expression in gills and intestine was influenced by Cu^{2+} exposure in a concentration-dependent manner, with a two-fold upregulation in the gills and a three-fold upregulation in the intestine following nine days of exposure to 100 mg/L Cu^{2+} [66]. The effect of Cu^{2+} on CA expression appears more complex if we consider that the metal in the same fish is also able to interfere with the regulation of CA expression by osmotic stress, disrupting the osmotic stress induction of the intestinal CAII. These results outline that CA plays a role in the combined effects of copper and osmotic stress on ion homeostasis and show how this enzyme represents a factor linking the physiological responses of the organism to multiple stresses.

In mussels, Caricato *et al.* [67] demonstrated that cadmium exposure (1.785 µM CdCl$_2$ for 14 days) enhances CA protein expression in digestive gland, as assessed by Western blotting analysis, CA activity measurement and immunofluorescence analysis in both laboratory and field experiments [67,68]. In basal conditions, digestive gland cells show a high CA specific activity, which can be functional to the well-developed lysosomal compartment of these cells. In fact CA, catalyzing the H$^+$ production from metabolic CO$_2$, can provide the H$^+$ necessary for the lysosome acidification. Therefore, the authors hypothesized that the observed metal-induced CA increased expression in mussel digestive gland is functionally related to the activation of the lysosomal system known to occur in pollutant-exposed organisms [69] and well described in mussel digestive gland [70].

The metal induction of CA expression arouses the question of how to interpret the different opposite effects that some metals can exert on CA, such as inhibition of the activity and upregulation of the gene expression. In mussel digestive gland, the IC_{50} value for Cd^{2+} inhibition of CA activity *in vitro* was in the millimolar range, while CA induction was observed following *in vivo* exposure to 1.785 μM $CdCl_2$ for two weeks [67]. In this case, it is possible to argue that the effect of Cd^{2+} on CA is dose-dependent with upregulation of CA expression at low concentrations and inhibition of the CA activity at higher concentrations. Moreover, the type of effect of metals on CA, induction *vs.* inhibition, is tissue specific. In fact, in the case of mussels, the same organisms exposed to 1.785 μM $CdCl_2$ for two weeks showed mantle CA significantly inhibited [57], while digestive gland CA significantly increased [67].

Although future work is needed to clarify this intriguing aspect and to deepen the research on the regulation of CA expression by metals, these results enrich the panorama of information about the effects of metals on CA, pointing to the importance of the concentration of exposure, the type of exposure and the metal and tissue specificity on the observed effects.

5. CA and Trace Metals: Applicative Insights and Perspectives

As emerged from the above described studies, the relationships between metals and CA are complex and multifaceted, metals being cofactors of CA, but also inhibitors of CA activity and modulators of CA expression. In this field, new applicative perspectives have been recently developed in the biotechnological and environmental fields, allowing translating new advances in basic science on the interaction between CA and metals into practice and novel applications.

5.1. CA Metallovariants

The ability of many divalent metal ions, such as Co^{2+}, Ni^{2+}, Mn^{2+}, Cu^{2+}, Cd^{2+} and Hg^{2+}, to easily bind to the three-histidine moiety within the CA active site in transmetallation experiments generated a number of CA metallovariants.

The interest in CA metallovariants goes beyond the study of the metal-protein interaction in the catalytic site structure. It allows probing the changes certain metals make to the function of the enzyme, including new catalytic reactions. This has been observed particularly in CA enzymes containing Co^{2+} and Mn^{2+} metal centers [71]. The natural function of CA is to catalyze the reversible hydration of CO_2 to HCO_3^-, but it is also known for its ability to catalyze the hydrolysis of esters with moderate enantioselectivity. Replacing zinc with manganese in the active site produced the manganese-substituted carbonic anhydrase (CA[Mn]) with peroxidase activity [72]. In the presence of bicarbonate and hydrogen peroxide, CA[Mn] catalyzed the efficient oxidation of *O*-dianisidine with K_{cat}/K_m comparable to that for horseradish peroxidase. CA[Mn] also catalyzed the moderately enantioselective epoxidation of olefins to epoxides. This enantioselectivity of CA[Mn] is similar to that observed in natural heme-based peroxidases, with the advantage that CA[Mn] avoids the formation of aldehyde side products [71].

Therefore, the ability of some transition metals to displace the native Zn^{2+} in the catalytic site discloses a CA catalytic promiscuity that has begun to be recognized as a valuable research and synthesis tool in potential biotechnological applications, leading to improvements in existing catalysts and providing novel synthesis pathways currently not available.

5.2. CA-Based Biosensors for Metal Ions

Pollution by trace metals is a world-wide problem, which raises concerns about potential effects on human health and the environment [73]. Mining and smelting, industrial and urban waste, wastewater discharges and shipping activity are the major anthropogenic sources of metals in the aquatic environment. Many sensitive and selective analytical methods for the detection of metal ions at low concentrations in environmental samples have been developed. In this field, great is the interest for simple, rapid and inexpensive alternatives to the classical analytical methods, able to provide a continuous analysis of metal concentration *in situ* and in real time. This interest is expressed by the development of biosensors for trace metal detection. Some of them are fluorescence-based biosensors for the determination of free metals in solution [74–76], such as Cu^{2+}, Co^{2+}, Zn^{2+}, Cd^{2+} and Ni^{2+} at concentrations down the picomolar range [76,77]. They utilize the affinity of the apoCA (in general variants of human CAII) for metals [76,78].

Therefore, the high affinity of CA for metals has driven the development of CA-based biosensing that has been shown to be a viable approach for determining certain divalent cations in environmental media, serving also as an archetype for other fluorescence-based biosensors [79].

5.3. CA-Based Ecotoxicological Biomarkers and Bioassays for Metal Pollution Assessment

In the last few decades, the use of effect-based methodologies, such as ecotoxicological bioassays and biomarkers, has received growing interest in environmental monitoring and assessment. Ecotoxicity bioassays are widely used for the assessment of environmental media quality, because they provide an integrated measurement of the various complex effects exerted by contaminants present in an environmental sample on the test biological system [80,81]. On the other hand, molecular and cellular biomarkers measured in exposed organism in the field can be helpful for detecting the occurrence of pollutant-induced stress syndrome in living organisms and gaining insight regarding the mechanisms causing the observed effects of chemicals [82,83]. Bioassays and biomarkers provide useful information about the degree of exposure to pollutants and the resulting effects on the organisms. In this regard, the use of bioassay/biomarkers provides early warning information useful for improving the processes of environmental risk assessment [84].

Recently, new perspectives on the potential use of CA as pollution biomarkers in environmental biomonitoring rose from the sensitivity of CA activity and expression to trace metals in several species.

Therefore, CA measurements have been included in multi-biomarker approaches on bioindicator organisms in field studies [85]. For example CA activity inhibition measurement in the bioindicator fish *Pimelodus maculatus* has been included in the multi-marker biomonitoring of three water reservoirs along the Paraiba do Sul River in an industrialized area of Brazil [86]. In corals, CA activity inhibition by metal exposure has been related to inhibition of coral growth through alteration of the calcification process, and it has been suggested as a potential biomarker of exposure to metal pollution [61].

The applicability of digestive gland CA activity and expression in a multimarker approach has been also demonstrated in the bioindicator organism *Mytilus galloprovincialis* [67,87] in two studies addressing the ecotoxicological biomonitoring and assessment of a coastal marine area in Southern Italy.

Branchial CA activity measured in the filter-feeding species *Crassostrea rhizophorae* was shown to be very responsive to coastal contamination in a recent study carried out in three human-impacted Brazilian estuaries [88]. Data were consistent with the usefulness of branchial CA in this species as a supporting biomarker for inexpensive and rapid analysis.

All of the available studies demonstrate the applicability of metal-induced CA alterations as pollution biomarkers. However, the successful use of CA as a biomarker in environmental biomonitoring requires a thorough knowledge of the specific metal-induced CA response in the specific bioindicator species used, because of the high species-specific and tissue-specific effects of metals on CA isoforms.

The *in vitro* inhibition of CA activity by metals has been applied by Lionetto *et al.* [57,89,90] to the development of an *in vitro* bioassay sensitive to the synergic effects of metals in a mixture and useful for the toxicity assessment of environmental aqueous samples [90]. This patented method has been recently applied to the monitoring of harbor sediments [68] and of reclaimed waste waters [91], showing a high agreement with standardized *in vivo* bioassays.

All of the studies carried out to date on the *in vitro* and *in vivo* sensitivity of CA to trace metals contribute to developing effect-based methodologies, such as CA-based pollution biomarkers and bioassays. They allow translating new advances in basic science into practice and novel applications in environmental biomonitoring (Figure 3).

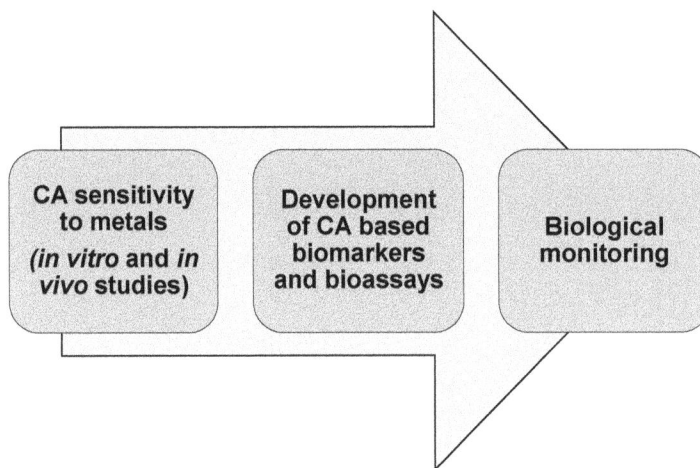

Figure 3. From *in vivo* and *in vivo* assessment of the sensitivity of CA to trace metals to environmental monitoring.

6. Conclusions

Although CA has been widely investigated in several aspects of its structure and functions, its complex relationships with metals have been only partially investigated to date, and some questions still remain to be clarified, such as, for example, the metal regulation of CA expression and its underlying mechanisms. This is an intriguing aspect of the research on this metalloenzyme that could raise new perspectives in the understanding of CA function and regulation.

The responses of CA activity and expression to metal exposure in animals need to be further characterized, and this could improve the potentiality of this enzyme for biotechnological and environmental applications.

Acknowledgments: Acknowledgments: This study was supported by PRIN (Progetti di Rilevate Interesse Nazionale) project 2010–2011 prot. 2010ARBLT7_005.

Author Contributions: Author Contributions: All authors contributed equally and extensively to this work. All authors in collaboration designed the study, analysed and discussed the literature, and wrote the manuscript.

Conflicts of Interest: Conflicts of Interest: The authors declare no conflict of interest.

References

1. Supuran, C.T. Carbonic anhydrase inhibitors. *Bioorg. Med. Chem. Lett.* **2010**, *20*, 3467–3474. [CrossRef] [PubMed]
2. Del Prete, S.; Vullo, D.; Fisher, G.M.; Andrews, K.T.; Poulsen, S.A.; Capasso, C.; Supuran, C.T. Discovery of a new family of carbonic anhydrases in the malaria pathogen *Plasmodium falciparum*—The η-carbonic anhydrases. *Bioorg. Med. Chem. Lett.* **2014**, *24*, 4389–4396. [CrossRef] [PubMed]
3. Ivanov, B.N.; Ignatova, L.K.; Romanova, A.K. Diversity in forms and functions of carbonic anhydrase in terrestrial higher plants. *Russ. J. Plant Physiol.* **2007**, *54*, 143–162. [CrossRef]
4. Cannon, G.C.; Heinhorst, S.; Kerfeld, C.A. Carboxysomal carbonic anhydrases: Structure and role in microbial CO$_2$ fixation. *BBA Proteins Proteom.* **2010**, *1804*, 382–392. [CrossRef] [PubMed]
5. Neri, D.; Supuran, C.T. Interfering with pH regulation in tumors as a therapeutic strategy. *Nat. Rev. Drug Discov.* **2011**, *10*, 767–777. [CrossRef] [PubMed]

6.   Supuran, C.T. Carbonic anhydrases: Novel therapeutic applications for inhibitors and activators. *Nat. Rev. Drug Discov.* **2008**, *7*, 168–181. [CrossRef] [PubMed]

7.   Kimber, M.S.; Pai, E.F. The active site architecture of *Pisum sativum* β-carbonic anhydrase is a mirror image of that of α-carbonic anhydrases. *EMBO J.* **2000**, *19*, 1407–1418. [CrossRef] [PubMed]

8.   Smith, K.S.; Ferry, J.G. Prokaryotic carbonic anhydrases. *FEMS Microbiol. Rev.* **2000**, *24*, 335–366. [CrossRef] [PubMed]

9.   Alber, B.E.; Ferry, J.G. A carbonic anhydrase from the archeon *Methanosarcina thermophile*. *Proc. Natl. Acad. Sci. USA* **1994**, *91*, 6909–6913. [CrossRef] [PubMed]

10. Roberts, S.B.; Lane, T.W.; Morel, F.M.M. Carbonic anhydrase in the marine diatom *Thalassiosira weissflogii* (Bacillariophyceae). *J. Phycol.* **1997**, *33*, 845–850. [CrossRef]

11. Tripp, B.C.; Smith, K.S.; Ferry, J.G. Carbonic anhydrase: New insights for an ancient enzyme. *J. Biol. Chem.* **2001**, *276*, 48615–48618. [CrossRef] [PubMed]

12. Xu, Y.; Supuran, C.T.; Morel, F.M.M. Cadmium-Carbonic Anhydrase. In *Encyclopedia of Inorganic and Bioinorganic Chemistry*; John Wiley & Sons: New York, NY, USA, 2011.

13. Supuran, C.T.; Scozzafava, A.; Casini, A. Carbonic anhydrase inhibitors. *Med. Res. Rev.* **2003**, *23*, 146–189. [CrossRef] [PubMed]

14. Supuran, C.T.; Scozzafava, A. Carbonic anhydrases as targets for medicinal chemistry. *Bioorg. Med. Chem. Lett.* **2007**, *15*, 4336–4350. [CrossRef] [PubMed]

15. Innocenti, A.; Scozzafava, S.; Parkkila, L.; Puccetti, G.; de Simone, G.; Supuran, C.T. Investigations of the esterase, phosphatase, and sulfatase activities of the cytosolic mammalian carbonic anhydrase isoforms I, II, and XIII with 4-nitrophenyl esters as substrates. *Bioorg. Med. Chem. Lett.* **2008**, *18*, 2267–2271. [CrossRef] [PubMed]

16. De Simone, G.; di Fiore, A.; Capasso, C.; Supuran, C.T. The zinc coordination pattern in the eta-carbonic anhydrase from *Plasmodium falciparum* is different from all other carbonic anhydrase genetic families. *Bioorg. Med. Chem. Lett.* **2015**, *25*, 1385–1389. [CrossRef] [PubMed]

17. Berg, J.M.; Tymoczko, J.L.; Stryer, L. *Biochemistry*, 7th ed.; W.H. Freeman: New York, NY, USA, 2012.

18. Williams, R.J.P. The biochemistry of zinc. *Polyhedron* **1987**, *6*, 61–69. [CrossRef]

19. Maret, W. Zinc biochemistry: From a single zinc enzyme to a key element of life. *Adv. Nutr.* **2013**, *4*, 82–91. [CrossRef] [PubMed]

20. Vahrenkamp, H. Why does nature use zinc—A personal view. *Dalton Trans.* **2007**, *42*, 4751–4759. [CrossRef] [PubMed]

21. Supuran, C.T. Structure-based drug discovery of carbonic anhydrase inhibitors. *J. Enzym. Inhib. Med. Chem.* **2012**, *27*, 759–772. [CrossRef] [PubMed]

22. MacAuley, S.R.; Zimmerman, S.A.; Apolinario, E.E.; Evilia, C.; Hou, Y.; Ferry, J.G.; Sowers, K.R. The archetype γ-class carbonic anhydrase (Cam) contains iron when synthesized *in vivo*. *Biochemistry* **2009**. [CrossRef] [PubMed]

23. Tripp, B.C.; Bell, C.B.; Cruz, F.; Krebs, C.; Ferry, J.G. A role for iron in an ancient carbonic anhydrase. *J. Biol. Chem.* **2004**, *279*, 21677–21677. [CrossRef] [PubMed]

24. Ferry, J.G. The gamma class of carbonic anhydrases. *Biochim. Biophys. Acta* **2010**, *1804*, 374–381. [CrossRef] [PubMed]

25. Wu, Y.; Zhao, X.; Li, P.; Huang, H. Impact of Zn, Cu, and Fe on the Activity of Carbonic Anhydrase of Erythrocytes in Ducks. *Biol. Trace Elem. Res.* **2007**, *118*, 227–232. [CrossRef] [PubMed]

26. Lane, T.W.; Morel, F.M.M. A biological function for cadmium in marine diatoms. *Proc. Natl. Acad. Sci. USA* **2000**, *97*, 4627–4631. [CrossRef] [PubMed]

27. Lane, T.W.; Saito, M.A.; George, G.N.; Pickering, I.J.; Prince, R.C.; Morel, F.M.M. Biochemistry: A cadmium enzyme from a marine diatom. *Nature* **2005**, *435*. [CrossRef] [PubMed]

28. Park, H.; McGinn, P.J.; Morel, F.M.M. Expression of cadmium carbonic anhydrase of diatoms in seawater. *Aquat. Microbial. Ecol.* **2008**, *51*, 183–193. [CrossRef]

29. Lionetto, M.G.; Lionetto, M.G.; Vilella, S.; Trischitta, F.; Cappello, M.S.; Giordano, M.E.; Schettino, T. Effects of $CdCl_2$ on electrophysiological parameters in the intestine of the teleost fish, *Anguilla anguilla*. *Aquat. Toxicol.* **1998**, *41*, 251–264.

30. Yee, D.; Morel, F.M.M. *In vivo* substitution of zinc by cobalt in carbonic anhydrase of a marine diatom. *Limnol. Oceanogr.* **1996**, *41*, 573–577. [CrossRef]

31. Mahon, B.P.; Pinar, M.A.; McKenna, R. Targeting Carbonic Anhydrase IX Activity and Expression. *Molecules* **2015**, *20*, 2323–2348. [CrossRef] [PubMed]

32. Dutta, S.; Goodsell, D. Carbonic anhydrase. Available online: http://pdb101.rcsb.org/motm/49 (accessed on 12 January 2016).

33. Håkansson, K.; Wehnert, A.; Liljas, A. X-ray analysis of metal-substituted human carbonic anhydrase II derivatives. *Acta Crystallogr. D* **1994**, *50*, 93–100. [CrossRef] [PubMed]

34. Kogut, K.A.; Rowlett, R.S. A comparison of the mechanisms of CO_2 hydration by native and Co^{2+}-substituted carbonic anhydrase II. *J. Biol. Chem.* **1987**, *262*, 16417–16424. [PubMed]

35. Marino, T.; Russo, N.; Toscano, M. A Comparative Study of the Catalytic Mechanisms of the Zinc and Cadmium Containing Carbonic Anhydrase. *J. Am. Chem. Soc.* **2005**, *127*, 4242–4253. [CrossRef] [PubMed]

36. Led, J.J.; Neesgaard, E. Paramagnetic Carbon-13 NMR Studies of the Kinetics and Mechanism of the HCO_3/CO_2 Exchange Catalyzed by Manganese(II) Human Carbonic Anhydrase I. *Biochemistry* **1987**, *26*, 183–192. [CrossRef] [PubMed]

37. Sven, L. Structure and mechanism of carbonic anhydrase. *Pharmacol. Ther.* **1997**, *74*, 1–20.

38. Hoffmann, K.M.; Samardzic, D.; van den Heever, K.; Rowlett, R.S. Co(II)-substituted *Haemophilus influenzae* b-carbonic anhydrase: Spectral evidence for allosteric regulation by pH and bicarbonate ion. *Arch. Biochem. Biophys.* **2011**, *511*, 80–87. [CrossRef] [PubMed]

39. Eriksson, E.A.; Kylsten, P.M.; Jones, A.T.; Liljas, A. Crystallographic studies of inhibitor binding sites in human carbonic anhydrase II; A penta coordinated binding of the SCN- ion to the zinc at high pH. *Proteins* **1988**, *4*, 283–293. [CrossRef] [PubMed]

40. Hogeback, J.; Schwarzer, M.; Wehe, C.A.; Sperlingab, M.; Karst, U. Investigating the adduct formation of organic mercury species with carbonic anhydrase and hemoglobin from human red blood cell hemolysate by means of LC/ESI-TOF-MS and LC/ICP-MS. *Metallomics* **2015**. [CrossRef] [PubMed]

41. Tu, C.; Wynns, G.C.; Silverman, D.N. Inhibition by cupric ions of ^{18}O exchange catalyzed by human carbonic anhydrase II. Relation to the interaction between carbonic anhydrase and hemoglobin. *J. Biol. Chem.* **1981**, *256*, 9466–9470. [PubMed]

42. Song, H.; Weitz, A.C.; Hendrich, M.P.; Lewis, E.A.; Emerson, J.P. Building reactive copper centers in human carbonic anhydrase II. *J. Biol. Inorg. Chem.* **2013**, *18*, 595–598. [CrossRef] [PubMed]

43. Nettles, W.L.; Song, H.; Farquhar, E.R.; Fitzkee, N.C.; Emerson, J.P. Characterization of the copper(II) binding sites in human carbonic anhydrase II. *Inorg. Chem.* **2015**, *54*, 5671–5680. [CrossRef] [PubMed]

44. Christensen, G.M.; Tucker, J.H. Effects of selected water toxicants on the *in vitro* activity of fish carbonic anhydrase. *Chem. Biol. Interact.* **1976**, *13*, 181–192. [CrossRef]

45. Lionetto, M.G.; Maffia, M.; Cappello, M.S.; Giordano, M.E.; Storelli, C.; Schettino, T. Effect of cadmium on carbonic anhydrase and Na⁺-K⁺-ATPase in eel, *Anguilla anguilla*, intestine and gills. *Comp. Biochem. Physiol. A* **1998**, *120*, 89–91. [CrossRef]

46. Lionetto, M.G.; Giordano, M.E.; Vilella, S.; Schettino, T. Inhibition of eel enzymatic activities by cadmium. *Aquat. Toxicol.* **2000**, *48*, 561–571. [CrossRef]

47. Ceyhun, S.B.; Şentürk, M.; Yerlikaya, E.; Erdoğan, O.; Küfrevioğlu, Ö.I.; Ekinci, D. Purification and characterization of carbonic anhydrase from the teleost fish *Dicentrarchus labrax* (European Seabass) liver and toxicological effects of metals on enzyme activity. *Environ. Toxicol. Pharmacol.* **2011**, *32*, 69–74. [CrossRef] [PubMed]

48. Kaya, E.D.; Söyüt, H.; Beydemir, S. Carbonic anhydrase activity from the gilthead sea bream (*Sparus aurata*) liver: The toxicological effects of heavy metals. *Environ. Toxicol. Pharmacol.* **2013**, *36*, 514–521. [CrossRef] [PubMed]

49. Soyut, H.; Beydemir, Ş.; Hisar, O. Effects of some metals on carbonic anhydrase from brains of rainbow trout. *Biol. Trace Elem. Res.* **2008**, *123*, 179–190. [CrossRef] [PubMed]

50. Vitale, A.M.; Monserrat, J.M.; Casthilo, P.; Rodriguez, E.M. Inhibitory effects of cadmium on carbonic anhydrase activity and ionic regulation of the estuarine crab, *Chasmagnathus granulata* (Decapoda, Grapsidae). *Comp. Biochem. Physiol. C* **1999**, *122*, 121–129. [CrossRef]

51. Skaggs, H.S.; Henry, R.P. Inhibition of carbonic anhydrase in the gills of two euryhaline crabs, *Callinectes sapidus* and *Carcinus maenas*, by heavy metals. *Comp. Biochem. Physiol. C* **2002**, *133*, 605–612. [CrossRef]

52. Ekinci, D.; Beydemir, Ş.; Küfrevioğlu, Ö.İ. *In vitro* inhibitory effects of some heavy metals on human erythrocyte carbonic anhydrases. *J. Enzym. Inhib. Med. Chem.* **2007**, *22*, 745–750. [CrossRef] [PubMed]

53. Kolayli, S.; Karahalil, F.; Sahin, H.; Dicer, B.; Supuran, C.T. Characterization and inhibition studies of an α-carbonic anhydrase from the endangered sturgeon species *Acipenser gueldenstaedti. J. Enzym. Inhib. Med. Chem.* **2011**, *26*, 895–900. [CrossRef] [PubMed]

54. Demirdağ, R.; Comakli, V.; Kuzu, M.; Yerlikaya, E.; Şentürk, M. Purification and characterization of carbonic anhydrase from Ağrı Balık Lake Trout Gill (*Salmo trutta* labrax) and effects of sulfonamides on enzyme activity. *J. Biochem. Mol. Toxicol.* **2013**, *29*, 278–282. [CrossRef] [PubMed]

55. Mercan, L.; Ekinci, D.; Supuran, C.T. Characterization of carbonic anhydrase from Turkish native "Gerze" chicken and influences of metal ions on enzyme activity. *J. Enzym. Inhib. Med. Chem.* **2014**, *29*, 773–776. [CrossRef] [PubMed]

56. Jebali, J.; Chouba, L.; Bannia, M.; Boussetta, H. Comparative study of the bioaccumulation and elimination of trace metals (Cd, Pb, Zn, Mn and Fe) in the digestive gland, gills and muscle of bivalve *Pinna nobilis* during a field transplant experiment. *J. Trace Elem. Med. Biol.* **2014**, *28*, 212–217. [CrossRef] [PubMed]

57. Lionetto, M.G.; Caricato, R.; Erroi, E.; Giordano, M.E.; Schettino, T. Potential application of carbonic anhydrase activity in bioassay and biomarker studies. *Chem. Ecol.* **2006**, *22*, S119–S125. [CrossRef]

58. Soto, M.; Ireland, M.P.; Marigómez, I. Changes in mussel biometry on exposure to metals: Implications in estimation of metal bioavailability in "Mussel-Watch" programmes. *Sci. Total Environ.* **2000**, *247*, 175–187. [CrossRef]

59. Brock, J.R.; Bielmyer, G.K. Metal accumulation and sublethal effects in the sea anemone, *Aiptasia pallida*, after waterborne exposure to metal mixtures. *Comp. Biochem. Physiol. C* **2013**, *158*, 150–158. [CrossRef] [PubMed]

60. Santini, O.; Chahbane, N.; Vasseur, P.; Frank, H. Effects of low-level copper exposure on Ca^{2+}-ATPase and carbonic anhydrase in the freshwater bivalve *Anodonta anatine. Toxicol. Environ. Chem.* **2011**, *93*, 1826–1837. [CrossRef]

61. Bielmyer, G.K.; Grosell, M.; Bhagooli, R.; Baker, A.C.; Langdon, C.; Gillette, P.; Capo, T.R. Differential effects of copper on three species of scleractinian corals and their algal symbionts (*Symbiodinium* spp.). *Aquat. Toxicol.* **2010**, *97*, 125–133. [CrossRef] [PubMed]

62. Gilbert, A.L.; Guzman, H.M. Bioindication potential of carbonic anhydrase activity in anemones and corals. *Mar. Pollut. Bull.* **2001**, *42*, 742–744. [CrossRef]

63. Khan, N.A.; Singh, S.; Anjum, N.A.; Nazar, R. Cadmium effects on carbonic anhydrase, photosynthesis, dry mass and antioxidative enzymes in wheat (*Triticum aestivum*) under lowand sufficient zinc. *J. Plant Interact.* **2008**, *3*, 31–37. [CrossRef]

64. Henkin, R.; Martin, B.M.; Agarwal, R. Efficacy of exogenous oral zinc in treatment of patients with carbonic anhydrase VI deficiency. *Am. J. Med. Sci.* **1999**, *318*, 392–405. [CrossRef]

65. Goto, T.; Shirakawa, H.; Furukawa, Y.; Komai, M. Decreased expression of carbonic anhydrase isozyme II, rather than of isozyme VI, in submandibular glands in long term zinc-deficient rats. *Br. J. Nutr.* **2008**, *99*, 248–253. [CrossRef] [PubMed]

66. De Polo, A.; Margiotta-Casaluci, L.; Lockyer1, A.E.; Scrimshaw, M.D. A new role for carbonic anhydrase 2 in the response of fish to copper and osmotic stress: Implications for multi-stressor studies. *PLoS ONE* **2014**, *9*. [CrossRef]

67. Caricato, R.; Lionetto, M.G.; Dondero, F.; Viarengo, A.; Schettino, T. Carbonic anhydrase activity in *Mytilus galloprovincialis* digestive gland: Sensitivity to heavy metal exposure. *Comp. Biochem. Physiol. C* **2010**, *152*, 241–247.

68. Lionetto, M.G.; Caricato, R.; Giordano, M.E.; Erroi, E.; Schettino, T. Carbonic anydrase and chemical pollutants: New applied perspectives. In Proceedings of the 65th Annual Meeting of the Italian Physiological Society, Anacapri, Italy, 28–30 September 2014.

69. Köhler, A. Lysosomal perturbations in fish liver as indicators for toxic effects of environmental pollution. *Comp. Biochem. Physiol.* **1991**, *100*, 123–127. [CrossRef]

70. Moore, M.N.; Simpson, M.G. Molecular and cellular pathology in environmental impact assessment. *Aquat. Toxicol.* **1992**, *22*, 313–322. [CrossRef]

71. Jing, Q.; Okrasa, K.; Kazlauskas, R.J. Manganese-Substituted α-Carbonic Anhydrase as an Enantioselective Peroxidase. *Top. Organomet. Chem.* **2009**, *25*, 45–61.

72. Okrasa, K.; Kazlauskas, R.J. Manganese-Substituted Carbonic Anhydrase as a New Peroxidase. *Chem. Eur. J.* **2006**, *12*, 1587–1596. [CrossRef] [PubMed]

73. De Mora, S.; Fowler, S.W.; Wyse, E.; Azemard, S. Distribution of heavy metals in marine bivalves, fish and coastal sediments in the Gulf and Gulf of Oman. *Mar. Pollut. Bull.* **2004**, *49*, 410–424. [CrossRef] [PubMed]

74. Zeng, H.H.; Thompson, R.B.; Maliwal, B.P.; Fones, G.R.; Moffett, J.W.; Fierke, C.A. Real-time determination of picomolar free Cu(II) in seawater using a fluorescence-based fiber optic biosensor. *Anal. Chem.* **2003**, *75*, 6807–6812. [CrossRef] [PubMed]

75. Bozym, R.; Hurst, T.K.; Westerberg, N.; Stoddard, A.; Fierke, C.A.; Frederickson, C.J.; Thompson, R.B. Determination of zinc using carbonic anhydrase-based fluorescence biosensors. *Methods Enzymol.* **2008**, *450*, 287–309. [PubMed]

76. Fierke, C.A.; Thompson, R.B. Fluorescence-based biosensing of zinc using carbonic anhydrase. *Biometals* **2001**, *14*, 205–222. [CrossRef] [PubMed]

77. Mei, Y.J.; Frederickson, C.J.; Giblin, L.J.; Weiss, J.H.; Medvedeva, Y.; Bentley, P.A. Sensitive and selective detection of zinc ions in neuronal vesicles using PYDPY1, a simple turn-on dipyrrin. *Chem. Commun.* **2011**, *47*, 7107–7109. [CrossRef] [PubMed]

78. McCall, K.A.; Fierke, C.A. Probing determinants of the metal ion selectivity in carbonic anhydrase using mutagenesis. *Biochemistry* **2004**, *43*, 3979–3986. [CrossRef] [PubMed]

79. Thompson, R.B.; Bozym, R.A.; Cramer, M.L.; Stoddard, A.K.; Westerberg, N.M.; Fierke, C.A. Chapter 6: Carbonic anhydrase-based biosensing of metal ions: Issue and future prospects. In *Fluorescence Sensors and Biosensors*; Thompson, R.B., Ed.; CRC Press: Boca Raton, FL, USA, 2005.

80. Costa, C.R.; Olivi, P.; Botta, C.M.R.; Espindola, E.L.G. Toxicity in aquatic environments: Discussion and evaluation methods. *Quim Nova* **2008**, *31*, 1820–1830. [CrossRef]

81. Manzo, S.; de Nicola, F.; de Luca Picione, F.; Maisto, G.; Alfani, A. Assessment of the effects of soil PAH accumulation by a battery of ecotoxicological tests. *Chemosphere* **2008**, *71*, 1937–1944. [CrossRef] [PubMed]

82. Forbes, V.E.; Palmqvist, A.; Bach, L. The use and misuse of biomarkers in ecotoxicology. *Environ. Toxicol. Chem.* **2006**, *25*, 272–280. [CrossRef] [PubMed]

83. Schettino, T.; Caricato, R.; Calisi, A.; Giordano, M.E.; Lionetto, M.G. Biomarker approach in marine monitoring and assessment: New insights and perspectives. *Open Environ. Sci.* **2012**, *6*, 20–27. [CrossRef]

84. Martínez-Gómez, C.; Vethaak, A.D.; Hylland, K.; Burgeot, T.; Khöler, A.; Lyons, B.P.; Thain, J.; Gubbins, M.J.; Davies, I.M. A guide to toxicity assessment and monitoring effects at lower levels of biological organization following marine oil spills in European waters. *J. Mar. Sci.* **2010**, *67*, 1105–1118. [CrossRef]

85. Lionetto, M.G.; Caricato, R.; Giordano, M.E.; Erroi, E.; Schettino, T. Carbonic anhydrase as pollution biomarker: An ancient enzyme with a new use. *Int. J. Environ. Res. Public Health* **2012**, *9*, 3965–3977. [CrossRef] [PubMed]

86. De Andrade Brito, I.; Freire, C.A.; Yamamoto, F.Y.; de Assis, H.C.S.; Souza-Bastos, L.R.; Cestari, M.M.; de Castilhos Ghisi, N.; Prodocimo, V.; Filipak Neto, F.; de Oliveira Ribeiro, C.A. Monitoring water quality in reservoirs for human supply through multi-biomarker evaluation in tropical fish. *J. Environ. Monit.* **2012**, *14*, 615–625. [CrossRef] [PubMed]

87. Caricato, R.; Lionetto, M.G.; Schettino, T. Studio di biomarkers in mitili (*Mytilus galloprovincialis*) traslocati in Mar Piccolo e in Mar Grande di Taranto. *Biol. Mar. Medit.* **2009**, *16*, 136–147.

88. Azevedo-Linhares, M.; Freire, C.A. Evaluation of impacted Brazilian estuaries using the native oyster *Crassostrea rhizophorae*: Branchial carbonic anhydrase as a biomarker. *Ecotoxicol. Environ. Saf.* **2015**, *122*, 483–489. [CrossRef] [PubMed]

89. Lionetto, M.G.; Caricato, R.; Erroi, E.; Giordano, M.E.; Schettino, T. Carbonic anhydrase based environmental bioassay. *Int. J. Environ. Anal. Chem.* **2005**, *85*, 895–903. [CrossRef]

90. Schettino, T.; Lionetto, M.G.; Erroi, E. Method for enzymatic assessment of the toxicity of environmental aqueous matrices. Patent WO 2009/135537, 2 March 2012.

91. Lionetto, M.G.; Caricato, R.; Calisi, A.; Giordano, M.E.; Erroi, E.; Schettino, T. Biomonitoring of water and soil quality: A case study of ecotoxicological methodology application to the assessment of reclaimed agroindustrial wastewaters used for irrigation. *Rend. Fis. Acc. Lincei* **2015**. [CrossRef]

International Journal of
Molecular Sciences

MDPI

Article

Effects of Protein-Iron Complex Concentrate Supplementation on Iron Metabolism, Oxidative and Immune Status in Preweaning Calves

Robert Kupczyński [1,*], Michał Bednarski [2], Kinga Śpitalniak [1] and Krystyna Pogoda-Sewerniak [1]

[1] Department of Environment Hygiene and Animal Welfare, Faculty of Biology and Animal Science, Wroclaw University of Environmental and Life Sciences, Chelmonskiego 38c, 51-630 Wroclaw, Poland; kinga.spitalniak@upwr.edu.pl (K.Ś.); krystyna.pogoda-sewerniak@upwr.edu.pl (K.P.-S.)
[2] Department of Epizootiology and Clinic of Birds and Exotic Animals, Faculty of Veterinary Medicine, Wroclaw University of Environmental and Life Sciences, pl. Grunwaldzki 45, 50-366 Wroclaw, Poland; michal.bednarski@upwr.edu.pl
* Correspondence: robert.kupczynski@upwr.edu.pl; Tel.: +48-71-320-5941

Received: 31 May 2017; Accepted: 7 July 2017; Published: 12 July 2017

Abstract: The objective of this study was to determine the effects of feeding protein-iron complex (PIC) on productive performance and indicators of iron metabolism, hematology parameters, antioxidant and immune status during first 35 days of a calf's life. Preparation of the complex involved enzymatic hydrolysis of milk casein (serine protease from *Yarrowia lipolytica* yeast). Iron chloride was then added to the hydrolyzate and lyophilizate. Calves were divided into treated groups: LFe (low iron dose) 10 g/day calf of protein-iron complex, HFe (height iron dose) 20 g/day calf, and control group. Dietary supplements containing the lower dose of concentrate had a significant positive effect on iron metabolism, while the higher dose of concentrate resulted in increase of total iron binding capacity (TIBC), saturation of transferrin and decrease of and unsaturated iron binding capacity (UIBC), which suggest iron overload. Additionally, treatment with the lower dose of iron remarkably increased the antioxidant parameters, mainly total antioxidant (TAS) and glutathione peroxidase activity (GPx). Higher doses of PIC were related to lower total antioxidant status. IgG, IgM, insulin, glucose, TNFα and IGF-1 concentration did not change significantly in either group after supplementation. In practice, the use of protein-iron complex concentrate requires taking into account the iron content in milk replacers and other feedstuffs.

Keywords: iron; calves; protein-iron complex; immunology; antioxidative status

1. Introduction

Iron plays an important role in the organisms of humans and animals, and its deficiency leads to numerous health conditions. The iron requirement of animals varies according to the age, sex and condition of the organism [1,2]. Young animals in the early stages of life are most susceptible to iron deficiency, however neonates do have some iron reserves in their body. Moreover, in the case of feeding calves with cow's milk, which has low iron concentration, rapid growth rates can lead to the development of temporary iron deficiency. This condition may be exacerbated by the immaturity of molecular mechanisms of iron absorption as in other animal species [2–4]. Transient iron deficiency in newborn calves most commonly manifests as anemia [2]. Presently, for young calves, the problem of iron deficiency is not as widespread as for e.g., piglets. This is due to the appropriate availability of this element in commercial milk replacers [5,6]. Young calves' iron requirements are estimated to be around 100 ppm, and are higher than in adult animals [7]. Iron supplementation has a significant effect on optimal calf growth rate, and can play a role in physical development and in hematopoiesis [5,8].

However, it should be noted that numerous studies-including Volker and Rotermund [7] above indicate that additional iron supplementation has beneficial effects on hematological parameters as well as on other parameters associated with iron metabolism [5]. Iron supplementation may play an important role in calf growth and resistance to infections [9]. Higher concentrations of Fe in the blood of calves are also associated with a higher concentration of Insulin-like Growth Factor 1 (IGF-1), a hormone that has major effects on growth and is responsible for metabolism [9,10]. On the other hand, excessive Fe content in the diet of ruminants can have toxic effects. They cause morphological damage to the intestine by increasing their permeability. High levels of iron in feed also negatively affect the absorption of other elements; this leads to oxidative stress at the cellular level, lipid peroxidation and increased gene expression of many enzymes considered to be antioxidants [10]. Cellular respiration is disturbed, and metabolic acidosis also occurs along with the associated consequences [11].

At the present time, iron supplementation is based primarily on inorganic compounds. However, these compounds can undergo oxidation and transform into insoluble forms [4,12]. For the purposes of animal nutrition and to increase the bioavailability of this element, research focuses on the use of chelates or proteinaceous iron preparations [13]. Casein proteins have very good iron binding properties, thereby decreasing their susceptibility to oxidation and therefore have high bioavailability [13–15]. Only biological trials can determine what dosing of iron is optimal for growth and development of animals with regards to hematological parameters, iron metabolism, antioxidant status or immune status, depending on the form of iron administration [2,5].

The objective of this study was to determine the effects of feeding protein–iron complex (PIC) on productive performance and indicators of iron metabolism, hematology parameters, antioxidant and immune status during first 35 days of a calf's life.

2. Results and Discussion

Casein protein concentrate is one of the most important byproducts of cheese manufacturing, and for this reason it can be used as a low-cost source of protein in food products. Iron-amino acid chelate was used for food fortification [16] as well as animal supplements [17]. Compared with amino acids, peptides are difficult to saturate with metals, although their absorption efficiency is higher. The absorption efficiency depends on the protein hydrolysis technology and the conditions of the active peptide iron chelate synthesis [18]. Analysis of the scanning electron microscopy microstructure indicates the presence of spherical particles of irregular pore size in the experimental formulation. The degree of degradation of spherical particles should be assessed as significant (Figure 1). Globules with smooth surfaces with a coarse appearance, characteristic of milk protein concentrate [19], did not occur in our study.

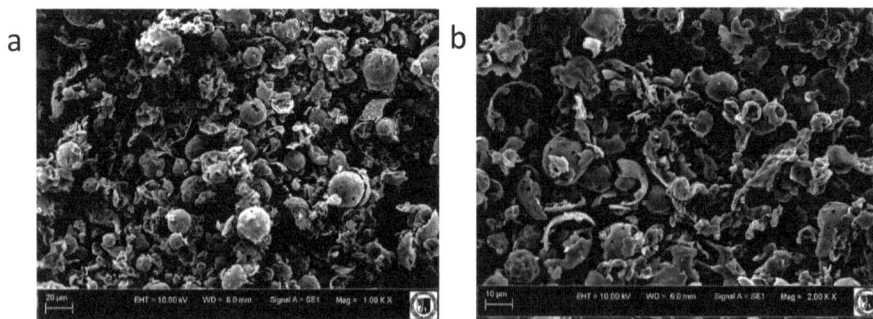

Figure 1. *Cont.*

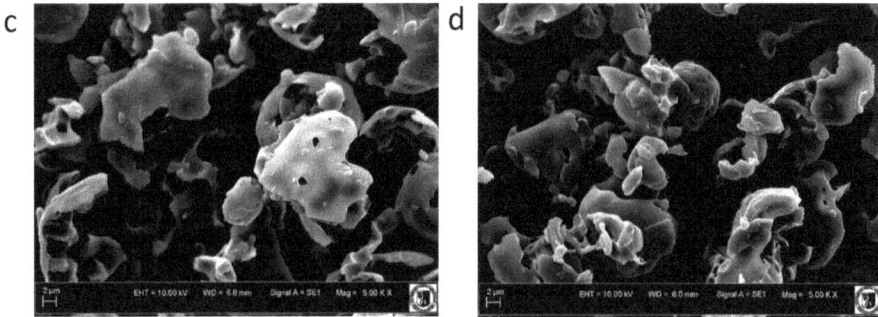

Figure 1. Microstructure of protein and protein-iron complex. View of the formulation at 20 and 10 μm particle size (**a**,**b**), protein fragmentation (**c**,**d**).

2.1. Hematological Parameters and Iron Metabolism

Iron is one of the important factors that influence growth, productivity, and animal immunity [9]. By feeding only cow's milk to young ruminant calves, iron intake is insufficient for normal erythropoiesis during the first month of life [8], which is manifested by reduced blood parameters, such as red blood cell (RBC), hematocrit (HCT) and hemoglobin (HGB) [20]. Volker and Rotermund [7] reported that oral administration of 100 mg iron per day, prevented anemia in calves. In turn, Mohri et al. [5] showed that higher of doses of 150 mg iron per day are required by growing calves. Generally, the iron requirements of these young ruminants are higher than those of mature ones and thought to be about 100–150 mg/kg of dry matter [21]. In addition, legislation in Europe-based on the concern for animal welfare-mandates the provision of solid feed to milk-fed veal calves [22].

Hematological examinations are a basic indicator of iron metabolism and hematopoiesis. In our study, iron supplementation did not significantly increase RBC and HGB concentration in comparison to the control group (Table 1). On the other hand, our investigations showed a statistically significant decrease ($p < 0.01$) of mean corpuscular volume (MCV) in the LFe (low iron dose) group of calves and an increase in mean corpuscular hemoglobin concentration (MCHC) (both experimental groups). No differences in Mean corpuscular hemoglobin (MCH) were detected between the groups. Mohri et al. [5,8] did not observe statistically significant differences in these parameters between calves during the first 28 days of life that received either oral or parenteral iron supplementation. The MCV decrease is associated with a decrease in erythrocyte volume during the first days of life [23], but curiously, Miltenburg et al. [24] reported that MCV was correlated with the administration of iron, which does not correspond to our results. In all studies, the MCV decreased in all groups, but the decrease was fastest in the groups where the PIC was administered.

Table 1. Mean values of hematological parameters in calf blood.

Item	Treatment			SEM	*p*-Value [1]		
	Control	Low Iron Dose (LFe)	Height Iron Dose (HFe)		D	T	D × T
WBC (G/L)	9.95	9.48	7.93	0.28	<0.01	0.11	0.88
RBC (T/L)	8.04	8.02	8.14	0.10	0.87	0.26	0.92
HGB (mmol/L)	6.01	6.21	6.25	0.09	0.60	0.25	0.79
HCT (L/L)	0.31	0.27	0.32	0.07	<0.01	0.37	0.41
PLT (G/L)	757.47	770.93	822.34	30.22	0.64	<0.01	0.89
MCV (fl)	38.39	33.64	39.29	0.28	<0.01	<0.01	0.14
MCH (fmol)	0.77	0.76	0.77	0.04	0.40	0.12	0.65
MCHC (mmol/L)	19.36	23.00	19.53	0.14	<0.01	<0.01	0.54

[1] Significant effect of experimental diet (D), time on diet (T), and their interaction (D × T); SEM-standard error of the means.

The concentration of Fe in the blood of young calves can vary by a wide range, from less than 10 μmol/L to as high as 30 μmol/L [2,25]. Moreover, over the first few days of life a progressive reduction in serum Fe concentration occurs, thus many authors point to the need for supplementation [2,20]. Primary iron deficiency is associated with a decrease of this trace element concentration in blood serum and elevated total iron binding capacity (TIBC) and unsaturated iron binding capacity (UIBC) [6]. In practice, iron deficiencies are not present when milk replacers are used, due to their high iron content. However, additional iron supplements may play an important role in calves' growth and resistance to infections [9]. For economic reasons, inorganic forms of Fe are generally used (iron sulfate). Current technology is heading towards the use of other iron compounds that are more readily absorbed, such as chelates or protein iron complex. Studies on monogastric animals indicated that the utilization of iron-amino acid chelate is nearly twice as efficient as Fe-sulfate [17].

In the current study, the highest serum iron concentrations were observed after administering a lower dose of PIC (Table 2). No statistically significant differences were observed in mean Fe concentration between groups for the entire study period, but there were differences in the ages of the animals at which blood samples were collected. In the LFe group the increase in Fe concentration was linear, whereas in the HFe (height iron dose) group Fe concentration only increased after the first week of PIC supplementation. The UIBC showed similar dependencies. The formulations used did not cause significant differences in TIBC and transferrin saturation. At a higher dose, especially after 2 weeks of supplementation, a rising tendency could be seen in these parameters. Transferrin concentration was significantly increased in both experimental groups and, over the course of the experiment, the increase was much higher in the HFe group than in the lower dose group. The highest transferrin concentrations were observed in calves on their 28th day of life. In the HFe group, apart from growth ($p < 0.01$) of transferrin concentration, the TIBC value increased as did the percentage of transferrin iron saturation. These results indicate some iron overload (HFe group), since virtually all circulating iron in the body is bound to transferrin [26]. However, mechanisms for the removal of Fe from the body were presumably sufficiently efficient that, despite such a significant increase in transferrin, Fe concentration was similar to the control group.

When compared with other rearing methods, the use of milk replacer leads to higher Fe values in calves and saturation of transferrin, and significantly lower TIBC [27]. In our study, the highest iron concentrations occurred in the LFe group (which received a lower PIC dose), whereas no significant differences were observed between the control group and the group with the highest PIC. Similar results as the LFe group were obtained by Mohri et al. [5]. However, the results from the HFe group are not consistent with the expectation that a higher iron dose in feed correlates with higher iron concentrations as Mohri et al. [5] demonstrated at similar doses of iron. Regardless, the results are not statistically significant. A similar tendency to our study was reported by Reece and Hotchkiss [27] where a slower growth of TIBC was observed in groups where iron was administered. Mohri et al. [5] observed the opposite trend: TIBC in calves receiving higher iron doses (150 mg/day) was significantly higher in comparison to the control group, and on the 28th day it was significantly lower. In this case, the changes were probably connected with forms of iron supplementation. Studies have also been conducted on administering iron parenterally. Providing Fe-dextran via this route for 2-day-old calves resulted in improved RBC parameters, increase of Fe in blood and weight gain during the first month of life in neonatal dairy calves [1]. High iron content in the diet of other species like rats resulted in decreased growth [28]. In our own studies, the amount of Fe contained in PIC did not significantly affect the rate of body weight gain.

Studies on animal models demonstrate that iron transporter proteins in the duodenum, liver and spleen are differentially regulated during developmental iron deficiency, and that early-life iron deficiency may cause long-term abnormalities in iron recycling from the spleen [29]. A strong negative correlation was found between Fe concentration and duodenal DMT1 (Fe duodenal divalent metal transporter-1) expression in high Fe rats, suggesting that the high body Fe stores were signaling reductions in intestinal Fe transport [30].

Table 2. Mean value of iron metabolism parameters.

Item	Treatment			SEM	p-Value [1]		
	Control	Low Iron Dose (LFe)	Height Iron Dose (HFe)		D	T	D × T
Iron (μmol/L)	15.06	16.59	14.70	0.57	0.35	0.01	0.88
UIBC (μmol/L)	6.68	5.47	6.15	0.46	0.62	<0.01	0.98
TIBC (μmol/L)	19.99	18.30	18.72	0.56	0.76	0.11	0.52
Transferrin saturation (%)	70.484	73.77	79.10	0.94	0.03	0.02	0.99
Transferrin (mg/mL)	2.041	3.59	5.53	0.42	<0.01	<0.01	0.01

[1] Significant effect of experimental diet (D), time on diet (T), and their interaction (D × T); SEM-standard error of the means.

2.2. Antioxidation Status, Biochemical and Immunologial Parameters

The Ganz and Nemeth [31] study suggests that excessive iron intake can lead to the production of free radicals and expose sensitive tissues to oxidative stress. The total antioxidant (TAS) value was significantly different between the groups (Table 3). The lowest mean for the entire study period was recorded in the HFe group, whereas the highest was in the control group. Statistically significant differences ($p < 0.01$) were found between the control group and the HFe group, and between the LFe group and the HFe group. These differences were apparent at the ages of 14, 28 and 35 days. Different dependencies were exhibited in glutathione peroxidase activity (GPx). The highest mean GPx activity was observed in the LFe group. At the beginning of the study, GPx activity was aligned between the groups. In the control group, the activity of this enzyme underwent a systematic reduction. Similar trends occurred in the experimental groups, albeit in the LFe group the reduction of activity was low. In the HFe group the activity decreased markedly, and the change was most apparent after one week of supplementation.

Table 3. TAS, GPx, SOD and concentration of selected biochemical and immunological parameters in calf blood.

Item	Treatment			SEM	p-Value [1]		
	Control	Low Iron Dose (LFe)	Height Iron Dose (HFe)		D	T	D × T
TAS (mmol/L)	1.15	0.91	0.86	0.22	<0.01	0.79	0.13
GPx (U/L)	59,716.22	60,833.10	50,422.23	324.11	0.01	0.01	0.88
SOD (U/mL)	1168.6	1041.8	1249.4	30.45	0.31	0.36	0.99
Insulin (ng/mL)	0.496	0.589	0.620	0.03	0.12	0.18	0.11
Glucose (mmol/L)	5.992	5.783	5.459	0.11	0.16	0.01	0.66
IGF-1 (ng/mL)	45.251	50.761	40.849	1.42	0.30	0.51	0.98
TNF-α (pg/mL)	103.76	108.79	95.891	1.14	0.49	0.15	0.71
IgG (mg/mL)	12.355	13.502	11.256	0.54	0.23	0.09	0.86
IgM (mg/mL)	0.70125	0.64937	0.78187	0.05	0.52	0.67	0.17

[1] Significant effect of experimental diet (D), time on diet (T), and their interaction (D × T); SEM-standard error of the means.

Total antioxidant assay allows an integrated antioxidant system to be evaluated, which includes all biological components that exhibit activity in realm of preventing excessive oxidation [32]. On the other hand, the determination of the major antioxidant enzymes superoxide dismutase (SOD) and GPx is aimed at evaluating the activity of intracellular antioxidants [33]. In the study conducted, on the last day of administering iron-containing compounds, the serum of experimental calves was lower in TAS, which may be attributable to a lower supply of free radicals.

The applied PIC was not confirmed to have any influence on SOD activity. However, there was a higher SOD activity in the HFe group compared to the LFe group. Iron overload in rats leads to a significant increase in catalase and SOD activity [34]. In the HFe group, higher SOD activity in the HFe group, which catalyzes the peroxide superoxide conversion reaction to hydrogen peroxide, may indicate a higher level of oxidative stress. Similar dependencies in cattle were found in other studies [35].

Long-term use of mineral blocks significantly increased serum levels of Fe, Mn, and Se, decreased the level of MDA, and increased GSH activity [36]. However, it is worth paying attention to the supply of selenium, which plays an important role in antioxidative processes. GSH-Px activity in the liver and heart was not affected by dietary Fe concentration [10], despite the administration of high doses of iron (750 mg of supplemental Fe/kg of dry matter).

No differences were found between groups regarding serum insulin and glucose levels, although at higher levels the functional additive that was used resulted in lower mean glucose and higher insulin levels. There were statistically significant ($p < 0.01$) differences in glucose concentration between the ages of the animals. In the HFe group the glucose concentration was significantly reduced, while in the control group and LFe the changes were not pronounced. In rat studies it was shown that glucose metabolism in adipose tissue appears to be affected by combination of: iron deficiency, excess through-interaction with adipocyte differentiation, tissue hyperplasia and hypertrophy, release of adipokines, lipid synthesis, and lipolysis [37]. Differences in glucose uptake in calves are largely due to the type of diet used. With milk replacers (MR) feed vs. colostrum or milk, insufficient glucose uptake was observed [38]. IGF-1 is a pleiotropic hormone exerting mitogenic and anti-apoptotic effects [39]. It is mainly produced in the liver, but can be mediated through the provision of energy, nutrients, minerals and vitamins, but also through the effects of non-nutritive factors [40]. In the case of calves born prematurely, there is a reduced thyroid status, as well as reduced IGF-1 concentrations in plasma [41]. Research on animal models and humans has revealed an important relationship between iron and IGF-1 [42,43]. Chronic iron deficiency leads to a decrease in hemoglobin in the blood, which results in a decrease in IGF-1 secretion in the liver [43]. Induced gestational-neonatal iron deficiency in rats indicates that early postnatal iron treatment of gestational iron deficiency reactivates the IGF system and promotes neurogenesis and differentiation in the hippocampus during a critical developmental period [44]. A reduction of IGF-1 secretion is also caused by iron overload [43].

Generally, calves exposed to iron deficiency anemia presented with a significant ($p < 0.01$) reduction in serum IGF-1 compared to calves with optimal iron concentration during the neonatal period [2]. Moderate IGF-1 serum concentrations were unaffected by dietary treatment (Table 3). A four-week supplementation regime (age 35 days) resulted in LFe iron level of 19.36, while in group III 16.59 µmol/L, with IGF-1 concentrations of 50.29 and 38.44 ng/mL, respectively. In the control group, Fe concentration was 17.80 µmol/L, with IGF-1 concentration of 43.26 ng/mL. These data indicate the optimal normalization of Fe concentration in the LFe group.

There exist dependencies between the level of feeding and the concentration of IGF-1 in the blood. Calves that are fed high protein and energy have a higher plasma concentration of IGF-1 than calves fed with a moderate or low diet [45]. In our studies, despite the fact that the supplementation was small, these dependencies were only confirmed in the LFe group. Concentrations of IGF-1 were affected by the amount of untreated calf milk replacer fed and were consistently higher in calves that were fed additional milk replacer [46]. Long-term iron supplementation in young dairy cattle results in an increase in IgG and IGF-1 concentration [47]. On the first day of our study, the IgG concentrations were measured and found to be equal in the groups, indicating that the passive immunity was adequate. Neither the dietary group nor the timing of blood sampling had any effect on serum IgG or IgM concentrations.

Infections can initiate an immune response, with concomitant production of cytokines such as tumor necrosis factor-α (TNF-α), interferon-γ (IFN-γ), interleukin-6 (IL-6), interleukin-10 (IL-10), and interleukin-1β (IL-1β) [46,48]. In practice, gastrointestinal infections are quite common [49]. This response is amplified by shipping stress followed by exposure to environmental pathogens. Animal feed has some effect on TNF-α concentration. Calves that had been fed once secreted less TNF-α in lipopolysaccharide stimulated whole blood cultures at 45 days of age compared with twice-fed calves, and these concentrations tended to persist through the immediate postweaning period [50]. A significant reduction in TNF-α was observed in adult ruminants that were fed with amino-acid-protected supplementation [51]. In our own studies, serum concentrations of TNF-α were unaffected by treatment (Table 3). In all of the studied calves, a decreasing trend was observed at day 14 in the serum TNF-α concentrations, followed by a subsequent increase. Clinical examinations as well as measured changes in IGF1; TNF-α and SOD activity were not significant, indicating that the calves did not exhibit inflammatory conditions that affected the formation of reactive oxygen species. Thus, an objective assessment of the impact of the dietary supplement could be made.

Total antioxidant activity decreased markedly after four weeks of supplementation of protein-iron complex, which may be attributable to a lower supply of free radicals or which suggests a decreased ability of preventing excessive oxidation. On the other hand, excessive iron intake can lead to the production of free radicals. The lower dose of protein-iron complex had a beneficial effect on the antioxidant status and the concentration of IGF-1 and iron in the blood at 35 days of age in comparison to the group with the higher dose of PIC.

3. Materials and Methods

3.1. Statement of Ethics

Studies have been conducted in accordance with the European Animal Welfare Guidelines and GLP. All the calves were handled in accordance with the regulations of the Polish Council on Animal Care and all the procedures for this trial were approved by the 2nd Local Ethical Committee for Experiments on Animals in Wroclaw (No. 63/2013).

3.2. Animals and Treatments

The study was carried on 30 Polish Holstein Friesian calves of the black-and-white variety. The animals were put into randomized groups, taking into account the age (7 days of age), body weight (ca. 40 ± 1.65 kg) and sex (50% females and 50% males in each group). The calves received a commercial milk replacer (Polmass, Bydgoszcz, Poland), and ad libitum granulated feed (Josera, Nowy Tomysl, Poland). Clean water was available at all times. Milk replacers were fed at 7:00 and 14:30 daily. During the first 21 days the calves received 6 L/day of MR, and then 8 L/day.

The calves were divided into: control group, fed with standard milk replacer ($n = 10$); experimental group (LFe), receiving a PIC additive in milk replacer at 10 g/day ($n = 10$); and HFe group, receiving PIC at 20 g/day ($n = 10$). The milk replacer formula contained 100 mg of iron in the form of iron sulfate; calves received an additional 27 mg (LFe) or 54 mg (HFe) of iron contained in protein-iron complex (PIC). Body mass of the calves was measured before morning feeding at 7, 14, 28, and 35 days of age. Dry matter intake was controlled in this study.

3.3. Process of Obtaining Protein-Iron Complex

Isoelectric casein was suspended in water and conditioned for 10 min at 60 °C, then cooled to 35 °C and basified to pH 8.0 using 5 mol NaOH (Merck, Darmstadt, Germany). In the next step, serine protease (enzyme preparation from *Yarrowia lipolytica* yeast) was introduced. Enzymatic hydrolysis was carried out for 3 h at 35 °C using 120 rpm stirring. The hydrolysis was interrupted by thermal deactivation of the enzyme at 80 °C for 20 min, then the hydrolyzate was cooled to 2 °C and a solution of Sigma-Aldrich iron (II) chloride (Buchs, Switzerland) was introduced. To a final concentration of 0.001 mol. binding was carried out under cooling for 12 h and then the resulting formulation was spray dried. The detailed process for obtaining the protein-iron complex is the subject of a patent application. The Fe contents were analyzed using an atomic absorption spectrophotometer Varian SpectrAA 220 (Agilent Technologies Inc., Palo Alto, Santa Clara, CA, USA) [52]. This analysis was used to determine the dose of PIC in treatment groups.

3.4. Scanning Electron Microscopy

The microstructure of protein-iron complex was examined by scanning electron microscopy using a Leo Zeiss 435 VP (Oberkochen, Germany), operating at 10 kV. Technical preparation of the samples included standard procedures for this type of microscopy, which were performed at Wroclaw University of Environmental and Life Sciences. Micrographs were recorded.

3.5. Clinical Observations and Sampling Procedures

During the study the calves were placed under clinical observation. Vitality, dehydration and fecal consistency were determined on the basis of a clinical trial and follow-up at 7, 14, 28, and 35 days of age. Blood samples were collected from all calves from the external jugular vein (*vena jugularis externa*) at 7, 14, 28, and 35 days of age. Seven days of age was treated as the starting point of the PIC (supplementation) experiment. For each sample, blood was drawn into a tube containing K2EDTA, a tube containing sodium heparin, and into a tube without anticoagulant (Sarstedt, Warsaw, Poland). The blood samples for serum and plasma were centrifuged at $3000 \times g$ for 10 min at a room temperature (2 h from collection), and the serum samples were frozen ($-20\ °C$) until the analysis.

3.6. Laboratory Analyses

Blood tests were performed to assess the effects of hematological, biochemical and immunological parameters. Analysis of hematological parameters was performed using an ABC Vet analyzer (Horiba ABX, Montpellier, France), which recorded parameters that included: red blood cells (RBC), white blood cells (WBC), hemoglobin (HGB), hematocrit (Hct), mean corpuscular volume (MCV), mean corpuscular hemoglobin (MCH), platelets (PLT) and mean corpuscular hemoglobin concentration (MCHC).

Iron was measured in serum by photometric test using Ferene Horiba ABX (Montpellier, France). Serum Unsaturated Iron Binding Capacity (UIBC) was determined by using a photometric method with ferrosine using reagents from Pointe Scientific (Canton, MI, USA). Serum Total Iron Binding Capacity (TIBC) was determined by using a photometric method, precipitating Fe^{3+} with calcium carbonate, using a BioMaxima (Lublin, Poland) precipitant, followed by a Horiba ABX (Montpellier, France) reagent for iron determination. Transferrin serum was labeled with the Bethyl (Montgomery, AL, USA) Bovine Transferrin ELISA Kit for testing immuno-enzymes. Transferrin saturation (TS) was calculated by TS = [Iron (μmol/L)]/[TIBC (μmol/L)] $\times$ 100 [L].

The laboratory analyses of blood serum were done using Pentra 400 biochemical analyzer Horiba ABX (Montpellier, France). The following parameters were estimated:

- glucose by oxidase method, reagents HORIBA ABX (Montpellier, France);
- glutathione peroxidase activity (GPx) by enzymatic method, Randox reagents Ransel RS (Crumlin, UK). The parameters determining the anti-oxidative status were also determined:
- Total antioxidant capacity (TAS) in serum by colorimetric method based on ABTS (2,2′-azine-di-[3-ethylbenzothiazoline sulfate]) method with peroxidase,
- glutathione peroxidase (GPx) in whole blood using enzymatic method,
- superoxide dismutase (SOD) in erythrocytes by the spectrophotometric, consisting of reaction with 2-(4-iodophenyl-3-(4-nitrophenol)-5-phenyltetrazoline chloride (I.N.T.)

These analyses were performed with the Pentra 400 biochemical analyzer by Horiba ABX (Montpellier, France), using reagents from Randox (Crumlin, Dunlin, Ireland).

Immunological parameters were determined in serum. Serum IgG immunoglobulin was assayed using a Bethyl Bovine IgG ELISA Kit. Serum IgM immunoglobulin was determined by Bovine IgM ELISA Kit, Bethyl (Montgomery, AL, USA). Tumor Necrosis Factor (TNF-α) was assayed in serum by the Bovine Tumor necrosis factor ELISA Kit from MyBio Source (San Diego, CA, USA). Serum IL-6 interleukin was assayed by the Bovine Interleukin-6 ELISA Kit immunoassay from MyBio Source (San Diego, CA, USA). Insulin-like growth factor 1 (IGF1) was assayed in serum using an enzyme-linked immunosorbent assay (IGF-1) ELISA Kit from MyBio Source (San Diego, CA, USA). Concentration of insulin in serum was determined by the ELISA method using the BioSource INS-EASIA Kit (BioSource Europe SA, Louvain-la-Neuve, Belgium). The above measurements were conducted using a Synergy fluorescence, luminescence and absorbance reader from BioTek Instruments (Winooski, VT, USA).

3.7. Statistical Analysis

Results obtained were subjected to statistical analysis using STATISTICA 10.0 software (Statistica, Tulsa, OK, USA). Data were analyzed using a general linear model for repeated measures ANOVA with dietary treatments (D) and sampling time (T) as fixed effects and their interactions (D × T) according to the model:

$$Y_{ijk} = \mu + \alpha_i + \beta_j + \alpha\beta_{ij} + \varepsilon_{ijk}$$

where Y_{ijk} is the dependent variable; μ is the overall mean; α_i-group (three groups); β_j-series of blood tests (1, 2, 3, 4); $\alpha\beta_{ij}$-group effect x series of tests; ε_{ijk}-random residual error.

Differences between treatment group means were analyzed for significance ($p < 0.05$) using the Duncan test. Production and health data were subjected to the nonparametric Wilcoxon test. The data are presented as average values and accompanied by standard error of the means.

4. Conclusions

The study has shown that the iron contained in the milk replacer feed only satisfies the basic iron requirements for erythropoiesis. Only the lower dose of protein-iron complex had a positive effect on the immune and antioxidant status. Further studies should focus on iron concentrations in the liver, as well as hepcidin-ferroportin axis and cytotoxicity, after adding PIC to the diets of calves.

Acknowledgments: Project POIG.01.03.01-02-080/12 "*Y. lipolytica* and *D. hansenii* yeast, enzymes and toxin killers used for preparation of preparations useful in industry and agrotechnics" co-financed by the European Regional Development Fund under the Operational Program Innovative Economy 2007–2013. Project supported by Wroclaw Centre of Biotechnology, programme The Leading National Research Centre (KNOW) for the years 2014–2018.

Author Contributions: Robert Kupczyński, Michał Bednarski and Kinga Śpitalniak contributed in equal manner to the design and execution of experiments, data analysis, writing, and revision of the work. Krystyna Pogoda-Sewerniak conducted laboratory analyses.

Conflicts of Interest: The authors declare no conflict of interest.

Abbreviations

PIC	Protein-iron complex
MR	Milk replacers
IGF-1	Insulin-like Growth Factor 1
SEM	Scanning electron microscopy
RBC	Red blood cell
WBC	White blood cell
HCT	Hematocrit
HGB	Hemoglobin
MCV	Mean corpuscular volume
MCH	Mean corpuscular hemoglobin
MCHC	Mean corpuscular hemoglobin concentration
DM	Dry matter
PLT	Platelets
TIBC	Total iron binding capacity
UIBC	Unsaturated iron binding capacity
TS	Transferrin saturation
LFe	Experimental group receiving low iron dose
HFe	Experimental group receiving height iron dose
DMT1	Duodenal divalent metal transporter-1
TAS	Total antioxidant capacity
GPx	Glutathione peroxidase activity
SOD	Superoxide dismutase

MDA	Malondialdehyde
GSH-Px	Erythrocyte glutathione peroxidase activity
TNF-α	Tumor necrosis factor-α
IFN-γ	Interferon-γ
IL	Interleukin
IgG	Immunoglobulin G
IgM	Immunoglobulin M

References

1. Bami, M.H.; Mohri, M.; Seifi, H.A.; Tabatabaee, A.A. Effects of parenteral supply of iron and copper on hematology, weight gain, and health in neonatal dairy calves. *Vet. Res. Commun.* **2008**, *32*, 553–561. [CrossRef] [PubMed]
2. Prodanović, R.; Kirovski, D.; Vujanac, I.; Dodovski, P.; Jovanović, L.; Šamanc, H. Relationship between serum iron and insulin-like growth factor-I concentrations in 10-day-old calves. *Acta Vet. BRNO* **2014**, *83*, 133–137. [CrossRef]
3. Miltenburg, G.A.J.; Wensing, T.H.; Breukink, H.J.; Marx, J.J.M. Mucosal uptake, mucosal transfer and retention of iron in veal calves. *Vet. Res. Commun.* **1993**, *17*, 209–217. [CrossRef] [PubMed]
4. Wienk, K.J.H.; Marx, J.J.M.; Beynen, A.C. The concept of iron bioavailability and its assessment. *Eur. J. Nut.* **1999**, *38*, 51–75. [CrossRef]
5. Mohri, M.; Sarrafzadeh, F.; Seifi, H.A.; Farzaneh, N. Effects of oral iron supplementation on some haematological parameters and iron biochemistry in neonatal dairy calves. *Comp. Clin. Pathol.* **2004**, *13*, 39–42. [CrossRef]
6. Jones, M.L.; Allison, R.W. Evaluation of the complete blood cell count. *Vet. Clin. Food Anim.* **2007**, *23*, 377–402. [CrossRef] [PubMed]
7. Volker, H.; Rotermund, L. Possibilities of oral iron supplementation for maintaining health status in calves. *Dtsch. Tierarztl. Wochenschr.* **2000**, *107*, 16–22. [PubMed]
8. Mohri, M.; Poorsina, S.; Sedaghat, R. Effects of parenteral supply of iron on RBC parameters, performance, and health in neonatal dairy calves. *Biol. Trace Elem. Res.* **2010**, *136*, 33–39. [CrossRef] [PubMed]
9. Atyabi, N.; Gharagozloo, F.; Nassiri, S.M. The necessity of iron supplementation for normal development of commercially reared suckling calves. *Comp. Clin. Pathol.* **2006**, *15*, 165–168. [CrossRef]
10. Hansen, S.L.; Ashwell, M.S.; Moeser, A.J.; Fry, R.S.; Knutson, M.D.; Spears, J.W. High dietary iron reduces transporters involved in iron and manganese metabolism and increases intestinal permeability in calves. *J. Dairy Sci.* **2010**, *93*, 656–665. [CrossRef] [PubMed]
11. Mendel, M.; Wiechetek, M. Iron poisoning in animals. *Med. Weter.* **2006**, *62*, 1357–1361.
12. Smith, J.E. Iron metabolism and its diseases. In *Clinical Biochemistry of Domestic Animals*, 4th ed.; Kaneko, J.J., Ed.; Academic Press Inc.: San Diego, CA, USA, 1989; p. 262.
13. Raja, B.K.; Jafri, E.S.; Dickson, D.; Acebron, A.; Cremonesi, P.; Fossati, G.; Simpson, R. Involvement of Iron (Ferric) Reduction in the Iron Absorption Mechanism of a Trivalent Iron-Protein Complex (Iron Protein Succinylate). *Pharmacol. Toxicol.* **2000**, *87*, 108–115. [CrossRef] [PubMed]
14. Sugiarto, M.; Ye, A.; Singh, H. Characterisation of binding of iron to sodium caseinate and whey protein isolate. *Food Chem.* **2009**, *114*, 1007–1013. [CrossRef]
15. Sugiarto, M.; Ye, A.; Taylor, W.M.; Singh, H. Milk protein-iron complexes: Inhibition of lipid oxidation in an emulsion. *Dairy Sci. Technol.* **2010**, *90*, 87–98. [CrossRef]
16. Layrisse, M.; García-Casal, M.N.; Solano, L.; Barón, M.A.; Arguello, F.; Llovera, D.; Ramírez, J.; Leets, I.; Tropper, E. Iron bioavailability in humans from breakfasts enriched with iron bis-glycine chelate, phytates and polyphenols. *J. Nutr.* **2000**, *130*, 2195–2199. [PubMed]
17. Ettle, T.; Schlegel, P.; Roth, F.X. Investigations on iron bioavailability of different sources and supply levels in piglets. *J. Anim. Physiol. Anim. Nutr.* **2008**, *92*, 35–43. [CrossRef] [PubMed]
18. Hellwig, M.; Geissler, S.; Peto, A.; Knütter, I.; Brandsch, M.; Henle, T. Transport of free and peptide-bound pyrraline at intestinal and renal epithelial cells. *J. Agric. Food Chem.* **2009**, *57*, 6474–6480. [CrossRef] [PubMed]
19. Shilpashree, B.G.; Arora, S.; Sharma, V.; Bajaj, R.K.; Tomar, S.K. Preparation of iron bound succinylated milk protein concentrate and evaluation of its stability. *Food Chem.* **2016**, *196*, 800–807. [CrossRef] [PubMed]

20. Bunger, U.; Kaphangst, P.; Fiebig, U.; Schonfelder, E.; Jentsch, D.; Ponge, J.; Furcht, G. Anaemia in male calves during rearing 4. Relations between birth weight, duration of trial and body weight gain while the calves were fed on colostrums, and the blood picture during weaning. *Arch. Tierernahr.* **1980**, *30*, 611–631. [PubMed]
21. Moosavian, H.R.; Mohri, M.; Seifi, H.A. Effects of parenteral over-supplementation of vitamin A and iron on hematology, iron biochemistry, weight gain, and health of neonatal dairy calves. *Food Chem. Toxicol.* **2010**, *48*, 1316–1320. [CrossRef] [PubMed]
22. Veissier, I.; Butterworth, A.; Bock, B.; Roe, E. European approaches to ensure good animal welfare. *Appl. Anim. Behav. Sci.* **2008**, *113*, 279–297. [CrossRef]
23. Jain, N.C. *Essentials of Veterinary Hematology*; Lea and Febiger: Philadelphia, PA, USA, 1993.
24. Miltenburg, G.A.J.; Wensing, T.; van Vliet, J.P.M.; Schuijt, G.; van de Broek, J.; Breukink, H.J. Blood hematology, plasma iron and tissue iron in dams in late gestation, at calving, and in veal calves at delivery and later. *J. Dairy Sci.* **1991**, *74*, 3086–3094. [CrossRef]
25. Knowles, T.G.; Edwards, J.E.; Bazeley, K.J.; Brown, S.N.; Butterworth, A.; Warriss, P.D. Changes in the blood biochemical and haematological profile of neonatal calves with age. *Vet. Rec.* **2000**, *147*, 593–598. [CrossRef] [PubMed]
26. Hatcher, H.C.; Singh, R.N.; Torti, F.M.; Torti, S.V. Synthetic and natural iron chelators: Therapeutic potential and clinical use. *Future Med. Chem.* **2009**, *1*, 1643–1670. [CrossRef] [PubMed]
27. Reece, W.O.; Hotchkiss, D.K. Blood studies and performance among calves reared by different methods. *J. Dairy Sci.* **1987**, *70*, 1601–1611. [CrossRef]
28. Zhang, H.; Gilbert, E.R.; Pan, S.; Zhang, K.; Ding, X.; Wang, J.; Qiufeng, Z.; Bai, S. Dietary iron concentration influences serum concentrations of manganese in rats consuming organic or inorganic sources of manganese. *Br. J. Nutr.* **2016**, *115*, 585–593. [CrossRef] [PubMed]
29. Oh, S.; Shin, P.K.; Chung, J. Effects of developmental iron deficiency and post-weaning iron repletion on the levels of iron transporter proteins in rats. *Nutr. Res. Pract.* **2015**, *9*, 613–618. [CrossRef] [PubMed]
30. Hansen, S.L.; Trakooljul, N.; Liu, H.C.; Moeser, A.J.; Spears, J.W. Iron transporters are differentially regulated by dietary iron, and modifications are associated with changes in manganese metabolism in young pigs. *J. Nutr.* **2009**, *139*, 1474–1479. [CrossRef] [PubMed]
31. Ganz, T.; Nemeth, E. Regulation of iron acquisition and iron distribution in mammals. *BBA Mol. Cell Res.* **2006**, *1763*, 690–699. [CrossRef] [PubMed]
32. Kusano, C.; Ferrari, B. Total Antioxidant Capacity: A biomarker in biomedical and nutritional studies. *J. Cell. Mol. Biol.* **2008**, *7*, 1–15.
33. McDowell, L.R.; Wilkinson, N.; Madison, R.; Felix, T. Vitamins and minerals functioning as antioxidants with supplementation considerations. In *Florida Ruminant Nutrition Symposium*; Best Western Gateway Grand: Gainesville, FL, USA, 2007; pp. 30–31.
34. Badria, F.A.; Ibrahim, A.S.; Badria, A.F.; Elmarakby, A.A. Curcumin attenuates iron accumulation and oxidative stress in the liver and spleen of chronic iron-overloaded rats. *PLoS ONE* **2015**, *10*, e0134156. [CrossRef] [PubMed]
35. Bernabucci, U.; Ronchi, B.; Lacetera, N.; Nardone, A. Influence of body condition on relationships between metabolic status and oxidative stress in periparturient dairy cows. *J. Dairy Sci.* **2005**, *88*, 2017–2026. [CrossRef]
36. Wang, H.; Liu, Z.; Huang, M.; Wang, S.; Cui, D.; Dong, S.; Li, S.; Qi, Z.; Liu, Y. Effects of long-term mineral block supplementation on antioxidants, immunity, and health of Tibetan sheep. *Biol. Trace Elem. Res.* **2016**, *172*, 326–335. [CrossRef] [PubMed]
37. Fernández-Real, J.M.; McClain, D.; Manco, M. Mechanisms linking glucose homeostasis and iron metabolism toward the onset and progression of type 2 diabetes. *Diabetes Care* **2015**, *38*, 2169–2176. [CrossRef] [PubMed]
38. Hammon, H.M.; Steinhoff-Wagner, J.; Flor, J.; Schönhusen, U.; Metges, C.C. Lactation Biology Symposium: Role of colostrum and colostrum components on glucose metabolism in neonatal calves. *J. Animal. Sci.* **2013**, *91*, 685–695. [CrossRef] [PubMed]
39. Annibalini, G.; Bielli, P.; De Santi, M.; Agostini, D.; Guescini, M.; Sisti, D.; Contarelli, S.; Brandi, G.; Villarini, A.; Stocchi, V.; et al. MIR retroposon exonization promotes evolutionary variability and generates species-specific expression of IGF-1 splice variants. *BBA-Gene Regul. Mech.* **2016**, *1859*, 757–768. [CrossRef] [PubMed]

40. Guilloteau, P.; Biernat, M.; Woliński, J.; Zabielski, R. Gut regulatory peptides and hormones of the small gut. In *Biology of the Intestine in Growing Animals*; Zabielski, R., Gregory, P.C., Weström, B., Eds.; Elsevier: Amsterdam, The Netherlands, 2002; Chapter 11; pp. 325–362.

41. Steinhoff-Wagner, J.; Görs, S.; Junghans, P.; Bruckmaier, R.M.; Kanitz, E.; Metges, C.C.; Hammon, H.M. Maturation of endogenous glucose production in preterm and term calves. *J. Dairy Sci.* **2011**, *94*, 5111–5123. [CrossRef] [PubMed]

42. Tran, P.V.; Fretham, S.J.; Wobken, J.; Miller, B.S.; Georgieff, M.K. Gestational-neonatal iron deficiency suppresses and iron treatment reactivates IGF signaling in developing rat hippocampus. *Am. J. Physiol. Endocrinol. Metab.* **2012**, *302*, 316–324. [CrossRef] [PubMed]

43. Soliman, A.T.; De Sanctis, V.; Yassin, M.; Adel, A. Growth and growth hormone–Insulin Like Growth Factor–I (GH-IGF-I) axis in chronic anemias. *Acta Biomed.* **2017**, *88*, 101–111. [CrossRef] [PubMed]

44. Tan, Y.; Liu, J.; Deng, Y.; Cao, H.; Xu, D.; Cu, F.L.; Lei, Y.Y.; Magdalou, J.; Wu, M.; Chen, L.; et al. Caffeine-induced fetal rat over-exposure to maternal glucocorticoid and histone methylation of liver IGF-1 might cause skeletal growth retardation. *Toxicol. Lett.* **2012**, *214*, 279–287. [CrossRef] [PubMed]

45. Brown, E.G.; VandeHaar, M.J.; Daniels, K.M.; Liesman, J.S.; Chapin, L.T.; Keisler, D.H.; Nielsen, M.W. Effect of increasing energy and protein intake on body growth and carcass composition of heifer calves. *J. Dairy Sci.* **2005**, *88*, 585–594. [CrossRef]

46. Quigley, J.D.; Wolfe, T.A.; Elsasser, T.H. Effects of additional milk replacer feeding on calf health, growth, and selected blood metabolites in calves. *J. Dairy Sci.* **2006**, *89*, 207–216. [CrossRef]

47. Cui, K.; Tu, Y.; Wang, Y.C.; Zhang, N.F.; Ma, T.; Diao, Q.Y. Effects of a limited period of iron supplementation on the growth performance and meat colour of dairy bull calves for veal production. *Anim. Prod. Sci.* **2016**, *57*, 778–784. [CrossRef]

48. Rodríguez, F.; González, J.F.; Arbelo, M.; Zucca, D.; Fernández, A. Cytokine expression in lungs of calves spontaneously infected with Mycoplasma bovis. *Vet. Res. Commun.* **2015**, *39*, 69–72. [CrossRef] [PubMed]

49. Bednarski, M.; Kupczyński, R.; Sobiech, P. Acid-base disorders in calves with chronic diarrhea. *Pol. J. Vet. Sci.* **2015**, *18*, 207–215. [CrossRef] [PubMed]

50. Hulbert, L.E.; Cobb, C.J.; Carroll, J.A.; Ballou, M.A. Effects of changing milk replacer feedings from twice to once daily on Holstein calf innate immune responses before and after weaning. *J. Dairy Sci.* **2011**, *94*, 2557–2565. [CrossRef] [PubMed]

51. Sun, F.; Cao, Y.; Cai, C.; Li, S.; Yu, C.; Yao, J. Regulation of Nutritional Metabolism in Transition Dairy Cows: Energy Homeostasis and Health in Response to Post-Ruminal Choline and Methionine. *PLoS ONE* **2016**, *11*, e0160659. [CrossRef] [PubMed]

52. Association of Official Analytical Chemists. *AOAC Official Methods of Analysis*, 15th ed.; Association of Official Analytical Chemists: Washington, DC, USA, 1990; p. 105.

International Journal of
Molecular Sciences

MDPI

Article

Administration of Zinc plus Cyclo-(His-Pro) Increases Hippocampal Neurogenesis in Rats during the Early Phase of Streptozotocin-Induced Diabetes

Bo Young Choi [1],[†], In Yeol Kim [1],[†], Jin Hee Kim [1], Bo Eun Lee [1], Song Hee Lee [1], A Ra Kho [1], Min Sohn [2] and Sang Won Suh [1],*

[1] Department of Physiology, College of Medicine, Hallym University, Chuncheon 24252, Korea; bychoi@hallym.ac.kr (B.Y.C.); inyeol@hallym.ac.kr (I.Y.K.); fate0710@hallym.ac.kr (J.H.K.); supsock1126@naver.com (B.E.L.); sshlee@hallym.ac.kr (S.H.L.); rnlduadkfk136@hallym.ac.kr (A.R.K.)
[2] Department of Nursing, Inha University, Incheon 22212, Korea; sohnmin@inha.ac.kr
* Correspondence: swsuh@hallym.ac.kr; Tel.: +82-10-8573-6364
† These authors contributed equally to this work.

Academic Editor: Reinhard Dallinger
Received: 24 November 2016; Accepted: 26 December 2016; Published: 1 January 2017

Abstract: The effects of zinc supplementation on hippocampal neurogenesis in diabetes mellitus have not been studied. Herein, we investigated the effects of zinc plus cyclo-(His-Pro) (ZC) on neurogenesis occurring in the subgranular zone of dentate gyrus after streptozotocin (STZ)-induced diabetes. ZC (27 mg/kg) was administered by gavage once daily for one or six weeks from the third day after the STZ injection, and histological evaluation was performed at 10 (early phase) or 45 (late phase) days after STZ injection. We found that the proliferation of progenitor cells in STZ-induced diabetic rats showed an increase in the early phase. Additionally, ZC treatment remarkably increased the number of neural progenitor cells (NPCs) and immature neurons in the early phase of STZ-induced diabetic rats. Furthermore, ZC treatment showed increased survival rate of newly generated cells but no difference in the level of neurogenesis in the late phase of STZ-induced diabetic rats. The present study demonstrates that zinc supplementation by ZC increases both NPCs proliferation and neuroblast production at the early phase of diabetes. Thus, this study suggests that zinc supplemented with a histidine/proline complex may have beneficial effects on neurogenesis in patients experiencing the early phase of Type 1 diabetes.

Keywords: zinc; zinc plus cyclo-(His-Pro); neurogenesis; hippocampus; diabetes; streptozotocin

1. Introduction

Type 1 diabetes is reported to make up approximately 5%–10% of the total diabetic population and represents a very significant risk to public health. Cognitive deficits recognized in type 1 diabetes patients include reduced information processing speeds [1,2] and worsening psychomotor function [1,3]. However, the mechanism by which type 1 diabetes patients develop cognitive dysfunction is not clear.

Several studies have demonstrated that new neurons are continuously produced in the rodent brain throughout life [4–6], both in the subgranular zone (SGZ) of the hippocampal dentate gyrus (DG) and subventricular zone (SVZ) of the lateral ventricle [7,8]. It is believed that hippocampal neurogenesis plays a role in learning and memory function [9,10]. Our previous studies have demonstrated that zinc, an essential trace element, is involved in hippocampal neurogenesis with or without brain injury. Our lab demonstrated that continuous free zinc release from degenerating DG cells may perpetually produce a signal that drives progenitor cell proliferation and aids the survival of neuroblasts following hypoglycemia [11], epilepsy [12], and traumatic brain injury [13]. In addition, we recently demonstrated that increasing hippocampal vesicular zinc by zinc supplemented with a histidine/proline complex (zinc plus cyclo-(His-Pro) (ZC)) promotes hippocampal neurogenesis under physiological conditions [14].

Therefore, the present study tested the hypothesis that ZC treatment can increase hippocampal neurogenesis in streptozotocin (STZ)-induced diabetic rats. We define the "early phase" as seven days after the STZ-induced diabetic condition and the "late phase" as 42 days after STZ-induced diabetic condition, in rats. We observed three interesting findings. First, progenitor cell proliferation in the hippocampus of diabetic rats showed an increase during the early phase. Second, zinc supplement provided by ZC administration in diabetic rats increased NPC proliferation rate and neuroblast production during the early phase. Third, zinc supplement by ZC administration in diabetic rats showed an increased survival rate of newly generated cells, but no difference in the level of neurogenesis during the late phase. Therefore, the present study suggests that zinc supplementation during the early phase of diabetes may have beneficial effects on hippocampal neurogenesis. However, zinc supplementation during the late phase produced no observable neurogenic effect in the hippocampus.

2. Results

2.1. ZC Treatment Does Not Affect Body Weight or Blood Glucose Level in Diabetic Rats

Weight change and blood glucose deregulation are the most prominent features of diabetes. Song et al. demonstrated that ZC administration improved body weight control in genetically diabetic rats [15] and decreased blood glucose levels in STZ-induced diabetic rats [16]. Thus, we examined the effects of ZC treatment on body weight and blood glucose level in rats that had undergone STZ-induced diabetes during our experimental period (i.e., early (10 days) or late (45 days) diabetic phase). Rats were given ZC (27 mg/kg, per os (PO)) once per day for one or six weeks. Compared to the sham-operated group, the STZ-injected group showed a decrease in body weight and a rapid increase in blood glucose levels. However, ZC treatment itself induced no weight or blood glucose changes, either in the sham-operated or in the STZ-injected group (Tables 1 and 2).

Table 1. Effect of zinc plus cyclo-(His-Pro) (ZC) on change in body weight of sham and streptozotocin (STZ)-induced diabetic rats. Values are means $\pm$ SE, n = 6–12 from each group; * $p < 0.05$, versus sham group.

Groups	Body Weight (g) 10 Days	
	Initial	Final
Sham + Vehicle (n = 12)	177.17 $\pm$ 1.89	224.75 $\pm$ 3.39
Sham + ZC (n = 7)	167.57 $\pm$ 5.36	227.29 $\pm$ 6.89
STZ + Vehicle (n = 12)	172.75 $\pm$ 3.08	165.92 $\pm$ 5.09 *
STZ + ZC (n = 10)	166.40 $\pm$ 1.56	162.00 $\pm$ 2.71 *

Groups	Body Weight (g) 45 Days	
	Initial	Final
Sham + Vehicle (n = 7)	133.14 $\pm$ 4.34	405.00 $\pm$ 7.51
Sham + ZC (n = 8)	127.00 $\pm$ 2.42	387.75 $\pm$ 9.66
STZ + Vehicle (n = 7)	133.20 $\pm$ 5.34	213.60 $\pm$ 24.41 *
STZ + ZC (n = 6)	132.60 $\pm$ 2.52	184.40 $\pm$ 13.08 *

Table 2. Effect of ZC on change in blood glucose level of sham and STZ-induced diabetic rats. Values are means ± SE, n = 6–12 from each group; * p < 0.05, versus sham group.

Groups	Blood Glucose Level (mg/dL)	
	10 Days	
	Initial	Final
Sham + Vehicle (n = 12)	116.92 ± 4.39	115.83 ± 4.04
Sham + ZC (n = 7)	117.29 ± 5.85	114.00 ± 4.48
STZ + Vehicle (n = 12)	112.75 ± 3.67	518.08 ± 17.24 *
STZ + ZC (n = 10)	115.90 ± 3.64	488.50 ± 34.20 *
Groups	**Blood Glucose Level (mg/dL)**	
	45 Days	
	Initial	**Final**
Sham + Vehicle (n = 7)	110.29 ± 3.63	90.57 ± 3.73
Sham + ZC (n = 8)	109.63 ± 4.74	87.00 ± 3.35
STZ + Vehicle (n = 7)	105.00 ± 3.78	551.60 ± 20.99 *
STZ + ZC (n = 6)	105.40 ± 5.10	562.80 ± 16.92 *

2.2. Short-Term ZC Treatment Increases the NPCs Proliferation in Diabetic Rats

To test whether ZC affects progenitor cell proliferation during the early phase of STZ-induced diabetes, rats were sacrificed at one week following daily ZC treatment with or without STZ injection. 5-Bromo-2-Deoxyuridine (BrdU) was intraperitoneally (IP) injected twice per day for four consecutive days from the sixth day following STZ injection. Cellular proliferation was assessed by BrdU and Ki67 immunohistochemistry. We found an increase in the number of cells labeled by both BrdU and Ki67 immunostaining in the SGZ of DG of animals that received STZ injection. In addition, ZC administration after STZ injection significantly increased the number of BrdU and Ki67 positive cells in the SGZ of DG (Figure 1).

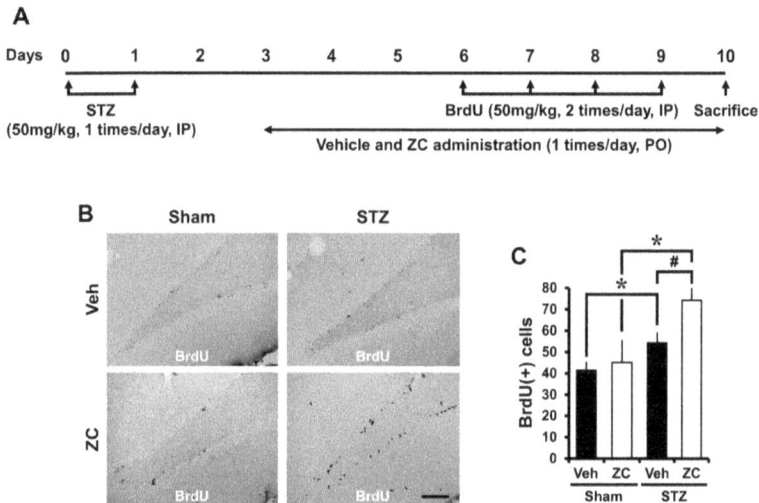

Figure 1. *Cont.*

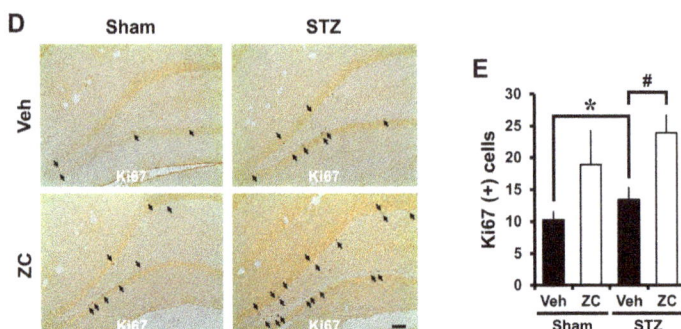

Figure 1. ZC increases proliferation of progenitor cells in the early phase of STZ-induced diabetic rats. (**A**) Experimental procedure in the early phase of STZ-induced diabetic rats. IP: intraperitoneal, PO: per os; (**B**) Bright field photomicrographs show BrdU (+) progenitor cells in the hippocampal dentate gyrus (DG). In the early phase, STZ-injected group showed a significant increase of BrdU (+) cells. ZC treatment by gavage for one week after STZ-induced hyperglycemia further increased BrdU (+) cells. Scale bar = 100 μm; (**C**) Bar graph represents the number of BrdU-positive cells in the subgranular zone (SGZ) of DG. Data are means ± SE, n = 7–12 from each group. * $p < 0.05$, versus vehicle-treated sham group; # $p < 0.05$, versus vehicle-treated STZ group; (**D**) Representative photomicrographs show Ki67 (+) cells in the hippocampal DG. The Ki67 (+) cells is indicated by a black arrow. In the early phase, the STZ-injected group showed a significant increase of Ki67 (+) cells. Short-term ZC treatment increased the number of Ki67 positive cells in STZ-induced diabetic rats. Scale bar = 100 μm; (**E**) Bar graph represents the number of Ki67 (+) cells in the SGZ of DG. Data are means ± SE, n = 7–12 from each group. * $p < 0.05$, versus vehicle-treated sham group; # $p < 0.05$, versus vehicle-treated STZ group.

Next, we determined whether theses progenitor cells are neural precursors. To this end, double immunofluorescence for BrdU and Nestin, an intermediate filament protein used to identify neural stem cells [17], was conducted. The number of BrdU and Nestin co-labeled cells in the DG was significantly increased after STZ injection. Additionally, the number of BrdU and Nestin co-labeled cells after daily ZC treatment in both the sham-operated and the STZ-injected group was also significantly increased in the DG compared to vehicle-treated rats (Figure 2).

2.3. Short-Term ZC Treatment Increases Neuroblast Production in Diabetic Rats

To assess whether ZC influences newly generated immature postmitotic neurons during the early phase of STZ-induced diabetes, rats were sacrificed one week after daily ZC treatment with or without STZ injection. Neuroblast production was evaluated by doublecortin (DCX) immunohistochemistry. In the sham-operated group, the number of DCX positive cells in the DG was similar between vehicle-treated rats and ZC-treated rats. The number of DCX positive cells was not different after STZ injection compared to the sham-operated group. However, ZC-treated rats showed a significantly higher number of DCX positive cells in the DG compared to vehicle-treated rats after STZ injection (Figure 3).

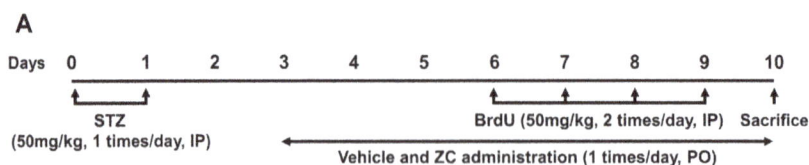

Figure 2. *Cont.*

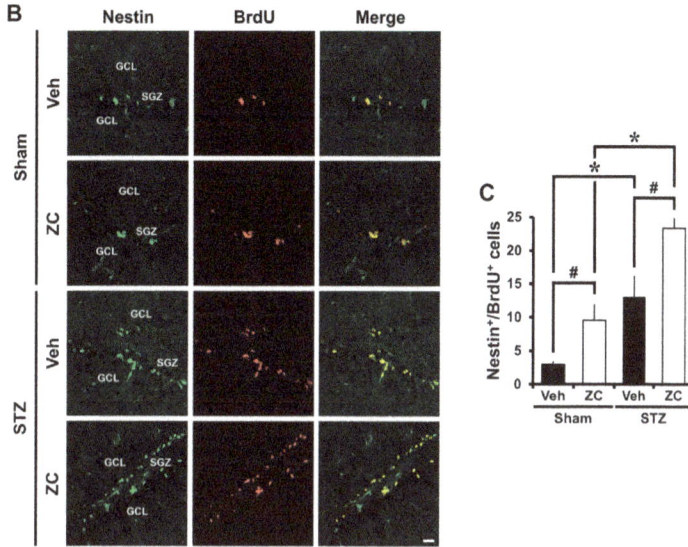

Figure 2. ZC increases neural progenitor cells (NPCs) proliferation in the early phase of STZ-induced diabetic rats. (**A**) Experimental procedure in the early phase of STZ-induced diabetic rats; (**B**) Fluorescent images show Nestin$^+$/BrdU$^+$ cells. In the early phase, the STZ-injected group showed a significant increase of Nestin$^+$/BrdU$^+$ cells. Short-term ZC treatment increased the number of Nestin$^+$/BrdU$^+$ cells in both the sham-operated and the STZ-injected group. Scale bar = 20 μm; (**C**) Bar graph represents the number of Nestin$^+$/BrdU$^+$ cells in the SGZ of DG. Data are means ± SE, n = 5–6 from each group. * $p < 0.05$, versus sham group; # $p < 0.05$, versus vehicle-treated group. GCL: granular cell layer; SGZ: subgranular zone.

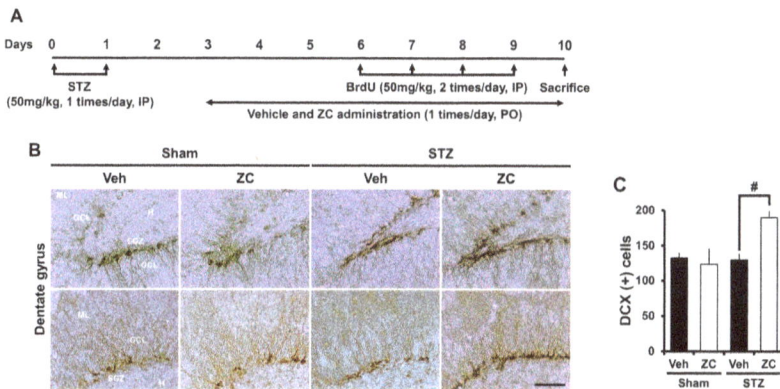

Figure 3. ZC increases DCX-positive cells in the early phase of STZ-induced diabetic rats. (**A**) Experimental procedure in the early phase of STZ-induced diabetic rats; (**B**) Photomicrographs show DCX (+) neuroblasts in the hippocampal DG. The STZ-injected group showed a similar level of DCX expression compared to the sham-operated group. However, ZC treatment by gavage for one week after STZ-induced hyperglycemia significantly increased DCX (+) cells. Scale bar = 100 μm; (**C**) Bar graph represents the number of DCX-positive cells in the SGZ/GCL. Data are means ± SE, n = 7–12 from each group. # $p < 0.05$, versus vehicle-treated STZ group. ML: molecular layer; GCL: granular cell layer; SGZ: subgranular zone; H: hilus.

2.4. Long-Term ZC Treatment Increases the Survival of BrdU Positive Cells in Diabetic Rat

To test whether ZC affects survival of newborn cells in the late phase of STZ-induced diabetes, rats were sacrificed six weeks after daily ZC treatment with or without STZ injection. Identical to the early phase, BrdU was intraperitoneally injected twice per day for four consecutive days from the sixth day following STZ injection. Surviving newborn cells, as detected by BrdU immunohistochemistry, were distributed throughout the entire granular cell layer (GCL). Some of those cells had already migrated into the GCL, whereas others were still localized in the SGZ. Although the progenitor cells proliferation rate was remarkably increased 10 days after STZ injection, the number of surviving newborn cells was decreased at 45 days after STZ injection. However, ZC-treated rats showed a significantly higher number of BrdU positive cells in the DG compared to vehicle-treated rats after STZ injection (Figure 4).

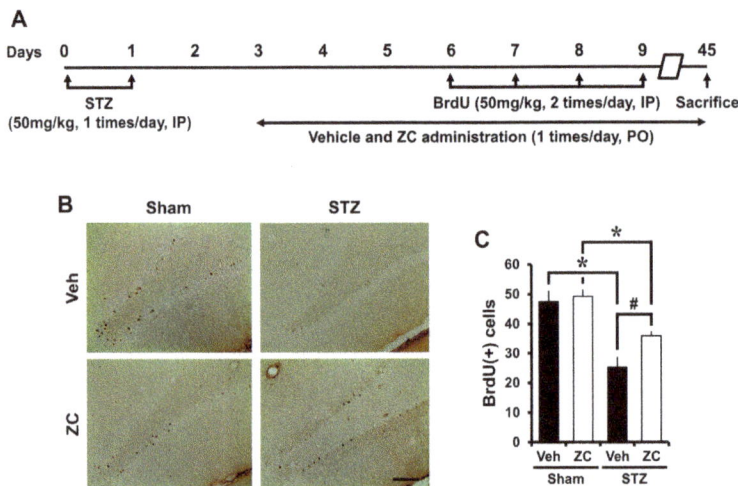

Figure 4. ZC increases survival of BrdU-positive cells in the late phase of STZ-induced diabetic rats. (**A**) Experimental procedure in the late phase of STZ-induced diabetic rats; (**B**) Bright field photomicrographs show surviving newborn cells. In the late phase, the STZ-injected group showed a significant decrease of BrdU (+) cells. However, ZC treatment by gavage for six weeks after STZ-induced hyperglycemia significantly increased BrdU (+) cells. Scale bar = 100 μm; (**C**) Bar graph represents BrdU-positive cells in the SGZ/GCL. Data are means ± SE, n = 6–8 from each group. * $p < 0.05$, versus sham group; # $p < 0.05$, versus vehicle-treated STZ group.

2.5. Long-Term ZC Treatment Does Not Affect the Neurogenesis in Diabetic Rats

We then investigated the phenotype of cells that have survived in the late phase of STZ-induced diabetes. The phenotypes of newly proliferated cells in the DG include neuron (BrdU$^+$/NeuN$^+$) and glia (BrdU$^+$/GFAP$^+$). Numerous BrdU positive cells were also positive for NeuN. The proportion and number of BrdU positive cells expressing NeuN was significantly decreased in the STZ-injected group compared to the sham-operated group. However, no significant differences in the proportion and number of BrdU positive cells expressing NeuN were found between vehicle- and ZC-treated rats under sham-operation or STZ injection. In addition, a significant difference was not observed in the number of BrdU positive cells expressing GFAP between the groups. However, the proportion of BrdU positive cells expressing GFAP was significantly increased in the STZ-injected group compared to the sham-operated group (Figure 5). These results suggest that ZC treatment has no effect on neurogenesis during the late phase of STZ-induced diabetes.

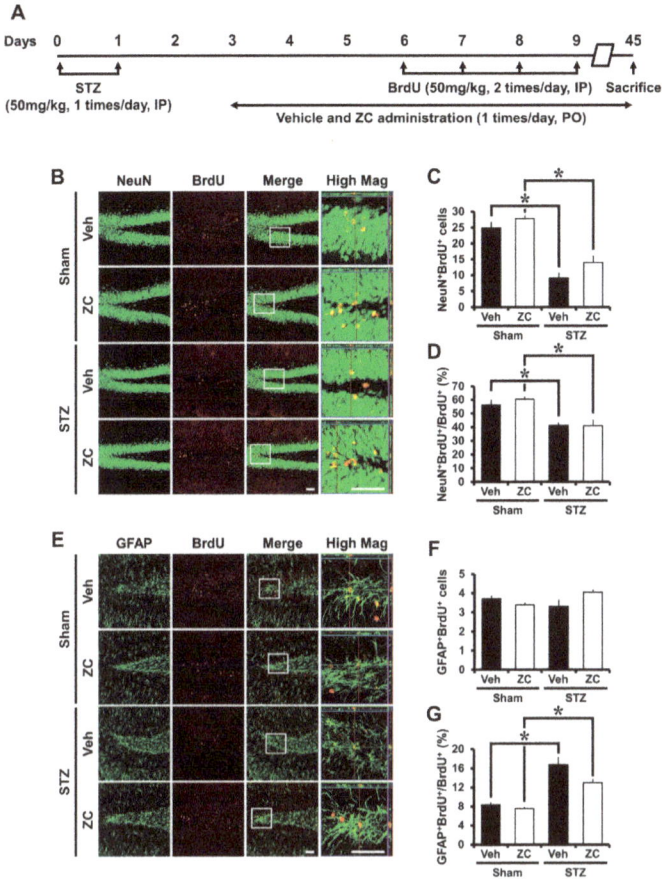

Figure 5. ZC does not affect neurogenesis in the late phase of STZ-induced diabetic rats. (**A**) Experimental procedure in the late phase of STZ-induced diabetic rats; (**B**) Representative images show double-labeling immunofluorescence of BrdU and NeuN positive cells in the hippocampal DG. STZ-injected group showed a significant decrease of BrdU$^+$/NeuN$^+$ cells. ZC treatment by gavage for six weeks after STZ-induced hyperglycemia showed no difference in the number of BrdU$^+$/NeuN$^+$ cells. Scale bar = 100 μm; (**C**) Bar graph represents the number of BrdU$^+$/NeuN$^+$ cells in the SGZ/GCL; (**D**) Bar graph show the proportion of BrdU$^+$/NeuN$^+$ cells in BrdU$^+$ cells of the SGZ/GCL. Data are means ± SE, *n* = 5–8 from each group. * *p* < 0.05, versus sham group; (**E**) Representative images show double-labeling immunofluorescence of BrdU and GFAP positive cells in the hippocampal DG. There were no differences among the groups in the number of BrdU$^+$/GFAP$^+$ cells. Scale bar = 100 μm; (**F**) Bar graph represents the number of BrdU$^+$/GFAP$^+$ cells in the SGZ/GCL; (**G**) Bar graph show the proportion of BrdU$^+$/GFAP$^+$ cells in BrdU$^+$ cells of the SGZ/GCL. Data are means ± SE, *n* = 5–8 from each group. * *p* < 0.05, versus sham group.

3. Discussion

The effects of zinc supplementation on hippocampal neurogenesis in diabetes mellitus have not been evaluated previously in rat. Hippocampal neurogenesis was assessed by BrdU, Ki67, and DCX immunohistochemistry at 10 (early phase) or 45 (late phase) days after STZ injection. In the present study, we found that daily zinc supplemented with a histidine/proline complex given during the early phase of diabetes increased progenitor cell proliferation and neuroblast production in diabetic rats. However, long-term zinc supplement in the late phase of diabetes showed no neurogenic effects in diabetic rats.

Patients suffering from uncontrolled diabetes suffer from various forms of cognitive impairment [1–3,18]. Several behavioral studies in diabetic rats have shown mixed results, possibly due to the different diabetes models used. For instance, different degrees or forms of stress arising from different behavioral tests may account for some variability. Larger behavioral problems are associated with the STZ-induced diabetic model than in non-diabetic rats or other diabetic animal models [19]. STZ-diabetic rodents can perform easy and basic tasks [20], however, if more-complicated tasks are employed, STZ-diabetic rats show apparent cognitive impairment [20]. As in chronic diabetic patients, a diverse number of cognitive functions worsen in diabetic rats. A reduction in conduction velocity was described initially in the peripheral nervous system and later in the central nervous system [21,22]. Visual and auditory problems arise at only three to four months after the onset of diabetes and steadily worsen thereafter [22,23]. The precise mechanism underlying cognitive dysfunction in diabetic patients remains unclear.

Synaptic plasticity is thought to supply a cellular basis for cognitive function in the brain [24]. In particular, the hippocampus is an important center for spatial memory in rodents. Several groups have demonstrated that hippocampal synaptic strength is weakened in diabetic rats, which leads to functional cognitive impairment. Accordingly, a reduction in synaptic strength has been reported in hippocampal slice experiments from STZ-induced diabetic rats [19]. Mounting data suggest that newly born cells from the subgranular zone of the hippocampus mature and functionally integrate into the DG. These newly developed cells display normal physiological parameters such as receptive membrane properties and action potentials, behaving like mature DG cells [25]. Newly generated neurons from the DG play a leading part in synaptic plasticity [26]. Reduction in the number or functional disintegration of newly generated cells worsens learning and memory [9]. Neurogenesis after transient cerebral ischemia not only leads to the replacing of injured cells, but also influences functional recovery [27,28]. Our previous study also demonstrated that hypoglycemia, seizure, or traumatic brain injury-induced cognitive impairment also display reduced neurogenesis in the hippocampus. We further suggested that divalent zinc is associated with injury-induced neurogenesis. Therefore, these results further support the concept that strategies for increasing endogenous neurogenesis may hold potential for the development of restorable therapy [11].

Ionic zinc is the second most plentiful transition element in the brain, following iron. Chelatable zinc is highly localized in the synaptic vesicle of mossy fiber terminals of the hippocampus and in the olfactory blub [29], areas where neurogenesis and neural migration actively occur in the adult brain [30]. Zinc ion is considered to be a biologically essential element for both systemic physiology and brain function. This ion is a constituent of over 1000 enzymes and many transcription factors, including zinc finger proteins. Thus, ionic zinc is involved in a truly broad diversity of cellular functions such as DNA synthesis and cell division [31]. Zinc also affects humoral control of cell division by pituitary growth hormone including nerve growth factor (NGF) [32] or insulin-like growth factor (IGF) [31]. Cerebellar granular cell division and migration was also impaired after severe zinc deficiency [33–35]. Reduced dietary zinc also impaired performance in short-term memory tasks [36]. The above evidence demonstrates that zinc is an essential element required for development, cell division, migration, and proliferation, and further suggests that this metal ion may have a vital role in cognitive function and neurogenesis [37].

Streptozotocin (STZ)-induced diabetic rats cannot secrete new insulin because of damage to the beta cells of the pancreatic islet. STZ administration induces a rapid breakdown of the beta cells, causes hyperglycemia [38], and produces metal ion dysmetabolism [39]. However, if the dose of STZ is not sufficient for complete destruction of the islet cells, new islet cells can be regenerated [40]. The zinc supplement compound cyclo-(His-Pro) (ZC) showed a reduction of blood glucose concentration in streptozotocin-induced diabetic rats, likely via stimulation of glucose uptake by the action of zinc on the b-subunit of insulin receptor [41,42]. Although ZC may have separate effects at the level of the insulin receptor or glucose transporter, it is very likely that they positively work on blood glucose control in diabetes patients via activating zinc metabolism [16]. It is not clear to what degree ZC's influence on blood glucose metabolism is directly linked to the action of zinc transporters, passive transport, or cation exchange [43]. Although previous studies showed a reduction of blood glucose levels by ZC administration, the present study showed no difference in blood glucose concentrations between the ZC treated group and the vehicle treated group. We cannot explain this difference. However, we can speculate that our STZ-induced model of diabetes may have been too severe, since over-administration of STZ not only destroys pancreas beta cells but may also inhibit insulin receptors [44] or glucose transporters [45] at high doses. This is one possibility to explain why ZC treatment in the present study had no ability to reduce blood glucose levels. Thus, the present study showed that ZC and vehicle-treated groups had similar blood glucose levels in the STZ-induced diabetic rat. Even though blood glucose levels were not affected by ZC administration, proliferation and differentiation were increased by ZC administration during the early phase of STZ-induced diabetes However, during the late phase neural differentiation was not increased by ZC administration. These results suggest that conditions that impose a state of chronic hyperglycemia may deleteriously affect neural differentiation. However, the exact mechanism should be confirmed by future studies.

Zinc influx and outflow is modulated by an active transport process that promotes homeostasis and prevents zinc toxicity. ZC enhances zinc transport from the intestinal lumen into the enterocytes [46]. ZC is a cyclic form of L-histidine and L-proline amino acid complex. Histidine and proline are highly distributed throughout the human body [45]. These amino acids are a permanent metabolite of both thyrotropin-releasing hormone [47] and histidine-proline-rich glycoproteins [48]. This glycoprotein is responsible for zinc transport from the intestine to body. The ZC motif, similar in sequence to specific zinc transporters, in the intestine facilitates zinc transport into human cells [49]. ZC is also found in high concentrations in the brain after administration [50]. Song et al. has reported that ZC is an efficient compound to increase zinc metabolism in the body [15]. Furthermore, our recent study demonstrated that zinc supplementation by ZC treatment increases hippocampal neurogenesis and levels of vesicular zinc [14]. Therefore, the present study suggests that ZC improves zinc absorption into the brain and has a beneficial effect on hippocampal neurogenesis.

4. Materials and Methods

4.1. Ethics Statement

This study was performed by the Guide for the Care and Use of Laboratory Animals of the National Institutes of Health (NIH). Use of animals in this study was approved by the Committee on Animal Use for Research and Education at Hallym University (Protocol # Hallym 2015-48).

4.2. Experimental Animals

Juvenile Sprague-Dawley male rats (90–100 g), aged four weeks, were purchased from DBL (DBL Co., Chungcheongbuk, Korea). The animals were kept in a temperature and humidity controlled room (22 ± 2 °C, 55% ± 5%, and a 12 h light: 12 h dark cycle), supplied with Purina diet (Purina, Gyeonggi, Korea) and water ad libitum. All animals were adapted for one week to avoid stress associated with transportation.

4.3. Rat Model of Type 1 Diabetes

For induction of type 1 diabetes, rats were intraperitoneally injected with streptozotocin (STZ, 50 mg/kg, IP) once per day for two consecutive days. STZ powder was dissolved in 0.1M sodium citrate buffer (pH = 4.5). In the present study we defined diabetes as overnight fasting tail blood glucose levels in excess of 200 mg/dL 24 h after the STZ injection. We found that most animals receiving STZ showed typical diabetes symptoms as seen in previous studies [51]. Tail blood glucose level was measured using a One Touch Basic Glucometer (Accu-Chek Active, Roche, Germany) [52].

4.4. Zinc Supplementation

For zinc supplementation, we used zinc plus cyclo-(His-Pro) (zinc plus CHP, ZC), formulated as a gel capsule containing 200 mg bovine prostate powder supplemented with 20 mg zinc [53]. ZC (27 mg/kg) was administered by gavage once daily for one or six weeks from the third day after the STZ injection, and then brains were harvested at 10 or 45 days, respectively. The vehicle-treated group was fed cyclo-(His-Pro) without zinc for the same periods. Animals were randomly divided into four groups to evaluate the effects of ZC treatment for early phase: (1) vehicle-treated sham group (Sham + Vehicle, $n = 12$); (2) ZC-treated sham group (Sham + ZC; $n = 7$); (3) vehicle-treated STZ group (STZ + Vehicle, $n = 12$); and (4) ZC-treated STZ group (STZ + ZC, $n = 10$). In the early phase, STZ-induced diabetic rats underwent hyperglycemia for seven days. Next, animals were randomly divided into four groups to evaluate the effects of ZC treatment for late phase: (1) vehicle-treated sham group (Sham + Vehicle, $n = 7$); (2) ZC-treated sham group (Sham + ZC; $n = 8$); (3) vehicle-treated STZ group (STZ + Vehicle, $n = 7$); and (4) ZC-treated STZ group (STZ + ZC, $n = 6$). In the late phase, STZ-induced diabetic rats underwent hyperglycemia for 42 days.

4.5. BrdU Labeling

To test the effects of zinc supplementation by ZC on neurogenesis in diabetic rats, 5-Bromo-2-Deoxyuridine (BrdU, 50 mg/kg; Sigma, St. Louis, MO, USA) was injected twice daily for four consecutive days starting the sixth day after the STZ injection. The rats were sacrificed at either 10 or 45 days after first STZ injection.

4.6. Brain Sections Preparation

To fix the brain, rats were anesthetized by overdose of intraperitoneal injection of urethane (1.5 g/kg, IP). For the blood wash out from the brain, 0.9% normal saline was perfused through heart for 10 min and then switched by 4% paraformaldehyde (PFA) in 0.1 M phosphate buffered saline (PBS, pH 7.4). Thirty minutes after PFA perfusion, the brains were taken out and post-fixed in the 4% PFA for one hour. After then, the brains were cryoprotected by 30% sucrose solution for overnight. When the brains were sank at the bottom, the entire brain was frozen by powered dry ice. After then, the brains were cut with Leica CM1850 cryostat (Leica Biosystems, Wetzlar, Germany) at 30 μm thickness.

4.7. Immunohistochemistry

Frozen sections were incubated in 0.6% H_2O_2 for 15 min at room temperature and washed three times with PBS. For immunohistochemical staining, mouse anti-BrdU (1:150, Roche, Basel, Switzerland), rabbit anti-Ki67 (1:1000, Novocastra, UK), or guinea pig anti-DCX (1:2000, Millipore, Billerica, MA, USA), diluted in PBS containing 0.3% normal chicken serum and 0.3%Triton X-100, were used as the primary antibodies and incubated overnight at 4 °C. The sections were washed three times for 10 min each with PBS, incubated in biotinylated anti-mouse, anti-rabbit or anti-guinea pig IgG (1:250, Vector Laboratories, Burlingame, CA, USA), and then avidin-biotinylated enzyme complex (ABC reagent, Vector Laboratories), and diluted 1:250 in the same solution as the primary antiserum. The immunoreactivity was revealed with 3,3'-diaminobenzidine (DAB, Sigma-Aldrich Co., St. Louis, MO, USA) in 0.01 M PBS buffer and mounted on the gelatin-coated slides.

4.8. Immunofluorescence Staining

For BrdU and Nestin double immunostaining in the early phase, sections were immersed with 2 N HCl at 37 °C for 90 min, neutralized two times for 10 min with 0.1 M sodium borate buffer, and incubated in a mixture of rat monoclonal anti-BrdU (1:150, Abcam, Cambridge, UK) and mouse monoclonal anti-Nestin (1:200, Abcam) for 2 h at room temperature. After rinse with PBS, the sections were incubated in a mixture of Alexa Fluor 488 donkey anti-mouse IgG (Nestin) and Alexa Fluor 594 donkey anti-rat IgG (BrdU) antibodies (1:250, Invitrogen, Grand Island, NY, USA) for 2 h at room temperature. To determine the phenotype of newly generated cells, double immunofluorescent labeling with BrdU and either NeuN or GFAP was performed in mice of the late phase (36 days after the last BrdU injection). Sections were incubated for 2 h in a mixture of mouse monoclonal anti-BrdU (1:150, Roche) and either rabbit polyclonal anti-NeuN (neuronal nuclei, 1:500, Millipore) or goat polyclonal anti-GFAP (glial fibrillary acidic protein, 1:200, Abcam) followed by 2 h of incubation in a mixture of Alexa Fluor 594 donkey anti-mouse IgG (BrdU) and either Alex Fluor 488 donkey anti-rabbit IgG (NeuN) or donkey anti-goat (GFAP) antibodies (1:250, Invitrogen) for 2 h at room temperature. Fluorescence signals were detected using a Zeiss LSM 710 confocal imaging system (Carl Zeiss, Oberkochen, Germany) with a sequential scanning mode for Alexa 488 and 594. Stacks of images (1024×1024 pixels) from consecutive slices of 0.9–1.2 μm in thickness were obtained by averaging eight scans per slice and were processed with ZEN 2010 (Carl Zeiss).

4.9. Quantification

To quantify BrdU, Ki67, and DCX-positive cells, sections were collected from 2.8 to 4.5 mm posterior to bregma and five coronal sections were analyzed from each animal. An experimenter masked to the treatment condition counted the number of BrdU, Ki67, and DCX-positive cells in the subgranular zone (SGZ) and granular cell layer (GCL) of dentate gyrus from both hemispheres. The mean numbers of BrdU, Ki67, and DCX-positive cells were used for statistical analyses. To analyze the phenotype of BrdU-positive cells, we determined whether BrdU-positive cells in the SGZ and GCL expressed Nestin, NeuN or GFAP with confocal microscopy. A double positive percentage was calculated as $BrdU^+/NeuN^+$ or $BrdU^+/GFAP^+$ cells for total BrdU-positive cells.

4.10. Data Analysis

Data are shown as mean + SEM. Statistical significance was assessed by analysis of variance (ANOVA) followed by the Student-Newman-Keuls *post-hoc* test. *p*-values less than 0.05 were considered statistically significant.

5. Conclusions

In the present study, we found that progenitor cell proliferation in the hippocampus of diabetic rats showed an increase in the early phase after STZ injection. Further, we found that ZC treatment significantly increased NPCs proliferation and neuroblast production in the early phase of STZ-induced diabetic rats. In addition, ZC treatment showed increased survival rate of newly generated cells, but no difference in the level of neurogenesis in the late phase of STZ-induced diabetic rats. Taken together, the present study suggests that zinc supplementation by ZC increases both NPCs proliferation and neuroblast production at the early phase of diabetes. Thus, this study suggests that zinc supplemented with a histidine/proline complex may have beneficial effects on neurogenesis in diabetic patients.

Acknowledgments: We thank Moon Ki Song (VA Greater Los Angles Healthcare System, Los Angeles, CA, USA) for generously donating the ZC with or without zinc. This work was supported by the Brain Research Program through the National Research Foundation of Korea (NRF) funded by the Ministry of Science, ICT & Future Planning (NRF-2016M3C7A1913844) to Sang Won Suh.

Author Contributions: Bo Young Choi and In Yeol Kim researched data, wrote/reviewed/edited manuscript; Jin Hee Kim, Bo Eun Lee, Song Hee Lee, and A Ra Kho researched data; Min Sohn conducted data analysis/reviewed/edited manuscript; Sang Won Suh contributed to discussion, wrote/reviewed/edited manuscript.

Conflicts of Interest: The authors declare no conflict of interest.

Abbreviations

ZC	Zinc plus cyclo-(His-Pro)
DG	Dentate gyrus
DCX	Doublecortin
STZ	Streptozotocin
GCL	Granular cell layer
SGZ	Subgranular zone
SVZ	Subventricular zone
NPCs	Neural progenitor cells
BrdU	5-Bromo-2-Deoxyuridine
PFA	Paraformaldehyde
PB	Phosphate buffer
ABC	Avidin-biotinylated enzyme complex
DAB	3,3′-Diaminobenzidine
NeuN	Neuronal nuclei
GFAP	Glial fibrillary acidic protein
IP	Intraperitoneal
PO	Per os
PBS	Phosphate-buffered saline

References

1. Brands, A.M.; Kessels, R.P.; Hoogma, R.P.; Henselmans, J.M.; van der Beek Boter, J.W.; Kappelle, L.J.; de Haan, E.H.; Biessels, G.J. Cognitive performance, psychological well-being, and brain magnetic resonance imaging in older patients with type 1 diabetes. *Diabetes* **2006**, *55*, 1800–1806. [CrossRef] [PubMed]
2. Ryan, C.M.; Williams, T.M.; Finegold, D.N.; Orchard, T.J. Cognitive dysfunction in adults with type 1 (insulin-dependent) diabetes mellitus of long duration: Effects of recurrent hypoglycaemia and other chronic complications. *Diabetologia* **1993**, *36*, 329–334. [CrossRef] [PubMed]
3. Weinger, K.; Jacobson, A.M.; Musen, G.; Lyoo, I.K.; Ryan, C.M.; Jimerson, D.C.; Renshaw, P.F. The effects of type 1 diabetes on cerebral white matter. *Diabetologia* **2008**, *51*, 417–425. [CrossRef] [PubMed]
4. Gould, E.; Reeves, A.J.; Fallah, M.; Tanapat, P.; Gross, C.G.; Fuchs, E. Hippocampal neurogenesis in adult old world primates. *Proc. Natl. Acad. Sci. USA* **1999**, *96*, 5263–5267. [CrossRef] [PubMed]
5. Eriksson, P.S.; Perfilieva, E.; Bjork-Eriksson, T.; Alborn, A.M.; Nordborg, C.; Peterson, D.A.; Gage, F.H. Neurogenesis in the adult human hippocampus. *Nat. Med.* **1998**, *4*, 1313–1317. [CrossRef] [PubMed]
6. Gage, F.H. Mammalian neural stem cells. *Science* **2000**, *287*, 1433–1438. [CrossRef] [PubMed]
7. Taupin, P.; Gage, F.H. Adult neurogenesis and neural stem cells of the central nervous system in mammals. *J. Neurosci. Res.* **2002**, *69*, 745–749. [CrossRef] [PubMed]
8. Abrous, D.N.; Koehl, M.; Le Moal, M. Adult neurogenesis: From precursors to network and physiology. *Physiol. Rev.* **2005**, *85*, 523–569. [CrossRef] [PubMed]
9. Shors, T.J.; Miesegaes, G.; Beylin, A.; Zhao, M.; Rydel, T.; Gould, E. Neurogenesis in the adult is involved in the formation of trace memories. *Nature* **2001**, *410*, 372–376. [CrossRef] [PubMed]
10. Feng, R.; Rampon, C.; Tang, Y.P.; Shrom, D.; Jin, J.; Kyin, M.; Sopher, B.; Miller, M.W.; Ware, C.B.; Martin, G.M.; et al. Deficient neurogenesis in forebrain-specific presenilin-1 knockout mice is associated with reduced clearance of hippocampal memory traces. *Neuron* **2001**, *32*, 911–926. [CrossRef]
11. Suh, S.W.; Fan, Y.; Hong, S.M.; Liu, Z.; Matsumori, Y.; Weinstein, P.R.; Swanson, R.A.; Liu, J. Hypoglycemia induces transient neurogenesis and subsequent progenitor cell loss in the rat hippocampus. *Diabetes* **2005**, *54*, 500–509. [CrossRef] [PubMed]
12. Kim, J.H.; Jang, B.G.; Choi, B.Y.; Kwon, L.M.; Sohn, M.; Song, H.K.; Suh, S.W. Zinc chelation reduces hippocampal neurogenesis after pilocarpine-induced seizure. *PLoS ONE* **2012**, *7*, e48543. [CrossRef] [PubMed]
13. Choi, B.Y.; Kim, J.H.; Kim, H.J.; Lee, B.E.; Kim, I.Y.; Sohn, M.; Suh, S.W. Zinc chelation reduces traumatic brain injury-induced neurogenesis in the subgranular zone of the hippocampal dentate gyrus. *J. Trace Elem. Med. Biol.* **2014**, *28*, 474–481. [CrossRef] [PubMed]

14. Choi, B.Y.; Kim, I.Y.; Kim, J.H.; Lee, B.E.; Lee, S.H.; Kho, A.R.; Sohn, M.; Suh, S.W. Zinc plus cyclo-(His-Pro) promotes hippocampal neurogenesis in rats. *Neuroscience* **2016**, *339*, 634–643. [CrossRef] [PubMed]
15. Song, M.K.; Rosenthal, M.J.; Song, A.M.; Uyemura, K.; Yang, H.; Ament, M.E.; Yamaguchi, D.T.; Cornford, E.M. Body weight reduction in rats by oral treatment with zinc plus cyclo-(His-Pro). *Br. J. Pharmacol.* **2009**, *158*, 442–450. [CrossRef] [PubMed]
16. Song, M.K.; Rosenthal, M.J.; Hong, S.; Harris, D.M.; Hwang, I.; Yip, I.; Golub, M.S.; Ament, M.E.; Go, V.L. Synergistic antidiabetic activities of zinc, cyclo (His-Pro), and arachidonic acid. *Metabolism* **2001**, *50*, 53–59. [CrossRef] [PubMed]
17. Frederiksen, K.; McKay, R.D. Proliferation and differentiation of rat neuroepithelial precursor cells in vivo. *J. Neurosci.* **1988**, *8*, 1144–1151. [PubMed]
18. Kodl, C.T.; Seaquist, E.R. Cognitive dysfunction and diabetes mellitus. *Endocr. Rev.* **2008**, *29*, 494–511. [CrossRef] [PubMed]
19. Gispen, W.H.; Biessels, G.J. Cognition and synaptic plasticity in diabetes mellitus. *Trends Neurosci.* **2000**, *23*, 542–549. [CrossRef]
20. Flood, J.F.; Mooradian, A.D.; Morley, J.E. Characteristics of learning and memory in streptozocin-induced diabetic mice. *Diabetes* **1990**, *39*, 1391–1398. [CrossRef] [PubMed]
21. Terada, M.; Yasuda, H.; Kikkawa, R.; Koyama, N.; Yokota, T.; Shigeta, Y. Electrophysiological study of dorsal column function in streptozocin-induced diabetic rats: Comparison with 2,5-hexanedione intoxication. *J. Neurol. Sci.* **1993**, *115*, 58–66. [CrossRef]
22. Biessels, G.J.; Cristino, N.A.; Rutten, G.J.; Hamers, F.P.; Erkelens, D.W.; Gispen, W.H. Neurophysiological changes in the central and peripheral nervous system of streptozotocin-diabetic rats—Course of development and effects of insulin treatment. *Brain* **1999**, *122*, 757–768. [CrossRef] [PubMed]
23. Morano, S.; Sensi, M.; di Gregorio, S.; Pozzessere, G.; Petrucci, A.F.; Valle, E.; Pugliese, G.; Caltabiano, V.; Vetri, M.; di Mario, U.; et al. Peripheral, but not central, nervous system abnormalities are reversed by pancreatic islet transplantation in diabetic lewis rats. *Eur. J. Neurosci.* **1996**, *8*, 1117–1123. [CrossRef] [PubMed]
24. Bliss, T.V.; Collingridge, G.L. A synaptic model of memory: Long-term potentiation in the hippocampus. *Nature* **1993**, *361*, 31–39. [CrossRef] [PubMed]
25. Song, H.J.; Stevens, C.F.; Gage, F.H. Neural stem cells from adult hippocampus develop essential properties of functional cns neurons. *Nat. Neurosci.* **2002**, *5*, 438–445. [CrossRef] [PubMed]
26. Van Praag, H.; Schinder, A.F.; Christie, B.R.; Toni, N.; Palmer, T.D.; Gage, F.H. Functional neurogenesis in the adult hippocampus. *Nature* **2002**, *415*, 1030–1034. [CrossRef] [PubMed]
27. Liu, J.; Solway, K.; Messing, R.O.; Sharp, F.R. Increased neurogenesis in the dentate gyrus after transient global ischemia in gerbils. *J. Neurosci.* **1998**, *18*, 7768–7778. [PubMed]
28. Raber, J.; Fan, Y.; Matsumori, Y.; Liu, Z.; Weinstein, P.R.; Fike, J.R.; Liu, J. Irradiation attenuates neurogenesis and exacerbates ischemia-induced deficits. *Ann. Neurol.* **2004**, *55*, 381–389. [CrossRef] [PubMed]
29. Perez-Clausell, J.; Danscher, G. Intravesicular localization of zinc in rat telencephalic boutons. A histochemical study. *Brain Res.* **1985**, *337*, 91–98. [CrossRef]
30. Ming, G.L.; Song, H. Adult neurogenesis in the mammalian central nervous system. *Annu. Rev. Neurosci.* **2005**, *28*, 223–250. [CrossRef] [PubMed]
31. MacDonald, R.S. The role of zinc in growth and cell proliferation. *J. Nutr.* **2000**, *130*, 1500S–1508S. [PubMed]
32. Stewart, G.R.; Frederickson, C.J.; Howell, G.A.; Gage, F.H. Cholinergic denervation-induced increase of chelatable zinc in mossy-fiber region of the hippocampal formation. *Brain Res.* **1984**, *290*, 43–51. [CrossRef]
33. Dvergsten, C.L.; Fosmire, G.J.; Ollerich, D.A.; Sandstead, H.H. Alterations in the postnatal development of the cerebellar cortex due to zinc deficiency. I. Impaired acquisition of granule cells. *Brain Res.* **1983**, *271*, 217–226. [CrossRef]
34. Sandstead, H.H.; Frederickson, C.J.; Penland, J.G. History of zinc as related to brain function. *J. Nutr.* **2000**, *130*, 496S–502S. [PubMed]
35. Golub, M.S.; Takeuchi, P.T.; Keen, C.L.; Gershwin, M.E.; Hendrickx, A.G.; Lonnerdal, B. Modulation of behavioral performance of prepubertal monkeys by moderate dietary zinc deprivation. *Am. J. Clin. Nutr.* **1994**, *60*, 238–243. [PubMed]

36. Keller, K.A.; Chu, Y.; Grider, A.; Coffield, J.A. Supplementation with L-histidine during dietary zinc repletion improves short-term memory in zinc-restricted young adult male rats. *J. Nutr.* **2000**, *130*, 1633–1640. [PubMed]

37. Suh, S.W.; Won, S.J.; Hamby, A.M.; Yoo, B.H.; Fan, Y.; Sheline, C.T.; Tamano, H.; Takeda, A.; Liu, J. Decreased brain zinc availability reduces hippocampal neurogenesis in mice and rats. *J. Cereb. Blood Flow Metab.* **2009**, *29*, 1579–1588. [CrossRef] [PubMed]

38. King, A.J. The use of animal models in diabetes research. *Br. J. Pharmacol.* **2012**, *166*, 877–894. [CrossRef] [PubMed]

39. Beltramini, M.; Zambenedetti, P.; Raso, M.; IbnlKayat, M.I.; Zatta, P. The effect of zn(II) and streptozotocin administration in the mouse brain. *Brain Res.* **2006**, *1109*, 207–218. [CrossRef] [PubMed]

40. Grossman, E.J.; Lee, D.D.; Tao, J.; Wilson, R.A.; Park, S.Y.; Bell, G.I.; Chong, A.S. Glycemic control promotes pancreatic β-cell regeneration in streptozotocin-induced diabetic mice. *PLoS ONE* **2010**, *5*, e8749. [CrossRef] [PubMed]

41. Ezaki, O. Iib group metal ions (Zn^{2+}, Cd^{2+}, Hg^{2+}) stimulate glucose transport activity by post-insulin receptor kinase mechanism in rat adipocytes. *J. Biol. Chem.* **1989**, *264*, 16118–16122. [PubMed]

42. Kimball, S.R.; Vary, T.C.; Jefferson, L.S. Regulation of protein synthesis by insulin. *Annu. Rev. Physiol.* **1994**, *56*, 321–348. [CrossRef] [PubMed]

43. Simons, T.J. Calcium-dependent zinc efflux in human red blood cells. *J. Membr. Biol.* **1991**, *123*, 73–82. [CrossRef] [PubMed]

44. Dominici, F.P.; Balbis, A.; Bartke, A.; Turyn, D. Role of hyperinsulinemia on hepatic insulin receptor concentration and autophosphorylation in the presence of high growth hormone levels in transgenic mice overexpressing growth hormone gene. *J. Endocrinol.* **1998**, *159*, 15–25. [CrossRef] [PubMed]

45. Szkudelski, T. The mechanism of alloxan and streptozotocin action in b cells of the rat pancreas. *Physiol. Res.* **2001**, *50*, 537–546. [PubMed]

46. Rosenthal, M.J.; Hwang, I.K.; Song, M.K. Effects of arachidonic acid and cyclo (His-Pro) on zinc transport across small intestine and muscle tissues. *Life Sci.* **2001**, *70*, 337–348. [CrossRef]

47. Kagabu, Y.; Mishiba, T.; Okino, T.; Yanagisawa, T. Effects of thyrotropin-releasing hormone and its metabolites, cyclo(his-pro) and trh-oh, on growth hormone and prolactin synthesis in primary cultured pituitary cells of the common carp, cyprinus carpio. *Gen. Comp. Endocrinol.* **1998**, *111*, 395–403. [CrossRef] [PubMed]

48. Morgan, W.T. Human serum histidine-rich glycoprotein. I. Interactions with heme, metal ions and organic ligands. *Biochim. Biophys. Acta.* **1978**, *535*, 319–333. [CrossRef]

49. Kambe, T.; Yamaguchi-Iwai, Y.; Sasaki, R.; Nagao, M. Overview of mammalian zinc transporters. *Cell. Mol. Life Sci.* **2004**, *61*, 49–68. [CrossRef] [PubMed]

50. Yamada, M.; Wilber, J.F. The distribution of histidyl-proline diketopiperazine (cyclo(His-Pro)) in discrete rat hypothalamic nuclei. *Neuropeptides* **1989**, *13*, 221–223. [CrossRef]

51. Choi, B.Y.; Kim, J.H.; Kim, H.J.; Yoo, J.H.; Song, H.K.; Sohn, M.; Won, S.J.; Suh, S.W. Pyruvate administration reduces recurrent/moderate hypoglycemia-induced cortical neuron death in diabetic rats. *PLoS ONE* **2013**, *8*, e81523. [CrossRef] [PubMed]

52. Aissaoui, A.; Zizi, S.; Israili, Z.H.; Lyoussi, B. Hypoglycemic and hypolipidemic effects of coriandrum sativum l. In meriones shawi rats. *J. Ethnopharmacol.* **2011**, *137*, 652–661. [CrossRef] [PubMed]

53. Song, M.K.; Rosenthal, M.J.; Naliboff, B.D.; Phanumas, L.; Kang, K.W. Effects of bovine prostate powder on zinc, glucose, and insulin metabolism in old patients with non-insulin-dependent diabetes mellitus. *Metabolism* **1998**, *47*, 39–43. [CrossRef]

International Journal of
Molecular Sciences

MDPI

Article

Successive Onset of Molecular, Cellular and Tissue-Specific Responses in Midgut Gland of *Littorina littorea* Exposed to Sub-Lethal Cadmium Concentrations

Denis Benito [1,†], Michael Niederwanger [2,†] (ORCID), Urtzi Izagirre [1], Reinhard Dallinger [2,*] and Manu Soto [1,*]

1   CBET Research Group, Research Centre for Experimental Marine Biology and Biotechnology (PiE-UPV/EHU), University of the Basque Country UPV/EHU, Areatza Pasalekua, 48620 Plentzia-Bizkaia, Basque Country, Spain; denis.benito@ehu.eus (D.B.); urtzi.izagirre@ehu.eus (U.I.)
2   Institute of Zoology and Center of Molecular Biosciences Innsbruck (CMBI), University of Innsbruck, Technikerstraße 25, A-6020 Innsbruck, Austria; michael.niederwanger@uibk.ac.at
*   Correspondence: reinhard.dallinger@uibk.ac.at (R.D.); manu.soto@ehu.eus (M.S.);
    Tel.: +43-512-507-51861 (R.D.); +34-946-015-512 (M.S.)
†   These authors contributed equally to this study.

Received: 4 July 2017; Accepted: 18 August 2017; Published: 22 August 2017

Abstract: Cadmium (Cd) is one of the most harmful metals, being toxic to most animal species, including marine invertebrates. Among marine gastropods, the periwinkle (*Littorina littorea*) in particular can accumulate high amounts of Cd in its midgut gland. In this organ, the metal can elicit extensive cytological and tissue-specific alterations that may reach, depending on the intensity of Cd exposure, from reversible lesions to pathological cellular disruptions. At the same time, *Littorina littorea* expresses a Cd-specific metallothionein (MT) that, due to its molecular features, expectedly exerts a protective function against the adverse intracellular effects of this metal. The aim of the present study was, therefore, to assess the time course of MT induction in the periwinkle's midgut gland on the one hand, and cellular and tissue-specific alterations in the digestive organ complex (midgut gland and digestive tract) on the other, upon exposure to sub-lethal Cd concentrations (0.25 and 1 mg Cd/L) over 21 days. Depending on the Cd concentrations applied, the beginning of alterations of the assessed parameters followed distinct concentration-dependent and time-dependent patterns, where the timeframe for the onset of the different response reactions became narrower at higher Cd concentrations compared to lower exposure concentrations.

Keywords: *Littorina littorea*; metallothionein induction; midgut gland tubules; digestive cells; basophilic cells; connective tissue calcium cells; lysosomes; lipofuscin

---

1. Introduction

Cadmium (Cd) is one of the most harmful metals, being considered as a cytotoxic [1], genotoxic [2,3] and carcinogenic agent [4]. It can provoke, moreover, radical stress [5,6] and exerts hormonal effects on eukaryotes [7,8]. Cd is also highly toxic to aquatic and marine animals [9–11] and is, therefore, of high relevance as a pollutant in the marine environment [12,13]. This is significant considering that, in spite of the normally rather low Cd concentrations in seawater [14,15], many marine species such as invertebrates and fish of upper trophic levels can accumulate high amounts of Cd in their tissues [16,17]. This may have important consequences for human seafood consumption [18,19], and may apply to heavily contaminated marine habitats to an even greater extent [20].

In particular, marine molluscs are able to accumulate metals from seawater reaching high tissue concentrations [21–24]. One of the marine organisms with an exceptionally high capability for Cd accumulation is the common periwinkle (*Littorina littorea*) [25,26]. This gastropod species lives on rocky shores of the intertidal zone of the North Atlantic Ocean, where it is intermittently exposed to adverse environmental conditions due to seasonal and daily fluctuations of water and oxygen supply, salinity, temperature, as well as the availability of mineral and metal ions [27,28]. *Littorina littorea* can resist these stressors by an adaptation of its metabolic pathways and its "cytoprotective repertoire" to the harsh conditions in its habitat [29,30]. In addition, *Littorina littorea* has apparently improved its fitness to survive by increasing its metal detoxification capacity [31,32].

Cd accumulation by *Littorina littorea* has been investigated from different approaches in order to determine the effects of sub-lethal exposure to this metal. Histopathological and histochemical analyses by means of autometallography and micro-analytical studies have been applied in order to better understand the toxicity mechanisms and the animal's responses to Cd at different levels of biological complexity [12,32–35]. On the other hand, several studies have shown that, at the biochemical and molecular levels, an important response reaction of *Littorina littorea* to Cd exposure and other environmental stressors involves upregulation of metallothionein (MT) [31,36]. The metal binding features and tertiary structure of the MT of *Littorina littorea* have recently been elucidated and thoroughly reported [37,38].

As already indicated above, one of the central organs for Cd metabolism and Cd-induced response reactions in *Littorina littorea* is its midgut gland [32,34,39,40]. Generally, the molluscan midgut gland is a lobed organ composed of an interdigitating mass of blind-ending tubules. These tubules are bound together by connective tissue and muscle fibers and are irrigated by blood from the visceral sinus. In addition, gonad follicles are located in the connective tissue among the digestive tubules within the visceral mass of the midgut gland. The tubules converge to form major ducts which eventually drain into the stomach. The number of ducts may differ in a species-specific manner. Overall, at least two cell types are present in the tubules: digestive and basophilic cells. Their overall morphology is quite similar among all molluscan species [41,42]. It has previously been shown that the morphology, integrity and relative frequency of digestive and basophilic cells in the midgut gland of *Littorina littorea* might be subjected to significant alterations upon exposure to toxicants or environmental stressors [32,39,43–46]. Cd, in particular, can cause cytological alterations that impair the integrity and functionality of the midgut gland of littorinid gastropods [33]. These alterations include, for example, shifts and relocations of glycogen reserves [47], changes and re-structuring of lysosomal compartments, alterations of cell morphology and composition in midgut gland epithelia [39], as well as modifications in the volume density of the basophilic cells of the digestive tubules [32]. In fact, several of these cell-specific alterations have occasionally been applied as biomarkers for environmental pollution in marine habitats [32,48].

Altogether, Cd-induced cytological alterations in the midgut gland of *Littorina littorea* on the one hand, and molecular and biochemical response mechanisms such as increased MT mRNA concentrations on the other, must be considered as concomitant processes which complement each other. Up until now, however, we have lacked a clear perception regarding how these different response levels may be interconnected. The temporal succession of cellular and molecular response reactions may provide a first hint to this kind of interdependence. The working hypothesis of the present study was, therefore, the assumption that the onset of response mechanisms in the midgut gland of *Littorina littorea* provoked by sub-lethal Cd concentrations may be detected at very short but differing time-courses and may disappear after a given period of recovery.

Hence, individuals of *Littorina littorea* were exposed to two different Cd concentrations ("low" and "high") at sub-lethal levels over a period of 21 days, followed by a recovery phase of 12 days. The general intention of our study was to follow and compare the time course of cellular and molecular response mechanisms to the Cd uptake in taken into the midgut gland of exposed snails, with a particular focus on the first 24 h of exposure.

2. Results

2.1. Cd Accumulation and Black Silver Deposits (BSD) in the Midgut Gland

Cd accumulation in the midgut gland of metal-exposed winkles (CdL "Cd Low", 0.25 mg Cd/L and CdH "Cd High", 1 mg Cd/L) was delayed in time, with significant differences from control values appearing at day 7 for CdH winkles, and for both metal treatment groups at the end of exposure (Figure 1).

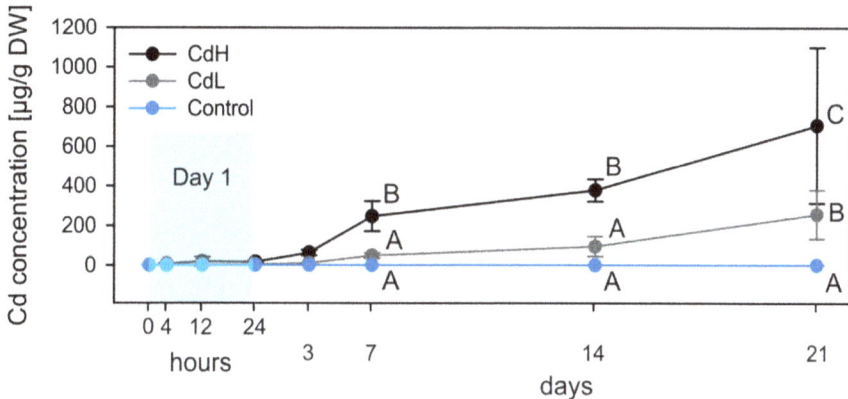

Figure 1. Cd accumulation in the digestive gland of the control (blue lines and symbols) and Cd-exposed winkles (grey line and symbols: CdL = 0.25 mg Cd /L; black line and symbols: CdH = 1 mg Cd /L). Mean values and standard deviations ($n = 4$) are shown. The significance of the accumulation curves (CdL and CdH) was confirmed by ANOVA ($p \leq 0.001$). The significant differences of values between Cd-exposed and control individuals at different time points (Holm–Sidak pairwise multiple comparison, significance level 0.05) are indicated by different letters. Sampling points during the first 24 h (Day 1) are under-laid in blue.

Concomitantly, metal detection by autometallography revealed only scarce black silver deposits (BSD) in the digestive cells of control winkles. Instead, they were mainly localized in the basal lamina (histological sense) of the digestive tubules (Figure 2A). After Cd exposure, BSD became more conspicuous, being evident in the basal lamina of the tubular epithelium and in lysosomes of tubular digestive cells, as well as in hemocytes of the interstitial connective tissue (Figure 2B–E). Moreover, BSD were also detected in the basal lamina of the adjacent stomach epithelium and in lysosomes of the stomach enterocytes (Figure 2D,E). During the period of recovery, the amount of BSD in the connective tissue cells of the midgut gland and in the digestive epithelia of the stomach declined again.

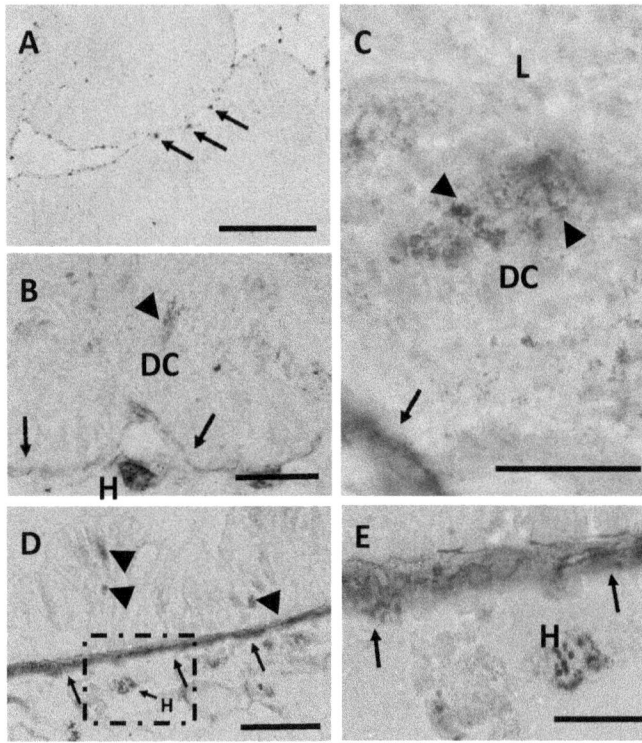

Figure 2. Autometallography staining of the midgut gland of winkles. (**A**) Controls, black silver deposits (BSD) in the basal lamina (arrows) of midgut gland tubuli (scale bar: 100 μm); (**B**) CdH (21 days), BSD in digestive cell lysosomes (arrowheads) and in the basal lamina (arrows) of midgut gland tubules. Note the presence of BSD within hemocytes (H) of the interstitial connective tissue (scale bar: 50 μm); (**C**) CdH (21 days), detail of BSD in digestive cell (DC) lysosomes (arrowheads) near the tubular lumen (L) and in the basal lamina (arrows) of a midgut gland tubule (scale bar: 20 μm); (**D**) CdH (21 days), BSD in the basal lamina of the epithelium of the stomach (arrows), in the lysosomes of epithelial cells (arrowheads) and in hemocytes (H) of the connective tissue (scale bar: 50 μm); (**E**) CdH (21days), inset of figure (**D**) (square), showing in detail BSD in hemocytes and in the basal lamina of the stomach epithelium (arrows).

2.2. Increase of Metallothionein mRNA Concentration

Until this point, we had only found one species of metallothionein (MT) mRNA in the midgut gland of *Littorina littorea*. However, there was a conspicuous allelic variability across individuals, with at least five different allelic MT variants, as seen in Figure 3A. Hence, the respective mRNAs and the translated amino acid sequences of the MT coding regions proved to be not completely identical to a previously published (reference) sequence from a study focusing on MT expression in periwinkles exposed to environmental stressors [31] (Figure 3A). For quantitative real-time PCR, the reference MT mRNA sequence was applied.

MT mRNA copy numbers in the midgut gland of control and Cd-exposed winkles are shown in Figure 3B. It appeared that the response of MT gene transcription was somewhat delayed with respect to the start of metal exposure, with a first significant peak of mRNA concentration at day 3 for the CdH group and at day 7 for the CdL group (Figure 3B). Moreover, while the intensity of Cd exposure (CdL or CdH) had an influence on the response time of the increase of MT mRNA concentration, the upper level of MT mRNA transcription reached during the experimental time course remained the same for both Cd exposure groups (CdL and CdH) (Figure 3B).

In fact, Cd concentration factors in the midgut gland of metal-exposed winkles increased until the end of exposure to about 250 times in the CdL group and to about 750 times in the CdH group (Figure 4A). At the same time, the increase in MT mRNA concentration rose from an initial value of about 1.7-fold to approximately 7.5-fold in both groups of metal-exposed winkles (Figure 4B), irrespective of whether exposed individuals had enriched the Cd in their midgut glands with lower (CdL) or much higher (CdH) concentration factors.

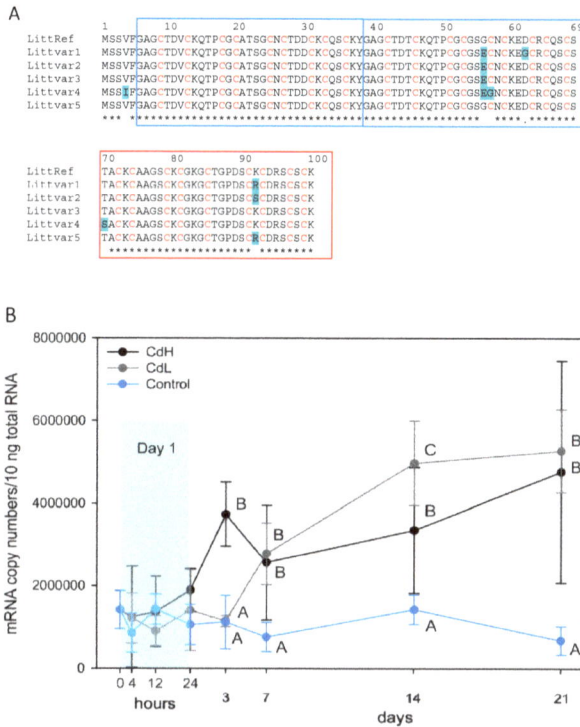

Figure 3. (A) Translated protein sequences of allelic variants of *Littorina littorea*, all of them caused by single nucleotide polymorphisms (SNPs) in the coding region of the respective gene. The sequence variants (Littvar1, Littvar2, Littvar3, Littvar4, Littvar5) are aligned with the first reported reference sequence (LittRef) of the *Littorina littorea* MT (GenBank Acc.Nr.: AY034179.1) [31]. Exchanged amino acid residues in the allelic variants are marked in blue. The tri-partite structure of the protein into different domains is indicated by blue (α domains) and red frames (β domain) [37]. The allelic variant sequences are available from GenBank under the acc. nrs.: KY963497 (Littvar 1), KY963498 (Littvar2), KY963499 (Littvar3), KY963500 (Littvar4) and KY963501 (Littvar5); (B) mRNA transcription of the Cd metallothionein (MT) gene of *Littorina littorea* in control animals (Control, blue line and symbols) and in winkles exposed to nominal Cd concentrations of 0.25 mg/L (CdL, grey line and symbols) and 1.00 mg/L (CdH, black line and symbols) over a period of 21 days. Mean values and standard deviations ($n = 4$) are shown. The significance of the mRNA transcription curves was confirmed by ANOVA ($p \le 0.001$). The significant differences of values between Cd-exposed and control individuals at different time points (Holm–Sidak pairwise multiple comparison, significance level 0.05) are indicated by different letters (A, B, C). Sampling points during the first 24 h (Day 1) are under-laid in blue.

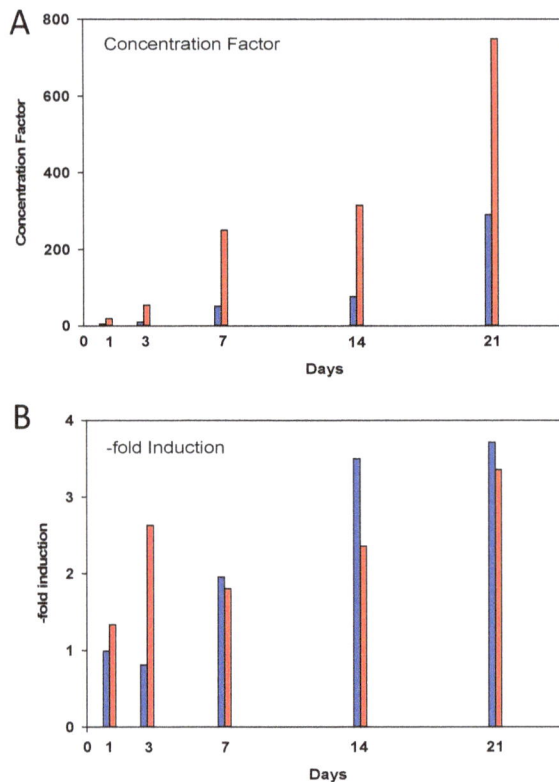

Figure 4. (**A**) Concentration factor for Cd in the midgut gland of *Littorina littorea*. Blue bars represent the low Cd exposure group (CdL), the red bars represent the high Cd exposure group (CdH). All values refer to respective control concentrations at the beginning of the experiment; (**B**) Fold induction factor for the increase of the MT mRNA concentration. Blue bars represent the low Cd exposure group (CdL). Red bars represent the high Cd exposure group (CdH). All values refer to the respective control mRNA copy numbers at the beginning of the experiment (set to 1).

2.3. Cd-Induced Response Patterns at the Cellular and Tissue Levels

2.3.1. Digestive Cells

The volume density of digestive cell lysosomes (Vv_{LYS}) in the midgut gland oscillated for all treatment groups tover the first three days of exposure (Figure 5A). Significant differences between Cd-exposed and control snails were observed after 7 days of exposure. From that point on, the V_{VLYS} values of winkles exposed to the higher Cd concentration (CdH)—but not those of the CdL group—remained significantly elevated over the whole period of exposure, without regeneration during the recovery period (Figure 5A). Upon microscopic examination, digestive cell lysosomes in the midgut gland of the CdH group appeared to be larger and aggregated in bigger clusters compared to the situation in the CdL and control groups, where lysosomes seemed to have a bigger diameter and to be aggregated in smaller groups, or did not appear in clusters at all. As shown by autometallography, the digestive cell lysososmes from Cd-exposed winkles contained dense signals of BSD, indicating metal loads (Figure 2B,C).

The digestive cell integrity in the midgut gland tubules of the control group remained stable over the entire the exposure period (Figure 5B). In contrast, the loss of digestive cell integrity increased significantly in Cd-exposed snails (CdL and CdH) from day 7 until day 21 of exposure, but returned to control levels during the recovery period (Figure 5B).

The presence of lipofuscin granules in digestive cells of the midgut gland tubular epithelium (Figure 6) increased gradually with increasing exposure time and Cd concentrations. Overall, lipofuscin was more abundant in the CdH group than in the CdL group. No changes were found during the experiment in the control group, in which the amount and intensity of lipofuscin staining remained stable.

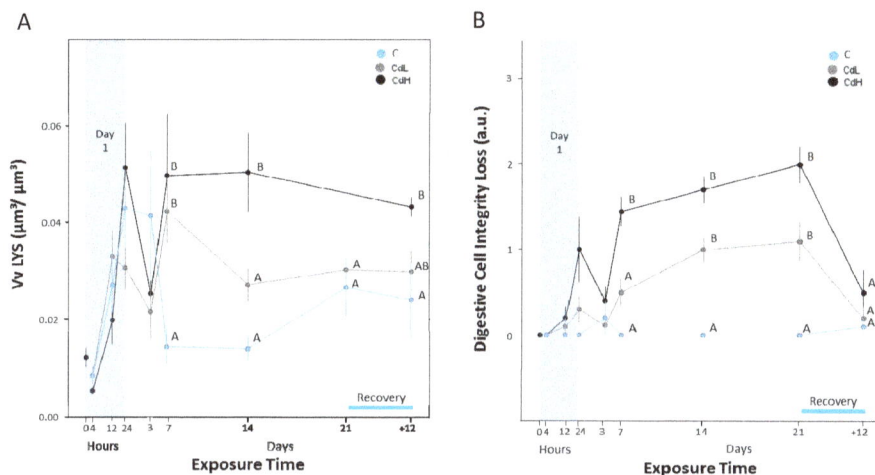

Figure 5. (A) Progression of volume density of lysosomes (Vv_{LYS}) in the digestive cells of midgut gland tubules of *Littorina littorea*. Means (symbols) and standard errors (bars) are shown. Different letter codes between single values of the same time point indicate statistically significant differences ($p < 0.05$); (B) Course of integrity loss of digestive cells (DCI) in the midgut gland tubular epithelium, expressed in arbitrary units (a.u.) (see Table 1 in the Materials and Methods section for explanation). Means (symbols) and standard errors (bars) are shown. Different letter codes between single values of the same time point indicate statistically significant differences ($p < 0.05$).

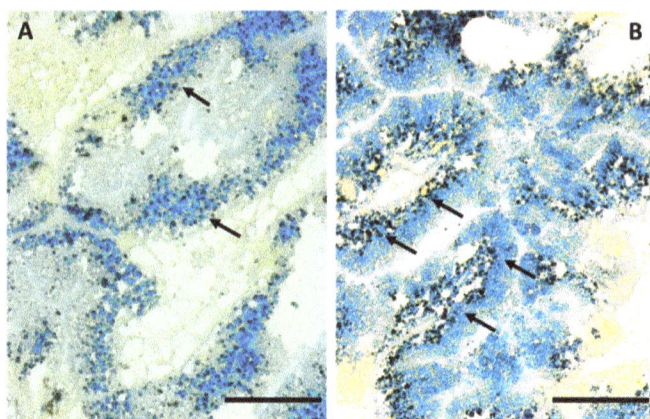

Figure 6. Lipofuscin staining in digestive cells of the tubular epithelium in the midgut gland of (A) a control periwinkle and; (B) an individual of the CdH group after 21 days of exposure. Scale bars: 200 µm; arrows indicate lipofuscin granules.

2.3.2. Basophilic Cells

Apart from a slight initial decrease, there was no significant change in the relative proportion of basophilic and digestive cells in the midgut gland tubuli of control periwinkles, expressed as the volume density of basophilic cells (V_{VBAS}). In contrast, both Cd concentrations (CdL and CdH) provoked a significant increase in the volume density of basophilic cells (V_{VBAS}) in metal-treated animals after only 12 h of exposure (Figure 7A). These values increased further until day 21 of exposure in both groups (CdL and CdH), leading to a clearly visible difference in the histological appearance (Figure 8) between the control (Figure 8A) and Cd-exposed individuals (Figure 8B).

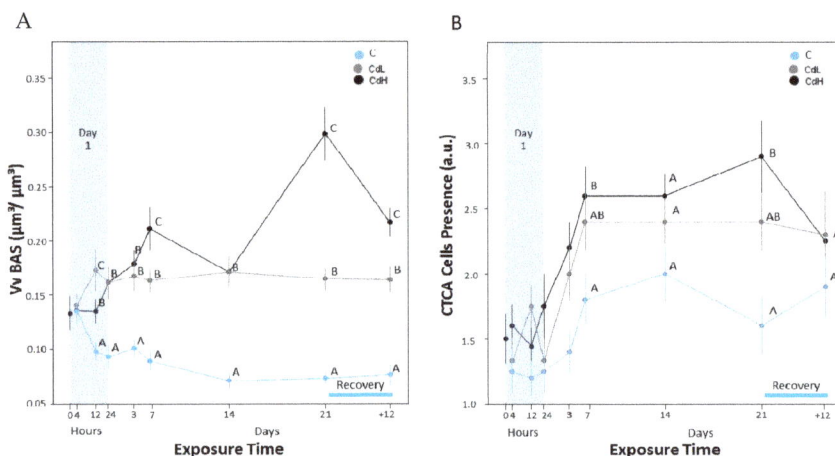

Figure 7. (**A**) Time course of volume density of basophilic cells (V_{VBAS}) in the midgut gland tubules of *Littorina littorea*. Means (symbols) and standard errors (bars) are shown. Different letter codes between single values of the same time point indicate statistically significant differences ($p < 0.05$); (**B**) Progression of the presence of connective tissue calcium cells (CTCA) in the midgut gland of periwinkles, expressed in arbitrary units (a.u.) (see Table 1 in the Materials and Methods section for explanation). Means (symbols) and standard errors (bars) are shown. Different letter codes between single values of the same time point indicate statistically significant differences ($p < 0.05$).

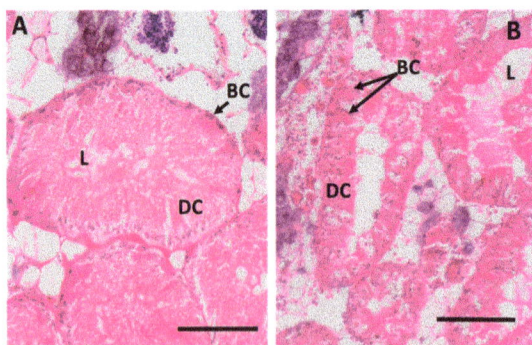

Figure 8. (**A**) Detailed view of the midgut gland tubules (hematoxylin-eosin staining) of a control periwinkle and (**B**) of a periwinkle of the CdH group after an exposure of 21 days. Note the drastic increase of the relative number of basophilic cells (arrows) and the enlarged lumen in (**B**), compared to control conditions (**A**). Scale bars: 200 µm; Abbreviations: L: lumen, DC: digestive cells, BC: basophilic cells.

During the period of recovery, the V_{VBAS} values in exposed groups (CdL and CdH) did not achieve the values of the control group any more, even though the value of the CdH group decreased slightly from day 21 of exposure till the end of the recovery period (Figure 7A).

The integrity of the basophilic cells showed the same trend for all treatment groups, as observed in the digestive cells (data not shown).

2.3.3. Connective Tissue Calcium Cells

The presence of connective tissue calcium cells in the midgut gland (Figure 7B) increased for all groups during the experiment, although this increase was higher in the Cd-exposed animals (CdL and CdH) than in the controls (Figure 7B). However, only values of the CdH group were significantly different from the control values after 7 and 21 days of metal exposure. After the period of recovery, values decreased in both Cd exposure groups (CdL and CdH), reaching control values again (Figure 7B). A microscopic inspection of histological midgut gland sections confirmed the increased presence of connective tissue calcium cells in Cd-exposed winkles (Figure 9).

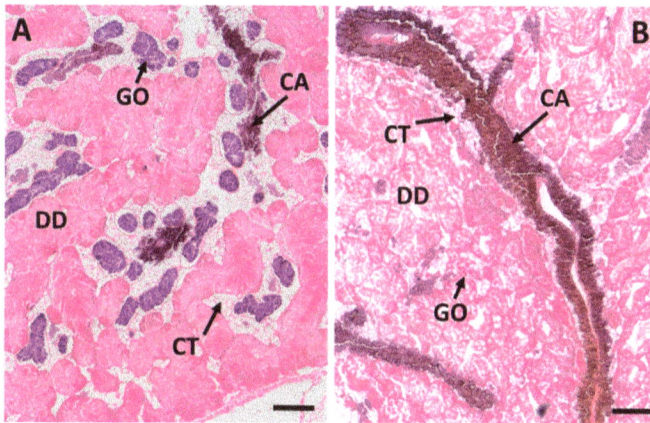

Figure 9. Overall view of the midgut gland/gonad complex stained with hematoxylin-eosin. (**A**) control periwinkle (male); (**B**) a periwinkle of the CdH group after 21 days of exposure (female). Scale bars: 200 µm; Abbreviations: DD: digestive tubular diverticula, CA: Connective Tissue Calcium cells, CT: connective tissue; GO: gonad tissue.

2.3.4. Midgut Gland Structure and Integrity

As shown in Figure 10 by quantitative parameters, the structural integrity in the midgut gland/gonad complex was subjected to apparent alterations over the course of the experiment. For example, the connective tissue to digestive tubular ratio (CTD) increased over the exposure time in the midgut gland of Cd-treated snails (CdL and CdH) (Figure 10A). Statistically significant differences between the Cd-exposed groups and the control group appeared from the seventh day on. During the recovery period, no return to control levels was noticed (Figure 10A). Concomitantly, the visibility of the tubular lumen in the midgut gland increased in all treatment groups during the experiment, including the control group (Figure 10B), although no statistically significant difference could be assessed between different treatments. Nevertheless, values were consistently higher in the Cd-exposed groups (CdL and CdH), and especially in the CdH group.

Upon microscopic observation of histological sections, the lumen visibility in the midgut gland of Cd-exposed winkles was evident when compared to control individuals (Figure 8). Moreover, the reduction of the digestive gland integrity was clearly visible, especially when compared with the digestive gland structure of control individuals (Figure 9).

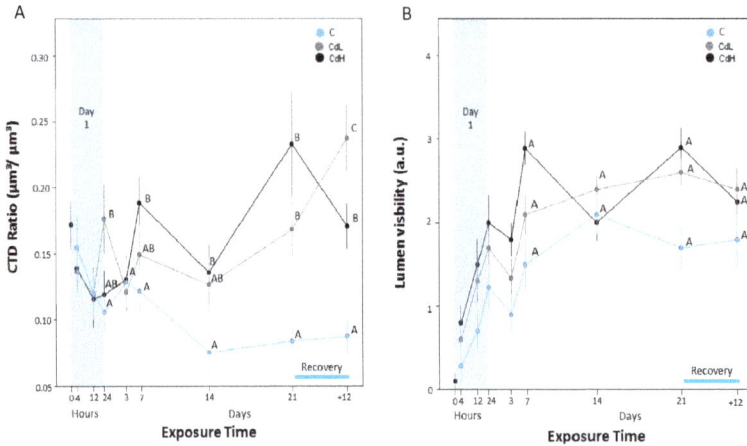

Figure 10. (**A**) Progression of structural integrity in the midgut gland/gonad complex of *Littorina littorea*, expressed by the connective tissue to digestive tubule ratio (CTDr). Means (symbols) and standard errors (bars) are shown. Different letter codes between single values of the same time point indicate statistically significant differences ($p < 0.05$); (**B**) Progression of lumen visibility of midgut gland tubules, expressed in arbitrary units (a.u.) (see Table 1 in the Materials and Methods section for explanation). Means (symbols) and standard errors (bars) are shown. Different letter codes between single values of the same time point indicate statistically significant differences ($p < 0.05$).

2.3.5. Integrity of Digestive Tract Epithelium

Apart from the midgut gland, a loss of tissue integrity due to Cd exposure was also observed in the digestive tract. When measured quantitatively, the structural integrity loss of epithelial cells of the digestive tract was pronounced, especially in the CdH-exposed winkles, when compared to control individuals (Figure 11). This was confirmed upon inspection of the digestive tract tissue condition in histological sections (Figure 12). Apart from a slight disintegration of the tissue structure, the digestive tract epithelium of Cd-exposed individuals also contained an increased density of mucocytes due to metal exposure (Figure 12A) compared to the intact structure of control winkles (Figure 12B).

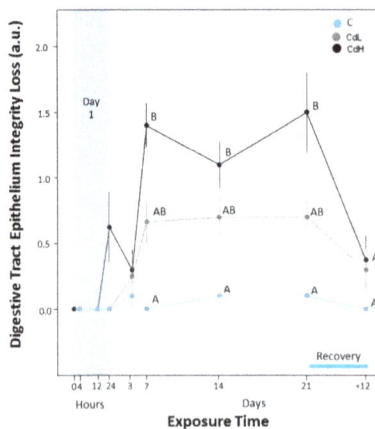

Figure 11. Course of the integrity loss of the epithelia of digestive tract (DTI), expressed in arbitrary units (a.u.) (see Table 1 in the Materials and Methods section for explanation). Means (symbols) and standard errors (bars) are shown. Different letter codes between single values of the same time point indicate statistically significant differences ($p < 0.05$).

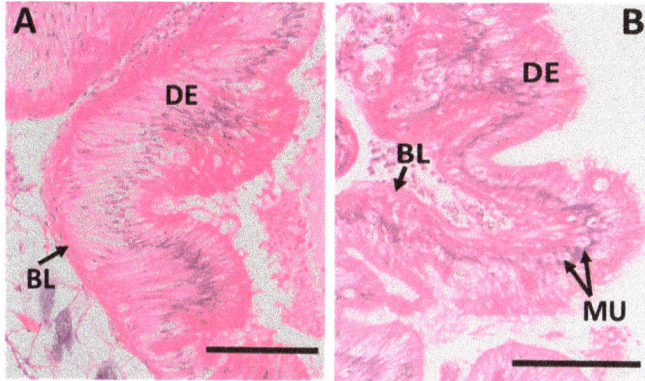

Figure 12. Sections of the stomach (digestive tract) stained with hematoxylin-eosin. (**A**) Control periwinkle; (**B**) Periwinkle of the CdH group after 21 days of exposure. Note the loss of integrity in the stomach epithelium (arrows) and the increase of mucocytes as a result of Cd-exposure. Scale bars: 200 μm; Abbreviations: DE: Digestive epithelia, BL: Basal lamina, MU: Mucocytes.

3. Discussion

3.1. Cellular Cadmium Accumulation and the Significance of MT mRNA Increase

The Cd accumulation capacity of *Littorina littorea* is huge, considering the rather low metal concentrations of the exposure medium in the present study, with nominal Cd concentrations of 0.25 mg/L (effective: 0.24 mg/L) in the CdL exposure and of 1.0 mg/L (effective: 0.85 mg/L) in the CdH exposure. Both concentrations are far below the known LC_{50} (96h) values for Cd in *Littorina littorea* [49,50], and are even slightly lower than sub-lethal Cd concentrations applied in some previous studies with this species [39,47]. In spite of this, one has to consider that real Cd concentrations are, even in seawater samples from contaminated habitats, about 1000 to 10,000 times lower than the experimental concentrations applied in the present study [14]. All the more remarkable is the resistance of *Littorina littorea* in its various response patterns to accumulated midgut gland Cd concentrations of more than 250 μg/g dry wt. in the low exposure group (CdL) and about 700 μg/g dry wt. in the high exposure group (CdH) (Figure 1). These concentrations are about 290 times (CdL) and up to 750 times (CdH), respectively, higher than the control levels detected in the midgut gland tissue of uncontaminated winkles (Figure 4A).

Evidently, a considerable proportion of the accumulated Cd is retained by cellular compartments. This is suggested by the results of autometallography (Figure 2) which shows a distinct increase of BSD spots in the basal lamina and digestive cells of midgut gland tubules, and in the connective tissue of the midgut gland/gonad complex. Cd exposure also led to marked BSD signals in the basal lamina of the stomach, in the lysosomes of the stomach epithelial cells, and in hemocytes of the connective tissue near the basal lamina of the stomach. Autometallography as a histological method was developed for the visualization of metal sulphide and selenide compounds in cellular structures [51]. A number of more recent studies have demonstrated that this method works excellently for the localization of metal deposits in metal-exposed marine molluscs [35], including *Littorina littorea* [32,52]. Autometallography fails, however, to demonstrate the presence of metal ions bound to cytosolic MT molecules [24], where metal ions such as Cd^{2+} are coordinated by the sulphur atoms of the protein's cysteine residues [53,54]. Only when these stable Cd-metallothionein complexes are semi-digested in the digestive cell lysosomes, exposing the bound Cd ions, or when these complexes are sufficiently concentrated in lysosomes, can autometallographed black silver deposits be formed to reveal the toxic metal [52]. Therefore, lysosomes can contain degradation products of MTs and serve as a final storage site of degraded MTs and, possibly, of other metal-binding proteins. In any case, the speciation of Cd is the same (Cd^{+2}) in the cytosol and in the lysosomal pool.

Yet, the participation of MT in the process of Cd accumulation and detoxification by *Littorina littorea* is essential [36,55], as also demonstrated by the present study. The increase of MT mRNA concentration in Cd-exposed periwinkles can be considered, therefore, as an efficient protective molecular response to the uptake intake of a metal ion that is highly toxic to most animal species at concentrations far below those observed in *Littorina littorea*. In fact, the MT protein of *Littorina littorea* possesses highly Cd-selective binding features [38]. Its Cd-binding capacity has been optimized through evolution by the addition of a third Cd-binding protein domain. The observed sequence variability caused by single nucleotide polymorphisms (SNPs) mostly affects non-critical amino acid positions (Figure 3A). Hence, it can be assumed that the overall performance of the MT as a cytosolic Cd complexing agent is not significantly affected. The tertiary structure of this protein has recently been elucidated by means of solution NMR, making the MT of *Littorina littorea* the first MT so far to have been shown to have a proven three-domain architecture [37].

Interestingly, the induction factors for the increase in MT mRNA concentration detected in the midgut gland of Cd-exposed winkles are, at about 6.6-fold, rather moderate (see Figure 4B) compared to those observed for MT genes in other marine species [56–58]. Moreover, it appeared that the upper limits of the increase of MT mRNA concentration in Cd-exposed winkles increased moderately with exposure time, but almost without significant differences between CdL and CdH-exposed individuals throughout (Figure 2B). This is also reflected by the pattern of increased MT induction values (Figure 4B). Instead, the rise of the MT mRNA concentration started faster in CdH than in CdL-exposed snails (Figure 2B). In the CdH group, the first significantly elevated values of MT mRNA concentration were observed after three days of metal exposure. This is a slightly delayed reaction compared with the early response of some cytological alterations, especially in the volume density of basophilic cells (V_{VBAS}), which was assessed only a few hours after the start of Cd exposure (Figure 7A). An obvious interpretation for this may be that, due to its high selectivity and capacity for Cd binding, the MT of *Littorina littorea* may already efficiently work at only moderate expression rates, maintaining its stress response capability in this way in an energy-saving mode [37]. In fact, the minimization of energy expenditure in metal-stressed invertebrates is often achieved by metabolic depression or other energy-saving strategies [59,60]. In such situations, the energetic demand for the upregulation of cellular protective mechanisms such as MT induction is preferably kept as low as possible [61]. The Cd-specific binding character and the extended loading capacity for additional Cd^{2+} ions due to the three-domain structure of the *Littorina littorea* MT may comply with these energy-sparing requirements, conferring to periwinkles an evolutionary advantage in coping with Cd stress [37]. This is all the more significant if we consider that an increase in MT mRNA concentration in *Littorina littorea* may also occur in response to environmental, non-metallic stressors [31].

Overall, the near concomitance of an increasing MT mRNA concentration with enormous alterations at the cellular level (see below) reinforces the impression that increased MT mRNA levels in the midgut gland of periwinkles apparently serves the purpose of protecting *Littorina littorea* from potentially adverse Cd effects. This interpretation is consistent with the view that, as well as in many other animal species, MTs may exert a protective role against the adverse effects of Cd^{2+} [62]. For *Littorina littorea*, this may be especially important during periods of increased Cd exposure via the external medium (as in the present study). The protective role of MT may also be important, however, under any condition that leads to significant cytological re-structuration or cellular wear in the midgut gland of winkles, with an increased potential for metal ions to leak out from their intracellular compartments [43]. This may occur due to the impact of non-metallic contaminants [44,63–65], the influence of fluctuating environmental stressors [31,66], or because of seasonal alterations of the physiological state of periwinkles [45,67].

3.2. Cadmium-Induced Cellular Response Patterns in the Midgut Gland

3.2.1. Digestive Cells

Cd exposure led to alterations in the shape and integrity of digestive cells in midgut gland tubules (Figure 5B). One of the reasons for this is that Cd-stressed digestive cells cast off their apical parts and become flatter [33]. In addition, the present study also shows that the relative number of digestive cells declined in favor of an increase of the volume density of basophilic cells (V_{VBAS}) (see Figure 7A). This seems to be in contrast with previous findings, where it was shown that Cd exposure of *Littorina littorea* stimulates digestive cell proliferation [32]. At the same time, however, Cd also causes an increase of digestive cell mortality. Overall, their number actually decreases in relation to basophilic cells, due to the fact that they die at a faster rate than they proliferate, as also observed in previous experiments [32]. In the present study, these alterations caused a drastic change of the midgut gland tubular morphology, with an increased visibility of the tubular lumen and a thinner surrounding epithelium consisting predominantly of basophilic cells (Figures 8 and 10). Generally, Cd is known for its stimulatory effects on cell proliferation in mammals [68], and even so in molluscs, including terrestrial [69,70] and marine gastropods like *Littorina littorea*, especially under the influence of pollutant stressors [32,39]. Occasionally, cell proliferation in gastropods has been observed to be accompanied by programmed cell death [70–72]. Both phenomena indicate that gastropod tissues can be subjected to extensive structural alterations in response to environmental stressors, documenting the huge plasticity of these animals in adapting to the fluctuating and sometimes adverse conditions of their environment [67,73].

Under Cd exposure, in particular, digestive cells assume another important metabolic task: their lysosomal system is apparently involved in Cd storage and detoxification. This is clearly demonstrated in our study by the increase of autometallographic BSD signals in lysosomes of the midgut gland digestive cells of Cd-exposed winkles (Figure 2B,C). The required increase of lysosomes for Cd sequestration was apparently achieved by an enlargement of the lysosomal volume in relation to the cytoplasmic space in digestive cells, as expressed quantitatively by the rising volume density of digestive cell lysosomes in Cd-exposed winkles (Figure 5A). This increased loading capacity of digestive cell lysosomes seems to be compromised by a decrease in the lysosomal membrane stability, as shown for *Littorina littorea* exposed to environmental stressors and polyaromatic hydrocarbons [43,63,65].

Overall, there are actually two relevant pools of Cd sequestration in the midgut gland of *Littorina littorea*: apart from the lysosomal compartment, the second pool of Cd sequestration is due to the binding of the metal to cytosolic MT, which in the present study is documented by the Cd-induced increase of the MT mRNA concentration (Figure 3). It is important to discover how these two metal pools (the cytosolic MT-related and the lysosomal one) interact. The data of the present study show that both pools react to the Cd load in a timely response between the 3rd and the 7th day after the start of Cd exposure (compare Figures 3B and 5A). Our interpretation is that, at the beginning of Cd exposure, most of the metal may first be sequestered by constitutive and de novo synthesized MT. At this stage of Cd accumulation, the MT-associated Cd pool may prevail. Due to ongoing MT turnover and degradation [36], however, a certain part of the liberated Cd ions would be bound to de novo synthesized MT, whereas a minor proportion of the metal (perhaps still attached to half-degraded and denatured MT) would end up in digestive cell lysosomes for final storage. From there, some Cd would leak out back into the cytoplasmic space due to lysosomal membrane destabilization (see above). Eventually, impaired Cd-loaded lysosomes would be discarded by fecal excretion, probably together with extruded digestive cells. The viability of this hypothesis was confirmed by studies in mussels, where metal-loaded lysosomes and cell debris were detected in the lumen of midgut gland tubules [24,35,74].

The increasing presence of lipofuscin granules in the digestive cells of Cd-exposed winkles (see Figure 6) may be the consequence of enhanced cellular proliferation and re-structuration due to

Cd exposure. This probably leads to an increment of autophagy and cell turnover, accompanied by impaired digestion within the lysosomal compartment. In other gastropod species too, Cd apparently stimulates the proliferation of endoplasmic reticulum and midgut gland tubular cells, followed by a concentration-dependent increase in the formation of residual bodies and lipofuscin granules [72].

3.2.2. Basophilic Cells

The results observed at the histological level show that some cellular alterations induced by Cd exposure in the present study developed faster than expected [33]. This holds particularly for the increase of the volume density of basophilic cells (Vv_{BAS}) in the midgut gland tubules of Cd-exposed periwinkles. This was, in the present study, the fastest responsive parameter among all observed variables, with significantly increased values only 12 h after the start of Cd treatment (Figure 7A). As a direct consequence of this increase, the tubular epithelium of the midgut gland undergoes a process of cell type replacement, which is considered to be a general biomarker for environmental stress in bivalves and gastropods [40,44,73,75,76]. Under control conditions, the majority of the midgut gland tubular epithelium is composed by digestive cells. Upon stress exposure, however, the proportion of digestive cells drops down, resulting in a gradual increase of the number of basophilic cells, which eventually may constitute the majority of the midgut gland tubular epithelium (see above). During the period of recovery, Vv_{BAS} levels remained significantly above control values, even though a slight (but statistically insignificant) tendency towards normalization could be observed [32]. In fact, one may expect that, in the course of a longer period of recovery, the original number of digestive cells may gradually be restored in the absence of Cd, giving also rise to a normalization of the Vv_{BAS} values [32]. The fact that periwinkles were forced to stay underwater for the entire course of Cd exposure could have caused, in the present study, an incomplete restoration of control values during the observed time span of recovery compared to previous studies [24,32,45].

3.2.3. Connective Tissue Calcium Cells

During Cd exposure, the number of connective tissue calcium cells increased slightly in response to Cd exposure, particularly in the CdH group, but returned to control levels after the period of recovery (Figure 7B). Connective tissue calcium cells are constituents of the mantle, foot, albumen gland and visceral complex (including the midgut gland) of gastropods and are closely related with blood vessels. They contain calcium carbonate, which is thought to regulate the calcium content, contributing to the ionic balance and the maintenance of haemolymph pH [77]. Metals such as Cd, Cu and Sr can also be incorporated into the mineral granules inside the connective tissue calcium cells [35]. Thus, in the present study, the slight increase of connective tissue calcium cell density in Cd-exposed winkles (Figures 7B and 9) is probably related to the presence of Cd in these cells. On the other hand, their calcium carbonate stores must remain disposable to rapid mobilization upon metabolic demand [77]. This would probably also lead to a rapid liberation of the trapped Cd^{2+} ions which, because of their similar ionic radii, may be confused with Ca^{2+} ions by cellular transport mechanisms [78] and Ca-dependent signalling pathways [68]. Hence, the inactivation of Cd^{2+} on behalf of calcium may be one of the important protective roles exerted by MT in the midgut gland tissue of *Littorina littorea*.

3.2.4. Cd-Induced Loss of Tissue Structure and Integrity

Clearly, Cd-induced alterations of cellular structure and integrity, as well as cell type-specific replacements, must have an impact on the appearance and integrity of the organs and tissues constituted by the respective cell types. For example, the structural degradation of digestive cells and their gradual replacement by basophilic cells in the midgut gland tubular epithelium of Cd-exposed winkles is apparently reflected by a loss of integrity at the tissue level. This can be quantified, as in the present study, by changes of the connective tissue to digestive tubule (CTD) ratio (see Figure 10A), or by alterations of the lumen visibility in midgut gland tubules (see Figure 10B). At least the first of these parameters—the CDT ratio—shows an obvious degradation in dependence of Cd exposure in both the CdL and CdH-treated winkles (Figure 10A). Similar results were reported from other studies in different mollusc species exposed to environmental stressors [79,80].

At the same time, as tissue alterations appeared in the midgut gland, a decrease of tissue integrity was also assessed in the epithelia of the digestive tract (Figure 11), which became clearly evident in histological preparations (Figure 12). As in the case of integrity loss in the tubular epithelium of the midgut gland, the decrease of the integrity of epithelial cells in the digestive tract of Cd-treated winkles was, in part, due to epithelial cells losing their height by discharging their apical ends under Cd exposure [39]. This is probably a reaction from the tissues to get rid of their Cd-loaded cells upon exposure [33]. The loss of digestive cells can lead to an impairment of digestion capability.

3.2.5. The Biomarker Potential of Assessed Parameters

Due to their short-termed capacity for being induced at sub-lethal concentrations of Cd, many of the parameters assessed in the present study may have a strong potential for their application as biomarkers in environmental monitoring [81]. This holds particularly for cellular biomarkers assessed in different midgut gland cell types and for sub-cellular biomarkers such as lysosomal impairment, most of which show a great responsiveness and rapidity in their patterns of reaction.

The midgut gland of molluscs is the crossroads for metabolic regulation, participating in the mechanisms of immune defense and the homeostatic regulation of the internal medium, as well as in the processes of stress response and metal detoxification. In fact, cellular endpoints such as changes in the lysosomal structure and lysosomal metal accumulation have previously been used with success to assess the effect of pollutants on molluscs [22,82].

The start of the MT mRNA increase depends, among other reasons, on the Cd concentration applied during exposure, showing that MT mRNA concentrations start to rise earlier at higher Cd concentrations in the CdH group compared to a somewhat delayed reaction in the CdL group (Figure 3B). In contrast, there was no clear relationship between Cd concentration of the exposure medium (CdL or CdH) or Cd concentration factors in the midgut gland of periwinkles on the one hand, and the values of transcriptional induction of the *MT* gene on the other (see Figure 4). This resembles previous findings reporting that the level of expressed MT protein in the midgut gland of Cd-exposed winkles was hardly affected by the intensity of Cd exposure [55], compromising to some extent the applicability of MT upregulation as a biomarker for Cd stress in *Littorina littorea*.

Taken together, the time patterns of the different molecular, cellular and tissue-specific response reactions to Cd exposure in *Littorina littorea* may very well provide a valuable biomarker battery approach for the purposes of biomonitoring (Figure 13). It appeared that the timeframe for the onset of the different response reactions became narrower at higher Cd concentrations (CdH) compared to the lower exposure (CdL). In this time-dependent succession, the first parameter which showed distinct reactions at both Cd concentrations (CdL and CdH) was, interestingly, the volume density of basophilic cells (V_{VBAS}) in the periwinkle's midgut gland, with significantly increased values after only 12 h of exposure (see Figure 13). This indicates that Cd-stressed periwinkles respond to the intake of the metal by starting the re-modelling of their cellular composition of the tubular tissue in the midgut gland at an early stage. The increase of the MT mRNA concentration follows next, prior to or together with the activation of the lysosomal compartment (Figure 13), suggesting that cytosolic inactivation of Cd by binding to MT and lysosomal Cd accumulation may be complementary processes of detoxification that go hand in hand, as explained above. All other cellular and tissue-specific response reactions to Cd exposure followed afterward.

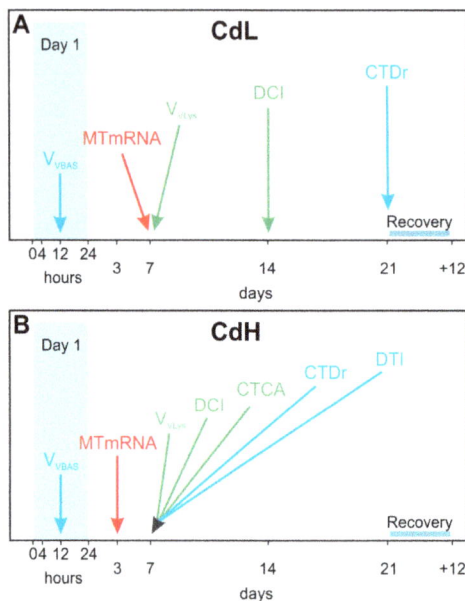

Figure 13. Time course of the onset of molecular (red), cellular (green) and tissue-specific alterations (blue) in *Littorina littorea* exposed to sub-lethal low (CdL) (**A**) and high (CdH) (**B**) Cd concentrations through 21 days of exposure, followed by a 12 days period of recovery. Abbreviations: MT mRNA: MT mRNA copy numbers, Vv_{LYS}: Volume Density of lysosomes in digestive cells of midgut gland tubules, DCI: Digestive cell integrity of midgut gland tubules; CTCA: presence of connective tissue calcium cells in the midgut gland, V_{VBAS}: Volume density of basophilic cells in the epithelium of midgut gland tubules, DTI: Digestive tract integrity. CTDr: Connective tissue to digestive tubule ratio.

In spite of their general validity as biomarkers, many of the cellular and molecular variables assessed in the present study may also respond to environmental stimuli not directly related to Cd stress. This has been shown, for example, for increasing MT mRNA concentrations in periwinkles exposed to anoxia and freezing [31], or for alterations of the lysosomal compartment in midgut gland digestive cells due to seasonal fluctuations or pollution impacts in the intertidal habitat of periwinkles. In the present study, too, the volume density of the lysosomal system (Vv_{LYS}) in the first three days of exposure must be more related to the experimental design than to the effects of Cd exposure (see Figure 5A), as suggested by the concomitant alteration of this variable in all exposure groups, independent of Cd concentration. A possible reason for this may be the fact that, in the present study, periwinkles were forced to remain immersed under water for the entire study. This condition may have given rise to alterations of the metabolism of periwinkles, with possible impacts on their digestive cycles, as previously reported for subtidal mussels [83]. Evidently, confounding factors like these must be observed when applying cellular and molecular variables of *Littorina littorea* for purposes of environmental monitoring.

4. Materials and Methods

4.1. Experimental Set-Up

Littorina littorea collected in Scrabster (Scotland) were purchased from a commercial dealer (Arrainko SL, Mercabilbao, Bilbao, Spain). Specimens with a maximum height of between 20–30 mm were selected and acclimated to experimentation conditions in a seawater flow through the system for a week before experimentation (Salinity: 33‰; Temperature: 17 °C). A group of 175 individuals was stated as the control group. Another two groups of 175 individuals each were subjected to nominal Cd concentrations (applied as $CdCl_2$) of 0.25 mg·Cd/L ("Cd Low", CdL) and to 1 mg·Cd/L ("Cd High", CdH) for 21 days, respectively. Real Cd concentrations measured in the seawater were 0.035 ± 0.005 mg/L for control conditions, 0.243 ± 0.006 mg/L for CdL exposure, and 0.852 ± 0.021 mg/L for CdH exposure (means ± standard deviations, $n = 9$). pH values in the seawater throughout the exposure were 7.72 ± 0.18 for control conditions, 7.76 ± 0.24 for CdL exposure, and 7.79 ± 0.21 for CdH exposure (means ± standard deviations, $n = 27$). The O_2 concentration in the seawater was constantly at ~4 mg/L.

During the experimentation period, winkles were maintained in 40 L of naturally sand-filtered well-seawater from a clean place (Plentzia, Bizkaia, Spain) and fed *ad libitum* with *Bifurcaria bifurcata*. Seawater and food were changed every second day, and a net was installed in the interface air-water of each tank to maintain periwinkles always underwater.

Winkles were sampled after 0, 4, 12 h and 1, 3, 7, 14 and 21 days of exposure. At the end of the exposure period, winkles were maintained for recovery in clean seawater for 12 days without Cd supply. 10 winkles per experimental group (controls, CdL and CdH) were sacrificed performed at each exposure time and after the recovery period. The midgut gland/gonad complex was dissected out and processed as described below to obtain the different endpoints.

4.2. mRNA Isolation, Allelic Variant Screening and Quantitative Real-Time PCR of the Reference Gene

Three Cd-exposed individuals were used for PCR and sequencing in order to confirm the primary protein structure of the reference MT of *Littorina littorea* (see below). An additional 20 individuals were used for the screening of allelic MT variants (see below).

For each experimental time point, four individuals of *Littorina littorea* were dissected on an ice-cooled stainless steel plate and total RNA was isolated from ~10 mg of homogenized (Precellys, Bertin Instruments, Montigny-le-Breonneux, France) hepatopancreatic tissue with the RNeasy®Plant Mini Kit (Qiagen, Venlo, The Netherlands) applying on-column DNase 1 digestion (Qiagen). RNA was screened for integrity visually on an agarose gel and quantified with the RiboGreen®RNA Quantification Kit from Molecular Probes (Invitrogen, Karlsruhe, Germany) on a VICTOR™X4 2030 Multilabel Reader (PerkinElmer, Waltham, MA, USA). First, strand cDNA was synthesized from 450 ng of total RNA with the Superscript® IV Reverse Transcriptase synthesis kit (Invitrogen, Life Technologies, Waltham, MA, USA) in a 20 µL approach for subsequent Real-time Detection PCR. The remaining tissue was processed further for Cd analysis as described below.

The primary structure of the *Littorina littorea* MT (GenBank Acc.Nr.: AY034179.1) [31] was confirmed by PCR and sequencing ($n = 3$). In a screening for allelic variation by sequencing 20 PCR—amplified and cloned individuals of *Littorina littorea*, 5 distinct isoforms ($n = 3$) were detected. PCR primers located in the untranslated region (sense 5′-CTGACGAGTGAACTGTTTTT-3′ and antisense 5′-GATGGGGAATGAGAAAATG-3′) were applied. The respective sequences were submitted to GenBank and are accessible under the acc. nrs.: KY963497 (allelic variant 1), KY963498 (allelic variant 2), KY963499 (allelic variant 3), KY963500 (allelic variant 4) and KY963501 (allelic variant 5).

Quantitative Real-time Detection PCR of the *Littorina littorea* MT was performed on a Quant studio 3 (Applied Biosystems, Thermo Fisher Scientific, Waltham, MA, USA) using Power SYBR Green (Applied Biosystems, Thermo Fisher Scientific, Waltham, MA, USA). The transcript with the defined amplicon length of 84 bp was amplified using the following concentrations and primers: *Littorina littorea* litt. sense, 900 nM; 5′-AATACGGAGCGGGTTGCA-3′ and *Litoorina littorea*. antisense, 900 nM; 5′-AGCGACAGTCCTCCTTACAGTTG-3′ applying the following protocol of 40 cycles: denaturation at 95 °C for 15 s, annealing and extension combined at 60 °C for 60 s. The 10 µL PCR reaction contained 1 µL of cDNA and 1× Power SYBR Green PCR master mix, 1× U-BSA and sense and antisense primer. Primers were designed using the Primer Express 3.0 software (Applied Biosystems) and optimal primer concentrations were assessed with a primer-matrix followed by dissociation curves. Calibration curves from amplicons were generated to determine C_q values (PCR efficiency ~92%) for copy number analysis using the Thermo Fisher Cloud Software, Version 1.0 (Life Technologies Corporation, Carlsbad, CA, USA).

4.3. Metal Analysis

Cd concentrations in the midgut gland tissues and seawater were assessed by flame atomic absorption spectrophotometry. After oven-drying the dissected midgut gland at 65 °C, the samples were pressure-digested in 2 mL tubes (Eppendorf, Hamburg, Germany) with a 1:1 mixture of nitric acid (Suprapure, Merck, Darmstadt, Germany) and deionized water in an aluminum oven covered with a heated lid at 69 °C until a clear solution was obtained. All samples were diluted to 2 mL with deionized water and Cd concentrations measured by graphite furnace atomic absorption spectrophotometry (model Z-8200, Hitachi, Tokyo, Japan). Standard metal solutions in 1% nitric acid were used for calibration. Accuracy of metal measurements of the midgut gland was verified with certified standard reference material (TORT-2, Lobster Hepatopancreas Reference Material for Trace Metals; National Research Council Canada).

4.4. Sample Processing for Microscopy

The midgut gland/gonad complex of control and exposed winkles was fixed in neutralized phosphate buffer with formaldehyde at 4%, dehydrated in a series of ethanol baths and paraffin, embedded using a Leica ASP3005 tissue processor and sectioned at 5 µm with a Leica RM2125RTS microtome for histopathological analysis (Section 2.3.3) and autometallographical staining (Section 2.3.4). Another portion of the midgut gland/gonad complex was dissected out, frozen in liquid nitrogen and stored at −80 °C, then sectioned at 8µm with a CM3050s Leica cryotome for histochemical analysis.

4.5. Histopathology

Paraffin sections (5 µm) were stained with haematoxylin-eosin (H/E) in order to analyze the integrity of the midgut gland, gonad and gills.

The volume density of basophilic cells (Vv_{BAS}) and the connective tissue area per digestive diverticula area ratio (CTD) were quantified by means of stereology as an indication of whether changes in cell-type composition and in the amount connective tissue occurred or not [24]. Counts were made in three randomly selected fields in one midgut gland slide per winkle (six winkles per sample). Slides were viewed at 40× objective (final magnification ~400×) using a drawing tube attached to a light microscope. A simplified version of the Weibel graticule multipurpose test system M-168 [83] was used, and hits on basophilic cells (b), digestive cells (d), diverticular lumens (l) and interstitial connective tissue (c) were recorded. Vv_{BAS} was calculated according to the Delesse's principle [83], as $V_{VBAS} = VBAS/VEP$, where VBAS is the volume of basophilic cells and VEP the volume of midgut gland epithelium. The CTD ratio was calculated as CTD = c / (b + d + l).

The general condition of the midgut gland was systematically analyzed under an Olympus BX61 microscope following a semi-quantitative approach. Briefly, the integrity of each organ was ranked with a value ranging from 0 to 4, where the highest (control) integrity was 0 and the highest possible damage was 4. The clear visualization of the limits of the lumen of the midgut gland tubule was ranked from 0–4, where 0 meant the impossibility of properly seeing the lumen under light microscope and 4 meant that the lumen was totally visible and their limits defined. The same rank (0–4) was used for semi-quantitative assessment of the presence of calcium cells in the interstitial connective tissue. The criteria used for the semi quantification of the histopathological endpoints are explained in Table 1.

Table 1. Criteria for the semi-quantification of the histopathological alterations observed in the midgut gland of *Littorina littorea* upon Cd exposure: Digestive cell integrity loss, Integrity loss of digestive tract epithelium, Lumen visibility of midgut gland tubules, Presence of Ca cells in midgut gland connective tissue. The appraisal of the apparent histopathological status is ranked in arbitrary units from 0 to 4, with 0 being the control value and 4 the worst possible status.

Endpoint	Score				
	0	1	2	3	4
Digestive cell integrity loss	Control integrity	Slightly lower cells	Lower cells slightly vacuolated epithelium	Low cells Vacuolated epithelium	Disintegrated tissue
Integrity loss of digestive tract epithelium	Control integrity	Slightly lower cells	Slightly lower cells slightly vacuolated epithelium	Slightly lower cells vacuolated epithelium	Disintegrated tissue
Lumen visibility of midgut gland tubules	<10% of the area of the tubule	<30% of the area of the tubule	<40% of the area of the tubule	<50% of the area of the tubule	>50% of the area of the tubule
Presence of Ca cells in the midgut gland connective tissue	<30% of the connective tissue	<45% of the connective tissue	<60% of the connective tissue	<85% of the connective tissue	>85% of the connective tissue

4.6. Autometallography

The intra-lysosomal accumulation of metals was determined in paraffin-embedded sections following the autometallography procedure of Danscher [51]. Paraffin sections (5 μm) were dewaxed in xylene and hydrated in decreasing ethanol degree baths, once hydrated slides were left in an oven at 37 °C overnight. Tissue sections were covered using temperate and homogenized commercial silver enhancement kit (BBI Solutions) solution (initiator and enhancer solution mixed in a 1:1 ratio) under safety light conditions following the product instructions. Sections were developed for 20 min and then washed for 2 min with tap water. The slides were mounted using Kaiser's glycerin gelatin. Metals were visualized as black silver deposits (BSD) using an Olympus light microscope. The quantification of the BSD extent (Volume density of BSD; Vv_{BSD}) by image analysis could not be performed because the size of the deposits was in many cases under the detection limit of the system.

4.7. Stereology of Digestive Cell Lysosomes

The histochemical activity of β-glucuronidase was demonstrated in unfixed cryotome sections as in [84]. Sections (8 μm) were cut in a CM3050 cryotome at a cabinet temperature of −25 °C, collected into warm glass slides and stored at −40 °C until required for staining. Sections were tempered at room temperature for 5 min and then transferred to the substrate incubation medium consisting of 22.4 mg of naphthol AS-BI-β-D-glucuronide dissolved in 0.96 mL sodium bicarbonate (50 mM) and made up to 80 mL with 0.1M acetate buffer (pH 4.5) containing 2.5% NaCl and 12 g of polyvinyl at a 20% (w/v) concentration as colloid stabilizer. Sections were incubated for 20 min at 37 °C in a water bath with constant agitation. Then slides were rinsed in 2.5% NaCl at 37 °C for 2 min and stained at room temperature for 10 min and under darkness conditions, with 1 mg/mL fast garnet GBC in 0.1 M phosphate buffer (pH 7.4) plus 2.5% NaCl. Sections were then fixed in Baker's calcium formol (4% formaldehyde, 1% calcium chloride, 2.5% sodium chloride) for 10 min at 4 °C and rinsed in distilled water. Finally, sections were gently washed in distilled water and mounted in Kaiser's gelatine.

A stereological procedure was applied to quantify the structure of the digestive cell lysosomes in periwinkles, using an image analysis system. The system consists of a B&W-CCD video camera, a Leitz Laborlux light microscope, a computer with video board and BMS software. An objective lens of 100× magnification was used. Binary images segregating lysosomes from digestive cell cytoplasm were obtained by the segmentation procedure, which was manually adjusted in the first measurement of a given section to correct slight differences in staining intensity between different sections. With the image analysis system, the lysosomal volume density ($Vv_{LYS} = VL/VC$) was generated, where V = volume, L = lysosomes and C = digestive cell cytoplasm. Five measurements were made per midgut gland. The stereological formulae included a correction factor for particles with an average diameter smaller than the section thickness [85]. Sample size was determined based on previous analyses of mean and standard deviation values of the four parameters, which at least resulted in a maintained constant for a sampling area over 16,000 μm^2 [86]. Since the total area of digestive cells scanned in each measurement was approximately 4000 μm^2, 5 measurements were made on one single section (total sampling area per mussel 20,000 μm^2).

4.8. Lipofuscin Determination

Lipofuscins were detected using the Schmorl method [87]. Lipofuscins are residual pigments stored in lysosomes, organelles in which the first detectable alterations caused by pollutants can be detected before any other effect on physiological parameters can be observed [88]. Sections (8 μm) were cut in a CM3050S Leica cryotome at a cabinet temperature of $-25\,^{\circ}C$, collected into warm glass slides and stored at $-40\,^{\circ}C$ until required for staining. First, slides were fixed in fixative as described above for 15 min at $4\,^{\circ}C$. After rinsing the sections in distilled water, there were immersed in the reaction medium containing 1% ferric chloride and 1% potassium ferricyanide in a ratio of 3:1 for 5 min. Then, the sections were rinsed in 1% acetic acid for 1 min. At last, sections were rinsed in distilled water and mounted with Kaiser's glycerin gelatin. The appearance of lipofuscins as bluish granular concretions was analyzed under Olympus light microscope.

4.9. Biometry

Biometric measurements were performed with periwinkles from D0 and D21 (Control, CdL and CdH). Whole animal and flesh weights were recorded; and in addition, the maximum length and width of the shells were measured with a caliper up to the nearest millimeter.

4.10. Statistics

The statistical analysis was carried out with the aid of the SPSS/PC+ statistical package V.22 (SPSS Inc., Microsoft Co.). For the lysosomal volume density, for the basophilic cell stereology and for the connective to diverticula ratio, one-way ANOVA and a subsequent Duncan's Posthoc test for multiple comparisons between pairs of mean values was applied ($p < 0.05$). For the semi-quantitative results obtained in the general histopathology, non-parametric Kruskal–Wallis tests were carried out comparing the variances between experimental groups ($p < 0.05$).

Data of qRT-PCR and metal analysis were statistically evaluated by Sigma Plot 12.5. For normal-distributed data, the *t*-test was applied whereas for data failing equal distribution the Holm-Sidak method was used. Statistical significance was set at $p \leq 0.05$. Additionally, an analysis of variance (ANOVA) was applied to test for significance of time-dependent variations of data ($p \leq 0.001$).

5. Conclusions

1. The exposure of periwinkles (*Littorina littorea*) to sub-lethal Cd concentrations (Cd Low, 0.25 and Cd High, 1 mg Cd/L) over 21 days provoked the successive induction of molecular, cellular and tissue-specific response reactions in the midgut gland and digestive tract of metal-exposed winkles. The assessed parameters were: increase of MT mRNA concentration, volume density of digestive cell lysosomes and lipofuscin formation in the midgut gland tubular epithelium, tubular digestive cell integrity, volume density of tubular basophilic cells, presence of connective tissue calcium cells, as well as tissue integrity of the midgut gland/gonad complex and of the digestive tract.

2. The beginning of alterations of the assessed parameters followed distinct concentration-dependent and time-dependent patterns, where the timeframe for the onset of the different response reactions became narrower at higher Cd concentrations (CdH) compared to lower exposure concentrations (CdL).

3. Interestingly, the first parameter that showed distinct reactions at both Cd concentrations (CdL and CdH) was the volume density of basophilic cells (V_{VBAS}) in the periwinkle's midgut

gland, with significantly increased values after only 12 h of exposure. This proves that Cd-stressed periwinkles respond to the intake of the metal by a very early re-modelling of the cellular composition in the tubular tissue of the midgut gland.

4. The increase of the MT mRNA concentration follows next, prior to or together with the activation of the lysosomal compartment of tubular digestive cells, suggesting that cytosolic inactivation of Cd by binding to MT and lysosomal Cd accumulation may be complementary processes of detoxification which interfere with each other. At the beginning of Cd exposure, most of the metal may first be sequestered by constitutive and de novo synthesized MT. At this stage of Cd accumulation, the MT-associated Cd pool may prevail. Due to ongoing MT turnover and degradation (Bebianno and Langston 1998), however, a certain part of liberated Cd ions would be bound to de novo synthesized MT, whereas a minor proportion of the metal (perhaps still attached to half-degraded and denatured MT) would end up for final storage in digestive cell lysosomes. Overall, the near concomitance of increasing MT mRNA transcription with enormous alterations at the cellular and tissue-specific levels reinforces the impression that the increasing MT mRNA concentrations in the midgut gland of periwinkles apparently serve the purpose of protecting *Littorina littorea* from potentially adverse effects induced by Cd^{2+} ions entering the cells or leaking out from impaired cellular structures.

5. An important response strategy regarding Cd stress in *Littorina littorea* is, apart from Cd sequestration by MT and lysosomes, the re-modelling of midgut gland tubules by cell replacement, where digestive cells die at a faster rate than they proliferate, implicating an increase of the number and volume density of basophilic cells, which prevail in tubules of Cd-stressed individuals. In addition to this, there is also a Cd-induced increase of the number of connective tissue calcium cells.

6. The specific Cd-induced molecular and cellular alterations in metal-stressed *Littorina littorea* may themselves be applied as biomarkers in environmental monitoring. Taken together, the time patterns of the different molecular, cellular and tissue-specific response reactions to Cd exposure in periwinkles may well provide a valuable biomarker battery approach.

Acknowledgments: This work was funded by the Austrian Science foundation (FWF) Project ref. I 1482-N28 (DACH) granted to Reinhard Dallinger.

Author Contributions: Reinhard Dallinger and Manu Soto designed and coordinated the study. Denis Benito, Michael Niederwanger and Urtzi Izagirre performed the experiments and the analyses. Denis Benito and Michael Niederwanger contributed equally to this study. Manu Soto and Reinhard Dallinger wrote the manuscript with the help of the other authors. All authors edited and approved the final version of the manuscript.

Conflicts of Interest: The authors declare no conflict of interest.

References

1. Hamada, T.; Tanimoto, A.; Sasaguri, Y. Apoptosis induced by cadmium. *Apoptosis* **1997**, *2*, 359–367. [CrossRef] [PubMed]
2. Bertin, G.; Averbeck, D. Cadmium: Cellular effects, modifications of biomolecules, modulation of DNA repair and genotoxic consequences (a review). *Biochimie* **2006**, *88*, 1549–1559. [CrossRef] [PubMed]
3. Giaginis, C.; Gatzidou, E.; Theocharis, S. DNA repair systems as targets of cadmium toxicity. *Toxicol. Appl. Pharmacol.* **2006**, *213*, 282–290. [CrossRef] [PubMed]
4. Waisberg, M.; Joseph, P.; Hale, B.; Beyersmann, D. Molecular and cellular mechanisms of cadmium carcinogenesis. *Toxicology* **2003**, *192*, 95–117. [CrossRef]
5. Liu, J.; Qu, W.; Kadiiska, M.B. Role of oxidative stress in cadmium toxicity and carcinogenesis. *Toxicol. Appl. Pharmacol.* **2009**, *238*, 209–214. [CrossRef] [PubMed]
6. Koutsogiannaki, S.; Franzellitti, S.; Fabbri, E.; Kaloyianni, M. Oxidative stress parameters induced by exposure to either cadmium or 17 β-estradiol on Mytilus galloprovincialis hemocytes. The role of signaling molecules. *Aquat. Toxicol.* **2014**, *146*, 186–195. [CrossRef] [PubMed]

7. Henson, M.C.; Chedrese, P.J. Endocrine Disruption by Cadmium, a Common Environmental Toxicant with Paradoxical Effects on Reproduction. *Exp. Biol. Med.* **2004**, *229*, 383–392. [CrossRef]

8. Byrne, C.; Divekar, S.D.; Storchan, G.B.; Parodi, D.A.; Martin, M.B. Cadmium—A metallohormone? *Toxicol. Appl. Pharmacol.* **2009**, *238*, 266–271. [CrossRef] [PubMed]

9. Ronald, E. Boron hazards to fish, wildlife, and invertebrates: A synoptic review. *Contam. Hazard. Rev.* **1990**, *85*, 1–20.

10. Zhu, J.Y.; Huang, H.Q.; Bao, X.D.; Lin, Q.M.; Cai, Z. Acute toxicity profile of cadmium revealed by proteomics in brain tissue of *Paralichthys olivaceus*: Potential role of transferrin in cadmium toxicity. *Aquat. Toxicol.* **2006**, *78*, 127–135. [CrossRef] [PubMed]

11. Parveen, N.; Shadab, G.G. Cytogenetic evaluation of cadmium chloride on *Channa punctatus*. *J. Environ. Biol.* **2012**, *33*, 663–666. [PubMed]

12. Marigomez, J.A.; Ireland, M.P. Accumulation, distribution and loss of cadmium in the marine prosobranch *Littorina littorea* (L.). *Sci. Total Environ.* **1989**, *78*, 1–12. [CrossRef]

13. Edwards, J.W.; Edyvane, K.S.; Boxall, V.A.; Hamann, M.; Soole, K.L. Metal levels in seston and marine fish flesh near industrial and metropolitan centres in South Australia. *Mar. Pollut. Bull.* **2001**, *42*, 389–396. [CrossRef]

14. Ferreira, A.C.; Costa, A.C.S.; Korn, M.G.A. Preliminary evaluation of the cadmium concentration in seawater of the Salvador City, Brazil. *Microchem. J.* **2004**, *78*, 77–83. [CrossRef]

15. Komjarova, I.; Blust, R. Comparison of liquid-liquid extraction, solid-phase extraction and co-precipitation preconcentration methods for the determination of cadmium, copper, nickel, lead and zinc in seawater. *Anal. Chim. Acta* **2006**, *576*, 221–228. [CrossRef] [PubMed]

16. Bargagli, R.; Nelli, L.; Ancora, S.; Focardi, S. Elevated cadmium accumulation in marine organisms from terra nova bay (antarctica). *Polar Biol.* **1996**, *16*, 513–520. [CrossRef]

17. Dietz, R.; Riget, F.; Cleemann, M.; Aarkrog, A.; Johansen, P.; Hansen, J.C. Comparison of contaminants from different trophic levels and ecosystems. *Sci. Total Environ.* **2000**, *245*, 221–231. [CrossRef]

18. Dural, M.; Göksu, M.Z.L.; Özak, A.A. Investigation of heavy metal levels in economically important fish species captured from the Tuzla lagoon. *Food Chem.* **2007**, *102*, 415–421. [CrossRef]

19. Bille, L.; Binato, G.; Cappa, V.; Toson, M.; Dalla Pozza, M.; Arcangeli, G.; Ricci, A.; Angeletti, R.; Piro, R. Lead, mercury and cadmium levels in edible marine molluscs and echinoderms from the Veneto Region (north-western Adriatic Sea—Italy). *Food Control* **2014**, *50*, 362–370. [CrossRef]

20. Jitar, O.; Teodosiu, C.; Oros, A.; Plavan, G.; Nicoara, M. Bioaccumulation of heavy metals in marine organisms from the Romanian sector of the Black Sea. *New Biotechnol.* **2015**, *32*, 369–378. [CrossRef] [PubMed]

21. Bryan, G.W. Brown seaweed, fucus vesiculosus, and the gastropod, Littorina littoralis, as indicators of trace-metal availability in estuaries. *Sci. Total Environ.* **1983**, *28*, 91–104. [CrossRef]

22. Soto, M. Simultaneous Quantification of Bioavailable Heavy Metals in Molluscs by Means of Cellular and Tissue Analysis. Implications for Monitoring Metal Pollution in Water Quality Assessment. Ph.D. Thesis, University of the Basque Country, Basque, Spain, 1995.

23. Regoli, F.; Nigro, M.; Orlando, E. Lysosomal and antioxidant responses to metals in the Antarctic scallop Adamussium colbecki. *Aquat. Toxicol.* **1998**, *40*, 375–392. [CrossRef]

24. Soto, M.; Zaldibar, B.; Cancio, I.; Taylor, M.G.; Turner, M.; Morgan, A.J.; Marigómez, I. Subcellular distribution of cadmium and its cellular ligands in mussel digestive gland cells as revealed by combined autometallography and X-ray microprobe analysis. *Histochem. J.* **2002**, *34*, 273–280. [CrossRef] [PubMed]

25. Marigomez, J.A.; Cajaraville, M.P.; Angulo, E. Cellular cadmium distribution in the common winkle, *Littorina littorea* (L.) determined by X-ray microprobe analysis and histochemistry. *Histochemistry* **1990**, *94*, 191–199. [CrossRef] [PubMed]

26. Nott, J.A.; Bebianno, M.J.; Langston, W.J.; Ryan, K.P. Cadmium in the gastropod *Littorina littorea*. *J. Mar. Biol. Assoc. U. K.* **1993**, *73*, 655–665. [CrossRef]

27. Klekowski, R. The influence of low salinity and desiccation on the survival, osmoregulation and water balance of *Littorina littorea* (L.) (*Prosobranchia*). *Pol. Arch. Hydrobiol.* **1963**, *11*, 241–250.

28. Churchill, T.A.; Storey, K.B. Metabolic responses to freezing and anoxia by the periwinkle *Littorina littorea*. *J. Therm. Biol.* **1996**, *21*, 57–63. [CrossRef]

29. Russell, E.L.; Storey, K.B. Anoxia and freezing exposures stimulate covalent modification of enzymes of carbohydrate metabolism in *Littorina littorea*. *J. Comp. Physiol. B* **1995**, *165*, 132–142. [CrossRef]

30. Storey, K.B.; Lant, B.; Anozie, O.O.; Storey, J.M. Metabolic mechanisms for anoxia tolerance and freezing survival in the intertidal gastropod, *Littorina littorea*. *Comp. Biochem. Physiol. A Mol. Integr. Physiol.* **2013**, *165*, 448–459. [CrossRef] [PubMed]

31. English, T.E.; Storey, K.B. Freezing and anoxia stresses induce expression of metallothionein in the foot muscle and hepatopancreas of the marine gastropod *Littorina littorea*. *J. Exp. Biol.* **2003**, *206*, 2517–2524. [CrossRef] [PubMed]

32. Zaldibar, B.; Cancio, I.; Marigomez, I. Reversible alterations in epithelial cell turnover in digestive gland of winkles (*Littorina littorea*) exposed to cadmium and their implications for biomarker measurements. *Aquat. Toxicol.* **2007**, *81*, 183–196. [CrossRef] [PubMed]

33. Marigomez, I. Aportaciones Cito-Histologicas a la Evaluación Ecotoxicologica de Niveles Subletales de Cadmio en el Medio Marino: Estudios de Laboratorio en el Gasterópodo Prosobranquio *Littorina littorea* (L.). Ph.D. Thesis, University of the Basque Country, Basque, Spain, 1989.

34. Marigomez, I.; Cajaraville, M.P.; Angulo, E. Histopathology of the digestive gland-gonad complex of the marine prosobranch *Littorina littorea* exposed to cadmium. *Dis. Aquat. Organ.* **1990**, *9*, 229–238. [CrossRef]

35. Marigomez, I.; Soto, M.; Cajaraville, M.P.; Angulo, E.; Giamberini, L. Cellular and subcellular distribution of metals in molluscs. *Microsc. Res. Tech.* **2002**, *56*, 358–392. [CrossRef] [PubMed]

36. Bebianno, M. Cadmium and metallothionein turnover in different tissues of the gastropod *Littorina littorea*. *Talanta* **1998**, *46*, 301–313. [CrossRef]

37. Baumann, C.; Beil, A.; Jurt, S.; Niederwanger, M.; Palacios, O.; Capdevila, M.; Atrian, S.; Dallinger, R.; Zerbe, O. Structural Adaptation of a Protein to Increased Metal Stress: NMR Structure of a Marine Snail Metallothionein with an Additional Domain. *Angew. Chem. Int. Ed. Engl.* **2017**, *56*, 4617–4622. [CrossRef] [PubMed]

38. Palacios, Ò.; Jiménez-Marti, E.; Niederwanger, M.; Gil-Moreno, S.; Zerbe, O.; Atrian, S.; Dallinger, R.; Capdevila, M. Analysis of metal-binding features of the wild type and two domain-truncated mutant variants of *Littorina littorea* metallothionein reveals its Cd-specific character. *Int. J. Mol. Sci.* **2017**, *18*, 1452. [CrossRef] [PubMed]

39. Vega, M.M.; Marigomez, J.A.; Angulo, E. Quantitative alterations in the structure of the digestive cell of *Littorina littorea* on exposure to cadmium. *Mar. Biol.* **1989**, *103*, 547–553. [CrossRef]

40. Marigómez, I.; Izagirre, U.; Lekube, X. Lysosomal enlargement in digestive cells of mussels exposed to cadmium, benzo[a]pyrene and their combination. *Comp. Biochem. Physiol. Part C Toxicol. Pharmacol.* **2005**, *141*, 188–193. [CrossRef] [PubMed]

41. Sumner, A.T. The distribution of some hydrolytic enzymes in the cells of the digestive gland of certain lamellibranchs and gastropods. *J. Zool.* **1969**, *158*, 277–291. [CrossRef]

42. Robinson, W.E. Assessment of bivalve intracellular digestion based on direct measurements. *J. Moll. Stud.* **1983**, *49*, 1–8. [CrossRef]

43. Moore, M.N.; Pipe, R.K.; Farrar, S.V. Lysosomal and microsomal responses to environmental factors in *Littorina littorea* from Sullom Voe. *Mar. Pollut. Bull.* **1982**, *13*, 340–345. [CrossRef]

44. Cajaraville, M.P.; Marigómez, J.A.; Angulo, E. Short-term toxic effects of 1-naphthol on the digestive gland-gonad complex of the marine prosobranch *Littorina littorea* (L): A light microscopic study. *Arch. Environ. Contam. Toxicol.* **1990**, *19*, 17–24. [CrossRef] [PubMed]

45. Marigomez, I.; Soto, M.; Etxeberria, M.; Angulo, E. Effects of size, sex, reproduction, and trematode infestation on the quantitative structure of digestive tubules in stressed winkles. In *Zoologische Jahrbücher. Abteilung für Anatomie und Ontogenie der Tiere*; Fischer, G., Ed.; Fischer: Jena, Germany, 1993; Volume 123, pp. 319–336.

46. Marigomez, I.; Cajaraville, M.P.; Soto, M.; Lekube, X. Cell-type replacement, a successful strategy of molluscs to adapt to chronic exposure to pollutants. *Cuad. Investig. Biol.* **1998**, *20*, 411–414.

47. Gil, J.M.; Marigomez, J.A.; Angulo, E. Histophysiology of polysaccharide and lipid reserves in various tissues of *Littorina littorea* exposed to sublethal concentrations of cadmium. *Comp. Biochem. Physiol. Part C Comp.* **1989**, *94*, 641–648. [CrossRef]

48. Cajaraville, M.P.; Bebianno, M.J.; Blasco, J.; Porte, C.; Sarasquete, C.; Viarengo, A. The use of biomarkers to assess the impact of pollution in coastal environments of the Iberian Peninsula: A practical approach. *Sci. Total Environ.* **2000**, *247*, 295–311. [CrossRef]

49. Kwamla Atupra, D. Determination of the Toxicity of Cadmium in Three Populations of *Littorina littorea* Taken from a Pollution and Salinity Gradient in the Scheldt Estuary (Mollusca: Gastropoda). Master's Thesis, Vrije Universiteit Brussel, Brussel, Belgium, 2001.

50. De Wolf, H.; Backeljau, T.; Blust, R. Sensitivity to cadmium along a salinity gradient in populations of the periwinkle, *Littorina littorea*, using time-to-death analysis. *Aquat. Toxicol.* **2004**, *66*, 241–253. [CrossRef] [PubMed]

51. Danscher, G. Autometallography. A new technique for light and electron microscopic visualization of metals in biological tissues (gold, silver, metal sulphides and metal selenides). *Histochemistry* **1984**, *81*, 331–335. [CrossRef] [PubMed]

52. Soto, M.; Cajaraville, M.P.; Angulo, E.; Marigómez, I. Autometallographic localization of protein-bound copper and zinc in the common winkle, *Littorina littorea*: A light microscopical study. *Histochem. J.* **1996**, *28*, 689–701. [CrossRef] [PubMed]

53. Kägi, J.H.R. Evolution, structure and chemical activity of class I metallothionein: An overview. In *Metallothionein*; Suzuki, K.T., Imura, N., Kimura, M., Eds.; Birkhauser Verlag: Basel, Switzerland, 1993; Volume III, pp. 29–55.

54. Dallinger, R.; Wang, Y.; Berger, B.; Mackay, E.A.; Kägi, J.H.R. Spectroscopic characterization of metallothionein from the terrestrial snail, Helix pomatia. *Eur. J. Biochem.* **2001**, *268*, 4126–4133. [CrossRef] [PubMed]

55. Bebianno, M.J.; Langston, W.J.; Simkiss, K. Metallothionein induction in Littorina-littorea (Mollusca, Prosobranchia) on exposure to cadmium. *J. Mar. Biol. Assoc. U. K.* **1992**, *72*, 329–342. [CrossRef]

56. Carginale, V.; Scudiero, R.; Capasso, C.; Capasso, A.; Kille, P.; di Prisco, G.; Parisi, E. Cadmium-induced differential accumulation of metallothionein isoforms in the Antarctic icefish, which exhibits no basal metallothionein protein but high endogenous mRNA levels. *Biochem. J.* **1998**, *332*, 475–481. [CrossRef] [PubMed]

57. Tom, M.; Chen, N.; Segev, M.; Herut, B.; Rinkevich, B. Quantifying fish metallothionein transcript by real time PCR for its utilization as an environmental biomarker. *Mar. Pollut. Bull.* **2004**, *48*, 705–710. [CrossRef] [PubMed]

58. Dondero, F.; Piacentini, L.; Banni, M.; Rebelo, M.; Burlando, B.; Viarengo, A. Quantitative PCR analysis of two molluscan metallothionein genes unveils differential expression and regulation. *Gene* **2005**, *345*, 259–270. [CrossRef] [PubMed]

59. Leung, K.M.Y.; Taylor, A.C.; Furness, R.W. Temperature-dependent physiological responses of the dogwhelk Nucella lapillus to cadmium exposure. *J. Mar. Biol. Assoc. U. K.* **2000**, *80*, S0025315400002472. [CrossRef]

60. Babczynska, A.; Wilczek, G.; Wilczek, P.; Szulinska, E.; Witas, I. Metallothioneins and energy budget indices in cadmium and copper exposed spiders Agelena labyrinthica in relation to their developmental stage, gender and origin. *Comp. Biochem. Physiol. Part C Toxicol. Pharmacol.* **2011**, *154*, 161–171. [CrossRef] [PubMed]

61. Sokolova, I.M.; Frederich, M.; Bagwe, R.; Lannig, G.; Sukhotin, A.A. Energy homeostasis as an integrative tool for assessing limits of environmental stress tolerance in aquatic invertebrates. *Mar. Environ. Res.* **2012**, *79*, 1–15. [CrossRef] [PubMed]

62. Klaassen, C.D.; Liu, J.; Choudhuri, S. METALLOTHIONEIN: An Intracellular Protein to Protect Against Cadmium Toxicity. *Annu. Rev. Pharmacol. Toxicol.* **1999**, *39*, 267–294. [CrossRef] [PubMed]

63. Pipe, R.K.; Moore, M.N. An ultrastructural study on the effects of phenanthrene on lysosomal membranes and distribution of the lysosomal enzyme β-glucuronidase in digestive cells of the periwinkle *Littorina littorea*. *Aquat. Toxicol.* **1986**, *8*, 65–76. [CrossRef]

64. Cajaraville, M.P.; Marigomez, J.A.; Angulo, E. A stereological survey of lysosomal structure alterations in *Littorina littorea* exposed to 1-naphthol. *Comp. Biochem. Physiol. Part C Comp. Pharmacol.* **1989**, *93*, 231–237. [CrossRef]

65. Lowe, D.M.; Moore, M.N.; Readman, J.W. Pathological reactions and recovery of hepatopancreatic digestive cells from the marine snail *Littorina littorea* following exposure to a polycyclic aromatic hydrocarbon. *Mar. Environ. Res.* **2006**, *61*, 457–470. [CrossRef] [PubMed]

66. Calvo-Ugarteburu, G.; Saez, V.; McQuaid, C.D.; Angulo, E. Validation of a planimetric procedure to quantify stress in *Littorina littorea* (Gastropoda: Mollusca): Is it independent of the reproductive cycle? In *Advances in Littorinid Biology, Proceedings of the Fourth International Symposium on Littorinid Biology, Roscoff, France, 19–25 September 1993*; Mill, P.J., McQuaid, C.D., Eds.; Springer Netherlands: Dordrecht, The Netherlands, 1995; pp. 37–44.

67. Marigómez, I.; Soto, M.; Angulo, E. Seasonal variability in the quantitative structure of the digestive tubules of *Littorina littorea*. *Aquat. Living Resour.* **1992**, *5*, 299–305. [CrossRef]

68. Beyersmann, D.; Hechtenberg, S. Cadmium, gene regulation, and cellular signalling in mammalian cells. *Toxicol. Appl. Pharmacol.* **1997**, *144*, 247–261. [CrossRef] [PubMed]

69. Zaldibar, B.; Cancio, I.; Soto, M.; Marigomez, I. Digestive cell turnover in digestive gland epithelium of slugs experimentally exposed to a mixture of cadmium and kerosene. *Chemosphere* **2007**, *70*, 144–154. [CrossRef] [PubMed]

70. Chabicovsky, M.; Klepal, W.; Dallinger, R. Mechanisms of cadmium toxicity in terrestrial pulmonates: Programmed cell death and metallothionein overload. *Environ. Toxicol. Chem.* **2004**, *23*, 648–655. [CrossRef] [PubMed]

71. Triebskorn, R.; Köhler, H.R. The impact of heavy metals on the grey garden slug, Deroceras reticulatum (Muller): Metal storage, cellular effects and semi-quantitative evaluation of metal toxicity. *Environ. Pollut.* **1996**, *93*, 327–343. [CrossRef]

72. Hödl, E.; Felder, E.; Chabicovsky, M.; Dallinger, R. Cadmium stress stimulates tissue turnover in Helix pomatia: Increasing cell proliferation from metal tolerance to exhaustion in molluscan midgut gland. *Cell. Tissue Res.* **2010**, *341*, 159–171. [CrossRef] [PubMed]

73. Marigómez, J.A.; Sáez, V.; Cajaraville, M.P.; Angulo, E. A planimetric study of the mean epithelial thickness (MET) of the molluscan digestive gland over the tidal cycle and under environmental stress conditions. *Helgoländer Meeresuntersuchungen* **1990**, *44*, 81–94. [CrossRef]

74. Cajaraville, M.P.; Robledo, Y.; Etxeberria, M.; Marigomez, I. Cellular biomarkers as useful tools in the biological monitoring of pollution: Molluscan digestive lysosomes. In *Cell Biology in Environmental Toxicology*; University of the Basque Country Press Service: Bilbao, Spain, 1995; pp. 29–55.

75. Lowe, D.M.; Clarke, K.R. Contaminant-induced changes in the structure of the digestive epithelium of Mytilus edulis. *Aquat. Toxicol.* **1989**, *15*, 345–358. [CrossRef]

76. Marigómez, I.; Soto, M.; Kortabitarte, M. Tissue-level biomarkers and biological effect of mercury on sentinel slugs Arion ater. *Arch. Environ. Contam. Toxicol.* **1996**, *31*, 54–62. [CrossRef] [PubMed]

77. Mason, A.Z.; Nott, J.A. The role of intracellular biomineralized granules in the regulation and detoxification of metals in gastropods with special reference to the marine prosobranch *Littorina littorea*. *Aquat. Toxicol.* **1981**, *1*, 239–256. [CrossRef]

78. Chmielowska-Bak, J.; Izbiańska, K.; Deckert, J. The toxic Doppelganger: On the ionic and molecular mimicry of cadmium. *Acta Biochim. Pol.* **2013**, *60*, 369–374. [PubMed]

79. Brooks, S.; Harman, C.; Soto, M.; Cancio, I.; Glette, T.; Marigomez, I. Integrated coastal monitoring of a gas processing plant using native and caged mussels. *Sci. Total Environ.* **2012**, *426*, 375–386. [CrossRef] [PubMed]

80. Rementeria, A.; Mikolaczyk, M.; Lanceleur, L.; Blanc, G.; Soto, M.; Schäfer, J.; Zaldibar, B. Assessment of the effects of Cu and Ag in oysters Crassostrea gigas (Thunberg, 1793) using a battery of cell and tissue level biomarkers. *Mar. Environ. Res.* **2016**, *122*, 11–22. [CrossRef] [PubMed]

81. McCarthy, J.F.; Shugart, L.R. Biological markers of environmental contamination. In *Biomarkers of Environmental Contamination*; McCarthy, J.F., Shugart, L.R., Eds.; Lewis Publishers: Chelsea, MI, USA, 1990; Volume 3, pp. 3–14.

82. Izagirre, U.; Ruiz, P.; Marigomez, I. Time-course study of the early lysosomal responses to pollutants in mussel digestive cells using acid phosphatase as lysosomal marker enzyme. *Comp. Biochem. Physiol. Part C Toxicol. Pharmacol.* **2009**, *149*, 587–597. [CrossRef] [PubMed]

83. Weibel, E.R. *Stereological Methods. Vol 1: Practical Methods for Biological Morphometr*; Academic Press: New York, NY, USA, 1979; p. 415.

84. Moore, M.N. Cytochemical demonstration of latency of lysosomal hydrolases in digestive cells of the common mussel, mytilus edulis, and changes induced by thermal stress. *Cell. Tissue Res.* **1976**, *175*, 279–287. [CrossRef] [PubMed]

85. Lowe, D.M.; Moore, M.N.; Clarke, K.R. Effects of oil on digestive cells in mussels: Quantitative alterations in cellular and lysosomal structure. *Aquat. Toxicol.* **1981**, *1*, 213–226. [CrossRef]
86. Etxeberria, M.; Sastre, I.; Cajaraville, M.P.; Marigomez, I. Digestive lysosome enlargement induced by experimental exposure to metals (Cu, Cd, and Zn) in mussels collected from a zinc-polluted site. *Arch. Environ. Contam. Toxicol.* **1994**, *27*, 338–345. [CrossRef]
87. Pearse, A.G.E. Pigments and pigment precursors. In *Histochemistry. Theoretical and Applied. Vol. 2: Analytical Technology AGE Pearse*, 4th ed.; Churchill Livingstone: Edinburgh, UK, 1985; pp. 874–928.
88. Regoli, F. Lysosomal responses as a sensitive stress index in biomonitoring heavy metal pollution. *Mar. Ecol. Prog. Ser.* **1992**, *84*, 63–69. [CrossRef]

International Journal of
Molecular Sciences

MDPI

Article

Challenging the *Metallothionein* (*MT*) Gene of *Biomphalaria glabrata*: Unexpected Response Patterns Due to Cadmium Exposure and Temperature Stress

Michael Niederwanger [ORCID], Martin Dvorak, Raimund Schnegg, Veronika Pedrini-Martha, Katharina Bacher, Massimo Bidoli and Reinhard Dallinger *

Institute of Zoology and Center of Molecular Biosciences Innsbruck (CMBI), University of Innsbruck, Technikerstrasse 25, A-6020 Innsbruck, Austria; michael.niederwanger@uibk.ac.at (M.N.); martin.dvorak@uibk.ac.at (M.D.); raimund.schnegg@uibk.ac.at (R.S.); Veronika.Pedrini-Martha@uibk.ac.at (V.P.-M.); csap9255@student.uibk.ac.at (K.B.); massimo.bidoli@student.uibk.ac.at (M.B.)
* Correspondence: reinhard.dallinger@uibk.ac.at; Tel.: +43-512-507-51861; Fax: +43-512-507-51899

Received: 14 July 2017; Accepted: 7 August 2017; Published: 11 August 2017

Abstract: Metallothioneins (MTs) are low-molecular-mass, cysteine-rich, metal binding proteins. In most animal species, they are involved in metal homeostasis and detoxification, and provide protection from oxidative stress. Gastropod MTs are highly diversified, exhibiting unique features and adaptations like metal specificity and multiplications of their metal binding domains. Here, we show that the MT gene of *Biomphalaria glabrata*, one of the largest *MT* genes identified so far, is composed in a unique way. The encoding for an MT protein has a three-domain structure and a C-terminal, Cys-rich extension. Using a bioinformatic approach involving structural and in silico analysis of putative transcription factor binding sites (TFBs), we found that this *MT* gene consists of five exons and four introns. It exhibits a regulatory promoter region containing three metal-responsive elements (MREs) and several TFBs with putative involvement in environmental stress response, and regulation of gene expression. Quantitative real-time polymerase chain reaction (qRT-PCR) data indicate that the *MT* gene is not inducible by cadmium (Cd) nor by temperature challenges (heat and cold), despite significant Cd uptake within the midgut gland and the high Cd tolerance of metal-exposed snails.

Keywords: metallothionein; metal binding domain; cadmium tolerance; heat shock; cold shock; *Biomphalaria glabrata*; *Gastropoda*; *Hygrophila*

1. Introduction

The adaptation process of animal species towards stressful conditions is achieved, among other methods, by relying on an array of "stress"-related gene encoding for proteins enabling protection against adverse environmental conditions, repair of damaged macromolecules, and detoxification of unwanted metal ions. Metallothioneins (MTs), in particular, belong to a superfamily of nonenzymatic metal binding proteins, occurring in bacteria and most eukaryotic kingdoms [1,2]. Typically, MTs are hydrophilic, low-molecular-mass proteins binding with high affinity to a number of transition metal ions such as Cd^{2+}, Zn^{2+} and Cu^+ [3–5]. One of their most distinctive properties is the high cysteine content (up to ~30%) with individual cysteine residues arranged in distinctive motifs (e.g., Cys-x-Cys, Cys-x-x-Cys) that form metal-binding clusters. Owing to their features, MTs are often involved in the homeostatic regulation and detoxification of metallic trace elements [6,7]. Moreover, MTs may also participate in the cellular protection from oxidative stress [8,9].

Gastropods thrive in marine, terrestrial and freshwater habitats and possess MTs with striking capabilities for protecting their hosts from metal stress [10–14] and adverse environmental conditions [15]. In fact, the primary structure of MTs from species of different gastropod clades is highly variable, suggesting a huge potential for evolutionary adaptation to different habitats and their varying environmental challenges [12,16].

Surprisingly, there is little knowledge available about the role and biological function of MTs in freshwater pulmonate snails [17,18], many of them belonging to the clade of *Hygrophila*. An important member of *Hygrophila* is *Biomphalaria glabrata*. This freshwater snail has attained a particular significance as an intermediary host of the trematode parasite *Schistosoma mansonii*, which leads to infections with schistosomiasis of millions of people worldwide every year [19]. Like many other species of *Hygropohila*, *Biomphalaria glabrata* can be considered as a (phylogenetically spoken) modern freshwater snail with a "lung", consisting of the mantle cavity. Species of this group carry an air bubble within this cavity from which oxygen enters the blood vessels directly by diffusion. This feature enables aquatic lung snails to remain under water over extended periods of time. The adaptive transition from marine to freshwater conditions may have evolved through a number of preadaptations in marine ancestors [20], confronting freshwater neocolonizers with the necessity to readjust their mechanisms for energy acquisition and osmotic regulation [21–24]. This may also have implications on their mechanisms for metal uptake and regulation [25,26].

Here we report the identification and annotation of an *MT* gene locus from the ram's horn snail, *Biomphalaria glabrata*, provide an analysis of its gene structure and promoter region in comparison with other gastropod *MT* genes, identify allelic variation and assess mRNA transcription in response to Cd and nonmetallic stressors.

2. Results and Discussion

2.1. Gene Map and Coding Sequence

The *MT* gene of *Biomphalaria glabrata* is more than 11,000 bp long. It consists of five exons and four introns with differing lengths (Figure 1A), the promoter region and the 3′ untranslated region. The translated coding region represented by the five exons (Figure 1B) spans 369 bp and was confirmed by sequencing of polymerase chain reaction (PCR)-amplified and cloned individual samples. Analysis of the gene by the software package of the transcription factor database (TRANSFAC) yielded a large number of putative transcription factor binding sites (TFBs), related to stress response and regulatory functions, interspersed throughout the whole gene sequence with theoretical binding sites for the TATA-binding protein (TBP), Cdx homeodomeain protein 1 (Cdx-1), MADS box transcription enhancer factor 2 (MEF2C), yeast activator protein 1 (YAP-1) and heat shock factors (HSFs) (for functional explanation, see Table 1) being the most abundant.

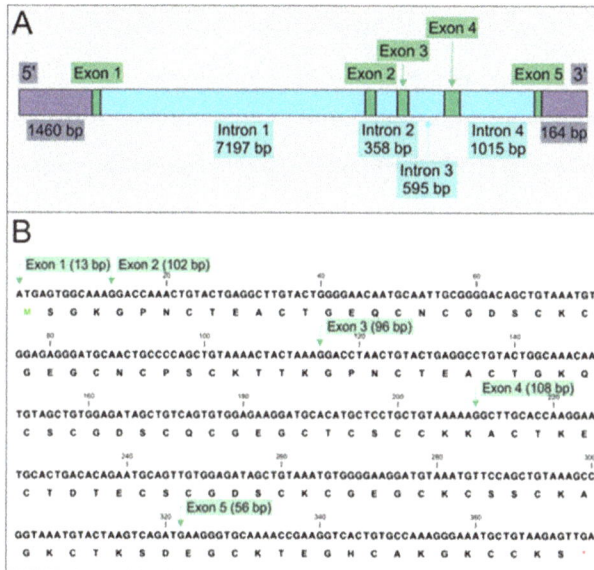

Figure 1. Structural organization of the *Metallothionein* (*MT*) gene from *Biomphalaria glabrata*. (**A**) Map showing the exon/intron structure of the gene. Positions and length of exons (green), introns (blue), the promoter at the 5′ end-and the 3′-untranslated region (UTR) (lilac) are specified by boxes above and underneath the gene map; (**B**) linked exons 1–5 (shaded in green) with starting points (green arrows) and length (in parentheses), coding for amino acid sequence of the MT protein (shown in one-letter amino acid code below base triplets).

2.2. Promoter Structure and Metal-Responsive Elements

Referring to the promoter analysis of the *CdMT* gene of *Helix pomatia* [16] and due to the abundance of putative Transcription Factor Binding Sites (TFBs), the promoter region of the *MT* gene of *Biomphalaria glabrata* was deliberately restricted to a length of 1460 bp (Figure 2). An analysis of this promoter region by the software package TRANSFAC indicated the presence of a large number of putative TFBs involved in stress response or transcriptional regulation (mostly with activating and enhancing functions) (Table 1). Expectedly, the promoter also contains three metal-responsive elements (MREs), considered to be the most characteristic TFBs of an *MT* gene. They are located in the proximal promoter region upstream from the start codon at positions −173 and −127 bp. Two of them consist of a bidirectional, palindromic sequence at −127 bp. No TATA box motif was detected by TRANSFAC. However, two sequence stretches were identified by us as potential TATA box-like motifs, located at positions −24 bp and −80 bp upstream of the start codon (Figure 2). Compared to the already characterized *MT* gene of *Helix pomatia* [16], the *MT* promoter of *Biomphalaria glabrata* differs from the former by the kind, number and allocation of putative TFBs. Moreover, the respective encoded MT protein shows a different number and organization of metal-binding domains, with different positions of cysteine residues within the primary structure compared to the *CdMT* of *Helix pomatia*, which suggesting specific adaptation strategies in the two species towards metal challenge and environmental stressors. In particular, the *Biomphalaria glabrata MT* gene promoter contains a proximal cluster of three MREs and in close vicinity to them also hosts putative binding sites for transcriptional activators and enhancers (Cdx-1, MEF-2C, prolamin box binding factor (PBF), and TBP) (Figure 3, Table 1).

```
         20              40              60              80
GTATTTTTAGTTCAACACATGGCACAGTCTCATTTATAAATACGATTATGTATATATGTTTCAAATATCAATAAGTATTATTTTATT
MEF-2C              MEF-2C                                              Cdx-1
     100            120             140             160
TAAATTAATAACTAATTATTCATTTATGTCTTTTTTTTCTCATAAAAGTGCGTCTACGGGAGGGATACGATTGGGTGATCACACCTA
     POU1F1
         180            200             220             240           260
CCGCCCTTCCAGGATTAGTTTGAATGGCCCTGCCAAAACCAATCAAAATATCTGAAAGAAACAAACTAACACGTAGGGCTCTTCAAT
                              HSFA2
         280            300                     320             340
TTAGGGAAATAACAAAAGAATGGAAAGAAGCATTTCATCTCACTGTTTAAAGTTGGTAGCATTAAAAATTTTTAAATACATTTTGAT
                      YAP-5
     360            380             400             420
AATAATATCCTTTAGCCACATTTTTCCCCAAACTATCTCCTATGACAAGGTCAACTTGGTTATCTTTTCATCACTTTTATCAGCTGA
     PPARgamma                                    GATA-1        GATA-1, TBP
     440            460             480             500           520
ATCTCGTGAGTGATGAGTCACATAGAAGTCCAAGTGGAGGGAAAAAAGCCACAATTAGAGAAGATCATTGAACTAAAAAGTATGTGA
            AP-1, AP-1, AP-1
         540            560             580             600
GACTGAGACCTCCAAGAAGCCAGAATAACGTCACAGAACCGTCAGAAACACGTCAAATTAAAAGAATAAATAATTTTGTTGTAGAAT
                      ATF        HSF        ATF                 Cdx-1, MEF-2C
     620            640             660             680
TTTGAGCTATATTTCCCTAACCCAAAACTAAAGGTCATAGTAACCGTAGTTTGACTGAAATCACTCGTAGATCTACGTTACGGTAGTT
                  HSFA2                                              p300
     700            720             740             760           780
CACTTATTGTGACTATAGGCTGAGGAGTCTGTGTGTGGGTAATGTAATTCTTGATGGACCATGTGGTGTTTTTTCTTCATATCTGT
                                              cMyc
         800            820             840             860
TGTGTTGTATTTGCATTTTTAACTAATGCGTCAAGTAATCTCACAATTTCAAAATAAAATAAAACGGCCTTTTTTCTCTTTTGTTTA
     POU2F1                          C/EBPgamma   Cdx-1, Cdx-1   PBF, PBF  PBF
         880            900             920             940
CGATGTAGTAAATCTATTTCTAAATTAGTAATCTTGGTCAACAGAAATGAAACAATTCTAAACACCTGATCATTTATAAGAAAAGAA
                 AP-3 YAP-1, YAP-1                                   HSF, HSF
     960            980            1,000           1,020         1,040
AATTCAACTCGTTTTGACACGCTCTACAATAAAATTCGATGCACTTTTAACTTATCGATTTTTTGAGTGATGCCCAGTCTATTTATA
     Cdx-1                                                           MEF-2C
     1,060          1,080           1,100           1,120
GTAACAAGGACAGTAAACGAGCCATTGCAAATGCTAATGACAAGTGACAAGGGTTAATCTAGTATTATACTGTTGTAGCTAGTGACA
     1,140          1,160           1,180           1,200
GAACTGTTTGTCTTACAGTTGTTTAAAAAAAGGTTTATTTTTCTAAAGAACAATAAAATACGTAGTTCTAAACTTCAAATCTATTTA
HSF                    Cdx-1, PBF, MEF-2C    HSF   Cdx-1         HSF
     1,220          1,240           1,260           1,280         1,300
GCCACATGGGTTTTCTTTTATGTCCATGAAAGTACAACGGACATCTGGATGGCCAGTGTTGAGACGACAATCCGTGTGCAAATTTAT
Cdx-1, TBP                                                          MRE
     1,320          1,340           1,360           1,380
CGCATGGCTGTTAGAATGATCTCGATTTTGGCCGTGTGCATATAACGGCCTTGGTTGGCAAGTAGGCTATATAAACCAGAGACTGTC
        HSFA2         MRE, MRE                              TATA
     1,400          1,420           1,440           1,460
TACTTTCTCTATTGAATTCCTCGTTGATCAAGAATATTTTATTTTAACAGTTCATAACAACAAACACCATG
     TATA Cdx-1                                           Start codon
```

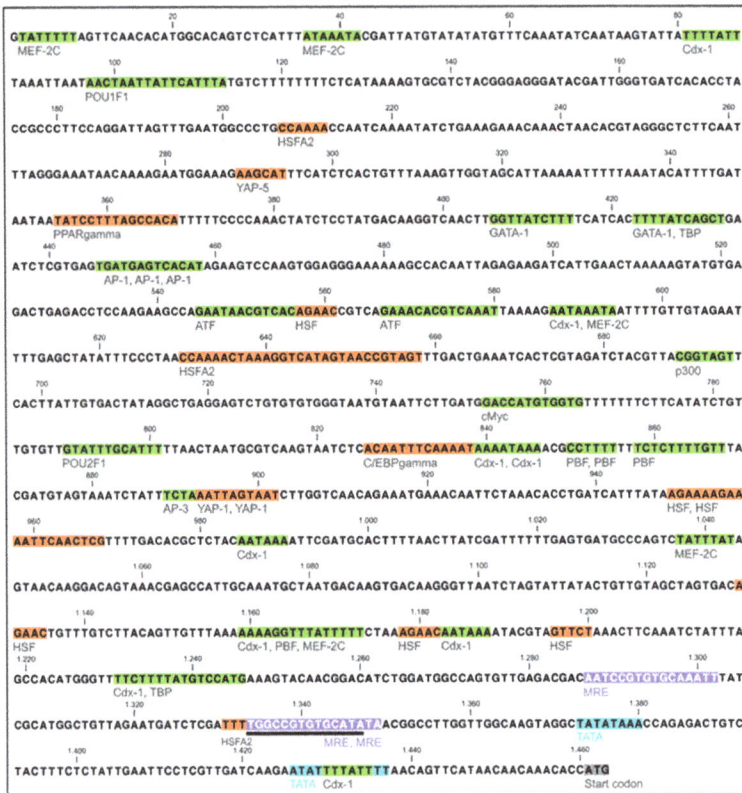

Figure 2. Nucleotide sequence of the promoter region (position 1 to 1460 bp upstream from the start codon), of the *MT* gene of *Biomphalaria glabrata*. Metal responsive elements (MREs) acting as potential binding sites for the presumed metal transcription factor I are highlighted in lilac. Potential transcription factor binding sites (TFBs) involved in stress response are marked in orange and hypothetical binding sites for regulatory transcription factors are marked in green. The putative TATA boxes are highlighted in blue and the start codon of the first exon is shaded in grey. Functional specification of TFBs identified using TRANSFAC: metal induction (lilac): MREs: aatccgtGTGCAaatt, tggccgtGTGCAtata, tggccgtGTGCAta (black line indicates palindromic MRE sequence of the MRE in opposite direction). Stress response (orange): HSF: GTTCT, AGAAC, agaaaattcaACTCG, AGAAAagaaaattca, TGAAGagaatatgcg; C/EBPγ: acaATTTCaaaat; PPARγ: tatCCTTTagccaca; YAP: aTTAGTaa, aAGCAT, agcaggATACGtaatgtatt. Transcriptional regulation (green): AP-1: atgAGTCAc, tgaTGAGTcacat; ATF: gaaacaCGTCAa, gaataaCGTCAc; Cdx-1: TTTATt, ttctTTTATgtccatg, TTTATt, aATAAA; cMyc: gaccATGTGGtg; GATA-1: ttTTTATCagc, ttTTTATCtct, gccctTATCAtttt; MEF-2C: atAAAATA, TATTTttt; POU2F1: aTTTGCattt; POU1F1a: aattATTCAt; p300: cgGTAGT, ACGTTcg; PBF: tctCTTTTgtt, CCTTTt; TBP: tTTTATgtcca, tTTTATcagct. For abbreviations and functional explanation of the single TFBs, see Table 1.

In contrast, the *CdMT* gene promoter of *Helix pomatia* contains four MREs, one of them at a distal position [16]. As in the case of *Helix pomatia*, the *MT* gene promoter of *Biomphalaria glabrata* also exhibits several theoretical TFBs for transcription factors involved in stress response and stress-related signaling (C/enhancer binding protein (C/EBPβ), heat shock element (HSE), peroxisome proliferator activated receptor (PPAR), and yeast activator protein (YAP)). A prominent feature of the *Biomphalaria glabrata MT* gene promoter is, moreover, its preponderant abundance of putative HSEs acting as binding sites for HSFs, suggesting that the *MT* gene of this species may perhaps be upregulated not only by metal stress, but also by nonmetallic stressors like heat shock. In this regard, it has been shown that some *MT* genes can be directly induced by stress-related transcription factors, including the heat shock transcription factor [27]. In particular, some *MT* genes may be activated by nonmetallic inducers, often through synergistic interaction of MREs with other stress-related TFBs such as antioxidant-responsive elements upon oxidative stress [28]. Overall, transcriptional upregulation by nonmetallic stressors like hypoxia, oxidative stress, physical stress and starvation, among others, is a common feature of many *MT* genes [16,29–32]. Interestingly, the expression of the heat shock protein 70 (HSP70) in *Biomphalaria glabrata* can be influenced by Cd exposure [33]. Our present study shows, however (see below), that the *MT* gene itself does not or only hesitantly react towards temperature-associated stress.

Table 1. Theoretical binding sites for transcription factors involved in stress-response and transcriptional regulation from the promoter region of *Biomphalaria glabrata*. Abbreviated terms indicate the binding sites and suggested functional context according to explanations in TRANSFAC databases (see materials and methods).

Response Elements and Binding Sites for Transcription Factors Involved in Stress Response		
Response element or binding site (abbreviation)	Explanation	Functional context
C/EBP	C/Enhancer binding protein	Glucocorticoid activation
HSE	Heat Shock element	Heat shock protein activation
MRE	Metal responsive element	Metal-induced expression
PPAR	Peroxisome proliferator activated receptor	Stress response
YAP	Yeast activator protein	Multidrug resistance in yeast [34]
Response Elements and Binding Sites for Transcription Factors Involved in Transcriptional Regulation		
Response element or binding site (abbreviation)	Explanation	Functional context
AP-1	Activator Protein 1	Nuclear transcriptional activator
ATF	Activating transcription factor	Transcriptional activator
cMyc	c-myc protein	Telomerase activator
GATA-1	GATA binding protein 1	Cell growth transcription factor
POU1F1a, POU2F1	Octamer-binding factors	Transcriptional activators
p300	E1A-associated 300-kDa protein	Transcriptional enhancer
PBF	Prolamin box binding factor	Transcriptional activator in plants
TBP	TATA-binding protein	Polymerase activator
Cdx-1	Cdx homeodomain protein	Activator
MEF-2C	MADS box transcription enhancer factor 2	Transcriptional enhancer

2.3. Domain Structure of the MT Protein

The translated primary structure suggests that the MT protein of *Biomphalaria glabrata* is a three-domain MT with three α domains and an additional, cysteine-containing C-terminal extension (Figures 1B and 3) [35]. The first N-terminal (α1) and the third C-terminal α-domain (α3) contain nine cysteine residues each, whereas the second N-terminal domain (α2) holds 10 cysteines, including a Cys–Cys double motif. The C-terminal extension holds five Cys residues, including again a double cysteine motif (Figure 3). Compared with the novel three-domain structure of an MT recently published for *Littorina littorea* [36] and primary structure data of the CdMT of *Helix pomatia* [12,16], it appears that the protein of *Biomphalaria glabrata* may be the first described three-domain MT of a gastropod species consisting, apart from the C-terminal extension, of three highly homologous α domains. In mammalian MT-2, two types of metal-binding domains can be distinguished (designated as α and β) that differ in their number of contained cysteine residues and consequently, in their metal binding stoichiometry [37]. Accordingly, the MT-2 α-domain contains eleven to twelve cysteine residues binding four divalent metal ions, and the β-domain includes nine cysteine residues binding three divalent metal ions.

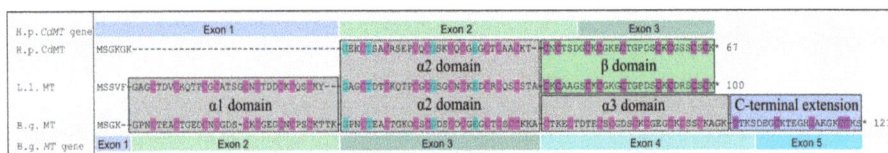

Figure 3. Alignment of the CdMT of *Helix pomatia* (H.p.) (GenBank acc. Nr.: AF399740.1), the MT of *Littorina littorea* (L.l.) (GenBank acc. Nr.: AY034179.1) and the MT of *Biomphalaria glabrata* (B.G.) (GenBank acc. Nr.: KT697617) displaying the amino acid sequence and suggested domain structure. Exon structures of the genes are shown for *Helix pomatia* (H.p. *CdMT* gene) (GenBank acc. Nr.: FJ755002.1) and for *Biomphalaria glabrata* (B.g. *MT* gene) (GenBank acc. Nr.: XM_013225031). Conserved cysteines are marked in pink. Conserved amino acid positions shared by all MTs are marked in blue. The α-domains are indicated by a grey box, β-domains by a green box and the C-terminal extension is marked by a blue box. Exon structures of the genes are indicated by colored boxes above (for H.p.) and underneath (for B.g.) the respective amino acid sequence. Additionally, the length is indicated at the end of each sequence.

While the mammalian (N-terminal) and the gastropod (C-terminal) MT β domains share a high degree of similarity [36], less similarity is found between the α domains of *Biomphalaria glabrata* MT and mammalian MT-2 (yielding, for very short sequence stretches, identities of 38–44% and expect values between 2.6 and 8.4). The *MT* gene of *Biomphalaria glabrata* achieved its extended length, apart from a cysteine containing C-terminal extension, presumably by modular triplication of an ancestral α-domain. As observed in a recent study about MT evolution in bivalves, various MT isoforms of the oyster, *Crassostrea virginica*, may have achieved their multi-domain structure by a series of exon and gene duplication events [38]. Upon comparison of protein and gene structures of the *MTs* from *Biomphalaria glabrata* and *Helix pomatia* (Figure 3), it appears that, at least partially, this hypothesis may also apply to the MT of *Biomphalaria glabrata*. The first two domains of this MT align nicely with the exon/intron structure of the gene (Figure 3) and suggest, exon-specific domain duplication. In contrast, the third α-domain does not reflect this hypothesis and may better be explained as a result of protein-specific domain duplication. No indications for exon-specific domain duplications were found in the *CdMT* gene structure of *Helix pomatia* (Figure 3). Hence, the domain structures of both MTs cannot be explained by exon duplication alone.

2.4. Allelic Variations

The coding region of the *Biomphalaria glabrata* MT shows a considerable degree of allelic variation caused by single nucleotide polymorphisms (SNPs) or mutations of single nucleotides. A screening of MT cDNA sequences from 20 individuals revealed the existence of four allelic variants that differed from the wild-type MT at three different nucleotide positions (133, 210 and 300) of the coding region (Figure 4). MT variant 1 shows a nucleotide mutation at position 210, changing the respective amino acid from lysine to asparagine. MT variant 2 shows SNPs at nucleotide positions 210 and 300, changing the respective amino acid position to either lysine or asparagine in the former, and the latter being silent. MT variant 3 differs from the wild-type MT due to a silent nucleotide mutation at position 300, and MT variant 4 shows two SNPs at nucleotide positions 210 and 300, identical to those in the MT variant 2, and beyond that a silent SNP at nucleotide position 133 (Figure 4). Since in many gastropod MTs, there is a clear relationship between primary structure and metal binding specificity, we assume that the variants of *Biomphalaria glabrata* MT may slightly differ with respect to their metal binding behavior [35].

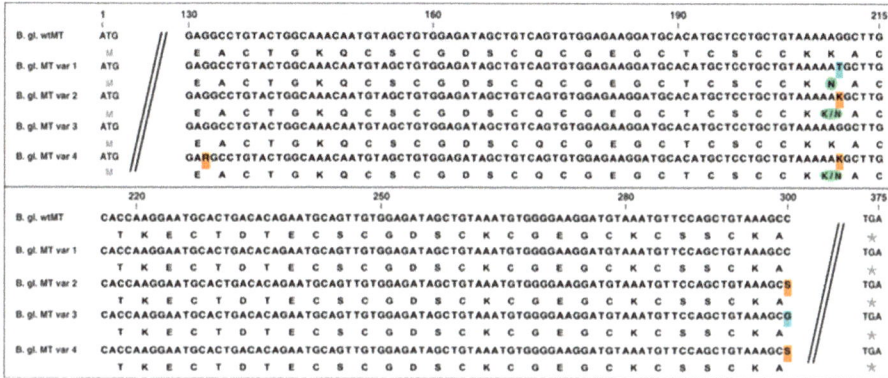

Figure 4. Alignment of the *Biomphalaria glabrata* wild-type MT (B. gl. wtMT) and its four allelic variants (B. gl. MT var 1–4). Parts of the sequence showing 100% coverage amongst all MTs were cut (two black lines). Nucleotide positions are indexed above the top sequence. Amino acids are displayed in single letter code underneath the respective base triplets and the start and stop codons are marked in grey. Nucleotide mutations are shaded in blue, single nucleotide polymorphisms (SNPs) are shaded in orange and the respective changes of amino acids are indicated by a green circle or ellipses. B. gl. MT var 1 shows a change at position 210 (G is replaced by T—changing the codon from lysine to asparagine). B. gl. MT var 2 shows SNPs at positions 210 (K: represents G or T-changing the codon to either lysine or asparagine) and 300 (S: represents G or C). B. gl. MT var 3 shows a shift at position 300 (C is replaced by G). B. gl. MT var 4 shows SNPs at positions 133 (R: represents A or G), 210 (K: represents G or T—changing the codon to either lysine or asparagine) and 300 (S: represents G or C).

2.5. Lacking Upregulation of the MT Gene Due to Cd Exposure

In contrast to several metal-specific MTs from other gastropod species [14,32,39–41], the *MT* gene in the midgut gland of *Biomphalaria glabrata* did not show significant transcriptional upregulation upon exposure to sublethal concentrations of Cd^{2+} (75 µg/L) (Figure 5A), in spite of the strong metal accumulation in this organ during exposure (Figure 5B), and no observed mortality in metal-exposed snails. Instead, it appeared that the constitutive expression of this gene was already highly elevated under control conditions, with values ranging from ~750,000 to ~2,000,000 copies/10 ng total RNA in unexposed individuals. Even the highly fluctuating transcription levels of MT mRNA during the first 8 days of exposure (with single values significantly above and below control levels on days 4 and 8, respectively) do not vary much the impression that the long-running trend of MT mRNA induction through the whole exposure period remains unaffected by Cd exposure (Figure 5A). In fact, the analysis of variance (ANOVA) test for the time-dependent mRNA expression curve of the Cd exposure group failed to indicate any significance and overall, the fluctuating time pattern of upregulation does not seem to reflect a typical induction pattern as observed for *MT* genes of other gastropod species [14,39,42].

As shown in Figure 5B, Cd accumulation in the midgut gland of *Biomphalaria glabrata* seems to reach some kind of saturation after Day 8 of the experiment, although the exposure was still persisting. This suggests that during this time, metal accumulation may go hand in hand with some degree of elimination, keeping the accumulation curve at a constant equilibrium. Unfortunately, we do not know the mechanisms of elimination in this case. However, it could be suggested that they may in part consist of cellular mechanisms of excretion, as observed in other gastropod species [43–45]. Overall, it can be concluded that despite Cd being accumulated strongly in the midgut gland tissue of metal-exposed snails (Figure 5B) with a concentration factor about 400 times above control levels, no concomitant upregulation of the MT mRNA could be observed. This might be because the constitutive expression of this gene was already high enough to inactivate Cd^{2+} ions entering the midgut gland cells. Since the MT of *Biomphalaria glabrata* is not Cd-specific at all [35], metal binding to the protein may in this case be achieved through exchange reactions at the protein's binding sites, for example with Cd^{2+} ions replacing Zn^{2+}. This also suggests that MT may not necessarily be the main actor for Cd detoxification in *Biomphalaria glabrata*. Alternatively or concomitantly, phytochelatines may be involved in Cd detoxification, as shown for Cd-exposed *Lymnaea stagnalis*, a close relative of *Biomphalaria glabrata* [46]. Moreover, diurnal variability of physiological activity and age-dependent *MT* gene expression, as shown for *MT* genes of the land snails *Cantareus aspersus* and *Helix pomatia*, may also play a role [32,47]. Interestingly, strains of *Biomphalaria glabrata* susceptible to infestation by the parasite *Schistosoma mansonii* have been shown to be more Cd-tolerant compared to parasite-resistant strains [48]. It is not known if the laboratory population of *Biomphalaria glabrata* used in the present study may represent a parasite-susceptible strain.

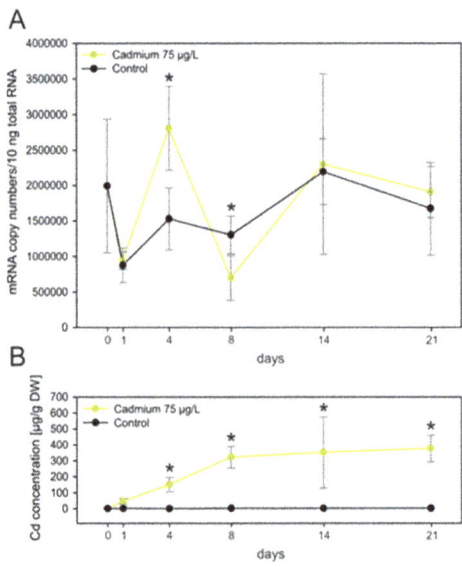

Figure 5. (**A**) *MT* induction pattern in the midgut gland of untreated individuals (black line) and individuals exposed to cadmium (75 µg/L) (yellow line) of *Biomphalaria glabrata* over a period of 21 days. Means and standard deviations are shown ($n = 4$). The course of MT mRNA transcription was tested by two-way analysis of variance (ANOVA) ($p \leq 0.05$) but was insignificant. Stars indicate significant differences at single time points according to multiple *t*-test comparisons; (**B**) cadmium concentration in the midgut gland of *Biomphalaria glabrata* from controls (black line) and exposed individuals (75 µg/L Cd) (yellow line). Means and standard deviations are shown ($n = 4$). Two-way ANOVA ($p \leq 0.05$) revealed significant differences between the two treatments. Stars indicate significant differences at single time points according to multiple *t*-test comparisons.

2.6. Transcriptional Response of the CdMT Gene Following Heat and Cold Shock Exposure

The presence of multiple heat shock elements (HSEs) as potential binding sites of heat shock factors within the promoter region (see Figure 2) suggested the possibility that the *MT* gene of *Biomphalaria glabrata* may respond to temperature-related stressors, e.g., heat or cold shock exposure (Figure 6) [49]. The ability of MTs to respond to both decreasing and increasing temperatures of exposure has already been reported [50,51] although the distinct mechanism remains unclear. Heat shock with an abrupt increase of the temperature by +7.5 °C applied through 8 days (Figure 6A,B) did not lead to a sudden response of the *MT* gene. Instead, a significant increase of *MT* upregulation could only be observed on day 8 of exposure (Figure 6A). The biological significance of this finding remains unclear, since mRNA upregulation of heat shock-responsive genes would normally occur much faster than in the present case [52].

In the same manner as sudden heat increase may be able to induce genes involved in stress response, a cold shock might as well be an inducer of these genes [49]. In our study, a temperature reduction by −15 °C (Figure 6C) through 24 h did not alter *MT* gene expression levels (Figure 6D). Rather, the mRNA copy numbers of snails subjected to the decreased temperature varied only slightly when compared to control individuals.

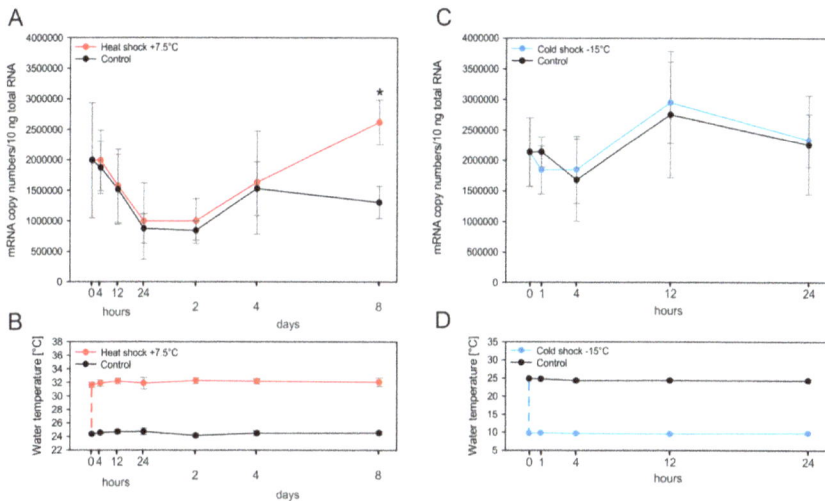

Figure 6. (**A**) *MT* induction and pattern in the midgut gland of *Biomphalaria glabrata* in untreated individuals (black line) and snails exposed to a sudden temperature increase (+7.5 °C) (red line), persisting through a period of 8 days. Means and standard deviations are shown (*n* = 4). The course of MT mRNA transcription levels was tested by two-way ANOVA ($p \leq 0.05$) but was insignificant. The star indicates a significant difference at a single time point according to multiple *t*-test comparisons; (**B**) water temperature profile for the duration of exposure. The dotted line indicates the heat shock at day 0; (**C**) *MT* induction pattern in the midgut gland of *Biomphalaria glabrata* in untreated individuals (black line) and animals exposed to a sudden temperature decrease (−15 °C) (blue line), through a period of 24 h. Means and standard deviations are shown (*n* = 4). The course of MT mRNA transcription levels was tested by two-way ANOVA ($p \leq 0.05$) but was insignificant. (**D**) water temperature profile for the duration of exposure. The dotted line indicates the cold shock at day 0.

3. Materials and Methods

3.1. Animals, Rearing Conditions and Experimental Set-Up

Individuals of *Biomphalaria glabrata* originated from a laboratory-grown culture at the Institute of Zoology in Innsbruck, where the snails were kept in freshwater aquarium tanks at 25 °C with a 12:12 h photoperiod. Snails were fed *ad libitum* with commercially available lettuce (*Lactuca sativa*) every third day. The trial was approved by the austrian science foundation (Fonds zur Förderung der wissenschaftlichen Forschung (FWF) (Project ref. I 1482-N28; 1 Jan 2014).

Prior to experiments, individuals of *Biomphalaria glabrata* were acclimatized for two weeks in reconstituted water (KCl 18 mg/L, $MgSO_4$ 190 mg/L, $NaHCO_3$ 98.5 mg/L, $CaCl_2$ 450 mg/L and NaCl 430 mg/L in milliQ water) at 25 °C and subsequently separated into different tanks. Three different exposure regimes were applied.

1. **Cadmium exposure:** 40 individuals were subjected to a nominal Cd concentration of 75 µg/L by adding CdCl$_2$ to the water. According to our own experiments and data from the literature, Cd LC-50 values at 96 h for *Biomphalaria glabrata* range between 0.1 mg/L [53] and 0.3 mg/L [54]. Forty control snails were kept in Cd-free water. Measured Cd concentrations in the water tanks were as follows (mean $\pm$ standard deviation, $n = 5$): Control, 0.24 $\pm$ 0.14 µg/L; Cd exposure, 63 $\pm$ 7.6 µg/L. Four snails of each group were sampled on days 1, 4, 8, 14 and 21.
2. **Heat shock experiment:** Post-acclimatization, 40 snails were directly subjected to tanks with an increased water temperature by 7.5 °C to exert a heat shock. Due to the expectation of a fast response, sampling occurred at 4 h, 12 h and at day 1. For detection of possible long-term effects, sampling was additionally extended to days 2, 4 and 8. Forty snails were kept in 25 °C water as a control and were sampled on day 0 and all other respective time points.
3. **Cold shock experiment:** Post-acclimatization, 35 snails were directly subjected to tanks with a water temperature decreased by 15 °C to exert a cold shock. Sampling was done at 1, 4, 12 and 24 h. 35 snails were kept in 25 °C water as a control and were sampled on day 0 and all other respective time points.

Throughout the experiment, the snails were fed with lettuce (*Lactuca sativa*) *ad libitum*. All sampled individuals were dissected and the midgut gland tissue was used for RNA isolation and tissue Cd analysis as described below.

3.2. Primary Structure of the Biomphalaria glabrata MT and Its Variants

The primary MT structure was identified by genome analysis using VectorBase in agreement with main proponents of the *Biomphalaria glabrata* genome project [55]. The complete gene structure was annotated and submitted to the national center for biotechnology information (NCBI) (GenBank acc. nr.: XM_013225031) (Supplementary Material). The respective mRNA sequence (KT697617) (including the 5′ and 3′ untranslated regions) was experimentally verified by sequencing rapid amplification of cDNA ends polymerase chain reaction (RACE-PCR)-amplified and cloned individuals as described below. 20 individuals of *Biomphalaria glabrata* were screened for allelic variation by sequencing PCR–amplified and cloned individuals, revealing four distinct allelic variants ($n = 3$). The respective sequences were submitted to GenBank and are accessible under the acc. nrs.: KY963493 (allelic variant 1), KY963494 (allelic variant 2), KY963495 (allelic variant 3) and KY963496 (allelic variant 4). Putative MT isoform sequences of *Biomphalaria glabrata* published earlier [14] could not be confirmed by genome analysis or PCR. They may have originated from cross-contamination during sample preparation. The respective sequences and accession numbers were therefore deleted from GenBank in agreement with the GenBank support team.

3.3. mRNA Isolation, Reverse Transcription, RACE-PCR and Sequencing

Total RNA was isolated from homogenized (Precellys, Bertin Instruments, Montigny-le-Bretonneux, France) midgut gland tissues of freshly dissected, untreated Ramshorn snails (*Biomphalaria glabrata*) originating from our laboratory culture at the University of Innsbruck, using the RNeasy Plant Mini Kit (QIAGEN, Venlo, The Netherlands) followed by DNase 1 digestion (Invitrogen, Thermo Fisher Scientific, Waltham, MA, USA). cDNA libraries of 23 individuals were generated using the Moloney murine leukemia virus reverse transcriptase (M-MLV RT) reverse transcriptase (Invitrogen, Thermo Fisher Scientific, Waltham, MA, USA). Three of these individuals were subjected to rapid amplification of cDNA ends (RACE-PCR) and the other 20 individuals were screened for allelic variation with the Titanium Taq PCR system using the following primers (located in the 5′ and 3′ untranslated regions): forward 5′-AAACACCATGAGTGGCAA-3′ and reverse 5′-CCACTCAACTCTTACAGC-3′ according to the recommended protocol, with a denaturation cycle of 95 °C for 1 min, followed by 30 cycles with 95 °C for 30 s, 50 °C for 50 s and 68 °C for 50 s, with a final extension step at 68 °C for an additional 3 min. RACE-PCR ($n = 3$) was applied using the SMARTer RACE 5′/3′ Kit (Takara Clontech, Shimogyo-ku, Kyoto, Japan) with the following primers (GSP1: 5′-ACTGAGGCTTGTACTGGGGA-3′, GSP2: 5′-TTGCATCCCTCTCCACATTTAC-3′) according to the recommended protocol. Amplified products were separated by a 1.5% agarose gel, stained with GelRed (Biotium Inc., Bay Area, CA, USA) and subsequently purified using the Qiaquick Gel Extraction Kit (QIAGEN, Venlo, The Netherlands). Cloning of PCR fragments was performed with TOPO TA Cloning Kit for Sequencing (Invitrogen, Thermo Fisher Scientific, Waltham, MA, USA). Plasmids were purified with the QIAprep Mini-Prep Kit (QIAGEN, Venlo, The Netherlands) and sequenced using the BigDye Terminator version 1.1 Cycle Sequencing Kit (Applied Biosystems, Thermo Fisher Scientific, Waltham, MA, USA). DNA Sequencing Analysis Software v5.2 (Applied Biosystems, Thermo Fisher Scientific, Waltham, MA, USA), CLC Workbench software (CLC Bio-Qiagen, Aarhus, Denmark) and ClustalW [56] were applied for sequence analysis.

3.4. Bioinformatic Analysis

Analysis of putative transcription factor binding sites (TFBs) for the *MT* gene of *Biomphalaria glabrata* was performed by means of the transcription factor database (TRANSFAC) software package (version 2014.4, BIOBASE, Wolfenbuettel, Germany) [57]. The program was run with default parameters except for the "matrix similarity" setting, where the parameter was set to 0.9.

3.5. mRNA Isolation, Reverse Transcription and Quantitative Real-Time Detection PCR

Individuals were dissected on an ice-cooled stainless steel plate and total RNA was isolated from ~10 mg of homogenized (Precellys, Bertin Instruments, Montigny-le-Bretonneux, France) midgut gland tissue with the RNeasy®Plant Mini Kit (QIAGEN, Venlo, The Netherlands) applying on-column DNase 1 digestion (QIAGEN, Venlo, The Netherlands). RNA was screened for integrity visually on an agarose gel and quantified with the RiboGreen®RNA Quantification Kit from Molecular Probes (Invitrogen, Thermo Fisher Scientific, Waltham, MA, USA) on a VICTOR™X4 2030 Multilabel Reader (PerkinElmer, Waltham, MA, USA). First strand cDNA was synthesized from 250 ng of total RNA with the Superscript® IV Reverse Transcriptase synthesis kit (Invitrogen, Thermo Fisher Scientific, Waltham, MA, USA) in a 20 μL approach for subsequent real-time detection PCR. The remaining tissue was processed further for Cd analysis as described below.

Quantitative real-time detection PCR of BglMT cDNA was performed with Power SYBR Green (Applied Biosystems, Thermo Fisher Scientific, Waltham, MA, USA) on a QuantStudio 3 (Applied Biosystems, Thermo Fisher Scientific, Waltham, MA, USA). The transcript with the defined amplicon length of 107 bp was amplified using the following concentrations and primers: sense primer: 900 nM; 5′-GCACTGACACAGAATGCAGTTG-3′ and antisense primer, 900 nM; 5′-TTTGCACCCTTCATCTGACTTAGT-3′ applying the following protocol of 40 cycles: denaturation at 95 °C for 15 s, annealing and extension combined at 60 °C for 60 s. The 10 μL PCR reaction contained 1 μL of cDNA and 1× Power SYBR Green PCR master mix, 1× U-BSA and sense and antisense primer. Primers were designed using the Primer Express 3.0 software (Applied Biosystems, Thermo Fisher Scientific, Waltham, MA, USA) and a primer matrix with subsequent dissociation curves was used to determine optimal primer concentrations. Calibration curves from amplicons were generated to determine C_q values for copy number analysis (PCR efficiency ~96%) using the Thermo Fisher Cloud Software, Version 1.0 (Life Technologies Corporation, Carlsbad, CA, USA).

3.6. Metal Analysis

Cd concentrations in the midgut gland tissues and the medium were assessed by means of atomic absorption spectrophotometry. After oven-drying of tissue aliquots at 65 °C and dry weight determination, samples were digested under pressure in 2 mL tubes (Eppendorf, Hamburg, Germany) with a 1:1 mixture of nitric acid (65%) (Suprapur, Merck, Darmstadt, Germany) and deionized water in an aluminum oven at 69 °C. After obtaining a clear solution, the samples were diluted with deionized water to 2 mL. Subsequently, Cd concentrations were measured with an atomic absorption spectrophotometer (model Z-8200, Hitachi, Tokyo, Japan). The system was calibrated with standard Cd solutions in 1% nitric acid and the accuracy of metal measurements of the midgut gland was verified using certified standard reference material (TORT-2, Lobster Hepatopancreas Reference Material for Trace Metals; National Research Council Canada, Ottawa, ON, Canada).

3.7. Statistical Methods

Data of qRT PCR and metal analysis were evaluated statistically by means of SigmaPlot 12.5. For normal-distributed data, the *t*-test was applied whereas for data failing equal distribution the Holm–Sidak method was used. Statistical significance was set at $p \leq 0.05$. Additionally, analysis of variance (ANOVA) was applied to test for significance of time-dependent variations of data ($p \leq 0.001$).

4. Conclusions

Our results show that *Biomphalaria glabrata* is apparently highly tolerant to Cd^{2+}, as demonstrated by the high metal concentrations accumulated in the snail's midgut gland. In spite of this, there is no upregulation of the *MT* gene due to Cd^{2+} exposure. Moreover, there is no increase of MT mRNA due to application of temperature shock exposures (heat and cold). It is assumed that the high constitutive expression of this MT may counteract metal stress by unspecific binding through metal exchange reactions at the protein's binding sites [35]; in a similar way, the high constitutive expression level may also contribute to heat and cold shock tolerance. Alternatively, and/or concomitantly, Cd detoxification in *Biomphalaria glabrata* may also be achieved through complexation by phytochelatins [46].

Supplementary Materials: Supplementary materials can be found at www.mdpi.com/1422-0067/18/8/1747/s1.

Acknowledgments: This work was funded by the Tiroler Wissenschaftsfonds (TWF) Project ref. 214457 granted to Michael Niederwanger and by the Fonds zur Förderung der wissenschaftlichen Forschng by the Austrian Science foundation (FWF) Project ref. I 1482-N28 (DACH) granted to Reinhard Dallinger.

Author Contributions: Reinhard Dallinger supervised and coordinated the research. Michael Niederwanger and Reinhard Dallinger conceived and designed experiments, analyzed the data and discussed the experimental results. Michael Niederwanger and Reinhard Dallinger identified and analyzed the *MT* gene structurally. Michael Niederwanger and Katharina Bacher performed cloning and sequencing of the allelic variants. Michael Niederwanger, Martin Dvorak, Raimund Schnegg and Veronika Pedrini-Martha performed mRNA induction studies. Michael Niederwanger, Martin Dvorak, Raimund Schnegg and Massimo Bidoli performed cadmium analysis of the tissue. Michael Niederwanger and Reinhard Dallinger were primarily responsible for writing the manuscript. All authors edited and approved the final version of the manuscript.

Conflicts of Interest: The authors declare no conflicts of interest.

References

1. Binz, P.-A.; Kägi, J.H.R. Metallothionein: Molecular evolution and classification. In *Metallothionein IV*; Klaassen, C.D., Ed.; Birkhäuser Basel: Basel, Switzerland, 1999; pp. 7–13.
2. Capdevila, M.; Atrian, S. Metallothionein protein evolution: A miniassay. *J. Biol. Inorg. Chem.* **2011**, *16*, 977–989. [CrossRef] [PubMed]
3. Kägi, J.H.; Kojima, Y. Chemistry and biochemistry of metallothionein. *Exp. Suppl.* **1987**, *52*, 25–61.
4. Andrews, G.K. Regulation of metallothionein gene expression. *Prog. Food Nutr. Sci.* **1990**, *14*, 193–258. [PubMed]
5. Kägi, J.H.R. Overview of *Metallothionein. Methods Enzymol.* **1991**, *205*, 613–626. [PubMed]
6. Klaassen, C.D.; Liu, J.; Choudhuri, S. Metallothionein: An intracellular protein to protect against cadmium toxicity. *Annu. Rev. Pharmacol. Toxicol.* **1999**, *39*, 267–294. [CrossRef] [PubMed]
7. Egli, D.; Domènech, J.; Selvaraj, A.; Balamurugan, K.; Hua, H.; Capdevila, M.; Georgiev, O.; Schaffner, W.; Atrian, S. The four members of the *Drosophila* metallothionein family exhibit distinct yet overlapping roles in heavy metal homeostasis and detoxification. *Genes Cells* **2006**, *11*, 647–658. [CrossRef] [PubMed]
8. Ghoshal, K.; Wang, Y.; Sheridan, J.F.; Jacob, S.T. Metallothionein induction in response to restraint stress. *J. Biol. Chem.* **1998**, *273*, 27904–27910. [CrossRef] [PubMed]

9. Baird, S.K.; Kurz, T.; Brunk, U.T. Metallothionein protects against oxidative stress-induced lysosomal destabilization. *Biochem. J.* **2006**, *394*, 275–283. [CrossRef] [PubMed]
10. Bebianno, M.J.; Serafim, M.A. P.; Rita, M.F. Involvement of metallothionein in cadmium accumulation and elimination in the clam *Ruditapes decussata*. *Bull. Environ. Contam. Toxicol.* **1994**, *53*, 726–732. [CrossRef] [PubMed]
11. Berger, B.; Dallinger, R.; Gehrig, P.; Hunziker, P.E. Primary structure of a copper-binding metallothionein from mantle tissue of the terrestrial gastropod *Helix pomatia* L. *J. Biochem.* **1997**, *328*, 219–224. [CrossRef]
12. Dallinger, R.; Berger, B.; Hunziker, P.; Kägi, J.H.R. Metallothionein in snail Cd and Cu metabolism. *Nature* **1997**, *143*, 831–833.
13. Hispard, F.; Schuler, D.; de Vaufleury, A.; Scheifler, R.; Badot, P.M.; Dallinger, R. Metal distribution and metallothionein induction after cadmium exposure in the Terrestrial Snail *Helix Aspersa*. *Environ. Toxicol.* **2008**, *27*, 1533–1542. [CrossRef] [PubMed]
14. Palacios, O.; Pagani, A.; Pérez-Rafael, S.; Egg, M.; Höckner, M.; Brandstätter, A.; Capdevila, M.; Atrian, S.; Dallinger, R. Shaping mechanisms of metal specificity in a family of metazoan metallothioneins: Evolutionary differentiation of mollusc metallothioneins. *BMC Biol.* **2011**, *9*, 4. [CrossRef] [PubMed]
15. English, T.E.; Storey, K.B. Freezing and anoxia stresses induce expression of metallothionein in the foot muscle and hepatopancreas of the marine gastropod *Littorina littorea*. *J. Exp. Biol.* **2003**, *206*, 2517–2524. [CrossRef] [PubMed]
16. Egg, M.; Höckner, M.; Brandstätter, A.; Schuler, D.; Dallinger, R. Structural and bioinformatic analysis of the Roman snail Cd-Metallothionein gene uncovers molecular adaptation towards plasticity in coping with multifarious environmental stress. *Mol. Ecol.* **2009**, *18*, 2426–2443. [CrossRef] [PubMed]
17. Gnatyshyna, L.L.; Fal'fushinskaya, G.I.; Golubev, O.P.; Dallinger, R.; Stoliar, O.B. Role of metallothioneins in adaptation of *Lymnaea stagnalis* (*Mollusca*: *Pulmonata*) to environment pollution. *Hydrobiol. J.* **2011**, *47*, 56–66. [CrossRef]
18. Martinez-Paz, P.; Morales, M.; Sanchez-Arguello, P.; Morcillo, G.; Martinez-Guitarte, J.L. Cadmium in vivo exposure alters stress response and endocrine-related genes in the freshwater snail *Physa acuta*. New biomarker genes in a new model organism. *Environ. Pollut.* **2017**, *220*, 1488–1497. [CrossRef] [PubMed]
19. Fenwick, A. The global burden of neglected tropical diseases. *Public Health* **2012**, *126*, 233–236. [CrossRef] [PubMed]
20. Sokolova, I.M.; Granovitch, A.I.; Berger, V.J.; Johannesson, K. Intraspecific physiological variability of the gastropod *Littorina saxatilis* related to the vertical shore gradient in the White and North Seas. *Mar. Biol.* **2000**, *137*, 297–308. [CrossRef]
21. Ferreira, M.V.; Alencastro, A.C.; Hermes-Lima, M. Role of antioxidant defenses during estivation and anoxia exposure in the freshwater snail *Biomphalaria tenagophila* (Orbigny, 1835). *Can. J. Zool.* **2003**, *81*, 1239–1248. [CrossRef]
22. Greenaway, P. Sodium regulation in the freshwter mollusc *Lymnaea stagnalis* (L.) (*gastropoda*: *pulmonata*). *J. Exp. Biol.* **1970**, *53*, 147–163. [PubMed]
23. Kefford, B.J.; Nugegoda, D. No evidence for a critical salinity threshold for growth and reproduction in the freshwater snail *Physa acuta*. *Environ. Pollut.* **2005**, *134*, 377–383. [CrossRef] [PubMed]
24. Oliveira-Filho, E.; Caixeta, N.; Simplício, N.; Sousa, S.; Aragão, T.; Muniz, D. Implications of water hardness in ecotoxicological assessments for water quality regulatory purposes: A case study with the aquatic snail *Biomphalaria glabrata* (Say, 1818). *Braz. J. Biol.* **2014**, *74*, 175–180. [CrossRef] [PubMed]
25. Dallinger, R.; Rainbow, P.S. *Ecotoxicology of Metals in Invertebrates*; CRC Press: Boca Raton, FL, USA, 1993.
26. Grosell, M.; Brix, K.V. High net calcium uptake explains the hypersensitivity of the freshwater pulmonate snail, *Lymnaea stagnalis*, to chronic lead exposure. *Aquat. Toxicol.* **2009**, *91*, 302–311. [CrossRef] [PubMed]
27. Tamai, K.T.; Liu, X.; Silar, P.; Sosinowski, T.; Thiele, D.J. Heat shock transcription factor activates yeast metallothionein gene expression in response to heat and glucose starvation via distinct signalling pathways. *Mol. Cell. Biol.* **1994**, *14*, 8155–8165. [CrossRef] [PubMed]
28. Davis, S.R.; Cousins, R.J. Metallothionein expression in animals: A physiological perspective on function. *J. Nutr.* **2000**, *130*, 1085–1088. [PubMed]
29. Sogawa, N.; Sogawa, C.A.; Fukuoka, H.; Mukubo, Y.; Yoneyama, T.; Okano, Y.; Furuta, H.; Onodera, K. The changes of hepatic metallothionein synthesis and the hepatic damage induced by starvation in mice. *Methods Find. Exp. Clin. Pharmacol.* **2003**, *25*, 601. [CrossRef] [PubMed]

30. Beattie, J.H.; Owen, H.L.H.; Wallace, S.M.; Arthur, J.R.; Kwun, I.-S.; Hawksworth, G.M.; Wallace, H.M. Metallothionein overexpression and resistance to toxic stress. *Toxicol. Lett.* **2005**, *157*, 69–78. [CrossRef] [PubMed]

31. Murphy, B.J.; Kimura, T.; Sato, B.G.; Shi, Y.; Andrews, G.K. Metallothionein induction by hypoxia involves cooperative interactions between metal-responsive transcription Factor-1 and hypoxia-inducible transcription Factor-1α. *Mol. Cancer Res.* **2008**, *6*, 483–490. [CrossRef] [PubMed]

32. Pedrini-Martha, V.; Niederwanger, M.; Kopp, R.; Schnegg, R.; Dallinger, R. Physiological, diurnal and stress-related variability of cadmium-metallothionein gene expression in *Land snails*. *PLoS ONE* **2016**, *11*, e0150442. [CrossRef] [PubMed]

33. Da Silva Cantinha, R.; Borrely, S.I.; Oguiura, N.; de Braganca Pereira, C.A.; Rigolon, M.M.; Nakano, E. HSP70 expression in *Biomphalaria glabrata* snails exposed to cadmium. *Ecotoxicol. Environ. Saf.* **2017**, *140*, 18–23. [CrossRef] [PubMed]

34. Nevitt, T.; Pereira, J.; Rodrigues-Pousada, C. YAP4 gene expression is induced in response to several forms of stress in *Saccharomyces cerevisiae*. *Yeast* **2004**, *21*, 1365–1374. [CrossRef] [PubMed]

35. Niederwanger, M.; Calatayud, S.; Zerbe, O.; Atrian, S.; Albalat, R.; Capdevila, M.; Palacios, Ò.; Dallinger, R. *Biomphalaria glabrata* Metallothionein: Lacking metal specificity of the protein and missing gene upregulation suggest metal sequestration by exchange instead of through selective binding. *Int. J. Mol. Sci.* **2017**, *18*, 1457. [CrossRef] [PubMed]

36. Baumann, C.; Beil, A.; Jurt, S.; Niederwanger, M.; Palacios, O.; Capdevila, M.; Atrian, S.; Dallinger, R.; Zerbe, O. Structural adaptation of a protein to increased metal stress: NMR structure of a marine snail metallothionein with an additional domain. *Angew. Chem. Int. Ed.* **2017**, *56*, 4617–4622. [CrossRef] [PubMed]

37. Braun, W.; Wagner, G.; Wörgötter, E.; Vasak, M.; Kägi, J.H. R.; Wüthrich, K. Polypeptide fold in the two metal clusters of metallothionein-2 by nuclear magnetic resonance in solution. *J. Mol. Biol.* **1986**, *187*, 125–129. [CrossRef]

38. Jenny, M.J.; Payton, S.L.; Baltzegar, D.A.; Lozier, J.D. Phylogenetic analysis of molluscan metallothioneins: Evolutionary insight from *Crassostrea virginica*. *J. Mol. Evol.* **2016**, *83*, 110–125. [CrossRef] [PubMed]

39. Bebianno, M.J.; Langston, W.J.; Simkiss, K. Metallothionein induction in *Littorina-littorea* (*Mollusca, Prosobranchia*) on exposure to cadmium. *J. Mar. Biol. Assoc. UK* **1992**, *72*, 329–342. [CrossRef]

40. Bebianno, M. Cadmium and metallothionein turnover in different tissues of the gastropod *Littorina littorea*. *Talanta* **1998**, *46*, 301–313. [CrossRef]

41. Roesijadi, G. Metallothionein induction as a measure of response to metal exposure in aquatic animals. *Environ. Health Perspect.* **1994**, *102*, 91–95. [CrossRef] [PubMed]

42. Benito, D.; Niederwanger, M.; Urtzi, I.; Dallinger, R.; Soto, M. Successive onset of molecular, cellular and tissue-specific responses in midgut gland of *Littorina littorea* exposed to sub-lethal cadmium concentrations. *Int. J. Mol. Sci.* **2017**, in press.

43. Phillips, D.J.H.; Rainbow, P.S. Strategies of trace metal sequestration in aquatic organisms. *Mar. Environ. Res.* **1989**, *28*, 207–210. [CrossRef]

44. Dallinger, R. Strategies of metal detoxification in terrestrial invertebrates. In *Ecotoxicology of Metals in Invertebrates*; CRC Press: Boca Raton, FL, USA, 1993; p. 245.

45. Rainbow, P. Trace metal concentrations in aquatic invertebrates: Why and so what? *Environ. Pollut.* **2002**, *120*, 497–507. [CrossRef]

46. Goncalves, S.F.; Davies, S.K.; Bennett, M.; Raab, A.; Feldmann, J.; Kille, P.; Loureiro, S.; Spurgeon, D.J.; Bundy, J.G. Sub-lethal cadmium exposure increases phytochelatin concentrations in the aquatic snail *Lymnaea stagnalis*. *Sci. Total Environ.* **2016**, *568*, 1054–1058.

47. Baurand, P.-E.; Dallinger, R.; Niederwanger, M.; Capelli, N.; Pedrini-Martha, V.; de Vaufleury, A. Differential sensitivity of snail embryos to cadmium: Relation to age and metallothionein gene expression. *Environ. Sci. Pollut. Res.* **2016**, *23*. [CrossRef] [PubMed]

48. Salice, C.J.; Anderson, T.A.; Roesijadi, G. Adaptive responses and latent costs of multigeneration cadmium exposure in parasite resistant and susceptible strains of a freshwater snail. *Ecotoxicology* **2010**, *19*, 1466–1475. [CrossRef] [PubMed]

49. Deane, E.E.; Woo, N.Y.S. Cloning and characterization of the *HSP70* multigene family from silver sea bream: Modulated gene expression between warm and cold temperature acclimation. *Biochem. Biophys. Res. Commun.* **2005**, *330*, 776–783. [CrossRef] [PubMed]

50. Beattie, J.H.; Black, D.J.; Wood, A.M.; Trayhurn, P. Cold-induced expression of the *metallothionein*-1 gene in brown adipose tissue of rats. *Am. J. Physiol. Regul. Integr. Comp. Physiol.* **1996**, *270*, R971–R977.

51. Piano, A.; Valbonesi, P.; Fabbri, E. Expression of cytoprotective proteins, heat shock protein 70 and metallothioneins, in tissues of *Ostrea edulis* exposed to heat and heavy metals. *Cell Stress Chaperones* **2004**, *9*, 134–142. [CrossRef] [PubMed]

52. Rinehart, J.P.; Yocum, G.D.; Denlinger, D.L. Developmental upregulation of inducible HSP70 transcripts, but not the cognate form, during pupal diapause in the flesh fly, *Sarcophaga crassipalpis*. *Insect Biochem. Mol. Biol.* **2000**, *30*, 515–521. [CrossRef]

53. Niederwanger, M.; Dallinger, R.; Stoliar, O.B.; Fal'fushinskaya, G.I. LC-50 Values for *Biomphalaria glabrata* upon Cd Exposure. Unpublished work, 2013.

54. Bellavere, C.; Gorbi, J. A comparative analysis of acute toxicity of chromium, copper and cadmium to *Daphnia magna, Biomphalaria glabrata*, and *Brachydanio rerio*. *Environ. Technol. Lett.* **1981**, *2*, 119–128. [CrossRef]

55. Adema, C.M.; Hillie, L.W.; Jones, C.S.; Loker, E.S.; Knight, M.; Minx, P.; Oliveira, G.; Raghavan, N.; Shedlock, A.; do Amaral, L.R.; et al. Whole genome analysis of a schistosomiasis-transmitting freshwater snail. *Nature* **2017**, *8*. [CrossRef] [PubMed]

56. Larkin, M.A.; Blackshields, G.; Brown, N.P.; Chenna, R.; Mcgettigan, P.A.; McWilliam, H.; Valentin, F.; Wallace, I.M.; Wilm, A.; Lopez, R.; et al. Clustal W and Clustal X version 2.0. *Bioinformatics* **2007**, *23*, 2947–2948. [CrossRef] [PubMed]

57. Matys, V. TRANSFAC(R) and its module TRANSCompel(R): Transcriptional gene regulation in eukaryotes. *Nucleic Acids Res.* **2006**, *34*, D108–D110. [CrossRef] [PubMed]

International Journal of
Molecular Sciences

MDPI

Article

Modulation of the PI3K/Akt Pathway and Bcl-2 Family Proteins Involved in Chicken's Tubular Apoptosis Induced by Nickel Chloride (NiCl$_2$)

Hongrui Guo [1], Hengmin Cui [1,2,*], Xi Peng [1,2], Jing Fang [1,2], Zhicai Zuo [1,2], Junliang Deng [1,2], Xun Wang [1,2], Bangyuan Wu [1], Kejie Chen [1] and Jie Deng [1]

[1] Key Laboratory of Animal Diseases and Environmental Hazards of Sichuan Province, Sichuan Agricultural University, Ya'an 625014, China; guohonrui@163.com (H.G.); pengxi197313@163.com (X.P.); fangjing4109@163.com (J.F.); zzcjl@126.com (Z.Z.); dengjl213@126.com (J.D.); wangxun99@163.com (X.W.); wubangyuan2008@163.com (B.W.); ckj930@126.com (K.C.); dengjl213@126.com (J.D.)

[2] College of Veterinary Medicine, Sichuan Agricultural University, Ya'an 625014, China

* Author to whom correspondence should be addressed; cui580420@sicau.edu.cn; Tel.: +86-835-288-2510; Fax: +86-835-288-2340.

Academic Editor: Reinhard Dallinger
Received: 17 August 2015; Accepted: 17 September 2015; Published: 23 September 2015

Abstract: Exposure of people and animals to environments highly polluted with nickel (Ni) can cause pathologic effects. Ni compounds can induce apoptosis, but the mechanism and the pathway of Ni compounds-induced apoptosis are unclear. We evaluated the alterations of apoptosis, mitochondrial membrane potential (MMP), phosphoinositide-3-kinase (PI3K)/serine-threonine kinase (Akt) pathway, and Bcl-2 family proteins induced by nickel chloride (NiCl$_2$) in the kidneys of broiler chickens, using flow cytometry, terminal deoxynucleotidyl transferase 2'-deoxyuridine 5'-triphosphate dUTP nick end-labeling (TUNEL), immunohistochemstry and quantitative real-time polymerase chain reaction (qRT-PCR). We found that dietary NiCl$_2$ in excess of 300 mg/kg resulted in a significant increase in apoptosis, which was associated with decrease in MMP, and increase in apoptosis inducing factor (AIF) and endonuclease G (EndoG) protein and mRNA expression. Concurrently, NiCl$_2$ inhibited the PI3K/Akt pathway, which was characterized by decreasing PI3K, Akt1 and Akt2 mRNA expression levels. NiCl$_2$ also reduced the protein and mRNA expression of anti-apoptotic Bcl-2 and Bcl-xL and increased the protein and mRNA expression of pro-apoptotic Bax and Bak. These results show that NiCl$_2$ causes mitochondrial-mediated apoptosis by disruption of MMP and increased expression of AIF and EndoG mRNA and protein, and that the underlying mechanism of MMP loss involves the Bcl-2 family proteins modulation and PI3K/Akt pathway inhibition.

Keywords: NiCl$_2$; apoptosis; PI3K/Akt pathway; Bcl-2; mitochondria

---

1. Introduction

Nickel (Ni) is one of the essential elements found in abundance in the earth's crust occurring at an average concentration of about 75 µg/g [1]. Ni and Ni compounds have many industrial and commercial uses, and the progress of industrialization has led to their increased release into ecosystems [2,3]. Ni is considered an essential element in animals, microorganisms and plants, and is a constituent of enzyme proteins and nucleic acid [1,4]. However, symptoms of toxicity can occur when too much Ni is taken up [3]. Ni is potentially hazardous to living organisms due to its genotoxicity, immunotoxicity, mutagenicity and cancinogenicity [5–7]. Exposure of workers to Ni compounds can produce adverse effects on their health, such as Ni allergy, contact dermatitis, lung fibrosis, cardiovascular diseases, kidney diseases, and cancer of the respiratory tract [8]. Many forms of Ni may induce carcinoma in human beings and animals [9–11]. The findings of Zheng et al. [12] show that oxidative stress and the mitochondrial pathway play important roles in nickel sulfate (NiSO$_4$)-induced apoptosis in *Carassius auratus* liver. Ni compounds can promote the generation of reactive oxygen species (ROS), interact directly or indirectly with nucleic acids and cause DNA damage [13]. It has been also suggested that nickel chloride (NiCl$_2$) can induce DNA damage indirectly through the formation of ROS [14–16]. Efremenko et al. [17] have reported that nickel sulfide (Ni$_3$S$_2$)-caused inflammation and proliferation in the lungs of rats. Our previous studies have shown that dietary NiCl$_2$ in excess of 300 mg/kg can cause immunotoxicity, oxidative damage and apoptosis in the kidneys, spleens, small intestines, and cecal tonsils of broiler chickens [18–28].

The phosphoinositide-3-kinase (PI3K)/serine-threonine kinase (Akt) signaling pathway plays a crucial role in cell growth and cell survival, and the pathway can be activated by many types of cellular stimuli or toxins [29]. Serine/threonine kinase Akt/PKB is the primary mediator of PI3K-initiated signaling. Akt, activated by PI3K, regulates cell survival through phosphorylation of a variety of downstream targets such as pro-apoptotic protein, transcription factors and another protein kinase [30,31]. The PI3K/Akt pathway can mediate cell-survival signals through the Bcl-2 family [4,32–34]. Among the Bcl-2 family proteins, Bcl-2 and Bcl-xL promote cell survival, while Bad, Bak, Bid, and Bax can induce cell death [35,36]. The Bcl-2 family of proteins, which is located on the mitochondrial membrane, can alter mitochondrial membrane permeability and trigger apoptosis [35,37,38], and Ni-induced apoptosis is reportedly associated with the PI3K/Akt pathway [4,33,34]. Wang et al. [39] suggest that nickel acetate induces cytotoxicity and apoptosis in HK-2 cells via ROS generation and that the mitochondria-mediated apoptotic signaling pathway is involved in the positive regulation of nickel acetate-induced renal cytotoxicity.

Although studies on apoptosis induced by Ni and Ni compounds have been reported, the mechanisms of Ni and Ni compounds-induced apoptosis are unclear. Therefore, the objective of this study was to determine potential mechanisms of NiCl$_2$-induced mitochondria-mediated apoptosis in kidneys of broiler chickens and the alteration of the PI3K/Akt pathway and Bcl-2 family proteins. We monitored apoptosis, the change of mitochondrial membrane potential (MMP), and the mRNA expression of apoptosis inducing factor (AIF) and of endonuclease G (EndoG). We also measured the PI3K/Akt pathway (mRNA expression levels of PI3K, Akt1, Akt2) and the Bcl-2 family of proteins (the protein and mRNA expression of anti-apoptotic Bcl-2 and Bcl-xL and pro-apoptotic Bax and Bak).

2. Results

2.1. Histopathological Changes in the Kidney

In Figures 1–4, NiCl$_2$ resulted in dose- and time-dependent histopathological changes in the kidney, including tubular granular degeneration, vacuolar degeneration, necrosis and apoptosis. In the granular and vacuolar degenerated tubular cells, tiny particles and small or large vacuoles appeared in the cytoplasm. Karyorrhexis, karyolysis and hypochromatosis appeared in the necrotic cells. In the apoptotic cells, cytoplasm was intensely eosinophilic, and nuclei were shrunken, dense, ring-shaped and crescentic. Apoptotic bodies were also observed.

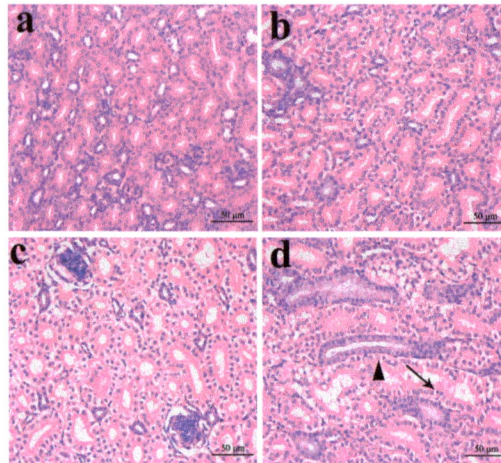

Figure 1. Histopathological changes in the kidney at 14 days of age. (**a**) Control group. No changes are observed (H·E × 400); (**b**) 300 mg/kg group. Tubular cells show slight granular degeneration (H·E × 400); (**c**) 600 mg/kg group. Tubular cells show granular degeneration and vacuolar degeneration (H·E × 400); and (**d**) 900 mg/kg group. Tubular cells show obvious granular and vacuolar degeneration. Also, few necrotic tubular cells (▲) and apoptotic tubular cells (↑) are observed (H·E × 400).

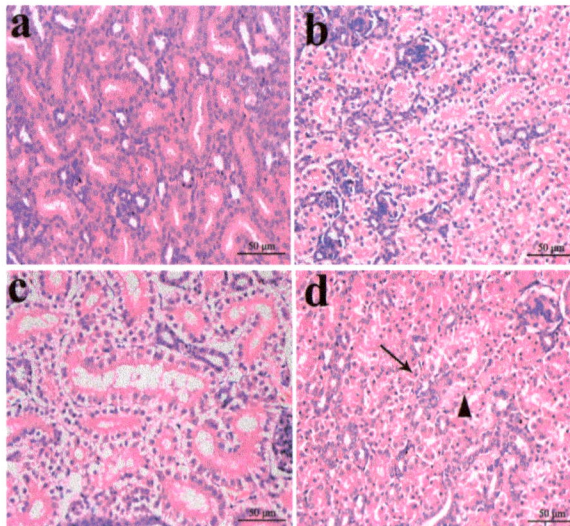

Figure 2. Histopathological changes in the kidney at 28 days of age. (**a**) Control group. No changes are observed (H·E × 400); (**b**) 300 mg/kg group. Tubular cells show granular degeneration (H·E × 400); (**c**) 600 mg/kg group. Tubular cells show obvious granular and vacuolar degeneration. Also, few necrotic tubular cells and apoptotic tubular cells are observed (H·E × 400); and (**d**) 900 mg/kg group. Tubular cells show marked granular and vacuolar degeneration. Also, some necrotic tubular cells (▲) and apoptotic tubular cells (↑) are observed (H·E × 400).

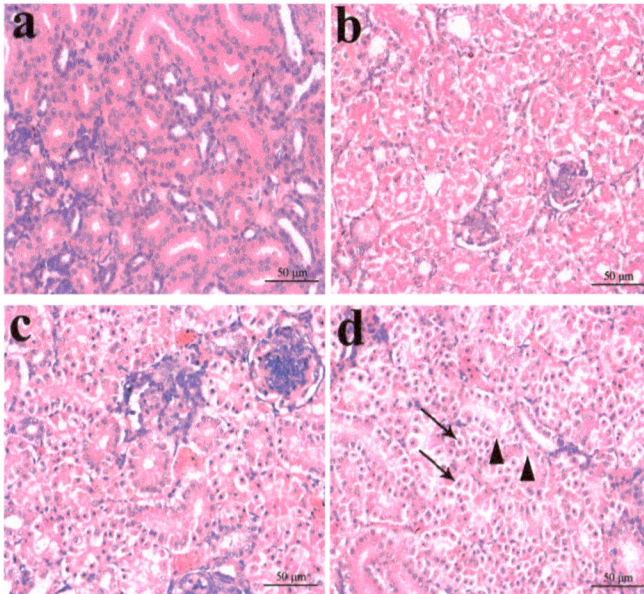

Figure 3. Histopathological changes in kidney at 28 days of age. (**a**) Control group. No changes are observed (H·E × 400); (**b**) 300 mg/kg group. Tubular cells show granular and vacuolar degeneration (H·E × 400); (**c**) 600 mg/kg group. Tubular cells show marked granular and vacuolar degeneration. Also, some necrotic tubular cells and apoptotic tubular cells are observed (H·E × 400); and (**d**) 900 mg/kg group. A large number of necrotic tubular cells (▲) and apoptotic tubular cells (↑) are observed (H·E × 400).

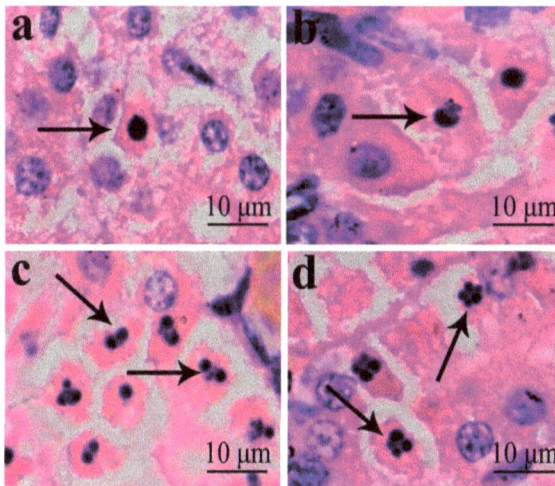

Figure 4. Morphological changes of apoptotic cells. (**a**) In the apoptotic cell, cytoplasm was intensely eosinophilic, and nucleus is shrunken and dense ring-shaped (↑). (H·E × 1000); and (**b**) In the apoptotic cell, nucleus is crescentic (↑). (H·E × 1000); (**c**) and (**d**) In the apoptotic cells, nuclei are cracked into two or multiple apoptotic bodies (↑). (H·E × 1000).

2.2. Effects of NiCl₂ on Apoptosis in the Kidney

The effects of dietary NiCl₂ on the apoptosis in the kidney were studied with TUNEL assay. The results presented in Figure 5, showed that the number of apoptotic cells was significantly greater ($p < 0.05$ or $p < 0.01$) in the 600 and 900 mg/kg groups at 14 days of age than in the control group. Apoptotic cells were significantly increased ($p < 0.05$ or $p < 0.01$) also in the three NiCl₂-treated groups at 28 to 42 days of age.

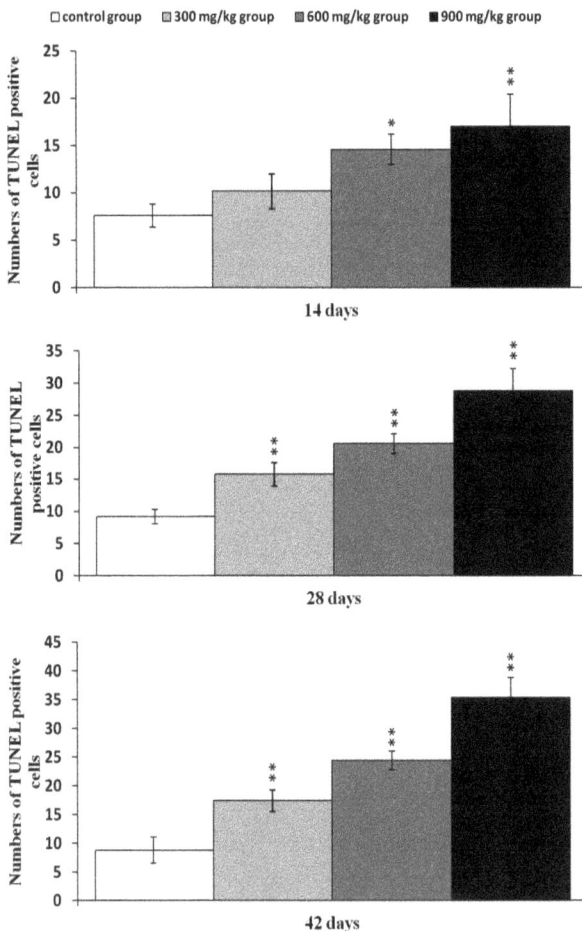

Figure 5. Changes of TUNEL-positive cells at 14, 28 and 42 days. Data are presented with the mean ± standard deviation ($n = 5 \times 5$); * $p < 0.05$, compared with the control group; ** $p < 0.01$, compared with the control group.

2.3. Effects of NiCl₂ on MMP, and AIF and EndoG Protein and mRNA Expression in the Kidney

To further confirm the role of mitochondria in NiCl₂-induced apoptosis, the changes of the MMP and release of AIF and EndoG protein from the mitochondria to the nucleus were examined.

As illustrated in Figure 6a,b, the results of the flow cytometry assay showed that NiCl₂ caused a significant loss ($p < 0.05$ or $p < 0.01$) in the MMP of the three NiCl₂-treated groups from 28 to 42 days of age and in the 600 and 900 mg/kg groups at 14 days of age as, compared with that in the control group.

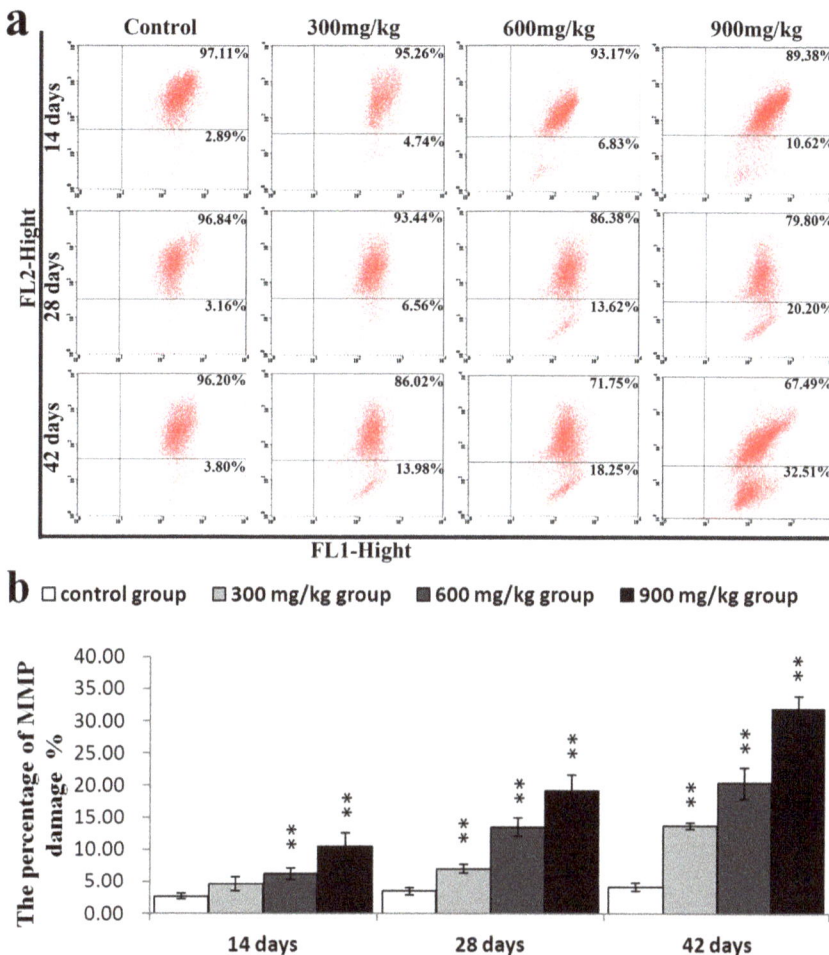

Figure 6. NiCl$_2$-induced mitochondrial dysfunction in the kidney. (**a**) Representative flow cytometric diagram of MMP analysis; and (**b**) The percentage of MMP damage. Data are presented with the mean $\pm$ standard deviation (n = 5); ** $p < 0.01$, compared with the control group.

In Figure 7, AIF and EndoG protein expression was significantly higher ($p < 0.05$ or $p < 0.01$) in the three NiCl$_2$-treated groups from 28 to 42 days of age and in the 600 and 900 mg/kg group at 14 days of age, than in the control group.

AIF mRNA expression was significantly higher ($p < 0.05$ or $p < 0.01$) in the three NiCl$_2$-treated groups from 28 to 42 days of age and in the 600 and 900 mg/kg group at 14 days of age, than in the control group. EndoG mRNA expression was significantly higher ($p < 0.05$ or $p < 0.01$) in the 600 and 900 mg/kg group from 14 to 42 days of age and in the 300 mg/kg group at 42 days of age than in the control group, as shown in Figure 8.

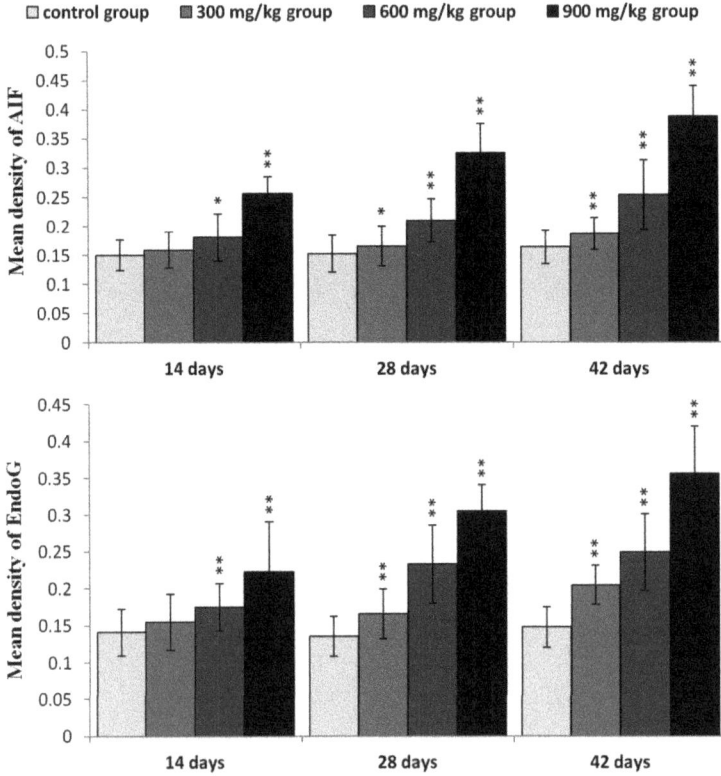

Figure 7. The protein expression levels of AIF and EndoG in the kidney. Data are presented with the mean ± standard deviation ($n = 5 \times 5$); * $p < 0.05$, compared with the control group; ** $p < 0.01$, compared with the control group.

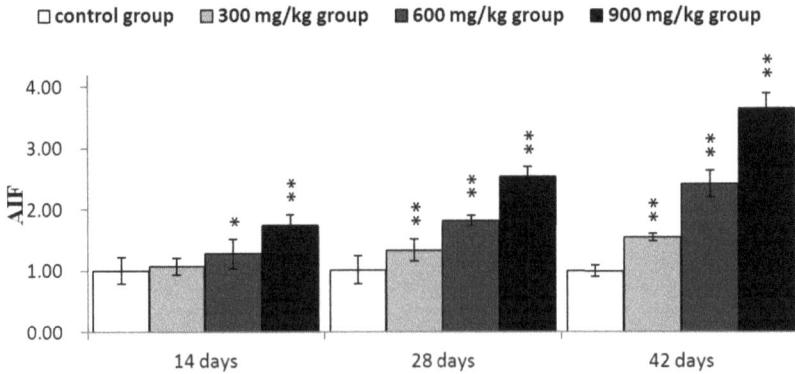

Figure 8. *Cont.*

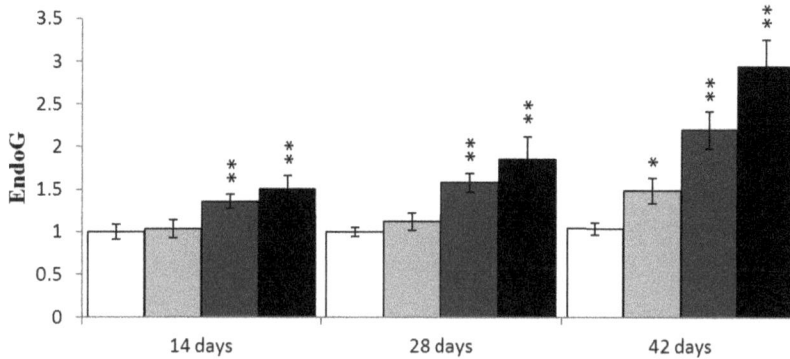

Figure 8. The mRNA expression levels of AIF and EndoG in the kidney. Data are presented with the mean ± standard deviation ($n = 5$); * $p < 0.05$, compared with the control group; ** $p < 0.01$, compared with the control group.

2.4. Effects of NiCl$_2$ on Phosphoinositide-3-Kinase (PI3K)/Serine-Threonine Kinase (Akt) Pathway in the Kidney

We investigated whether PI3K/Akt was involved in NiCl$_2$-mediated apoptosis. As shown in Figure 9, PI3K mRNA expression was significantly decreased ($p < 0.05$ or $p < 0.01$) in the 600 and 900 mg/kg groups from 14 to 42 days of age and in the 300 mg/kg group at 42 days of age. The mRNA expression of Akt1 and Akt2 was significantly lower ($p < 0.05$ or $p < 0.01$) in the three NiCl$_2$-treated groups from 28 to 42 days of age and in the 600 and 900 mg/kg groups at 14 days of age than in the control group.

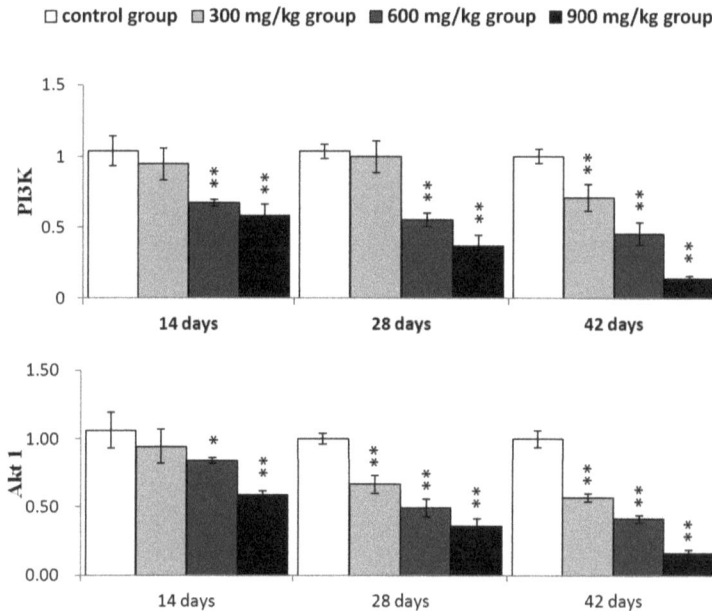

Figure 9. *Cont.*

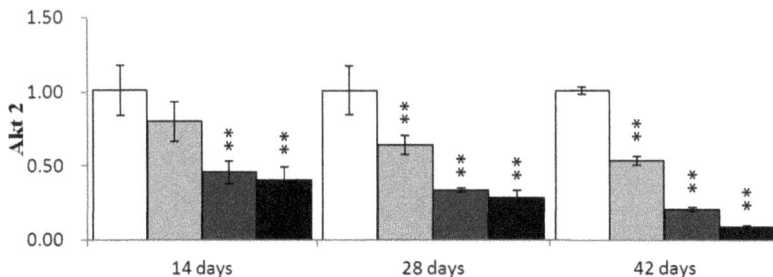

Figure 9. The mRNA expression levels of PI3K, Akt1 and Akt2 in the kidney. Data are presented with the mean ± standard deviation ($n = 5$); * $p < 0.05$, compared with the control group; ** $p < 0.01$, compared with the control group.

2.5. Expression of NiCl$_2$ on Bcl-2 Family Protein and mRNA Expression in the Kidney

It has been suggested that Bcl-2 can be a crucial mediator downstream of PI3K/Akt signaling. And Bcl-2 famly proteins have been shown to regulate the MMP. Therefore, we examined the effect of NiCl$_2$ treatment on Bcl-2 family proteins in the kidney.

In Figure 10, Bcl-2 protein expression was significantly decreased ($p < 0.05$ or $p < 0.01$) in the three NiCl$_2$-treated groups from 28 to 42 days of age and in the 900 mg/kg group at 14 days of age when compared with those in the control group. Bcl-xL protein expression was significantly decreased ($p < 0.05$ or $p < 0.01$) in the 900 mg/kg group at 14 days of age, in the 600 and 900 mg/kg groups at 48 days of age and in the three NiCl$_2$-treated groups at 42 days of age. The protein expression of Bax and Bak was significantly increased ($p < 0.05$ or $p < 0.01$) in the three NiCl$_2$-treated groups from 28 to 42 days of age. And, Bax protein expression was significantly increased ($p < 0.05$ or $p < 0.01$) in the 600 and 900 mg/kg groups at 14 days of age when compared with those in the control group.

The protein expression of Bax/Bcl-2 ratio was significantly higher ($p < 0.05$ or $p < 0.01$) in the three NiCl$_2$-treated groups from 28 to 42 days of age and in the 900 mg/kg groups from 14 days of age than in the control group, as shown in Figure 11.

In Figure 12, Bcl-2 mRNA expression was significantly decreased ($p < 0.05$ or $p < 0.01$) in the three NiCl$_2$-treated groups from 28 to 42 days of age and in the 900 mg/kg group at 14 days of age. Bcl-xL mRNA expression was significantly lower ($p < 0.05$ or $p < 0.01$) in the three NiCl$_2$-treated groups from 42 days of age and in the 900 mg/kg group from 14 to 28 days of age than that in the control group. The mRNA expression of Bax and Bak was significantly increased ($p < 0.05$ or $p < 0.01$) in the three NiCl$_2$-treated groups from 28 to 42 days of age and in the 600 and 900 mg/kg groups at 14 days of age.

The Bax/Bcl-2 ratio was significantly higher ($p < 0.05$ or $p < 0.01$) in the three NiCl$_2$-treated groups from 28 to 42 days of age and in the 600 and 900 mg/kg groups from 14 days of age than in the control group, as shown in Figure 13.

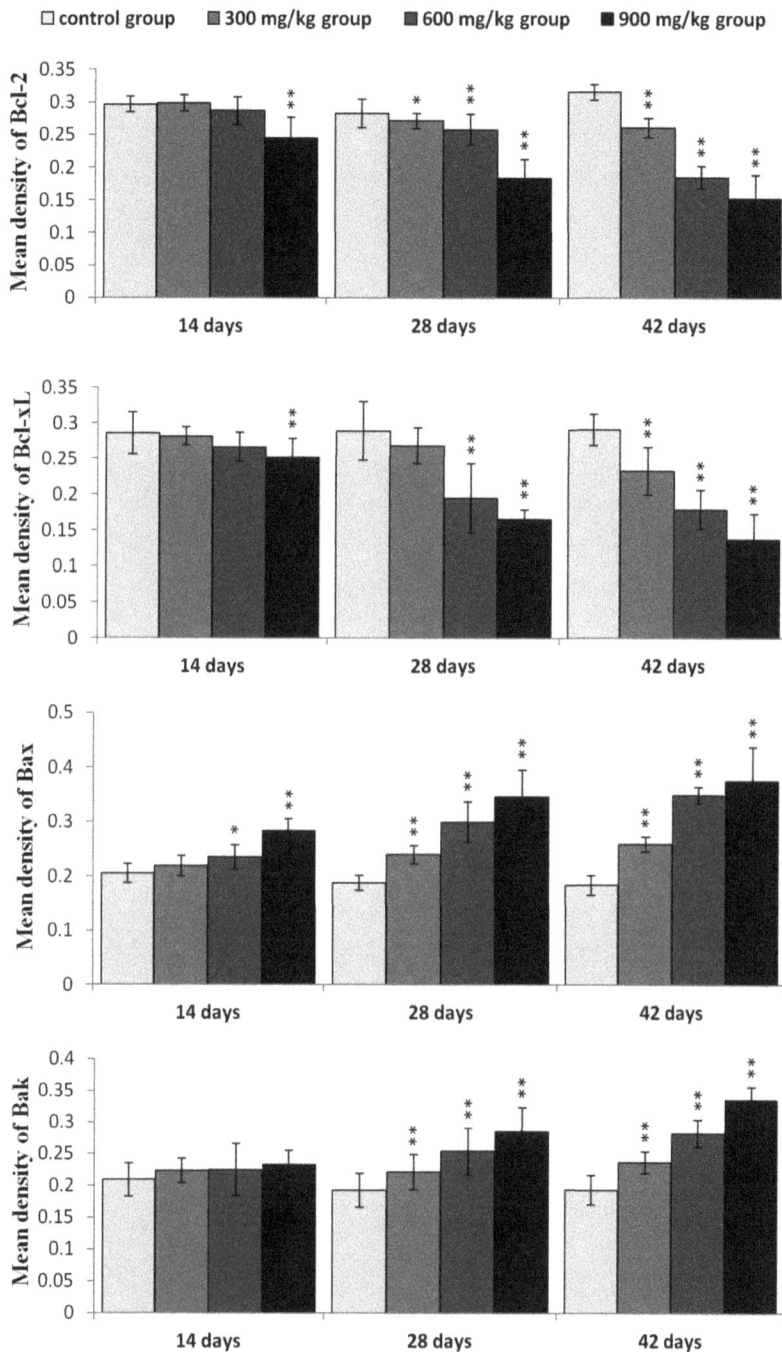

Figure 10. The protein expression levels of Bcl-2, Bcl-xL, Bax and Bak in the kidney. Data are presented with the mean $\pm$ standard deviation ($n = 5 \times 5$); * $p < 0.05$, compared with the control group; ** $p < 0.01$, compared with the control group.

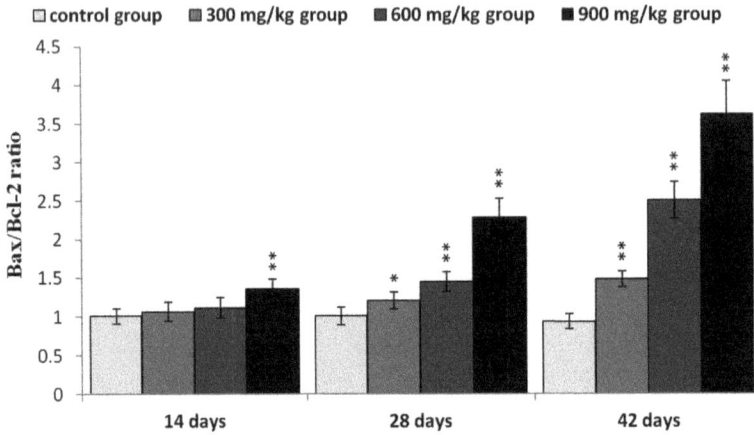

Figure 11. The ratio of Bax/Bcl-2 protein expression in the kidney. Data are presented with the mean $\pm$ standard deviation ($n = 5 \times 5$); * $p < 0.05$, compared with the control group; ** $p < 0.01$, compared with the control group.

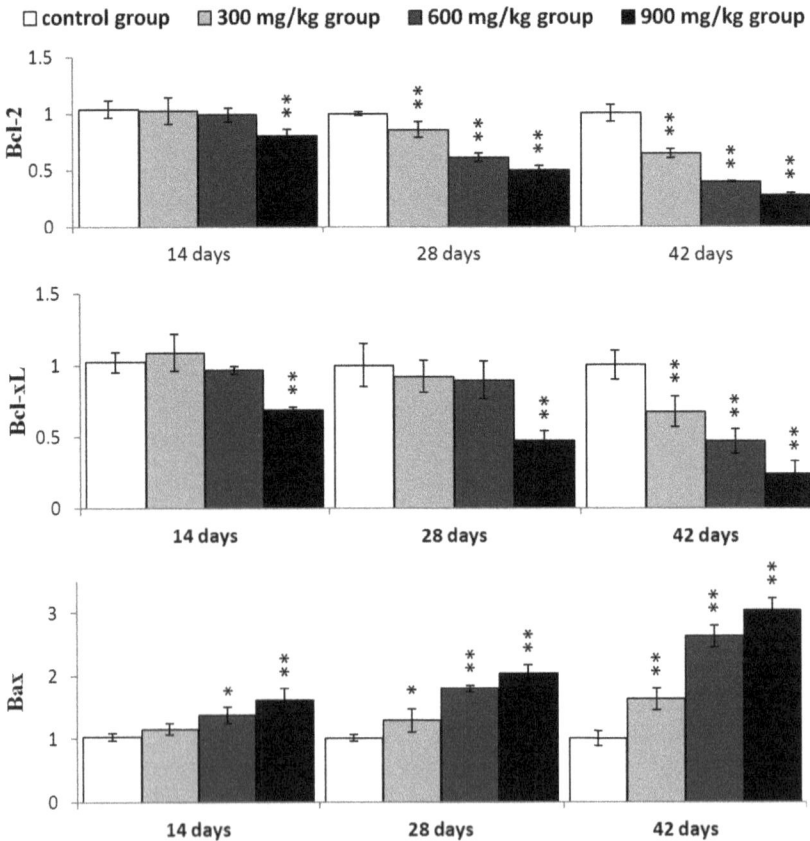

Figure 12. *Cont.*

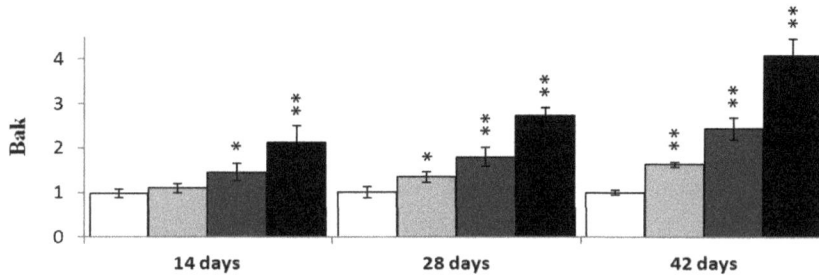

Figure 12. The mRNA expression levels of Bcl-2, Bcl-xL, Bax and Bak in the kidney. Data are presented with the mean ± standard deviation ($n = 5$); * $p < 0.05$, compared with the control group; ** $p < 0.01$, compared with the control group.

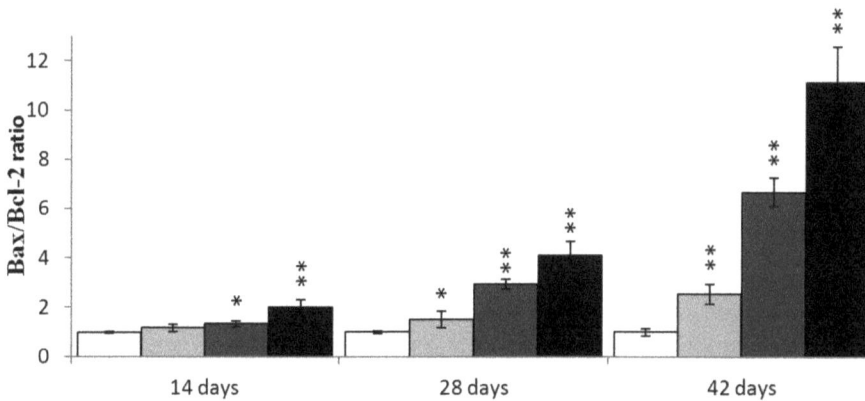

Figure 13. The ratio of Bax/Bcl-2 mRNA expression in the kidney. Data are presented with the mean ± standard deviation ($n = 5$); * $p < 0.05$, compared with the control group; ** $p < 0.01$, compared with the control group.

3. Discussion

This study explores the molecular control pathways of dietary $NiCl_2$-induced apoptosis in the kidney of young chickens. We found consistent evidence that dietary $NiCl_2$ in excess of 300 mg/kg had adverse effects on the kidney cells. The results showed that $NiCl_2$ significantly increased the TUNEL-postive cells, which were regarded as the apoptotic cells. The histopathological changes also showed that $NiCl_2$ increased apoptosis in the kidney. Our previous studies have proved that $NiCl_2$ induced apoptosis in the thymus, cecal tonsil and spleen [18,21,40]. Also, our findings are in agreement with the the the results of Ma et al. and Hossain et al. [41,42] who demonstrate that nickel nanowires (Ni NWs) induce apoptosis in Hela cells and human pancreatic adenocarcinoma cells. Also, Gathwan et al. [43] have reported that $NiCl_2$ can cause apoptotis in liver of male mice.

Mitochondria play an important role in the regulation of cell apoptosis [35]. Changes in the MMP are considered an early event in apoptosis and many proapoptotic proteins can be released from the mitochondria into the cytoplasm when the MMP is damaged [44]. We investigated whether mitochondria were involved in $NiCl_2$-induced apoptosis in broilers. As shown by flow cytometry analysis, $NiCl_2$ reduced the MMP in the kidney. Moreover, AIF and EndoG protein and mRNA expression were increased after disruption of the MMP. AIF and EndoG are thought to be mitochondrial cysteine proteases, whose release can be blocked by Bcl-2. After the MMP is damaged, AIF and EndoG can translocate from mitochondria to the nucleus, and there cause DNA fragmentation and cleavage of genomic DNA without being activated by caspases [45,46]. The data here shows that $NiCl_2$ induces apoptosis through a mitochondria-mediated pathway in the kidney. Previous studies show that nickel acetate can leads to apoptosis via a mitochondrial-mediated pathway [39]. Also, Zhao et al. [47] have reported that nickel nanoparticles (Ni NPs) and nickel fine particles induce changes in the MMP, and up-regulate and mobilize AIF from mitochondria to cytoplasm in JB6 cells.

Recently, it has been reported that regulation of mitochondrial-mediated apoptosis requires the involvement of the Bcl-2 family proteins [48]. The anti-apoptotic members of the Bcl-2 family proteins tend to stabilize the barrier function of mitochondrial membranes, whereas pro-apoptotic members of Bcl-2 family proteins destabilize this function [36]. Loss of MMP is a prerequisite for mitochondrial-mediated apoptosis as it is associated with the reshuffling of Bcl-2 family proteins [49]. The ratio between anti- and pro-apoptotic proteins is a determinant of tissue homeostasis because it influences the sensitivity of cells to apoptosis [35,50]. In this study, $NiCl_2$ increased pro-apoptotic Bax and Bak protein and mRNA expression, and concomitantly decreased anti-apoptotic Bcl-2, Bcl-xL protein and mRNA expression, which caused a significant increase in the Bax/Bcl-2 ratio, and then promoted apoptosis. Our previous studies have also shown that $NiCl_2$ increases the mRNA expression levels of Bax and decreases the mRNA expression levels of Bcl-2 in the thymus, cecal tonsil and spleen [18,21,40].

The PI3K/Akt pathway also plays a role in the regulation of Bcl-2 family proteins, which are believed to be important targets for anti-apoptosis [29,51]. The activation of PI3K/Akt pathway leads to increase in Bcl-2, Bcl-xL expression, and decrease in Bad, Bax expression [30,52–54]. We have investigated whether the PI3K/Akt pathway is involved in $NiCl_2$-induced mitochondria-mediated apoptosis. The results showed that dietary $NiCl_2$ reduced the expression of PI3K and Akt mRNA, implying that inhibition of the PI3K/Akt pathway is one mechanism of $NiCl_2$-induced apoptosis. This implication is consistent with the report of Liu et al. [55] that $NiCl_2$ induces apoptosis and decreases PI3K, p-Akt and Bcl-2 protein expression and increases Bax protein expression in the liver of Kunming mice. Our results are also in agreement with the results of Wang et al. [39] in which nickel acetate-induced apoptosis is characterized by decreasing the protein expression of Bcl-2 and Bcl-xL and increasing the protein expression of Bad, Bcl-Xs, Bax, cytochrome c and caspases 9, 3 and 6 in human proximal tubule cells. Pan et al. [56] suggest that Ni_3S_2 can down-regulate several anti-apoptotic proteins, including Bcl-2 and Bcl-xL in human bronchial epithelial (BEAS-2B) cells. Contrary to our results and the above-mentioned references, Shi [54] suggests that $NiCl_2$ up-regulates expression of Akt, Bcl-2, and Bcl-$_X$L in BEAS-2B cells.

Based on the results of our study and the discussion above, the mechanism of $NiCl_2$-caused mitochondria-mediated apoptosis in the kidney is summarized in Figure 14.

$NiCl_2$ modulates the Bcl-2 family proteins and inhibits the PI3K/Akt pathway. This action is followed by MMP disruption, which increases AIF and EndoG protein and mRNA expression. The highly expressed AIF and EndoG translocate from mitochondria to the nucleus and there causes DNA damage, which finally leads to renal-cell apoptosis.

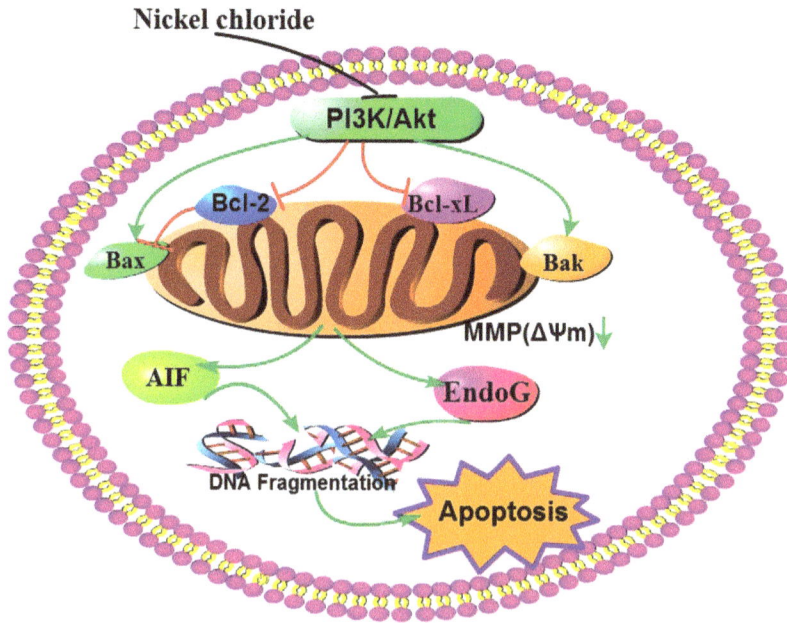

Figure 14. Schematic diagram of NiCl$_2$-caused mitochondria-mediated apoptosis.

4. Experimental Section

4.1. Animals and Treatment

Two hundred and eighty one-day-old healthy broilers (Chia Tai Group, Wenjiang, Sichuan, China) were divided into four groups. There were seventy broilers in each group. Broilers were housed in cages with electrical heaters, and provided with water as well as under-mentioned experimental diets *ad libitum* for 42 days. The growth cycle of commercial broilers is about 42 days, after which they are used for consumption. In this rapid growth period food consumption is high, and broilers will easily be affected by diet containing metal pollutants (such as Ni). The aim of our study is to evaluate the effect of dietary NiCl$_2$ on the broilers in this period of rapid growth.

To observe the time-dependent dynamic change, we chose three time points (14, 28 and 42 days of age) for examining histopathological injury, the alterations of apoptosis, mitochondrial membrane potential (MMP), apoptotic protein expression and mRNA expression levels.

In this study, a corn-soybean basal diet formulated by the National Research Council [57] was the control diet. NiCl$_2$ (NiCl$_2$·6H$_2$O, ChengDu Kelong Chemical Co., Ltd., Chengdu, China) was mixed into the corn-soybean basal diet to produce the experimental diets containing 300, 600 and 900 mg/kg NiCl$_2$, respectively.

The basis of doses (300, 600 and 900 mg/kg NiCl$_2$) selection: Ling and Leach reported that dietary NiCl$_2$ concentrations of 300 mg/kg and over resulted in significant reduction in growth rate. Mortality and anemia were observed in chicks receiving 1100 mg/kg nickel [58]. Weber and Reid found a significant growth reduction at 700 mg/kg NiSO$_4$ and nickel acetate and over [59]. Chicks fed more than 250–300 mg/kg Ni in the diet exhibited depressed growth and reduced feed intake [60]. Bersenyi et al. [61] reported that supplementation of 500 mg/kg NiCl$_2$ reduced weight gain (by 10%), feed intake (by 4%) and worse feed conversion efficiency (FCE) (by 5%) in growing broiler cockerels. According to the above-mentioned research results and our preliminary experiment, we chose the doses of 300, 600 and 900 mg/kg NiCl$_2$ in this study for observing the does-dependent changes.

Our experiments involving the use of broilers, and all experimental procedures were approved by Animal Care and Use Committee, Sichuan Agricultural University (Approval No: 2012-024).

4.2. Histopathological Examination of Kidney

Five chickens in each group were humanely killed at 14, 28 and 42 days of age. Kidneys were removed, fixed in 4% paraformaldehyde, dehydrated in ethanol and embedded in paraffin. Serial slices at 5 mm thickness were prepared and stained with haematoxylin and eosin (H&E), and examined by light microscopy.

4.3. Detection of Renal Apoptosis by TUNEL

Five broilers in each group were humanely sacrificed at 14, 28, and 42 days of age. Kidneys were removed, fixed in 4% paraformaldehyde, dehydrated in ethanol and embedded in paraffin.

TUNEL analysis was carried out according to the manual of In Situ Cell Death Detection Kit (Cat: 11684817980, Roche, Mannheim, Germany). Briefly, tissue sections (5 µm thick) were rehydrated in a series of xylene and ethanol solutions and then rinsed in ddH_2O, digested with 50 µL proteinase K (diluted in Tris·HCl pH 7.8) for 15 min, and incubated with 3% H_2O_2 in methanol for 15 min at room temperature to inactivate endogenous peroxidase. The sections were transferred to a reaction mixture containing biotin-dUTP terminal deoxynucleotidyl and incubated in a humidified chamber for 1 h at 37 °C, followed by washing in phosphate buffer saline (PBS), pH 7.2–7.4. Sections were incubated in Converter-POD (HRP) for 30 min at 37 °C. Reaction product was visualized with DAB kit (AR1022, Boster, Wuhan, China). After final washing in ddH_2O, slices were lightly counterstained with hematoxylin, dehydrated in ethanol, cleared in xylene and mounted.

Cells were observed with light microscopy (Olympus, Shimadzu, Japan). The nuclei of apoptotic cells containing DNA strand breaks were stained brown. The TUNEL positive cells (apoptotic cells) were counted by use of a computer-supported imaging system connected to a light microscope with an objective magnification of 1000×. Apoptotic cells were quantified by use of Image-Pro Plus 5.1 (Madia Cybernetics, Bethesda, MD, USA) image analysis software. Five sections in each chicken and five fields in each section were measured and averaged.

4.4. Detection of Mitochondrial Membrane Potential (ΔΨm) in the Kidney by Flow Cytometry

Mitochondrial membrane potential was measured with a mitochondrial membrane potential detection kit (Cat: 551302, Lot: 3242965, BD, Franklin lakes, NJ, USA).

At 14, 28 and 42 days of age, five broilers in each group were humanely killed, and the kidneys were immediately removed and ground to form a cell suspension, which was filtered through a 300-mesh nylon screen. The cells were washed twice with ice-cold PBS (pH 7.2–7.4) and suspended in PBS at a concentration of 1×10^6 cells/mL. One milliliter of the cell suspension was transferred to a 5-mL culture tube. The collected cells were incubated with JC-1 (5,5′,6,6′-tetra-chloro-1,1′,3,3′-tetra-ethylbenzimidalyl-carbocyanineiodide) working solution for 15 min in a 37 °C, 5% CO_2 incubator. The staining solution was removed and the cells were washed twice with JC-1 staining buffer. The cell-associated fluorescence was measured with flow cytometry (BD, Franklin Lakes, NJ, USA). Normal ΔΨm produces red fluorescence for JC-1 aggregates, but loss of the ΔΨm results in the disaggregation of JC-1 and produces green fluorescence.

4.5. Detection of Protein Expression in the Kidney by Immunohistochemistry

Five chickens in each group were humanely sacrificed for gross examination at 14, 28 and 42 days of age. Kidneys were collected and fixed in 10% neutral buffered formalin, and then processed, trimmed, and embedded in paraffin wax.

The method used was that described by Wu et al. [21]. Tissue slices were dewaxed in xylene, rehydrated through a graded series of ethanol solutions, washed in distilled water and PBS and endogenous peroxidase activity was blocked by incubation with 3% H_2O_2 in methanol for 15 min.

The sections were subjected to antigen retrieval procedure by microwaving in 0.01 M sodium citrate buffer pH 6.0. Additional washing in PBS was performed before 30 min of incubation at 37 °C in 10% normal goat serum (Boster). The slices were incubated overnight at 4 °C with anti-Bax (1:400) (Cat: 14796, Cell Signaling Technology, Danvers, MA, USA); anti-Bak (1:400) (Cat: 12105, Cell Signaling Technology); anti-Bcl-xL (1:300) (Cat: 2764, Cell Signaling Technology, anti-Bcl-2 (1:400) (Cat: 15071, Cell Signaling Technology); anti-AIF(1:200) (Cat: sc-9416, Santa Cruz Biotechnology, Dallas, TX, USA); anti-EndoG (1:100), (Cat: orb6003, Biorbyt, San Francisco, CA, USA). After washing in PBS, the slices were exposed to 1% biotinylated goat anti-mouse IgG secondary antibody (Boster) for 1 h at 37 °C, and then incubated with strept avidin-biotin complex (SABC; Boster) for 30 min at 37 °C. To visualize the immunoreaction, sections were immersed in diaminobenzidine hydrochloride (DAB; Boster). The slices were monitored microscopically and stopped by immersion in distilled water, as soon as brown staining was visible. Slices were lightly counterstained with hematoxylin, dehydrated in ethanol, cleared in xylene and mounted.

The protein expression levels were measured using a computer-supported imaging system connected to a light microscope (OlympusAX70) with an objective magnification of 400×. The intensity of staining for each protein was quantified using Image-pro Plus 5.1 (Madia Cybernetics, Bethesda, MD, USA). Each group was measured in five sections and each section was measured using five views, and averaged.

4.6. Detection of mRNA Expression in the Kidney by qRT-PCR

The kidneys from five chickens in each group were taken at 14, 28, and 42 days of age and stored in liquid nitrogen. They kidneys were homogenized in liquid nitrogen with a mortar and pestle.

As described [21], total RNA was extracted from the frozen kidney powders with RNAiso Plus (9108/9109, Takara, Dalian, China) according to the manufacturer's protocol. Next, cDNA was synthesized with a Prim-Script™ RT reagent Kit (RR047A, Takara, Japan) according to the manufacturer's protocol. The cDNA product was used as a template for qRT-PCR analysis. Sequences for target genes were obtained from the NCBI database. Oligonucleotide primers were designed by use of Primer 5 software and synthesized at Takara (Dalian, China), as shown in Table 1.

Table 1. A list of primers in qRT-PCR analysis of mRNA expression of the apoptotic proteins.

Gene Symbol	Accession Number	Primer	Primer Sequence (5'–3')	Product Size (bp)	Tm (°C)
PI3K	NM001004410	F	CGGATGTTGCCTTACGGTTGT	162	58
		R	GTTCTTGTCCTTGAGCCACTGAT		
Akt1	AF039943	F	TGATGGCACATTCATTGGCTAC	122	58
		R	TGTTTGGTTTAGGTCGTTCTGTCT		
Akt2	AF181260	F	CCGAAGTGCTGGAGGACAAC	115	60
		R	CGCTCGTGGTCCTGGTTGTA		
Bcl-2	NM205339	F	GATGACCGAGTACCTGAACC	114	61
		R	CAGGAGAAATCGAACAAAGGC		
Bax	XM422067	F	TCCTCATCGCCATGCTCAT	69	62
		R	CCTTGGTCTGGAAGCAGAAGA		
Bak	NM001030920	F	TCTACCAGCAAGGCATCACGG	122	60
		R	ATCGAGTGCAGCCACCCATC		
Bcl-xL	GU230783	F	ATGAGTTTGAGCTGAGGTACCGG	150	59
		R	AGAAGAAAGCCACGATGCGC		
AIF	NM001007490	F	CTGGGTCCTGATGTGGGCTAT	123	58
		R	TGTCCCTGACTGCTCTGTTGC		
EndoG	XM415487	F	TGCCTGGAATAACCTTGAGAAATAC	170	61
		R	TGAAGAAATGGGTAGGGACGG		
β-actin	L08165	F	TGCTGTGTTCCCATCTATCG	178	62
		R	TTGGTGACAATACCGTGTTCA		

All qRT-PCR were performed by use of the SYBR® Premix Ex Taq™ II system (DRR820A, Takara, Japan) with on a Model C1000 Thermal Cycler (Bio Rad, Hercules, CA, USA).

Chicken β-actin expression was used as an internal reference housekeeping gene. Gene expression values from control group subsamples at 14, 28, and 42 days of age were used to calibrate gene expression in subsamples from corresponding experimental subsamples. All data output from the qRT-PCR experiments were analyzed by use of the $2^{-\Delta\Delta Ct}$ method [62].

4.7. Statistical Analysis

The significance of difference among the four groups of broiler chicks was assessed with variance analysis, and results were presented as mean ± standard deviation (M ± SD). The variation was measured by use of one-way analysis of variance (ANOVA) test of SPSS 16.0 for windows. $p < 0.05$ was considered statistical significance.

5. Conclusions

In chicks, dietary $NiCl_2$ in excess of 300 mg/kg was found to cause mitochondrial-mediated apoptosis by disruption of MMP and increased expression of AIF and EndoG mRNA and protein. These results indicate that the underlying mechanism of MMP disruption involves Bcl-2 family protein modulation and PI3K/Akt pathway inhibition.

Acknowledgments: The study was supported by the program for Changjiang scholars and innovative research team in university (IRT 0848) and the Shuangzhi project of Sichuan Agricultural University (03570327; 03571189).

Author Contributions: Hongrui Guo and Hengmin Cui designed the experiments. Hongrui Guo, Bangyuan Wu, Kejie Chen and Jie Deng carried out the experiments. Hongrui Guo, Hengmin Cui, Xi Peng, Jing Fang, Zhicai Zuo, Junliang Deng, Xun Wang, Bangyuan Wu, Kejie Chen and Jie Deng analyzed and interpreted data. Hongrui Guo and Hengmin Cui concluded the scientific findings and wrote and revised the manuscript.

Conflicts of Interest: The authors declare no conflict of interest.

References

1. Poonkothai, M.; Vijayavathi, B.S. Nickel as an essential element and a toxicant. *Int. J. Environ. Sci.* **2012**, *1*, 285–288.
2. Pasanen, K.; Pukkala, E.; Turunen, A.W.; Patama, T.; Jussila, I.; Makkonen, S.; Salonen, R.O.; Verkasalo, P.K. Mortality among population with exposure to industrial air pollution containing nickel and other toxic metals. *J. Occup. Environ. Med.* **2012**, *54*, 583–591. [CrossRef] [PubMed]
3. Cempel, M.; Nikel, G. Nickel: A review of its sources and environmental toxicology. *Pol. J. Environ. Stud.* **2006**, *15*, 375–382.
4. Lu, H.; Shi, X.; Costa, M.; Huang, C. Carcinogenic effect of nickel compounds. *Mol. Cell. Biochem.* **2005**, *279*, 45–67. [CrossRef] [PubMed]
5. Kubrak, O.I.; Husak, V.V.; Rovenko, B.M.; Poigner, H.; Kriews, M.; Abele, D.; Lushchak, V.I. Antioxidant system efficiently protects goldfish gills from Ni^{2+}-induced oxidative stress. *Chemosphere* **2013**, *90*, 971–976. [CrossRef] [PubMed]
6. Alarifi, S.; Ali, D.; Alakhtani, S.; Al Suhaibani, E.S.; Al-Qahtani, A.A. Reactive oxygen species-mediated DNA damage and apoptosis in human skin epidermal cells after exposure to nickel nanoparticles. *Biol. Trace Elem. Res.* **2014**, *157*, 84–93. [CrossRef] [PubMed]
7. Das, K.K.; Das, S.N.; Dhundasi, S.A. Nickel, its adverse health effects & oxidative stress. *Indian J. Med. Res.* **2008**, *128*, 412–425. [PubMed]
8. Uddin, A.N.; Burns, F.J.; Rossman, T.G.; Chen, H.; Kluz, T.; Costa, M. Dietary chromium and nickel enhance UV-carcinogenesis in skin of hairless mice. *Toxicol. Appl. Pharm.* **2007**, *221*, 329–338. [CrossRef] [PubMed]
9. Costa, M.; Klein, C.B. Nickel carcinogenesis, mutation, epigenetics, or selection. *Environ. Health Persp.* **1999**, *107*, A438–A439. [CrossRef]
10. Denkhausa E, S.K. Nickel essentiality, toxicity, and carcinogenicity. *Crit. Rev. Oncol. Hematol.* **2002**, *42*, 35–36. [CrossRef]
11. Goodman, J.E.; Prueitt, R.L.; Dodge, D.G.; Thakali, S. Carcinogenicity assessment of water-soluble nickel compounds. *Crit. Rev. Toxicol.* **2009**, *39*, 365–417. [CrossRef] [PubMed]

12. Zheng, G.H.; Liu, C.M.; Sun, J.M.; Feng, Z.J.; Cheng, C. Nickel-induced oxidative stress and apoptosis in Carassius auratus liver by JNK pathway. *Aquat. Toxicol.* **2014**, *147*, 105–111. [CrossRef] [PubMed]

13. Salnikow, K.; Costa, M. Epigenetic mechanisms of nickel carcinogenesis. *J. Environ. Pathol. Toxicol. Oncol.* **1999**, *19*, 307–318.

14. Chen, C.-Y.; Lin, T.-K.; Chang, Y.-C.; Wang, Y.-F.; Shyu, H.-W.; Lin, K.-H.; Chou, M.-C. Nickel(II)-induced oxidative stress, apoptosis, G2/M arrest, and genotoxicity in normal rat kidney cells. *J. Toxicol. Environ. Health A* **2010**, *73*, 529–539. [CrossRef] [PubMed]

15. Şaplakoğlu, U.; İşcan, M.; İşcan, M. DNA single-strand breakage in rat lung, liver and kidney after single and combined treatments of nickel and cadmium. *Mutat. Res.* **1997**, *394*, 133–140. [CrossRef]

16. Dally, H.; Hartwig, A. Induction and repair inhibition of oxidative DNA damage by nickel (II) and cadmium (II) in mammalian cells. *Carcinogenesis* **1997**, *18*, 1021–1026. [CrossRef] [PubMed]

17. Efremenko, A.Y.; Campbell, J.L., Jr.; Dodd, D.E.; Oller, A.R.; Clewell, H.J., 3rd. Time- and concentration-dependent genomic responses of the rat airway to inhaled nickel subsulfide. *Toxicol. Appl. Pharmacol.* **2014**, *279*, 441–454. [CrossRef] [PubMed]

18. Tang, K.; Guo, H.; Deng, J.; Cui, H.; Peng, X.; Fang, J.; Zuo, Z.; Wang, X.; Wu, B.; Li, J.; Yin, S. Inhibitive effects of nickel chloride (NiCl$_2$) on thymocytes. *Biol. Trace Elem. Res.* **2015**, *164*, 242–252. [CrossRef] [PubMed]

19. Wu, B.; Cui, H.; Peng, X.; Fang, J.; Zuo, Z.; Deng, J.; Wang, X.; Huang, J. Toxicological effects of nickel chloride on the cytokine mRNA expression and protein levels in intestinal mucosal immunity of broilers. *Environ. Toxicol.* **2014**. [CrossRef] [PubMed]

20. Wu, B.; Cui, H.; Peng, X.; Fang, J.; Zuo, Z.; Deng, J.; Huang, J. toxicological effects of nickel chloride on IgA$^+$ B cells and sIgA, IgA, IgG, IgM in the intestinal mucosal immunity in broilers. *Int. J. Environ. Res. Public Health* **2014**, *11*, 8175–8192. [CrossRef] [PubMed]

21. Wu, B.; Cui, H.; Peng, X.; Fang, J.; Zuo, Z.; Deng, J.; Huang, J. Dietary nickel chloride induces oxidative stress, apoptosis and alters Bax/Bcl-2 and caspase-3 mRNA expression in the cecal tonsil of broilers. *Food Chem. Toxicol.* **2014**, *63*, 18–29. [CrossRef] [PubMed]

22. Wu, B.; Cui, H.; Peng, X.; Fang, J.; Zuo, Z.; Deng, J.; Huang, J. Analysis of the Toll-like receptor 2-2 (TLR2-2) and TLR4 mRNA expression in the intestinal mucosal immunity of broilers fed on diets supplemented with nickel chloride. *Int. J. Environ. Res. Public Health* **2014**, *11*, 657–670. [CrossRef] [PubMed]

23. Tang, K.; Li, J.; Yin, S.; Guo, H.; Deng, J.; Cui, H. Effects of nickel chloride on histopathological lesions and oxidative damage in the thymus. *Health* **2014**, *6*, 2875. [CrossRef]

24. Huang, J.; Cui, H.; Peng, X.; Fang, J.; Zuo, Z.; Deng, J.; Wang, X.; Wu, B. Effect of dietary nickel chloride on splenic immune function in broilers. *Biol. Trace Elem. Res.* **2014**, *159*, 183–191. [CrossRef] [PubMed]

25. Huang, J.; Cui, H.; Peng, X.; Fang, J.; Zuo, Z.; Deng, J.; Wang, X.; Wu, B. Downregulation of TLR4 and 7 mRNA expression levels in broiler's spleen caused by diets supplemented with nickel chloride. *Biol. Trace Elem. Res.* **2014**, *158*, 353–358. [CrossRef] [PubMed]

26. Guo, H.; Wu, B.; Cui, H.; Peng, X.; Fang, J.; Zuo, Z.; Deng, J.; Wang, X.; Deng, J.; Yin, S.; Li, J.; Tang, K. NiCl$_2$-down-regulated antioxidant enzyme mRNA expression causes oxidative damage in the broiler's kidney. *Biol. Trace Elem. Res.* **2014**, *162*, 288–295. [CrossRef] [PubMed]

27. Wu, B.; Cui, H.; Peng, X.; Fang, J.; Zuo, Z.; Huang, J.; Luo, Q.; Deng, Y.; Wang, H.; Liu, J. Changes of the serum cytokine contents in broilers fed on diets supplemented with nickel chloride. *Biol. Trace Elem. Res.* **2013**, *151*, 234–239. [CrossRef] [PubMed]

28. Wu, B.; Cui, H.; Peng, X.; Fang, J.; Zuo, Z.; Deng, J.; Huang, J. Dietary nickel chloride induces oxidative intestinal damage in broilers. *Int. J. Environ. Res. Public Health* **2013**, *10*, 2109–2119. [CrossRef] [PubMed]

29. Porta, C.; Figlin, R.A. Phosphatidylinositol-3-kinase/Akt signaling pathway and kidney cancer, and the therapeutic potential of phosphatidylinositol-3-kinase/Akt inhibitors. *J. Urol.* **2009**, *182*, 2569–2577. [CrossRef] [PubMed]

30. Franke, T.F.; Hornik, C.P.; Segev, L.; Shostak, G.A.; Sugimoto, C. PI3K/Akt and apoptosis: Size matters. *Oncogene* **2003**, *22*, 8983–8998. [CrossRef] [PubMed]

31. Liu, C.M.; Ma, J.Q.; Sun, Y.Z. Puerarin protects rat kidney from lead-induced apoptosis by modulating the PI3K/Akt/eNOS pathway. *Toxicol. Appl. Pharmacol.* **2012**, *258*, 330–342. [CrossRef] [PubMed]

32. Datta, S.R.; Dudek, H.; Tao, X.; Masters, S.; Fu, H.; Gotoh, Y.; Greenberg, M.E. Akt phosphorylation of BAD couples survival signals to the cell-intrinsic death machinery. *Cell* **1997**, *91*, 231–41. [CrossRef]

33. Harris, G.K.; Shi, X. Signaling by carcinogenic metals and metal-induced reactive oxygen species. *Mutat. Res.* **2003**, *533*, 183–200. [CrossRef] [PubMed]

34. Rana, S.V. Metals and apoptosis: Recent developments. *J. Trace Elem. Med. Biol.* **2008**, *22*, 262–84. [CrossRef] [PubMed]

35. Martinou, J.C.; Youle, R.J. Mitochondria in apoptosis: Bcl-2 family members and mitochondrial dynamics. *Dev. Cell* **2011**, *21*, 92–101. [CrossRef] [PubMed]

36. Cory, S.; Adams, J.M. The Bcl2 family: Regulators of the cellular life-or-death switch. *Nat. Rev. Cancer* **2002**, *2*, 647–656. [CrossRef] [PubMed]

37. Gross, A.; McDonnell, J.M.; Korsmeyer, S.J. Bcl-2 family members and the mitochondria in apoptosis. *Gene Dev.* **1999**, *13*, 1899–1911. [CrossRef] [PubMed]

38. Budihardjo, I.; Oliver, H.; Lutter, M.; Luo, X.; Wang, X. Biochemical pathways of caspase activation during apoptosis. *Annu. Rev. Cell Dev. Biol.* **1999**, *15*, 269–290. [CrossRef] [PubMed]

39. Wang, Y.F.; Shyu, H.W.; Chang, Y.C.; Tseng, W.C.; Huang, Y.L.; Lin, K.H.; Chou, M.C.; Liu, H.L.; Chen, C.Y. Nickel (II)-induced cytotoxicity and apoptosis in human proximal tubule cells through a ROS- and mitochondria-mediated pathway. *Toxicol. Appl. Pharmacol.* **2012**, *259*, 177–186. [CrossRef] [PubMed]

40. Huang, J.; Cui, H.; Peng, X.; Fang, J.; Zuo, Z.; Deng, J.; Wu, B. The association between splenocyte apoptosis and alterations of Bax, Bcl-2 and Caspase-3 mRNA expression, and oxidative stress induced by dietary nickel chloride in broilers. *Int. J. Environ. Res. Public Health* **2013**, *10*, 7310–7326. [CrossRef] [PubMed]

41. Ma, C.; Song, M.; Zhang, Y.; Yan, M.; Zhang, M.; Bi, H. Nickel nanowires induce cell cycle arrest and apoptosis by generation of reactive oxygen species in HeLa cells. *Toxicol. Rep.* **2014**, *1*, 114–121. [CrossRef]

42. Hossain, M.Z.; Kleve, M.G. Nickel nanowires induced and reactive oxygen species mediated apoptosis in human pancreatic adenocarcinoma cells. *Int. J. Nanomed.* **2011**, *6*, 1475. [CrossRef] [PubMed]

43. Gathwan, K.H.; Al-Karkhi, I.H.T.; Jaffar al-Mulla, E.A. Hepatic toxicity of nickel chloride in mice. *Res. Chem. Intermed.* **2012**, *39*, 2537–2542. [CrossRef]

44. Vaux, D.L. Apoptogenic factors released from mitochondria. *Biochim. Biophys. Acta* **2011**, *1813*, 546–550. [CrossRef] [PubMed]

45. Susin, S.A.; Zamzami, N.; Castedo, M.; Hirsch, T.; Marchetti, P.; Macho, A.; Daugas, E.; Geuskens, M.; Kroemer, G. Bcl-2 inhibits the mitochondrial release of an apoptogenic protease. *J. Exp. Med.* **1996**, *184*, 1331–41. [CrossRef] [PubMed]

46. Li, L.Y.; Wang, X.L.X. Endonuclease G is an apoptotic DNase when released from mitochondria. *Nature* **2001**, *412*, 95–99. [CrossRef] [PubMed]

47. Zhao, J.; Bowman, L.; Zhang, X.; Shi, X.; Jiang, B.; Castranova, V.; Ding, M. Metallic nickel nano- and fine particles induce JB6 cell apoptosis through a caspase-8/AIF mediated cytochrome c-independent pathway. *J. Nanobiotechnol.* **2009**, *7*, 2. [CrossRef] [PubMed]

48. Brunelle, J.K.; Letai, A. Control of mitochondrial apoptosis by the Bcl-2 family. *J. Cell Sci.* **2009**, *122*, 437–441. [CrossRef] [PubMed]

49. Steinbach, J.P.; Weller, M. Apoptosis in gliomas: Molecular mechanisms and therapeutic implications. *J. Neurooncol.* **2004**, *70*, 247–256. [CrossRef] [PubMed]

50. Korsmeyer, S.J. BCL-2 gene family and the regulation of programmed cell death. *Cancer Res.* **1999**, *59*, 1693s–1700s. [CrossRef]

51. Osaki, M.; Oshimura, M.A.; Ito, H. PI3K-Akt pathway: Its functions and alterations in human cancer. *Apoptosis* **2004**, *9*, 667–676. [CrossRef] [PubMed]

52. Aziz, M.H.; Nihal, M.; Fu, V.X.; Jarrard, D.F.; Ahmad, N. Resveratrol-caused apoptosis of human prostate carcinoma LNCaP cells is mediated via modulation of phosphatidylinositol 3′-kinase/Akt pathway and Bcl-2 family proteins. *Mol. Cancer Ther.* **2006**, *5*, 1335–1341. [CrossRef] [PubMed]

53. Hu, L.; Sun, Y.; Hu, J. Catalpol inhibits apoptosis in hydrogen peroxide-induced endothelium by activating the PI3K/Akt signaling pathway and modulating expression of Bcl-2 and Bax. *Eur. J. Pharmacol.* **2010**, *628*, 155–163. [CrossRef] [PubMed]

54. Pan, J.J.; Chang, Q.S.; Wang, X.; Son, Y.O.; Liu, J.; Zhang, Z.; Bi, Y.Y.; Shi, X. Activation of Akt/GSK3beta and Akt/Bcl-2 signaling pathways in nickel-transformed BEAS-2B cells. *Int. J. Oncol.* **2011**, *39*, 1285–1294. [PubMed]

55. Liu, C.M.; Zheng, G.H.; Ming, Q.L.; Chao, C.; Sun, J.M. Sesamin protects mouse liver against nickel-induced oxidative DNA damage and apoptosis by the PI3K-Akt pathway. *J. Agric. Food Chem.* **2013**, *61*, 1146–1154. [CrossRef] [PubMed]

56. Pan, J.; Chang, Q.; Wang, X.; Son, Y.; Zhang, Z.; Chen, G.; Luo, J.; Bi, Y.; Chen, F.; Shi, X. Reactive oxygen species-activated Akt/ASK1/p38 signaling pathway in nickel compound-induced apoptosis in BEAS 2B cells. *Chem. Res. Toxicol.* **2010**, *23*, 568–577. [CrossRef] [PubMed]

57. National Research Council (NRC). *Nutrient Requirements of Poultry*, 9th ed.; National Academy Press: Washington, DC, USA, 1994.

58. Ling, J.; Leach, R. Studies on nickel metabolism: Interaction with other mineral elements. *Poult. Sci.* **1979**, *58*, 591–596. [CrossRef] [PubMed]

59. Weber, C.W.; Reid, B.L. Nickel toxicity in growing chicks. *J. Nutr.* **1968**, *95*, 612–616. [PubMed]

60. Szilagyi, M.; Szentmihalyi, S.; Anke, M. Changes in some of the biochemical parameters in Ni and Mo deficient animals [goat, sheep, pig, chicken, rat]. *Proc. (Hungary).* 1981, Volume 1, pp. 257–260, International System for Agricultural Science and Technology. Available online: http://agris.fao.org/agris-search/search.do?recordID=HU8200908 (accessed on 24 April 2015).

61. Bersényi, A.; Fekete, S.G.; Szilágyi, M.; Berta, E.; Zöldág, L.; Glávits, R. Effects of nickel supply on the fattening performance and several biochemical parameters of broiler chickens and rabbits. *Acta Vet. Hung.* **2004**, *52*, 185–197. [CrossRef] [PubMed]

62. Livak, K.J.; Schmittgen, T.D. Analysis of relative gene expression data using real-time quantitative PCR and the $2^{-\Delta\Delta Ct}$ method. *Methods* **2001**, *25*, 402–408. [CrossRef] [PubMed]

International Journal of
Molecular Sciences

MDPI

Review

Cadmium Protection Strategies—A Hidden Trade-Off?

Adolf Michael Sandbichler and Martina Höckner *

University of Innsbruck, Institute of Zoology, Technikerstraße 25, 6020 Innsbruck, Austria;
adolf.sandbichler@uibk.ac.at
* Correspondence: martina.hoeckner@uibk.ac.at; Tel.: +43-512-507 (ext. 51878); Fax: +43-512-507-2930

Academic Editor: Reinhard Dallinger
Received: 28 November 2015; Accepted: 18 January 2016; Published: 21 January 2016

Abstract: Cadmium (Cd) is a non-essential transition metal which is introduced into the biosphere by various anthropogenic activities. Environmental pollution with Cd poses a major health risk and Cd toxicity has been extensively researched over the past decades. This review aims at changing the perspective by discussing protection mechanisms available to counteract a Cd insult. Antioxidants, induction of antioxidant enzymes, and complexation of Cd to glutathione (GSH) and metallothionein (MT) are the most potent protective measures to cope with Cd-induced oxidative stress. Furthermore, protection mechanisms include prevention of endoplasmic reticulum (ER) stress, mitophagy and metabolic stress, as well as expression of chaperones. Pre-exposure to Cd itself, or co-exposure to other metals or trace elements can improve viability under Cd exposure and cells have means to reduce Cd uptake and improve Cd removal. Finally, environmental factors have negative or positive effects on Cd toxicity. Most protection mechanisms aim at preventing cellular damage. However, this might not be possible without trade-offs like an increased risk of carcinogenesis.

Keywords: cadmium; protection; detoxification; antioxidants; chelation; metallothionein; hormesis; oxidative stress; mitochondrial damage; nuclear response factor 2 signaling

1. Introduction

Over the last decades, several studies have looked into the toxic effects of cadmium (Cd) at cellular and organismic levels to assess the risk of increasing environmental pollution by heavy metals. Cd is a known carcinogenic and immunotoxic heavy metal. An estimated 30,000 tons of Cd are released into the environment each year. Cd is highly persistent in the environment and also enters the food chain [1]. Cd toxicity is mainly based on so-called ionic mimicry which is defined by the replacement of elements like calcium (Ca^{2+}) and trace elements like zinc by Cd^{2+} [2]. This can lead to protein mis- or unfolding and malfunction and eventually cause endoplasmic reticulum (ER) stress and cell death [3].

The induction of oxidative stress appears to be another indicator of the damaging mechanism of Cd as shown by a considerable body of evidence. This is caused indirectly, as Cd is not a redox active metal, through the depletion of the cells' major antioxidants and direct interference with active centers of the electron transport chain [4–6]. We discuss the protective measures employed at the cellular and organismic level when confronted with Cd. With this change in perspective from "what is damaged" to "how detrimental effects can be overcome or even bypassed", this review discusses protection strategies against Cd insult. We focus on the effect of Cd at the cellular level including results from *in vivo* studies where novel defense mechanisms are presented but detailed cellular explanations have yet to be found.

Available defense strategies against Cd are grouped according to their underlying mechanisms. These include antioxidant defense, mitochondrial protection, metal chelation, prevention of macromolecular damage, cytoskeletal rearrangements, hormetic response, co-exposure to other metals or trace elements, reduced uptake of Cd, removal of Cd, and toxicity of Cd altered by environmental factors.

In the current review, we summarize the variety of protective responses against Cd insult which are based on highly diverse mechanisms. However, when implemented, most of these defense strategies contain trade-offs like anti-apoptotic effects and risk of carcinogenesis.

2. Results and Discussion

2.1. Protection via Antioxidants

Cd is not able to produce radicals in Fenton type chemistry. Nonetheless, it induces oxidative stress through a multifaceted mechanism including the reduction of antioxidative defense and the production of reactive oxygen species (ROS) by mitochondrial damage (see Section 2.2).

Upon entry into the cell, Cd forms complexes with thiol residues from the tripeptide-reduced glutathione (GSH), the main intracellular antioxidative substance. GSH complexation with Cd^{2+} (termed GS-Cd) is considered a first line of defense since it prevents the heavy metal from causing further damage and in some cases enables active removal through specialized transporters (see Section 2.9) [7–9]. Due to the reduction of free GSH levels by Cd^{2+} binding, the cells redox balance is shifted to a more oxidized state and antioxidative defense is impaired. Interestingly, only recently a study on rat proximal tubule cells has shown the induction of GSH synthase subunit genes. As a protective response to Cd intoxication, GSH synthase recycles oxidized glutathione [10]. The same study also tested for chronic effects *in vivo* and found elevated gene expression for catalase (CAT), mitochondrial superoxide dismutase 2 (SOD), glutathione peroxidase 4, and peroxiredoxin 2 after daily subcutaneous Cd injections.

A second important redox system besides GSH/oxidized glutathione (GSSG) is the thioredoxin (Trx) system. The central enzyme Trx reductase (TrxR), a selenoprotein which recuperates reduced Trx using nicotinamide adenine dinucleotide phosphate (NADPH), can be induced by Cd to evoke a protective response. In bovine arterial endothelial cells, such Cd-induced expression of TrxR isoform 1 was mediated by nuclear response factor 2 (Nrf2) which binds to an antioxidative response element (ARE) in the promotor region of TrxR1 [11].

Other examples for the induction of antioxidative enzymes via ARE binding of Nrf2 include hemeoxygenase-1 and glutamate-cysteine ligase [12] or SOD [13].

Different natural compounds and phytochemicals have protective potential in Cd intoxication (Table 1). Many of the compounds tested are referred to as "natural antioxidants" but actually function as activators of Nrf2 leading to the upregulation of the antioxidant machinery [14]. Given these observations it is not surprising that Nrf2 signaling is believed to be an important regulator of cellular resistance to oxidants [15]. Indeed, upregulation of Nrf2 has also been shown to have negative effects: A growing body of evidence finds that cancer cells employ this mechanism to raise their resistance to oxidative stress, reprogram metabolism, and sustain cell proliferation [14]. Interestingly, Cd itself has only weak genotoxic effects but secondary carcinogenic effects and tissue damage can occur by way of oxidative stress [6,16–18]. Such carcinogenic damage can be reduced by a number of natural antioxidants (Table 1). However, if this includes Nrf2 activation, short-term amelioration of Cd-induced ROS may lead to carcinogenic effects in the long term. Ultimately, due to the direct inhibition of DNA repair enzymes such as the human 8-oxoguanine DNA *N*-glycosylase (hOGG1) by Cd, the carcinogenic potential of Cd is even potentiated by DNA changes [18,19].

Table 1. Protective natural compounds and phytochemicals against Cd intoxication.

Substance	Source	[Cd]/Duration/Experimental Animal	References
Curcumin [a,b]	Turmeric (*Curcuma longa* L.)	24 h Cd exposure, *in vivo*, rodents *In vitro*, human airway epithelial cells	[20–23]
Ginger	Ginger (*Zingiber officinale*)	200 mg/kg b.w., 12 weeks, *in vivo*, rabbits	[24]
Resveratrol [b]	Polyphenol from skin of grapes (*Vitis vinifera*)	7 mg/kg b.w., 24 h exposure, *in vivo*, mice	[21]
Physalis extract	*Physalis peruviana* L.	6.5 mg/kg b.w., 5 days, *in vivo*, rats	[25]
Grapefruit juice [a]	Grapefruit	1.5 mg/kg b.w., from day 7 of gestation until day 17 of pregnancy, *in vivo*, mice	[26]
Garlic extract or Allicin [b]	Garlic	5 or 10 ppm, 45 days, *in vivo*, Freshwater catfish (*Clarias batrachus*)	[27]
Royal jelly [a]	from Honey bees	2 mg/kg b.w., 6–7 weeks, *in vivo*, mice	[28]
Spirulina [a]	Micro-algae spirulina (*Arthrospira maxima*)	1.5 mg/kg b.w., 1 time Cd challenge, *in vivo*, pregnant mice; 3.5 mg/kg b.w., 1 time Cd intraperitoneal dose, *in vivo*, rats	[29,30]
Farnesol [a]	Isoprenoid from aromatic plants	5 mg/kg b.w., 1 time Cd, *in vivo*, mice	[31]
Theaflavin	Polyphenol from black tea (*Camellia sinensis*)	0.4 mg/kg b.w., once a day, for 5 weeks, *in vivo*, rats	[32]
Taxifolin	Bioflavonoid from conifers	100 μM Cd, *in vivo*, Zebrafish (*Danio rerio*)	[33]
Quercetin	Bioflavonoid from apples and onions	4 mg/kg b.w. for 2 weeks, *in vivo*, mice; 1.2 mg Cd/kg/day, 5 times/week during nine weeks, *in vivo*, rats 5 μM, *in vitro*, in cultured granulosa cells from chicken ovarian follicles	[34–36]
Naringenin	Bioflavonoid from grapefruit	5 mg/kg, orally for 4 weeks, *in vivo*, rats	[37]
Rosemary extract [b]	*Rosmarinus officinalis* L.	30 mg/kg b.w., 5 consecutive days/week for 8 weeks, *in vivo*, rats	[38]
Catechin [a,b]	Polyphenol from Green tea (*Camellia sinensis*)	50 ppm *ad libitum*, 20 weeks, *in vivo*, rats	[39]
Sulforaphane [a,b]	Isothiocyanate from cruciferous vegetables	*In vitro* in human hepatocytes and *in vivo* in mice; 0.2 mg/kg, 15 days, *in vivo*, rats	[40,41]

[a] shown to prevent Cd-induced genotoxic effects; [b] suspected to induce nuclear response factor 2 (Nrf2) signaling [14,42]; b.w. body weight.

Numerous studies have already shown the protective role of hormones like melatonin [43–45], antioxidative vitamins [27,40,46–49], and antioxidants such as *N*-acetylcysteine (NAC) [50–54].

In the following, we show different effects of two antioxidants, ascorbic acid (vitamin C, VC) and NAC, on Cd-impaired cell survival in a zebrafish embryonic fibroblast cell line (Z3) in order to assess the potential and putative differences of VC and NAC in the recovery from Cd-induced oxidative stress. Z3 cells were serum-deprived by incubation in Hank's buffered salt solution (HBSS), which is known to induce ROS and eventually lead to apoptosis [55]. In fact, cell density in HBSS-treated cells but also cells treated with cell culture media lacking fetal bovine serum (FBS) was decreased compared to cells incubated with complete media (Figure 1).

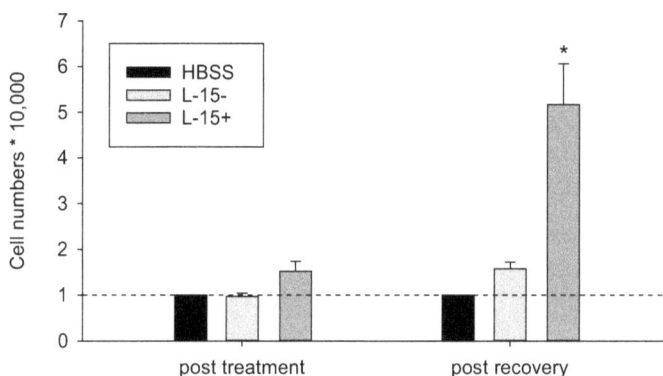

Figure 1. Cell density assay (Hoechst 33342) with Z3 zebrafish cells in control experiments using different culture media. Cell numbers were measured once after the 18 h treatment and once after the 6 h recovery period. L-15−: L-15 complete media without FBS L-15+: L-15 complete media. Cell numbers were normalized to 10,000 cells of the Hank's buffered salt solution (HBSS) control. Statistical analysis was performed prior to data normalization using a *t*-test. Groups were compared to HBSS treatment (* $p \leqslant 0.05$). Values are mean $\pm$ standard error from 3 biological replicates.

There is no general agreement on the effect of NAC on Cd toxicity since several studies suggest a protective effect on cell viability, e.g., in rat hepatocytes, when cells were co-exposed to Cd and NAC [53]. However, another study observes a cumulative toxic effect of NAC and Cd. In human HaCaT cells, only pre-treatment with NAC restored Cd-induced cell death which led the authors to the conclusion that Cd and NAC might form complexes with one another or with the culture media [56]. In Leydig cells, NAC pre-treatment also revealed decreased cell death via the reduction of oxidative damage [57], and in HepG2 cells, Cd-induced apoptosis could be reduced by NAC-dependent upregulation of catalase [58]. Another study reveals that NAC changes the expression of cytokines and chemokines and suggests that the immunomodulatory effect protects against Cd toxicity [59]. Studies on the protection mechanisms of NAC reveal that NAC increases phosphorylated p38 MAPK by decreasing the ROS level in a human osteosarcoma cell line [60]. Similarly, it has been found that in zebrafish embryo NAC protects against msh6 inhibition which is part of the DNA mismatch repair, most likely also by decreasing ROS [61].

Original data included in the present review article reveal that NAC is able to restore cell numbers of Z3 zebrafish cells upon HBSS starvation and Cd exposure (Figure 2). The experiments were conducted in HBSS to overcome putative problems caused by the formation of complexes between Cd^{2+} and components of the cell culture media, as stated above. We, therefore, conclude that NAC protects against Cd-induced oxidative stress via its antioxidant capacity affecting cellular mechanisms which might differ between cell types and tissues.

Interestingly, VC was, in contrast to NAC, not able to restore cell numbers upon HBSS starvation in Z3 zebrafish cells (Figure 2). Co-exposure to HBSS, $CdCl_2$, and VC even caused a cumulative toxic effect further decreasing Z3 cell numbers (Figure 2C). Preparation of the treatment solutions in HBSS excludes complex formation with cell culture media components, so we suggest that VC and $CdCl_2$ form compounds with higher toxicity than Cd alone or that VC increases or facilitates Cd uptake by Z3 cells. As shown before, Cd is responsible for δ-aminolevulinate dehydratase enzyme inhibition in rat lung and VC even increased the inhibiting effect [62]. However, VC has also been shown to attenuate germ cell apoptosis by protecting against ER stress and unfolded protein response (UPR) in mouse testes [63]. According to another study, VC inhibits lipid peroxidation in rat testes [46]. VC has also been shown to protect against Cd-induced renal injuries [64] and to reduce Cd accumulation in liver and kidney of catfish [27].

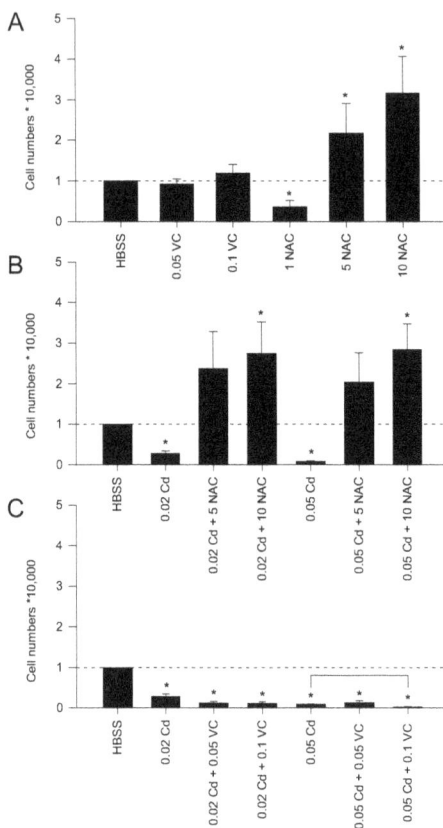

Figure 2. Cell density assay (Hoechst 33342) with Z3 zebrafish cells. (**A**) Effect of vitamin C (VC) and *N*-acetylcysteine (NAC) on HBSS incubated cells; (**B**) Recovery from HBSS and Cd treatment using NAC; (**C**) Recovery from HBSS and Cd treatment using VC. Cell numbers were normalized to 10,000 cells of the HBSS control. Statistical analysis was performed prior to data normalization using a *t*-test. Exposures were compared to HBSS treatment (* $p \leqslant 0.05$). Square bracket indicates statistical significance from comparison of normalized data. Values are mean ± standard error from 3 biological replicates.

Taken together, the present results from a zebrafish cell line and many previous studies show that NAC and VC have different effects on Cd toxicity. The impact of NAC and VC might, therefore, be cell type and tissue-specific; underlying mechanisms, however, remain to be resolved.

In conclusion, it can be stated that a major protection mechanism against Cd-induced ROS can be found in the induction and upregulation of the intrinsic antioxidative machinery. Indeed, several studies employ the idea of assaying the induction of oxidative response as a biomarker for Cd contamination, for example in Nile tilapia [65] and bivalves [66–68]. Furthermore, it can be postulated that many different means able to reduce oxidative stress will also ameliorate Cd-induced toxicity although this may be cell and tissue-specific and might also have negative effects.

2.2. Mitochondrial Protection Counteracts Cd Insult

A well-studied detrimental effect of Cd exposure is mitochondrial damage due to increased ROS levels but also deviations in Ca^{2+} homeostasis [69–71]. Since mitochondria are important Ca^{2+}

stores inside the cell, Cd^{2+} leads to a competitive inhibition of calcium translocation and homeostasis. Consequently, ROS and Ca^{2+} disturbance lead to numerous changes in the mitochondrial status including the reduction of oxidative phosphorylation, depolarization of mitochondrial membrane potential ($\Delta\Psi_m$), increase of superoxide and decrease of ATP production [72], and, ultimately, to mitophagy [73,74]. The mitochondria-specific increase in ROS was shown to be caused by direct interaction of Cd with the electron transport chain (ETC) [75].

The exact sequence of mitochondrial degradation caused by Cd has been demonstrated along different lines. Some studies show Cd-induced opening of the mitochondrial permeability transition pore (mPTP) [75–77]. According to another study on rat proximal tubule cells, Cd^{2+} enters mitochondria via the mitochondrial calcium uniporter (MCU) and induces mPTP-independent swelling of mitochondria [78]. However, both mitochondrial dysfunctions caused by Cd, mPTP and Cd entry via the MCU, lead to the release of cytochrome C from the intermembrane space into the cytosol, an important step in the initiation of apoptosis.

In summary, mitochondria represent a central target for Cd-induced toxicity and different means of mitochondrial protection against Cd toxicity apply. When considering, for example, normal mitochondrial turnover which is tightly controlled by fission and fusion rates of mitochondrial fragments, we can postulate that prevention of stress-induced mitochondrial fragmentation should protect mitochondria against Cd. Indeed, a recent study has shown that silencing a central fission-promoting protein (Drp1) reduces Cd-induced mitophagy [79].

Clearly, Cd toxicity in mitochondria is often based on oxidative stress and most of the afore mentioned detrimental effects such as depolarization of $\Delta\Psi_m$, mPTP, swelling or fission would also occur under ROS stress, for example as a result of hypoxia/reoxigenation. Therefore, antioxidative intervention is able to protect mitochondria against Cd insult. Both, pyruvate, known to protect against oxidative stress [72], and melatonin, known for its effects on free radical scavenging [80], have been shown to directly protect mitochondria. Examples of natural antioxidative substances or substances activating antioxidative defense under Cd exposure are listed in Table 1.

While different pathways of Cd-induced mitochondrial damage have been studied in recent years, it remains unknown how cells protect themselves and their mitochondria against heavy metal insult. Remarkably, several protective options exist. A recent study has found a positive induction of mitochondrial biogenesis and mitochondrial DNA content after acute Cd toxicity in rat proximal tubule cells as well as after chronic exposure *in vivo* [10]. The authors also found a distinct upregulation of anti-apoptotic genes with chronic exposure. This result may indicate an attempt to counteract Cd-induced apoptosis triggered by mitochondria and/or ER. Another study identified an upregulation of the mitochondrial NADP+-dependent enzyme isocitrate dehydrogenase to ameliorate oxidative stress by providing NADPH which serves as a reduction equivalent for the regeneration of GSH [81].

Another protective effect involving mitochondria can be observed in the Cd-induced downregulation of metabolism itself. Dogwhelks, aquatic gastropods, cope with a 20-day Cd exposure by metabolic depression. This physiological adaptation is a common response to intermittent hypoxia but it could also be shown to occur under Cd stress. The authors claim this response to be a strategy to minimize Cd^{2+} uptake and meet the extra energy demand for detoxification [82]. It is tempting to ask whether this is merely an effect of oxidative stress and mitochondrial damage or truly a protection mechanism. Indications can be drawn from a recent publication studying energy utilization of mitochondria in the freshwater crab *Sinopotamon henanense*. These experiments show that mitochondria respond to acute Cd exposure with an upregulation of energy production (higher levels of $\Delta\Psi_m$, NADH/NAD+ and ATP/ADP ratio) to cope with the energy demand of cellular defense mechanisms such as metallothionein (MT) production. However, with increasing exposure time a decline of energy production accompanied by excessive mitochondrial impairment was observed [83]. Consequently, it can be stated that mitochondrial energetic homeostasis is a fundamental requirement for successful Cd defense [83,84] but long-term countermeasures may depend on a balanced mitochondrial turnover with the risk of emphasizing anti-apoptotic signaling.

2.3. Protection by Metal Chelation

One of the major detoxification mechanisms protecting the cell from Cd-induced damage is the direct binding of Cd^{2+} to metal chelators. Among the most important and well-studied Cd^{2+}-binding proteins are MTs [85]. MTs occur throughout the animal kingdom and are involved in diverse cellular tasks including antioxidative functions [86,87]. However, their main responsibility is the homeostasis and detoxification of metals. Several MT isoforms have been described, the numbers differ within species with 12, *i.e.*, most, being present in mammals. The first *in vivo* Cd^{2+} binding studies using mouse MT1 were performed in the late 90s suggesting that domain duplication events in MTs might have evolved to not only function in trace metal homeostasis but also to cope with toxic metals like Cd [88]. In terrestrial gastropods, the evolution of a MT isoform showing an extraordinary Cd^{2+}-binding specificity was observed [89]. A recent study on mammalian MT1A revealed that the domain-specific order of the binding reaction and not the binding affinities account for the binding of zinc or Cd^{2+} [90], whereas it had already been shown that MT1 is more significantly sequestering Cd^{2+} than MT2 [91]. Several examples show that MT isoforms evolved to take over isoform-specific functions like Cd detoxification in mollusks [92,93], sea urchins [94], *Drosophila* [95], *C. elegans* [96], and *Tetrahymena* species [97,98]. However, due to its angiogenic, anti-apoptotic and proliferative functions, MT upregulation has been connected with poor prognosis and increased chemotherapeutic resistance [99,100] in some types of cancer.

Combined with the fact that MT gene expression can be directly induced by Cd, it can be stated that this mechanism presents one of the most efficient and prominent protection strategies against Cd. In vertebrates and insects, the metal transcription factor 1 (MTF-1) is responsible for MT induction. In the presence of Cd, MT-bound zinc is replaced by Cd^{2+} which is then able to activate MTF-1 [101]. Then again, except for insects, the MT activation mechanisms in invertebrates might be regulated [102] differently, in earthworms probably via the cAMP response element (CRE)-binding protein [103].

MTs are mainly expressed in the liver where Cd-MT complexes are formed. A thorough overview of structural characterization and binding affinities of Cd^{2+} in MTs can be found elsewhere [104]. Initially, Cd-MT complexes are stored in lysosomes, but are released into the bloodstream once liver cells die off. In colonic epithelial cells the uptake of Cd-MT complexes and their translocation to lysosomes lead to a decrease of systemic Cd toxicity [105]. However, Cd-MT complexes might still bear the risk of cellular damage. This has been shown in a study using a rat ADP ribosylation factor 1 (Arf1) mutant (Arf1 is involved in late endosome/lysosome trafficking) which decreased Cd toxicity in renal cells probably by attenuating the release of Cd^{2+} from degraded MT1 complexes into the cytosol [106]. The kidney is also known to be severely affected by Cd exposure. According to a recent study, Cd^{2+} causes hyperpermeability and hence disrupts the endothelial cell barrier in the glomerulus [107]. Due to its low molecular weight, the Cd-MT complex is filtered out at the glomerulus and is incorporated into proximal tubular cells. Subsequently, this can lead to kidney injuries. However, if the receptor responsible for Cd-MT incorporation is inhibited, Cd-MT-induced toxicity is reduced in the kidney [108] (see Section 2.8). Taken together, MT is pivotal in the protection against Cd-induced toxicity but also plays a central role in the systemic cycling of Cd and may hold carcinogenic potential due to its diverse functions.

Glutathione, which has already been described as an antioxidant, also acts as a metal-chelating agent able to bind Cd [109]. In addition, GSH is involved in cellular removal of Cd and is discussed later. Phytochelatins (PCs), which are formed from condensation of glutathione molecules, have recently been discovered in invertebrate species [110] and are also believed to function as a Cd detoxification system [111]. In contrast to Cd-MT, Cd-PC complexes taken up with the food have been shown to not co-localize with lysosomes [105] which might hint at different storage and excretion routes of PCs and MTs.

Cd chelation via MT, GSH, and PC represents a highly efficient detoxification system. However, a putative degradation of the metal-protein complex may lead to a repeated release of toxic Cd ions.

2.4. Protection against Macromolecular Damage

The endoplasmic reticulum (ER) is the major Ca^{2+} store inside the cell. It is, therefore, not surprising that Cd intoxication involves ER stress by altering Ca^{2+} homeostasis [112]. Moreover, the ER is the site for protein folding and refolding, which also play a major role in Cd toxicity. Since Cd^{2+} has a similar hydration radius like Ca^{2+}, it enters the cell through Ca^{2+} channels, interacts with Ca^{2+} pumps in the ER membrane and damages the ER upon entry [113]. Furthermore, Cd^{2+} is structurally very similar to essential trace elements like zinc. This ionic mimicry is responsible for protein misfolding or malfunction. Therefore, the ER is not only challenged directly by altered Ca^{2+} levels but also by an increase of damaged proteins. The cellular response to ER stress can involve adaptive mechanisms which protect the cell against stress or can lead to Cd-induced apoptosis. Several proteins have been found to be involved in mediating between cell survival and cell death. However, the point of no return has not yet been identified [114]. An indicator of ER stress is the upregulation of the unfolded protein response (UPR) which can activate pro-survival signals or induce apoptotic cell death. Several types of tumors depend on this mechanism, because several branches of the UPR positively affect cell transformation and tumor aggressiveness [115]. A strategy to reduce macromolecular damage causing ER stress and subsequent UPR is the expression of chaperones like Grp78. Grp78 is located in the ER and is known to be induced upon Cd exposure to prevent protein unfolding or misfolding as shown in LLC-PK1 renal epithelial cells [116].

Regarding ER stress and Cd intoxication, it could also be shown that once again Nrf2 [117], ubiquitin ligase FBXO6 [118] as well as ascorbic acid [63], a well-known antioxidant, attenuate Cd-induced ER stress. In concordance with the latter, the prevention of ER stress in Cd-resistant cells is responsible for cell survival via the activation of p38 and the induction of autophagy [119].

The heat shock response represents a general protective mechanism against environmental stress and specifically against Cd exposure via an increased expression of heat shock proteins (HSPs). HSPs represent cytosolic chaperones involved in protein folding and in the antioxidant response. The protective role of HSPs in Cd toxicity might be exerted via ROS scavenging [120]. So far, a time-dependent induction of HSPs upon Cd exposure has been revealed [121]. However, Cd-induced reduction of FcHsp70 was observed in the Chinese shrimp *Fenneropenaeus chinensis* [122], the Pacific oyster *Crassostrea gigas* [123], and in a human myeloid cell line [124]. In addition, the mRNA and protein level of HSPs can also differ as shown in the cyprinid fish *Tanichthys albonubes* [125].

In summary, the prevention and repair of molecular damage presents one of the major cellular tasks to maintain or re-establish homeostasis upon Cd exposure. In this context, ER stress prevention is an important protection mechanism in the short-term response to Cd administration but also bears the potential risk of carcinogenesis.

2.5. Cd Resistance and Cytoskeletal Rearrangements

As stated in the previous section, ionic mimicry, the competitive replacement of calcium ions by Cd^{2+}, is a highly toxic mechanism for many cellular processes [2] such as the regulation of cytoskeletal elements through polymerization of the actin cytoskeleton [126]. Cd exposure has been shown to cause oxidation of peptidyl-cysteines in proteins regulating the actin skeleton [127] and epigenetic methylation of actin and myosin promotor regions in chinese hamster ovary cells [128]. Further studies have found F-actin depolymerization and apoptosis to be another effect of Cd^{2+}—the chronological order of events is, however, still unknown [129]. Also, increased amounts of microtubules and microfilaments are able to protect a mouse cell line from Cd-induced damage by increasing the level of protein sulfhydryls. In the cytoskeletal and cytosolic fraction of Cd-resistant cells, the basal level of protein sulfhydryl groups was elevated. These cells show no cytoskeletal rearrangements upon Cd stress in contrast to parental cells [130]. Interestingly, in Cd-resistant rat lung epithelial cells, cytokeratins were upregulated, most likely to prevent Cd-induced apoptosis—a change in keratin expression is a highly probable protective response to long-term Cd exposure [131]. The involvement of Cd in malignant transformation of an immortalized cell line and the involvement of keratin was confirmed later [132]. Concluding, this protective mechanism also holds a potential trade-off in the form of carcinogenic transformation.

2.6. Protection against Cd by Cd—Hormetic Responses

Many terms have been used to describe beneficial dose-response relationships: hormesis, preconditioning, cross-resistance or adaptive protection. However, it has been suggested, that these phenomena all describe the same principle, namely the plasticity of biological processes and systems to adapt and respond to different kinds of stressors [133]. A simplified description of hormesis is the opposite dose-response relationship at low *versus* higher concentrations of a toxicant [134]. Accordingly, at low dosages, heavy metals can have a beneficial effect on the organism. A review of the mechanisms responsible for hormesis suggests that, regardless of the actual mechanisms involved, the intensity of the response is a measure of biological plasticity [135,136]. We, therefore, discuss the literature on the mechanisms underlying this biological plasticity to Cd exposure and its protective effects.

Cd has been shown to stimulate cell proliferation in zebrafish liver cells and to decrease the percentage of apoptotic cells by a change in expression of growth factors and DNA repair genes. Genomic instability might then, however, contribute to Cd-induced carcinogenesis [137]. Hormesis also induces other effects like the increase in cellular metabolic activity as shown in mouse fibroblast cells upon exposure to low levels of Cd which also coincided with an increased production of stress proteins like HSPs and MTs [138]. A study using HaCaT cells reveals that the proliferative response to low metal concentrations needs NADPH oxidase (NOX) stimulation which is activated by endogenous factors [139].

Hormetic effects of Cd were mainly studied at the organismic level. In adult rainbow trout (*Oncorrhynchus mykiss*), for example, chronic exposure to low dietary amounts of Cd decreases the toxic effect of waterborne Cd [140]. In mice, HSP70 and its activating heat shock factor 1 (HSF1) take over a major role in the protection and preconditioning to Cd administration [141]. In earthworms, hormetic effects upon Cd exposure affect antioxidant enzymes by increasing the activity of CAT and SOD [142]. Hormesis has also been described as a species-specific phenomenon. While exposure to small amounts of metals increased the rate of growth and reproduction in one species of snails, another species did not display any signs of hormesis [143].

It has also been shown that Cd induces cross-resistance to other metals like zinc [144] and manganese [145] or oxidative stress as shown in V97 Chinese hamster fibroblasts [9]. However, the cross-resistance effect does not seem to be bidirectional since stressors like oxidative stress can render cells more prone to a Cd challenge [146].

However, the beneficial effect of hormesis may not come without trade-offs. The exposure to dead spores causes longevity but also leads to reduced immune functions [147]. An additional stressor (depleted uranium) in the presence of radiation hormesis leads to an even higher toxicity (increased apoptosis) than the additional stressor alone would have caused [148]. It is important to note that the very ability of preconditioning can be deactivated by Cd exposure as demonstrated in a recent study in rats. The latter effect was attributed to the inhibition of hypoxia-inducible factor 1a (Hif1a) stabilization and the promotion of Hif1a degradation [149]. However, other authors show a clear induction of the Hif1a/vascular endothelial growth factor signaling axis by Cd [150].

In conclusion, beneficial effects derived from hormesis or hormesis-like phenomena should be critically reviewed especially when discussing the outcome at the organismic level.

2.7. Protective Effect by Co-Exposure to Other Metals or Trace Elements

Pre-exposure or co-exposure to other elements such as copper, selenium, zinc, and manganese has a protective effect on Cd toxicity. For copper, the protective effect of co-exposure to Cd has been shown, for example, in mice [151]. However, the cellular mechanism behind this effect remains unknown. For the trace element selenium, several studies have found a wide-spread beneficial effect on antioxidant status and lipid peroxidation *in vivo* when co-exposed or pre-exposed to Cd [152–154].

Remarkably, selenium shows similar protective effects on mitochondrial dysfunction as the classical antioxidant NAC in LLC-PK1 cells [155]. Based on a follow-up study, the same authors conclude that selenium reduces oxidative stress-induced mitochondrial apoptosis [156]. Similar results for selenium have been obtained in chicken splenic lymphocytes exposed to Cd [157]. For zinc, *in vivo* studies in rats show direct antioxidant effects which alleviated Cd oxidative stress [158] as well as genotoxicity [159]. In addition, zinc is also known to induce MT in adult zebrafish [13] or in Madin–Darby bovine kidney cells [160]. Similarly, in mice, the protective effect of manganese pre-exposure has been connected to antioxidative effects, induction of MT and protection of Ca^{2+} homeostasis [161].

All things considered, the reduction of Cd-induced oxidative stress may be the main protective effect caused by co-exposure to trace elements and other metals. Additionally, the co-induction of MT represents an important protective function (see Section 2.3). Recent studies have established yet another protective mechanism: By competing with Cd^{2+} uptake via shared transport mechanisms, Mn^{2+} and Zn^{2+} as well as Fe^{2+} and Ca^{2+} can significantly reduce or inhibit the entry of Cd^{2+} [162]. In the following section we focus on the reduced uptake of Cd^{2+} as a protective mechanism.

2.8. Protection by Reduced Uptake of Cd

Due to its high hydrophilicity, Cd has to enter cells via active or passive transport proteins such as receptors, transporters and pores or receptor-mediated endocytosis (RME) of Cd^{2+} bound to MT (Cd-MT) [163]. Cd^{2+} often uses uptake routes intended for essential divalent ions such as Ca^{2+}, Fe^{2+}, Zn^{2+}, or Mn^{2+}. Consequently, downregulation of transport proteins is an important protective mechanism for cells, especially for long-term resistance against the heavy metal. One approach to study this mechanism is to use Cd-resistant cell lines and to delineate their mode of Cd^{2+} transport because reduced uptake of Cd^{2+} has been shown to be an important feature of Cd-resistant cells. In the case of mouse embryonic cells, this resistance occurs due to a downregulation of transport systems such as the zinc transporter, divalent metal transporter, and voltage-dependent Ca^{2+} channels [145]. According to another study, in MT 1 and 2 knock-out cells, long-term Cd resistance is acquired by downregulation of T-type Ca^{2+} channels [164]. Finally, also for RME of Cd-MT, an important entry pathway of Cd^{2+} in mammalian kidney, studies indicate a protective mechanism by downregulation of kidney cell surface receptors such as cubilin in a rat model with subchronic exposure [165] and megalin in proximal tubule cells [166,167]. Originally, these experiments addressed Cd-induced proteinuria, the impaired reabsorption of proteins from the proximal tubule due to Cd intoxication. Interestingly, this impairment also represents a protective mechanism against additional Cd-MT uptake with obvious organismic trade-offs.

These studies are important examples for the protection of cells against Cd. The variety of different transport systems involved in Cd movement across the cell membrane as shown by several excellent reviews [162,163,168,169] may include many more protective pathways.

2.9. Protection through Removal of Cd

The phenomenon of multidrug resistance was first identified in tumor cell lines which developed resistance to chemotherapeutic treatments. Central to this resistance is the induction of multidrug resistance protein 1. Also known as P-glycoprotein (P-gp), this ATP-dependent transmembrane transporter belonging to the ATP-binding cassette (ABC) class of transmembrane proteins is responsible for pumping cytotoxic substances out of the cell. For example, with prolonged exposure time, a study on proximal tubule cells observed a reduction in Cd-associated apoptosis which was due to a four-fold upregulation of the drug efflux pump multidrug resistance P-gp [170]. The signal for the induction of the pump after Cd exposure was transduced via oxygen radicals and could be prevented by antioxidant intervention. As mentioned above, once inside the cell, Cd^{2+} readily binds to thiol groups of GSH. Therefore, when GS-Cd is removed by P-gp, GSH equivalents also leave the cell. In this respect, complexation of GSH with Cd^{2+} and the resulting efflux from the cell might again represent a way of immediate cellular protection with the inevitably adverse long-term effects of lower GSH levels.

Interestingly, a study on Cd-resistant zebrafish cells (ZF4-Cd) connects the cells' resistance to an upregulation of multidrug resistance-associated protein (MRP) transport activity, higher rates of Cd removal, elevated expression of other ABC class proteins, and increased content of cellular GSH [171]. It is apparent that upregulation of GSH production is a protective mechanism which serves cells not only as an antioxidant but also protects them as a mediator for Cd removal. By blocking GSH synthesis with buthionine sulfoximine (BSO), a study on proximal tubule cells shows that Cd efflux depends on GSH. This study identifies a novel exit route for GSH and GS-Cd in the ABC family member cystic fibrosis conductance regulator (CFTR), a chloride channel. The authors propose a dual response model involving the CFTR in which low Cd intoxication might be resolved by direct removal of GS-Cd. Higher Cd concentrations might lead to severe GSH depletion with decreased ability of the cell to scavenge Cd-induced ROS, ultimately leading to apoptosis [172].

The environmental equivalent to multidrug resistance has been described as multixenobiotic resistance (MXR). This process has predominantly been observed in aquatic organisms where different anthropogenic contaminants are able to induce the P-gp transporter in order to develop a cellular defense mechanism [173,174]. A similar MXR response towards Cd contamination has been found in aquatic mollusks [175–178] and fish [179]. Natural variation in abiotic factors can also alter Cd-toxicity. This will be addressed in the next section.

2.10. Toxicity of Cd by Altered Environmental Factors

A set concentration of Cd in the environment of an organism can greatly vary in its effects under different abiotic conditions such as temperature, oxygen partial pressure, or salinity. For example, a study in Dogwhelk (*Nucella lapillus*) shows that Cd toxicity is positively correlated to temperature. As part of the protective response, metabolism is reduced and higher energy requirements needed for the stress response are met by using internal glycogen stores [82]. In the oyster, Cd damage is also reduced at lower temperatures leading to higher levels of activity of the antioxidative enzyme aconitase [180,181].

This type of response usually involves lower mitochondrial metabolic flux and ATP turnover at lower temperatures, resulting in a weaker toxicological damage in the presence of Cd. As highlighted in Section 2.2, energetic homeostasis is an important prerequisite for successfully handling Cd toxicity. Interestingly, organisms undergoing thermal acclimation respond better to concurrent toxicological challenges [181–184].

Co-exposure to hypoxia has been shown to increase the tissue accumulation of Cd in freshwater clams (*Corbicula fluminea*) but also to increase protection by MT induction. However, the combined exposure may at best have a compensatory effect on overall viability [176]. The low oxygen tension leads to increased ventilatory activity with the result of enhancing the Cd bioaccumulation rate [185].

Several studies also investigate the impact of ion content and salinity on Cd toxicity. In the gastropod *N. lapillus*, the response to low salinity levels includes altered Cd accumulation and MT expression [186]. Studies on trout gill Cd^{2+} uptake show that hard water (with more Ca^{2+} ions) protects against Cd^{2+} uptake and toxicity [187]. However, a considerable number of studies have found conflicting results for dissolved ions and salinity and a general rule of effect does not apply to different experimental situations. An attempt to include all relevant water chemistry parameters able to interact with metal toxicity has been made for daphnids and fish in the form of the biotic ligand model (BLM) [188]. In green algae, the BLM shows that Cd^{2+} uptake and toxicity are reduced upon calcium, zinc and cobalt exposure; these elements obviously influence Cd toxicity in aquatic environments [189].

Consequently, when using the responses of biomarkers to project Cd intoxication, it is necessary to consider the influence of different abiotic factors [190].

3. Experimental Section

3.1. Cell Culture

An adherent embryonic fibroblast zebrafish cell line (Z3) [191] was used for exposure experiments. The cells were grown in cell culture flasks to 80% confluency in Leibovitz 15 (L-15, Thermo Fisher Scientific, Carlsbad, CA, USA) complete media supplemented with 15% fetal bovine serum (FBS), L-glutamine, penicillin-streptomycin, and gentamycin. After trypsination, cells were seeded into 96-well plates and left for attachment at 25 °C overnight. The following day cells were washed once with HBSS (Thermo Fisher Scientific) and incubated with 200 μL of the treatment solutions for 18 h followed by a recovery period of 6 h (200 μL of HBSS without treatments).

3.2. Treatments

Cells were treated with two different antioxidants NAC (5, 10 mM) (Roth, Karlsruhe, Germany) and L-ascorbic acid (0.05, 0.1 mM) (Roth) as well as in combination with $CdCl_2$ (20 μM, 50 μM) (Sigma-Aldrich, St. Louis, MO, USA). All treatments were prepared in sterile HBSS containing Ca^{2+} and Mg^{2+} with pH adjusted to 7.6. We also included controls treated with L-15 complete media (L-15+) and with L-15 media lacking FBS (L-15−).

3.3. Cell Density Assay

After one washing step with HBSS, cell density was immediately measured after the recovery period or, for control experiments, after the treatment period using a fluorescent dye (Hoechst 33342) in a plate reader (Victor X4, Perkin Elmer, Waltham, MA, USA) according to standard procedures described previously [192]. For blank correction, the dye solution without cells was used. Absolute cell numbers were calculated according to a previously prepared standard curve. The antioxidant stock and working solutions were freshly prepared prior to each treatment in HBSS (pH 7.6). All experiments were performed using six technical repeats and a minimum of three biological replicates.

3.4. Statistical Analysis

Data were normalized to 10,000 cells of the HBSS treatment to overcome seeding-related differences in cell numbers in the biological replicates. Statistical analysis using *t*-tests was, however, performed prior to data normalization. All groups were compared to the HBSS exposure group. Significance level was set to $p \leqslant 0.05$. Normalized data were used to reveal the cumulative toxicity of the 50 μM $CdCl_2$ and 0.1 mM VC co-exposure compared to the 50 μM $CdCl_2$-treated cells.

4. Conclusions

Cd is introduced into the environment largely by human activities. On the cellular and organismic levels, several mechanisms can be adopted to cope with Cd and protect against Cd-induced toxicity.

Perhaps the most prominent protection strategy is the prevention of oxidative stress which is one of the major mechanisms by which Cd exerts its toxicity. It can be postulated that many different means able to reduce oxidative stress will also ameliorate Cd-induced toxicity. However, alteration in cellular redox balance can have negative effects like an increased risk of carcinogenesis. Mitochondrial energetic homeostasis is a fundamental requirement for successful Cd defense but long-term countermeasures may depend on a balanced mitochondrial turnover bearing the risk of enhancing anti-apoptotic signaling. The prevention of cellular damage by free Cd^{2+} via metal chelation seems to be a perfect short-term detoxification strategy. Storage and degradation of, e.g., Cd-MT complexes in lysosomes, however, bear the risk of releasing free Cd^{2+} into the cytosol after cell death. ER stress prevention appears to be another highly important protection mechanism in short-term responses to Cd administration. Again, this process potentially leads to carcinogenesis by inducing cell survival pathways. Cytoskeletal rearrangements have also been shown to protect against Cd toxicity, but might also be responsible for carcinogenic transformation. Due to the presence of trade-offs, hormesis or hormesis-like phenomena reducing Cd-induced cellular damage must be critically reviewed, especially when discussing the outcome at an organismic level. Protection via reduced Cd uptake might involve impaired reabsorption. The improved removal of Cd bears the risk of an increased loss of essential proteins leading to negative side-effects.

Antioxidants [193] and Cd chelation [194,195] have been proposed as a therapeutic approach to Cd intoxication. The risk of side-effects should, however, not be underestimated.

It can, therefore, be concluded that Cd protection or Cd detoxification strategies that prevent cellular damage seldom come without trade-offs like, primarily, an increased risk of carcinogenesis. However, an impressive cellular machinery has evolved across the animal kingdom and can be adopted to cope with Cd insult and other anthropogenic stressors in natural habitats.

Acknowledgments: Acknowledgments: We would like to thank Johannes Schibler for conducting preliminary experiments leading to the present work and Gerda Ludwig for critical reading of the manuscript.

Author Contributions: Author Contributions: Adolf Michael Sandbichler contributed to the preparation of the manuscript, the experimental setup and performance of the laboratory experiments. Martina Höckner contributed to the preparation of the manuscript, the experimental setup and performance of the laboratory experiments.

Conflicts of Interest: Conflicts of Interest: The authors declare no conflict of interest.

References

1. Järup, L.; Akesson, A. Current status of Cd as an environmental health problem. *Toxicol. Appl. Pharmacol.* **2009**, *238*, 201–208. [CrossRef] [PubMed]
2. Choong, G.; Liu, Y.; Templeton, D.M. Interplay of Calcium and cadmium in mediating cadmium toxicity. *Chem. Biol. Interact.* **2014**, *211*, 54–65. [CrossRef] [PubMed]
3. Gardarin, A.; Chédin, S.; Lagniel, G.; Aude, J.-C.; Godat, E.; Catty, P.; Labarre, J. Endoplasmic reticulum is a major target of cadmium toxicity in yeast. *Mol. Microbiol.* **2010**, *76*, 1034–1048. [CrossRef] [PubMed]
4. Stohs, S.J.; Bagchi, D. Oxidative mechanisms in the toxicity of metal ions. *Free Radic. Biol. Med.* **1995**, *18*, 321–336. [CrossRef]
5. Valko, M.; Morris, H.; Cronin, M.T.D. Metals, toxicity and oxidative stress. *Curr. Med. Chem.* **2005**, *12*, 1161–1208. [CrossRef] [PubMed]
6. Stohs, S.J.; Bagchi, D.; Hassoun, E.; Bagchi, M. Oxidative mechanisms in the toxicity of chromium and cadmium ions. *J. Environ. Pathol. Toxicol. Oncol.* **2000**, *19*, 201–213. [CrossRef] [PubMed]
7. Singhal, R.K.; Anderson, M.E.; Meister, A. Glutathione, a first line of defense against cadmium toxicity. *FASEB J.* **1987**, *1*, 220–223. [PubMed]
8. Rana, S.V.S.; Singh, R. Influence of antioxidants on metallothionein-mediated protection in cadmium-fed rats. *Biol. Trace Elem. Res.* **2002**, *88*, 71–78. [CrossRef]
9. Chubatsu, L.S.; Gennari, M.; Meneghini, R. Glutathione is the antioxidant responsible for resistance to oxidative stress in V79 Chinese hamster fibroblasts rendered resistant to cadmium. *Chem. Biol. Interact.* **1992**, *82*, 99–110. [CrossRef]
10. Nair, A.R.; Lee, W.-K.; Smeets, K.; Swennen, Q.; Sanchez, A.; Thévenod, F.; Cuypers, A. Glutathione and mitochondria determine acute defense responses and adaptive processes in cadmium-induced oxidative stress and toxicity of the kidney. *Arch. Toxicol.* **2015**, *89*, 2273–2289. [CrossRef] [PubMed]
11. Sakurai, A.; Nishimoto, M.; Himeno, S.; Imura, N.; Tsujimoto, M.; Kunimoto, M.; Hara, S. Transcriptional regulation of thioredoxin reductase 1 expression by cadmium in vascular endothelial cells: Role of NF-E2-related factor-2. *J. Cell. Physiol.* **2005**, *203*, 529–537. [CrossRef] [PubMed]
12. Chen, J.; Shaikh, Z.A. Activation of Nrf2 by cadmium and its role in protection against cadmium-induced apoptosis in rat kidney cells. *Toxicol. Appl. Pharmacol.* **2009**, *241*, 81–89. [CrossRef] [PubMed]
13. Arini, A.; Gourves, P.Y.; Gonzalez, P.; Baudrimont, M. Metal detoxification and gene expression regulation after a Cd and Zn contamination: An experimental study on *Danio rerio*. *Chemosphere* **2015**, *128*, 125–133. [CrossRef] [PubMed]
14. Huang, Y.; Li, W.; Su, Z.-Y.; Kong, A.-N.T. The complexity of the Nrf2 pathway: Beyond the antioxidant response. *J. Nutr. Biochem.* **2015**, *26*, 1401–1413. [CrossRef] [PubMed]

15. Ma, Q. Role of Nrf2 in Oxidative Stress and Toxicity. *Annu. Rev. Pharmacol. Toxicol.* **2013**, *53*, 401–426. [CrossRef] [PubMed]
16. Liu, J.; Qu, W.; Kadiiska, M.B. Role of oxidative stress in cadmium toxicity and carcinogenesis. *Toxicol. Appl. Pharmacol.* **2009**, *238*, 209–214. [CrossRef] [PubMed]
17. Cuypers, A.; Plusquin, M.; Remans, T.; Jozefczak, M.; Keunen, E.; Gielen, H.; Opdenakker, K.; Nair, A.R.; Munters, E.; Artois, T.J.; *et al.* Cadmium stress: An oxidative challenge. *BioMetals* **2010**, *23*, 927–940. [CrossRef] [PubMed]
18. Bertin, G.; Averbeck, D. Cadmium: Cellular effects, modifications of biomolecules, modulation of DNA repair and genotoxic consequences (a review). *Biochimie* **2006**, *88*, 1549–1559. [CrossRef] [PubMed]
19. Bravard, A.; Campalans, A.; Vacher, M.; Gouget, B.; Levalois, C.; Chevillard, S.; Radicella, J.P. Inactivation by oxidation and recruitment into stress granules of hOGG1 but not APE1 in human cells exposed to sub-lethal concentrations of cadmium. *Mutat. Res.* **2010**, *685*, 61–69. [CrossRef] [PubMed]
20. Eybl, V.; Kotyzová, D.; Bludovská, M. The effect of curcumin on cadmium-induced oxidative damage and trace elements level in the liver of rats and mice. *Toxicol. Lett.* **2004**, *151*, 79–85. [CrossRef] [PubMed]
21. Eybl, V.; Kotyzova, D.; Koutensky, J. Comparative study of natural antioxidants—Curcumin, resveratrol and melatonin—In cadmium-induced oxidative damage in mice. *Toxicology* **2006**, *225*, 150–156. [CrossRef] [PubMed]
22. Daniel, S.; Limson, J.L.; Dairam, A.; Watkins, G.M.; Daya, S. Through metal binding, curcumin protects against lead- and cadmium-induced lipid peroxidation in rat brain homogenates and against lead-induced tissue damage in rat brain. *J. Inorg. Biochem.* **2004**, *98*, 266–275. [CrossRef] [PubMed]
23. Rennolds, J.; Malireddy, S.; Hassan, F.; Tridandapani, S.; Parinandi, N.; Boyaka, P.N.; Cormet-Boyaka, E. Curcumin regulates airway epithelial cell cytokine responses to the pollutant cadmium. *Biochem. Biophys. Res. Commun.* **2012**, *417*, 256–261. [CrossRef] [PubMed]
24. Baiomy, A.A.; Mansour, A.A. Genetic and histopathological responses to cadmium toxicity in rabbit's kidney and liver: Protection by Ginger (*Zingiber officinale*). *Biol. Trace Elem. Res.* **2015**. in press.
25. Abdel Moneim, A.E.; Bauomy, A.A.; Diab, M.M.S.; Shata, M.T.M.; Al-Olayan, E.M.; El-Khadragy, M.F. The protective effect of *Physalis peruviana* L. against cadmium-induced neurotoxicity in rats. *Biol. Trace Elem. Res.* **2014**, *160*, 392–399. [CrossRef] [PubMed]
26. Argüelles, N.; Alvarez-González, I.; Chamorro, G.; Madrigal-Bujaidar, E. Protective effect of grapefruit juice on the teratogenic and genotoxic damage induced by cadmium in mice. *J. Med. Food* **2012**, *15*, 887–893. [CrossRef] [PubMed]
27. Kumar, P.; Prasad, Y.; Patra, A.K.; Ranjan, R.; Swarup, D.; Patra, R.C.; Pal, S. Ascorbic acid, garlic extract and taurine alleviate cadmium-induced oxidative stress in freshwater catfish (*Clarias batrachus*). *Sci. Total Environ.* **2009**, *407*, 5024–5030. [CrossRef] [PubMed]
28. Cavuşoğlu, K.; Yapar, K.; Yalçin, E. Royal jelly (honey bee) is a potential antioxidant against cadmium-induced genotoxicity and oxidative stress in albino mice. *J. Med. Food* **2009**, *12*, 1286–1292. [CrossRef] [PubMed]
29. Argüelles-Velázquez, N.; Alvarez-González, I.; Madrigal-Bujaidar, E.; Chamorro-Cevallos, G. Amelioration of cadmium-produced teratogenicity and genotoxicity in mice given *Arthrospira maxima* (Spirulina) Treatment. *Evid. Based Complement. Altern. Med.* **2013**, *2013*, 1–8. [CrossRef] [PubMed]
30. Paniagua-Castro, N.; Escalona-Cardoso, G.; Hernández-Navarro, D.; Pérez-Pastén, R.; Chamorro-Cevallos, G. Spirulina (*Arthrospira*) protects against cadmium-induced teratogenic damage in mice. *J. Med. Food* **2011**, *14*, 398–404. [CrossRef] [PubMed]
31. Jahangir, T.; Khan, T.H.; Prasad, L.; Sultana, S. Alleviation of free radical mediated oxidative and genotoxic effects of cadmium by farnesol in Swiss albino mice. *Redox Rep.* **2005**, *10*, 303–310. [CrossRef] [PubMed]
32. Wang, W.; Sun, Y.; Liu, J.; Wang, J.; Li, Y.; Li, H.; Zhang, W.; Liao, H. Protective effect of theaflavins on cadmium-induced testicular toxicity in male rats. *Food Chem. Toxicol.* **2012**, *50*, 3243–3250. [CrossRef] [PubMed]
33. Krishnan, M.; Jayaraj, R.L.; Jagatheesh, K.; Elangovan, N. Taxifolin mitigates oxidative DNA damage *in vitro* and protects zebrafish (*Danio rerio*) embryos against cadmium toxicity. *Environ. Toxicol. Pharmacol.* **2015**, *39*, 1252–1261.
34. Jia, Y.; Lin, J.; Mi, Y.; Zhang, C. Quercetin attenuates cadmium-induced oxidative damage and apoptosis in granulosa cells from chicken ovarian follicles. *Reprod. Toxicol.* **2011**, *31*, 477–485. [CrossRef] [PubMed]

35. Vicente-Sánchez, C.; Egido, J.; Sánchez-González, P.D.; Pérez-Barriocanal, F.; López-Novoa, J.M.; Morales, A.I. Effect of the flavonoid quercetin on cadmium-induced hepatotoxicity. *Food Chem. Toxicol.* **2008**, *46*, 2279–2287. [CrossRef] [PubMed]

36. Bu, T.; Mi, Y.; Zeng, W.; Zhang, C. Protective effect of quercetin on cadmium-induced oxidative toxicity on germ cells in male mice. *Anat. Rec.* **2011**, *294*, 520–526. [CrossRef] [PubMed]

37. Renugadevi, J.; Prabu, S.M. Cadmium-induced hepatotoxicity in rats and the protective effect of naringenin. *Exp. Toxicol. Pathol.* **2010**, *62*, 171–181. [CrossRef] [PubMed]

38. Sakr, S.A.; Bayomy, M.F.; El-Morsy, A.M. Rosemary extract ameliorates cadmium-induced histological changes and oxidative damage in the liver of albino rat. *J. Basic Appl. Zool.* **2015**, *71*, 1–9. [CrossRef]

39. Choi, J.-H.; Rhee, I.-K.; Park, K.-Y.; Park, K.-Y.; Kim, J.-K.; Rhee, S.-J. Action of green tea catechin on bone metabolic disorder in chronic cadmium-poisoned rats. *Life Sci.* **2003**, *73*, 1479–1489. [CrossRef]

40. Jahan, S.; Khan, M.; Ahmed, S.; Ullah, H. Comparative analysis of antioxidants against cadmium induced reproductive toxicity in adult male rats. *Syst. Biol. Reprod. Med.* **2014**, *60*, 28–34. [CrossRef] [PubMed]

41. Wang, W.; He, Y.; Yu, G.; Li, B.; Sexton, D.W.; Wileman, T.; Roberts, A.A.; Hamilton, C.J.; Liu, R.; Chao, Y.; *et al.* Sulforaphane protects the liver against CdSe quantum dot-induced cytotoxicity. *PLoS ONE* **2015**, *10*, e0138771.

42. Su, Z.-Y.; Shu, L.; Khor, T.O.; Lee, J.H.; Fuentes, F.; Kong, A.-N.T. A perspective on dietary phytochemicals and cancer chemoprevention: Oxidative stress, Nrf2, and epigenomics. *Top. Curr. Chem.* **2013**, *329*, 133–162. [PubMed]

43. El-Sokkary, G.H.; Nafady, A.A.; Shabash, E.H. Melatonin administration ameliorates cadmium-induced oxidative stress and morphological changes in the liver of rat. *Ecotoxicol. Environ. Saf.* **2010**, *73*, 456–463. [CrossRef] [PubMed]

44. Kim, C.Y.; Lee, M.J.; Lee, S.M.; Lee, W.C.; Kim, J.S. Effect of melatonin on cadmium-induced hepatotoxicity in male Sprague–Dawley rats. *Tohoku J. Exp. Med.* **1998**, *186*, 205–213. [CrossRef] [PubMed]

45. Pi, H.; Xu, S.; Reiter, R.J.; Guo, P.; Zhang, L.; Li, Y.; Li, M.; Cao, Z.; Tian, L.; Xie, J.; *et al.* SIRT3-SOD2-mROS-dependent autophagy in cadmium-induced hepatotoxicity and salvage by melatonin. *Autophagy* **2015**, *11*, 1037–1051. [CrossRef] [PubMed]

46. García, M.T.A.; González, E.L.M. Natural antioxidants protect against cadmium-induced damage during pregnancy and lactation in rats' pups. *J. Food Sci.* **2010**, *75*, T18–T23. [CrossRef] [PubMed]

47. Ognjanović, B.I.; Pavlović, S.Z.; Maletić, S.D.; Zikić, R.V.; Stajn, A.S.; Radojicić, R.M.; Saicić, Z.S.; Petrović, V.M. Protective influence of vitamin E on antioxidant defense system in the blood of rats treated with cadmium. *Physiol. Res.* **2003**, *52*, 563–570. [PubMed]

48. Novelli, J.; Novelli, E.L.B.; Manzano, M.A.; Lopes, A.M.; Cataneo, A.C.; Barbosa, L.L.; Ribas, B.O. Effect of α-tocopherol on superoxide radical and toxicity of cadmium exposure. *Int. J. Environ. Health Res.* **2000**, *10*, 125–134. [CrossRef]

49. El-Sokkary, G.H.; Awadalla, E.A. The protective role of Vitamin C against cerebral and pulmonary damage induced by cadmium chloride in male adult albino rat. *Open Neuroendocrinol. J.* **2011**, *4*, 1–8. [CrossRef]

50. Liu, T.; He, W.; Yan, C.; Qi, Y.; Zhang, Y. Roles of reactive oxygen species and mitochondria in cadmium-induced injury of liver cells. *Toxicol. Ind. Health* **2011**, *27*, 249–256. [PubMed]

51. Abe, T.; Yamamura, K.; Gotoh, S.; Kashimura, M.; Higashi, K. Concentration-dependent differential effects of N-acetyl-L-cysteine on the expression of HSP70 and metallothionein genes induced by cadmium in human amniotic cells. *Biochim. Biophys. Acta* **1998**, *1380*, 123–132. [CrossRef]

52. Odewumi, C.O.; Badisa, V.L.D.; Le, U.T.; Latinwo, L.M.; Ikediobi, C.O.; Badisa, R.B.; Darling-Reed, S.F. Protective effects of N-acetylcysteine against cadmium-induced damage in cultured rat normal liver cells. *Int. J. Mol. Med.* **2010**, *27*, 1193–1205. [CrossRef] [PubMed]

53. Wang, J.; Zhu, H.; Liu, X.; Liu, Z. N-acetylcysteine protects against cadmium-induced oxidative stress in rat hepatocytes. *J. Vet. Sci.* **2014**, *15*, 485–493. [CrossRef] [PubMed]

54. Wispriyono, B.; Matsuoka, M.; Igisu, H.; Matsuno, K. Protection from cadmium cytotoxicity by N-acetylcysteine in LLC-PK1 cells. *J. Pharmacol. Exp. Ther.* **1998**, *287*, 344–351. [PubMed]

55. Wu, C.-A.; Chao, Y.; Shiah, S.-G.; Lin, W.-W. Nutrient deprivation induces the Warburg effect through ROS/AMPK-dependent activation of pyruvate dehydrogenase kinase. *Biochim. Biophys. Acta* **2013**, *1833*, 1147–1156. [CrossRef] [PubMed]

56. Nzengue, Y.; Steiman, R.; Garrel, C.; Lefèbvre, E.; Guiraud, P. Oxidative stress and DNA damage induced by cadmium in the human keratinocyte HaCaT cell line: Role of glutathione in the resistance to cadmium. *Toxicology* **2008**, *243*, 193–206. [CrossRef] [PubMed]

57. Khanna, S.; Mitra, S.; Lakhera, P.C.; Khandelwal, S. *N*-acetylcysteine effectively mitigates cadmium-induced oxidative damage and cell death in Leydig cells *in vitro*. *Drug Chem. Toxicol.* **2015**, *39*, 74–80. [CrossRef] [PubMed]

58. Oh, S.-H.; Lim, S.-C. A rapid and transient ROS generation by cadmium triggers apoptosis via caspase-dependent pathway in HepG2 cells and this is inhibited through *N*-acetylcysteine-mediated catalase upregulation. *Toxicol. Appl. Pharmacol.* **2006**, *212*, 212–223. [CrossRef] [PubMed]

59. Odewumi, C.O.; Latinwo, L.M.; Ruden, M.L.; Badisa, V.L.D.; Fils-Aime, S.; Badisa, R.B. Modulation of cytokines and chemokines expression by NAC in cadmium chloride treated human lung cells. *Environ. Toxicol.* **2015**. [CrossRef] [PubMed]

60. Hu, K.-H.; Li, W.-X.; Sun, M.-Y.; Zhang, S.-B.; Fan, C.-X.; Wu, Q.; Zhu, W.; Xu, X. Cadmium Induced Apoptosis in MG63 Cells by Increasing ROS, Activation of p38 MAPK and Inhibition of ERK 1/2 Pathways. *Cell. Physiol. Biochem.* **2015**, *36*, 642–654. [CrossRef] [PubMed]

61. Hsu, T.; Huang, K.-M.; Tsai, H.-T.; Sung, S.-T.; Ho, T.-N. Cadmium (Cd)-induced oxidative stress down-regulates the gene expression of DNA mismatch recognition proteins MutS homolog 2 (MSH2) and MSH6 in zebrafish (*Danio rerio*) embryos. *Aquat. Toxicol.* **2013**, *126*, 9–16. [CrossRef] [PubMed]

62. Luchese, C.; Zeni, G.; Rocha, J.B.T.; Nogueira, C.W.; Santos, F.W. Cadmium inhibits δ-aminolevulinate dehydratase from rat lung *in vitro*: Interaction with chelating and antioxidant agents. *Chem. Biol. Interact.* **2007**, *165*, 127–137. [CrossRef] [PubMed]

63. Ji, Y.-L.; Wang, Z.; Wang, H.; Zhang, C.; Zhang, Y.; Zhao, M.; Chen, Y.-H.; Meng, X.-H.; Xu, D.-X. Ascorbic acid protects against cadmium-induced endoplasmic reticulum stress and germ cell apoptosis in testes. *Reprod. Toxicol.* **2012**, *34*, 357–363. [CrossRef] [PubMed]

64. Manna, P.; Sinha, M.; Sil, P.C. Taurine plays a beneficial role against cadmium-induced oxidative renal dysfunction. *Amino Acids* **2009**, *36*, 417–428. [CrossRef] [PubMed]

65. Almeida, J.; Diniz, Y.; Marques, S.F.; Faine, L.; Ribas, B.; Burneiko, R.; Novelli, E.L. The use of the oxidative stress responses as biomarkers in Nile tilapia (*Oreochromis niloticus*) exposed to *in vivo* cadmium contamination. *Environ. Int.* **2002**, *27*, 673–679. [CrossRef]

66. Geret, F.; Serafim, A.; Bebianno, M.J. Antioxidant enzyme activities, metallothioneins and lipid peroxidation as biomarkers in *Ruditapes decussatus*? *Ecotoxicology* **2003**, *12*, 417–426. [CrossRef] [PubMed]

67. Cossu, C.; Doyotte, A.; Jacquin, M.C.; Babut, M.; Exinger, A.; Vasseur, P. Glutathione reductase, selenium-dependent glutathione peroxidase, glutathione levels, and lipid peroxidation in freshwater bivalves, *Unio tumidus*, as biomarkers of aquatic contamination in field studies. *Ecotoxicol. Environ. Saf.* **1997**, *38*, 122–131. [CrossRef] [PubMed]

68. Doyotte, A. Antioxidant enzymes, glutathione and lipid peroxidation as relevant biomarkers of experimental or field exposure in the gills and the digestive gland of the freshwater bivalve *Unio tumidus*. *Aquat. Toxicol.* **1997**, *39*, 93–110. [CrossRef]

69. Thévenod, F.; Lee, W.-K. Cadmium and cellular signaling cascades: Interactions between cell death and survival pathways. *Arch. Toxicol.* **2013**, *87*, 1743–1786. [CrossRef] [PubMed]

70. Gobe, G.; Crane, D. Mitochondria, reactive oxygen species and cadmium toxicity in the kidney. *Toxicol. Lett.* **2010**, *198*, 49–55. [CrossRef] [PubMed]

71. Cannino, G.; Ferruggia, E.; Luparello, C.; Rinaldi, A.M. Cadmium and mitochondria. *Mitochondrion* **2009**, *9*, 377–384. [CrossRef] [PubMed]

72. Poteet, E.; Winters, A.; Xie, L.; Ryou, M.-G.; Liu, R.; Yang, S.-H. *In vitro* protection by pyruvate against cadmium-induced cytotoxicity in hippocampal HT-22 cells. *J. Appl. Toxicol.* **2014**, *34*, 903–913. [CrossRef] [PubMed]

73. Wei, X.; Qi, Y.; Zhang, X.; Qiu, Q.; Gu, X.; Tao, C.; Huang, D.; Zhang, Y. Cadmium induces mitophagy through ROS-mediated PINK1/Parkin pathway. *Toxicol. Mech. Methods* **2014**, *24*, 504–511. [CrossRef] [PubMed]

74. Pi, H.; Xu, S.; Zhang, L.; Guo, P.; Li, Y.; Xie, J.; Tian, L.; He, M.; Lu, Y.; Li, M.; *et al.* Dynamin 1-like-dependent mitochondrial fission initiates overactive mitophagy in the hepatotoxicity of cadmium. *Autophagy* **2013**, *9*, 1780–1800. [CrossRef] [PubMed]

75. Belyaeva, E.A.; Sokolova, T.V.; Emelyanova, L.V.; Zakharova, I.O. Mitochondrial electron transport chain in heavy metal-induced neurotoxicity: Effects of cadmium, mercury, and copper. *Sci. World J.* **2012**, *2012*, 1–14. [CrossRef] [PubMed]

76. Dorta, D.J.; Leite, S.; deMarco, K.C.; Prado, I.M.R.; Rodrigues, T.; Mingatto, F.E.; Uyemura, S.A.; Santos, A.C.; Curti, C. A proposed sequence of events for cadmium-induced mitochondrial impairment. *J. Inorg. Biochem.* **2003**, *97*, 251–257. [CrossRef]

77. Li, M.; Xia, T.; Jiang, C.S.; Li, L.J.; Fu, J.L.; Zhou, Z.C. Cadmium directly induced the opening of membrane permeability pore of mitochondria which possibly involved in cadmium-triggered apoptosis. *Toxicology* **2003**, *194*, 19–33. [CrossRef]

78. Lee, W.-K.; Bork, U.; Gholamrezaei, F.; Thévenod, F. Cd^{2+}-induced cytochrome c release in apoptotic proximal tubule cells: Role of mitochondrial permeability transition pore and Ca^{2+} uniporter. *Am. J. Physiol. Ren. Physiol.* **2005**, *288*, F27–F39. [CrossRef] [PubMed]

79. Xu, S.; Pi, H.; Chen, Y.; Zhang, N.; Guo, P.; Lu, Y.; He, M.; Xie, J.; Zhong, M.; Zhang, Y.; *et al.* Cadmium induced Drp1-dependent mitochondrial fragmentation by disturbing Calcium homeostasis in its hepatotoxicity. *Cell Death Dis.* **2013**, *4*, e540. [CrossRef] [PubMed]

80. Guo, P.; Pi, H.; Xu, S.; Zhang, L.; Li, Y.; Li, M.; Cao, Z.; Tian, L.; Xie, J.; Li, R.; *et al.* Melatonin Improves mitochondrial function by promoting MT1/SIRT1/PGC-1 α-dependent mitochondrial biogenesis in cadmium-induced hepatotoxicity *in vitro. Toxicol. Sci.* **2014**, *142*, 182–195. [CrossRef] [PubMed]

81. Kil, I.S.; Shin, S.W.; Yeo, H.S.; Lee, Y.S.; Park, J.-W. Mitochondrial NADP+-dependent isocitrate dehydrogenase protects cadmium-induced apoptosis. *Mol. Pharmacol.* **2006**, *70*, 1053–1061. [CrossRef] [PubMed]

82. Leung, K.M.Y.; Taylor, A.C.; Furness, R.W. Temperature-dependent physiological responses of the dogwhelk *Nucella lapillus* to cadmium exposure. *J. Mar. Biol. Assoc. UK* **2000**, *80*, 647–660. [CrossRef]

83. Yang, J.; Liu, D.; He, Y.; Wang, L. Mitochondrial energy metabolism in the hepatopancreas of freshwater crabs (*Sinopotamon henanense*) after cadmium exposure. *Environ. Sci. Process. Impacts* **2015**, *17*, 156–165. [CrossRef] [PubMed]

84. Chen, C.-Y.; Zhang, S.-L.; Liu, Z.-Y.; Tian, Y.; Sun, Q. Cadmium toxicity induces ER stress and apoptosis via impairing energy homeostasis in cardiomyocytes. *Biosci. Rep.* **2015**, *35*, e00214. [CrossRef] [PubMed]

85. Andersen, O. Chelation of cadmium. *Environ. Health Perspect.* **1984**, *54*, 249–266. [CrossRef] [PubMed]

86. Viarengo, A.; Burlando, B.; Ceratto, N.; Panfoli, I. Antioxidant role of metallothioneins: A comparative overview. *Cell. Mol. Biol.* **2000**, *46*, 407–417. [PubMed]

87. Sato, M.; Kondoh, M. Recent studies on metallothionein: Protection against toxicity of heavy metals and oxygen free radicals. *Tohoku J. Exp. Med.* **2002**, *196*, 9–22. [CrossRef] [PubMed]

88. Cols, N.; Romero-Isart, N.; Bofill, R.; Capdevila, M.; Gonzàlez-Duarte, P.; Gonzàlez-Duarte, R.; Atrian, S. *In vivo* copper- and cadmium-binding ability of mammalian metallothionein beta domain. *Protein Eng.* **1999**, *12*, 265–269. [CrossRef] [PubMed]

89. Palacios, O.; Pagani, A.; Pérez-Rafael, S.; Egg, M.; Höckner, M.; Brandstätter, A.; Capdevila, M.; Atrian, S.; Dallinger, R. Shaping mechanisms of metal specificity in a family of metazoan metallothioneins: Evolutionary differentiation of mollusc metallothioneins. *BMC Biol.* **2011**, *9*. [CrossRef] [PubMed]

90. Pinter, T.B.J.; Irvine, G.W.; Stillman, M.J. Domain Selection in Metallothionein 1A: Affinity-controlled mechanisms of zinc binding and cadmium exchange. *Biochemistry* **2015**, *54*, 5006–5016. [CrossRef] [PubMed]

91. Jara-Biedma, R.; González-Dominguez, R.; García-Barrera, T.; Lopez-Barea, J.; Pueyo, C.; Gómez-Ariza, J.L. Evolution of metallotionein isoforms complexes in hepatic cells of *Mus musculus* along cadmium exposure. *Biometals* **2013**, *26*, 639–650. [CrossRef] [PubMed]

92. Palacios, O.; Pérez-Rafael, S.; Pagani, A.; Dallinger, R.; Atrian, S.; Capdevila, M. Cognate and noncognate metal ion coordination in metal-specific metallothioneins: The *Helix pomatia* system as a model. *J. Biol. Inorg. Chem.* **2014**, *19*, 923–935. [CrossRef] [PubMed]

93. Höckner, M.; Stefanon, K.; de Vaufleury, A.; Monteiro, F.; Pérez-Rafael, S.; Palacios, O.; Capdevila, M.; Atrian, S.; Dallinger, R. Physiological relevance and contribution to metal balance of specific and non-specific Metallothionein isoforms in the garden snail, *Cantareus aspersus. Biometals* **2011**, *24*, 1079–1092. [CrossRef] [PubMed]

94. Tomas, M.; Domènech, J.; Capdevila, M.; Bofill, R.; Atrian, S. The sea urchin metallothionein system: Comparative evaluation of the SpMTA and SpMTB metal-binding preferences. *FEBS Open Biol.* **2013**, *3*, 89–100. [CrossRef] [PubMed]

95. Egli, D.; Domènech, J.; Selvaraj, A.; Balamurugan, K.; Hua, H.; Capdevila, M.; Georgiev, O.; Schaffner, W.; Atrian, S. The four members of the *Drosophila* metallothionein family exhibit distinct yet overlapping roles in heavy metal homeostasis and detoxification. *Genes Cells* **2006**, *11*, 647–658. [CrossRef] [PubMed]

96. Höckner, M.; Dallinger, R.; Stürzenbaum, S.R. Nematode and snail metallothioneins. *J. Biol. Inorg. Chem.* **2011**, *16*, 1057–1065. [CrossRef] [PubMed]

97. Domènech, J.; Bofill, R.; Tinti, A.; Torreggiani, A.; Atrian, S.; Capdevila, M. Comparative insight into the Zn(II)-, Cd(II)- and Cu(I)-binding features of the protozoan *Tetrahymena pyriformis* MT1 metallothionein. *Biochim. Biophys. Acta* **2008**, *1784*, 693–704. [CrossRef] [PubMed]

98. Wang, Q.; Xu, J.; Chai, B.; Liang, A.; Wang, W. Functional comparison of metallothioneins MTT1 and MTT2 from *Tetrahymena thermophila*. *Arch. Biochem. Biophys.* **2011**, *509*, 170–176. [CrossRef] [PubMed]

99. Eckschlager, T.; Adam, V.; Hrabeta, J.; Figova, K.; Kizek, R. Metallothioneins and cancer. *Curr. Protein Pept. Sci.* **2009**, *10*, 360–375. [CrossRef] [PubMed]

100. Pedersen, M.Ø.; Larsen, A.; Stoltenberg, M.; Penkowa, M. The role of metallothionein in oncogenesis and cancer prognosis. *Prog. Histochem. Cytochem.* **2009**, *44*, 29–64. [CrossRef] [PubMed]

101. Günther, V.; Lindert, U.; Schaffner, W. The taste of heavy metals: Gene regulation by MTF-1. *Biochim. Biophys. Acta* **2012**, *1823*, 1416–1425. [CrossRef] [PubMed]

102. Höckner, M.; Stefanon, K.; Schuler, D.; Fantur, R.; de Vaufleury, A.; Dallinger, R. Coping with cadmium exposure in various ways: The two helicid snails *Helix pomatia* and *Cantareus aspersus* share the metal transcription factor-2, but differ in promoter organization and transcription of their Cd-metallothionein genes. *J. Exp. Zool. A Ecol. Genet. Physiol.* **2009**, *311*, 776–787. [CrossRef] [PubMed]

103. Höckner, M.; Dallinger, R.; Stürzenbaum, S.R. Metallothionein gene activation in the earthworm (*Lumbricus rubellus*). *Biochem. Biophys. Res. Commun.* **2015**, *460*, 537–542. [CrossRef] [PubMed]

104. Freisinger, E.; Vašák, M. Cadmium in metallothioneins. *Met. Ions Life Sci.* **2013**, *11*, 339–371. [PubMed]

105. Langelueddecke, C.; Lee, W.-K.; Thévenod, F. Differential transcytosis and toxicity of the hNGAL receptor ligands cadmium-metallothionein and cadmium-phytochelatin in colon-like Caco-2 cells: Implications for *in vivo* cadmium toxicity. *Toxicol. Lett.* **2014**, *226*, 228–235. [CrossRef] [PubMed]

106. Wolff, N.A.; Lee, W.-K.; Thévenod, F. Role of Arf1 in endosomal trafficking of protein-metal complexes and cadmium-metallothionein-1 toxicity in kidney proximal tubule cells. *Toxicol. Lett.* **2011**, *203*, 210–218. [CrossRef] [PubMed]

107. Li, L.; Dong, F.; Xu, D.; Du, L.; Yan, S.; Hu, H.; Lobe, C.G.; Yi, F.; Kapron, C.M.; Liu, J. Short-term, low-dose cadmium exposure induces hyperpermeability in human renal glomerular endothelial cells. *J. Appl. Toxicol.* **2015**. [CrossRef] [PubMed]

108. Onodera, A.; Tani, M.; Michigami, T.; Yamagata, M.; Min, K.-S.; Tanaka, K.; Nakanishi, T.; Kimura, T.; Itoh, N. Role of megalin and the soluble form of its ligand RAP in Cd-metallothionein endocytosis and Cd-metallothionein-induced nephrotoxicity *in vivo*. *Toxicol. Lett.* **2012**, *212*, 91–96. [CrossRef] [PubMed]

109. Delalande, O.; Desvaux, H.; Godat, E.; Valleix, A.; Junot, C.; Labarre, J.; Boulard, Y. Cadmium-glutathione solution structures provide new insights into heavy metal detoxification. *FEBS J.* **2010**, *277*, 5086–5096. [CrossRef] [PubMed]

110. Liebeke, M.; Garcia-Perez, I.; Anderson, C.J.; Lawlor, A.J.; Bennett, M.H.; Morris, C.A.; Kille, P.; Svendsen, C.; Spurgeon, D.J.; Bundy, J.G. Earthworms produce phytochelatins in response to arsenic. *PLoS ONE* **2013**, *8*, e81271.

111. Hall, J.; Haas, K.L.; Freedman, J.H. Role of MTL-1, MTL-2, and CDR-1 in mediating cadmium sensitivity in Caenorhabditis elegans. *Toxicol. Sci.* **2012**, *128*, 418–426. [CrossRef] [PubMed]

112. Hirano, T.; Ueda, H.; Kawahara, A.; Fujimoto, S. Cadmium toxicity on cultured neonatal rat hepatocytes: Biochemical and ultrastructural analyses. *Histol. Histopathol.* **1991**, *6*, 127–133. [PubMed]

113. Biagioli, M.; Pifferi, S.; Ragghianti, M.; Bucci, S.; Rizzuto, R.; Pinton, P. Endoplasmic reticulum stress and alteration in Calcium homeostasis are involved in cadmium-induced apoptosis. *Cell Calcium* **2008**, *43*, 184–195. [CrossRef] [PubMed]

114. Gorman, A.M.; Healy, S.J.M.; Jäger, R.; Samali, A. Stress management at the ER: Regulators of ER stress-induced apoptosis. *Pharmacol. Ther.* **2012**, *134*, 306–316. [CrossRef] [PubMed]

115. Luo, B.; Lee, A.S. The critical roles of endoplasmic reticulum chaperones and unfolded protein response in tumorigenesis and anti-cancer therapies. *Oncogene* **2013**, *32*, 805–818. [CrossRef] [PubMed]
116. Liu, F.; Inageda, K.; Nishitai, G.; Matsuoka, M. Cadmium induces the expression of Grp78, an endoplasmic reticulum molecular chaperone, in LLC-PK1 renal epithelial cells. *Environ. Health Perspect.* **2006**, *114*, 859–864. [CrossRef] [PubMed]
117. Liu, J.; Wu, K.C.; Lu, Y.-F.; Ekuase, E.; Klaassen, C.D. Nrf2 protection against liver injury produced by various hepatotoxicants. *Oxid. Med. Cell. Longev.* **2013**, *2013*. [CrossRef] [PubMed]
118. Du, K.; Takahashi, T.; Kuge, S.; Naganuma, A.; Hwang, G.-W. FBXO6 attenuates cadmium toxicity in HEK293 cells by inhibiting ER stress and JNK activation. *J. Toxicol. Sci.* **2014**, *39*, 861–866. [CrossRef] [PubMed]
119. Lim, S.-C.; Hahm, K.-S.; Lee, S.-H.; Oh, S.-H. Autophagy involvement in cadmium resistance through induction of multidrug resistance-associated protein and counterbalance of endoplasmic reticulum stress WI38 lung epithelial fibroblast cells. *Toxicology* **2010**, *276*, 18–26. [CrossRef] [PubMed]
120. Gaubin, Y.; Vaissade, F.; Croute, F.; Beau, B.; Soleilhavoup, J.-P.; Murat, J.-C. Implication of free radicals and glutathione in the mechanism of cadmium-induced expression of stress proteins in the A549 human lung cell-line. *Biochim. Biophys. Acta* **2000**, *1495*, 4–13. [CrossRef]
121. Liu, H.; He, J.; Chi, C.; Shao, J. Differential *HSP70* expression in *Mytilus coruscus* under various stressors. *Gene* **2014**, *543*, 166–173. [CrossRef] [PubMed]
122. Luan, W.; Li, F.; Zhang, J.; Wen, R.; Li, Y.; Xiang, J. Identification of a novel inducible cytosolic *Hsp70* gene in Chinese shrimp *Fenneropenaeus chinensis* and comparison of its expression with the cognate *Hsc70* under different stresses. *Cell Stress Chaperones* **2010**, *15*, 83–93. [CrossRef] [PubMed]
123. Boutet, I.; Tanguy, A.; Rousseau, S.; Auffret, M.; Moraga, D. Molecular identification and expression of heat shock cognate 70 (*HSC70*) and heat shock protein 70 (*HSP70*) genes in the Pacific oyster *Crassostrea gigas*. *Cell Stress Chaperones* **2003**, *8*, 76–85. [CrossRef]
124. Vilaboa, N.E.; Calle, C.; Pérez, C.; de Blas, E.; García-Bermejo, L.; Aller, P. cAMP increasing agents prevent the stimulation of heat-shock protein 70 (*HSP70*) gene expression by cadmium chloride in human myeloid cell lines. *J. Cell Sci.* **1995**, *108 Pt 8*, 2877–2883. [PubMed]
125. Jing, J.; Liu, H.; Chen, H.; Hu, S.; Xiao, K.; Ma, X. Acute effect of copper and cadmium exposure on the expression of heat shock protein 70 in the Cyprinidae fish *Tanichthys albonubes*. *Chemosphere* **2013**, *91*, 1113–1122. [CrossRef] [PubMed]
126. Wang, Z.; Templeton, D.M. Cellular factors mediate cadmium-dependent actin depolymerization. *Toxicol. Appl. Pharmacol.* **1996**, *139*, 115–121. [CrossRef] [PubMed]
127. Go, Y.-M.; Orr, M.; Jones, D.P. Actin cytoskeleton redox proteome oxidation by cadmium. *Am. J. Physiol. Lung Cell. Mol. Physiol.* **2013**, *305*, L831–L843. [CrossRef] [PubMed]
128. Colon Rodriguez, I.; Negron Berrios, J. Effects of cadmium on epigenetics of cytoskeletal genes in CHO cells. *FASEB J.* **2015**, *29*, 884.47.
129. Templeton, D.M.; Liu, Y. Effects of cadmium on the actin cytoskeleton in renal mesangial cells. *Can. J. Physiol. Pharmacol.* **2013**, *91*, 1–7. [CrossRef] [PubMed]
130. Li, W.; Kagan, H.M.; Chou, I.N. Alterations in cytoskeletal organization and homeostasis of cellular thiols in cadmium-resistant cells. *Toxicol. Appl. Pharmacol.* **1994**, *126*, 114–123. [CrossRef] [PubMed]
131. Lau, A.T.Y.; Chiu, J.-F. The possible role of cytokeratin 8 in cadmium-induced adaptation and carcinogenesis. *Cancer Res.* **2007**, *67*, 2107–2113. [CrossRef] [PubMed]
132. Somji, S.; Garrett, S.H.; Toni, C.; Zhou, X.D.; Zheng, Y.; Ajjimaporn, A.; Sens, M.A.; Sens, D. A Differences in the epigenetic regulation of MT-3 gene expression between parental and Cd^{+2} or As^{+3} transformed human urothelial cells. *Cancer Cell Int.* **2011**, *11*. [CrossRef] [PubMed]
133. Calabrese, E.J. Converging concepts: Adaptive response, preconditioning, and the Yerkes–Dodson Law are manifestations of hormesis. *Ageing Res. Rev.* **2008**, *7*, 8–20. [CrossRef] [PubMed]
134. Hoffmann, G.R. A perspective on the scientific, philosophical, and policy dimensions of hormesis. *Dose Response* **2009**, *7*, 1–51. [CrossRef] [PubMed]
135. Calabrese, E.J.; Blain, R.B. The hormesis database: The occurrence of hormetic dose responses in the toxicological literature. *Regul. Toxicol. Pharmacol.* **2011**, *61*, 73–81. [CrossRef] [PubMed]
136. Calabrese, E.J. Hormetic mechanisms. *Crit. Rev. Toxicol.* **2013**, *43*, 580–606. [CrossRef] [PubMed]
137. Chen, Y.Y.; Zhu, J.Y.; Chan, K.M. Effects of cadmium on cell proliferation, apoptosis, and proto-oncogene expression in zebrafish liver cells. *Aquat. Toxicol.* **2014**, *157*, 196–206. [CrossRef] [PubMed]

138. Damelin, L.H.; Vokes, S.; Whitcutt, J.M.; Damelin, S.B.; Alexander, J.J. Hormesis: A stress response in cells exposed to low levels of heavy metals. *Hum. Exp. Toxicol.* **2000**, *19*, 420–430. [CrossRef] [PubMed]
139. Mohammadi-Bardbori, A.; Rannug, A. Arsenic, cadmium, mercury and nickel stimulate cell growth via NADPH oxidase activation. *Chem. Biol. Interact.* **2014**, *224*, 183–188. [CrossRef] [PubMed]
140. Chowdhury, M.J.; Pane, E.F.; Wood, C.M. Physiological effects of dietary cadmium acclimation and waterborne cadmium challenge in rainbow trout: Respiratory, ionoregulatory, and stress parameters. *Comp. Biochem. Physiol. C Toxicol. Pharmacol.* **2004**, *139*, 163–173. [CrossRef] [PubMed]
141. Wirth, D.; Christians, E.; Li, X.; Benjamin, I.J.; Gustin, P. Use of HSF1(−/−) mice reveals an essential role for HSF1 to protect lung against cadmium-induced injury. *Toxicol. Appl. Pharmacol.* **2003**, *192*, 12–20. [CrossRef]
142. Zhang, Y.; Shen, G.; Yu, Y.; Zhu, H. The hormetic effect of cadmium on the activity of antioxidant enzymes in the earthworm Eisenia fetida. *Environ. Pollut.* **2009**, *157*, 3064–3068. [CrossRef] [PubMed]
143. Lefcort, H.; Freedman, Z.; House, S.; Pendleton, M. Hormetic effects of heavy metals in aquatic snails: Is a little bit of pollution good? *Ecohealth* **2008**, *5*, 10–17. [CrossRef] [PubMed]
144. Banjerdkij, P.; Vattanaviboon, P.; Mongkolsuk, S. Cadmium-induced adaptive resistance and cross-resistance to zinc in Xanthomonas campestris. *Curr. Microbiol.* **2003**, *47*, 260–262. [CrossRef] [PubMed]
145. Fujishiro, H.; Kubota, K.; Inoue, D.; Inoue, A.; Yanagiya, T.; Enomoto, S.; Himeno, S. Cross-resistance of cadmium-resistant cells to manganese is associated with reduced accumulation of both cadmium and manganese. *Toxicology* **2011**, *280*, 118–125. [CrossRef] [PubMed]
146. Sengupta, S.; Bhattacharyya, N.P. Oxidative stress-induced cadmium resistance in Chinese hamster V79 cells. *Biochem. Biophys. Res. Commun.* **1996**, *228*, 267–271. [CrossRef] [PubMed]
147. Mcclure, C.D.; Zhong, W.; Hunt, V.L.; Chapman, F.M.; Hill, F.V.; Priest, N.K. Hormesis results in trade-offs with immunity. *Evolution* **2014**, *68*, 2225–2233. [CrossRef] [PubMed]
148. Ng, C.Y.P.; Pereira, S.; Cheng, S.H.; Adam-Guillermin, C.; Garnier-Laplace, J.; Yu, K.N. Combined effects of depleted uranium and ionising radiation on zebrafish embryos. *Radiat. Prot. Dosim.* **2015**, *167*, 1–5. [CrossRef] [PubMed]
149. Belaidi, E.; Beguin, P.C.; Levy, P.; Ribuot, C.; Godin-Ribuot, D. Prevention of HIF-1 activation and iNOS gene targeting by low-dose cadmium results in loss of myocardial hypoxic preconditioning in the rat. *Am. J. Physiol. Heart Circ. Physiol.* **2008**, *294*, H901–H908. [CrossRef] [PubMed]
150. Jing, Y.; Liu, L.Z.; Jiang, Y.; Zhu, Y.; Guo, N.L.; Barnett, J.; Rojanasakul, Y.; Agani, F.; Jiang, B.H. Cadmium increases HIF-1 and VEGF expression through ROS, ERK, and AKT signaling pathways and induces malignant transformation of human bronchial epithelial cells. *Toxicol. Sci.* **2012**, *125*, 10–19. [CrossRef] [PubMed]
151. Li, D.; Katakura, M.; Sugawara, N. Improvement of acute cadmium toxicity by pretreatment with copper salt. *Bull. Environ. Contam. Toxicol.* **1995**, *54*, 878–883. [CrossRef] [PubMed]
152. El-Sharaky, A.S.; Newairy, A.A.; Badreldeen, M.M.; Eweda, S.M.; Sheweita, S.A. Protective role of selenium against renal toxicity induced by cadmium in rats. *Toxicology* **2007**, *235*, 185–193. [CrossRef] [PubMed]
153. Ognjanović, B.I.; Marković, S.D.; Pavlović, S.Z.; Zikić, R.V.; Stajn, A.S.; Saicić, Z.S. Effect of chronic cadmium exposure on antioxidant defense system in some tissues of rats: Protective effect of selenium. *Physiol. Res.* **2008**, *57*, 403–411.
154. Liu, L.; Yang, B.; Cheng, Y.; Lin, H. Ameliorative effects of selenium on cadmium-induced oxidative stress and endoplasmic reticulum stress in the chicken kidney. *Biol. Trace Elem. Res.* **2015**, *167*, 308–319. [CrossRef] [PubMed]
155. Zhou, Y.-J.; Zhang, S.-P.; Liu, C.-W.; Cai, Y.-Q. The protection of selenium on ROS mediated-apoptosis by mitochondria dysfunction in cadmium-induced LLC-PK$_1$ cells. *Toxicol. Vitro* **2009**, *23*, 288–294. [CrossRef] [PubMed]
156. Wang, Y.; Wu, Y.; Luo, K.; Liu, Y.; Zhou, M.; Yan, S.; Shi, H.; Cai, Y. The protective effects of selenium on cadmium-induced oxidative stress and apoptosis via mitochondria pathway in mice kidney. *Food Chem. Toxicol.* **2013**, *58*, 61–67. [CrossRef] [PubMed]
157. Liu, S.; Xu, F.; Yang, Z.; Li, M.; Min, Y.; Li, S. Cadmium-induced injury and the ameliorative effects of selenium on chicken splenic lymphocytes: Mechanisms of oxidative stress and apoptosis. *Biol. Trace Elem. Res.* **2014**, *160*, 340–351. [CrossRef] [PubMed]

158. Brzóska, M.M.; Rogalska, J. Protective effect of zinc supplementation against cadmium-induced oxidative stress and the RANK/RANKL/OPG system imbalance in the bone tissue of rats. *Toxicol. Appl. Pharmacol.* **2013**, *272*, 208–220. [CrossRef] [PubMed]

159. Coogan, T.P.; Bare, R.M.; Waalkes, M.P. Cadmium-induced DNA strand damage in cultured liver cells: Reduction in cadmium genotoxicity following zinc pretreatment. *Toxicol. Appl. Pharmacol.* **1992**, *113*, 227–233. [CrossRef]

160. Zhang, D.; Liu, J.; Gao, J.; Shahzad, M.; Han, Z.; Wang, Z.; Li, J.; Sjölinder, H. Zinc supplementation protects against cadmium accumulation and cytotoxicity in Madin–Darby bovine kidney cells. *PLoS ONE* **2014**, *9*, e103427. [CrossRef] [PubMed]

161. Eybl, V.; Kotyzová, D. Protective effect of manganese in cadmium-induced hepatic oxidative damage, changes in cadmium distribution and trace elements level in mice. *Interdiscip. Toxicol.* **2010**, *3*, 68–72. [CrossRef] [PubMed]

162. Himeno, S.; Yanagiya, T.; Fujishiro, H. The role of zinc transporters in cadmium and manganese transport in mammalian cells. *Biochimie* **2009**, *91*, 1218–1222. [CrossRef] [PubMed]

163. Thévenod, F. Catch me if you can! Novel aspects of cadmium transport in mammalian cells. *Biometals* **2010**, *23*, 857–875. [CrossRef] [PubMed]

164. Leslie, E.M.; Liu, J.; Klaassen, C.D.; Waalkes, M.P. Acquired cadmium resistance in metallothionein-I/II(−/−) knockout cells: Role of the T-type Calcium channel Cacnα1G in cadmium uptake. *Mol. Pharmacol.* **2006**, *69*, 629–639. [CrossRef] [PubMed]

165. Santoyo-Sánchez, M.P.; Pedraza-Chaverri, J.; Molina-Jijón, E.; Arreola-Mendoza, L.; Rodríguez-Muñoz, R.; Barbier, O.C. Impaired endocytosis in proximal tubule from subchronic exposure to cadmium involves angiotensin II type 1 and cubilin receptors. *BMC Nephrol.* **2013**, *14*. [CrossRef] [PubMed]

166. Gena, P.; Calamita, G.; Guggino, W.B. Cadmium impairs albumin reabsorption by down-regulating megalin and ClC5 channels in renal proximal tubule cells. *Environ. Health Perspect.* **2010**, *118*, 1551–1556. [CrossRef] [PubMed]

167. Wolff, N.A.; Abouhamed, M.; Verroust, P.J.; Thévenod, F. Megalin-dependent internalization of cadmium-metallothionein and cytotoxicity in cultured renal proximal tubule cells. *J. Pharmacol. Exp. Ther.* **2006**, *318*, 782–791. [CrossRef] [PubMed]

168. Jenkitkasemwong, S.; Wang, C.-Y.; Mackenzie, B.; Knutson, M.D. Physiologic implications of metal-ion transport by ZIP14 and ZIP8. *Biometals* **2012**, *25*, 643–655. [CrossRef] [PubMed]

169. Thévenod, F.; Wolff, N.A. Iron transport in the kidney: Implications for physiology and cadmium nephrotoxicity. *Metallomics* **2016**. [CrossRef] [PubMed]

170. Thevenod, F.; Friedmann, J.M.; Katsen, A.D.; Hauser, I.A. Up-regulation of multidrug resistance P-glycoprotein via nuclear factor-κB activation protects kidney proximal tubule cells from cadmium- and reactive oxygen species-induced apoptosis. *J. Biol. Chem.* **2000**, *275*, 1887–1896. [CrossRef] [PubMed]

171. Long, Y.; Li, Q.; Wang, Y.; Cui, Z. MRP proteins as potential mediators of heavy metal resistance in zebrafish cells. *Comp. Biochem. Physiol. Part C Toxicol. Pharmacol.* **2011**, *153*, 310–317. [CrossRef] [PubMed]

172. L'hoste, S.; Chargui, A.; Belfodil, R.; Duranton, C.; Rubera, I.; Mograbi, B.; Poujeol, C.; Tauc, M.; Poujeol, P. CFTR mediates cadmium-induced apoptosis through modulation of ROS level in mouse proximal tubule cells. *Free Radic. Biol. Med.* **2009**, *46*, 1017–1031. [CrossRef] [PubMed]

173. Bard, S.M. Multixenobiotic resistance as a cellular defense mechanism in aquatic organisms. *Aquat. Toxicol.* **2000**, *48*, 357–389. [CrossRef]

174. Ferreira, M.; Costa, J.; Reis-Henriques, M.A. ABC transporters in fish species: A review. *Front. Physiol.* **2014**, *5*, 266. [CrossRef] [PubMed]

175. Ivanina, A.; Sokolova, I. Effect of cadmium exposure on P-glycoprotein expression and activity in eastern oysters, *Crassostrea virginica*. *FASEB J.* **2008**, *22*, 757.11. [CrossRef] [PubMed]

176. Legeay, A.; Achard-Joris, M.; Baudrimont, M.; Massabuau, J.-C.; Bourdineaud, J.-P. Impact of cadmium contamination and oxygenation levels on biochemical responses in the Asiatic clam *Corbicula fluminea*. *Aquat. Toxicol.* **2005**, *74*, 242–253. [CrossRef] [PubMed]

177. Achard, M. Induction of a multixenobiotic resistance protein (MXR) in the Asiatic clam *Corbicula fluminea* after heavy metals exposure. *Aquat. Toxicol.* **2004**, *67*, 347–357. [CrossRef] [PubMed]

178. Eufemia, N.A.; Epel, D. Induction of the multixenobiotic defense mechanism (MXR), P-glycoprotein, in the mussel *Mytilus californianus* as a general cellular response to environmental stresses. *Aquat. Toxicol.* **2000**, *49*, 89–100. [CrossRef]

179. Zucchi, S.; Corsi, I.; Luckenbach, T.; Bard, S.M.; Regoli, F.; Focardi, S. Identification of five partial ABC genes in the liver of the Antarctic fish Trematomus bernacchii and sensitivity of ABCB1 and ABCC2 to Cd exposure. *Environ. Pollut.* **2010**, *158*, 2746–2756. [CrossRef] [PubMed]

180. Sanni, B.; Cherkasov, A.; Sokolova, I.M. Mitochondrial aconitase is sensitive to oxidative stress induced by cadmium and elevated temperatures but not protected by uncoupling proteins in eastern oysters *Crassostrea virginica*. *FASEB J.* **2007**, *21*, A819–d–820.

181. Cherkasov, A.S.; Biswas, P.K.; Ridings, D.M.; Ringwood, A.H.; Sokolova, I.M. Effects of acclimation temperature and cadmium exposure on cellular energy budgets in the marine mollusk *Crassostrea virginica*: Linking cellular and mitochondrial responses. *J. Exp. Biol.* **2006**, *209*, 1274–1284. [CrossRef] [PubMed]

182. Sokolova, I.M. Cadmium effects on mitochondrial function are enhanced by elevated temperatures in a marine poikilotherm, *Crassostrea virginica* Gmelin (Bivalvia: Ostreidae). *J. Exp. Biol.* **2004**, *207*, 2639–2648. [CrossRef] [PubMed]

183. Ivanina, A.V.; Taylor, C.; Sokolova, I.M. Effects of elevated temperature and cadmium exposure on stress protein response in eastern oysters *Crassostrea virginica* (Gmelin). *Aquat. Toxicol.* **2009**, *91*, 245–254. [CrossRef] [PubMed]

184. Vergauwen, L.; Knapen, D.; Hagenaars, A.; Blust, R. Hypothermal and hyperthermal acclimation differentially modulate cadmium accumulation and toxicity in the zebrafish. *Chemosphere* **2013**, *91*, 521–529. [CrossRef] [PubMed]

185. Tran, D.; Boudou, A.; Massabuau, J.-C. How water oxygenation level influences cadmium accumulation pattern in the Asiatic clam *Corbicula fluminea*: A laboratory and field study. *Environ. Toxicol. Chem.* **2001**, *20*, 2073–2080. [CrossRef] [PubMed]

186. Leung, K.M.Y.; Svavarsson, J.; Crane, M.; Morritt, D. Influence of static and fluctuating salinity on cadmium uptake and metallothionein expression by the dogwhelk *Nucella lapillus* (L.). *J. Exp. Mar. Biol. Ecol.* **2002**, *274*, 175–189. [CrossRef]

187. Pascoe, D.; Evans, S.A.; Woodworth, J. Heavy metal toxicity to fish and the influence of water hardness. *Arch. Environ. Contam. Toxicol.* **1986**, *15*, 481–487. [CrossRef] [PubMed]

188. Niyogi, S.; Wood, C.M. Biotic ligand model, a flexible tool for developing site-specific water quality guidelines for metals. *Environ. Sci. Technol.* **2004**, *38*, 6177–6192. [CrossRef] [PubMed]

189. Lavoie, M.; Fortin, C.; Campbell, P.G.C. Influence of essential elements on cadmium uptake and toxicity in a unicellular green alga: The protective effect of trace zinc and cobalt concentrations. *Environ. Toxicol. Chem.* **2012**, *31*, 1445–1452. [CrossRef] [PubMed]

190. Vidal, M.-L.; Bassères, A.; Narbonne, J.-F. Influence of temperature, pH, oxygenation, water-type and substrate on biomarker responses in the freshwater clam *Corbicula fluminea* (Müller). *Comp. Biochem. Physiol. C Toxicol. Pharmacol.* **2002**, *132*, 93–104. [CrossRef]

191. Pando, M.P.; Pinchak, A.B.; Cermakian, N.; Sassone-Corsi, P. A cell-based system that recapitulates the dynamic light-dependent regulation of the vertebrate clock. *Proc. Natl. Acad. Sci. USA* **2001**, *98*, 10178–10183. [CrossRef] [PubMed]

192. Sandbichler, A.M.; Aschberger, T.; Pelster, B. A method to evaluate the efficiency of transfection reagents in an adherent zebrafish cell line. *BioRes. Open Access* **2013**, *2*, 20–27. [CrossRef] [PubMed]

193. Brzóska, M.M.; Borowska, S.; Tomczyk, M. Antioxidants as a Potential Preventive and Therapeutic Strategy for Cadmium. *Curr. Drug Targets* **2015**, *16*. [CrossRef]

194. Ivanova, J.; Gluhcheva, Y.; Arpadjan, S.; Mitewa, M. Effects of cadmium and monensin on renal and cardiac functions of mice subjected to subacute cadmium intoxication. *Interdiscip. Toxicol.* **2014**, *7*, 111–115. [CrossRef] [PubMed]

195. Smith, S.W. The role of chelation in the treatment of other metal poisonings. *J. Med. Toxicol.* **2013**, *9*, 355–369. [CrossRef] [PubMed]

International Journal of
Molecular Sciences

MDPI

Review

Possible Immune Regulation of Natural Killer T Cells in a Murine Model of Metal Ion-Induced Allergic Contact Dermatitis

Kenichi Kumagai [1,2], Tatsuya Horikawa [3,†], Hiroaki Shigematsu [1,2,†], Ryota Matsubara [1,2],
Kazutaka Kitaura [2], Takanori Eguchi [2,4], Hiroshi Kobayashi [2,5], Yasunari Nakasone [1],
Koichiro Sato [1], Hiroyuki Yamada [6], Satsuki Suzuki [7], Yoshiki Hamada [1] and Ryuji Suzuki [2,*]

[1]   Department of Oral and Maxillofacial Surgery, School of Dental Medicine, Tsurumi University,
       2-3-1 Tsurumi, Tsurumi-ku, Yokohama 230-8501, Japan; kumagai-kenichi@tsurumi-u.ac.jp (K.Ku.);
       shigematsu-h@tsurumi-u.ac.jp (H.S.); matsubara-ryota@tsurumi-u.ac.jp (R.M.);
       from7ritoyou@gmail.com (Y.N.); sato-ki@tsurumi-u.ac.jp (K.S.); hamada-y@tsurumi-u.ac.jp (Y.H.)
[2]   Department of Rheumatology and Clinical Immunology, Clinical Research Center for Rheumatology and
       Allergy, Sagamihara National Hospital, National Hospital Organization, 18-1 Sakuradai, Minami-ku,
       Sagamihara 252-0392, Japan; k-kitaura@sagamihara-hosp.gr.jp (K.Ki.); fhb19830419@yahoo.co.jp (T.E.);
       hirok0614@yahoo.co.jp (H.K.)
[3]   Department of Dermatology, Nishi-Kobe Medical Center, 5-7-1 Kojidai, Kobe 651-2273, Japan;
       thorikaw@med.kobe-u.ac.jp
[4]   Department of Oral and Maxillofacial Surgery, Toshiba Rinkan Hospital, 7-9-1 Kamitsuruma, Minami-ku,
       Sagamihara 252-0385, Japan
[5]   Department of Oral and Maxillofacial Surgery, Shonan Tobu Hospital, 500 Nishikubo,
       Chigasaki 253-0083, Japan
[6]   Division of Oral Maxillofacial Surgery, Department of Reconstructive Oral and Maxillofacial Surgery, Iwate
       Medical University School of Dentistry, Morioka, Iwate 020-8505, Japan; yamadah@iwate-med.ac.jp
[7]   Section of Biological Sciences, Research Center for Odontology, The Nippon Dental University School of Life
       Dentistry at Tokyo, 1-9-20 Fujimi, Chiyoda-ku, Tokyo 102-8159, Japan; satsukis@tky.ndu.ac.jp
*   Correspondance: r-suzuki@sagamihara-hosp.gr.jp; Tel.: +81-42-742-8311; Fax: +81-42-742-7990
†   These authors contributed equally to this work.

Academic Editor: Reinhard Dallinger
Received: 20 November 2015; Accepted: 7 January 2016; Published: 12 January 2016

Abstract: Metal often causes delayed-type hypersensitivity reactions, which are possibly mediated by accumulating T cells in the inflamed skin, called irritant or allergic contact dermatitis. However, accumulating T cells during development of a metal allergy are poorly characterized because a suitable animal model is unavailable. We have previously established novel murine models of metal allergy and found accumulation of both metal-specific T cells and natural killer (NK) T cells in the inflamed skin. In our novel models of metal allergy, skin hypersensitivity responses were induced through repeated sensitizations by administration of metal chloride and lipopolysaccharide into the mouse groin followed by metal chloride challenge in the footpad. These models enabled us to investigate the precise mechanisms of the immune responses of metal allergy in the inflamed skin. In this review, we summarize the immune responses in several murine models of metal allergy and describe which antigen-specific responses occur in the inflamed skin during allergic contact dermatitis in terms of the T cell receptor. In addition, we consider the immune regulation of accumulated NK T cells in metal ion–induced allergic contact dermatitis.

Keywords: metal allergy; delayed-type hypersensitivity; natural killer T cells; metal ion; T cell receptor

1. Introduction

Metal allergy is categorized as a delayed-type hypersensitivity (DTH) reaction, and the number of patients with metal allergy has increased because metal is increasingly used in jewelry, surgical instruments, and dental restorations [1]. Traditionally, nickel (Ni), cobalt (Co), palladium (Pd), and chromium (Cr) have been reported as causal metals of allergic contact dermatitis (ACD).

Metal ions are thought to form highly defined geometricall but reversible coordination complexes with partner molecules within the body, thereby becoming antigens. Metal allergy is characterized by the recruitment of lymphocytes and inflammatory cells, including T cells, dendritic cells, granulocytes, and macrophages, to the site of allergic inflammation. The immune response of infiltrating T cells during the elicitation phase of DTH is thought to be autoreactive reaction [2], yet their antigen specificity for each metal has not been determined.

In this review, we summarize the immune responses in several murine models of metal allergy and which antigen-specific responses occur in the inflamed skin during ACD in terms of the T cell receptor (TCR). In addition, we consider the immune regulation of accumulated natural killer (NK) T cells in metal ion-induced ACD.

2. Establishment of Murine Models for Metal Allergy

Most studies of metal allergies have been performed *in vitro* using sensitized human lymphocytes, and it is believed that T cells are essential for metal allergy, as well as contact hypersensitivity to classical haptens [2]. However, the mechanism through which pathogenic T cells at the sites of allergic inflammation contribute to the development of metal allergy has not been explored. To examine the involvement of these accumulated T cells in metal allergy, the establishment of suitable animal models is desired.

Previous studies have generated several murine models of metal allergy in the inflamed ear [3,4]. They examined the effect of lipopolysaccharide (LPS) on allergies to Ni and other metals in mice, and concluded that LPS is a major inducer of metal allergies and potently promotes such allergies via innate immunity and histidine decarboxylase (HDC) induction in cells. LPS, a component of the cell walls of many Gram-negative bacteria, signals through a complex consisting of CD14, toll-like receptor (TLR) 4, and MD-2 [5], leading to secretion of pro-inflammatory cytokines and potent activation of the innate immune system [6]. LPS stimulates innate immunity via TLR4 and stimulates dendritic cells expressing TLRs, inducing maturation and modifying adaptive immunity. Recently, Schmidt *et al.* reported that Ni directly stimulates human TLR4, but not mouse TLR4, which may be crucial for the development of contact allergy [7].

In addition to ACD, skin exposure to metal without primary sensitization produces irritant contact dermatitis (ICD). ICD is a non-specific inflammatory dermatitis mainly mediated by the direct toxicity of chemicals into the skin, which triggers the activation of the innate immune system. ACD corresponds to a delayed-type hypersensitivity response, and skin inflammation is caused by antigen-specific T cells. However, the precise immune response both of the ICD and ACD has not been elucidated because of their coexistence with the DTH reaction. Previous murine models of metal allergy have not been used to investigate the differences between ICD and ACD, and have not revealed the histopathological features of the DTH response, such as spongiosis and liquefaction degeneration of the epidermis caused by intercellular edema and lymphocytic infiltration in the epidermis. Therefore, previous metal allergic murine models have not provided complete evidence of the metal ion-specific immune response.

Based on these previous reports, our novel murine models of allergies against metals such as Pd, Ni, and Cr were induced through sensitization by administration of metal chlorides and LPS into the mouse groin followed by challenge with several kinds of metal chlorides in the footpad of mice [8–10]. In our metal allergic mouse models, we found significant differences between ICD and ACD. The footpad swelling was measured every day after the challenge with metal chloride injection into the footpad. The footpad swelling reached maximum at one day after challenge in all mice. The footpad swelling at one day after challenge in ICD and ACD mice was similar. The footpad swelling was reduced at seven days after challenge in ICD mice. However, the footpad swelling continued for 14 days after challenge in ACD mice [10]. We found that both types of contact dermatitis could not be differentiated by the macroscopic appearance because footpad swelling was the same in both mouse models at one day after challenge. However, at seven days after challenge, inflammatory cells accumulated around the superficial venular bed and extended into the epidermis in ACD mice. Furthermore, the epidermal keratinocytes were partially separated, creating spongiotic dermatitis.

Regarding the skin pathophysiology in ACD, the inflammatory process is caused by a type-IV cell-mediated reaction occurring in a sensitized individual after renewed contact with the antigen. In the acute stage of ACD, the spongiosis and vesicle formation are mainly observed histologically. In subacute eczema, acanthosis increases with the formation of a parakeratotic cornified layer in inflamed skin. The epithelial ridges become elongated and broadened, and hyperkeratosis is evident in chronic eczema. The infiltration of dermal lymphocytes and vascular dilatations can be found at all stages. The pathological features in our metal ion-induced ACD mice at seven days after challenge resemble these features. Therefore, our model enables investigation of the precise immune response of metal-allergic inflamed skin.

In terms of the LPS interaction with metal allergy, we identified the accumulation of T cells in the footpads of metal chloride– and LPS-sensitized, metal chloride–challenged mice compared with those of sensitized mice without LPS administration. This finding may suggest that signals from TLR4 induced by LPS stimuli might be required for the induction of metal allergy. Thus, we speculate that LPS is indispensable for the establishment of metal allergy in mice and plays an important role as an adjuvant to induce metal-specific T cells in the inflamed skin of metal ion–induced ACD mice.

Our studies demonstrated successful establishment of a murine model of metal ion-induced ACD with appropriate pathological reactions that reflect the metal ion-specific immune response. In addition, we identified the allergen-specific positive T cells in metal ion-induced ACD mice.

3. Immune Responses in Inflamed Metal ACD in Terms of the TCR

We next focused on the accumulation of metal-specific T cells in the inflamed skin. Similar to contact hypersensitivity to classical haptens, the infiltration of lymphocytes into the sites of allergic inflammation is essential for mediating metal allergy. In addition to T cells, mast cells, NK cells, and granulocytes have been shown to be involved in metal-induced inflammation in other animal models [3,4,11,12]. However, the involvement of the antigen specificity and diversity of pathogenic T cells in the development of metal allergy remains unclear.

During an immune response to metal ion–associated antigens, the clonal T cells expansion is observed due to the antigen-specific immune response. T cells bearing TCRs recognize antigens in the form of peptide fragments in relation with major histocompatibility complex (MHC) class I and II molecules on antigen-presenting cells. The high specificity of T cells is determined by the TCRs displayed on their surface, which are heterodimers of an α- and β-chain (Va and Vb) or a γ- and δ-chain (V gamma and V delta). We previously developed an adaptor ligation-mediated polymerase chain reaction (AL-PCR) method that allows TCR repertoires to be defined based on the expression levels of transcripts, even when only small numbers of cells are available [13]. This method enables amplification of all variable regions of the rearranged TCR genes through PCR cycles without skewing. Applying this method to a microplate hybridization assay is simple and reproducible, and enables rapid analysis of TCR repertoires in several diseases in humans and mice [14–16]. Random insertions of non-germinal element (N) nucleotides or deletions of nucleotides have been identified in the VN (D) NJ junction region, designated as the complementary determining region 3 (CDR3), and are thought to be responsible for the recognition of antigenic peptide content [17,18]. Thus, any specific recognition of antigens by CDR3 will lead to clonal expansion of T cells. Because CDR3 has different sequences and lengths, it is possible to analyze the diversity of TCRs using a CDR3-size spectratyping method that provides a rapid scan of all TCR Va and Vb-region transcripts grouped according to the utilized V-region gene and chain length [19].

Metal ions can induce the proliferation of human T cells *in vitro*, and limited TCR repertoires have been observed in human T cells from patients with metal allergies [20–22]. However, TCR analysis has not been performed in the previous murine models of metal allergy. We hypothesized that restricted TCR usage may reflect the prolonged exposure of the host immune system to putative metal ion–associated antigens. To elucidate the immune responses to metal ion–associated antigens, we examined the characteristics of the T cells within allergic tissue specimens in our models in terms of TCR repertoires and CDR3-size spectratyping. Recently, we successfully identified each metal-specific TCR repertoire in the inflamed skin at the elicitation phase in Pd-, Ni-, and Cr-allergic mice [8–10]. Unexpectedly, the TCR repertoire was found to be specific: TCR Va14Ja18 and Vb8.2 were commonly used in BALB/c and C57BL/6 mice by Ni-induced allergic skin. TCR Va14Ja18 and Vb8-2 are a major subset of mouse invariant NK T (iNK T) cells, and we found that iNK T cells accumulated in the inflamed skin of Ni-allergic mice. In Cr allergy, the infiltrating T cells from Cr-induced ACD mice expressed CD4+ and used a specific TCR repertoire expressing TCR Va11-1, Va14-1, Vb8-2, and Vb14-1. In Ni and Cr allergies, we have identified iNK T cells in lymphocytic infiltrates at a high frequency during the elicitation phase. Thus, we suggest that metal-specific T cells driven by invariant NK T cells might contribute to the pathogenesis of metal allergy.

4. The Possible Role of NK T Cells in Metal Allergy

NK T cells (also called iNK T cells and Va14 iNK T cells) were originally defined as T cells that expressed CD161c and other receptors typical of NK cells. This definition has subsequently been refined to specify a subpopulation of T cells that express a TCR with an invariant Va14Ja18 TCR Va chain paired with a restricted subset of TCR Vb chains [23]. NK T cells recognize glycolipids (microbial derived or self-glycolipids) presented in the context of CD1d (a MHC class I–like molecule) [24]. After activation, NK T cells produce Th1- and Th2-type molecules such as interferon (IFN)-γ and interleukin-4 to regulate adaptive immune responses [25]. Mature NK T cells in mice are mainly CD4 single-positive or CD4/CD8 double-negative cells. NK T cells predominantly express Vb8.2 joined with a bias of a particular Jb usage including Jb2.1, Jb2.5, or Jb2.7 [26,27].

Recently, NK T cells were also identified in the lesional skin of contact dermatitis (including Ni allergy) in humans [28]. Thus, NK T cells may be involved in the pathogenesis of contact hypersensitivity reactions, possibly by recognizing endogenous antigens such as self-lipids. Water-soluble Ni salts can pass through the ion channels of cell membranes or can be taken up by cells along with other molecules such as proteins and amino acids, so Ni can be incorporated readily into the body [29]. We hypothesized that Ni allergy can be triggered by intracellular Ni that can combine with self-lipids on cellular membranes. We therefore speculate that Ni-modified self-lipids may activate NK T cells in the lesions of Ni-induced dermatitis in our model. A previous study suggested a role of NK T cells in the sensitization phase of contact hypersensitivity (CHS) by promoting the survival and maturation of dendritic cells in draining lymph nodes [30]. The authors indicated that the dendritic cell–NK T cell interaction has a pivotal role in the sensitization phase of CHS. Regarding the cross-reactivity of metal allergy, a previous study suggested that, among the tested ultrapure metals (Ni, Pd, Co, Cr, Cu, and Au), only Ni and Pd cross-reacted [31]. These results may imply that NK T cells participate in the cross-reactivity of metal allergy. A previous study showed that the Th1 subset of NK T cells has a dominant and critical role in allergy. In addition, the immune response in Ni allergy was skewed towards a Th1 response with a minimal Th2 response [32]. The Th1 subset of NK T cells may expand locally in response to Ni.

Based on these findings, we summarized the schematic mechanism of metal ion–induced ACD and the possible role of NK T cells in both sensitization and elicitation phases (Figure 1a,b). We propose that NK T cells act as an early amplification step in the innate immune response in the sensitization phase, and bridge to the adaptive immunity in the elicitation phase. We thus speculate that the regulation of excessive NK T cell immune responses might be a new diagnostic or therapeutic target for metal allergy.

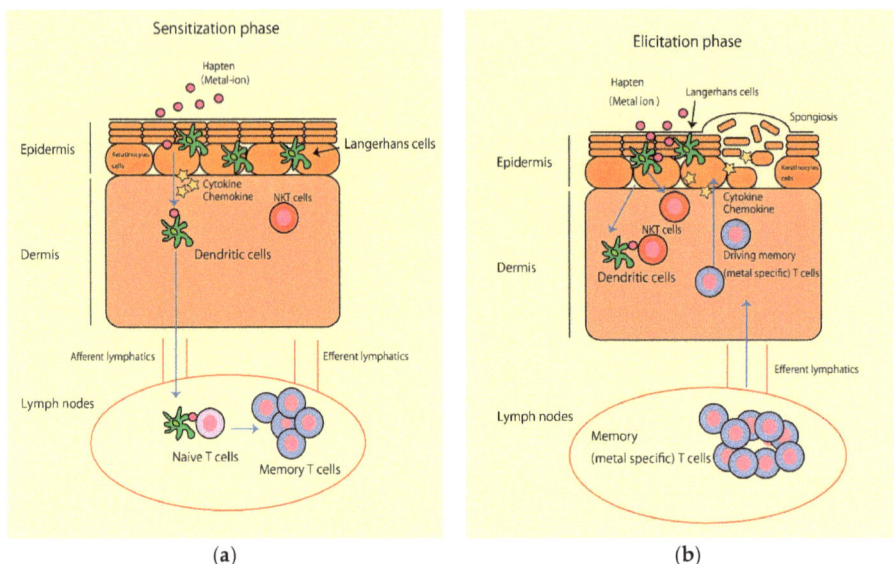

Figure 1. Schematic mechanism of metal ion–induced allergic contact dermatitis. (**a**) Sensitization phase in metal allergy. Step 1: Metal ions form complexes with partner molecules within the body, thereby becoming antigens. A hapten (metal ion) combines with a native protein and activates keratinocytes (KCs), cutaneous Langerhans cells (LCs), and dermal dendritic cells (DCs) through the innate immune system; Step 2: Activated DCs capture antigens, mature, and migrate to the regional lymph nodes via afferent lymphatics; Step 3: Migrated DCs present antigens to naive T cells in draining lymph nodes. NK T cells affect DC functions and regulate the excessive immune response; (**b**) Elicitation phase in metal allergy. Step 1: KCs are activated by re-exposure to haptens and produce various cytokines and chemokines that activate endothelial cells and draining memory metal-specific T cells; Step 2: Infiltrated metal-specific effector T cells are activated and produce proinflammatory cytokines and chemokines that activate KCs and induce further inflammatory cell infiltration; Step 3: NK T cells may regulate the excessive acquired immune response caused by metal-specific effector T cells.

5. Conclusions and Future Direction

Our novel mouse model is useful for understanding the pathological roles of T cells in metal ion allergy, and the regulation of NK T cells might be involved in the immune responses of metal allergy. Further studies using this mouse model of metal allergy will contribute to the diagnosis of metal allergy in terms of the regulation of NK T cells, as well as to new treatments to control metal-specific T cells.

Acknowledgments: Acknowledgments: This work was supported by JSPS KAKENHI Grant-in-Aid for Scientific Research C Grant Number 15K11329.

Author Contributions: Author Contributions: Equal contributors to this work: Kenichi Kumagai, Tatsuya Horikawa, Hiroaki Shigematsu. Conceived and designed the review: Kenichi Kumagai, Hiroaki Shigematsu, Tatsuya Horikawa, Satsuki Suzuki, Ryuji Suzuki. Contributed reagents/materials/analysis tools: Hiroaki Shigematsu, Takanori Eguchi, Hiroshi Kobayashi, Ryota Matsubara, Yasunari Nakasone, Koichiro Sato, Hiroyuki Yamada, Kazutaka Kitaura. Wrote the manuscript: Kenichi Kumagai, Tatsuya Horikawa, Hiroaki Shigematsu. Coordinated the study as the principal investigator: Ryuji Suzuki, Yoshiki Hamada.

Conflicts of Interest: Conflicts of Interest: The authors state no conflict of interest.

References

1. Raap, U.; Stiesch, M.; Reh, H.; Kapp, A.; Werfel, T. Investigation of contact allergy to dental metals in 206 patients. *Contact Dermatitis* **2009**, *60*, 339–343. [CrossRef] [PubMed]
2. Kalish, R.S. Recent developments in the pathogenesis of allergic contact dermatitis. *Arch. Dermatol.* **1991**, *127*, 1558–1563. [CrossRef] [PubMed]
3. Sato, N.; Kinbara, M.; Kuroishi, T.; Kimura, K.; Iwakura, Y.; Ohtsu, H.; Sugawara, S.; Endo, Y. Lipopolysaccharide promotes and augments metal allergies in mice, dependent on innate immunity and histidine decarboxylase. *Clin. Exp. Allergy* **2007**, *37*, 743–751. [CrossRef] [PubMed]
4. Kinbara, M.; Sato, N.; Kuroishi, T.; Takano-Yamamoto, T.; Sugawara, S.; Endo, Y. Allergy-inducing nickel concentration is lowered by lipopolysaccharide at both the sensitization and elicitation steps in a murine model. *Br. J. Dermatol.* **2011**, *164*, 356–362. [CrossRef] [PubMed]
5. Nagarajan, N.A.; Kronenberg, M. Invariant NKT cells amplify the innate immune response to lipopolysaccharide. *J. Immunol.* **2007**, *178*, 2706–2713. [CrossRef] [PubMed]
6. Hazlett, L.D.; Li, Q.; Liu, J.; McClellan, S.; Du, W.; Barrett, R.P. NKT cells are critical to initiate an inflammatory response after Pseudomonas aeruginosa ocular infection in susceptible mice. *J. Immunol.* **2007**, *179*, 1138–1146. [CrossRef] [PubMed]
7. Schmidt, M.; Raghavan, B.; Müller, V.; Vogl, T.; Fejer, G.; Tchaptchet, S.; Keck, S.; Kalis, C.; Nielsen, P.J.; Galanos, C.; *et al.* Crucial role for human Toll-like receptor 4 in the development of contact allergy to nickel. *Nat. Immunol.* **2010**, *11*, 814–819. [CrossRef] [PubMed]
8. Kobayashi, H.; Kumagai, K.; Eguchi, T.; Shigematsu, H.; Kitaura, K.; Kawano, M.; Horikawa, T.; Suzuki, S.; Matsutani, T.; Ogasawara, K.; *et al.* Characterization of T cell receptors of Th1 cells infiltrating inflamed skin of a novel murine model of palladium-induced metal allergy. *PLoS ONE* **2013**, *8*, e76385.
9. Eguchi, T.; Kumagai, K.; Kobayashi, H.; Shigematsu, H.; Kitaura, K.; Suzuki, S.; Horikawa, T.; Hamada, Y.; Ogasawara, K.; Suzuki, R. Accumulation of invariant NKT cells into inflamed skin in a novel murine model of nickel allergy. *Cell Immunol.* **2013**, *284*, 163–171. [CrossRef] [PubMed]
10. Shigematsu, H.; Kumagai, K.; Kobayashi, H.; Eguchi, T.; Kitaura, K.; Suzuki, S.; Horikawa, T.; Matsutani, T.; Ogasawara, K.; Hamada, Y.; *et al.* Accumulation of metal-specific T cells in inflamed skin in a novel murine model of chromium-induced allergic contact dermatitis. *PLoS ONE* **2014**, *9*, e85983. [CrossRef] [PubMed]
11. Ishihara, K.; Goi, Y.; Hong, J.J.; Seyama, T.; Ohtsu, H.; Wada, H.; Ohuchi, K.; Hirasawa, N. Effects of nickel on eosinophil survival. *Int. Arch. Allergy Immunol.* **2009**, *149*, 57–60. [CrossRef] [PubMed]
12. Kim, J.Y.; Huh, K.; Lee, K.Y.; Yang, J.M.; Kim, T.J. Nickel induces secretion of IFN-gamma by splenic natural killer cells. *Exp. Mol. Med.* **2009**, *41*, 288–295. [CrossRef] [PubMed]
13. Matsutani, T.; Yoshioka, T.; Tsuruta, Y.; Iwagami, S.; Suzuki, R. Analysis of TCRAV and TCRBV repertoires in healthy individuals by microplate hybridization assay. *Hum. Immunol.* **1997**, *56*, 57–69. [CrossRef]
14. Gotoh, A.; Hamada, Y.; Shiobara, N.; Kumagai, K.; Seto, K.; Horikawa, T.; Suzuki, R. Skew in T cell receptor usage with polyclonal expansion in lesions of oral lichen planus without hepatitis C virus infection. *Clin. Exp. Immunol.* **2008**, *154*, 192–201. [CrossRef] [PubMed]
15. Kumagai, K.; Hamada, Y.; Gotoh, A.; Kobayashi, H.; Kawaguchi, K.; Horie, A.; Yamada, H.; Suzuki, S.; Suzuki, R. Evidence for the changes of antitumor immune response during lymph node metastasis in head and neck squamous cell carcinoma. *Oral Surg. Oral Med. Oral Pathol. Oral Radiol. Endod.* **2010**, *110*, 341–350. [CrossRef] [PubMed]
16. Kitaura, K.; Fujii, Y.; Hayasaka, D.; Matsutani, T.; Shirai, K.; Nagata, N.; Lim, C.K.; Suzuki, S.; Takasaki, T.; Suzuki, R.; *et al.* High clonality of virus-specific T lymphocytes defined by TCR usage in the brains of mice infected with West Nile virus. *J. Immunol.* **2011**, *15*, 3919–3930. [CrossRef] [PubMed]
17. Engel, I.; Hedrick, SM. Site-directed mutations in the VDJ junctional region of a T cell receptor beta chain cause changes in antigenic peptide recognition. *Cell* **1988**, *12*, 473–484. [CrossRef]
18. Davis, M.M.; Bjorkman, P.J. T-Cell antigen receptor genes and T cell recognition. *Nature* **1988**, *4*, 334–395.
19. Puisieux, I.; Even, J.; Pannetier, C.; Jotereau, F.; Favrot, M.; Kourilsky, P. Oligoclonality of tumor-infiltrating lymphocytes from human melanomas. *J. Immunol.* **1994**, *153*, 2807–2818. [PubMed]
20. Hashizume, H.; Seo, N.; Ito, T.; Takigawa, M.; Yagi, H. Promiscuous interaction between gold-specific T cells and APCs in gold allergy. *J. Immunol.* **2008**, *181*, 8096–8102. [CrossRef] [PubMed]

21. Budinger, L.; Neuser, N.; Totzke, U.; Merk, H.F.; Hertl, M. Preferential usage of TCR-Vbeta17 by peripheral and cutaneous T cells in nickel-induced contact dermatitis. *J. Immunol.* **2001**, *167*, 6038–6044. [CrossRef] [PubMed]
22. Silvennoinen-Kassinen, S.; Karvonen, J.; Ikaheimo, I. Restricted and individual usage of T cell receptor β-gene variables in nickel-induced CD4+ and CD8+ cells. *Scand. J. Immunol.* **1998**, *48*, 99–102. [CrossRef] [PubMed]
23. Taniguchi, M.; Harada, M.; Kojo, S.; Nakayama, T.; Wakao, H. The regulatory role of Valpha14 NKT cells in innate and acquired immune response. *Annu. Rev. Immunol.* **2003**, *21*, 483–513. [CrossRef] [PubMed]
24. Sieling, P.A. CD1-Restricted T cells: T cells with a unique immunological niche. *Clin. Immunol.* **2000**, *96*, 3–10. [CrossRef] [PubMed]
25. Bendelac, A.; Rivera, M.N.; Park, S.H.; Roark, J.H. Mouse CD1-specific NK1 T cells: Development, specificity, and function. *Annu. Rev. Immunol.* **1997**, *15*, 535–562. [CrossRef] [PubMed]
26. Fowlkes, B.J.; Kruisbeek, A.M.; Ton-That, H.; Weston, M.A.; Coligan, J.E.; Schwartz, R.H.; Pardoll, D.M. A novel population of T-cell receptor alpha betabearing thymocytes which predominantly expresses a single V β gene family. *Nature* **1987**, *329*, 251–254. [CrossRef] [PubMed]
27. Ceredig, R.; Lynch, F.; Newman, P. Phenotypic properties, interleukin 2 production, and developmental origin of a "mature" subpopulation of Lyt-2L3T4-mouse thymocytes. *Proc. Natl. Acad. Sci. USA* **1987**, *84*, 8578–8582. [CrossRef] [PubMed]
28. Gober, M.D.; Fishelevich, R.; Zhao, Y.; Unutmaz, D.; Gaspari, A.A. Human natural killer T cells infiltrate into the skin at elicitation sites of allergic contact dermatitis. *J. Investig. Dermatol.* **2008**, *128*, 1460–1469. [CrossRef] [PubMed]
29. Lin, X.; Costa, M. Transformation of human osteoblasts to anchorage-independent growth by insoluble nickel particles. *Environ. Health Perspect.* **1994**, *102*, 289–292. [CrossRef] [PubMed]
30. Shimizuhira, C.; Otsuka, A.; Honda, T. Natural killer T cells are essential for the development of contact hypersensitivity in BALB/c mice. *J. Investig. Dermatol.* **2014**, *134*, 2709–2718. [CrossRef] [PubMed]
31. Kinbara, M.; Nagai, Y.; Takano-Yamamoto, T.; Sugawara, S.; Endo, Y. Cross-reactivity among some metals in a murine metal allergy model. *Br. J. Dermatol.* **2011**, *165*, 1022–1029. [CrossRef] [PubMed]
32. Almogren, A.; Adam, M.H.; Shakoor, Z.; Gadelrab, M.O.; Musa, H.A. Th1 and Th2 cytokine profile of CD4 and CD8 positive peripheral blood lymphocytes in nickel contact dermatitis. *Cent. Eur. J. Immunol.* **2013**, *38*, 100–106. [CrossRef]

International Journal of
Molecular Sciences

MDPI

Article

Fexofenadine Suppresses Delayed-Type Hypersensitivity in the Murine Model of Palladium Allergy

Ryota Matsubara [1,2], Kenichi Kumagai [1,2,*,†], Hiroaki Shigematsu [1,2], Kazutaka Kitaura [2], Yasunari Nakasone [1,2], Satsuki Suzuki [3], Yoshiki Hamada [1] and Ryuji Suzuki [2,*,†]

[1]   Department of Oral and Maxillofacial Surgery, School of Dental Medicine, Tsurumi University,
       2-3-1 Tsurumi, Tsurumi-ku, Yokohama 230-8501, Japan; matsubara-ryota@tsurumi-u.ac.jp (R.M.);
       shigematsu-h@tsurumi-u.ac.jp (H.S.); nakasone-yasunari@tsurumi-u.ac.jp (Y.N.);
       hamada-y@tsurumi-u.ac.jp (Y.H.)
[2]   Department of Rheumatology and Clinical Immunology, Clinical Research Center for Rheumatology and
       Allergy, Sagamihara National Hospital, National Hospital Organization, 18-1 Sakuradai, Minami-ku,
       Sagamihara 252-0392, Japan; k-kitaura@sagamihara-hosp.gr.jp
[3]   Section of Biological Science, Research Center for Odontology, The Nippon Dental University School of
       Life Dentistry at Tokyo, 1-9-20 Fujimi, Chiyoda-ku, Tokyo 102-8159, Japan; satsukis@tky.ndu.ac.jp
*   Correspondence: kumagai-kenichi@tsurumi-u.ac.jp (K.K.); r-suzuki@sagamihara-hosp.gr.jp (R.S.);
       Tel.: +81-45-581-1001 (K.K) +81-42-742-8311 (R.S.); Fax: +81-45-573-9599 (K.K); +81-42-742-7990 (R.S.)
†   These authors contributed equally to this work.

Received: 17 May 2017; Accepted: 20 June 2017; Published: 25 June 2017

Abstract: Palladium is frequently used in dental materials, and sometimes causes metal allergy. It has been suggested that the immune response by palladium-specific T cells may be responsible for the pathogenesis of delayed-type hypersensitivity in study of palladium allergic model mice. In the clinical setting, glucocorticoids and antihistamine drugs are commonly used for treatment of contact dermatitis. However, the precise mechanism of immune suppression in palladium allergy remains unknown. We investigated inhibition of the immune response in palladium allergic mice by administration of prednisolone as a glucocorticoid and fexofenadine hydrochloride as an antihistamine. Compared with glucocorticoids, fexofenadine hydrochloride significantly suppressed the number of T cells by interfering with the development of antigen-presenting cells from the sensitization phase. Our results suggest that antihistamine has a beneficial effect on the treatment of palladium allergy compared to glucocorticoids.

Keywords: metal allergy; palladium; anti histamine; fexofenadine hydrochloride; corticosteroid

1. Introduction

Metal allergy is an inflammatory disease categorized as a delayed-type hypersensitivity (DTH) reaction, and is thought to be caused by the release of ions that function as haptens from metal materials [1]. Adverse skin reactions to metal ions, such as intractable dermatitis, pustulosis palmaris et plantaris, and incompatibility reactions to metal-containing biomaterials are serious problems [2–4]. It has been suggested that metal allergy is associated with the infiltration of lymphocytes into sites of allergic inflammation [5–8]. Among metals in biomaterials, palladium (Pd), which is frequently used in industry, jewelry, surgical instruments, dental implants [9], and dental materials (it is a common constituent of dental restorative alloys for crowns and bridges) [10,11], causes metal allergy [12]. The incidence of patients sensitized to Pd has increased in recent years [13–15].

Metal-induced DTH is driven by T cell sensitization to metal ions. T cells are largely responsible for the development of metal allergy in mice and humans [16,17]. We have previously generated a novel murine model of Pd allergy [6] and found an accumulation of Pd-specific T cells that exert cytotoxic effects and secrete inflammatory mediators to produce skin reactions at 7 days after the final challenge. It has been suggested that antihistamine drugs are effective for treatment of Pd-induced allergic contact dermatitis (ACD) mice within 24 h after the challenge [18]. However, the precise mechanism of inhibition for effective treatment of Pd allergy is unclear during the long period after the challenge.

Glucocorticoids and antihistamine drugs are the first selection for symptomatic treatment of contact dermatitis [19]. Although glucocorticoids are effective for most inflammatory skin disorders, their use is limited by local and systemic side effects [20]. Therefore, clinicians have to balance the benefits of treatment with glucocorticoids or antihistamines against the potential long-term detrimental effects on the DTH.

In the present study, we examined inhibition of the immune response in the Pd-induced murine model by administration of prednisolone as a glucocorticoid and fexofenadine hydrochloride as an antihistamine.

2. Results

2.1. Footpad Swelling in Pd-Induced Allergic Mice Administered with Fexofenadine or Prednisolone

All experimental protocols were determined as described in the Materials and Methods (Figure 1, Tables 1 and 2). To address how inflamed skin was inhibited by fexofenadine or prednisolone, we employed Pd-induced allergic contact dermatitis (ACD) mice (Figure 2). To determine the dose of fexofenadine and prednisolone, we have converted the approximate dose of human dosage into our Pd-induced ACD mice. In the preliminary experiment, the dose of fexofenadine (5 mg/kg) showed suppression equal to that shown by the dose of fexofenadine (50 mg/kg) in Pd-induced ACD mice (Figure S1A). Otherwise, the dose of prednisolone (5 mg/kg) showed the suppression of footpad swelling, but the dose of prednisolone (10 mg/kg) was more effective for suppression after each challenge (Figure S1B). Pd-induced ACD mice treated with fexofenadine (5 mg/kg, and 10 mg/kg) and prednisolone (5 mg/kg, and 10 mg/kg) did not show weight loss and poor fur condition, however, the Pd-induced ACD mice treated with prednisolone (30 mg/kg, and 50 mg/kg) showed weight loss, and poor fur condition (Figure S2). Thus, we determined the dose of fexofenadine and prednisolone as 10 mg/kg.

In all groups, the peak of footpad swelling was observed at 24 h after challenge. Next, we examined whether administration of fexofenadine or prednisolone affected the footpad swelling. Administration of fexofenadine and prednisolone at 10 mg/kg/d significantly suppressed the increase in footpad swelling compared with untreated ACD mice (Figure 2). The administration of fexofenadine led to significant inhibition of the increase in footpad swelling compared with prednisolone. Regarding the administration of fexofenadine, the fexofenadine administration before challenge only (Group B) reduced the responses to a lesser degree than the fexofenadine administration before both sensitization and challenge (Group A). With respect to the administration of prednisolone, the prednisolone administration before challenge only (Group D) reduced the responses to a lesser degree than the fexofenadine administration before sensitization (Group C). As shown in a representative photograph, clear swelling and redness of the footpad were observed at 7 days after the last challenge, which were suppressed by fexofenadine administration (Figure 3).

Table 1. Experimental groups of contact dermatitis to Pd-induced allergy.

BALB/cAJcl	Sensitization	Challenge for Elicitation
ACD	+	+
ICD	−	+
Saline	+	−

ACD: allergic contact dermatitis; ICD: irritant contact dermatitis.

Table 2. Experimental groups of suppression of contact dermatitis to Pd-induced allergy by oral administration of fexofenadine or prednisolone.

Group	Sensitization	Challenge for Elicitation
A	F *	F
B	–	F
C	F	P **
D	–	P

* Fexofenadine hydrochloride; ** Prednisolone.

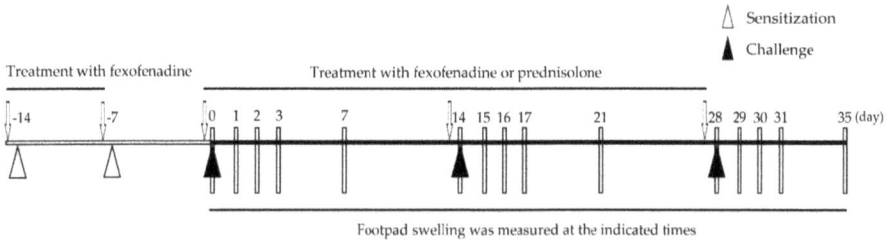

Footpad swelling was measured at the indicated times

Figure 1. Schedule of the sensitization and elicitation of palladium (Pd)-induced allergic mice and oral administration of fexofenadine or prednisolone. The start of the first challenge was defined as day 0. Sensitization using palladium was performed every week throughout the experimental period from day –14 to day 0. The challenge for elicitation using Pd was performed every 2 weeks throughout the experimental period from day 0 to day 35. Arrows show when fexofenadine hydrochloride and prednisolone were orally administered at 10 mg/kg at 1 h before sensitization or challenge. Between challenges, the bar indicates the measurement day of left and right footpads. Footpad swelling was measured at 28, 29, 30, 31, and 35 days. All mice were sacrificed at day 35, and the footpads were taken as samples.

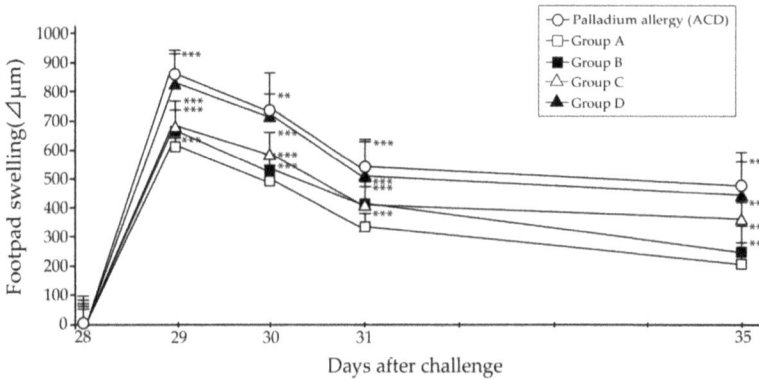

Figure 2. Effect of fexofenadine hydrochloride and prednisolone on footpad swelling with Pd induced allergy. Fexofenadine and prednisolone were orally administered to Pd-induced ACD mice at 1 h before sensitization or challenge. Furthermore, Pd-induced ACD mice received fexofenadine hydrochloride or prednisolone, and the mice were divided into four groups as shown in Table 2. ○, Pd-induced ACD; □, Group A; ■, Group B; △, Group C; ▲, Group D. Bars and error bars indicate the mean + standard deviation (SD). ** $p < 0.01$ is considered as very significant, and *** $p < 0.001$ is considered as extremely significant.

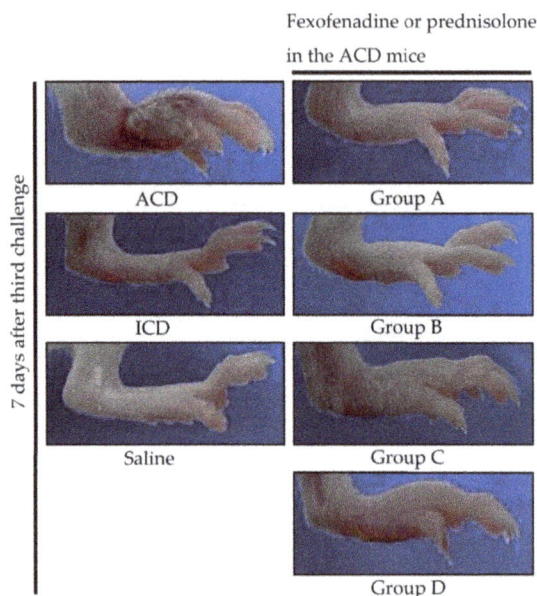

Figure 3. Macroscopic findings in Pd-induced ACD mice administered fexofenadine or prednisolone. Photographs show footpad swelling of a representative mouse at 7 days after the last challenge of Pd. In comparison with the footpad of Pd allergy mice, in terms of the redness and swelling, the feet of the groups A, B, C, and D showed decreases; in particular groups A and B showed dominant decreases.

2.2. Histological and Immunohistochemical Analyses of F4/80 and T cell Markers in Footpads of Pd-Induced Allergic Mice Administered with Fexofenadine or Prednisolone

To verify whether antigen-presenting cells (APCs) and T cells infiltrated into the site of inflamed skin, we analyzed the footpad skin of each mouse by immunohistochemistry. Hematoxylin and eosin (H&E) staining showed epithelial acanthosis, and epidermal spongiosis and liquefaction degeneration of the epithelial basal layer infiltrated with a dense mononuclear cells in the epithelial basal layer and upper dermis of ACD mice (Figure 4A,D,E). In particular, epidermal keratinocytes were separated partially (Figure 4A). The inflammatory reaction in the footpads was diminished in mice that received only fexofenadine (Figure 4B,C). Immunohistochemical staining showed that cluster of differentiation (CD) 3-positive T cells and F4/80-positive cells predominantly existed in the epithelial basal layer and upper dermis of ACD mice (Figure 4F,K). CD3-positive T cells were mainly present in the epithelial basal layer and upper dermis of mice that received prednisolone at a higher degree than in mice that received fexofenadine (Figure 4F–J). F4/80-positive cells had diminished in the epithelial basal layer of mice administered with fexofenadine compared with prednisolone (Figures 4K–O and 5). Furthermore, administration of fexofenadine from sensitization significantly suppressed F4/80-positive cells in the epithelial basal layer (Figures 4L and 5). In contrast, prednisolone did not suppress the infiltrating F4/80-positive cells (Figures 4O and 5).

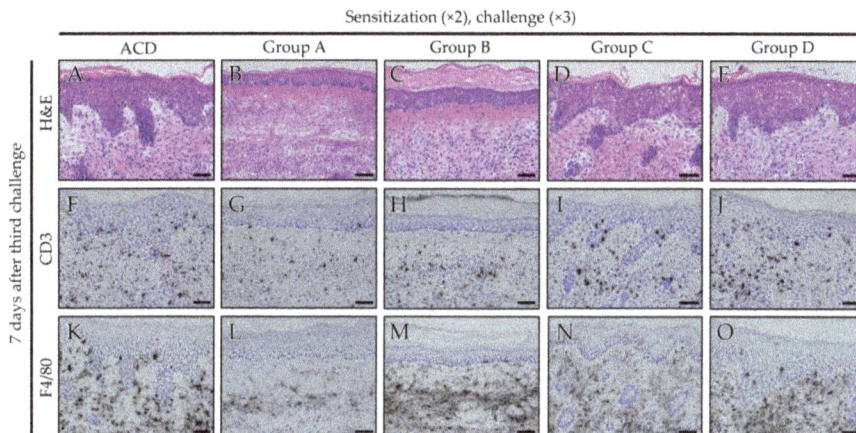

Figure 4. Histopathology and immunohistochemical analyses of accumulated T cells and antigen-presenting cells (APCs) in Pd-induced ACD mice treated with fexofenadine or prednisolone. Histopathology and immunohistochemical analyses of monoclonal antibody (mAb) that binds to a surface molecule on mature macrophages and dendritic cells (F4/80-positive cells) and cluster of differentiation (CD) 3-positive T cells in footpad tissues. Frozen footpad tissue sections were stained with hematoxylin and eosin (H&E) (**A–E**) and anti-CD3 (**F–J**) and anti-F4/80 (**K–O**) antibodies at 7 days after the last challenge. Scale bar = 100 µm.

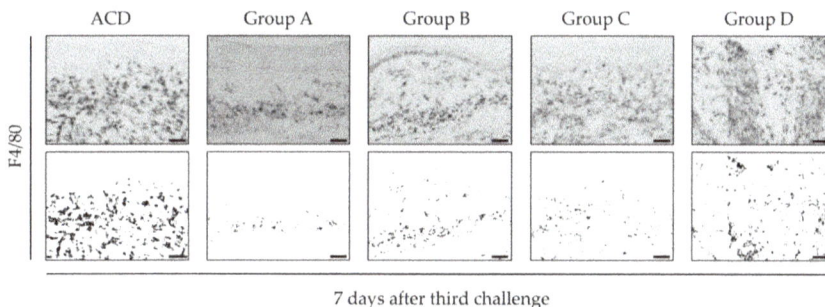

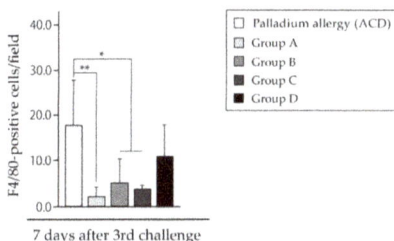

Figure 5. Imaging calculation of F4/80-positive cells. Measurements were made in six different regions of each sample, and mean values were used for statistical analysis. The figure show the representative photos of each group. The number of F4/80-positive cells at 7 days after the last challenge per target area was counted using ImageJ software with Java-based color deconvolution (v. 1.41). Bars and error bars indicate the mean + standard deviation (SD). * $p < 0.05$ is considered as significant, and ** $p < 0.01$ is considered as very significant. Scale bar = 100 µm.

2.3. Expression Levels of T Cell Markers, Related Cytokines, and APC-Derived Signals Induced by Fexofenadine and Prednisolone

We investigated the expression levels of T cell markers, related cytokines, and APC-derived signals by quantitative polymerase chain reaction (qPCR). Cytokine expression at 7 days after challenge was measured in footpads of Pd-challenged mice. In mice that received fexofenadine, CD4 levels were significantly lower than in ACD mice. However, CD8 levels in skin were not significantly different (Figure 6A). Notably, mice that received fexofenadine before both sensitization and challenge had suppressed messenger RNA (mRNA) expression levels of APC-derived signals (Figure 6B). Prednisolone administration before challenge only did not affect T cell suppression. We also examined the expression levels of proinflammatory cytokine interleukin (IL)-1β, histidine decarboxylase (HDC), T helper type (Th) 1 cytokines (tumor necrosis factor (TNF)-α, interferon (IFN)-γ, and IL-12), Th2 cytokines (IL-4 and IL-5), and Th1/Th2 cytokine imbalance (IFN-γ/IL-4 and IFN-γ/IL-5). Mice that received fexofenadine before both sensitization and challenge had suppressed mRNA expression levels of IL-1β, HDC, and TNF-α (Figure 7A,B). Furthermore, mice that received only fexofenadine had suppressed mRNA expression levels of IL-4 and IL-5 (Figure 7C). However, there was no significant differences in the Th1/Th2 cytokine imbalance (Figure 7D). In contrast, mice that received prednisolone did not have suppressed mRNA expression levels of these T cell and APC markers.

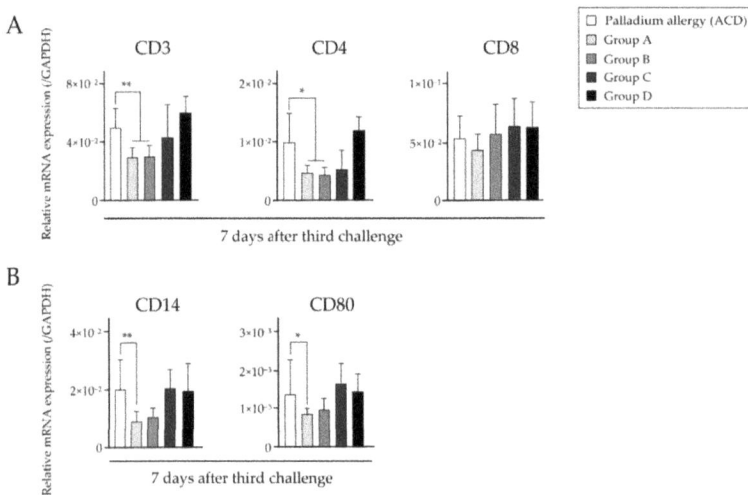

Figure 6. Effects of fexofenadine hydrochloride on mRNA expression levels indicating T cell phenotypes and T cell-related markers in Pd-induced ACD mice. Messenger RNA (mRNA) expression levels of CD3, CD4, CD8 (**A**), CD14, and CD80 (**B**) in footpads were assessed at 7 days after challenge. Glyceraldehyde-3-phosphate dehydrogenase (GAPDH) gene expression was used as an internal control. Bars and error bars indicate the mean + standard deviation (SD). Statistical significance was tested by the unpaired Mann Whitney test. * $p < 0.05$ is considered as significant and ** $p < 0.01$ is considered as very significant.

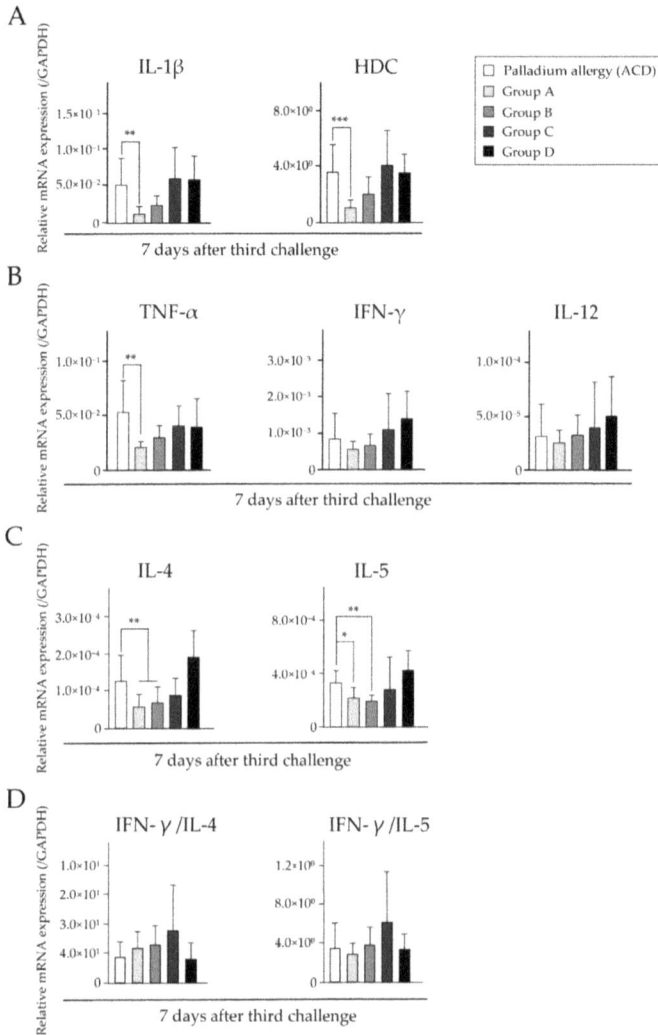

Figure 7. Effects of fexofenadine hydrochloride on the mRNA expression levels of T cell-related cytokines and chemokines in Pd-induced ACD mice. mRNA expression of (**A**) proinflammatory cytokine interleukin (IL) -1β, Histidine decarboxylase (HDC), (**B**) T helper type (Th) 1 cytokines (tumor necrosis factor (TNF) -α, Interferon (IFN) -γ, and IL-12), (**C**) T helper type (Th) 2 cytokines (IL-4 and IL-5), and (**D**) Th1/Th2 cytokine imbalance (IFN-γ/IL-4 and IFN-γ/IL-5) are shown. GAPDH gene expression was used as an internal control. Bars and error bars indicate the mean + standard deviation (SD). Statistical significance was tested by the unpaired Mann Whitney test. * $p<0.05$ is considered as significant, ** $p < 0.01$ is considered as very significant, and *** $p < 0.001$ is considered as extremely significant. TNF: tumor necrosis factor; HDC: histidine decarboxylase; IFN: interferon; IL: interleukin.

3. Discussion

In this study, we demonstrated that fexofenadine inhibited immune responses of Pd allergy compared with glucocorticoids. Furthermore, the administration of fexofenadine hydrochloride during the sensitization phase significantly suppressed the number of T cells by interfering with the development of APCs in the elicitation phase.

It has been suggested that antihistamine drugs interfere with T cell-related inflammatory molecules at various points of the DTH immune cascade [21]. In our previous study, administration of antihistamine suppressed footpad swelling in the Pd allergy mouse model [18]. However, the precise mechanism of immunosuppression in Pd allergy remained unknown. In the present study, we investigated the suppressive effect of antihistamine in terms of the timing of administration compared with prednisolone, and found new aspects of antihistamine in metal allergy.

Metal allergy has two phases in the cutaneous hypersensitivity response: sensitization and elicitation. During the sensitization phase, cutaneous APCs take up and process antigens, and then migrate to regional lymph nodes where they activate T cells with consequent production of memory T cells, which localize in the dermis. In the elicitation phase, subsequent exposure to the sensitization chemical leads to antigen presentation to memory T cells in the dermis [22,23]. ACD is based on DTH reactions that involve antigen presentation by APCs and the T cell response [24]. Sensitization of metal-reactive T cells requires disruption of the barrier function and interaction of metal ions with major histocompatibility complex/peptide complexes presented by APCs to naive T cells [25]. Activation of APCs is essential for the establishment of sensitization with haptens [26]. Fexofenadine has been reported to suppress APC functions concerning skin immunity [27], whereas prednisolone did not. Thus, our results suggest that fexofenadine may act on APCs in DTH.

DTH led to spongiosis in the inflamed footpad skin and epithelial hyperplasia. Moreover, T cells are recruited to allergic sites in ACD lesions. In the present study, the administration of fexofenadine suppressed the development of CD3-positive T cells and F4/80-positive cells in the Pd-induced ACD mice (Figures 4 and 5). In addition, apparent spongiosis changes, edema, and epithelial hyperplasia almost remained unaffected. The activated T cells cause edema and epidermal spongiosis [26]. These findings indicate that CD3-positive T cells were inhibited by fexofenadine in response to Pd in the skin of Pd-induced ACD mice. Local redness and swelling of footpads were followed by an increase in T cell numbers [28], which were also macroscopically found after fexofenadine administration. Therefore, fexofenadine was capable of inhibiting Pd allergy just prior to sensitization or elicitation (Figure 4G,H). F4/80 is a monoclonal antibody (mAb) that binds to a surface molecule on mature macrophages and dendritic cells. It is used to explore the role of epidermal and dermal cells as APCs during the induction of DTH in mice [29]. This study indicated that F4/80-positive cells were inhibited by fexofenadine administration before sensitization in response to Pd in the skin of Pd-induced ACD mice (Figure 4L). Fexofenadine administration before sensitization significantly suppressed the development of both CD3-positive T cells and F4/80-positive cells. A previous study has reported that olopatadine administration before sensitization reduces the DTH reaction to a lesser degree than olopatadine administration before challenge only [30].

We further investigated the cytokine expression profiles in the footpads of Pd-induced ACD mice. A previous study has suggested that Th1-type cytokines or Th1 and Th2 cytokines are preferentially produced in response to Pd [6,12]. It has been suggested that fexofenadine suppresses the infiltration of lymphocytes and Th2 cytokine production [31]. Fexofenadine treatment has prevented the development of allergy in challenge even in sensitized mice [31]. In this study, fexofenadine reduced the expression levels of CD3, CD4, IL-4, and IL-5. Thus, our results were consistent with previous studies. The hypersensitivity allergic inflammatory response involves histamine and Th2 cells. Histamine is synthesized by a catalytic enzyme called HDC. Mast cells and basophils are the most well-described cellular sources of histamine, but dendritic cells and T cells can also express HDC upon stimulation [32]. During the sensitization phase of DTH, cutaneous APCs migrate into skin-draining lymph nodes. It is known that the development of DTH is based on APC maturation and presentation of antigens to naive T cells in lymph nodes [33]. The maturation and migration of APCs are mainly regulated by TNF-α on mast cells [34]. Furthermore, mast cells and mast cell-associated TNF-α are supposed to be important contributors to the migration of hapten-bearing APCs from the initial stages of the sensitization phase of DTH [35]. Histamine not only alters APC and T cell functions, but also inhibits CD8 T cell proliferation [31], thereby deviating the immunoregulatory effects of the immune response to a Th2 response. Histamine enhances the secretion of Th2 cytokines, such as IL-4 and IL-5, and inhibits the production of Th1 cytokines such as IFN-γ and IL-12 [36]. Th2 cytokines were suppressed because fexofenadine inhibits histamine. Hence, treatment with fexofenadine prevented the increases in IL-4 and IL-5 levels, but IFN-γ levels were hardly affected (Figure 7B,C). The Th1/Th2 cytokine imbalance with a predominance of Th2 cytokines is known to be crucial for the pathogenesis of hypersensitivity allergic diseases [21].Fexofenadine controls host immune responses by suppressing Th2 responses, and also affects the imbalance of Th1/Th2 cytokines (Figure 7D). Therefore, modulation of the Th1/Th2 cytokine imbalance is a promising strategy for the treatment of hypersensitivity allergic diseases.

Current symptomatic treatment for metal allergy employs antihistamines or general immune suppressants such as corticosteroids [37]. Corticosteroids have various inhibitory actions in the immune system as well as harmful effects. Both immediate hypersensitivity and DTH reactions are rare in patients treated with systemic corticosteroids. However, delayed-type reactions to systemically administered steroids may present as a generalized dermatitis [38]. Although steroids are used as first-line therapy for allergic diseases, they can induce ACD in sensitized patients [39]. The therapeutic use of corticosteroids requires a careful balance between helping the patient by reducing inflammatory manifestations of the disease and causing harm from the toxic side-effects [40]. For this reason, allergic disease is often treated in combination with other drugs to keep the doses and toxic effects to a minimum. Fexofenadine hydrochloride effectively inhibits histamine-induced cutaneous wheals more than prednisolone [41,42]. Histamine is the major mediator of acute inflammatory and immediate hypersensitivity responses, and has been suggested to affect chronic inflammation [43]. Fexofenadine suppresses not only histamine, but also APC functions concerning skin immunity [27]. Histamine plays a modulatory role in the cortical arousal system mainly through the H1 receptor (H1R). The fexofenadine inhibited H1R occupancy in the human brain than the other antihistamine, therefore, fexofenadine hydrochloride has high safety compared with other antihistamines [44,45]. For allergic responses, the best treatment is avoidance of the allergen, which is not always possible. Although the unwanted immune responses that occur in allergy present somewhat difficult problems, the therapeutic goal in all cases is to inhibit the harmful immune response and thus avoid damage to tissues or disruption of their function [46]. We elucidated immune suppression of fexofenadine in Pd allergy, and demonstrated the inhibitory activities of APCs and infiltrating T cells in Pd allergy. Our data may indicate the clinical utility of fexofenadine as a candidate therapeutic drug for Pd allergy.

4. Materials and Methods

4.1. Ethics Statement

This study was performed in strict accordance with the recommendations in the Guidelines for Care and Use of Laboratory Animals of Tsurumi University and the Clinical Research Center of Sagamihara National Hospital, Japan. All animal experiments were performed according to the relevant ethical requirements with approval from the committees for animal experiments at Tsurumi University (approval number 27A061, issued on 9 June 2015) and the Clinical Research Center for Rheumatology and Allergy, Sagamihara National Hospital (approval number H22-2010-1, issued on 31 March 2010). All surgeries were performed under three types of mixed anesthetic agents, and all efforts were made to minimize suffering.

4.2. Animals

BALB/cAJcl mice (4-week-old females, each experimental group included six mice, average weight = 18.3 g) were purchased from CLEA Japan (Tokyo, Japan). During the study period, all mice remained in good health, and they were assigned randomly to various groups. Animals were acclimated for at least 7 days before experimental use. Mice at 11–12 weeks of age were used for experiments, and their weights ranged from 22.7 to 25.0 g (average 24.1 g). All mice were kept in plastic cages with a lid made of stainless steel wire at our conventional animal facility that maintained the temperature at 19–23°C and humidity at 30–70% with a 12-h day/night cycle. Food and water were available ad libitum.

4.3. Reagents

PdCl$_2$ (>99% pure) was purchased from Wako Pure Chemical Industries (Osaka, Japan). Lipopolysaccharide (LPS) from *Escherichia coli* (O55:B5) prepared by phenol–water extraction was purchased from Sigma (St Louis, MO, USA). Prednisolone (>99% pure) was purchased from Sigma-Aldrich Japan (Tokyo, Japan). Fexofenadine hydrochloride (>98% pure) was purchased from Tokyo Chemical Industry Co., Ltd. (Tokyo Japan). PdCl$_2$, prednisolone, fexofenadine hydrochloride, and LPS were dissolved in sterile saline.

4.4. Anesthetic Agents

Medetomidine hydrochloride was purchased from Nippon Zenyaku Kogyo Co., Ltd. (Fukushima, Japan). Midazolam was purchased from Sandoz (Tokyo, Japan). Butorphanol tartrate was purchased from Meiji Seika Pharma Co., Ltd. (Tokyo, Japan). These anesthetics were kept at room temperature (RT). Three types of mixed anesthetic agents were prepared with medetomidine hydrochloride at a dose of 0.3 mg/kg, midazolam at a dose of 4 mg/kg, and butorphanol tartrate at a dose of 5 mg/kg. The concentration ratio of the three types of mixed anesthetic agents was determined by a previous study [47]. A total of 0.75 mL medetomidine hydrochloride was mixed with 2 mL midazolam and 2.50 mL butorphanol tartrate, and adjusted to a volume of 19.75 mL with sterile saline. All agents were diluted in sterile saline and stored at 4 °C in the dark. The mixed anesthetic agents were administered to mice at a volume of 0.01 mL/g of body weight.

4.5. Experimental Protocol

Based on previous reports [6,48], we established the experimental protocols (Figure 1). Each experimental group of mice was separated into seven sets with each set consisting of six randomly chosen mice. All experiments were carried out in another room after transfer from the animal holding room.

Sensitization: A total of 125 µL of 10 mM PdCl$_2$ and 10 µg/mL LPS in sterile saline was injected twice at an interval of 7 days via the intradermal (i.d.) route into the left and right groin of mice (250 µL each). At 7 days after the second sensitization, mice were challenged for the first time.

Challenge for elicitation: At day 7 after the second sensitization, non-sensitized mice (ICD mice) or sensitized mice (ACD mice) were challenged for elicitation with 25 µL of 10 mM PdCl$_2$ without LPS in sterile saline into the left and right footpad by i.d. injection under anesthesia with three types of mixed anesthetic agents. Mice sensitized with Pd plus LPS and then challenged with sterile saline were used as the control (Table 1). After weighing the mice, the appropriate volumes of the three types of mixed anesthetic agents were administered by intraperitoneal injection into the lower left or right quadrant of each mouse under manual restraint.

4.6. Oral Administration of Fexofenadine Hydrochloride and Prednisolone

Pd allergy was induced in mice as described above. On the day of each challenge, each drug (fexofenadine hydrochloride, or prednisolone) was orally administered to each ACD mouse once daily at 1 h before challenge. In some experimental ACD mouse groups, the dose of fexofenadine hydrochloride was orally administered at 1 h before sensitization. The mice received 10 mg/kg fexofenadine hydrochloride or prednisolone via a gastric tube. The dose of fexofenadine hydrochloride or prednisolone were 0.01 mg/g of weight in 100 µL of sterile saline. The approximate dose indicated that we converted from murine weight for humans. Control and ICD mice received the same volume of sterile saline alone. ACD mice were divided into four groups: Group A, fexofenadine hydrochloride was orally administered at 1 h before sensitization and challenge; Group B, fexofenadine hydrochloride was orally administered at 1 h before challenge only; Group C, fexofenadine hydrochloride was orally administered at 1 h before sensitization and prednisolone was orally administered at 1 h before challenge; and Group D, prednisolone was orally administered at 1 h before challenge only (Table 2).

4.7. Measurement of Allergic Footpad Swelling

Footpad swelling was measured before challenge and at 24 h, 48 h, and 72 h, and 1 week after challenge using a Peacock dial thickness gauge (Ozaki MFG Co. Ltd., Tokyo, Japan). The difference in footpad thickness before and after challenge was recorded. All procedures were performed by the same operator.

4.8. Immunohistochemistry

Footpads were obtained from Pd-induced ACD mice, fexofenadine hydrochloride-treated mice, and prednisolone-treated mice for histology and immunohistochemical analyses. Tissue samples were fixed with 4% paraformaldehyde-lysine-periodate overnight at 4 °C. After washing with phosphate buffered saline (PBS), fixed tissues were soaked in 5% sucrose/PBS for 1 h, 15% sucrose/PBS for 3 h, and then 30% sucrose/PBS overnight at 4 °C. Tissue samples were embedded in Tissue Mount (Chiba Medical, Saitama, Japan) and snap-frozen in a mixture of acetone and dry ice. Frozen sections were cut into 6-μm-thick cryosections that were air dried on poly-L-lysine-coated glass slides. For histological analyses, the cryosections were stained with H&E. For immunohistochemical analyses, antigen retrieval was performed, and the cryosections were stained with anti-mouse F4/80 (1:1000; Cl-A3-1, Abcam, Cambridge, UK) and anti-CD3 (1:500; SP7, Abcam, Cambridge, UK) mAbs. The F4/80 monoclonal antibody has been used to detect mouse macrophages populations in a wide range of footpad tissue. Non-specific binding of mAbs was blocked by incubation of the sections in PBS containing 5% normal goat rabbit serum, 0.025% Triton X-100 (Wako Pure Chemicals, Osaka, Japan), and 5% bovine serum albumin (Sigma-Aldrich) for 30 min at RT. The sections were incubated with primary mAbs for 1 h at RT. After washing three times with PBS for 5 min each, intrinsic peroxidase was quenched using 3% H_2O_2 in methanol. After soaking the sections in distilled water, they were washed twice and then incubated with a secondary antibody (biotinylated goat anti-hamster immunoglobulin G (IgG) or biotinylated rabbit anti-rat IgG) for 1 h at RT. After washing three times, the sections were treated with Vectastain ABC Reagent (Vector Laboratories, Burlingame, CA, USA) for 30 min at RT, followed by 3,3-diaminobenzidine (DAB) staining (0.06% DAB and 0.03% H_2O_2 in 0.1 M Tris-HCl, pH 7.6; Wako Pure Chemicals, Osaka, Japan). The tissue sections were counterstained with hematoxylin to visualize nuclei.

Images of F4/80-positive cells in immunostained sections were obtained using a BX 51 microscope (Olympus Optical, Tokyo, Japan) at ×200 magnification and stored in TIF format (4080 × 3072 resolution, 95 dpi). Selected tissue regions were delineated and subtracted from the respective layer area. Images were then thresholded to highlight the stained areas, but not the respective isotype controls. The highlighted section was analyzed and presented as a fraction of the selected region. To obtain representative results, measurements were made in six different regions of each sample, and mean values were used for statistical analysis. The number of positively immunostained cells per target area was counted using ImageJ software with Java-based color deconvolution (v. 1.41, US National Institutes of Health).

4.9. RNA Extraction and cDNA Synthesis

Fresh footpads were obtained from mice and immediately soaked in RNAlater RNA Stabilization Reagent (Qiagen, Hilden, Germany). Total RNA from footpads and spleens was extracted using the RNeasy Lipid Tissue Mini Kit (Qiagen, Hilden, Germany) according to the manufacturer's instructions.

cDNA was synthesized from DNA-free RNA using the PrimeScript™ RT reagent Kit (Takara Bio, Tokyo, Japan) according to the manufacturer's instructions.

4.10. qPCR

The expression levels of immune response-related genes, including T cell-related CD antigens, cytokines, and cytotoxic granules, were measured by qPCR using the Bio-Rad CFX96 system (Bio-Rad, Hercules, CA, USA). Specific primers for GAPDH, CD3, CD4, CD8, CD14, CD80, IL-1β, IFN-γ, TNF-α, IL-4, IL-5, IL-12, and HDC have been described previously [7,49,50]. Freshly isolated total RNA from the footpads of mice was converted to complementary DNA (cDNA). The PCR consisted of 5 μL SsoFast™ EvaGreen® Supermix (Bio-Rad), 3.5 μL RNase/DNase-free water, 0.5 μL of 5 μM primer mix, and 1 μL cDNA in a final volume of 10 μL. Cycling conditions were as follows: 30 s at 95 °C followed by 45 cycles of 1 s at 95 °C and 5 s at 60 °C. At the end of each program, melting curve analysis was performed from 65 °C to 95 °C to confirm the homogeneity of PCR products. All assays were repeated three times, and mean values were used to calculate gene expression levels. Five 10-fold serial dilutions of each standard transcript were used to determine the absolute quantification, specification, and amplification efficiency of each primer set. Standard transcripts were generated by in vitro transcription of the corresponding PCR product in a plasmid. The nucleotide sequences were confirmed by DNA sequencing using the CEQ8000 Genetic Analysis System (Beckman Coulter, Fullerton, CA, USA). Their quality and concentration were validated using an Agilent DNA 7500 Kit in an Agilent 2100 Bioanalyzer (Agilent, Santa Clara, CA, USA). GAPDH gene expression was used as an internal control. The expression levels of each target gene were normalized to GAPDH expression.

4.11. Statistical Analysis

The statistical significance of differences between mean values of each experimental group was analyzed using the Mann Whitney test by GraphPad Prism 5 software for Windows (GraphPad Software, Inc., San Diego, CA, USA). A p-value of less than 0.05 was considered as significant, a p-value of less than 0.01 was considered as highly significant, and a p-value of less than 0.001 was considered as extremely significant.

5. Conclusions

Our results show that the interference with the development of antigen-presenting cells by fexofenadine has the beneficial effect on the treatment of palladium allergy compared to prednisolone.

Supplementary Materials: The following are available online at www.mdpi.com/1422-0067/18/7/1357/s1.

Acknowledgments: This work was supported by the Japan Society for the Promotion of Science KAKENHI Grant-in-Aid for Scientific Research C Grant No. 15K11329; and by the Japan Society for the Promotion of Science KAKENHI Grant-in-Aid for Scientific Research B Grant No. 16K20443.

Author Contributions: Ryota Matsubara, Kenichi Kumagai and Ryuji Suzuki conceived and designed the experiments; Ryota Matsubara, Hiroaki Shigematsu, Satsuki Suzuki and Ryuji Suzuki performed the experiments; Ryota Matsubara, Kazutaka Kitaura and Yasunari Nakasone analyzed the data; Ryota Matsubara contributed reagents/materials/analysis tools; Ryota Matsubara, Kenichi Kumagai and Yoshiki Hamada wrote the paper.

Conflicts of Interest: The authors declare no conflict of interest.

References

1. Raap, U.; Stiesch, M.; Reh, H.; Kapp, A.; Werfel, T. Investigation of contact allergy to dental metals in 206 patients. *Contact Dermat.* **2009**, *60*, 339–343. [CrossRef] [PubMed]
2. Budinger, L.; Hertl, M. Immunologic mechanisms in hypersensitivity reactions to metal ions: An overview. *Allergy* **2000**, *55*, 108–115. [CrossRef] [PubMed]
3. Nakamura, K.; Imakado, S.; Takizawa, M.; Adachi, M.; Sugaya, M.; Wakugawa, M.; Asahina, A.; Tamaki, K. Exacerbation of pustulosis palmaris et plantaris after topical application of metals accompanied by elevated levels of leukotriene B4 in pustules. *J. Am. Acad. Dermatol.* **2000**, *42*, 1021–1025. [CrossRef] [PubMed]

4.   Hanafusa, T.; Yoshioka, E.; Azukizawa, H.; Itoi, S.; Tani, M.; Kira, M.; Katayama, I. Systemic allergic contact dermatitis to palladium inlay manifesting as annular erythema. *Eur. J. Dermatol.* **2012**, *22*, 697–698. [PubMed]

5.   Eguchi, T.; Kumagai, K.; Kobayashi, H.; Shigematsu, H.; Kitaura, K.; Suzuki, S.; Horikawa, T.; Hamada, Y.; Ogasawara, K.; Suzuki, R. Accumulation of invariant NKT cells into inflamed skin in a novel murine model of nickel allergy. *Cell. Immunol.* **2013**, *284*, 163–171. [CrossRef] [PubMed]

6.   Kobayashi, H.; Kumagai, K.; Eguchi, T.; Shigematsu, H.; Kitaura, K.; Kawano, M.; Horikawa, T.; Suzuki, S.; Matsutani, T.; Ogasawara, K.; et al. Characterization of T cell receptors of Th1 cells infiltrating inflamed skin of a novel murine model of palladium-induced metal allergy. *PLoS ONE* **2013**, *8*, e76385. [CrossRef] [PubMed]

7.   Shigematsu, H.; Kumagai, K.; Kobayashi, H.; Eguchi, T.; Kitaura, K.; Suzuki, S.; Horikawa, T.; Matsutani, T.; Ogasawara, K.; Hamada, Y.; et al. Accumulation of metal-specific T cells in inflamed skin in a novel murine model of chromium-induced allergic contact dermatitis. *PLoS ONE* **2014**, *9*, e85983. [CrossRef] [PubMed]

8.   Kumagai, K.; Horikawa, T.; Shigematsu, H.; Matsubara, R.; Kitaura, K.; Eguchi, T.; Kobayashi, H.; Nakasone, Y.; Sato, K.; Yamada, H.; et al. Possible Immune Regulation of Natural Killer T Cells in a Murine Model of Metal Ion-Induced Allergic Contact Dermatitis. *Int. J. Mol. Sci.* **2016**, *17*, 87. [CrossRef] [PubMed]

9.   Chow, M.; Botto, N.; Maibach, H. Allergic contact dermatitis caused by palladium-containing dental implants. *Dermatitis* **2014**, *25*, 273–274. [CrossRef] [PubMed]

10.   Kawano, M.; Nakayama, M.; Aoshima, Y.; Nakamura, K.; Ono, M.; Nishiya, T.; Nakamura, S.; Takeda, Y.; Dobashi, A.; Takahashi, A.; et al. NKG2D+ IFN-γ+ CD8+ T cells are responsible for palladium allergy. *PLoS ONE* **2014**, *9*, e86810. [CrossRef] [PubMed]

11.   Muris, J.; Feilzer, A.J.; Kleverlaan, C.J.; Rustemeyer, T.; van Hoogstraten, I.M.; Scheper, R.J.; von Blomberg, B.M. Palladium-induced Th2 cytokine responses reflect skin test reactivity. *Allergy* **2012**, *67*, 1605–1608. [CrossRef] [PubMed]

12.   Stejskal, V.; Reynolds, T.; Bjorklund, G. Increased frequency of delayed type hypersensitivity to metals in patients with connective tissue disease. *J. Trace Elem. Med. Biol.* **2015**, *31*, 230–236. [CrossRef] [PubMed]

13.   Faurschou, A.; Menne, T.; Johansen, J.D.; Thyssen, J.P. Metal allergen of the 21st century—A review on exposure, epidemiology and clinical manifestations of palladium allergy. *Contact Dermat.* **2011**, *64*, 185–195. [CrossRef] [PubMed]

14.   Larese Filon, F.; Uderzo, D.; Bagnato, E. Sensitization to palladium chloride: A 10-year evaluation. *Am. J. Contact Dermat.* **2003**, *14*, 78–81. [CrossRef] [PubMed]

15.   Muris, J.; Goossens, A.; Goncalo, M.; Bircher, A.J.; Gimenez-Arnau, A.; Foti, C.; Bruze, M.; Andersen, K.E.; Rustemeyer, T.; Feilzer, A.J.; et al. Sensitization to palladium in Europe. *Contact Dermat.* **2015**, *72*, 11–19. [CrossRef] [PubMed]

16.   Niiyama, S.; Tamauchi, H.; Amoh, Y.; Terashima, M.; Matsumura, Y.; Kanoh, M.; Habu, S.; Komotori, J.; Katsuoka, K. Th2 immune response plays a critical role in the development of nickel-induced allergic contact dermatitis. *Int. Arch. Allergy Immunol.* **2010**, *153*, 303–314. [CrossRef] [PubMed]

17.   Sinigaglia, F.; Scheidegger, D.; Garotta, G.; Scheper, R.; Pletscher, M.; Lanzavecchia, A. Isolation and characterization of Ni-specific T cell clones from patients with Ni-contact dermatitis. *J. Immunol.* **1985**, *135*, 3929–3932. [PubMed]

18.   Iguchi, N.; Takeda, Y.; Sato, N.; Ukichi, K.; Katakura, A.; Ueda, K.; Narushima, T.; Higuchi, S.; Ogasawara, K. The antihistamine olopatadine regulates T cell activation in palladium allergy. *Int. Immunopharmacol.* **2016**, *35*, 70–76. [CrossRef] [PubMed]

19.   Tamura, T.; Matsubara, M.; Takada, C.; Hasegawa, K.; Suzuki, K.; Ohmori, K.; Karasawa, A. Effects of olopatadine hydrochloride, an antihistamine drug, on skin inflammation induced by repeated topical application of oxazolone in mice. *Br. J. Dermatol.* **2004**, *151*, 1133–1142. [CrossRef] [PubMed]

20.   Baeck, M.; Goossens, A. Systemic contact dermatitis to corticosteroids. *Allergy* **2012**, *67*, 1580–1585. [CrossRef] [PubMed]

21.   Okamoto, T.; Iwata, S.; Ohnuma, K.; Dang, N.H.; Morimoto, C. Histamine H1-receptor antagonists with immunomodulating activities: Potential use for modulating T helper type 1 (Th1)/Th2 cytokine imbalance and inflammatory responses in allergic diseases. *Clin. Exp. Immunol.* **2009**, *157*, 27–34. [CrossRef] [PubMed]

22.   Kenneth, M.; Paul, T.; Mark, W. *Janeway's Immune biology*, 7th ed.; Garland Science: New York, NY, USA, 2008; p. 587.

23. Saito, M.; Arakaki, R.; Yamada, A.; Tsunematsu, T.; Kudo, Y.; Ishimaru, N. Molecular Mechanisms of Nickel Allergy. *Int. J. Mol. Sci.* **2016**, *17*, 202. [CrossRef] [PubMed]

24. Schmidt, M.; Raghavan, B.; Muller, V.; Vogl, T.; Fejer, G.; Tchaptchet, S.; Keck, S.; Kalis, C.; Nielsen, P.J.; Galanos, C.; et al. Crucial role for human Toll-like receptor 4 in the development of contact allergy to nickel. *Nat. Immunol.* **2010**, *11*, 814–819. [CrossRef] [PubMed]

25. McKee, A.S.; Fontenot, A.P. Interplay of innate and adaptive immunity in metal-induced hypersensitivity. *Curr. Opin. Immunol.* **2016**, *42*, 25–30. [CrossRef] [PubMed]

26. Honda, T.; Egawa, G.; Grabbe, S.; Kabashima, K. Update of immune events in the murine contact hypersensitivity model: Toward the understanding of allergic contact dermatitis. *J. Investig. Dermatol.* **2013**, *133*, 303–315. [CrossRef] [PubMed]

27. Sugita, K.; Kabashima, K.; Tokura, Y. Fexofenadine downmodulates antigen-presenting ability of murine epidermal Langerhans cells. *J. Dermatol. Sci.* **2008**, *49*, 88–91. [CrossRef] [PubMed]

28. Moulon, C.; Vollmer, J.; Weltzien, H.U. Characterization of processing requirements and metal cross-reactivities in T cell clones from patients with allergic contact dermatitis to nickel. *Eur. J. Immunol.* **1995**, *25*, 3308–3315. [CrossRef] [PubMed]

29. Kurimoto, I.; Grammer, S.F.; Shimizu, T.; Nakamura, T.; Streilein, J.W. Role of F4/80+ cells during induction of hapten-specific contact hypersensitivity. *Immunology* **1995**, *85*, 621–629. [PubMed]

30. Tokura, Y.; Kobayashi, M.; Ito, T.; Takahashi, H.; Matsubara, A.; Takigawa, M. Anti-allergic drug olopatadine suppresses murine contact hypersensitivity and downmodulates antigen-presenting ability of epidermal Langerhans cells. *Cell. Immunol.* **2003**, *224*, 47–54. [CrossRef] [PubMed]

31. Gelfand, E.W.; Cui, Z.H.; Takeda, K.; Kanehiro, A.; Joetham, A. Fexofenadine modulates T-cell function, preventing allergen-induced airway inflammation and hyperresponsiveness. *J. Allergy Clin. Immunol.* **2002**, *110*, 85–95. [CrossRef] [PubMed]

32. Ferstl, R.; Akdis, C.A.; O'Mahony, L. Histamine regulation of innate and adaptive immunity. *Front. Biosci.* **2012**, *17*, 40–53. [CrossRef]

33. Shimizuhira, C.; Otsuka, A.; Honda, T.; Kitoh, A.; Egawa, G.; Nakajima, S.; Nakashima, C.; Watarai, H.; Miyachi, Y.; Kabashima, K. Natural killer T cells are essential for the development of contact hypersensitivity in BALB/c mice. *J. Investig. Dermatol.* **2014**, *134*, 2709–2718. [CrossRef] [PubMed]

34. Otsuka, A.; Kubo, M.; Honda, T.; Egawa, G.; Nakajima, S.; Tanizaki, H.; Kim, B.; Matsuoka, S.; Watanabe, T.; Nakae, S.; et al. Requirement of interaction between mast cells and skin dendritic cells to establish contact hypersensitivity. *PLoS ONE* **2011**, *6*, e25538. [CrossRef] [PubMed]

35. Suto, H.; Nakae, S.; Kakurai, M.; Sedgwick, J.D.; Tsai, M.; Galli, S.J. Mast cell-associated TNF promotes dendritic cell migration. *J. Immunol.* **2006**, *176*, 4102–4112. [CrossRef] [PubMed]

36. Packard, K.A.; Khan, M.M. Effects of histamine on Th1/Th2 cytokine balance. *Int. Immunopharmacol.* **2003**, *3*, 909–920. [CrossRef]

37. Kenneth, M.; Paul, T.; Mark, W. *Janeway's Immune biology*, 7th ed.; Garland Science: New York, NY, USA, 2008; p. 580.

38. Browne, F.; Wilkinson, S.M. Effective prescribing in steroid allergy: Controversies and cross-reactions. *Clin. Dermatol.* **2011**, *29*, 287–294. [CrossRef] [PubMed]

39. Schlapbach, C.; Simon, D. Update on skin allergy. *Allergy* **2014**, *69*, 1571–1581. [CrossRef] [PubMed]

40. Greaves, M.W.; Tan, K.T. Chronic urticaria: Recent advances. *Clin. Rev. Allergy Immunol.* **2007**, *33*, 134–143. [CrossRef] [PubMed]

41. Inoue, T.; Katoh, N.; Kishimoto, S.; Matsunaga, K. Inhibitory effects of oral prednisolone and fexofenadine on skin responses by prick tests with histamine and compound 48/80. *J. Dermatol. Sci.* **2002**, *30*, 180–184. [CrossRef]

42. Bernstein, D.I.; Schoenwetter, W.F.; Nathan, R.A.; Storms, W.; Ahlbrandt, R.; Mason, J. Efficacy and safety of fexofenadine hydrochloride for treatment of seasonal allergic rhinitis. *Ann. Allergy Asthma Immunol.* **1997**, *79*, 443–448. [CrossRef]

43. Jutel, M.; Akdis, M.; Akdis, C.A. Histamine, histamine receptors and their role in immune pathology. *Clin. Exp. Allergy* **2009**, *39*, 1786–1800. [CrossRef] [PubMed]

44. Yanai, K.; Tashiro, M. The physiological and pathophysiological roles of neuronal histamine: An insight from human positron emission tomography studies. *Pharmacol. Ther.* **2007**, *113*, 1–15. [CrossRef] [PubMed]

45. Hiraoka, K.; Tashiro, M.; Grobosch, T.; Maurer, M.; Oda, K.; Toyohara, J.; Ishii, K.; Ishiwata, K.; Yanai, K. Brain histamine H1 receptor occupancy measured by PET after oral administration of levocetirizine, a non-sedating antihistamine. *Expert Opin. Drug Saf.* **2015**, *14*, 199–206. [CrossRef] [PubMed]
46. Kenneth, M.; Paul, T.; Mark, W. *Janeway's Immune biology*, 7th ed.; Garland Science: New York, NY, USA, 2008; p. 655.
47. Kawai, S.; Takagi, Y.; Kaneko, S.; Kurosawa, T. Effect of three types of mixed anesthetic agents alternate to ketamine in mice. *Exp. Anim.* **2011**, *60*, 481–487. [CrossRef] [PubMed]
48. Tamura, T.; Matsubara, M.; Hasegawa, K.; Ohmori, K.; Karasawa, A. Olopatadine hydrochloride suppresses the rebound phenomenon after discontinuation of treatment with a topical steroid in mice with chronic contact hypersensitivity. *Clin. Exp. Allergy* **2005**, *35*, 97–103. [CrossRef] [PubMed]
49. Fujii, Y.; Kitaura, K.; Nakamichi, K.; Takasaki, T.; Suzuki, R.; Kurane, I. Accumulation of T-cells with selected T-cell receptors in the brains of Japanese encephalitis virus-infected mice. *Jpn. J. Infect. Dis.* **2008**, *61*, 40–48. [PubMed]
50. Kitaura, K.; Fujii, Y.; Hayasaka, D.; Matsutani, T.; Shirai, K.; Nagata, N.; Lim, C.K.; Suzuki, S.; Takasaki, T.; Suzuki, R.; et al. High clonality of virus-specific T lymphocytes defined by TCR usage in the brains of mice infected with West Nile virus. *J. Immunol.* **2011**, *187*, 3919–3930. [CrossRef] [PubMed]

International Journal of
Molecular Sciences

MDPI

Article

TRAV7-2*02 Expressing CD8+ T Cells Are Responsible for Palladium Allergy

Yuri Takeda [1,2], Yoshiko Suto [1], Koyu Ito [1], Wataru Hashimoto [2], Tadashi Nishiya [3],
Kyosuke Ueda [4], Takayuki Narushima [4], Tetsu Takahashi [2] and Kouetsu Ogasawara [1,*]

[1] Department of Immunobiology, Institute of Development, Aging and Cancer,
    Tohoku University 4-1 Seiryo-machi, Aoba-ku, Sendai 980-8575, Japan; yuri.takeda@dent.tohoku.ac.jp (Y.T.);
    yosshi.snoopy.11r25@gmail.com (Y.S.); Koyu.ito.c6@tohoku.ac.jp (K.I.)
[2] Department of Oral and Maxillofacial Surgery, Graduate School of Dentistry, Tohoku University,
    4-1 Seiryo-machi, Aoba-ku, Sendai, Miyagi 980-8575, Japan; wataru-thk@umin.ac.jp (W.H.);
    tetsu@dent.tohoku.ac.jp (T.T.)
[3] Department of Pharmacology, School of Pharmaceutical Sciences, Ohu University, 31-1 Misumido,
    Tomitamachi, Koriyama, Fukushima 963-8611, Japan; t-nishiya@pha.ohu-u.ac.jp
[4] Department of Materials Processing, Graduate School of Engineering, Tohoku University, 6-6-02 Aza Aoba,
    Aramaki, Aoba-ku, Sendai, Miyagi 980-8579, Japan; ueda@material.tohoku.ac.jp (K.U.);
    narut@material.tohoku.ac.jp (T.N.)
*   Correspondence: immunobiology@m.tohoku.ac.jp; Tel.: +81-22-717-8579; Fax: +81-22-717-8822

Academic Editor: Reinhard Dallinger
Received: 7 March 2017; Accepted: 26 May 2017; Published: 31 May 2017

Abstract: While metallic biomaterials have led to an improvement in the quality of life, metal allergies, especially to palladium (Pd), has caused a recent increase in allergic patients. Metal allergy is known to be a T cell-mediated delayed-type hypersensitivity (DTH); however, the pathogenic T cell subsets and the specific T cell receptor (TCR) have not been identified. Therefore, we attempted to identify the pathogenic T cells responsible for Pd allergy. We found that activating CD8+ T cells significantly increased and that the TRAV (TCRα variable) 7-2*02 chain skewed in Pd allergic mice. Furthermore, adoptive transfer experiments revealed that in vitro-cultured Pd-stimulated antigen presenting cells (APCs) function as memory APCs with recipient mice developing Pd allergy and that the frequency of TRAV7-2*02 increases the same as conventional Pd allergic mice. In contrast, neither proliferation of CD8+ T cells nor increasing of TRAV7-2*02 was observed in major histocompatibility complex I (MHC I)-deficient Pd-APCs transferred to mice. Taken together, we revealed that TRAV7-2*02-expressing CD8+ T cells are the pathogenic T cells for the development of Pd allergy. We also identified the CDR3 consensus motif of pathogenic TCRs as CAAXSGSWQLIF in TRAV7-2*02/TRAJ (TCRα junction)22*01 positive cells. These results suggest that the specific TCRs represent novel targets for the development of diagnostics and treatments for metal allergy.

Keywords: animal models; autoimmunity; T cells; biomaterial(s); metal allergy; T cell receptor (TCR)

1. Introduction

Metal is superior to other materials in hardness, strength, durability, and workability and, therefore, is broadly useful in many fields. Palladium (Pd), in particular, has a wide array of uses, including as an extremely important metallic biomaterial for dentistry in the reconstruction of occlusions during dental restorations. However, metallic biomaterials sometimes cause metal allergy [1], and the incidence of Pd allergy, in particular, has increased recently [2]. However, the pathological mechanism of metal allergy has not been clarified. Clinical symptoms of metal allergy include intractable diseases, such as dermatitis, inflammation of oral mucosa, oral lichen planus, and stomatitis.

Metal allergy is categorized as a delayed-type hypersensitivity (DTH) reaction, which is a T cell-mediated disease with onset at 24 hours after exposure to the causal metal [3,4]. The pathogenic mechanism of metal allergy includes the function of metal ions as haptens that exert antigenicity and induce pathogenic T cells, which cause metal allergy. T cells are classified into two broad subsets, CD4+ T cells and CD8+ T cells. CD4+ T cells binding major histocompatibility complex II (MHC II) on presenting cells and play roles as helper and regulatory T cells by producing cytokines. CD8+ T cells recognize MHC I and function as proinflammatory T cells that directly kill infected or transformed cells. It has been reported that both CD8+ and CD4+ T cells are required for DTH [3]. Thus, both types of cells [5,6] were generally thought to be responsible for Pd allergy; however, controversy remains regarding the role of pathogenic T cell subsets in DTH.

Various cytokines control the activation or inhibition of immune responses that comprise the host biological defense system against foreign antigens. IFN-γ, the most important cytokine in metal allergy, is an inflammatory cytokine that enhances immune responses, including Th1 responses and cytotoxic activity [6–8]. Previously, we reported that there was a decrease in ear swelling in IFN-γ-deficient mice (IFN-γ$^{-/-}$ mice) compared to wild-type (WT) mice after Pd challenge [6]; however, in a previous study the distribution of T cell subsets was not examined in IFN-γ$^{-/-}$ mice after Pd challenge.

In order to establish acquired immunity, the TCR recognizes antigenic peptides present in the context of MHC, which leads to T cell activation [9,10]. To obtain vast repertoire diversity, the TCR α and β genes undergo complex rearrangement of variable (V), diverse (D), and junction (J) region gene segments resulting in the generation of a specific TCR from 1×10^{18} possible clonotypes. Examination of peripheral blood mononuclear cells obtained from patients with metal allergy showed that metal ions can induce proliferation of human T cells in vitro and limited TCRs are utilized in human T cells from metal allergy patients [11]. However, the specific TCR present on pathogenic T cells of patients with Pd allergy has not previously been identified. Therefore, we attempted to identify the T cell subsets responsible for Pd allergy and investigated whether a specific TCR exists on these pathogenic T cells.

2. Results

2.1. Activated T Cells Accumulate in Regional Lymph Nodes during Pd Allergy

To examine the accumulation of activated T cells in regional lymph nodes, we analyzed the T cell populations present in submandibular lymph node (SLN) cells of Pd-allergic mice by flow cytometry. To this end, Pd + lipopolysaccharide (LPS)-sensitized WT mice (sensitized-WT mice) and phosphate buffered saline (PBS)-injected WT mice (unsensitized-WT mice) were challenged by injection of Pd solution into ear auricles and ear swelling was assessed (Figure S1A). Following Pd challenge, ear swelling of sensitized-WT mice was significantly increased compared to that of unsensitized-WT mice (** $p < 0.01$). Moreover, WT mice sensitized with Pd alone or LPS alone and then challenged with Pd showed no evidence of ear swelling [6]. Both CD8+ and CD4+ T cells accumulated in ear auricles of sensitized WT mice (Figure S1B). Since the ears were most swollen at 24 h post Pd challenge, we assessed the T cell population in the SLNs at this time and found no difference in CD3+ T cell numbers between unsensitized-WT mice and naïve mice. In contrast, there was a significant increase in CD3+ T cells in the SLNs of sensitized-WT mice (** $p < 0.01$) (Figure 1A). Although CD4+ T cells appeared to be slightly increased in the SLN of sensitized WT mice, this difference was not statistically significant (Figure 1A). In contrast, CD8+ T cells were significantly increased in the SLN of sensitized WT mice compared to unsensitized WT (* $p < 0.05$) or naïve mice (** $p < 0.01$) (Figure 1A). We next measured IFN-γ production by SLN T cells. Previously, we showed that IFN-γ producing CD8+ T cells are responsible for Pd allergy [6]. However, in the previous study we analyzed IFN-γ production in the limited condition of T cells enriched by sequential adoptive transfer in Pd allergic mice. Therefore, we investigated which T cells are activated in a conventional model of Pd allergy. Interestingly, not only CD8+ T cells, but also CD4+ T cells produced IFN-γ in sensitized-WT mice (** $p < 0.01$) (Figure 1B,C).

Since ear auricles were not swollen in response to Pd challenge of IFN-$\gamma^{-/-}$ mice (Figure 1D), we assessed CD8$^+$ and CD4$^+$ T cells in the SLN cells of sensitized-IFN-$\gamma^{-/-}$ mice 24 h after Pd challenge and found no increase in CD8$^+$ T cells (* $p < 0.05$) compared to that seen in sensitized-WT mice (Figure 1E). These results suggest that activation of CD8$^+$ T cells, rather than CD4$^+$ T cells, enhances the development of Pd allergy.

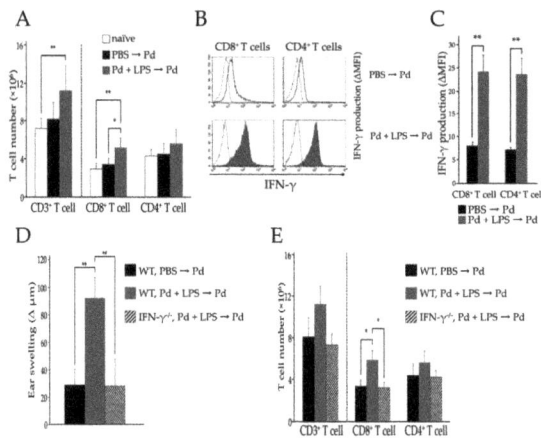

Figure 1. Number of T cells in each subset (**A,E**), IFN-γ production (**B,C**), and ear swelling (**D**) are represented at 24 h post-Pd challenge among naïve, unsensitized-WT (WT, PBS → Pd), sensitized-WT (WT, Pd + LPS → Pd), and sensitized-IFN-$\gamma^{-/-}$ mice (WT, Pd + LPS → Pd), (n = 4–5). Values are means ± Standard deviation (SD). * $p < 0.05$ and ** $p < 0.01$. Similar results were obtained in two independent experiments.

2.2. The Pathogenic TCR Repertoire in Pd Allergic Mice

To identify the specific TCRs that contribute to Pd allergy, we sequenced the entire TCR repertoire present in the SLNs of Pd allergic mice. For this, SLN cells were collected 24 h after Pd challenge and the TCRα and TCRβ chain repertoires were examined. The results showed a clear skewing of TCR repertoires in SLN cells of sensitized-WT mice (Figure 2A,B). These repertoires were selected with a greater than 1% frequency in sensitized-WT mice. For the TCRα chain, frequencies of TRAV7-2*02, TRAV8-1*03, TRAV9N-2*01, and TRAV14N-1*01 were significantly increased in sensitized-WT mice (** $p < 0.01$) compared to that in unsensitized WT mice, (Figure 2A). Interestingly, among all repertoires, the frequency of TRAV7-2*02 was found to be greatly increased. For the TCRβ chain, frequencies of TRBV (TCRβ variable) 3*01 (** $p < 0.01$), TRBV13-2*01 (* $p < 0.05$), and TRBV26*01 (* $p < 0.05$) were significantly increased in sensitized-WT mice (Figure 2B), and TRBV13-2*01 was the most frequently represented of the TRBV repertoire. To examine whether these specific TRAV and TRBV repertoires are expressed on both CD8$^+$ and CD4$^+$ T cells, we purified CD8$^+$ and CD4$^+$ T cells from SLN cells of Pd allergic mice and examined their TCR repertories (Figure S2). Although TRAV7-2*02, TRAV8-1*03, TRAV9N-2*01, and TRAV14N-1*01 TCRα chains were expressed on CD4$^+$ T cells in sensitized-WT mice, only TRAV7-2*02 was expressed on CD8$^+$ T cells (Figure 2C). Furthermore, the frequency of TRAV7-2*02 in sensitized-WT mice was 10 times higher than that in unsensitized-WT mice (Figure 2C). In addition, TRAV8-1*03, TRAV9N-2*01, and TRAV14N-1*01 were not detected on CD8$^+$ T cells (Figure 2C). For the TCRβ chain, TRBV13-2*01 not found to be significantly increased on either CD4$^+$ or CD8$^+$ T cells in sensitized-WT mice compared to that in unsensitized-WT mice (Figure 2D). Thus, Pd allergy-specific TCRβ chain was not identified. These results suggest that the TCRα chain, but not the TCRβ chain, is skewed on CD8$^+$ T cells in Pd allergy. Moreover, TRAV7-2*02 is likely the Pd allergy-specific TCR.

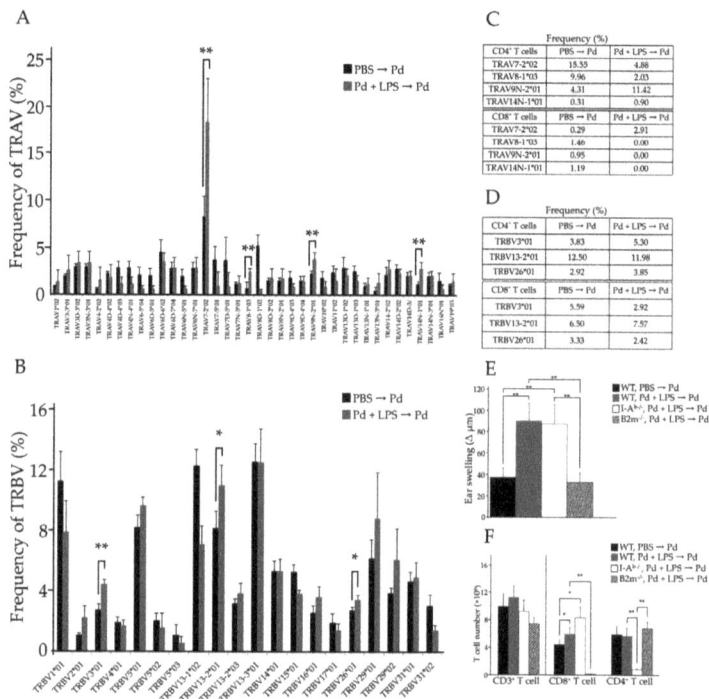

C

Frequency (%)

CD4+ T cells	PBS → Pd	Pd + LPS → Pd
TRAV7-2*02	18.33	4.88
TRAV8-1*03	9.96	2.03
TRAV9N-2*01	4.31	11.42
TRAV14N-1*01	0.31	0.90
CD8+ T cells	PBS → Pd	Pd + LPS → Pd
TRAV7-2*02	0.29	2.91
TRAV8-1*03	1.46	0.00
TRAV9N-2*01	0.95	0.00
TRAV14N-1*01	1.19	0.00

D

Frequency (%)

CD4+ T cells	PBS → Pd	Pd + LPS → Pd
TRBV3*01	3.83	5.30
TRBV13-2*01	12.50	11.98
TRBV26*01	2.92	3.85
CD8+ T cells	PBS → Pd	Pd + LPS → Pd
TRBV3*01	5.59	2.92
TRBV13-2*01	6.50	7.57
TRBV26*01	3.33	2.42

Figure 2. The frequency of TRAV (**A**) and TRBV (**B**) repertoires were compared between unsensitized-WT mice and sensitized-WT mice ($n = 4$); (**C,D**) The skewing of TCR repertoires in CD8+ and CD4+ T cells was analyzed. Ear swelling (**E**) and the number of T cells in each subset (**F**) were compared among unsensitized-WT mice, sensitized-WT mice, sensitized-I-A$^{b-/-}$ mice, and sensitized-B2m$^{-/-}$ mice ($n = 4$–5). Values are means ± SD. * $p < 0.05$ and ** $p < 0.01$. Similar results of ear swelling (**E**) and T cell numbers (**F**) were obtained in two independent experiments.

2.3. MHC I and Activated CD8+ T Cells Are Critical for Pd Allergy

To investigate whether MHC I or II plays a more important role in Pd allergy, we induced Pd allergy in C57BL/6-deficient β2m-microglobulin (B2m) mice (B2m$^{-/-}$ mice) and C57BL/6-deficient MHC II *H2-Ab1* (I-A^b) mice (I-A$^{b-/-}$ mice). Since B2m constitutes a part of the MHC I molecule, MHC I is not expressed on the cell surface in these mice and, thus, CD8+ T cells are impaired, while in I-A$^{b-/-}$ mice CD4+ T cells are impaired. Compared to unsensitized-WT mice, ear auricles were significantly more swollen in sensitized I-A$^{b-/-}$ mice at 24 h after Pd challenge (** $p < 0.01$). However, ear swelling was not observed in B2m$^{-/-}$ mice following Pd challenge (Figure 2E). In addition, CD4+ T cells were not significantly increased in SLNs of B2m$^{-/-}$ mice (Figure 2F), while the number of CD8+ T cells in SLNs of I-A$^{b-/-}$ mice was greatly increased (** $p < 0.01$) (Figure 2F). Furthermore, TRAV7-2*02 was highly expressed on SLN cells from Pd-sensitized I-A$^{b-/-}$ and WT mice (data not shown). Therefore, MHC I is essential and activated CD8+ T cells are important for the development of Pd allergy.

2.4. Adoptive Transfer of In Vitro Prepared Pd-Treated APCs Induces Pd Allergy

We examined whether Pd allergy could develop in mice with adoptively-transferred Pd-stimulated APCs. To this end, APCs were prepared from bone marrow cells and cultured in mouse macrophage colony stimulating factor (mM-CSF) containing medium. After seven days in culture, CD11b[+] F4/80[+] APCs were sorted and stimulated with Pd + LPS (Pd-APCs) or LPS (LPS-APCs) for 24 h. Flow cytometric analysis revealed that these APCs also express the co-stimulatory molecules CD80, CD86, and CD40 (Figure S3A). To assess their effects on Pd sensitization in mice, these APCs were adoptively transferred into naïve WT mice. Seven days after the transfer, recipient mice were challenged with Pd injection into ear auricles. Ear swelling in the mice given Pd-APCs was significantly increased (** $p < 0.01$) compared to that of mice given LPS-APCs (Figure 3A). Further, CD8[+] and CD4[+] T cells were significantly increased in SLNs of mice that received Pd-APCs (Figure 3B). These results indicate that APCs cultured with Pd + LPS in vitro function as the driving force in the development of Pd allergy. In addition, the TCR repertoire in mice given Pd-APCs is skewed in a manner similar to that seen in Pd-sensitized WT mice, with significant increases in TRAV7-2*02 expression, but not TRBV expression, starting 24 h after Pd challenge (Figure 3C,D). Therefore, Pd-memory APCs lead to Pd allergy, and the TCRα chain, rather than TCRβ chain, is involved in the development of Pd allergy.

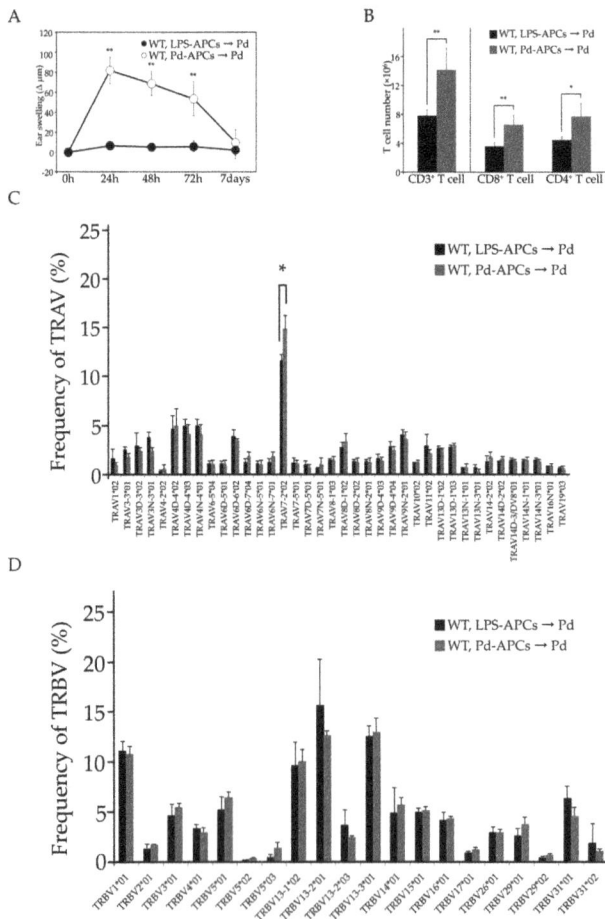

Figure 3. Ear swelling (**A**), number of T cells in each subset (**B**), TRAV (**C**), and TRBV (**D**) repertoires were examined after Pd challenge of mice given WT LPS-APCs or WT Pd-APCs ($n = 4$). Values are means ± SD. * $p < 0.05$ and ** $p < 0.01$. Similar results of ear swelling (**A**) and T cell numbers (**B**) were obtained in two independent experiments.

2.5. MHC I-Dependent CD8+ T Cell Activation Is Essential for the Onset of Pd Allergy

Neither MHC I molecules or CD8+ T cells function in B2m$^{-/-}$ mice as development of CD8+ T cells is impaired in the absence of this molecule. Thus, we investigated whether MHC I-deficient Pd-APCs derived from B2m$^{-/-}$ mice are able to induce Pd allergy. To this end, we prepared WT Pd-APCs and B2m$^{-/-}$ Pd-APCs and adoptively transferred them into naïve WT mice. Seven days later, these mice were challenged with Pd injection and ear thickness was measured from 24 h after the challenge. Ear swelling of B2m$^{-/-}$ Pd-APC recipient mice was significantly less (** $p < 0.01$) than that of mice given WT Pd-APCs (Figure 4A). Neither CD4+ nor CD8+ T cells increased significantly in mice given B2m$^{-/-}$ Pd-APCs compared to those given WT Pd-APCs (** $p < 0.01$) (Figure 4B). Moreover, the frequency of TRAV7-2*02 was not increased in mice given B2m$^{-/-}$ Pd-APCs compared to mice given WT Pd-APCs (** $p < 0.01$) (Figure 4C). These results indicate that TRAV7-2*02 expressing CD8+ T cells are activated and proliferate in a manner dependent on MHC I, which leads to Pd allergy.

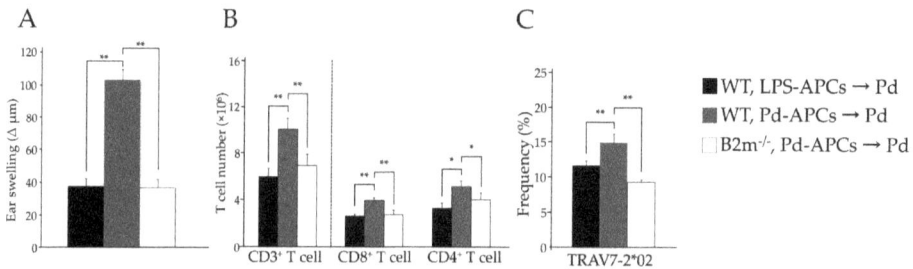

Figure 4. Ear thickness (**A**), number of cells in each T cell subset (**B**), and frequency of TRAV7-2*02 (**C**) were compared among mice given WT LPS-APCs, WT Pd-APCs, or B2m$^{-/-}$ Pd-APCs ($n = 4$) followed by Pd challenge. Values are means $\pm$ SD. * $p < 0.05$ and ** $p < 0.01$. Similar results of (**A,B**) were obtained in two independent experiments.

2.6. Identification of CDR3 Amino Acid Sequences in Pd Allergy

We assessed the skewing of J segment representation in TRAV7-2*02 during Pd allergy. Twenty-four hours from Pd challenge, TRAJ6*01, TRAJ15*01, TRAJ22*01, TRAJ26*01, TRAJ27*01, TRAJ31*01, and TRAJ37*01 within the TRAV7-2*02 repertoire were expressed in all Pd-sensitized WT mice ($n = 4$), with the frequency of TRAJ22*01 being markedly increased, comparatively (Table 1). CDR3 is known to be the hypervariable region that specifically recognizes the antigen-MHC complex [12,13]. Therefore, we also examined the CDR3 amino acid sequence of TRAV7-2*02/TRAJ22*01 and identified a consensus frame, CAAXSGSWQLIF (X = 1 to 4 amino acids), in all TRAV7-2*02/TRAJ22*01 (Table 2). In addition, the CDR3 amino acids common to TRAV7-2*02/TRAJ22*01 were expressed in Pd-APCs and transferred to mice at 24 h after Pd challenge (Table 3). Taken together, these results indicate that TRAV7-2*02/TRAJ22*01 is responsible for the induction of Pd allergy.

Table 1. Frequency of TRAJ repertoires bearing TRAV7-2*02.

TRAV7-2*02 Bearing TRAJ	Frequency (%)
TRAJ6*01	6.8
TRAJ15*01	10.6
TRAJ22*01	25.5
TRAJ26*01	4.6
TRAJ27*01	4.8
TRAJ31*01	4.6
TRAJ37*01	7.1

Representative data are shown. Similar results were observed in other sensitized WT mice ($n = 4$).

Table 2. The CDR3 amino acid sequences of TRAV7-2*02/TRAJ22*01 in Pd sensitized-WT mice.

	TRAV	CDR3			TRAJ	Frequency (%)
		3′ V-Region	N	5′ J-Region		
Mouse 1	TRAV7-2*02	CAAR	G	SGSWQLIF	TRAJ22*01	22.1
	TRAV7-2*02	CAA	T	SGSWQLIF	TRAJ22*01	20.1
	TRAV7-2*02	CA	AGA	SSGSWQLIF	TRAJ22*01	20.1
	TRAV7-2*02	CAAS	I	SGSWQLIF	TRAJ22*01	14.9
	TRAV7-2*02	CA	AA	SSGSWQLIF	TRAJ22*01	14.9
	TRAV7-2*02	CA	ARA	SSGSWQLIF	TRAJ22*01	7.8
Mouse 2	TRAV7-2*02	CAA	SHA	SSGSWQLIF	TRAJ22*01	39.3
	TRAV7-2*02	CA	AR	SSGSWQLIF	TRAJ22*01	39.3
	TRAV7-2*02	CAA	IA	SSGSWQLIF	TRAJ22*01	21.4
Mouse 3	TRAV7-2*02	CA	AR	SSGSWQLIF	TRAJ22*01	69.0
	TRAV7-2*02	CAA	KIP	GSWQLIF	TRAJ22*01	31.0
Mouse 4	TRAV7-2*02	CA	AA	SSGSWQLIF	TRAJ22*01	33.3
	TRAV7-2*02	CAAS	L	SSGSWQLIF	TRAJ22*01	42.8
	TRAV7-2*02	CAAS	P	SSGSWQLIF	TRAJ22*01	23.9

We identified a consensus frame, CAAXSGSWQLIF (X = 1 to 4 amino acids) (n = 4).

Table 3. The CDR3 amino acid sequences of TRAV7-2*02/TRAJ22*01 in mice given WT Pd-APCs.

	TRAV	CDR3			TRAJ	Frequency (%)
		3′ V-Region	N	5′ J-Region		
Mouse 1	TRAV7-2*02	CAA	T	SSGSWQLIF	TRAJ22*01	30.8
	TRAV7-2*02	CAA	S	SGSWQLIF	TRAJ22*01	23.1
	TRAV7-2*02	CAAS	PA	SSGSWQLIF	TRAJ22*01	15.4
	TRAV7-2*02	CAA	R	SSGSWQLIF	TRAJ22*01	15.4
	TRAV7-2*02	CA	AR	SGSWQLIF	TRAJ22*01	15.4
Mouse 2	TRAV7-2*02	CA	AA	SSGSWQLIF	TRAJ22*01	30.8
	TRAV7-2*02	CAAS	A	SGSWQLIF	TRAJ22*01	23.1
	TRAV7-2*02	CAAS	R	SSGSWQLIF	TRAJ22*01	15.4
	TRAV7-2*02	CA	AR	SSGSWQLIF	TRAJ22*01	15.4
	TRAV7-2*02	CA	AQ	SSGSWQLIF	TRAJ22*01	15.4
Mouse 3	TRAV7-2*02	CAAS	I	SSGSWQLIF	TRAJ22*01	50.0
	TRAV7-2*02	CAAS	I	SGSWQLIF	TRAJ22*01	50.0
Mouse 4	TRAV7-2*02	CAA	T	SSGSWQLIF	TRAJ22*01	25.9
	TRAV7-2*02	CAAS	MA	SGSWQLIF	TRAJ22*01	25.9
	TRAV7-2*02	CAAS	I	SSGSWQLIF	TRAJ22*01	18.5
	TRAV7-2*02	CAA	GS	SSGSWQLIF	TRAJ22*01	14.8
	TRAV7-2*02	CAA	SP	SGSWQLIF	TRAJ22*01	14.8

Similarly, the CDR3 amino acid sequences of TRAV7-2*02/TRAJ22*01 are shown (n = 4).

3. Discussion

In this study we demonstrate the skewing of specific TCRs during the development of Pd allergy and the contribution of CD8[+] T cells, rather than CD4[+] T cells, to the development of this allergy in mice. Moreover, we identify that TRAV7-2*02 expressing CD8[+] T cells mainly function as pathogenic T cells. Furthermore, using adoptive transfer experiments we show that APCs stimulated in vitro with Pd function as metal memory APCs in vivo and induce sensitization for metal allergy when transferred. Indeed, skewing of TRAV7-2*02 was observed in the Pd allergic WT mice that received in vitro derived Pd-APCs. These results indicate that Pd memory APCs were generated in vitro and induce the pathogenic T cell activation seen during the Pd allergic response.

Although there has previously been controversy regarding this, we found that activated CD8[+] T cells, rather than CD4[+] T cells, play a key role as proinflammatory T cells in Pd allergy (Figure 1).

Twenty-four hours after Pd challenge, both CD8$^+$ and CD4$^+$ T cells became activated and produced IFN-γ (Figure 1B,C). Thus, we assessed CD8$^+$ and CD4$^+$ T cells in the SLNs of sensitized-IFN-$\gamma^{-/-}$ mice 24 hours after Pd challenge and found no increase in CD8$^+$ T cells (* $p < 0.05$) over that seen in sensitized-WT mice (Figure 1E). Therefore, CD4$^+$ T cells are likely involved only in exacerbating Pd allergy, because IFN-γ production by CD4$^+$ T cells increased in sensitized-WT mice. However, while the ears of sensitized-I-A$^{b-/-}$ mice and sensitized-WT mice were swollen, those of sensitized-B2m$^{-/-}$ mice were not (Figure 2E). In addition, CD8$^+$ T cells were significantly increased in sensitized-I-A$^{b-/-}$ mice without the cooperation of CD4$^+$ T cell activation (Figure 2F). These results indicate that Pd allergy is caused by antigen presentation by MHC I to CD8$^+$ T cells; however, Pd allergy does not occur following the MHC II-CD4$^+$ T cell interaction. Thus, both MHC I expression and activation of CD8$^+$ T cells are essential for the development of Pd allergy. Consistent with our results, CD8$^+$ T cells have been shown to function as effector cells in DTH responses [14].

It has been reported that PdCl$_2$ is sufficiently reliable reagent for development of Pd allergy and that solution of PdCl$_2$ injection induces T cell dependent allergic diseases and in mice [6,15,16]. Therefore, we used PdCl$_2$ for these experiments and referred the appropriate dose of PdCl$_2$ [6,15,16]. Interestingly, since Na$_2$[PdCl$_4$] has been shown to be effective in human patch testing [17], Na$_2$[PdCl$_4$] may be more useful reagent than PdCl$_2$ in Pd allergic mouse model. Future experiments will be required for applying animal experiments.

LPS is a component of microbes that, when recognized by immune cells, induces MHC expression on APCs (Figure S3B). Thus, it has been suggested that this microbial component plays a role in the immunological response leading to metal allergy [18]. Moreover, it is likely that the resident flora contributes to the progression of metal allergy.

The predominant TCRα chain expressed during Pd allergy is TRAV7-2*02/TRAJ22*01. While TRAV7-2*02 was expressed on both subsets of T cells (Figure 2C,D), increased frequencies of TRAV7-2*02 expressing CD8$^+$ T cells were observed in sensitized-I-A$^{b-/-}$ allergic mice (data not shown). Therefore, adoptive transfer experiments were performed to assess whether TRAV7-2*02 expressing CD8$^+$ T cells are important for Pd allergy. Interestingly, we found ear swelling and increases in CD8$^+$ T cells in mice given Pd-APCs at a similar level to conventional Pd allergic mice (Figure 3A,B). Furthermore, the TRAV7-2*02 repertoire was found to be significantly represented in mice given Pd-APCs after Pd challenge (Figure 3C). In contrast, the frequency of TRAV7-2*02 was not increased in mice given B2m$^{-/-}$ Pd-APCs compared to mice given WT Pd-APCs (Figure 4C). Consistent with the lack of a response by TRAV7-2*02 expressing CD8$^+$ T cells, Pd allergy did not develop in mice given B2m$^{-/-}$ Pd-APCs (Figure 4). Therefore, these results indicate that CD8$^+$ T cells play a role as pathogenic T cells for the development of Pd allergy, and that TRAV7-2*02 expressing CD8$^+$ T cells are activated by Pd-stimulated APCs expressing MHC I. Accordingly, our findings indicate the presence of a specific TCR in metal allergy. Furthermore, the adoptive transfer experiments indicate that APCs stimulated in vitro with Pd function to sensitize recipient naïve mice. Therefore, it is possible that adoptive transfer experiments may be useful for further study of antigen presentation in metal allergy.

The skewed TCRs identified in Pd allergy were TRAV7-2*02/TRAJ22*01. As for the TCRβ chain, the TRBV13 family has been reported to be frequently expressed in various mouse disease models including in NOD mice [19] and in an induced arthritis model [20]. These results imply that TCRβ chains exhibit flexible binding capacities against pathogenic antigens. However, in Pd allergy there was actually more frequent skewing of the TCRα chain than of the TCRβ chain, and only Pd-specific TCRα chains were detected. Consistent with our results, TRAV has a high specificity for recognizing nickel bound to peptide in humans [21,22]. Therefore, the Pd binding peptide may be restricted to the TCRα chain. Moreover, we found that TRAJ22*01 was the most commonly-expressed J segment (Table 1), and we identified the CDR3 consensus frame, CAAXSGSWQLIF, in TRAV7-2*02/TRAJ22*01 (Table 2). Taken together, it can be concluded that TRAV7-2*02/TRAJ22*01 expressing CD8$^+$ T cells function as pathogenic T cells in Pd allergy. Based on our results, transgenic mice expressing TRAV7-2*02/TRAJ22*01 on CD8$^+$ T cells should be highly-sensitive Pd allergy animal models and would be a valuable tool to explore the molecular mechanisms underlying the pathogenesis of metal allergy. Furthermore, the response of TRAV7-2*02 expressing CD8$^+$ T cells could be assessed as an indicator of metal safety and, thus, our findings may be helpful for development of innovative biomaterials.

4. Materials and Methods

4.1. Ethics Statement

All mice were maintained under specific pathogen-free conditions, and used according to the guidelines of the Institutional Animal Care and Use Committee established at Tohoku University. The project identification code is 2016AcA-016 (date of approval: 25 March 2016).

4.2. Mice

C57BL/6 mice were obtained from CLEA Japan (Tokyo, Japan). B2m$^{-/-}$ mice or IFN-$\gamma^{-/-}$ mice were obtained from the Jackson Laboratory (Bar Harbor, ME, USA). I-A$^{b-/-}$ mice were kindly provided by Diane Mathis (Harvard Medical School, Boston, MA, USA).

4.3. Antibodies and Reagents

Monoclonal antibodies (mAbs) for immunohistochemistry were purchased from BD Biosciences (San Jose, CA, USA). All mAbs for flow cytometry were purchased from BioLegend (San Diego, CA, USA). LPS from *Escherichia coli* (O55:B5) [15,23], phorbol myristate acetate (PMA), ionomycin, and propidium iodide (PI) were purchased from Sigma-Aldrich (St. Louis, MO, USA). PdCl$_2$ was purchased from Wako Pure Chemical Industries, Ltd. (Osaka, Japan).

4.4. Induction of Pd Allergy

Induction of Pd allergy was previously described [6]. Briefly, for sensitization mice were injected twice intraperitoneally with 250 μL of 10 mM PdCl$_2$ containing 10 μg/mL LPS in PBS at an interval of seven days. LPS was used as an adjuvant [15]. As a control, mice were injected with PBS only (unsensitized). Seven days after the second sensitization, mice were challenged by intradermal injection of 20 μL of 0.5 mM PdCl$_2$ into each ear. Ear swelling was measured before and after the challenge using a Peacock dial thickness gauge (Ozaki MFG Co., Ltd., Tokyo, Japan) [6,15,23,24].

4.5. Histological Analysis

To identify the T cells present in ear auricles of WT mice with Pd allergy, frozen ear auricles were sliced and immunostained with anti-mouse CD4 mAb (H129.19), and anti-mouse CD8a mAb (53-6.7). CD4$^+$ and CD8$^+$ T cells were visualized by staining with 3,3′-diaminobenzidine (DAB) chromogen, and these sections were counter-stained with hematoxylin. The DAB signals were detected with an Olympus IX81 microscope, an Olympus DP71 CCD camera (Olympus, Tokyo, Japan), and Lumina Vision software (Mitani Corporation, Fukui, Japan). The scale bar indicates 100 μm.

4.6. Adoptive Transfer of Bone Marrow-Derived Antigen-Presenting Cells

To differentiate APCs, mouse bone marrow cells were cultured in RPMI1640 complete medium with 10% FBS and 10% CMG14-12 culture supernatant containing mM-CSF [25,26]. On day 7, the collected APCs were treated with 0.2 mM PdCl$_2$ and 10 ng/mL LPS (Pd-APCs) or 10 ng/mL LPS (LPS-APCs, as a control), and were cultured for 24 h. On day 8, these APCs were washed twice in PBS to completely remove the supernatant, and the harvested Pd-APCs or LPS-APCs (1 × 10^6 cells per mouse) were adoptively transferred via intravenous injection into naïve C57BL/6 mice for sensitization (Figure S4). Seven days after the adoptive transfer, recipient mice were challenged by intradermal injection using the same procedure described above.

Int. J. Mol. Sci. **2017**, *18*, 1162

4.7. Flow Cytometric Analysis of Cell Populations

SLN cells were pretreated with anti-CD16 and CD32 mAbs (2.4G2) to block Fc receptors, and then stained with the following specific mAbs: anti-CD3ε (145-2C11), anti-CD4 (GK1.5), anti-CD8a (53-6.7), anti-F4/80 (CI:A3-1), anti-CD11b (M1/70), anti-CD80 (16-10A1), anti-CD86 (GL-1), anti-CD40 (3/23), anti-H-2K^b (AF6-88.5), and isotype-matched controls. SLN cells were washed twice, and were then stained with PI followed by analysis on a FACSCanto II (BD Biosciences). Live cells were identified based on characteristic forward and side scatter and by exclusion of PI. Cell number was calculated using flow cytometric analysis. Number of cells in each subset of T cells was calculated by the following procedure: CD3$^+$ T cell numbers = total SLN cells numbers × CD3ε positive T cells population (%), and CD8$^+$ (or CD4$^+$) T cell numbers = total SLN cells numbers × CD3ε and CD8a (or CD4) double-positive T cell population (%). To identify IFN-γ producing T cells, SLN cells were isolated 24 hours after Pd challenge, and then were incubated with 20 ng/mL PMA plus 0.5 µg/mL ionomycin for four hours. Cells were then stained with anti-CD4, anti-CD8a, and anti-IFN-γ (XMG1.2) mAbs according to manufacturer's instructions and analyzed using a FACSCanto II.

4.8. T Cell Repertoire Sequence Analysis Using a Next-Generation Sequencer

To investigate the skewing of TCRs in Pd allergy, we analyzed the TCR repertoire using a next generation sequencer. Briefly, total RNA was prepared from the SLN 24 hours after Pd challenge [16]. Complementary DNA was synthesized from total RNA and TCR chains were amplified using adaptor ligation-mediated PCR. The specific PCR primers were 5'-AGG TGA AGC TTG TCT GGT TGC TC-3' (TCRα) and 5'-TGC AAT CTC TGC TTT TGA TGG CTC-3' (TCRβ). Then, using the PCR products as templates, TCR sequences were analyzed using the 454 GS junior+ system (Roche Applied Science, Indianapolis, IN, USA) according to the manufacturer's protocol. Alignments among approximately 100,000 sequences/run were performed with IMGT/V-QUEST (http://www.imgt.org).

4.9. Statistics

Student's *t*-test was used for analysis of differences and values of $p < 0.05$ or 0.01 were considered statistically significant [6,27].

5. Conclusions

In conclusion, we show that MHC I-dependent CD8$^+$ T cells activate and proliferate as pathogenic T cells and we identified TRAV7-2*02/TRAJ22*01 as the specific TCR in Pd allergic mice. Furthermore, Pd-APCs prepared in vitro can function as memory APCs and activate pathogenic T cells in recipient mice. In addition, the CDR3 consensus frame of pathogenic TCRs is CAAXSGSWQLIF in TRAV7-2*02/TRAJ22*01. Thus, these specific TCRs are promising, novel targets for generating improved diagnostics and treatments for Pd allergy.

Supplementary Materials: Supplementary materials can be found at www.mdpi.com/1422-0067/18/6/1162/s1.

Acknowledgments: We thank Risako Suzuki, Toru Kawakami, and Yoshihiro Yamaguchi for their helpful suggestions, and Madoka Itabashi, Masashi Arai, Jyuri Ohura, Wakana Ito, and Megumi Takahashi for their technical assistance. This work was supported by JSPS KAKENHI (grant numbers 15K11192, 15K11103, 16H05530, 16H06497 (Kouetsu Ogasawara), 15K19075 (Koyu Ito)) and by projects for the promotion of the indigenous creation and development of innovative medical devices in the Tohoku area (Kouetsu Ogasawara).

Author Contributions: Yuri Takeda designed and performed the experiments, analyzed the data, and wrote the manuscript. Yoshiko Suto, Koyu Ito, Wataru Hashimoto, Tadashi Nishiya, Kyosuke Ueda, Takayuki Narushima, and Tetsu Takahashi provided technical support and discussed the experimental strategy. Kouetsu Ogasawara contributed to the conception, wrote the manuscript, and supervised experiments.

Conflicts of Interest: The authors declare no conflict of interest.

Abbreviations

Pd	Palladium
DTH	Delayed-type hypersensitivity
TCR	T cell receptor
TRAV	TCRα variable
APCs	Antigen presenting cells
MHC	Major histocompatibility complex
TRAJ	TCRα junction
WT	Wild-type
SLN	Submandibular lymph node
LPS	Lipopolysaccharide
PBS	Phosphate buffered saline
SD	Standard deviation
TRBV	TCRβ variable
B2m	β2m-Microgloblin
B2m$^{-/-}$ mice	C57BL/6 deficient β2m-microgloblin mice
I-A$^{b-/-}$ mice	C57BL/6 deficient MHC II *H2-Ab1* (I-A^b) mice
mM-CSF	Mouse macrophage colony stimulating factor
mAbs	Monoclonal antibodies
PMA	Phorbol myristate acetate
PI	Propidium iodide

References

1. Raap, U.; Stiesch, M.; Reh, H.; Kapp, A.; Werfel, T. Investigation of contact allergy to dental metals in 206 patients. *Contact Dermat.* **2009**, *60*, 339–343. [CrossRef] [PubMed]
2. Faurschou, A.; Menne, T.; Johansen, J.D.; Thyssen, J.P. Metal allergen of the 21st century—A review on exposure, epidemiology and clinical manifestations of palladium allergy. *Contact Dermat.* **2011**, *64*, 185–195. [CrossRef] [PubMed]
3. Kaplan, D.H.; Igyarto, B.Z.; Gaspari, A.A. Early immune events in the induction of allergic contact dermatitis. *Nat. Rev. Immunol.* **2012**, *12*, 114–124. [CrossRef] [PubMed]
4. Schmidt, M.; Goebeler, M. Immunology of metal allergies. *J. Dtsch. Dermatol. Ges.* **2015**, *13*, 653–660. [CrossRef] [PubMed]
5. Muris, J.; Feilzer, A.J.; Kleverlaan, C.J.; Rustemeyer, T.; van Hoogstraten, I.M.; Scheper, R.J.; von Blomberg, B.M. Palladium-induced Th2 cytokine responses reflect skin test reactivity. *Allergy* **2012**, *67*, 1605–1608. [PubMed]
6. Kawano, M.; Nakayama, M.; Aoshima, Y.; Nakamura, K.; Ono, M.; Nishiya, T.; Nakamura, S.; Takeda, Y.; Dobashi, A.; Takahashi, A.; et al. NKG2D$^+$ IFN-γ$^+$ CD8$^+$ T cells are responsible for palladium allergy. *PLoS ONE* **2014**, *9*, e86810. [CrossRef] [PubMed]
7. Chang, J.T.; Wherry, E.J.; Goldrath, A.W. Molecular regulation of effector and memory T cell differentiation. *Nat. Immunol.* **2014**, *15*, 1104–1115. [CrossRef] [PubMed]
8. Ogasawara, K.; Yoshinaga, S.K.; Lanier, L.L. Inducible costimulator costimulates cytotoxic activity and IFN-gamma production in activated murine NK cells. *J. Immunol.* **2002**, *169*, 3676–3685. [CrossRef] [PubMed]
9. Rossjohn, J.; Gras, S.; Miles, J.J.; Turner, S.J.; Godfrey, D.I.; McCluskey, J. T cell antigen receptor recognition of antigen-presenting molecules. *Annu. Rev. Immunol.* **2015**, *33*, 169–200. [CrossRef] [PubMed]
10. Castro, C.D.; Luoma, A.M.; Adams, E.J. Coevolution of T-cell receptors with MHC and non-MHC ligands. *Immunol. Rev.* **2015**, *267*, 30–55. [CrossRef] [PubMed]
11. Hashizume, H.; Seo, N.; Ito, T.; Takigawa, M.; Yagi, H. Promiscuous interaction between gold-specific T cells and APCs in gold allergy. *J. Immunol.* **2008**, *181*, 8096–8102. [CrossRef] [PubMed]
12. Hennecke, J.; Wiley, D.C. T cell receptor-MHC interactions up close. *Cell* **2001**, *104*, 1–4. [CrossRef]
13. Garcia, K.C.; Adams, E.J. How the T cell receptor sees antigen—A structural view. *Cell* **2005**, *122*, 333–336. [CrossRef] [PubMed]

14. Honda, T.; Egawa, G.; Grabbe, S.; Kabashima, K. Update of immune events in the murine contact hypersensitivity model: Toward the understanding of allergic contact dermatitis. *J. Investig. Dermatol.* **2013**, *133*, 303–315. [CrossRef] [PubMed]

15. Sato, N.; Kinbara, M.; Kuroishi, T.; Kimura, K.; Iwakura, Y.; Ohtsu, H.; Sugawara, S.; Endo, Y. Lipopolysaccharide promotes and augments metal allergies in mice, dependent on innate immunity and histidine decarboxylase. *Clin. Exp. Allergy* **2007**, *37*, 743–751. [CrossRef] [PubMed]

16. Kobayashi, H.; Kumagai, K.; Eguchi, T.; Shigematsu, H.; Kitaura, K.; Kawano, M.; Horikawa, T.; Suzuki, S.; Matsutani, T.; Ogasawara, K.; et al. Characterization of T cell receptors of Th1 cells infiltrating inflamed skin of a novel murine model of palladium-induced metal allergy. *PLoS ONE* **2013**, *8*, e76385. [CrossRef] [PubMed]

17. Muris, J.; Kleverlaan, C.J.; Feilzer, A.J.; Rustemeyer, T. Sodium tetrachloropalladate (Na$_2$[PdCl$_4$]) as an improved test salt for palladium allergy patch testing. *Contact Dermat.* **2008**, *58*, 42–46. [CrossRef] [PubMed]

18. Schmidt, M.; Raghavan, B.; Muller, V.; Vogl, T.; Fejer, G.; Tchaptchet, S.; Keck, S.; Kalis, C.; Nielsen, P.J.; Galanos, C.; et al. Crucial role for human Toll-like receptor 4 in the development of contact allergy. *Nat. Immunol.* **2010**, *11*, 814–819. [CrossRef] [PubMed]

19. Toivonen, R.; Arstila, T.P.; Hanninen, A. Islet-associated T-cell receptor-beta CDR sequence repertoire in prediabetic NOD mice reveals antigen-driven T-cell expansion and shared usage of VbetaJbeta TCR chains. *Mol. Immunol.* **2015**, *64*, 127–135. [CrossRef] [PubMed]

20. Haqqi, T.M.; Anderson, G.D.; Banerjee, S.; David, C.S. Restricted heterogeneity in T-cell antigen receptor V β gene usage in the lymph nodes and arthritic joints of mice. *Proc. Natl. Acad. Sci. USA* **1992**, *89*, 1253–1255. [CrossRef] [PubMed]

21. Lu, L.; Vollmer, J.; Moulon, C.; Weltzien, H.U.; Marrack, P.; Kappler, J. Components of the ligand for a Ni++ reactive human T cell clone. *J. Exp. Med.* **2003**, *197*, 567–574. [CrossRef] [PubMed]

22. Gamerdinger, K.; Moulon, C.; Karp, D.R.; Van Bergen, J.; Koning, F.; Wild, D.; Pflugfelder, U.; Weltzien, H.U. A new type of metal recognition by human T cells: Contact residues for peptide-independent bridging of T cell receptor and major histocompatibility complex by nickel. *J. Exp. Med.* **2003**, *197*, 1345–1353. [CrossRef] [PubMed]

23. Shigematsu, H.; Kumagai, K.; Kobayashi, H.; Eguchi, T.; Kitaura, K.; Suzuki, S.; Horikawa, T.; Matsutani, T.; Ogasawara, K.; Hamada, Y.; et al. Accumulation of metal-specific T cells in inflamed skin in a novel murine model of chromium-induced allergic contact dermatitis. *PLoS ONE* **2014**, *9*, e85983. [CrossRef] [PubMed]

24. Kinbara, M.; Nagai, Y.; Takano-Yamamoto, T.; Sugawara, S.; Endo, Y. Cross-reactivity among some metals in a murine metal allergy model. *Br. J. Dermatol.* **2011**, *165*, 1022–1029. [CrossRef] [PubMed]

25. Takeshita, S.; Kaji, K.; Kudo, A. Identification and characterization of the new osteoclast progenitor with macrophage phenotypes being able to differentiate into mature osteoclasts. *J. Bone Miner. Res.* **2000**, *15*, 1477–1488. [CrossRef] [PubMed]

26. Nishiya, T.; DeFranco, A.L. Ligand-regulated chimeric receptor approach reveals distinctive subcellular localization and signaling properties of the Toll-like receptors. *J. Biol. Chem.* **2004**, *279*, 19008–190017. [CrossRef] [PubMed]

27. Pennino, D.; Eyerich, K.; Scarponi, C.; Carbone, T.; Eyerich, S.; Nasorri, F.; Garcovich, S.; Traidl-Hoffmann, C.; Albanesi, C.; Cavani, A. IL-17 amplifies human contact hypersensitivity by licensing hapten nonspecific Th1 cells to kill autologous keratinocytes. *J. Immunol.* **2010**, *184*, 4880–4888. [CrossRef] [PubMed]

International Journal of
Molecular Sciences

MDPI

Review

Molecular Mechanisms of Nickel Allergy

Masako Saito, Rieko Arakaki, Akiko Yamada, Takaaki Tsunematsu, Yasusei Kudo and Naozumi Ishimaru *

Department of Oral Molecular Pathology, Institute of Biomedical Sciences,
Tokushima University Graduate School, 3-18-15 Kuramoto Tokushima 770-8504, Japan;
m.saito@tokushima-u.ac.jp (M.S.); arakaki.r@tokushima-u.ac.jp (R.A.); aki.yamada@tokushima-u.ac.jp (A.Y.);
tsunematsu@tokushima-u.ac.jp (T.T.); yasusei@tokushima-u.ac.jp (Y.K.)
* Correspondence: ishimaru.n@tokushima-u.ac.jp; Tel./Fax: +81-88-633-7464

Academic Editor: Reinhard Dallinger
Received: 15 January 2016; Accepted: 29 January 2016; Published: 2 February 2016

Abstract: Allergic contact hypersensitivity to metals is a delayed-type allergy. Although various metals are known to produce an allergic reaction, nickel is the most frequent cause of metal allergy. Researchers have attempted to elucidate the mechanisms of metal allergy using animal models and human patients. Here, the immunological and molecular mechanisms of metal allergy are described based on the findings of previous studies, including those that were recently published. In addition, the adsorption and excretion of various metals, in particular nickel, is discussed to further understand the pathogenesis of metal allergy.

Keywords: metal allergy; Ni; DTH; DC; T cell; TLR; TSLP

---

1. Introduction

Contact dermatitis is usually caused by external exposure of the skin to allergens, such as metals, chemicals, and plants. Metal allergy is an inflammatory disease categorized as a delayed-type hypersensitivity (DTH) reaction. Humans come in contact with various metals daily. For example, metal alloys are widely used in costume jewelry, dental materials, or glasses. Although many individuals develop a metal allergy, the precise molecular mechanism underlying this allergy remains unknown.

Some metals cause contact allergic reactions categorized as type IV DTH, in which skin inflammation is mediated by hapten-specific T cells [1,2]. In this review, the cellular and molecular mechanisms identified by basic and clinical studies on metal allergy are described. In addition, the adsorption and excretion of metals in the human body and useful animal models for investigating metal allergy are reviewed. Furthermore, the adsorption and excretion of metals in the body are discussed. Finally, the pathogenesis of metal allergy is described with respect to the potential molecular mechanisms of this immune response.

2. Metal Allergy

Metals, such as gold (Au), silver (Ag), mercury (Hg), nickel (Ni), titanium (Ti), chromium (Cr), copper (Cu), and cobalt (Co) are ubiquitous in our environment and are widely used in costume jewelry, coins, mobile phones, and dental materials. Approximately, 10%–15% of the human population suffers from contact hypersensitivity to metals [1,2]. This allergy is considerably more common in women than in men, with an approximate population frequency of 10% in women *vs.* 2% in men [3,4]. Clinically, metal allergy is related to the cause of contact dermatitis, pustulosis palmoplantaris, lichen planus, dyshidrotic eczema, and burning mouth syndrome [5–8]. Moreover, patients with autoimmune conditions, including systemic lupus erythematosus, rheumatoid arthritis, and Sjögren's syndrome, have an increased frequency of metal allergy [9].

A previous study indicated that nickel (II) sulfate has the highest sensitization rate and affects approximately 15% of the population, followed by cobalt chloride and potassium dichromate, which approximately 5% and 3% of the population, respectively [10]. Nickel allergy is the most common [2,11], and clinically important condition that is becoming a threat to public health [12,13]. The use of nickel alloys is common in dentistry, and high concentrations of nickel can be found in food. Nickel-casting alloys are cheap and have favorable physical properties but are prone to corrosion in the oral environment [14]. Metal allergy is mainly diagnosed by patch testing. Several reports have demonstrated that the removal of causal metal can successfully improve allergic symptoms. Therefore, in addition to the metal concentration, a special quality of metal seems to be important for the pathogenesis of metal allergy [15–17].

Nickel ions released from various alloys are potent allergens or haptens that can trigger skin inflammation [18–20]. They penetrate the skin and activate epithelial cells that produce various cytokines or chemokines. The reaction follows complex immune responses that involve the activation of antigen-presenting cells (APCs) and T cells [21–23]. Some cytokines activate APCs, such as Langerhans cells (LCs) or dendritic cells (DCs). Activated APCs migrate to the draining lymph nodes where they present the allergens or haptens to naive CD4-positive T cells. Subsequent re-exposure to the same allergen or hapten would lead to the activation of hapten-specific T-cells, which subsequently enter the bloodstream and produce visible signs of hypersensitivity at 48 to 72 h after allergen or hapten exposure [24]. However, the precise molecular mechanisms that mediate the interactions between epithelial and immune cells in nickel allergy remain unknown.

3. Animal Models and Molecular Mechanism of Metal Allergy

Many researchers have used animal models to investigate nickel allergy by administrating adjuvants. For example, nickel chloride (II) is administered twice into mice in combination with adjuvants, such as incomplete Freund's adjuvant and complete Freund's adjuvant and ear swelling is evaluated for DTH after 48 h [25].

3.1. Keratinocytes and APCs in Ni Allergy Models

Nickel penetrates the skin tissue and activates keratinocytes, leading to the release of certain cytokines such as interleukin (IL)-1β and tumor necrosis factor alpha. Subsequently, nickel attaches to the major histocompatibility complex (MHC) molecules on LCs and DCs that are upregulated by the cytokines from the surrounding keratinocytes. These cytokines control the expression of E-cadherin and chemokines, including matrix metalloproteinase-9, secondary lymphoid tissue chemokine (SLC), and macrophage inflammatory protein-3β, that are produced by the APCs [26–29]. Subsequently, the APCs migrate to draining lymph nodes where they present these haptens to naive T cells. Re-exposure to the same hapten induces a hypersensitive reaction in an effector phase at the site of exposure (Figure 1).

Several studies have shown that the activation of p38 mitogen-activated protein kinase in dermal DCs is required to trigger a T cell-mediated immune response in a mouse model of nickel allergy [30–33]. Ni-activated epithelial DCs or LCs exhibit the upregulation of CD80, CD83, CD86, and MHC class II [30]. Moreover, nickel plays an important role in the maturation and activation of immature LCs or DCs in the skin via phosphorylated MAP kinase kinase 6 (MKK6) [31–34]. Therefore, Ni-stimulated DCs prime activate T cells to induce skin inflammation at the site of exposure to nickel. However, the injection of short interfering (si) RNAs targeting *MKK6* prevents a hypersensitive reaction after Ni immunization in a mouse model, suggesting that manipulating *MKK6* in DCs might be a good therapeutic strategy for nickel allergy [25].

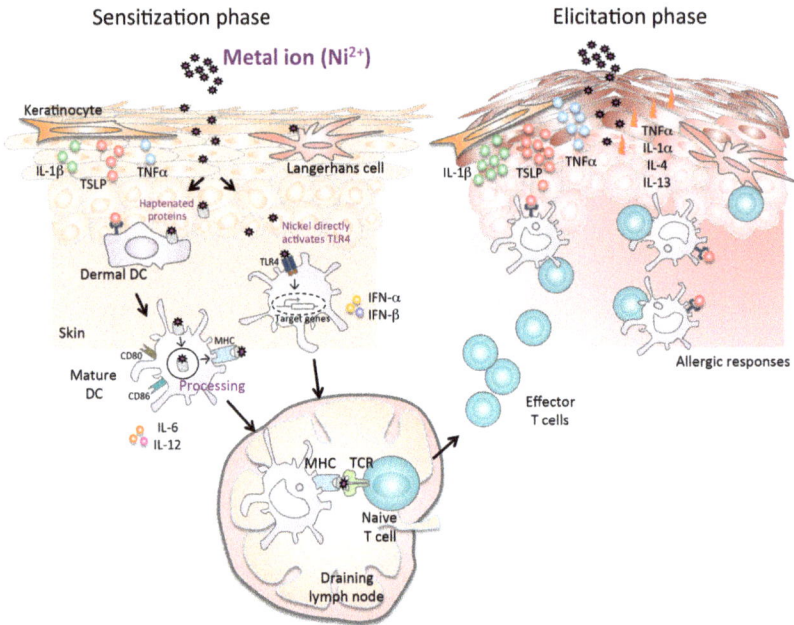

Figure 1. A complex mechanism of metal allergy. The sensitization phase begins after nickel exposure to the skin. Nickel penetration into the skin results in the production of proinflammatory cytokines (TNF-α and IL-1β), TSLP, and chemokines, which induce activation and migration of haptenated protein-loaded epidermal and dermal DCs through afferent lymph to the draining lymph nodes. Particularly in humans, nickel directly activates the TLR4 pathway in DCs. In the draining lymph nodes, haptenated-peptide presentation results in the proliferation, activation and subsequent differentiation of hapten-specific T cells. Secretion of cytokines in the draining lymph nodes during the sensitization phase contributes to efficient hapten-specific T cell activation, proliferation, and differentiation. At the end of this phase, primed specific T cells migrate out of the lymph nodes to the skin. In the elicitation phase, the subsequent application of the same hapten leads to uptake by cells, which is presented to the recirculating hapten-specific T cells. The activated T cells produce inflammatory cytokines and chemokines at the site of exposure that promote an allergic reaction, leading to the development of characteristic skin lesions.

3.2. Critical Role of Toll-Like Receptor 4 in Ni Allergy

Human toll-like receptor (TLR) 4 has been shown to play a crucial role in the development of contact allergy to nickel [35]. TLR4-deficient mice expressing transgenic human TLR4 developed contact hypersensitivity to nickel, whereas those expressing mouse TLR4 did not [35]. Although the cell type contributing to a TLR4-mediated allergic reaction has not been identified, immune cells such as DCs, macrophages, and endothelial cells were found to be associated with the allergic reaction to Ni via TLR4 [35]. Ni-induced activation of TLR4 leads to the activation of nuclear factor (NF)-κB, p38, and interferon regulatory factor 3, resulting in the induction of multiple proinflammatory cytokines that trigger an allergic response. These findings explain why Ni^{2+}, but not other contact allergens, directly triggers NF-κB-dependent activation of human DCs [35]. Furthermore, a recent study suggested that other metals, including cobalt and palladium, induce IL-8 production in HEK293 cells via TLR4/MD2 [36]. Lipopolysaccharides are an important inducer of nickel allergy and enhance the allergic response in TLR4-mutant mice [37]. However, TLR4 signaling by keratinocytes controls wound healing by inducing CCL5 expression [38]. Keratinocytes are known to produce danger signal-induced cytokines or chemokines in the skin tissue [39]. A recent study demonstrated that ionized gold is recognized by TLR3. Epithelial TLR3 plays a crucial role in the localized irritation reactivity to gold in the skin and mucosa. Therefore, in addition to gold, nickel, copper and mercury salts may activate an innate immune response in keratinocytes [40]. Taken together, these findings reveal a mechanism of skin contact allergy development and might contribute to the elucidation of novel therapeutic strategies such as those based on interference with distinct immune detention pathways.

3.3. Thymic Stromal Lymphopoietin and Its Receptor in Ni Allergy

Using a mouse model, a recent study showed that the increased expression of the thymic stromal lymphopoietin (TSLP) receptor (TSLPR) on DCs plays a key role in triggering an allergic response to nickel [41]. In this mouse model of nickel allergy, DCs in ear tissues were activated via TSLPR signaling induced by keratinocyte-derived TSLP. Furthermore, DTH reactions in mice with Ni-induced allergy were reduced significantly by the injection of a Tslp–siRNA combined with atelocollagen into the ear skin of the ear [41]. These results suggest that nickel allergy is triggered by a TSLP/TSLPR-mediated interaction between epithelial and immune cells. TSLP is produced by keratinocytes, the tonsil crypt epithelium, and bronchial epithelial cells [42,43]. Furthermore, TSLP induces allergic inflammatory reactions in patients with asthma and atopic dermatitis [42,43].

Although numerous patients develop allergic symptoms against various metals, experimental animal models of nickel allergy have been widely used to elucidate the molecular mechanisms of metal allergy. Allergic diseases are multifactorial disorders caused by various factors, such as genetics and the environment in addition to exposure to metals. Moreover, little is known regarding the molecular or cellular mechanisms underlying haptenization of metal allergens during an allergic reaction.

CD25$^+$ T cells isolated from peripheral blood of human nickel-allergy patients demonstrated a limited or no capacity to suppress metal-specific CD4$^+$ and CD8$^+$ T cell responses. In contrast, CD4$^+$CD25$^+$ T cells from peripheral blood of non-allergic subjects strongly regulate immune responses to nickel in a cytokine-independent, cell-contact-dependent mechanism. These results indicate that in healthy individuals CD25$^+$ Treg can control the activation of both naive and effector nickel-specific T cells [44,45].

Further studies on animal models might reveal the precise mechanism by which metal allergy promote the clinical application of new therapeutic strategies.

4. Adsorption and Excretion of Metals

Understanding how metals are metabolized in the body is one of the key factors in better elucidating the process of developing a metal allergy [46]. The accumulation of metals in the body is influenced by exposure time, absorption medium, tissue distribution, and metal excretion. Nickel has been well studied among the metals associated with allergy. The biological half-life of nickel is estimated to range from 17 to 39 h and 20 to 34 h in the urine and plasma, respectively [47]. The model allows the precise prediction of the state and extent of exposure, which is affected by varying concentrations of metals in the atmosphere [47].

The investigation using the excised human skin showed that Ni ions are detected to penetrate the skin using a very sensitive method to quantify the amount of nickel permeating to the skin [48–50]. Although the permeation process is slow with a lag time of approximately 50 h, the rate using aqueous nickel chloride is increased compared with that in aqueous nickel sulfate [50]. Thus, the selection of nickel salt is an important consideration when conducting a skin patch test for detecting nickel permeation to the skin [50].

Absorption of Ni via the gastrointestinal tract by diet remarkably affects the bioavailability of nickel in the body; approximately 25% of nickel ingested in drinking water after an over-night fast is absorbed from the intestine and excreted in the urine, whereas only 1% of nickel ingested is absorbed [51]. The compartmental model and kinetic parameters decrease the uncertainty of toxicological assessments of human exposures to Ni via drinking water and food [51].

The other well-studied example is cobalt. Water-soluble cobalt salts are rapidly absorbed from the small intestine, although the bioavailability of cobalt is limited and highly variable [52]. Cobalt uptake occurs substantially through the lungs following inhalation and cobalt oxide in dust, and welding fumes leads to the systemic dissemination of ultrafine particles via the lymph and vascular system, releasing soluble cobalt ions [52]. After a single dose of cobalt to humans, the concentration of cobalt in the serum is initially high but decreases rapidly by the tissue uptake, primarily by the liver and kidney combined with urinary and fecal excretion. Renal excretion is rapid but decreases over the first few days, followed by a slow phase lasting for several weeks. Therefore, the metal is sustained in the tissues for several years [53,54]. During the first 24 h, 40% cobalt is eliminated and approximately 70% is eliminated after a week. However, one month later approximately 20% and one year later approximately 10% remain [55].

The adsorption and excretion of metals in the human body are also controlled by genetic factors. A genome-wide association study (GWAS) demonstrated single nucleotide polymorphisms associated with whole blood levels of metals [56]. Eleven metals and trace elements including aluminum, cadmium, cobalt, copper, chromium, mercury, manganese, molybdenum, nickel, lead, and zinc, were evaluated in a cohort of 949 individuals by using mass spectrometry. In addition, DNA samples were also genotyped. This GWAS analysis revealed that two regions, 4q24 and 1q41, are associated with serum magnesium levels; these regions encode a protein involved in manganese and zinc transport, SLC39A8 and SLC30A10, respectively. These data revealed metabolic pathways of metals and suggested that different subsets of individuals are more susceptible to metal toxicity [56].

5. Conclusions and Perspectives

The incidence of allergic diseases has been increasing worldwide. The pathogenesis and mechanisms of the allergic response is highly complex, and many patients develop refractory disease. Because metal allergy is caused by materials used in products that are common in our daily life, chances of triggering the onset of allergic reactions are high. The clinical symptoms of metal allergy include rashes, swelling, and pain. Molecular pathogenesis of a metal allergy suggests that excess responses to metals occur via the complicated process of the interactions among the immune system, epithelial barrier, and homeostatic mechanism. The unique features, adsorption, and the excretion of metals in the human body complicate the pathogenesis and symptoms of metal allergy. Molecular mechanisms of metal allergy need to be determined to develop novel therapeutic strategies. Analysis and characterization of the precise mechanisms could have clinical implications leading to the development of new diagnostic or treatment methods for metal allergy.

Acknowledgments: Acknowledgments: This work was partly supported by Grants-in-Aids for the Ministry of Scientific Research from the Ministry of Education, Culture, Sports, Science and Technology of Japan (No.15K15676).

Author Contributions: Author Contributions: All authors participated in developing the ideas presented in this manuscript, and researching the literature. Masako Saito and Naozumi Ishimaru wrote the paper.

Conflicts of Interest: Conflicts of Interest: The authors declare no conflicts of interest.

References

1. Loh, J.; Fraser, J. Metal-derivatized major histocompatibility complex: Zeroing in on contact hypersensitivity. *J. Exp. Med.* **2003**, *197*, 549–552. [CrossRef] [PubMed]
2. Budinger, L.; Hertl, M. Immunologic mechanisms in hypersensitivity reactions to metal ions: An overview. *Allergy* **2000**, *55*, 108–115. [CrossRef] [PubMed]
3. Peltonen, L. Nickel sensitivity in the general population. *Contact Dermat.* **1979**, *5*, 27–32. [CrossRef]
4. Nielsen, N.H.; Menne, T. Allergic contact sensitization in an unselected Danish population. *Acta Derm. Venereol.* **1992**, *72*, 456–460. [PubMed]
5. Pigatto, P.D.; Guzzi, G. Systemic allergic dermatitis syndrome caused by mercury. *Contact Dermat.* **2008**, *59*, 66.

6. Yoshihisa, Y.; Shimizu, T. Metal allergy and systemic contact dermatitis: An overview. *Dermatol. Res. Pract.* **2012**, *2012*, 749561. [CrossRef] [PubMed]

7. Yokozeki, H.; Katayama, I.; Nishioka, K.; Kinoshita, M.; Nishiyama, S. The role of metal allergy and local hyperhidrosis in the pathogenesis of pompholyx. *J. Dermatol.* **1992**, *19*, 964–967. [CrossRef] [PubMed]

8. Song, H.; Yin, W.; Ma, Q. Allergic palmoplantar pustulosis caused by cobalt in cast dental crowns: A case report. *Oral Surg. Oral Med. Oral Pathol. Oral Radiol. Endod.* **2011**, *111*, e8–e10. [CrossRef] [PubMed]

9. Stejskal, V.; Reynolds, T.; Bjorklund, G. Increased frequency of delayed type hypersensitivity to metals in patients with connective tissue disease. *J. Trace Elem. Med. Biol.* **2015**, *31*, 230–236. [CrossRef] [PubMed]

10. Mahler, V.; Geier, J.; Schnuch, A. Current trends in patch testing—new data from the German Contact Dermatitis Research Group (DKG) and the Information Network of Departments of Dermatology (IVDK). *JDDG* **2014**, *12*, 583–592. [CrossRef] [PubMed]

11. Garner, L.A. Contact dermatitis to metals. *Dermatol. Ther.* **2004**, *17*, 321–327. [CrossRef] [PubMed]

12. Peiser, M.; Tralau, T.; Heidler, J.; Api, A.M.; Arts, J.H.; Basketter, D.A.; English, J.; Diepgen, T.L.; Fuhlbrigge, R.C.; Gaspari, A.A.; *et al.* Allergic contact dermatitis: Epidemiology, molecular mechanisms, *in vitro* methods and regulatory aspects. Current knowledge assembled at an international workshop at BfR, Germany. *Cell. Mol. Life Sci.* **2012**, *69*, 763–781. [CrossRef] [PubMed]

13. Schram, S.E.; Warshaw, E.M.; Laumann, A. Nickel hypersensitivity: A clinical review and call to action. *Int. J. Dermatol.* **2010**, *49*, 115–125. [CrossRef] [PubMed]

14. Wataha, J.C.; Drury, J.L.; Chung, W.O. Nickel alloys in the oral environment. *Expert Rev. Med. Devices* **2013**, *10*, 519–539. [CrossRef] [PubMed]

15. Matsuzaka, K.; Mabuchi, R.; Nagasaka, H.; Yoshinari, M.; Inoue, T. Improvement of eczematous symptoms after removal of amalgam-like metal in alveolar bone. *Bull. Tokyo Dent. Coll.* **2006**, *47*, 13–17. [CrossRef] [PubMed]

16. Laeijendecker, R.; Dekker, S.K.; Burger, P.M.; Mulder, P.G.; Van Joost, T.; Neumann, M.H. Oral lichen planus and allergy to dental amalgam restorations. *Arch. Dermatol.* **2004**, *140*, 1434–1438. [CrossRef] [PubMed]

17. Yaqob, A.; Danersund, A.; Stejskal, V.D.; Lindvall, A.; Hudecek, R.; Lindh, U. Metal-specific lymphocyte reactivity is downregulated after dental metal replacement. *Neuro Endocrinol. Lett.* **2006**, *27*, 189–197. [PubMed]

18. Kapsenberg, M.L.; Wierenga, E.A.; Stiekema, F.E.; Tiggelman, A.M.; Bos, J.D. Th1 lymphokine production profiles of nickel-specific CD4+T-lymphocyte clones from nickel contact allergic and non-allergic individuals. *J. Investig. Dermatol.* **1992**, *98*, 59–63. [CrossRef] [PubMed]

19. Mortz, C.G.; Lauritsen, J.M.; Bindslev-Jensen, C.; Andersen, K.E. Prevalence of atopic dermatitis, asthma, allergic rhinitis, and hand and contact dermatitis in adolescents. The Odense Adolescence Cohort Study on Atopic Diseases and Dermatitis. *Br. J. Dermatol.* **2001**, *144*, 523–532. [CrossRef] [PubMed]

20. Thierse, H.J.; Gamerdinger, K.; Junkes, C.; Guerreiro, N.; Weltzien, H.U. T cell receptor (TCR) interaction with haptens: Metal ions as non-classical haptens. *Toxicology* **2005**, *209*, 101–107. [CrossRef] [PubMed]

21. Curtis, A.; Morton, J.; Balafa, C.; MacNeil, S.; Gawkrodger, D.J.; Warren, N.D.; Evans, G.S. The effects of nickel and chromium on human keratinocytes: Differences in viability, cell associated metal and IL-1alpha release. *Toxicol. In Vitro* **2007**, *21*, 809–819. [CrossRef] [PubMed]

22. Larsen, J.M.; Bonefeld, C.M.; Poulsen, S.S.; Geisler, C.; Skov, L. IL-23 and T(H)17-mediated inflammation in human allergic contact dermatitis. *J. Allergy Clin. Immunol.* **2009**, *123*, 486–492. [CrossRef] [PubMed]

23. Sebastiani, S.; Albanesi, C.; Nasorri, F.; Girolomoni, G.; Cavani, A. Nickel-specific CD4(+) and CD8(+) T cells display distinct migratory responses to chemokines produced during allergic contact dermatitis. *J. Investig. Dermatol.* **2002**, *118*, 1052–1058. [CrossRef] [PubMed]

24. Steinman, R.M.; Pack, M.; Inaba, K. Dendritic cells in the T-cell areas of lymphoid organs. *Immunol. Rev.* **1997**, *156*, 25–37. [CrossRef] [PubMed]

25. Watanabe, M.; Ishimaru, N.; Ashrin, M.N.; Arakaki, R.; Yamada, A.; Ichikawa, T.; Hayashi, Y. A novel DC therapy with manipulation of MKK6 gene on nickel allergy in mice. *PLoS ONE* **2011**, *6*, e19017. [CrossRef] [PubMed]

26. Roake, J.A.; Rao, A.S.; Morris, P.J.; Larsen, C.P.; Hankins, D.F.; Austyn, J.M. Dendritic cell loss from nonlymphoid tissues after systemic administration of lipopolysaccharide, tumor necrosis factor, and interleukin 1. *J. Exp. Med.* **1995**, *181*, 2237–2247. [CrossRef] [PubMed]

27. Lore, K.; Sonnerborg, A.; Spetz, A.L.; Andersson, U.; Andersson, J. Immunocytochemical detection of cytokines and chemokines in Langerhans cells and *in vitro* derived dendritic cells. *J. Immunol. Methods* **1998**, *214*, 97–111. [CrossRef]

28. Riedl, E.; Stockl, J.; Majdic, O.; Scheinecker, C.; Rappersberger, K.; Knapp, W.; Strobl, H. Functional involvement of E-cadherin in TGF-beta 1-induced cell cluster formation of *in vitro* developing human Langerhans-type dendritic cells. *J. Immunol.* **2000**, *165*, 1381–1386. [CrossRef] [PubMed]

29. Geissmann, F.; Dieu-Nosjean, M.C.; Dezutter, C.; Valladeau, J.; Kayal, S.; Leborgne, M.; Brousse, N.; Saeland, S.; Davoust, J. Accumulation of immature Langerhans cells in human lymph nodes draining chronically inflamed skin. *J. Exp. Med.* **2002**, *196*, 417–430. [CrossRef] [PubMed]

30. Villadangos, J.A.; Cardoso, M.; Steptoe, R.J.; van Berkel, D.; Pooley, J.; Carbone, F.R.; Shortman, K. MHC class II expression is regulated in dendritic cells independently of invariant chain degradation. *Immunity* **2001**, *14*, 739–749. [CrossRef]

31. Verhasselt, V.; Buelens, C.; Willems, F.; De Groote, D.; Haeffner-Cavaillon, N.; Goldman, M. Bacterial lipopolysaccharide stimulates the production of cytokines and the expression of costimulatory molecules by human peripheral blood dendritic cells: Evidence for a soluble CD14-dependent pathway. *J. Immunol.* **1997**, *158*, 2919–2925. [PubMed]

32. Kyriakis, J.M. Life-or-death decisions. *Nature* **2001**, *414*, 265–266. [CrossRef] [PubMed]

33. Arrighi, J.F.; Rebsamen, M.; Rousset, F.; Kindler, V.; Hauser, C. A critical role for p38 mitogen-activated protein kinase in the maturation of human blood-derived dendritic cells induced by lipopolysaccharide, TNF-alpha, and contact sensitizers. *J. Immunol.* **2001**, *166*, 3837–3845. [CrossRef] [PubMed]

34. Jorgl, A.; Platzer, B.; Taschner, S.; Heinz, L.X.; Hocher, B.; Reisner, P.M.; Gobel, F.; Strobl, H. Human Langerhans-cell activation triggered *in vitro* by conditionally expressed MKK6 is counterregulated by the downstream effector RelB. *Blood* **2007**, *109*, 185–193. [CrossRef] [PubMed]

35. Schmidt, M.; Raghavan, B.; Muller, V.; Vogl, T.; Fejer, G.; Tchaptchet, S.; Keck, S.; Kalis, C.; Nielsen, P.J.; Galanos, C.; *et al.* Crucial role for human Toll-like receptor 4 in the development of contact allergy to nickel. *Nat. Immunol.* **2010**, *11*, 814–819. [CrossRef] [PubMed]

36. Rachmawati, D.; Bontkes, H.J.; Verstege, M.I.; Muris, J.; von Blomberg, B.M.; Scheper, R.J.; van Hoogstraten, I.M. Transition metal sensing by Toll-like receptor-4: Next to nickel, cobalt and palladium are potent human dendritic cell stimulators. *Contact Dermat.* **2013**, *68*, 331–338. [CrossRef] [PubMed]

37. Sato, N.; Kinbara, M.; Kuroishi, T.; Kimura, K.; Iwakura, Y.; Ohtsu, H.; Sugawara, S.; Endo, Y. Lipopolysaccharide promotes and augments metal allergies in mice, dependent on innate immunity and histidine decarboxylase. *Clin. Exp. Allergy* **2007**, *37*, 743–751. [CrossRef] [PubMed]

38. Suga, H.; Sugaya, M.; Fujita, H.; Asano, Y.; Tada, Y.; Kadono, T.; Sato, S. TLR4, rather than TLR2, regulates wound healing through TGF-beta and CCL5 expression. *J. Dermatol. Sci.* **2014**, *73*, 117–124. [CrossRef] [PubMed]

39. Reche, P.A.; Soumelis, V.; Gorman, D.M.; Clifford, T.; Liu, M.; Travis, M.; Zurawski, S.M.; Johnston, J.; Liu, Y.J.; Spits, H.; *et al.* Human thymic stromal lymphopoietin preferentially stimulates myeloid cells. *J. Immunol.* **2001**, *167*, 336–343. [CrossRef] [PubMed]

40. Rachmawati, D.; Buskermolen, J.K.; Scheper, R.J.; Gibbs, S.; von Blomberg, B.M.; van Hoogstraten, I.M. Dental metal-induced innate reactivity in keratinocytes. *Toxicol. In Vitro* **2015**, *30*, 325–330. [CrossRef] [PubMed]

41. Ashrin, M.N.; Arakaki, R.; Yamada, A.; Kondo, T.; Kurosawa, M.; Kudo, Y.; Watanabe, M.; Ichikawa, T.; Hayashi, Y.; Ishimaru, N. A critical role for thymic stromal lymphopoietin in nickel-induced allergy in mice. *J. Immunol.* **2014**, *192*, 4025–4031. [CrossRef] [PubMed]

42. Ying, S.; O'Connor, B.; Ratoff, J.; Meng, Q.; Mallett, K.; Cousins, D.; Robinson, D.; Zhang, G.; Zhao, J.; Lee, T.H.; *et al.* Thymic stromal lymphopoietin expression is increased in asthmatic airways and correlates with expression of Th2-attracting chemokines and disease severity. *J. Immunol.* **2005**, *174*, 8183–8190. [CrossRef] [PubMed]

43. Yoo, J.; Omori, M.; Gyarmati, D.; Zhou, B.; Aye, T.; Brewer, A.; Comeau, M.R.; Campbell, D.J.; Ziegler, S.F. Spontaneous atopic dermatitis in mice expressing an inducible thymic stromal lymphopoietin transgene specifically in the skin. *J. Exp. Med.* **2005**, *202*, 541–549. [CrossRef] [PubMed]

44. Cavani, A. Breaking tolerance to nickel. *Toxicology* **2005**, *209*, 119–121. [CrossRef] [PubMed]

45. Cavani, A.; Nasorri, F.; Ottaviani, C.; Sebastiani, S.; De Pita, O.; Girolomoni, G. Human CD25+ regulatory T cells maintain immune tolerance to nickel in healthy, nonallergic individuals. *J. Immunol.* **2003**, *171*, 5760–5768. [CrossRef] [PubMed]

46. Christensen, J.M. Human exposure to toxic metals: Factors influencing interpretation of biomonitoring results. *Sci. Total Environ.* **1995**, *166*, 89–135. [CrossRef]

47. Tossavainen, A.; Nurminen, M.; Mutanen, P.; Tola, S. Application of mathematical modelling for assessing the biological half-times of chromium and nickel in field studies. *Br. J. Ind. Med.* **1980**, *37*, 285–291. [CrossRef] [PubMed]

48. Albohn, H. Comparative epicutaneous tests with contact allergens and croton oil in 5 different body rions. *Z. Haut Gechlechtskr.* **1966**, *40*, 118–124. (In Gemany).

49. Fullerton, A.; Menne, T.; Hoelgaard, A. Patch testing with nickel chloride in a hydrogel. *Contact Dermat.* **1989**, *20*, 17–20. [CrossRef]

50. Fullerton, A.; Andersen, J.R.; Hoelgaard, A.; Menne, T. Permeation of nickel salts through human skin *in vitro. Contact Dermat.* **1986**, *15*, 173–177. [CrossRef]

51. Sunderman, F.W., Jr.; Hopfer, S.M.; Sweeney, K.R.; Marcus, A.H.; Most, B.M.; Creason, J. Nickel absorption and kinetics in human volunteers. *Proc. Soc. Exp. Biol. Med.* **1989**, *191*, 5–11. [CrossRef] [PubMed]

52. Leggett, R.W. The biokinetics of inorganic cobalt in the human body. *Sci. Total Environ.* **2008**, *389*, 259–269. [CrossRef] [PubMed]

53. Lauwerys, R.; Lison, D. Health risks associated with cobalt exposure—An overview. *Sci. Total Environ.* **1994**, *150*, 1–6. [CrossRef]

54. Mosconi, G.; Bacis, M.; Vitali, M.T.; Leghissa, P.; Sabbioni, E. Cobalt excretion in urine: Results of a study on workers producing diamond grinding tools and on a control group. *Sci. Total Environ.* **1994**, *150*, 133–139. [CrossRef]

55. Simonsen, L.O.; Harbak, H.; Bennekou, P. Cobalt metabolism and toxicology—A brief update. *Sci. Total Environ.* **2012**, *432*, 210–215. [CrossRef] [PubMed]

56. Ng, E.; Lind, P.M.; Lindgren, C.; Ingelsson, E.; Mahajan, A.; Morris, A.; Lind, L. Genome-wide association study of toxic metals and trace elements reveals novel associations. *Hum. Mol. Genet.* **2015**, *24*, 4739–4745. [CrossRef] [PubMed]

International Journal of
Molecular Sciences

MDPI

Article

Methylmercury Uptake into BeWo Cells Depends on LAT2-4F2hc, a System L Amino Acid Transporter

Christina Balthasar [1], Herbert Stangl [2], Raimund Widhalm [1] [ID], Sebastian Granitzer [1,3], Markus Hengstschläger [1] and Claudia Gundacker [1,*]

[1] Center for Pathobiochemistry and Genetics, Institute of Medical Genetics, Währinger Strasse 10, Medical University of Vienna, 1090 Wien, Vienna, Austria; christina.balthasar@gmail.com (C.B.); raimund.widhalm@meduniwien.ac.at (R.W.); Sebastian.Granitzer@kl.ac.at (S.G.); markus.hengstschlaeger@meduniwien.ac.at (M.H.)

[2] Center for Pathobiochemistry and Genetics, Institute of Medical Chemistry, Währinger Strasse 10, Medical University of Vienna, 1090 Wien, Vienna, Austria; herbert.stangl@meduniwien.ac.at

[3] Karl Landsteiner University of Health Sciences, Dr.-Karl-Dorrek-Straße 30, 3500 Krems an der Donau, Austria

* Correspondence: claudia.gundacker@meduniwien.ac.at; Tel.: +43-1-40160-56503

Received: 2 June 2017; Accepted: 1 August 2017; Published: 8 August 2017

Abstract: The organic mercury compound methylmercury (MeHg) is able to target the fetal brain. However, the uptake of the toxicant into placental cells is incompletely understood. MeHg strongly binds to thiol-S containing molecules such as cysteine. This MeHg-L-cysteine exhibits some structural similarity to methionine. System L plays a crucial role in placental transport of essential amino acids such as leucine and methionine and thus has been assumed to also transport MeHg-L-cysteine across the placenta. The uptake of methylmercury and tritiated leucine and methionine into the choriocarcinoma cell line BeWo was examined using transwell assay and small interfering (si)RNA mediated gene knockdown. Upon the downregulation of large neutral amino acids transporter (LAT)2 and 4F2 cell-surface antigen heavy chain (4F2hc), respectively, the levels of [^{3}H]leucine in BeWo cells are significantly reduced compared to controls treated with non-targeting siRNA ($p < 0.05$). The uptake of [^{3}H]methionine was reduced upon LAT2 down-regulation as well as methylmercury uptake after 4F2hc silencing ($p < 0.05$, respectively). These findings suggest an important role of system L in the placental uptake of the metal. Comparing the cellular accumulation of mercury, leucine, and methionine, it can be assumed that (1) MeHg is transported through system L amino acid transporters and (2) system L is responsible for the uptake of amino acids and MeHg primarily at the apical membrane of the trophoblast. The findings together can explain why mercury in contrast to other heavy metals such as lead or cadmium is efficiently transported to fetal blood.

Keywords: methylmercury; leucine; methionine; human placenta; *SLC7A5*; *SLC7A8*; *SLC3A2*; Forskolin

1. Introduction

One essential function of the human placenta is to accomplish the exchange of nutrients, gases, and metabolites between the mother and the fetus. The placental tissue is a barrier separating the maternal and fetal blood stream. Any substance that crosses the maternal-fetal interface from the maternal to the fetal side has to pass the outer syncytiotrophoblast (STB), the underlying cytotrophoblast (CTB), and the fetal endothelial cells (FECs). The initially complete cytotrophoblast layer becomes discontinuous as pregnancy progresses, resulting in just two continuous layers (STB and FECs) in the term placenta.

The human placenta is highly vulnerable to toxicants, including heavy metals [1–3]. It is evident that mercury crosses the placenta, accumulates in placental tissue, and passes onto the fetal blood and fetal organs [4–6]. An active transport across the human placenta has been assumed as cord blood mercury levels are, on average, almost twice the levels of maternal blood [7]. In a recent study, we showed system L transporters to be involved in methylmercury uptake into human placental cells, i.e., human primary trophoblasts and the human choriocarcinoma cell line BeWo [8].

The organic mercury compound methylmercury (MeHg) is a toxicant well known to target neurodevelopment. The toxicokinetics of mercury are determined by its high affinity to sulfhydryl groups. Methylmercury is present in the body as water-soluble complexes, mainly, if not exclusively, attached to the sulphur atom of thiol ligands such as L-cysteine, glutathione, or metallothioneins [9]. MeHg-L-cysteine, at least its amino acid component, is recognized by system L transporters [10]. These are heterodimers comprised of a light chain (large neutral amino acids transporter, LAT1 or LAT2) covalently bound by a disulfide bridge to a heavy subunit (4F2 cell-surface antigen heavy chain, 4F2hc) (syn. CD98). The solute carriers LAT1-4F2hc (*SLC7A5-SLC3A2*) and LAT2-4F2hc (*SLC7A8-SLC3A2*) are obligatory exchangers (1:1 stoichiometry), transporting large branched and aromatic neutral amino acids, including leucine and methionine. LAT2 also transports some smaller amino acids. The tissue distribution and cellular localization of LAT1 suggests that its main role is to transport amino acids into proliferating cells and across some barriers, including the placenta, while LAT2 is rather involved in the efflux step of transepithelial amino acid transport [11,12]. Previous studies indicate that these amino acid transporters also transport methylmercury across membranes of various cell types [13–16] and the rat placenta [17].

We aimed to analyse the uptake of methylmercury into placental cells in a setting as close as possible to the in vivo situation. Polarized cells take up and efflux molecules through both their basal and apical surfaces when grown on transwell supports. BeWo cells, notably the clone 24, build confluent, polarized monolayers, forming tight-junctions and microvilli on the apical side [18], enabling the study of the uptake, intracellular transport, and release of molecules on both their basal and apical membranes. In addition, the trophoblast-derived cells still display characteristics of human primary trophoblast cells, including the secretion of human chorionic gonadotropin (hCG). BeWo cells are therefore widely used to study that transport of nutrients, drugs, pathogens, and immunoglobulins across the STB [19].

The study goals were: (1) to establish the role of LAT1, LAT2 and 4F2hc in MeHg uptake into BeWo cells by comparing MeHg uptake with uptake of [^{3}H]methionine and [^{3}H]leucine into BeWo cells upon LAT1, LAT2 and 4F2hc silencing, respectively, and (2) to determine flux of the substrates in relation to transporter localization.

2. Results

2.1. Pre-Experiments

The BeWo cell culture protocol was optimized to study the uptake of MeHg, [^{3}H]methionine, and [^{3}H]leucine into placental cells in transwell plates (Figure 1A,B). Immunoblots (LAT1 and 4F2hc) and quantitative polymerase chain reaction (qPCR) (LAT2) confirmed efficient siRNA mediated knockdown under these conditions (Figure 1C). It has to be noted that, upon LAT1 downregulation, 4F2hc expression is also markedly reduced.

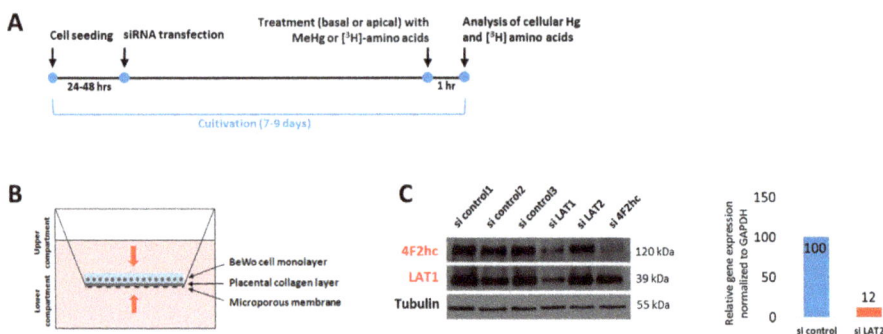

Figure 1. Experimental protocol and knockdown efficiency. (**A**) siRNA-mediated gene knockdown was conducted between 24 h and 48 h after seeding. BeWo cells were exposed to 2 μM MeHg or [^{3}H] amino acids around day 8 for one hour. To control for monolayer permeability, Lucifer Yellow (experiments on MeHg) and [^{14}C]mannitol (experiments on amino acids) were used (**B**) Transwell assay to study the uptake of MeHg and [^{3}H] amino acids into BeWo cells. (**C**) Confirmation of knockdown efficiency on protein level by immunoblotting (LAT1, 4F2hc) and on mRNA level by qPCR (LAT2) from one representative experiment. All three commercially available LAT2 antibodies (Table 1) were proven unable to detect the protein [20].

First we conducted a time course experiment to compare the uptake of MeHg and the amino acids; BeWo cells were exposed to two MeHg doses (0.9 μM, 2.0 μM) and to 1 μCi/mL tritium-labeled methionine and leucine, respectively, for 10 min, 30 min, and 60 min (Figure 2A,B). The 0.9 μM dose was selected as a multiple of 0.03 μM MeHg, which is equivalent to about 6 μg/L, representing physiological concentration [8]. As expected, mercury levels were twofold higher in cells exposed to 2.0 μM MeHg than in those exposed to 0.90 μM MeHg. Compared to the rather continuous uptake of amino acids, the uptake of methylmercury, particularly the 2 μM dosage, occurred in a non-linear manner. It was strongest during the first ten minutes, followed by a steady uptake for further 20 min, and then uptake increased significantly again. The incubation time for all following experiments was set at 60 min. At that time point, the cells had accumulated detectable concentrations of all substrates (Figure 2A,B). As shown in Figure 2C,D, BeWo cells exposed to 2 μM MeHg (i.e., 180 μg/L) for one hour accumulate mercury to non-cytotoxic levels as far as indicated by unchanged cell numbers (Figure 2E).

2.2. Forskolin-Induced BeWo Cell Fusion Increases LAT1 and LAT2 Expression

The Forskolin-induced fusion of BeWo cells was used to analyze changes of system L expression during differentiation (Figure 3A). Both LAT1 and LAT2 expression significantly increased upon 48 h of Forskolin treatment in relation to dimethyl sulfoxide (DMSO) treated controls, an effect confirmed by immunoblotting for LAT1 (Figure 3B).

2.3. LAT2 and 4F2hc Downregulation Reduces Mercury Uptake into BeWo Cells

Adding MeHg to apical compartments upon LAT2 and 4F2hc silencing resulted in significantly decreased mercury content of the BeWo cells (76% and 58%, respectively) in relation to the controls (Figure 4A). No such effect could be detected when methylmercury was added to the basal compartment (data not shown). The basal to apical permeability determined by Lucifer Yellow paracellular transport was $5.2 \pm 1.7\%$ ($n = 8$) and was approximately twice as high as that from apical to basal ($3.4 \pm 1.3\%$, $n = 8$). The ratio thus, as expected, correlated to the ratio of basal to apical volumes of media (2:1).

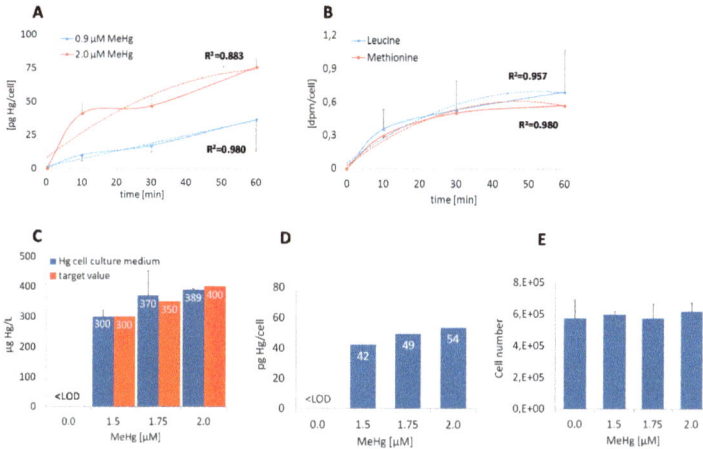

Figure 2. Time-dependent uptake of MeHg, [^{3}H]methionine, and [^{3}H]leucine. (**A**) BeWo cells were treated with 0.9 μM and 2 μM MeHg and (**B**) with 1 μCi/mL [^{3}H]leucine (100 Ci/ mmol) and 1 μCi/mL [^{3}H]methionine (84.5 mCi/mmol) for 10 min, 30 min, and 60 min in transwell plates. Data represent mean values ± SD from two independent experiments. (**C**) The apical addition of 0.0, 1.5 μM, 1.75 μM, and 2.0 μM MeHg (target values in red) to cell culture medium leads to the dose-dependent increase of total mercury concentrations in BeWo cells (**D**) after one hour at (**E**) unchanged cell numbers. (**C, E**) The data represent mean values ± SD from one experiment made in triplicate. (**D**) Prior to mercury analysis, BeWo cell aliquots were pooled.

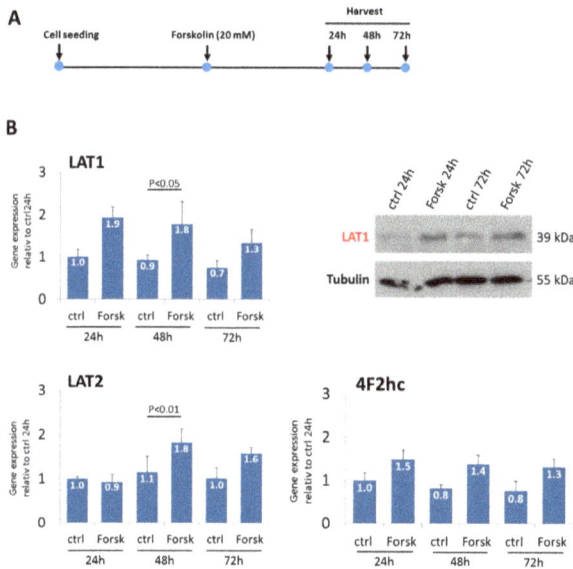

Figure 3. Expression of LAT1, LAT2, and 4F2hc under Forskolin treatment. (**A**) BeWo cells were treated with 20 mM Forskolin when about 50% confluent and harvested after 24, 48, and 72 h. (**B**) Gene expression of LAT1, LAT2, and 4F2hc over time. Increased LAT1 levels upon Forskolin treatment were confirmed via immunoblotting. One representative Western Blot is shown. Data are mean values ± SD from one experiment based on four replicates. Results from ANOVA are given when $p < 0.05$. ctrl: Control; Forsk: Forskolin.

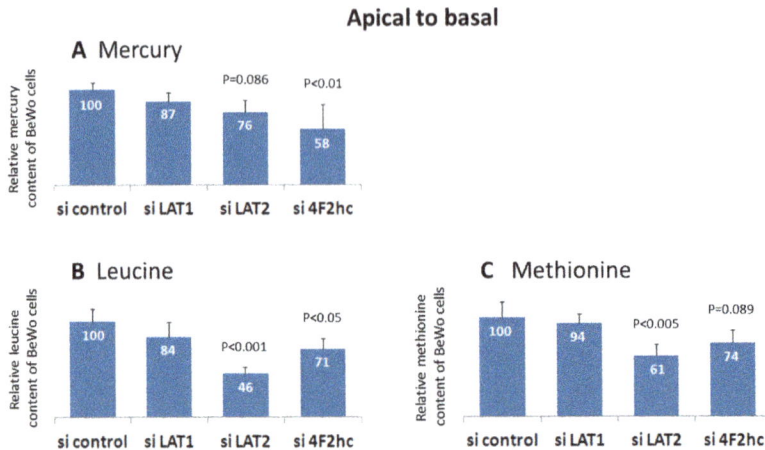

Figure 4. Uptake of MeHg, [^{3}H]methionine, and [^{3}H]leucine upon system L subunit silencing. Relative cellular contents of (**A**) total mercury, (**B**) [^{3}H]leucine, and (**C**) [^{3}H]methionine in BeWo cells after LAT1, LAT2, and 4F2hc silencing. MeHg, [^{3}H]leucine, and [^{3}H]methionine were added to the apical chamber of the transwell. Data are mean values ± SD from three independent experiments; results from ANOVA are given when $p < 0.1$.

2.4. LAT2 and 4F2hc Downregulation Reduces Methionine and Leucine Uptake into BeWo Cells

LAT2 and 4F2hc downregulation resulted in the significantly reduced uptake of leucine (46% and 71%, respectively) and methionine (61% and 74%, respectively) when amino acids were added to the apical chamber (Figure 4B,C). No such effect was seen when the amino acids were added to the basal compartment (data not shown). In LAT1 downregulated BeWo cells, a trend for lower leucine uptake was observed. The permeability determined by paracellular mannitol transport was 2.1 ± 0.5% ($n = 6$) in experiments examining apical to basal leucine transport. The basal to apical permeability was 4.5 ± 0.9%. With regard to methionine transport, apical to basal permeability was 2.2 ± 0.4% ($n = 6$), and basal to apical permeability was 5.0 ± 1.3% ($n = 6$). The ratio of permeability thus, as expected, correlated with the ratio from apical to basal volumes of media (1:2).

3. Discussion

The concept of a placenta barrier suggests that a placental cell, first and foremost the STB, is able to distinguish between essential nutrients that have to be transported to the fetal blood stream and unwanted substances that should not reach the fetal circulation. It is, however, evident that the toxicant mercury in the form of MeHg-L-cysteine is recognized by system L when expressed at the blood brain barrier or in *Xenopus laevis* eggs (e.g., [13,14]). The question arose whether the toxicant is transported in the same way as amino acids across the human placenta.

While placental amino acid transport is comparatively well understood [11,21], our knowledge on placental mercury transport still is incomplete. The aim of the present study was to address the role of placental system L amino acid transporters in MeHg uptake into BeWo cells, a trophoblast transport model endogenously expressing system L. It has to be noted that BeWo cells are mostly mononuclear (if not stimulated to fuse in vitro) and thereby model the undifferentiated trophoblast rather than the syncytiotrophoblast. As human primary trophoblast cells start to differentiate rapidly after plating and form syncytia in a discontinuous manner [22], they are rarely used in transwell studies. However, in a recent report, a validated model of a confluent human primary trophoblast monolayer has been proposed [23].

Previous findings [8,9,13,14,24–26] suggest that MeHg transport across barriers depends on cysteine, is stereo-selective (MeHg is transported in presence of L-cysteine but not in presence of D-cysteine), and is carrier-mediated by system L and system $b^{0,+}$. In vitro demethylation to mercuric mercury is implausible as, in humans, MeHg is slowly metabolized to inorganic mercury, predominantly by the intestinal microflora at a rate of about 1% of the body burden per day and to some extent also in phagocytic cells. It is therefore to be expected that most of the MeHg added to cell culture medium is present as monovalent cation (CH_3Hg^+) rapidly bound to ligands due to the high affinity of mercury ions to sulfhydryl group-containing molecules. Although it is likely that a substantial part of MeHg is bound to cysteine in the extracellular space as well as to glutathione in the cytosol, the respective amounts of MeHg compounds present in the extra- and intracellular compartments have not been quantified so far. It has to be noted that the amino acid composition of the serum (FCS) we added to cell culture medium is not provided by the manufacturer.

In order to demonstrate the inhibition of MeHg transport after the inactivation of system L subunits, we studied the transport of methionine, leucine, and methylmercury in parallel. Methionine was included because MeHg-L-cysteine structurally mimics the amino acid part of the molecule [10], while leucine specifically reflects system L activity as the amino acid is mainly transported by system L [11]. In this work, we provide evidence that MeHg is transported across BeWo cells through system L amino acid transporters.

3.1. Dose and Time Dependent Uptake into BeWo Cells

BeWo cells accumulate mercury to levels directly proportional to MeHg dosages. The mercury levels do not reach equilibrium during the first hour of exposure. Primary human trophoblast cells reach a steady state in mercury accumulation after about four hours [8]. Both methionine and leucine reach a steady state at around 30 to 60 min. The findings are in accordance with the time course of histidine uptake through LAT1 [27]. Simmons-Willis et al. [14] suggested MeHg-cysteine to be a better substrate for system L than endogenous amino acids, as they observed higher V_{max} values for MeHg-L-cysteine than for methionine transport through LAT1 and LAT2. Nonetheless, this finding cannot explain why BeWo cells still accumulate mercury while amino acid levels are already in a steady state. System L activity in combination with other amino acid uniporters/exchangers obviously tightly regulates intracellular methionine and leucine levels, while MeHg, once in the cell, dissociates from cysteine to bind to other intracellular ligands, e.g., glutathione or metallothionein [28], and thus no longer is under the control of amino acid transporters.

3.2. LAT1, LAT2, and 4F2hc Downregulation Does Not Affect BeWo Cell Number

The validation of siRNA-mediated gene knockdown by immunoblotting showed that the silencing of LAT1 results in the downregulation of 4F2hc. LAT2 knockdown could be confirmed on the mRNA level as commercial LAT2 antibodies (Table 1) were shown to be unable to detect the target protein [20].

We observed BeWo cell numbers to remain unaffected by the downregulation of any of the system L subunits (data not shown). In addition, no effects on cell morphology such as shrinking could be observed by visual inspection (inverted light microscope). Amino acid supply is crucial for cell growth and proliferation [29]. Gene targeting of *Slc3a2* (4F2hc) in conventional knockout mice is embryonically lethal as it is obligatory for murine embryogenesis [30]. 4F2hc was shown to play a role in tumorigenesis in renal cancer cell lines [31] and in the skin homeostasis of *Slc3a2* conditional knockout mice [32]. A global homozygous knockout of *Slc7a5* (LAT1) in mice was also embryonically lethal. The heterozygous *Slc7a5* knockout animals, however, had no overt phenotype, suggesting that its function in mTOR-S6K signaling is sufficiently compensated by, for instance, LAT2 [33]. A loss of LAT1 results in tumor growth inhibition [34]. In contrast to 4F2hc and LAT1, the *Slc7a8* (LAT2) knockout mouse did not apparently differ from the wild type mouse, apart from a mild aminoaciduria [35]. Overall, we conclude that our experimental model (i.e., a transfection period of five to seven days; Figure 1A) does not mimic the long-term effects of system L subunit inhibition.

3.3. Forskolin-Induced BeWo Cell Fusion Increases LAT1 and LAT2 Expression

Our observation that Forskolin induces the up-regulation of LAT1 is in accordance with previous reports in BeWo cells [36]. Elevated levels of LAT1 might be essential for the formation of the syncytiotrophoblast since a recent study has demonstrated impairments in trophoblastic fusion in LAT1 knockdown mice as well as in LAT1 deficient BeWo cells [37].

3.4. LAT2 and 4F2hc Downregulation Reduces Uptake of Methylmercury, Leucine and Methionine into BeWo Cells

BeWo cells accumulate significantly less mercury (76% and 58%) upon LAT2 and 4F2hc silencing relative to controls. This finding is, in principle, in accordance with other reports showing the transporter subunits to be involved in MeHg uptake into rat brain [13], C6 rat glioma cells [38], B35 rat neurons [16], rat placenta [17], *Xenopus laevis* oocytes [14], and Chinese hamster ovary cells [15]. In our previous study on BeWo cells cultivated in conventional dishes, we observed a similar reduction of mercury uptake upon LAT2 silencing (75%) but no such effect upon 4F2hc silencing. Moreover, we found the strongest effect on cellular mercury when LAT1 was down-regulated [8]. It remains unclear whether these discrepancies emerge from the different methods of cell culturing (conventional dishes versus transwell), leading to differences in cell morphology and polarity [39], or from the different MeHg treatments (0.90 µM in our previous work versus 2 µM in the present study).

BeWo cells responded to the down-regulation of LAT2 and 4F2hc with significantly reduced uptake of methionine (61% and 74%) and leucine (46% and 71%), whereas LAT1 silencing had no apparent effect (94% and 84%). The latter observation is in accordance with a previous report from Gaccioli et al. [40] in human primary trophoblast cells. The less pronounced effect of LAT2 knockdown on leucine uptake in human primary trophoblast cells (uptake reduced to 87% relative to controls) compared to BeWo cells (reduced to 46%; Figure 4B) might be explained by the circumstance that human primary trophoblasts are hard to transfect.

In transwell experiments with BeWo cells, LAT1 knockdown had no significant effect on that transport of methylmercury, leucine, and methionine, although we observed a trend for lowered uptake here as well (Figure 4). In *X. laevis* oocytes, LAT1 was shown to transport the substrate faster than LAT2 (V_{max} of 286 vs.75) but also to have a lower affinity to MeHg-L-cysteine than LAT2 (K_m of 98 µM vs. 64 µM) [14]. To our knowledge, no other studies exist in which mercury and amino acid uptake have been directly compared. In lung cancer cells, Dann et al. [34] observed a significant reduction of methionine levels (to about a third relative to controls) upon the downregulation of LAT1. Nicklin et al. [29] found both LAT1 and 4F2hc silencing to exert the same effect on leucine transport (reduction to a third) into HeLa cells.

3.5. Uptake of MeHg, Leucine and Methionine in Relation to Transporter Localization

Most of the so far available data suggest that system L transporters, LAT1, LAT2, and 4F2hc, respectively, localize primarily to the apical side of the STB [41–43]. Our data based on RNAi validated antibodies [20] showed the heavy chain 4F2hc to be localized at both STB plasma membranes (apical and basolateral). Moreover, we found the light chains, LAT1 and particularly LAT2, localized in intracellular vesicular structures of the STB [8]. Our data indicate that LAT2-4F2hc exerts its function predominantly at the apical side of trophoblast cells, as the uptake of mercury, leucine, and methionine remained unaffected by system L inactivation when substrates were added to the basal chamber.

4. Materials and Methods

4.1. Cell Culture

BeWo cells (clone 24), a kind gift from Dr. Isabella Ellinger (Medical University of Vienna), were cultured at 37 °C and 95% air/5% CO_2 in DMEM high glucose (Life Technologies, Carlsbad, CA, USA), supplemented with 10% FBS Good (Pan Biotech, Aidenbach, Germany) and 1% GlutaMAX

(Life Technologies). The cells were detached from culture dishes with trypsin/ethylenediaminetetraacetic acid (EDTA). The cell number was determined with a CASY cell counter and analyzer (CASY® Model TTC 45/60/150, Innovatis Technologies Inc., Woodbridge, VA, USA).

4.2. Transwell Studies

BeWo cells were seeded onto permeable Transwell inserts (12-well polycarbonate membrane with 0.4 μm pore size, Corning Inc., Corning, NY, USA) that had been coated with human placental collagen (50 μg/cm^2; Bornstein and Traub Type IV, Sigma-Aldrich Corporation, St. Louis, MO, USA) according to the protocol of Bode et al. [22] at a density of 2.5×10^4 cells/well. The cells were transiently transfected 24 h to 48 h post seeding with non-targeting and specific siRNA targeting *SLC7A5*, *SLC7A8*, and *SLC3A2* (encoding LAT1, LAT2, and 4F2hc) (GE Dharmacon, Lafayette, CO, USA) using Lipofectamine RNAiMax (Life Technologies) as described by Rosner et al. [44], with the minor modification of employing only $\frac{1}{4}$ of the original transfection reagent amount. Thereafter the cells were cultivated until a confluent monolayer was formed (around eight days after seeding).

BeWo cells were treated with 2 μM MeHg (aqueous CH$_3$HgCl) (Alfa Aesar, Haverhill, MA, USA) for one hour, added to medium at the apical (0.5 mL) or basal (1 mL) side of the transwell. The paracellular permeability of each well was determined concomitantly to MeHg transport by adding 100 μM Lucifer Yellow (CH Dilithium Salt, Sigma-Aldrich) to the target compartment, while the opposing compartment was filled with cell culture medium only. Lucifer Yellow's fluorescence was measured in black 96-well plates (Corning) in a microplate reader (BioTek Instruments, Winooski, VT, USA) using a 485 ± 20 nm excitation and 528 ± 20 nm emission filter.

In the same way as for MeHg, BeWo cells were treated with 1 μCi/mL [^{3}H]Leucine (100 Ci/mmol, Hartmann Analytik, Braunschweig, Germany) or 1 μCi/mL [^{3}H]Methionine (84.5 mCi/mmol, Perkin Elmer Inc., Waltham, MA, USA) in Hank's Balanced Salt Solution (HBSS) (Sigma-Aldrich), supplemented with 0.2 μCi/mL [^{14}C]mannitol (56.8 mCi/mmol, Perkin Elmer) to test for cell permeability. After incubation, the cells were washed three times with ice cold Hank's balanced salt solution (HBSS), followed by lysis in 250 μL NaOH (1 mol/L, Merck & Co., Kenilworth, NJ, USA). 200 μL of solubilized cells were added to 5 mL liquid scintillation fluid (Ultima Gold, Perkin Elmer), and the radioactivity of each cell lysate sample was determined by scintillation counting (TRI-Carb 2800 TR, Perkin Elmer).

4.3. Forskolin Treatment

BeWo cells were cultured until 50% confluency on 60 mm dishes (Corning). At this point, the cells were either incubated with Forskolin (20 mM) or DMSO as a control. Cells were harvested after 24 h, 48 h, and 72 h. Changes in gene expression were analysed by quantitative PCR (qPCR) and Immunoblotting.

4.4. RNA Isolation, cDNA Synthesis and Quantitative PCR

Total RNA was isolated using TRI Reagent®(Sigma), according to the manufacturer's instructions. RNA was reverse transcribed with a Go-Script Reverse Transcription System (Promega, Madison, WI, USA) using random hexamer primers. Gene expression was analyzed using a Taq Man Expression System (Applied Biosystems, Foster City, CA, USA) in an Applied Biosystems StepOnePlus™ Real-Time PCR System, according to the manufacturer's protocol. The cDNA was diluted 1:11, and 2 μL was used as a template in a 15 μL reaction. Glyceraldehyde 3-phosphate dehydrogenase (GAPDH) and TATA-box binding protein (TBP) were used as reference genes. The employed primers were Hs00794796 m1 (SLC7A8), Hs99999905_m1 (GAPDH), and Hs00427620_m1 (TBP). In Forskolin experiments, we used the primers Hs00185826_1 (SLC7A5), Hs00247916_m1 (LAT2), and Hs00374243_m1 (SLC3A2), and as reference gene we used Hs0082473_m1 (Ubiquitin C).

4.5. Protein Extraction and Immunoblotting

The cells were lysed in RIPA (Radioimmunoprecipitation assay) buffer (50 mM Tris, pH 7.6, 150 mM NaCl, 1% Triton, 0.1% SDS, 0.5% sodium deoxycolate), supplemented with 2 mg/mL aprotinin, 0.3 mg/mL benzamidin chloride, 2 mg/mL leupeptin, and 10 mg/mL trypsin inhibitor (Sigma). The protein samples were separated using SDS-PAGE and transferred to nitrocellulose membranes. Blots were blocked for 1 h in 5% nonfat dry milk in tris-buffered saline containing 0.1% Tween 20 (TBST), followed by incubation in 5% bovine serum albumin (BSA)/TBST containing the primary antibody overnight at 4 °C. Thereafter, blots were washed and incubated with corresponding secondary horseradish peroxidase (HRP)-conjugated antibodies. The enhanced chemiluminescence method (Pierce™ ECL western blotting substrate, Thermo Fisher Scientific, Waltham, MA, USA) was used to visualize the signals. For a list of the employed primary and secondary antibodies, see Table 1.

Table 1. List of the primary and secondary antibodies used in immunoblotting.

Product Name	Company, Article No. (Dilution)
anti-4F2hc rabbit polyclonal antibody	Cell Signaling, #13180 (1:1000)
anti-LAT1 rabbit polyclonal antibody	Cell Signaling, #5347 (1:1000)
anti-LAT2 mouse monoclonal antibody	OriGene, TA500513S (1:500)
anti-LAT2 rabbit polyclonal antibody	Santa Cruz, sc-133726 (1:100)
anti-LAT2 rabbit polyclonal antibody	ImmunoGlobe, 0142-10 (1:1000)
anti-α-Tubulin mouse monoclonal antibody	Merck, 05-829 (1:5000)
mouse IgG-heavy and light chain antibody	Bethyl, A90-116P (1:10,000)
rabbit IgG-heavy and light chain antibody	Bethyl, A120-101P (1:10,000)

Cell Signaling Technology: Danvers, MA, USA; OriGene: Rockville, ML, USA; Santa Cruz Biotechnology: Dallas, TX, USA; ImmunoGlobe: Himmelstadt, Germany; Bethyl: Montgomery, TX, USA.

4.6. Analysis of Mercury

The samples and reference material were acid-digested with nitric acid (69%; Suprapur®; Carl Roth, Karlsruhe, Germany) in a microwave oven (MARS6, CEM Corporation, Matthews, NC, USA). The samples, stabilized with HCl, were stored at 4 °C for up to three days and diluted in a ratio of 1:2.5 before they were analyzed for total mercury content by cold vapour atomic fluorescence spectroscopy (CV-AFS) (Mercur Plus, Analytik Jena AG, Jena, Germany). Quality control was achieved by measuring blank test solutions (limit of detection was 0.024 µg/L) and reference materials (Seronorm Trace Elements Urine L-2, 210705, LOT 1011645). The mercury levels of the reference material (30.6 ± 7.2 µg/L; $n = 19$) lay well within the certified range (23.8–55.8 µg/L). All samples were measured in duplicate by the working curve method (RSD < 15%).

4.7. Statistics and Software

Data represent mean values $\pm$ SD (standard deviation). Regression lines and coefficients of determination were made with MS Excel. ANOVA was applied for the comparison of group differences, followed by a Bonferroni test to correct for multiple testing. We used IBM SPSS Statistics 24 (IBM, Armonk, NY, USA) and set the critical significance level at $\alpha = 0.05$.

5. Conclusions

The present study is the first one in which uptake of methylmercury, leucine, and methionine were examined upon the knockdown of system L subunits LAT1, LAT2, and 4F2hc in parallel and in a setting as close as possible to the in vivo situation. The direction and magnitude of the effects are comparable. Altogether, the findings clearly indicate that LAT2-4F2hc is a significant contributor to methylmercury uptake into placental cells. The findings support the assumption that methylmercury (in the extracellular environment most likely present as MeHg-L-cysteine) is accidentally taken up into the human cytotrophoblast because the compound resembles essential amino acids. The 'mimicry' can explain why mercury in contrast to other heavy metals such as lead or cadmium is efficiently transported to the fetal blood.

Acknowledgments: The study was supported by NFB (Niederösterreichische Forschungs und Bildungsgesellschaft) (Project LS10-26 and LSC15-014). We thank Marie-Christine Giuffrida for technical support.

Author Contributions: Christina Balthasar, Herbert Stangl, and Claudia Gundacker conceived and designed the experiments; Christina Balthasar, Herbert Stangl, Sebastian Granitzer, and Raimund Widhalm performed the experiments; Christina Balthasar, Herbert Stangl, Sebastian Granitzer, and Claudia Gundacker analysed the data; Markus Hengstschläger and Herbert Stangl contributed reagents/materials/analysis tools; and Claudia Gundacker and Raimund Widhalm wrote the paper.

Conflicts of Interest: The authors declare no conflict of interest.

References

1. Gundacker, C.; Hengstschläger, M. The role of the placenta in fetal exposure to heavy metals. *Wien. Med. Wochenschr.* **2012**, *162*, 201–206. [CrossRef] [PubMed]
2. Laine, J.E.; Ray, P.; Bodnar, W.; Cable, P.H.; Boggess, K.; Offenbacher, S.; Fry, R.C. Placental cadmium levels are associated with increased preeclampsia risk. *PLoS ONE* **2015**, *10*, e0139341. [CrossRef] [PubMed]
3. St-Pierre, J.; Fraser, M.; Vaillancourt, C. Inhibition of placental 11beta-hydroxysteroid dehydrogenase type 2 by lead. *Reprod. Toxicol.* **2016**, *65*, 133–138. [CrossRef] [PubMed]
4. Murcia, M.; Ballester, F.; Enning, A.M.; Iñiguez, C.; Valvi, D.; Basterrechea, M.; Rebagliato, M.; Vioque, J.; Maruri, M.; Tardon, A.; et al. Prenatal mercury exposure and birth outcomes. *Environ. Res.* **2016**, *151*, 11–20. [CrossRef] [PubMed]
5. Gundacker, C.; Fröhlich, S.; Graf-Rohrmeister, K.; Eibenberger, B.; Jessenig, V.; Gicic, D.; Prinz, S.; Wittmann, K.J.; Zeisler, H.; Vallant, B.; et al. Perinatal lead and mercury exposure in Austria. *Sci. Total Environ.* **2010**, *408*, 5744–5749. [CrossRef] [PubMed]
6. Ask, K.; Akesson, A.; Berglund, M.; Vahter, M. Inorganic mercury and methylmercury in placentas of swedish women. *Environ. Health Perspect.* **2002**, *110*, 523–526. [CrossRef] [PubMed]
7. Stern, A.H.; Smith, A.E. An assessment of the cord blood: Maternal blood methylmercury ratio: Implications for risk assessment. *Environ. Health Perspect.* **2003**, *111*, 1465–1470. [CrossRef] [PubMed]
8. Straka, E.; Ellinger, I.; Balthasar, C.; Scheinast, M.; Schatz, J.; Szattler, T.; Bleichert, S.; Saleh, L.; Knöfler, M.; Zeisler, H.; et al. Mercury toxicokinetics of the healthy human term placenta involve amino acid transporters and ABC transporters. *Toxicology* **2016**, *340*, 34–42. [CrossRef] [PubMed]
9. Clarkson, T.W. The three modern faces of mercury. *Environ. Health Perspect.* **2002**, *110*, 11–23. [CrossRef] [PubMed]
10. Hoffmeyer, R.E.; Singh, S.P.; Doonan, C.J.; Ross, A.R.S.; Hughes, R.J.; Pickering, I.J.; George, G.N. Molecular mimicry in mercury toxicology. *Chem. Res. Toxicol.* **2006**, *19*, 753–759. [CrossRef] [PubMed]
11. Jansson, T. Amino acid transporters in the human placenta. *Pediatr. Res.* **2001**, *49*, 141–147. [CrossRef] [PubMed]
12. Verrey, F. System L: Heteromeric exchangers of large, neutral amino acids involved in directional transport. *Pflugers Arch.* **2003**, *445*, 529–533. [CrossRef] [PubMed]
13. Kerper, L.E.; Ballatori, N.; Clarkson, T.W. Methylmercury transport across the blood-brain barrier by an amino acid carrier. *Am. J. Physiol. Regul. Integr. Comp. Physiol.* **1992**, *262*, R761–R765.
14. Simmons-Willis, T.A.; Koh, A.S.; Clarkson, T.W.; Ballatori, N. Transport of a neurotoxicant by molecular mimicry: The methylmercury-L-cysteine complex is a substrate for human l-type large neutral amino acid transporter (lat) 1 and lat2. *Biochem. J.* **2002**, *367*, 239–246. [CrossRef] [PubMed]
15. Yin, Z.; Jiang, H.; Syversen, T.; Rocha, J.B.; Farina, M.; Aschner, M. The methylmercury-L-cysteine conjugate is a substrate for the l-type large neutral amino acid transporter. *J. Neurochem.* **2008**, *107*, 1083–1090. [CrossRef] [PubMed]
16. Heggland, I.; Kaur, P.; Syversen, T. Uptake and efflux of methylmercury in vitro: Comparison of transport mechanisms in C6, B35 and RBE4 cells. *Toxicol. In Vitro* **2009**, *23*, 1020–1027. [CrossRef] [PubMed]
17. Kajiwara, Y.; Yasutake, A.; Adachi, T.; Hirayama, K. Methylmercury transport across the placenta via neutral amino acid carrier. *Arch Toxicol.* **1996**, *70*, 310–314. [CrossRef] [PubMed]

18. Prouillac, C.; Lecoeur, S. The role of the placenta in fetal exposure to xenobiotics: Importance of membrane transporters and human models for transfer studies. *Drug Metab. Dispos.* **2010**, *38*, 1623–1635. [CrossRef] [PubMed]

19. Ellinger, I.; Schwab, M.; Stefanescu, A.; Hunziker, W.; Fuchs, R. IgG transport across trophoblast-derived BeWo cells: A model system to study IgG transport in the placenta. *Eur. J. Immunol.* **1999**, *29*, 733–744. [CrossRef]

20. Ellinger, I.; Chatuphonprasert, W.; Reiter, M.; Voss, A.; Kemper, J.; Straka, E.; Scheinast, M.; Zeisler, H.; Salzer, H.; Gundacker, C. Don't trust an(t)ybody—Pitfalls during investigation of candidate proteins for methylmercury transport at the placental interface. *Placenta* **2016**, *43*, 13–16. [CrossRef] [PubMed]

21. Lewis, R.M.; Brooks, S.; Crocker, I.P.; Glazier, J.; Hanson, M.A.; Johnstone, E.D.; Panitchob, N.; Please, C.P.; Sibley, C.P.; Widdows, K.L.; et al. Review: Modelling placental amino acid transfer—From transporters to placental function. *Placenta* **2013**, *34*, S46–S51. [CrossRef] [PubMed]

22. Bode, C.; Jin, H.; Rytting, E.; Silverstein, P.; Young, A.; Audus, K. In vitro models for studying trophoblast transcellular transport. *Methods Mol. Med.* **2006**, *122*, 225–239. [PubMed]

23. Huang, X.; Lüthi, M.; Ontsouka, E.C.; Kallol, S.; Baumann, M.U.; Surbek, D.V.; Albrecht, C. Establishment of a confluent monolayer model with human primary trophoblast cells: Novel insights into placental glucose transport. *Mol. Hum. Reprod.* **2016**, *22*, 442–456. [CrossRef] [PubMed]

24. Clarkson, T.; Vyas, J.; Ballatori, N. Mechanisms of mercury disposition in the body. *Am. J. Ind. Med.* **2007**, *50*, 757–764. [CrossRef] [PubMed]

25. Clarkson, T.W.; Magos, L. The toxicology of mercury and its chemical compounds. *Crit. Rev. Toxicol.* **2006**, *36*, 609–662. [CrossRef] [PubMed]

26. Ballatori, N. Transport of toxic metals by molecular mimicry. *Environ. Health Perspect.* **2002**, *110*, 689–694. [CrossRef] [PubMed]

27. Napolitano, L.; Scalise, M.; Galluccio, M.; Pochini, L.; Albanese, L.M.; Indiveri, C. LAT1 is the transport competent unit of the LAT1/CD98 heterodimeric amino acid transporter. *Int. J. Biochem. Cell Biol.* **2015**, *67*, 25–33. [CrossRef] [PubMed]

28. Rooney, J.P.K. The role of thiols, dithiols, nutritional factors and interacting ligands in the toxicology of mercury. *Toxicology* **2007**, *234*, 145–156. [CrossRef] [PubMed]

29. Nicklin, P.; Bergman, P.; Zhang, B.; Triantafellow, E.; Wang, H.; Nyfeler, B.; Yang, H.; Hild, M.; Kung, C.; Wilson, C.; et al. Bidirectional transport of amino acids regulates mtor and autophagy. *Cell* **2009**, *136*, 521–534. [CrossRef] [PubMed]

30. Tsumura, H.; Suzuki, N.; Saito, H.; Kawano, M.; Otake, S.; Kozuka, Y.; Komada, H.; Tsurudome, M.; Ito, Y. The targeted disruption of the CD98 gene results in embryonic lethality. *Biochem. Biophys. Res. Commun.* **2003**, *308*, 847–851. [CrossRef]

31. Poettler, M.; Unseld, M.; Braemswig, K.; Haitel, A.; Zielinski, C.C.; Prager, G.W. CD98hc (SLC3A2) drives integrin-dependent renal cancer cell behavior. *Mol. Cancer Res.* **2013**, *12*, 169. [CrossRef] [PubMed]

32. Boulter, E.; Estrach, S.; Errante, A.; Pons, C.; Cailleteau, L.; Tissot, F.; Meneguzzi, G.; Féral, C.C. CD98hc (SLC3A2) regulation of skin homeostasis wanes with age. *J. Exp. Med.* **2013**, *210*, 173–190. [CrossRef] [PubMed]

33. Poncet, N.; Mitchell, F.E.; Ibrahim, A.F.M.; McGuire, V.A.; English, G.; Arthur, J.S.C.; Shi, Y.-B.; Taylor, P.M. The catalytic subunit of the system L1 amino acid transporter (SLC7A5) facilitates nutrient signalling in mouse skeletal muscle. *PLoS ONE* **2014**, *9*, e89547. [CrossRef] [PubMed]

34. Dann, S.G.; Ryskin, M.; Barsotti, A.M.; Golas, J.; Shi, C.; Miranda, M.; Hosselet, C.; Lemon, L.; Lucas, J.; Karnoub, M.; et al. Reciprocal regulation of amino acid import and epigenetic state through Lat1 and EZH2. *EMBO J.* **2015**, *34*, 1773–1785. [CrossRef] [PubMed]

35. Braun, D.; Wirth, E.K.; Wohlgemuth, F.; Reix, N.; Klein, M.O.; Grüters, A.; Köhrle, J.; Schweizer, U. Aminoaciduria, but normal thyroid hormone levels and signalling, in mice lacking the amino acid and thyroid hormone transporter SLC7A8. *Biochem. J.* **2011**, *439*, 249–255. [CrossRef] [PubMed]

36. Dalton, P.; Christian, H.C.; Redman, C.W.G.; Sargent, I.L.; Boyd, C.A.R. Differential effect of cross-linking the CD98 heavy chain on fusion and amino acid transport in the human placental trophoblast (Bewo) cell line. *Biochim. Biophys. Acta* **2007**, *1768*, 401–410. [CrossRef] [PubMed]

37. Ohgaki, R.; Ohmori, T.; Hara, S.; Nakagomi, S.; Kanai-Azuma, M.; Kaneda-Nakashima, K.; Okuda, S.; Nagamori, S.; Kanai, Y. Essential roles of L-type amino acid transporter 1 in syncytiotrophoblast development by presenting fusogenic 4f2hc. *Mol. Cell. Biol.* **2017**, *37*, e00427-16. [CrossRef] [PubMed]
38. Zimmermann, L.T.; Santos, D.B.; Naime, A.A.; Leal, R.B.; Dórea, J.G.; Barbosa, F. Jr.; Aschner, M.; Rocha, J.B.T.; Farina, M. Comparative study on methyl- and ethylmercury-induced toxicity in C6 glioma cells and the potential role of LAT-1 in mediating mercurial-thiol complexes uptake. *Neurotoxicology* **2013**, *38*, 1–8. [CrossRef] [PubMed]
39. Baker, B.M.; Chen, C.S. Deconstructing the third dimension—How 3D culture microenvironments alter cellular cues. *J. Cell Sci.* **2012**, *125*, 3015–3024. [CrossRef] [PubMed]
40. Gaccioli, F.; Aye, I.L.M.H.; Roos, S.; Lager, S.; Ramirez, V.I.; Kanai, Y.; Powell, T.L.; Jansson, T. Expression and functional characterisation of system l amino acid transporters in the human term placenta. *Reprod. Biol. Endocrinol.* **2015**, *13*, 57. [CrossRef] [PubMed]
41. Ayuk, P.T.Y.; Sibley, C.P.; Donnai, P.; D'Souza, S.; Glazier, J.D. Development and polarization of cationic amino acid transporters and regulators in the human placenta. *Am. J. Physiol. Cell Physiol.* **2000**, *278*, C1162–C1171. [PubMed]
42. Kudo, Y.; Boyd, C.A.R. Characterisation of L-tryptophan transporters in human placenta: A comparison of brush border and basal membrane vesicles. *J. Physiol.* **2001**, *531*, 405–416. [CrossRef] [PubMed]
43. Okamoto, Y.; Sakata, M.; Ogura, K.; Yamamoto, T.; Yamaguchi, M.; Tasaka, K.; Kurachi, H.; Tsurudome, M.; Murata, Y. Expression and regulation of 4f2hc and hlat1 in human trophoblasts. *Am. J. Physiol. Cell Physiol.* **2002**, *282*, C196–C204. [PubMed]
44. Rosner, M.; Siegel, N.; Fuchs, C.; Slabina, N.; Dolznig, H.; Hengstschläger, M. Efficient siRNA-mediated prolonged gene silencing in human amniotic fluid stem cells. *Nat. Protocols* **2010**, *5*, 1081–1095. [CrossRef] [PubMed]

International Journal of
Molecular Sciences

Review

Cadmium Handling, Toxicity and Molecular Targets Involved during Pregnancy: Lessons from Experimental Models

Tania Jacobo-Estrada [1] , Mitzi Santoyo-Sánchez [2], Frank Thévenod [3] and Olivier Barbier [2,*]

1   Departamento de Sociedad y Política Ambiental, CIIEMAD, Instituto Politécnico Nacional,
    30 de Junio de 1520 s/n, La Laguna Ticomán, Ciudad de México 07340, Mexico; tjacoboe@ipn.mx
2   Departamento de Toxicología, Centro de Investigación y de Estudios Avanzados del Instituto Politécnico
    Nacional, Av. Instituto Politécnico Nacional 2508, Gustavo A. Madero, San Pedro Zacatenco,
    Ciudad de México 07360, Mexico; santoyomitzi@gmail.com
3   Department of Physiology, Pathophysiology & Toxicology and ZBAF (Centre for Biomedical Education and
    Research), Faculty of Health-School of Medicine, Witten/Herdecke University, Stockumer Str 12 (Thyssenhaus),
    D 58453 Witten, Germany; frank.thevenod@uni-wh.de
*   Correspondence: obarbier@cinvestav.mx; Tel.: +52-1-(55)-5747-3800

Received: 29 June 2017; Accepted: 18 July 2017; Published: 22 July 2017

Abstract: Even decades after the discovery of Cadmium (Cd) toxicity, research on this heavy metal is still a hot topic in scientific literature: as we wrote this review, more than 1440 scientific articles had been published and listed by the PubMed.gov website during 2017. Cadmium is one of the most common and harmful heavy metals present in our environment. Since pregnancy is a very particular physiological condition that could impact and modify essential pathways involved in the handling of Cd, the prenatal life is a critical stage for exposure to this non-essential element. To give the reader an overview of the possible mechanisms involved in the multiple organ toxic effects in fetuses after the exposure to Cd during pregnancy, we decided to compile some of the most relevant experimental studies performed in experimental models and to summarize the advances in this field such as the Cd distribution and the factors that could alter it (diet, binding-proteins and membrane transporters), the Cd-induced toxicity in dams (preeclampsia, fertility, kidney injury, alteration in essential element homeostasis and bone mineralization), in placenta and in fetus (teratogenicity, central nervous system, liver and kidney).

Keywords: cadmium; pregnancy; multiple organs toxicity; fetus; placenta

1. Introduction

Cadmium (Cd) is one of the most common and harmful transition metals present in our environment. Unfortunately, this non-essential element is toxic at very low doses and non-biodegradable with a very long biological half-life. Kjellström and Nordberg were the first researchers to work on long term persistence of Cd body loads and to evaluate the half-life of Cd, establishing a range of half-times from 6 to 38 years for the human kidney and 4 to 19 years for the human liver [1].

Whereas the molecular and cellular mechanisms of Cd toxicity are studied in great detail, as well as the toxicokinetics and toxicodynamics of Cd [2], many other aspects need to be elucidated considering the impact of Cd exposure. Numerous parameters are crucial to consider during the risk assessment of toxicants and environmental pollutants since they could intrinsically modify the absorption, distribution, metabolism, and excretion of these xenobiotics by the host organism. Hence, the genetic background, the epigenetic modifications, the diet, a physio-pathological condition, the individual behavior, and the co-exposure to other xenobiotics must be considered [3,4].

A unique physiological condition that could impact and modify essential pathways involved in the handling of Cd (and other environmental pollutants) is pregnancy [5,6]. With regard to the mother, during gestation, most organs and systems (renal, cardiovascular, blood, gastrointestinal, respiratory and endocrine) show multi-faceted and progressive changes in their anatomy and functions; e.g., ventilation is significantly increased (amplifying the contact with airborne molecules and particulate matter), total body water and plasma volume increase (modifying the volume of distribution and lowering the concentrations of proteins and elements in the body fluids), glomerular filtration rate increases (accelerating the proximal uptake of small filtered binding-proteins and/or their excretion through urine), gastrointestinal motility decreases whereas gastric pH increases (modifying the absorption), etc. [6]. Added to this situation, it is also necessary to consider the appearance of a new and temporary organ, the placenta, whose function participates importantly in the specific toxicity towards the fetus. If the placenta regulates blood flow, plays the role of a transport barrier, and metabolizes chemicals, it can also be a target for xenobiotics-induced toxicity [5,7,8]. Regarding the fetus, in utero/prenatal life corresponds to a challenging period due the rapid changes that occur during development. From preimplantation to organogenesis and finally birth, molecular expression, cell differentiation, structures and functions of the new organs undergo profound and dynamic modifications [9]. Thus, the nature of the fetus as a target of toxicity and the molecular pathways involved are constantly changing too [5,10].

Over the years, epidemiological studies have demonstrated the critical role of prenatal exposure to cadmium in the development of human individuals and its impact on public health. Thus, birth outcomes related to Cd exposure, such as alteration of Apgar 5-min score, birth weight, risk of being small-for-gestational age [11–13], association with poorer cognition [14] or epigenetic modifications [15,16], among others, have been evidenced in populations from all over the world.

To give the reader an overview of the possible mechanisms involved in the multiple organ toxic effects in fetuses and dams after exposure to Cd during pregnancy, we decided to compile some of the most relevant experimental studies and summarize the advances in this field during the last decades.

2. General Information about Cd

2.1. Mode of Action (MOA) of Cd Toxicity

As reviewed before [17–21], the toxic form of Cd is the ionized form Cd^{2+}. Conjugates and bound forms of filtered Cd are not toxic by themselves but the divalent form released from the complexes is responsible for the cellular toxicity altering the mitochondrial activity (through outer membrane rupture and uncoupling of respiration), provoking oxidative stress and apoptosis, interacting with transporter and ion channels, among others. Cadmium and divalent metals are well-known inducers of Ca^{2+} mobilization with a potency order of $Cd^{2+} > Co^{2+}\ Ni^{2+} > Fe^{2+} > Mn^{2+}$ [22]. Moreover, Cd^{2+} can compete with Zn^{2+} and Ca^{2+} transport and alters the uptake and cellular homeostasis of these important ligands, enzyme-cofactor and second messenger [23]. Thus, Cd^{2+} can be responsible for interfering with Ca^{2+} and Zn^{2+} pathways and, consequently, disrupting signaling, transport, metabolism and cell fate [24].

The main MOA of Cd toxicity is the induction of cell death through oxidative stress [25]. Ca^{2+} and reactive oxygen species (ROS), which are molecular triggers controlling cell function and activate effectors (Apoptosis signal-regulating kinase 1-c-Jun N-terminal kinase/p38, calpains, caspases and ceramides), are also capable to cause irreversible damages to organelles, such as mitochondria and endoplasmic reticulum (ER). Interestingly, localized ROS/Ca^{2+} levels play a role as second messengers triggering mechanisms such as cellular adaptation and survival via the signal transduction (extracellular signal-regulated kinases-1/2, phosphoinositide-3-kinase–protein kinase AKt) and the transcriptional regulation (redox effector factor-1 Ref1- nuclear factor erythroid 2-related factor 2 Nrf2, nuclear factor kappa-light-chain-enhancer of activated B cells NF-κB, wnt, activator protein 1 AP-1, bestrophin-3). Of course, many different proteins and processes mediated by ROS/Ca^{2+}

signaling (metallothionein, B-cell lymphoma 2 proteins, ubiquitin-proteasome, ER stress associated with unfolded proteins, autophagy, cell cycle regulation) can induce dual responses including death or survival, in accordance to the cell type and the conditions of exposure [20,24].

2.2. Sources of Exposure

Cadmium is a transition metal (group II b) with eight stable isotopes that was discovered by the German chemist Strohmeier in 1817, because of the study of some of the impurities of zinc carbonate [26].

This metal is found in nature in many forms. The most common are: elemental cadmium (Cd^0), and as cadmium carbonate, chloride, oxide, sulfate and sulfide salts. Considering natural sources, Cd is widely distributed in the earth's crust. Rock wear and erosion result in the release of Cd. Volcanic activity on marine and terrestrial surfaces also contributes to its release [27]. Considering anthropogenic sources, Cd is released from human-related activities. Cadmium is generally obtained as a by-product of the zinc concentrates. The zinc: cadmium ratios in typical minerals range from 200:1 to 400:1. This metal may be produced in a secondary manner by the recycling of batteries, copper-cadmium alloys and powders from electric arc furnaces. It is estimated that the production of secondary Cd accounts for approximately 20% of the total production of metallic Cd. It is used in the manufacture of various products in common use, being the nickel-cadmium batteries the article that consumes most of the world's production of this element (90%). A percentage of Cd is also used in the manufacture of pigments, coatings, stabilizers for plastics, non-ferrous alloys, electroplating, photovoltaic devices, etc. These activities related to the extraction of metal and the manufacture, use and disposal of Cd products, as well as agriculture, have the potential to release Cd into the environment and to be sources of exposure for humans and other animals [3,28].

The route of exposure impacts importantly Cd absorption: inhalation, that occurs mainly through tobacco smoke, approximates 25% (range 5–50%), whereas the oral route, through contaminated water and food (offal and seafood), was estimated at 5% (range 1–10%) [3].

Smoking one cigarette increases blood concentration by approximately 0.1–0.2 µg Cd per liter because each cigarette contains from 1 to 2 µg of Cd [1,3,29,30]; nevertheless, Cd content can vary depending on the origin of the tobacco leaves: in studies on Mexican cigarettes, it was found that each cigarette contained from 2.5 to 2.8 µg of Cd [31,32].

Considering a specific route of exposure for newborns, the presence of Cd in breast milk has been evidenced before [33,34] but recent findings once more highlighted the importance of this route since mouse pups showed a significant Cd exposure due to lactation after dams inhaled cadmium oxide (CdO) nanoparticles, opening the discussion about breast feeding as the major source of Cd exposure after birth and the clinical debate over lactation versus formula feeding in exposed populations [35].

Even if Cd concentrations in ambient air are generally considered as low [2], anthropogenic activities release large amounts of Cd, and air pollution also appears as a substantial source of exposure, especially in urban areas where the concentration of this metal ranges from 2 to 15 ng per cubic meter. However, some reports describe that Cd concentration in fine particulate matter with 2.5 micrometers or less in size ($PM_{2.5}$) can be higher, for instance, in Mexico City and its metropolitan area, levels of 35 to 40 ng per cubic meter have been detected previously [3,36,37].

3. Cadmium Distribution during Pregnancy

During pregnancy, both inhalational and oral Cd absorption increase [35,38,39]. This effect can be explained by the physiological changes that take place during this stage, like increased respiratory rate, decreased gastrointestinal motility and decreased gastric emptying [40]. In addition, the overexpression of receptors and transporters in the gut due to high nutrient demand [41] can promote Cd absorption too. Cadmium accumulates in the lung or gut depending on the route of exposure; then, it is distributed to the liver, kidneys, placenta, mammary glands, uterus and fetus [42–45] and it can be excreted into the milk [46,47].

It is believed that the absorption of Cd in the intestine is facilitated by different transporters, such as the Divalent Metal Transporter-1 (DMT-1), calcium channels, amino acid transporters, and by endocytosis of the cadmium-metallothionein (CdMT) complex [19,48].

3.1. Role of Diet

The diet is an important source of Cd. A variety of foodstuff may contain important amounts of Cd, for example potatoes, wheat, seafood (mollusks and crustaceans), offal, spinach, cereal bars, chocolate, soy and grains [3,49]. Therefore, during pregnancy, diet is the main source of Cd exposure.

Nutritional status can influence the absorption and distribution of Cd. Deficiency of essential metals is related to an increase in oral Cd absorption of up to 10 times [50,51], because Cd uses different essential metal transporters as entry pathways. For instance, iron and/or calcium deficiency, result in remarkable increases of Cd accumulation in liver and kidneys [51].

On the opposite side, a diet high in essential micronutrients can decrease oral Cd absorption [52] or its toxic effects [53]. Antioxidants [54,55] and probiotics [56,57] have also been proposed to decrease Cd toxicity.

Thus, the influence of the nutritional status on increases of absorption and accumulation of Cd is evident so that malnutrition and gestation are risk factors for the toxic effects of Cd. If they occur in combination, the risk of the absorption and accumulation of Cd increases and therefore toxic effects may be potentiated.

3.2. The Role of DMT-1 in Cd Distribution during Pregnancy

The presence of the transporter DMT-1 in the gut is very important for iron absorption, and its expression is dependent of iron body levels [58,59]. Cadmium absorption increases when there is an iron deficiency, and this effect has been related with increased DMT-1 expression in the intestine [38,52,60].

During the gestational stage, there is a bigger iron demand because this metal is necessary for fetal development [61]. This necessary iron is obtained from iron stores, and in answer to depletion of iron stores during late pregnancy, the expression of DMT1 and ferroportin-1 (FPN1) increases in the gut to obtain iron from the diet [61]. Although FPN1 is not involved in Cd transport [62], Leazer et al. observed an increase of Cd absorption during late pregnancy that may be related to an increase of DMT-1 in this period [38]. The iron movement during gestation could be related to Cd movement too. In the late stage of pregnancy and lactation period, iron is transferred to the fetus and mammary glands to be secreted into the milk [61]. Nakamura et al. suggested a role for DMT1 in Cd transfer to the fetus [43] and it will be discussed in Section 4.

3.3. Role of Metallothionein in Cd Distribution

Metallothionein (MT) is a low molecular weight protein that is important for zinc and copper homeostasis. It is ubiquitous, and each MT molecule can bind seven atoms of Cd [63].

Brako et al. [47] using knockout (k.o.) mice for isoforms 1 and 2 of MT (MT-I/MT-II k.o.) observed that, during early pregnancy, Cd absorption and distribution is independent of MT. Meanwhile, during the late gestation and lactation stages, MT increases in the gut and restricts the movement of Cd to the blood but it does not inhibit total Cd absorption. Therefore, in the absence of MT (MT-I/MT-II k.o.) Cd concentrations in the blood, liver, placenta, mammary glands and fetus are higher than in normal mice [47].

On the other hand, MT is important for Cd redistribution during late pregnancy. Chan and Cherian (1993) exposed Sprague-Dawley female rats to eight daily injections (s.c.) of 1.0 mg Cd/kg as CdCl$_2$ during 2 weeks. Then, 1.5 weeks later, the rats mated with an untreated male of the same strain. Later, the rats were sacrificed on gestational days (GD) 1, 7, 14 and a day before delivery. They observed that during the middle and late pregnancy, Cd liver concentrations decreased, and increased in the placenta and kidneys, meanwhile in non-pregnant rats the amount of Cd in liver and kidneys showed no apparent changes [64]. This Cd redistribution could be explained by MT redistribution, since they observed the same pattern with MT levels during this time of gestation [47,64]. Placenta has been showed to highly express megalin in cytotrophoblasts during first and third trimester [65]: megalin is a transmembrane receptor which mediates uptake and trafficking of many ligands including MT in kidney and other tissues [66–68]. However, other transporters that participate in receptor-mediated endocytosis of MT could be also involved [69,70]. These findings can be relevant for women in reproductive age who have been exposed chronically to Cd because this behavior increases the risk of Cd-induced toxicity in the kidneys and placenta during gestation.

4. Toxic Effects in Dams

There are many papers that evaluate the effects of Cd exposure during pregnancy, but most of them only assess toxic effects in the offspring and forget about dams. The toxic effects that have been observed in dams (Table 1) are dependent on Cd dose, exposure route, and the period of exposure i.e., before pregnancy, early pregnancy, late pregnancy and/or lactation period. However, Cd exposure during pregnancy has been associated with various effects on kidney, blood pressure and alterations of essential elements in the body, the latter being the cause of other pathologies in the mothers and fetuses.

Table 1. Cadmium toxicity in dams.

Outcome	Reference
Lower maternal weight gain.	[45,71]
Increase of systolic blood pressure and increased proteinuria associated with preeclampsia.	[72–74]
Abnormal glucocorticoid synthesis; induction of angiotensin II type 1-receptor-agonist autoantibodies (AT1-AA) and activation of complement component 5 (C5).	[73–75]
Kidney injury reported by excretion of Kim-1 into urine; increase of blood urea nitrogen; changes in the morphology of proximal tubules, like irregularly shaped nuclei and presence of vacuoles. Glomeruloendotheliosis and infiltrated inflammatory cells.	[2,29,52]
Decrease of progesterone and estradiol in placenta and plasma and decreased steroidogenic enzymes in reproductive organs. Increased uterine weight.	[45,76–78]
Alteration in the levels of essential elements in the body.	[39,43,44,79,80]
Decreased bone mineral density.	[81,82]

4.1. Preeclampsia

Preeclampsia (PE) is a specific syndrome of gestation, characterized by increased blood pressure and proteinuria after the 20th week of gestation. The causes could be genetic, modifications in the vascular endothelium induced by lifestyle factors, or immunological disorders [83]. Using in vivo models, Cd exposure has been associated to preeclampsia development with increased systolic pressure and proteinuria [72–74]. The mechanisms suggested for preeclampsia induction by Cd exposure are: placental damage caused by oxidative DNA damage [65,72], and high levels of corticosterone in plasma, consequence of placental alterations (downregulation of 11β-hydroxysteroid dehydrogenase, 11β-HSD2) [73,75]. Other mechanisms possibly related to the development of preeclampsia are immunological disorders. Zhang et al. observed an increased activation of complement component 5 (C5) in preeclamptic patients, determined by the increase in the concentration of the fragment C5 (C5a) in blood samples, which is a marker for complement activation [74]. The complement activation has been considered to be partially responsible for the development of preeclampsia and kidney injury [84]. In the same study, Wistar rats were exposed to 0.125 mg of Cd/kg/day intraperitoneally from GD9–14, as a model of preeclampsia, which was confirmed by an increase in systolic blood pressure and proteinuria. With this model, they observed increased levels of C5a and angiotensin II type 1-receptor-agonist autoantibodies (AT1-AA) induced by Cd exposure. When the rats were treated with an antagonist of C5a receptor (0.5 mg of PMX53/kg i.p. starting from 1 day before

Cd administration), the increases of pressure and protein excretion were minor. Moreover, losartan treatment prevented the Cd-induced increase in C5a levels. Thus, the authors suggested that Cd induce C5 activation through AT1-AA via AT1 receptor and therefore the preeclampsia characteristics [67,74].

However, all the studies linking Cd to preeclampsia have a disadvantage, they use a route of Cd exposure that is not environmentally relevant (intraperitoneal). Future studies with more relevant routes of Cd exposure are needed.

4.2. Kidney Damage

The kidneys are an important target of chronic Cd exposure [3]. During pregnancy, Cd is mobilized from the liver to the placenta and kidneys [64], also the expression of MT in the gut, favors the movement of Cd to the kidneys [47]. Therefore, kidneys are in great risk of experiencing Cd toxicity during pregnancy. Blum et al. exposed pregnant CD-1 mice to Cd by inhalation (CdO nanoparticles, 230 μg/m^3) from GD4.5 to 16.5, and they observed the excretion of kidney injury molecule 1 (Kim-1) into the urine [35]. Kim-1 is an early biomarker of proximal tubule kidney injury that changes before creatinine [85], the most common biomarker for kidney damage. Chan and Cherian (1993), exposed rats to Cd two weeks before mating and observed kidney damage until lactation stage, described by an increase of blood urea nitrogen (BUN) [64]. In addition, changes in renal morphology have been reported, such as glomerular endotheliosis, infiltration of inflammatory cells [72], irregularly shaped nuclei and presence of vacuoles [35].

Due to the redistribution and increased Cd absorption during gestation, more studies are needed to assess renal damage in dams exposed to Cd.

4.3. Lesser Incidence of Pregnancy

Animal studies showed that Cd exposure during gestation prevented implantation or caused less implantation sites, more resorptions and less number of fetuses (see Section 6), which is translated into a lesser incidence of pregnancy [45,71,86]. During pregnancy, the uterus and the placenta are Cd targets. It is possible that Cd decreases implantation due to a disturbance of the activity of implantation-related enzymes: cathepsin-D and alkaline phosphatase [76]. Also, it can be due to Cd-induced blastocyst death after implantation, caused by decreased levels of progesterone triggered by a lesser number of trophoblast cells [45,76,77,87].

Cd is known as an endocrine disruptor [88]. As it will be stated in the subsequent section, progesterone and estrogen are necessary to maintain pregnancy. Cadmium decreases hormones levels during pregnancy, which in turn can be caused by a decrease of steroidogenic enzymes [76–78]. In addition, complement activation have been related to dysregulation of angiogenic factors and spontaneous abortion [89]. During pregnancy, Cd exposure increases C5a [74]. Higher maternal plasma concentration of C5a has been associated with miscarriage and fetal death [90,91]; therefore, the Cd-induced increase in C5a levels during pregnancy, could explain the lesser incidence of pregnancy; nevertheless, studies to clarify this relationship are necessary.

Even if hormone levels do not affect the maintenance of pregnancy [76], low doses of Cd may cause effects on the offspring (intrauterine exposure), which should be evaluated.

4.4. Altered Levels of Essential Elements in the Body

Since Cd is a divalent metal ion, it is absorbed using transporters for essential metal ions. During gestational stage, gastrointestinal absorption increases, and the transport of essential elements increases too. Also during this stage, Cd exposure has been associated with altered concentrations of micronutrients in internal organs. For instance, iron levels in liver, kidneys, placenta and fetus decrease [39,44]; zinc levels increase in liver and decrease in kidneys and placenta [44,64,92], and copper levels show contrasting behaviors in liver and kidneys [51,64]. Although there are differences in the route of Cd administration, all studies conclude that Cd induced lower micronutrient concentrations in the fetuses.

The alteration in the concentrations of essential metals could be explained by Cd competition for receptors and transporters in the gastrointestinal tract, thus decreasing the absorption of nutrients [51]. Similarly, the induction of proteins such as MT in liver, kidney and gut could increase the concentrations of metals, such as as zinc and copper [47,63]. In addition, Cd reduces the activity of enzymes such as ceruloplasmin during gestation [79]. Ceruloplasmin is an important protein for copper transport from the liver to other organs [93] and plays a crucial role in copper transport from the dam to the fetus [94,95]. This effect could explain the accumulation of copper in liver and/or kidney and its decreased distribution to the fetus.

4.5. Decreased Bone Mineral Density

Cd may influence bone turnover by disruption of calcium absorption due to competition for their transporters in the intestine, or as consequence of kidney injury [2,30]. There are many animal studies about Cd toxicity during pregnancy; however, almost none of them investigated calcium levels. Trottier et al. reported that Cd inhalation (50 μg/m^3) during late gestation, decreased placental calcium levels after 5 days of exposure [80], suggesting that Cd interferes with calcium metabolism and/or transport. Another study showed that Cd promotes a loss of calcium from dam's skeleton during gestation and lactation stages that was induced by a calcium deficient diet [81]. The known interference between Cd and calcium, in addition to the high calcium demand during gestation and lactation, suggests a risk of the mother and fetus to develop osteomalacia or osteoporosis due to Cd exposure but more studies are needed.

5. Transport of Cd across the Placenta

The placenta is the organ responsible for enabling the transference of nutrients from the mother to the developing fetus, as well as the secretion of hormones that play a major role in the maintenance of pregnancy. At some point, this organ was considered an effective barrier that protected the fetus from toxic agents, but now, it is well known that its capacity to block toxicants from reaching the fetus is somewhat limited.

In this regard, the expression of MT-I and MT-II in the placenta restrains the passage of Cd to the fetus [47,96]. For instance, the levels of Cd in pups of MT-I/MT-II knockout (k.o.) mice exposed to this metal in utero were 100-fold higher than wild type mice [47]. Nevertheless, although the placental expression of both isoforms of MT can be induced by oral Cd exposure [43], it seems that the ability of this protein to block its transfer from the dam to the fetus is partial or can be surpassed when large amounts of Cd are present, which leads to the diffusion of this metal across the placenta.

The molecular mechanism by which Cd reaches the developing fetus remains to be established. Some studies suggest the participation of metal transporters, such as DMT-1, Zrt/Irt-like protein 14 (ZIP-14) and zinc transporter 2 (ZnT2) [38,43].

The placental expression of DMT-1 increases in a time-dependent manner over the progression of gestation and is higher than in other dam organs like large and small intestine, liver, and kidney [38]. Additionally, the placental gene expression of metal transporters can be induced in Wistar rats by an oral exposure to increasing doses of Cd (0, 1, 2 and 5 mg Cd/kg) from 3 weeks before mating and until GD20 [43]. Taking this into account, and the fact that these transporters lack substrate specificity (their substrates include iron, manganese, cobalt and cadmium, among others), it is quite possible that they facilitate the diffusion of Cd across the placenta. Still, more studies are needed to fully comprehend the extent of participation of these proteins.

In addition to the aforementioned transporters, TRPV6 (transient receptor potential cation channel subfamily V member 6), also known as calcium transporter type 1 (CaT1), could play a role in Cd diffusion. A study by Moreau et al. [97] using primary cultures of human placenta showed its expression in cytotrophoblasts where it contributed to calcium uptake. This finding suggested its participation in calcium maternal-fetal transfer, which was later corroborated with TRPV6 k.o. mice, whose fetuses had lower serum calcium levels and mineral content [98]. TRPV6, however, is not specific for calcium uptake, in vitro studies have demonstrated its ability to transport other elements, such as zinc, lanthanum, gadolinium and Cd [99,100], which raises the possibility of this channel contributing to Cd passage across the placenta. Studies that approach this issue are still needed.

Similarly to TRPV6, it is possible that the membrane receptor megalin plays a role in Cd transfer. Megalin is known to participate in the reabsorption of the Cd-MT complex in proximal tubules [66] and it is present in cytotrophoblasts as well as syncytiotrophoblasts of human [65] and mice placenta [101] where it is thought to contribute to the transfer of nutrients to the fetus. So it is plausible that this receptor mediates, to some extent, Cd endocytosis and subsequent transport across the placenta; nevertheless, this speculation needs to be confirmed by experimental studies.

Moreover, the use of Cd in the core of nanoparticles with biomedical applications poses new challenges in terms of assessing Cd toxicity. Because of their small size, nanoparticles can reach further than larger compounds, so it is of primary importance to conduct studies that evaluate the rate and amount of placental accumulation and transfer of cadmium nanoparticles (Cd-NPs). Although some studies start to address this issue [87], there is a long way to go to understand the molecular mechanisms involved in the transfer of Cd-NPs.

5.1. Placental Susceptibility to Cd

The placenta is one of the primary targets of Cd toxicity. Placental effects of this metal have been described in several animal species as early as the mid-60s. Some of the effects reported by those early studies include placental necrosis, vascular congestion, hemorrhage, decreased utero-placental blood flow and leukocyte infiltration [102–104]. The progress in laboratory techniques in recent years has allowed the evaluation of Cd-induced structural and functional alterations in the placenta at a molecular level. Table 2 summarizes some of Cd placental effects reported in the contemporary literature.

Table 2. Cadmium toxicity in placenta.

Outcome	Reference
Reduction of placental weight.	[77,87,105]
Decreased levels of total proteins, RNA, total lipids, cholesterol and glycogen.	[76]
Altered mRNA and protein levels, as well as activity of steroidogenic enzymes: Decreased: 3β-Hydroxysteroid dehydrogenase (HSD), 17β-HSD and 11β-HSD2. Increased: CYP11A1, CYP11B1 and CYP21.	[73,76]
Altered production of hormones necessary for the maintenance of pregnancy.	[76,77,87]
Placental atrophy, and swelling, vacuolization, deformation and death of trophoblasts due to apoptosis and necrosis in the junctional and labyrinthine zones of the placenta. Increased mRNA levels of molecules involved in apoptosis (p.53 and Bax).	[77,87,96,106,107]
DNA fragmentation in the junctional zone.	[77]
Decreased Zn and Cu placental concentration and altered metal transporters expression.	[38,43,87,96]
Altered MT mRNA and protein expression.	[43,107]
Oxidative stress and endoplasmic reticulum stress.	[87,106]
Reduced inner space of maternal and fetal blood vessels in the labyrinth layer. Thickening in the media vessel walls.	[73,105,106]
Higher corticosterone levels.	[73,75]
Decreased mRNA and protein levels of glucose transporter 3 (GLUT3). Hypermethylation of GLUT3 promoter region (E19.5).	[108]
Increased placental mRNA and protein levels of DNA methyltransferase 3-like (DNMT3L) and DNA methyltransferase 3 β (DNMT3B).	[108]
Lower protein levels of ATP-binding cassette (ABC) ABCG2 and ABCB4 transporters.	[109]
Decreased mRNA and protein levels of placental proton-coupled folate transporter (PCFT).	[105]

The experimental designs used to assess Cd placental toxicity are very diverse. Most studies administered this metal by routes that are not relevant for human exposure, and those seem to have the most severe outcomes in placental structure and/or function, even at low doses. However, studies that used oral administration, which is the main route of exposure for the non-smoking population, also caused alterations that can affect placental functions and that, ultimately, could result in fetal damage [43,75,96,108].

For example, a higher rate of trophoblastic apoptosis/necrosis is the most reported result of Cd intoxication no matter the route, length or period of exposure [77,87,96,106,107]. Interestingly, this effect appears to be dependent on the cellular type, because the four types of trophoblasts have different sensitivities to Cd. Yamagishi et al. reported that this metal showed the highest affinity for cytotrophoblasts but the most severely injured cells were spongiotrophoblasts [107]. The reason for this, however, is not clear yet.

Trophoblastic death can lead, partially, to some of the other Cd effects, such as altered endocrine function because trophoblasts in the junctional zone are responsible for the synthesis of hormones like progesterone, estradiol and placental lactogens [76,77,87]. This, in turn, could affect the maintenance of the pregnant state.

The precise mechanism by which Cd exerts its harmful effects in the placenta has not been fully unraveled, but the high vulnerability of this organ may derive, at least in part, from the ability of the trophoblasts to synthetize MT, which leads to Cd accumulation. Also, recent evidence showed that intragastric exposure to Cd 20 days prior to mating, resulted in lower protein levels of the transporters ABCG2 and ABCB4 in the placenta. These transporters protect the fetus from xenobiotics by transferring them, against their concentration gradient, to the maternal circulation [109]. Thus, a lower presence of these molecules could also increase Cd accumulation in the placenta. Once accumulated, it can induce oxidative stress, DNA fragmentation and endoplasmic reticulum stress [77,87,106].

Knowing the importance of the placenta in the progression of pregnancy and the development of the fetus, placental toxicity plays a pivotal role in Cd fetal effects. Section 6 will discuss such outcomes.

5.2. Cadmium Distribution in the Fetus

Cadmium transfer to the fetus is limited, but once this metal has reached the fetal body, it is of great importance to know how it gets distributed because it gives us the opportunity to estimate the most vulnerable organs and the magnitude of its effects. Unfortunately, most studies measure Cd fetal burden as a whole [44,75,86,108,110] which can only tell us whether this metal reached the fetus or not.

Despite that, some authors show that Cd can build up in the liver and kidney of Wistar rat fetuses (GD20) and pups (PND1) after an oral exposure to 0.5 and 5 mg Cd/kg through gavage [34,96]. Cadmium levels in both organs were comparable in each experimental design, which suggests a similar expression of MT or other metal-binding proteins in fetal liver and kidney. Nevertheless, as postnatal development progresses, liver Cd concentration peaks around PND14, after which the levels start to decrease. Meanwhile, the kidneys start to accumulate this metal as time goes by. Altogether, these findings point at a redistribution of Cd from the liver to the kidney like the one that happens in adult individuals [34].

After an exposure through inhalation or a subcutaneous injection, Cd experiences a similar distribution. For instance, it accumulated in the liver, brain and heart of Hartley guinea pig fetuses exposed to a dose of 0.05 mg Cd/m^3, for 4 h a day from GD35 to 40 [80]. Also, the exposure to 17.4 mg Cd/m^3 for 2 h from GD8–20 significantly increased the levels of this metal in the kidney of Wistar rat fetuses [42]. Additionally, exposure to Cd acetate by a subcutaneous injection from 7 days prior to mating and until GD20 or PND21 also leads to hepatic Cd build up in rat fetuses and pups [111].

These studies suggest that Cd distribution in the fetal body is quite similar to that found in adult animals probably because the pattern of MT synthesis, distribution is alike [112]. More importantly, differences in the animal species, type of salt used, route and length of exposure do not seem to affect significantly its fetal distribution.

6. Effects of Cd in the Fetal Organism

Fetal effects of Cd intoxication have been known for many years and they span a wide variety of organs. Table 3 lists the main outcomes of Cd gestational exposure in the recent literature.

Table 3. Cadmium toxicity in fetuses and pups.

Outcome	Reference
Lower fetal body weight, length and head diameter.	[42,73,75,77,87,105,106,108,110,113–116]
Higher number of resorptions, dead fetuses and post-implantation losses.	[77,86,87,105,107,113,117]
Increased apoptosis and decreased proliferation in the mesenchyme of limb buds of embryos.	[118]
Embryos with lower morphological score, somites number, DNA content, yolk sac diameter and cephalic length.	[119]
Higher percentage of embryos with an open neural tube and altered expression of genes related to the development of the Central Nervous System (CNS) and cell cycle arrest.	[113,119]
Increased levels of malondialdehyde (MDA) and myeloperoxidase (MPO) and decreased levels of Glutathione (GSH), Superoxide dismutase (SOD) and catalase (CAT) in embryos, placenta, fetal kidneys (except MPO), and fetal liver.	[86,110,119]
Fetal symmetrical kidneys, and renal cavitation and damage (tubular necrosis and degeneration, presence of hyaline cylinders in tubules, and proteinaceous material in the renal pelvis).	[42,120]
Delayed chondrogenesis that leads to decreased ossification of head bones, sternebrae and cervical vertebrae. Further, Cd causes fused ribs and vertebrae. Absence or lesser number of vertebrae, skull bones, ribs, tail bones, metacarpal and metatarsal bones, and phalanges.	[86,117,118]
Increased frequency of abnormalities such as cleft palate, unilateral anophtalmia, microphtalmia, hypoplasic lungs, genital anomalies, vein deformation, postaxial forelimb ectrodactyly (predominantly right-sided and with the loss of digit 5), clubfoot, polydactyly, anencephaly, exencephaly, encephalomeningocele, micrognathia, exophthalmos, tail deformity, amelia, brachygnathia, omphalocele, anotia, hemoperitoneum, brain edema and undifferentiated limbs.	[86,87,105,106,113,117,118,120,121]
Higher Cd levels in blood, liver and kidneys in the offspring at postnatal days (PND) 0 through 60.	[34,42,80,96,111,122]
Diminished weight gain of the offspring from PND0 to 21.	[122]
Lower concentration of Zn, Fe and Cu in fetal liver, and Ca levels in fetal kidney.	[111,123]
Lower activities of hepatic estradiol metabolizing enzymes (17-β-hydroxysteroid and UDP glucoronyl transferase) in fetuses (GD20) and pups (PND21).	[111]
Decreased DNA and glycogen hepatic content at PND21.	[111]
At PND21, lower activities of alkaline phosphatase, acid phosphatase and Na$^+$/K$^+$ ATPase in kidney tissue.	[122]
Reduced anogenital index in pups at PND1 and 21.	[114]
Delayed hair appearance, testicular descent, palmar grasp and negative geotaxis in pups.	[114]

Like placental toxicity, the mechanism of fetal toxicity remains elusive. Multiple studies show contrasting evidence in terms of maternal transfer of Cd, which raises the question of whether fetal toxicity is consequence of Cd action per se or of placental toxicity. Because the mother-fetus binomial has a very fine balance and cannot be separated, it is most likely that toxic effects in the fetuses are the result of both Cd fetal distribution and accumulation, and placental alterations.

Placental changes play a critical role in prenatal toxicity because they modify its ability to produce hormones, as well as its function in oxygen and nutrient transfer. For instance, placental hormonal synthesis is not only of primary importance for pregnancy sustainment, but it can also have detrimental effects in dams and fetuses. Cadmium exposure of rats during the gestational period increases the levels of corticosterone in the placenta and plasma of dams and fetuses possibly due to a decreased activity of placental 11β-HSD2 [73,75]. High cortisone levels have been related to the presence intrauterine growth restriction (IUGR), which, at the same time, can induce perinatal morbidity and the development of certain diseases in adult life due to fetal programming [75].

In addition, Cd-induced deficiency in the maternal transfer of nutrients like glucose, and essential elements, such as copper, calcium, sodium, potassium, zinc and iron [44,80,108,111,123], which are necessary for an optimal development, resulted in reduced body weight and fetal death of the offspring of rats and guinea pigs and are two of the most reported effects of this heavy metal (Table 3). This reduced transfer may be due to a decrease in the protein levels of transporters [108] and competition between Cd and other metals for their transporters in both the intestine and placenta of pregnant rats. In addition, the reduction in the lumen of the blood vessels in the labyrinth region of the placenta (Table 2) can interfere with nutrient transfer because the area of exchange is significantly smaller.

6.1. Teratogenicity

Cd is a well-established teratogenic agent. Its effects were reported 30 years ago, and even though the type of effects may vary depending on the dose, route and length of exposure, they have been reproduced by many groups, which underlines the danger of Cd gestational exposure and stresses the importance of finding ways to counteract those outcomes. In this sense, some studies found that co-exposure with the amino acid glycine can revert Cd-induced malformations due to the antioxidant activity of this molecule (evidenced by a decrease in lipid peroxidation levels) [86,119].

The main teratogenic effects caused by Cd are summarized in Table 3 and they include cleft palate, tail deformity, delayed chondrogenesis and postaxial forelimb ectrodactyly. The mechanism of toxicity behind these outcomes is not entirely deciphered, although some studies have shed some light on some of them. For example, postaxial forelimb ectrodactyly can be the result of a higher rate of apoptosis in limb buds, decreased expression of molecules involved in limb patterning, such as Fgf4, and decreased polarizing activity of the limb bud [118,121]. Interestingly, this outcome shows a right limb predominance when the exposure occurs on GD9, the reason why this happens is unknown. Also, altered chondrogenesis may be caused by a lower expression of Sox9 in the zeugopods of mouse limb buds [118].

6.2. Central Nervous System

Neural tube defects comprise a series of congenital diseases that affect the brain, spine and spinal cord. Their etiology is poorly understood, but it is thought that genetic, nutritional and environmental factors, or a combination of all, can induce them [124]. In this regard, when administered in the early organogenesis (during neurulation), Cd can induce open neural tubes that can result in anencephaly, exencephaly and encephalomeningocele (Table 3). Considering that folic acid is a vitamin necessary for a normal development, these Cd-induced defects may be the result of a lower folate content in embryos, triggered by a decreased expression of folate transporters in the placenta [45]. Additionally, direct actions of this metal may also contribute since the ex vivo exposure of mice embryos to 1 μM of $CdCl_2$ for 48 h caused a higher incidence of open neural tubes. This outcome correlated with the presence of oxidative stress that was counteracted by glycine co-exposure [119]. Furthermore, evidence shows that Cd can alter the expression of between 20 and 30 genes involved in CNS development in two strains of mice (Swiss and C57BL/6J) as early as 12 h after exposure. More importantly, a higher rate of cases of mice with exencephaly suggested that the C57BL/6J strain has a higher sensitivity to the action of this metal, although the reason for this is not known [113].

In the late organogenesis, Cd can also cross the blood-brain barrier and accumulate in the brain of the offspring after a single exposure [80]. Once there, it can cause brain edema [87] and, possibly, long-term effects.

6.3. Liver

The liver is one of the main organs where Cd builds up and as such, is the target of its toxicity. The exposure of Wistar rats to drinking water containing 10 mg/kg $CdCl_2$ during gestation causes degenerative changes in hepatocytes of fetuses at term, such as dilation of the smooth endoplasmic reticulum, slightly damaged nuclear membranes, and mitochondrial dilation [125]. In addition, a higher dose of Cd (50 ppm $CdCl_2$) changes the hepatic mRNA and protein levels of the glucocorticoid receptor due to modifications of the level of methylation on its promoter region. These alterations changed the expression of phosphoenolpyruvate carboxykinase and acyl-CoA oxidase that are involved in carbohydrate and lipid metabolism, respectively. Interestingly, all modifications found were sex-dependent [116], which suggests that Cd has a differential mechanism of toxicity.

Similarly, subcutaneous administration of Cd decreases the activity of estradiol metabolizing enzymes in fetuses, and 21-day old rat pups [111].

6.4. Kidney

The deleterious effects of Cd in the kidney of adult individuals, and the molecular mechanisms behind them are the subject of a vast amount of studies. The fetal kidney as a target of this metal, however, is far understudied in comparison.

Cd can reach and accumulate in fetal kidney [34,42,96]; thus, it can cause structural alterations and damage just like in fully developed kidneys. Accordingly, Roman et al. conducted a study that showed that a single intraperitoneal injection on GD10 with 5 mg $CdCl_2$/kg induces significant alterations in proximal and distal tubules, as well as glomerular edema in the offspring [126]. More recently, our group found that inhalational exposure to Cd from GD8 to GD20 provoked tubular necrosis and degeneration, among other effects, in the kidneys of Wistar rat fetuses (GD21). Such alterations correlated with an increase in the levels of early kidney injury biomarkers in amniotic fluid samples that are composed of fetal urine [42]. Similarly, a previous study presented evidence that repeated doses of Cd increased β2-microglobulin excretion and lower activity of γ-glutamyl transferase, alkaline phosphatase and N-acetylglucosaminidase in the urine of 3-day old rat pups, which suggested a reduced number of nephrons or a delayed tubular maturation [115]. All these renal outcomes could be the result of Cd-induced oxidative stress caused by lower levels of glutathione (GSH), superoxide dismutase (SOD) and catalase (CAT) [110]; nevertheless, further studies are required to corroborate these conclusions.

In addition, there is increasing evidence that links low birth weight with low nephron endowment [127]. Therefore, it is plausible that even if Cd does not reach fetal kidneys, it can cause low nephron endowment with subsequent alterations of renal structure and function, such as glomerulosclerosis.

Lastly, whether the renal effects caused by Cd gestational exposure are permanent or not, is still on debate because there is contrasting evidence that supports either possibility [34,115]. Discrepancies found may be the result of differences in the experimental models used, such as the period and length of exposure. Therefore, it is of great importance to keep on studying the renal effects of an in utero exposure to Cd since this organ is particularly vulnerable to this heavy metal as a result of its long embryonic development and its ability to accumulate Cd.

7. Conclusions

This review of the current literature illustrates how exposure to Cd during pregnancy deeply impacts the health of the fetuses and mothers and underlines the complexity of the multi-organs toxicity induced by this heavy metal (Figure 1).

In addition, it is important to highlight that many facets of Cd toxicity remain unclear and that some dogmas related to the physiological handling of Cd are still discussed and need to be clarified. For example, our vision of the central role of metallothionein as the main carrier of Cd in blood stream

Int. J. Mol. Sci. **2017**, *18*, 1590

or as the main source of delivery of Cd to proximal tubule epithelial cells (through receptor mediated endocytosis), could change in the coming years due to the evidence of the major role of other specific metal-binding proteins, such as transferrin, ferritin and Neutrophil gelatinase-associated lipocalin (NGAL), or new receptors, such as Lipocalin-2 (24p3/NGAL), involved in its binding and transport as well as the recent discovery of novel entry pathways for Cd [4,69,128].

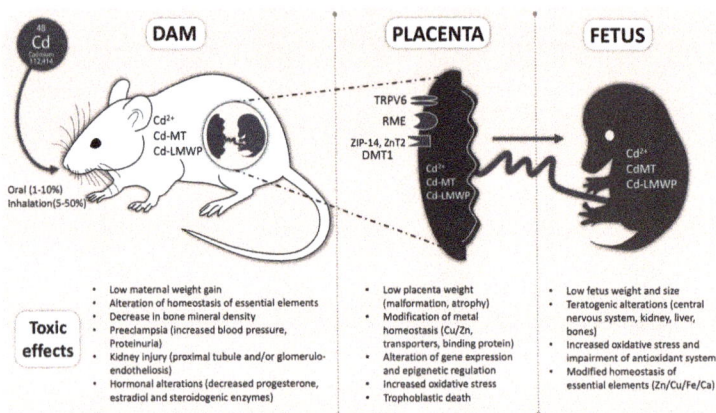

Figure 1. General scheme of the toxic effects of Cadmium (Cd) exposure in dam, placenta and fetus. MT: Metallothionein; LWMP: Low Molecular Weight Proteins; RME: Receptor-Mediated Endocytosis (e.g., Megalin or 24p3 Receptor); TRPV6: Transient Receptor Potential Cation Channel Subfamily V Member 6; DMT1: Divalent Metal Transporter-1; ZIP-14: Zrt/Irt-like Protein 14; ZnT2: Zinc Transporter 2.

Acknowledgments: This review is the result of a collaboration between research groups from Mexico and Germany and is supported by a joint grant Conacyt-BMBF (Conacyt 267755; BMBF 01DN16039).

Author Contributions: All authors contributed equally to the writing and reviewing of this work.

Conflicts of Interest: The authors declare no conflict of interest. The founding sponsors had no role in the design of the study; in the collection, analyses, or interpretation of data; in the writing of the manuscript, and in the decision to publish the results.

References

1. Kjellstrom, T.; Nordberg, G.F. A kinetic model of cadmium metabolism in the human being. *Environ. Res.* **1978**, *16*, 248–269. [CrossRef]

2. Jarup, L.; Akesson, A. Current status of cadmium as an environmental health problem. *Toxicol. Appl. Pharmacol.* **2009**, *238*, 201–208. [CrossRef] [PubMed]

3. ATSDR. *U.S.Toxicological Profile for Cadmium*; Department of Health and Human Sevices, Public Health Service, Centers for Disease Control Atlanta; ATSDR: Atlanta, GA, USA, 2012. Available online: https://www.atsdr.cdc.gov/toxprofiles/tp.asp?id=48&tid=15 (accessed on 29 June 2017).

4. Goyer, R.; Clarkson, T. Toxics effects of metals. In *Casarett and Doull's Toxicology: The Basic Science of Poisons*; Klaassen, C.D., Ed.; McGraw-Hill, Health Professions Division: New York, NY, USA, 2013.

5. Rogers, J. Developmental toxicology. In *Casarett and Doull's Toxicology: The Basic Science of Poisons*; Klaassen, C.D., Ed.; McGraw-Hill, Health Professions Division: New York, NY, USA, 2013.

6. Costantine, M.M. Physiologic and pharmacokinetic changes in pregnancy. *Front. Pharmacol.* **2014**, *5*, 65. [CrossRef] [PubMed]

7. Gundacker, C.; Hengstschlager, M. The role of the placenta in fetal exposure to heavy metals. *Wien. Med. Wochenschr.* **2012**, *162*, 201–206. [CrossRef] [PubMed]

8. Gundacker, C.; Neesen, J.; Straka, E.; Ellinger, I.; Dolznig, H.; Hengstschlager, M. Genetics of the human placenta: Implications for toxicokinetics. *Arch. Toxicol.* **2016**, *90*, 2563–2581. [CrossRef] [PubMed]

9. Little, M.H. Improving our resolution of kidney morphogenesis across time and space. *Curr. Opin. Genet. Dev.* **2015**, *32*, 135–143. [CrossRef] [PubMed]

10. Marciniak, A.; Patro-Malysza, J.; Kimber-Trojnar, Z.; Marciniak, B.; Oleszczuk, J.; Leszczynska-Gorzelak, B. Fetal programming of the metabolic syndrome. *Taiwan J. Obstet. Gynecol.* **2017**, *56*, 133–138. [CrossRef] [PubMed]

11. Al-Saleh, I.; Shinwari, N.; Mashhour, A.; Rabah, A. Birth outcome measures and maternal exposure to heavy metals (lead, cadmium and mercury) in Saudi Arabian population. *Int. J. Hyg. Environ. Health* **2014**, *217*, 205–218. [CrossRef] [PubMed]

12. Salihu, H.M.; Wilson, R.E. Epidemiology of prenatal smoking and perinatal outcomes. *Early Hum. Dev.* **2007**, *83*, 713–720. [CrossRef] [PubMed]

13. Luo, Y.; McCullough, L.E.; Tzeng, J.Y.; Darrah, T.; Vengosh, A.; Maguire, R.L.; Maity, A.; Samuel-Hodge, C.; Murphy, S.K.; Mendez, M.A.; et al. Maternal blood cadmium, lead and arsenic levels, nutrient combinations, and offspring birthweight. *BMC Public Health* **2017**, *17*, 354. [CrossRef] [PubMed]

14. Sanders, A.P.; Claus Henn, B.; Wright, R.O. Perinatal and childhood exposure to cadmium, manganese, and metal mixtures and effects on cognition and behavior: A review of recent literature. *Curr. Environ. Health Rep.* **2015**, *2*, 284–294. [CrossRef] [PubMed]

15. Appleton, A.; Jackson, B.; Karagas, M.; Marsit, C. Prenatal exposure to neurotoxic metals is associated with increased placental glucocorticoid receptor DNA methylation. *Epigenetics* **2017**. [CrossRef] [PubMed]

16. Vilahur, N.; Vahter, M.; Broberg, K. The epigenetic effects of prenatal cadmium exposure. *Curr. Environ. Health Rep.* **2015**, *2*, 195–203. [CrossRef] [PubMed]

17. Johri, N.; Jacquillet, G.; Unwin, R. Heavy metal poisoning: The effects of cadmium on the kidney. *Biometals* **2010**, *23*, 783–792. [CrossRef] [PubMed]

18. Barbier, O.; Jacquillet, G.; Tauc, M.; Cougnon, M.; Poujeol, P. Effect of heavy metals on, and handling by, the kidney. *Nephron Physiol.* **2005**, *99*, 105–110. [CrossRef] [PubMed]

19. Thevenod, F. Catch me if you can! Novel aspects of cadmium transport in mammalian cells. *Biometals* **2010**, *23*, 857–875. [CrossRef] [PubMed]

20. Thevenod, F.; Lee, W.K. Cadmium and cellular signaling cascades: Interactions between cell death and survival pathways. *Arch. Toxicol.* **2013**, *87*, 1743–1786. [CrossRef] [PubMed]

21. Thevenod, F.; Lee, W.K. Toxicology of cadmium and its damage to mammalian organs. *Met. Ions Life Sci.* **2013**, *11*, 415–490. [PubMed]

22. Smith, J.B.; Dwyer, S.D.; Smith, L. Cadmium evokes inositol polyphosphate formation and calcium mobilization. Evidence for a cell surface receptor that cadmium stimulates and zinc antagonizes. *J. Biol. Chem.* **1989**, *264*, 7115–7118. [PubMed]

23. Yang, H.; Shu, Y. Cadmium transporters in the kidney and cadmium-induced nephrotoxicity. *Int. J. Mol. Sci.* **2015**, *16*, 1484–1494. [CrossRef] [PubMed]

24. Thevenod, F. Cadmium and cellular signaling cascades: To be or not to be? *Toxicol. Appl. Pharmacol.* **2009**, *238*, 221–239. [CrossRef] [PubMed]

25. Kukongviriyapan, U.; Apaijit, K.; Kukongviriyapan, V. Oxidative stress and cardiovascular dysfunction associated with cadmium exposure: Beneficial effects of curcumin and tetrahydrocurcumin. *Tohoku J. Exp. Med.* **2016**, *239*, 25–38. [CrossRef] [PubMed]

26. Fassett, D. *Metallic Contaminants and Human Health*; Academic Press: Cambridge, MA, USA, 1972.

27. Cadmium. *Background and National Experience with Reducing Risk*; Organisation for Economic Co-Operation and Development: Paris, France, 1995; Available online: http://www.oecd.org/officialdocuments/publicdisplaydocumentpdf/?doclanguage=en&cote=ocde/gd(94)97 (accessed on 29 June 2017).

28. Emsley, J. *Nature's Building Blocks: An a-z Guide to the Elements*; Oxford University Press: Oxford, UK, 2011; Volume 48, p. 22.

29. Kosanovic, M.; Jokanovic, M.; Jevremovic, M.; Dobric, S.; Bokonjic, D. Maternal and fetal cadmium and selenium status in normotensive and hypertensive pregnancy. *Biol. Trace Elem. Res.* **2002**, *89*, 97–103. [CrossRef]

30. Jarup, L.; Berglund, M.; Elinder, C.G.; Nordberg, G.; Vahter, M. Health effects of cadmium exposure—A review of the literature and a risk estimate. *Scand. J. Work Environ. Health* **1998**, *24*, 1–51. [PubMed]

31. Saldivar, L.; Luna, M.; Reyes, E.; Soto, R.; Fortoul, T.I. Cadmium determination in mexican-produced tobacco. *Environ. Res.* **1991**, *55*, 91–96. [CrossRef]

32. Elinder, C.G.; Kjellstrom, T.; Lind, B.; Linnman, L.; Piscator, M.; Sundstedt, K. Cadmium exposure from smoking cigarettes: Variations with time and country where purchased. *Environ. Res.* **1983**, *32*, 220–227. [CrossRef]

33. Rebelo, F.M.; Caldas, E.D. Arsenic, lead, mercury and cadmium: Toxicity, levels in breast milk and the risks for breastfed infants. *Environ. Res.* **2016**, *151*, 671–688. [CrossRef] [PubMed]

34. Jacquillet, G.; Barbier, O.; Rubera, I.; Tauc, M.; Borderie, A.; Namorado, M.C.; Martin, D.; Sierra, G.; Reyes, J.L.; Poujeol, P.; et al. Cadmium causes delayed effects on renal function in the offspring of cadmium-contaminated pregnant female rats. *Am. J. Physiol. Ren. Physiol.* **2007**, *293*, F1450–F1460. [CrossRef] [PubMed]

35. Blum, J.L.; Edwards, J.R.; Prozialeck, W.C.; Xiong, J.Q.; Zelikoff, J.T. Effects of maternal exposure to cadmium oxide nanoparticles during pregnancy on maternal and offspring kidney injury markers using a murine model. *J. Toxicol. Environ. Health* **2015**, *78*, 711–724. [CrossRef] [PubMed]

36. Chow, J.C.; Watson, J.G.; Edgerton, S.A.; Vega, E. Chemical composition of PM$_{2.5}$ and PM$_{10}$ in mexico city during winter 1997. *Sci. Total Environ.* **2002**, *287*, 177–201. [CrossRef]

37. Guerra, R.; Vera-Aguilar, E.; Uribe-Ramirez, M.; Gookin, G.; Camacho, J.; Osornio-Vargas, A.R.; Mugica-Alvarez, V.; Angulo-Olais, R.; Campbell, A.; Froines, J.; et al. Exposure to inhaled particulate matter activates early markers of oxidative stress, inflammation and unfolded protein response in rat striatum. *Toxicol. Lett.* **2013**, *222*, 146–154. [CrossRef] [PubMed]

38. Leazer, T.M.; Liu, Y.; Klaassen, C.D. Cadmium absorption and its relationship to divalent metal transporter-1 in the pregnant rat. *Toxicol. Appl. Pharmacol.* **2002**, *185*, 18–24. [CrossRef] [PubMed]

39. Mikolić, A.; Schonwald, N.; Piasek, M. Cadmium, iron and zinc interaction and hematological parameters in rat dams and their offspring. *J. Trace Elem. Med. Biol.* **2016**, *38*, 108–116. [CrossRef] [PubMed]

40. Moya, J.; Phillips, L.; Sanford, J.; Wooton, M.; Gregg, A.; Schuda, L. A review of physiological and behavioral changes during pregnancy and lactation: Potential exposure factors and data gaps. *J. Expo. Sci. Environ. Epidemiol.* **2014**, *24*, 449–458. [CrossRef] [PubMed]

41. Astbury, S.; Mostyn, A.; Symonds, M.E.; Bell, R.C. Nutrient availability, the microbiome, and intestinal transport during pregnancy. *Appl. Physiol. Nutr. Metab.* **2015**, *40*, 1100–1106. [CrossRef] [PubMed]

42. Jacobo-Estrada, T.; Cardenas-Gonzalez, M.; Santoyo-Sanchez, M.; Parada-Cruz, B.; Uria-Galicia, E.; Arreola-Mendoza, L.; Barbier, O. Evaluation of kidney injury biomarkers in rat amniotic fluid after gestational exposure to cadmium. *J. Appl. Toxicol.* **2016**, *36*, 1183–1193. [CrossRef] [PubMed]

43. Nakamura, Y.; Ohba, K.; Ohta, H. Participation of metal transporters in cadmium transport from mother rat to fetus. *J. Toxicol. Sci.* **2012**, *37*, 1035–1044. [CrossRef] [PubMed]

44. Mikolić, A.; Piasek, M.; Sulimanec Grgec, A.; Varnai, V.M.; Stasenko, S.; Kralik Oguić, S. Oral cadmium exposure during rat pregnancy: Assessment of transplacental micronutrient transport and steroidogenesis at term. *J. Toxicol. Sci.* **2015**, *35*, 508–519. [CrossRef] [PubMed]

45. Blum, J.L.; Xiong, J.Q.; Hoffman, C.; Zelikoff, J.T. Cadmium associated with inhaled cadmium oxide nanoparticles impacts fetal and neonatal development and growth. *Toxicol. Sci.* **2012**, *126*, 478–486. [CrossRef] [PubMed]

46. Petersson Grawe, K.; Oskarsson, A. Cadmium in milk and mammary gland in rats and mice. *Arch. Toxicol.* **2000**, *73*, 519–527. [CrossRef] [PubMed]

47. Brako, E.E.; Wilson, A.K.; Jonah, M.M.; Blum, C.A.; Cerny, E.A.; Williams, K.L.; Bhattacharyya, M.H. Cadmium pathways during gestation and lactation in control versus metallothoinein 1,2-knockout mice. *Toxicol. Sci.* **2003**, *71*, 154–163. [CrossRef] [PubMed]

48. Vesey, D.A. Transport pathways for cadmium in the intestine and kidney proximal tubule: Focus on the interaction with essential metals. *Toxicol. Lett.* **2010**, *198*, 13–19. [CrossRef] [PubMed]

49. Satarug, S.; Vesey, D.A.; Gobe, G.C. Health risk assessment of dietary cadmium intake: Do current guidelines indicate how much is safe? *Environ. Health Perspect.* **2017**, *125*, 284–288. [CrossRef] [PubMed]

50. Min, K.S.; Ueda, H.; Tanaka, K. Involvement of intestinal calcium transporter 1 and metallothionein in cadmium accumulation in the liver and kidney of mice fed a low-calcium diet. *Toxicol. Lett.* **2008**, *176*, 85–92. [CrossRef] [PubMed]

51. Min, K.S.; Sano, E.; Ueda, H.; Sakazaki, F.; Yamada, K.; Takano, M.; Tanaka, K. Dietary deficiency of calcium and/or iron, an age-related risk factor for renal accumulation of cadmium in mice. *Biol. Pharm. Bull.* **2015**, *38*, 1557–1563. [CrossRef] [PubMed]

52. Suzuki, T.; Momoi, K.; Hosoyamada, M.; Kimura, M.; Shibasaki, T. Normal cadmium uptake in microcytic anemia *mk/mk* mice suggests that DMT1 is not the only cadmium transporter in vivo. *Toxicol. Appl. Pharmacol.* **2008**, *227*, 462–467. [CrossRef] [PubMed]

53. Jacquillet, G.; Barbier, O.; Cougnon, M.; Tauc, M.; Namorado, M.C.; Martin, D.; Reyes, J.L.; Poujeol, P. Zinc protects renal function during cadmium intoxication in the rat. *Am. J. Physiol. Ren. Physiol.* **2006**, *290*, F127–F137. [CrossRef] [PubMed]

54. Renugadevi, J.; Prabu, S.M. Naringenin protects against cadmium-induced oxidative renal dysfunction in rats. *Toxicology* **2009**, *256*, 128–134. [CrossRef] [PubMed]

55. Eybl, V.; Kotyzova, D.; Koutensky, J. Comparative study of natural antioxidants-curcumin, resveratrol and melatonin- in cadmium-induced oxidative damage in mice. *Toxicology* **2006**, *225*, 150–156. [CrossRef] [PubMed]

56. Zhai, Q.; Narbad, A.; Chen, W. Dietary strategies for the treatment of cadmium and lead toxicity. *Nutrients* **2015**, *7*, 552–571. [CrossRef] [PubMed]

57. Zhai, Q.; Tian, F.; Zhao, J.; Zhang, H.; Narbad, A.; Chen, W. Oral administration of probiotics inhibits absorption of the heavy metal cadmium by protecting the intestinal barrier. *Appl. Environ. Microbiol.* **2016**, *82*, 4429–4440. [CrossRef] [PubMed]

58. Kuhn, L.C. Iron regulatory proteins and their role in controlling iron metabolism. *Metallomics* **2015**, *7*, 232–243. [CrossRef] [PubMed]

59. Shawki, A.; Anthony, S.R.; Nose, Y.; Engevik, M.A.; Niespodzany, E.J.; Barrientos, T.; Ohrvik, H.; Worrell, R.T.; Thiele, D.J.; Mackenzie, B. Intestinal DMT1 is critical for iron absorption in the mouse but is not required for the absorption of copper or manganese. *Am. J. Physiol. Gastrointest. Liver Physiol.* **2015**, *309*, G635–G647. [CrossRef] [PubMed]

60. Kim, D.W.; Kim, K.Y.; Choi, B.S.; Youn, P.; Ryu, D.Y.; Klaassen, C.D.; Park, J.D. Regulation of metal transporters by dietary iron, and the relationship between body iron levels and cadmium uptake. *Arch. Toxicol.* **2007**, *81*, 327–334. [CrossRef] [PubMed]

61. Gao, G.; Liu, S.Y.; Wang, H.J.; Zhang, T.W.; Yu, P.; Duan, X.L.; Zhao, S.E.; Chang, Y.Z. Effects of pregnancy and lactation on iron metabolism in rats. *BioMed Res. Int.* **2015**, *2015*, 105325. [CrossRef] [PubMed]

62. Mitchell, C.J.; Shawki, A.; Ganz, T.; Nemeth, E.; Mackenzie, B. Functional properties of human ferroportin, a cellular iron exporter reactive also with cobalt and zinc. *Am. J. Physiol. Cell Physiol.* **2014**, *306*, C450–C459. [CrossRef] [PubMed]

63. Klaassen, C.D.; Liu, J.; Choudhuri, S. Metallothionein: An intracellular protein to protect against cadmium toxicity. *Annu. Rev. Pharmacol. Toxicol.* **1999**, *39*, 267–294. [CrossRef] [PubMed]

64. Chan, H.M.; Cherian, M.G. Mobilization of hepatic cadmium in pregnant rats. *Toxicol. Appl. Pharmacol.* **1993**, *120*, 308–314. [CrossRef] [PubMed]

65. Storm, T.; Christensen, E.I.; Christensen, J.N.; Kjaergaard, T.; Uldbjerg, N.; Larsen, A.; Honore, B.; Madsen, M. Megalin is predominantly observed in vesicular structures in first and third trimester cytotrophoblasts of the human placenta. *J. Histochem. Cytochem.* **2016**, *64*, 769–784. [CrossRef] [PubMed]

66. Klassen, R.B.; Crenshaw, K.; Kozyraki, R.; Verroust, P.J.; Tio, L.; Atrian, S.; Allen, P.L.; Hammond, T.G. Megalin mediates renal uptake of heavy metal metallothionein complexes. *Am. J. Physiol. Ren. Physiol.* **2004**, *287*, F393–F403. [CrossRef] [PubMed]

67. Wolff, N.A.; Abouhamed, M.; Verroust, P.J.; Thevenod, F. Megalin-dependent internalization of cadmium-metallothionein and cytotoxicity in cultured renal proximal tubule cells. *J. Pharmacol. Exp. Ther.* **2006**, *318*, 782–791. [CrossRef] [PubMed]

68. Onodera, A.; Tani, M.; Michigami, T.; Yamagata, M.; Min, K.S.; Tanaka, K.; Nakanishi, T.; Kimura, T.; Itoh, N. Role of megalin and the soluble form of its ligand RAP in Cd-metallothionein endocytosis and Cd-metallothionein-induced nephrotoxicity in vivo. *Toxicol. Lett.* **2012**, *212*, 91–96. [CrossRef] [PubMed]

69. Thevenod, F.; Wolff, N.A. Iron transport in the kidney: Implications for physiology and cadmium nephrotoxicity. *Metallomics* **2016**, *8*, 17–42. [CrossRef] [PubMed]

70. Langelueddecke, C.; Roussa, E.; Fenton, R.A.; Thevenod, F. Expression and function of the lipocalin-2 (24p3/NGAL) receptor in rodent and human intestinal epithelia. *PLoS ONE* **2013**, *8*, e71586. [CrossRef] [PubMed]

71. Samuel, J.B.; Stanley, J.A.; Princess, R.A.; Shanthi, P.; Sebastian, M.S. Gestational cadmium exposure-induced ovotoxicity delays puberty through oxidative stress and impaired steroid hormone levels. *J. Med. Toxicol.* **2011**, *7*, 195–204. [CrossRef] [PubMed]

72. Zhang, X.; Xu, Z.; Lin, F.; Wang, F.; Ye, D.; Huang, Y. Increased oxidative DNA damage in placenta contributes to cadmium-induced preeclamptic conditions in rat. *Biol. Trace Elem. Res.* **2016**, *170*, 119–127. [CrossRef] [PubMed]

73. Wang, F.; Zhang, Q.; Zhang, X.; Luo, S.; Ye, D.; Guo, Y.; Chen, S.; Huang, Y. Preeclampsia induced by cadmium in rats is related to abnormal local glucocorticoid synthesis in placenta. *Reprod. Biol. Endocrinol.* **2014**, *12*. [CrossRef] [PubMed]

74. Zhang, Q.; Huang, Y.; Zhang, K.; Huang, Y.; Yan, Y.; Wang, F.; Wu, J.; Wang, X.; Xu, Z.; Chen, Y.; et al. Cadmium-induced immune abnormality is a key pathogenic event in human and rat models of preeclampsia. *Environ. Pollut.* **2016**, *218*, 770–782. [CrossRef] [PubMed]

75. Ronco, A.M.; Urrutia, M.; Montenegro, M.; Llanos, M.N. Cadmium exposure during pregnancy reduces birth weight and increases maternal and foetal glucocorticoids. *Toxicol. Lett.* **2009**, *188*, 186–191. [CrossRef] [PubMed]

76. Nampoothiri, L.P.; Gupta, S. Biochemical effects of gestational coexposure to lead and cadmium on reproductive performance, placenta, and ovary. *J. Biochem. Mol. Toxicol.* **2008**, *22*, 337–344. [CrossRef] [PubMed]

77. Lee, C.K.; Lee, J.T.; Yu, S.J.; Kang, S.G.; Moon, C.S.; Choi, Y.H.; Kim, J.H.; Kim, D.H.; Son, B.C.; Lee, C.H.; et al. Effects of cadmium on the expression of placental lactogens and pit-1 genes in the rat placental trophoblast cells. *Mol. Cell. Endocrinol.* **2009**, *298*, 11–18. [CrossRef] [PubMed]

78. Piasek, M.; Laskey, J.W. Acute cadmium exposure and ovarian steroidogenesis in cycling and pregnant rats. *Reprod. Toxicol.* **1994**, *8*, 495–507. [CrossRef]

79. Chmielnicka, J.; Sowa, B. Cadmium interaction with essential metals (Zn, Cu, Fe), metabolism metallothionein, and ceruloplasmin in pregnant rats and fetuses. *Ecotoxicol. Environ. Saf.* **1996**, *35*, 277–281. [CrossRef] [PubMed]

80. Trottier, B.; Athot, J.; Ricard, A.C.; Lafond, J. Maternal-fetal distribution of cadmium in the guinea pig following a low dose inhalation exposure. *Toxicol. Lett.* **2002**, *129*, 189–197. [CrossRef]

81. Wang, C.; Brown, S.; Bhattacharyya, M.H. Effect of cadmium on bone calcium and ^{45}Ca in mouse dams on a calcium-deficient diet: Evidence of itai-itai-like syndrome. *Toxicol. Appl. Pharmacol.* **1994**, *127*, 320–330. [CrossRef] [PubMed]

82. Otha, H.; Ichikawa, M.; Seki, Y. Effects of cadmium intake on bone metabolism of mothers during pregnancy and lactation. *Tohoku J. Exp. Med.* **2002**, *196*, 33–42.

83. Pauli, J.M.; Repke, J.T. Preeclampsia: Short-term and long-term implications. *Obstet. Gynecol. Clin. N. Am.* **2015**, *42*, 299–313. [CrossRef] [PubMed]

84. Qing, X.; Redecha, P.B.; Burmeister, M.A.; Tomlinson, S.; D'Agati, V.D.; Davisson, R.L.; Salmon, J.E. Targeted inhibition of complement activation prevents features of preeclampsia in mice. *Kidney Int.* **2011**, *79*, 331–339. [CrossRef] [PubMed]

85. Vaidya, V.S.; Ferguson, M.A.; Bonventre, J.V. Biomarkers of acute kidney injury. *Annu. Rev. Pharmacol. Toxicol.* **2008**, *48*, 463–493. [CrossRef] [PubMed]

86. Paniagua-Castro, N.; Escalona-Cardoso, G.; Chamorro-Cevallos, G. Glycine reduces cadmium-induced teratogenic damage in mice. *Reprod. Toxicol.* **2007**, *23*, 92–97. [CrossRef] [PubMed]

87. Zhang, W.; Yang, L.; Kuang, H.; Yang, P.; Aguilar, Z.P.; Wang, A.; Fu, F.; Xu, H. Acute toxicity of quantum dots on late pregnancy mice: Effects of nanoscale size and surface coating. *J. Hazard. Mater.* **2016**, *318*, 61–69. [CrossRef] [PubMed]

88. Hofer, N.; Diel, P.; Wittsiepe, J.; Wilhelm, M.; Degen, G.H. Dose- and route-dependent hormonal activity of the metalloestrogen cadmium in the rat uterus. *Toxicol. Lett.* **2009**, *191*, 123–131. [CrossRef] [PubMed]

89. Girardi, G.; Yarilin, D.; Thurman, J.M.; Holers, V.M.; Salmon, J.E. Complement activation induces dysregulation of angiogenic factors and causes fetal rejection and growth restriction. *J. Exp. Med.* **2006**, *203*, 2165–2175. [CrossRef] [PubMed]

90. Lynch, A.M.; Salmon, J.E. Dysregulated complement activation as a common pathway of injury in preeclampsia and other pregnancy complications. *Placenta* **2010**, *31*, 561–567. [CrossRef] [PubMed]
91. Denny, K.J.; Coulthard, L.G.; Finnell, R.H.; Callaway, L.K.; Taylor, S.M.; Woodruff, T.M. Elevated complement factor C5a in maternal and umbilical cord plasma in preeclampsia. *J. Reprod. Immunol.* **2013**, *97*, 211–216. [CrossRef] [PubMed]
92. Piasek, M.; Blanusa, M.; Kostial, K.; Laskey, J.W. Low iron diet and parenteral cadmium exposure in pregnant rats: The effects on trace elements and fetal viability. *Biometals* **2004**, *17*, 1–14. [CrossRef] [PubMed]
93. Linder, M.C.; Wooten, L.; Cerveza, P.; Cotton, S.; Shulze, R.; Lomeli, N. Copper transport. *Am. J. Clin. Nutr.* **1998**, *67*, 965S–971S. [PubMed]
94. Lee, S.H.; Lancey, R.; Montaser, A.; Madani, N.; Linder, M.C. Ceruloplasmin and copper transport during the latter part of gestation in the rat. *Proc. Soc. Exp. Biol. Med.* **1993**, *203*, 428–439. [CrossRef] [PubMed]
95. Chu, Y.L.; Sauble, E.N.; Cabrera, A.; Roth, A.; Ackland, M.L.; Mercer, J.F.; Linder, M.C. Lack of ceruloplasmin expression alters aspects of copper transport to the fetus and newborn, as determined in mice. *Biometals* **2012**, *25*, 373–382. [CrossRef] [PubMed]
96. Nakamura, Y.; Ohba, K.; Suzuki, K.; Ohta, H. Health effects of low-level cadmium intake and the role of metallothionein on cadmium transport from mother rats to fetus. *J. Toxicol. Sci.* **2012**, *37*, 149–156. [CrossRef] [PubMed]
97. Moreau, R.; Daoud, G.; Bernatchez, R.; Simoneau, L.; Masse, A.; Lafond, J. Calcium uptake and calcium transporter expression by trophoblast cells from human term placenta. *Biochim. Biophys. Acta* **2002**, *1564*, 325–332. [CrossRef]
98. Suzuki, Y.; Kovacs, C.S.; Takanaga, H.; Peng, J.B.; Landowski, C.P.; Hediger, M.A. Calcium channel TRPV6 is involved in murine maternal-fetal calcium transport. *J. Bone Miner. Res.* **2008**, *23*, 1249–1256. [CrossRef] [PubMed]
99. Kovacs, G.; Danko, T.; Bergeron, M.J.; Balazs, B.; Suzuki, Y.; Zsembery, A.; Hediger, M.A. Heavy metal cations permeate the TRPV6 epithelial cation channel. *Cell Calcium* **2011**, *49*, 43–55. [CrossRef] [PubMed]
100. Kovacs, G.; Montalbetti, N.; Franz, M.C.; Graeter, S.; Simonin, A.; Hediger, M.A. Human TRPV5 and TRPV6: Key players in cadmium and zinc toxicity. *Cell Calcium* **2013**, *54*, 276–286. [CrossRef] [PubMed]
101. Burk, R.F.; Olson, G.E.; Hill, K.E.; Winfrey, V.P.; Motley, A.K.; Kurokawa, S. Maternal-fetal transfer of selenium in the mouse. *FASEB J.* **2013**, *27*, 3249–3256. [CrossRef] [PubMed]
102. Pařízek, J. Vascular changes at sites of oestrogen biosynthesis produced by parenteral injection of cadmium salts: The destruction of the placenta by cadmium salts. *J. Reprod. Fertil.* **1964**, *7*, 263–265. [CrossRef] [PubMed]
103. Pařízek, J. The peculiar toxicity of cadmium during pregnancy–an experimental "toxaemia of pregnancy" induced by cadmium salts. *J. Reprod. Fertil.* **1965**, *9*, 111–112. [CrossRef] [PubMed]
104. Levin, A.A.; Miller, R.K. Fetal toxicity of cadmium in the rat: Decreased utero-placental blood flow. *Toxicol. Appl. Pharmacol.* **1981**, *58*, 297–306. [CrossRef]
105. Zhang, G.B.; Wang, H.; Hu, J.; Guo, M.Y.; Wang, Y.; Zhou, Y.; Yu, Z.; Fu, L.; Chen, Y.H.; Xu, D.X. Cadmium-induced neural tube defects and fetal growth restriction: Association with disturbance of placental folate transport. *Toxicol. Appl. Pharmacol.* **2016**, *306*, 79–85. [CrossRef] [PubMed]
106. Wang, Z.; Wang, H.; Xu, Z.M.; Ji, Y.L.; Chen, Y.H.; Zhang, Z.H.; Zhang, C.; Meng, X.H.; Zhao, M.; Xu, D.X. Cadmium-induced teratogenicity: Association with ROS-mediated endoplasmic reticulum stress in placenta. *Toxicol. Appl. Pharmacol.* **2012**, *259*, 236–247. [CrossRef] [PubMed]
107. Yamagishi, Y.; Furukawa, S.; Tanaka, A.; Kobayashi, Y.; Sugiyama, A. Histological localization of cadmium in rat placenta by LA-ICP-MS. *J. Toxicol. Pathol.* **2016**, *29*, 279–283. [CrossRef] [PubMed]
108. Xu, P.; Wu, Z.; Xi, Y.; Wang, L. Epigenetic regulation of placental glucose transporters mediates maternal cadmium-induced fetal growth restriction. *Toxicology* **2016**, *372*, 34–41. [CrossRef] [PubMed]
109. Liu, L.; Zhou, L.; Hu, S.; Zhou, S.; Deng, Y.; Dong, M.; Huang, J.; Zeng, Y.; Chen, X.; Zhao, N.; et al. Down-regulation of ABCG2 and ABCB4 transporters in the placenta of rats exposed to cadmium. *Oncotarget* **2016**, *7*, 38154–38163. [CrossRef] [PubMed]
110. Enli, Y.; Turgut, S.; Oztekin, O.; Demir, S.; Enli, H.; Turgut, G. Cadmium intoxication of pregnant rats and fetuses: Interactions of copper supplementation. *Arch. Med. Res.* **2010**, *41*, 7–13. [CrossRef] [PubMed]

111. Pillai, A.; Gupta, S. Effect of gestational and lactational exposure to lead and/or cadmium on reproductive performance and hepatic oestradiol metabolising enzymes. *Toxicol. Lett.* **2005**, *155*, 179–186. [CrossRef] [PubMed]

112. Webb, M.; Cain, K. Functions of metallothionein. *Biochem. Pharmacol.* **1982**, *31*, 137–142. [CrossRef]

113. Robinson, J.F.; Yu, X.; Hong, S.; Griffith, W.C.; Beyer, R.; Kim, E.; Faustman, E.M. Cadmium-induced differential toxicogenomic response in resistant and sensitive mouse strains undergoing neurulation. *Toxicol. Sci.* **2009**, *107*, 206–219. [CrossRef] [PubMed]

114. Couto-Moraes, R.; Felicio, L.F.; Bernardi, M.M. Post-partum testosterone administration does not reverse the effects of perinatal exposure to cadmium on rat offspring development. *J. Appl. Toxicol.* **2010**, *30*, 233–241. [CrossRef] [PubMed]

115. Saillenfait, A.M.; Payan, J.P.; Brondeau, M.T.; Zissu, D.; de Ceaurriz, J. Changes in urinary proximal tubule parameters in neonatal rats exposed to cadmium chloride during pregnancy. *J. Appl. Toxicol.* **1991**, *11*, 23–27. [CrossRef] [PubMed]

116. Castillo, P.; Ibanez, F.; Guajardo, A.; Llanos, M.N.; Ronco, A.M. Impact of cadmium exposure during pregnancy on hepatic glucocorticoid receptor methylation and expression in rat fetus. *PLoS ONE* **2012**, *7*, e44139. [CrossRef] [PubMed]

117. Díaz, M.d.C.; González, N.; Gómez, S.; Quiroga, M.; Najle, R.; Barbeito, C. Effect of a single dose of cadmium on pregnant wistar rats and their offspring. *Reprod. Domest. Anim.* **2014**, *49*, 1049–1056. [CrossRef] [PubMed]

118. Liao, X.; Lee, G.S.; Shimizu, H.; Collins, M.D. Comparative molecular pathology of cadmium- and all-trans-retinoic acid-induced postaxial forelimb ectrodactyly. *Toxicol. Appl. Pharmacol.* **2007**, *225*, 47–60. [CrossRef] [PubMed]

119. Paniagua-Castro, N.; Escalona-Cardoso, G.; Madrigal-Bujaidar, E.; Martinez-Galero, E.; Chamorro-Cevallos, G. Protection against cadmium-induced teratogenicity in vitro by glycine. *Toxicol. In Vitro* **2008**, *22*, 75–79. [CrossRef] [PubMed]

120. Salvatori, F.; Talassi, C.B.; Salzgeber, S.A.; Spinosa, H.S.; Bernardi, M.M. Embryotoxic and long-term effects of cadmium exposure during embryogenesis in rats. *Neurotoxicol. Teratol.* **2004**, *26*, 673–680. [CrossRef] [PubMed]

121. Scott, W.J., Jr.; Schreiner, C.M.; Goetz, J.A.; Robbins, D.; Bell, S.M. Cadmium-induced postaxial forelimb ectrodactyly: Association with altered sonic hedgehog signaling. *Reprod. Toxicol.* **2005**, *19*, 479–485. [CrossRef] [PubMed]

122. Antonio Garcia, T.; Corredor, L. Biochemical changes in the kidneys after perinatal intoxication with lead and/or cadmium and their antagonistic effects when coadministered. *Ecotoxicol. Environ. Saf.* **2004**, *57*, 184–189. [CrossRef]

123. Kuriwaki, J.; Nishijo, M.; Honda, R.; Tawara, K.; Nakagawa, H.; Hori, E.; Nishijo, H. Effects of cadmium exposure during pregnancy on trace elements in fetal rat liver and kidney. *Toxicol. Lett.* **2005**, *156*, 369–376. [CrossRef] [PubMed]

124. Kondo, A.; Kamihira, O.; Ozawa, H. Neural tube defects: Prevalence, etiology and prevention. *Int. J. Urol.* **2009**, *16*, 49–57. [CrossRef] [PubMed]

125. Oğuz, E.O.; Abban, G.; Kutlubay, R.; Turgut, S.; Enli, Y.; Erdogan, D. Transmission electron microscopy study of the effects of cadmium and copper on fetal rat liver tissue. *Biol. Trace Elem. Res.* **2007**, *115*, 127–135. [CrossRef] [PubMed]

126. Roman, T.R.N.; De Lima, E.G.; Azoubel, R.; Batigália, F. Renal morphometry of fetuses rats treated with cadmium. *Int. J. Morphol.* **2004**, *22*, 231–236. [CrossRef]

127. Chong, E.; Yosypiv, I.V. Developmental programming of hypertension and kidney disease. *Int. J. Nephrol.* **2012**, *2012*, 760580. [CrossRef] [PubMed]

128. Langelueddecke, C.; Roussa, E.; Fenton, R.A.; Wolff, N.A.; Lee, W.K.; Thevenod, F. Lipocalin-2 (24p3/neutrophil gelatinase-associated lipocalin (NGAL)) receptor is expressed in distal nephron and mediates protein endocytosis. *J. Biol. Chem.* **2012**, *287*, 159–169. [CrossRef] [PubMed]

International Journal of
Molecular Sciences

MDPI

Review

Unravelling the Role of Metallothionein on Development, Reproduction and Detoxification in the Wall Lizard *Podarcis sicula*

Rosaria Scudiero [1],* (iD), Mariailaria Verderame [1], Chiara Maria Motta [1] and Palma Simoniello [2]

[1] Department of Biology, University Federico II, Via Mezzocannone 8, 80134 Napoli, Italy;
   ilaria.verderame@unina.it (M.V.); chiaramaria.motta@unina.it (C.M.M.)
[2] Department of Sciences and Technology, University Parthenope, Centro Direzionale, Isola C4, 80143 Napoli,
   Italy; palma.simoniello@uniparthenope.it
* Correspondence: rosaria.scudiero@unina.it; Tel.: +39-0812535217; Fax: +39-0812535035

Received: 27 May 2017; Accepted: 16 July 2017; Published: 19 July 2017

Abstract: Metallothioneins (MTs) are an evolutionary conserved multigene family of proteins whose role was initially identified in binding essential metals. The physiological role of MT, however, has been revealed to be more complex than expected, since not only are MTs able to bind to toxic heavy metals, but many isoforms have shown specialized and alternative functions. Within this uncertainty, the information available on MTs in non-mammalian vertebrates, particularly in neglected tetrapods such as the reptiles, is even more scant. In this review, we provide a summary of the current understanding on metallothionein presence and function in the oviparous lizard *Podarcis sicula*, highlighting the results obtained by studying *MT* gene expression in most representative adult and embryonic tissues. The results demonstrate that in adults, cadmium induces MT transcription in a dose- and tissue-specific manner. Thus, the MT mRNAs appear, at least in some cases, to be an unsuitable tool for detecting environmental ion contamination. In early embryos, maternal RNAs sustain developmental needs for MT protein until organogenesis is well on its way. At this time, transcription starts, but again in a tissue- and organ-specific manner, suggesting an involvement in alternative roles. In conclusion, the spatiotemporal distribution of transcripts in adults and embryos definitively confirms that MT has deserved the title of elusive protein.

Keywords: *Podarcis sicula* metallothionein; gene expression; in situ hybridisation; embryonic development; cadmium; oogenesis; spermatogenesis

---

1. Introduction

Since its discovery in 1957 from horse kidney as a cadmium-binding protein [1], metallothionein (MT) functions have been associated with metal micronutrients homeostasis and heavy metals detoxification [2]. Despite many studies aimed at validating the use of MT as a good biomarker of heavy metal contamination, increasing evidence demonstrates the involvement of this protein in many normal or pathological cellular processes, so that MT has deserved the title of elusive protein [3,4].

Found in all the eukaryotic taxa [5], MT has been extensively studied in invertebrates and vertebrates, both at the evolutionary and structural levels [6–10]. Among vertebrates, MT studies are particularly numerous on mammals, while within the lower vertebrates, the MTs have been studied in greater detail only in fish, in relation to their detoxifying capacity and the significant metal pollution of the aquatic environment [6,11,12].

Metallothioneins belong to a multigenic family. In mammals, four tandemly clustered genes (*MT1* to *MT4*) have been identified, with *MT1* and *MT2* being present in almost all tissues, and *MT3* and *MT4* being more abundant in neuronal and epithelial cells, respectively [13,14]. In addition, *MT1* has undergone further duplication events that have resulted in many duplicate isoforms (13 in humans, from *MT1a* to *MT1m*) [15].

At first, MTs were considered structurally conserved among vertebrates; recent studies however have demonstrated a greater complexity than expected, resulting in a certain confusion in the current nomenclature of the different MT types [16,17]. Indeed, mammalian MT4 and avian MT1 are the most ancestral isoforms in tetrapods, the MT1 and MT2 isoforms do not have a monophyletic origin, and finally, mammalian MT3 clusters form a sister group of the amphibian MT clade [18]. In the latter, an atypical MT has recently been identified, characterised by a lower number of cysteine residues (17 instead of 20) and by two Cys–Cys–Cys motifs [19].

Although a plethora of studies have attempted to clarify MT evolution in vertebrates [15–21], little attention has been dedicated to the MTs in non-mammalian vertebrates, particularly in neglected tetrapods such as reptiles. Therefore, to shed some light on the presence and function of this protein in adult and embryos of an amniotic egg-laying tetrapod, we have undertaken studies using as experimental model the Italian wall lizard *Podarcis sicula* (Figure 1).

Figure 1. Adult specimens of *Podarcis sicula* during copulation.

Here, we provide a summary of the current understanding on MTs' presence and function, highlighting the results obtained by studying its expression by molecular and cytological approaches.

2. *P. sicula* Metallothionein

The cDNA encoding the *P. sicula* MT was firstly isolated from the liver (GenBank accession number AJ609541) [22]. The protein (UniProt Q708T3) is made of 63 amino acids with 20 cysteines arranged in a fashion typical of vertebrates MT1 and MT2; *P. sicula* MT shows about 80% identity and 85% similarity with both mammalian and avian MTs (Figure 2). To date, no other isoforms have been found, not in other tissues (including the brain), or after cadmium induction [20,23].

Figure 2. Graphical representation of amino acidic sequence similarity among rat, chicken, and lizard metallothioneins (MTs).

3. Metallothionein in *P. sicula* Adult Tissues

Gene expression analyses performed on tissues of *P. sicula* adult specimens collected in rural areas show that MT transcripts are present in all of the five tissues examined (Table 1). Interestingly, the highest amount of transcript is found in the brain, followed by the two detoxifying organs, liver and kidney, then the ovary and the gut (brain > liver > kidney > ovary > gut) [23]. In this regard, we should point out that there is only one MT isoform in lizard brain; therefore it can be postulated that its abundance would permit those functions that in mammalian brains are accomplished by MT1/MT2 and MT3.

Table 1. Abundance of metallothionein transcript and protein in *P. sicula* tissues, under natural conditions and after in vivo Cd-exposure.

Tissue	Control mRNA/Protein	Acute Cd-Exposure mRNA/Protein	Chronic Cd-Exposure mRNA/Protein
Brain	+++/n.d.	No change/n.d.	No change/n.d.
Liver	++/+++	+3-fold/+2-fold	No change/+2-fold
Kidney	++/n.d.	+4-fold/n.d.	+4-fold/n.d.
Ovary	+/not present	+3-fold/+++	+4-fold/++
Gut	+/n.d.	n.d./n.d.	+30-fold/n.d.

n.d. = not determined. MT relative abundance : + < ++ < +++.

To test short- and long-term Cd effects on MT expression, adult lizards were exposed to acute or chronic treatments [23–26]. In the former case, a single intra-peritoneal Cd administration (2 μg/g body mass) increases MT transcript content (Table 1), in concomitance with the accumulation of Cd ions [23]. The only exception is represented by the brain, in which the MT mRNA content remains unchanged. We have already demonstrated that both intra-peritoneally and dietary administered cadmium crosses the blood-brain barrier and accumulates in the brain [23]. Hence, it is evident that the lack of induction is not due to the absence of Cd ions. However, since the analysis is carried out on the whole organ, it cannot be excluded that up and down regulations occurred in different brain districts, leaving the total measured concentration unaltered. This possibility is now under investigation by cytological analyses.

In the liver, a threefold increase in both MT transcript and protein is observed (Table 1); however, this Cd-induction does not alter MT mRNA localisation. MT transcripts in fact are detected exclusively in the large Kupffer cells and in monocytes in vessels, whereas the hepatocytes remain completely unstained, both under natural conditions and after Cd-treatment [26].

In the ovary, a single intra-peritoneal Cd-treatment gives rise to a threefold increase in MT transcript content after two days from administration. In this case, the cytological data demonstrate that transcript localisation significantly changes, moving from small follicular stem cells to large nurse cells [27].

The response to oral cadmium uptake is also different in the various tissues examined. The chronic dietary treatment (1 μg/g body mass, every second day, for 60 days) induces *MT* gene expression in the gut, kidney, and ovary, whereas it is ineffective in the liver and brain (Table 1). Possibly, the high constitutive amount of MT transcript present in these organs blocks any further increase of *MT* gene transcription.

Taken together, data indicate that under these experimental conditions, in *P. sicula* the relationship between Cd accumulation and MT transcript induction/concentration is not always predictable, depending on the organ and/or on the exposure route. The conclusion, therefore, is that monitoring MT mRNA is not necessarily a suitable tool for detecting environmental cadmium contamination. This conclusion is reinforced by the observation, in invertebrates and vertebrates, that the translational inhibition of MT mRNA transcripts occurs. Determining in parallel the changes in MT protein concentration might partly prevent misinterpretations in monitoring studies.

4. Metallothionein in *P. sicula* Reproductive Organs

4.1. Metallothionein in the Ovary

P. sicula is an oviparous species, characterised by a seasonal reproductive cycle with low gonadal activity in autumn-winter and maximal gonadal activity in spring-summer, coinciding, in females, with two or three ovulatory waves [28]. In this period, due to the rising temperature, typical morpho-physiological modifications occur in the ovary; these include the increase in the number and size of follicles, and the onset of vitellogenesis. At the end of the ovulatory period, the ovary enters a fall stasis until November, when a temporary recrudescence occurs. No vitellogenic follicles are produced, and the ovary enters a new stasis that lasts until the next spring [29].

Beside yolk, during oogenesis the oocytes store RNAs and organelles [30], and micronutrients such as copper and zinc [22] to support the early stages of embryo development. The rate of metals uptake, in particular, is clearly synchronized throughout the stages of maturation, since their increases correlate closely; at the end of oocyte maturation, the copper and zinc content in *Podarcis* eggs is similar, suggesting comparable requirements for both metals during embryogenesis [22].

Interestingly enough, we found that in *P. sicula* oocytes and eggs both zinc and copper ions are associated with high molecular mass metal-binding proteins, whilst no metal is bound to low molecular mass proteins. The MTs are therefore absent throughout the ovary, in germ cells as in somatic follicle cells. However, MT transcripts are present: Northern blot analysis in fact demonstrates that they are expressed constitutively in the ovary, in all periods of the ovarian cycle, and that they accumulate during the reproductive period in ovaries containing large vitellogenic follicles approaching ovulation [22]. By using the in situ hybridisation technique, we have clarified where the MT mRNA is localised. In small previtellogenic follicles (<150 μm diameter), it localises only in the oocyte's cytoplasm and not in the small stem cells forming the follicular epithelium. Later, in mid previtellogenesis (follicles 400–1400 μm diameter), the small stem cells, and to a lesser extent the differentiated pyriform cells, contain the MT transcripts. In even larger previtellogenic follicles (1.5–2 mm), the MT mRNA are localised in the small cells, and to a lesser extent in the regressing pyriform cells. In vitellogenic follicles (>2 mm diameter), pyriform cells are no longer present, and all the MT mRNA is localised in the cytoplasm of the small cells [27,31]. In the oocyte cytoplasm, the hybridisation signal is undetectable, indicating at first that no MT mRNA is present. However, the oocyte volume is very large, and the transcripts are dispersed and probably masked by the large yolk platelets (Figure 3). Their presence in fact is confirmed by molecular techniques [22]. The origin of the oocyte transcripts is also intriguing. The oocyte nucleus is silent since the early diplotene stage; however, the large pyriform cells synthesize RNAs, organelles, and cytoplasm that transfer into the oocyte cytoplasm via intercellular bridges at the end of previtellogenesis, before regressing [30].

The absence of protein indicates that the egg transcripts in *Podarcis* are stored to meet the future needs of the growing embryo. In effect, the MT mRNAs are still present in the discoblastulae [31]. Though in other species translation is resumed soon after fertilization [32], an asynchronous presence of transcripts and protein is rather common, and described, for example, in the embryos of the Roman snail *Helix pomatia* [33].

The induction of MT synthesis in ovaries following in vivo exposure to cadmium indicates that the translation of blocked MT mRNA can be triggered if an anomalous increase in metal content occurs [22]. This event is accompanied not only by an expected increase in MT expression (to restore the transcript pool), but, more significantly, by a change in transcript localisation, from small stem cells to differentiated pyriform cells [27]. The metal therefore seems to induce changes also in the function of the ovarian MTs.

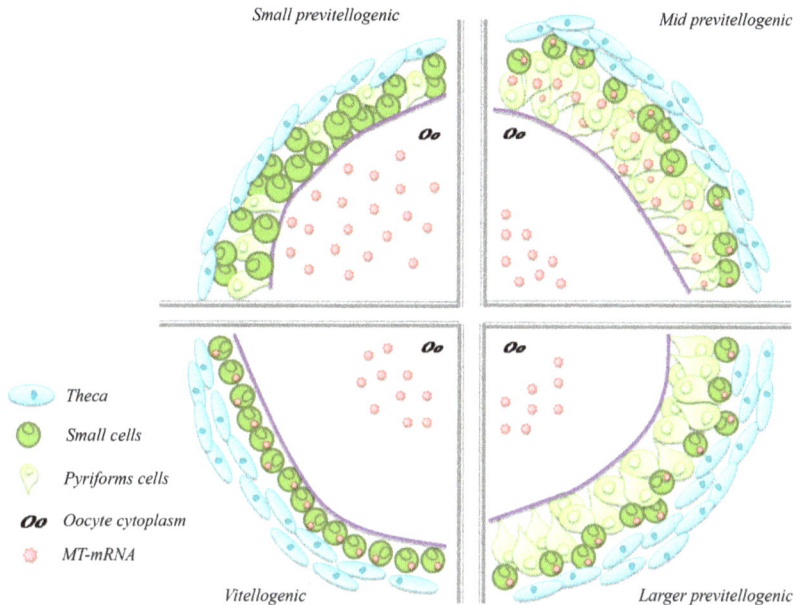

Figure 3. Schematic representation of the MT mRNA localisation in *P. sicula* follicles during oogenesis.

4.2. Metallothionein in the Testis

The presence of MT in the testis of vertebrates has been a matter of debate for a long time, and for many years authors have reported a substantial lack of the protein [34,35]. Today, both MT expression and synthesis in testis has been widely demonstrated, even though cell localisation is still contradictory [36–40]. Evidence exists on MT localisation in both somatic and germ cells, but in line with what occurs in the ovary, the localisation changes over time, probably depending on the state of the gonad. In mouse and rat testes, high levels of MT transcripts are not always accompanied to MT protein co-localisation [41], as demonstrated for the *Podarcis* ovary.

Due to the lack of information, we decided to investigate the presence and the localisation of MT transcript and protein in the testis of *P. sicula*, in two representative phases of the annual reproductive cycle, to highlight the role of this protein during spermatogenesis. The *P. sicula* testis shows a tubular structure, like mammals. During the reproductive period, in spring-summer, the seminiferous epithelium contains germ cells in all of the stages of maturation, from spermatogonia to spermatozoa, and many sperms fill the lumen of the tubules. This period is followed by: (1) a late-summer stasis, characterised by the block of spermatogenesis; (2) an autumnal resumption, favoured by the increase in circulating testosterone, with the renewal of a defective spermatogenesis without spermiation; (3) a winter stasis; (4) an early spring resumption [42,43].

We performed in situ hybridisation and immunocytochemical analyses on *P. sicula* testis in the reproductive period and autumnal resumption. The results demonstrate the presence of MT transcript and protein in testicular cells; however, MT expression and synthesis are different in the two phases of the reproductive cycle [44].

In the mating period, spermatogonia, primary and secondary spermatocytes, and spermatids contain both MT transcript and protein, whereas spermatozoa show only the protein. This finding suggests that the MTs are translated in the earlier stages and the protein is stored, at least partly, for spermatozoa (Figure 4).

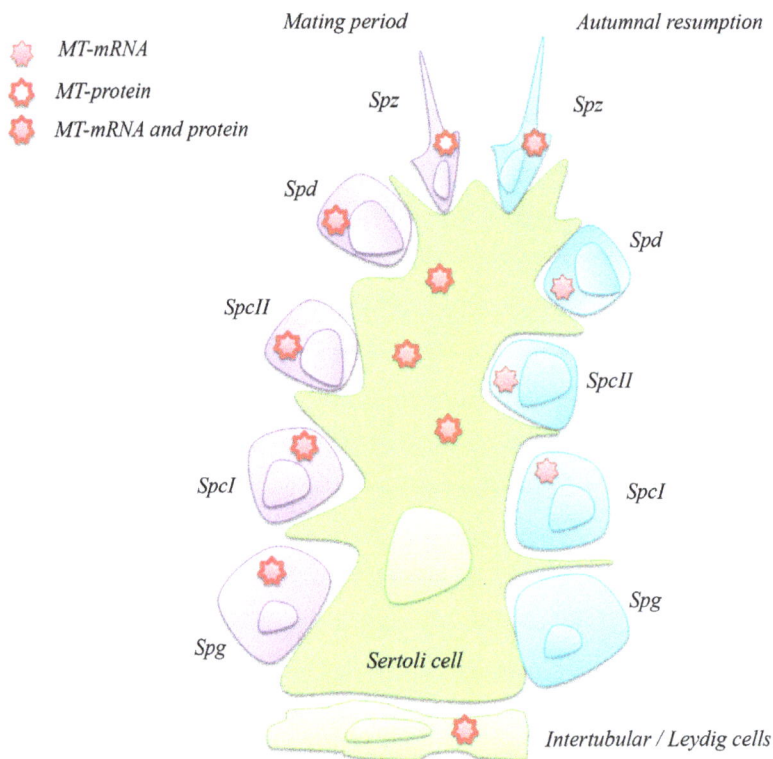

Figure 4. Schematic representation of MT mRNA and protein localisation in *P. sicula* testis during two stages of reproductive cycle. Spg, spermatogonia; SpcI, primary spermatocytes; SpcII, secondary spermatocytes; Spd, spermatids; Spz, spermatozoa.

During the autumnal resumption, MT mRNA is present in primary and secondary spermatocytes, spermatids, and spermatozoa, but not in spermatogonia; conversely, the protein is evident only in spermatozoa (Figure 4). The spermatogonia are in a resting phase, the recruitment of new gametes being unnecessary; the spermatocytes and spermatids, probably left from the previous spermatogenetic wave, are also quiescent but maintain their MT mRNA untranslated after having depleted the MT protein store. The spermatozoa, almost certainly left from the past reproductive episode, maintain both transcripts and protein, a condition recalling that of summer spermatids. It would be interesting to analyse the ultrastructure of these cells to demonstrate their true maturation stage.

What is clear from these data is that many regulatory signals exist between transcription and the production of a functional protein, and that they can be conveniently studied in *Podarcis*, an animal model possessing a single MT protein isoform.

In Sertoli and intertubular cells, including the Leydig cells, MT transcript and protein are always present (Figure 4), a fact not surprising when it is considered that these two cell types are involved in the endocrine control of reproduction.

The presence of the MT protein in all spermatogenetic stages during the reproductive period may indicate a key role of MT in germ cells' maturation, in particular in zinc distribution in the seminiferous epithelium. It is known that zinc ions, interacting with spermatozoa, are involved in sperm motility and viability, and the deficiency of this micronutrient leads to the atrophy of germ cells and a failure of sperm spawning [45,46]. In addition, a reversible metal exchange between MT and the zinc finger motifs of the estrogen receptors has been demonstrated; hence, the presence of MT in all spermatogenic cells during the mating period may ensure the proper zinc availability required for the correct functionality of the estrogen receptors [47].

5. Metallothionein in *P. sicula* Development

5.1. Metallothionein Expression in P. sicula Embryos under Natural Conditions

Despite it being demonstrated that null-MT mice are viable and reproduced normally [48], under natural conditions *MT* genes are precociously expressed in embryos, thus reinforcing the idea that MTs play critical functions during embryonic development. In mammals, MT transcripts are present as maternal messengers stored in eggs, and the embryonic *MT* genes are among the first to be transcribed, first in blastocyst and then in organs such as the liver and kidney [49,50].

As described above, in *P. sicula* eggs, MT mRNA accumulates abundantly; as the next step, we decided to investigate by in situ hybridisation the spatiotemporal changes in transcript localisation throughout development, from egg deposition to hatching (about 50 days later). Freshly laid eggs were incubated in a terrarium maintained at natural conditions of humidity and temperature, and embryos collected at regular time intervals [31].

In *Podarcis* blastula, collected when the fertilised eggs are still in oviducts, MT transcripts of maternal origin are present in both embryonic and extra-embryonic areas.

Later, at the beginning of organogenesis (5–10 days post deposition (pd)), MT transcripts are present in the cytoplasm of the developing neurons of the central nervous system and in the undifferentiated retina. Transcripts are also present in mesodermal derivates such as somites and mesenchyme, but not in the kidney or in endodermal derivates like the liver and gut (Figure 5).

At 20 days from deposition, the MT-mRNA distribution slightly changes, and from this moment on the embryonic gene is expressed, at least in some regions. In the brain, for example, where the typical anatomical districts are recognisable, transcripts are present in the ventricular zones of telencephalon and diencephalon, in the developing optic cortex and in several nuclei of mesencephalon and medulla oblongata, in the ependymal cells of all vesicles, and in the horns of the spinal cord. In the retina, the MT mRNA is present in the ganglion cells and in the inner and outer nuclear layers' cells. In the trunk, only the renal glomeruli show the presence of the MT transcript, whereas in the renal tubules, and lung, liver, and gut mucosa they are still absent (Figure 5).

At 40 days pd, MT transcripts simultaneously disappear from the telencephalon, diencephalon, mesencephalon, and ganglion layers of the differentiating retina and the grey matter of the spinal cord. However, they remain in the nuclei of the medulla oblongata, in the granular and molecular layers of the cerebellum, and appear, for the first time, in the white matter of the spinal cord. This shift in MT localisation in the central nervous system could be explained by a consumption of the pool of maternal mRNAs and/or a temporary silencing of the MT embryonic gene. In the trunk region, the MT mRNA is present in the renal glomeruli and for the first time in the lung parenchyma, whereas it is still absent in the liver and gut mucosa and the kidney tubules (Figure 5).

Immediately before hatching, at 50–55 days pd, the distribution of MT transcripts undergoes further relevant changes. Regarding the central nervous system, they appear in the cortical areas of the telencephalon and mesencephalon, in several basal nuclei, and in ependymal cells. Their distribution in the medulla oblongata and in the spinal cord does not change with respect to the previous stage. In the fully differentiated retina, the MT mRNA is present in the outer and inner nuclear layers, and again in the ganglion cells layer. In the other organs, transcripts are observed for the first time in the gut mucosal cells and in the liver, but not yet in the kidney tubules (Figure 5).

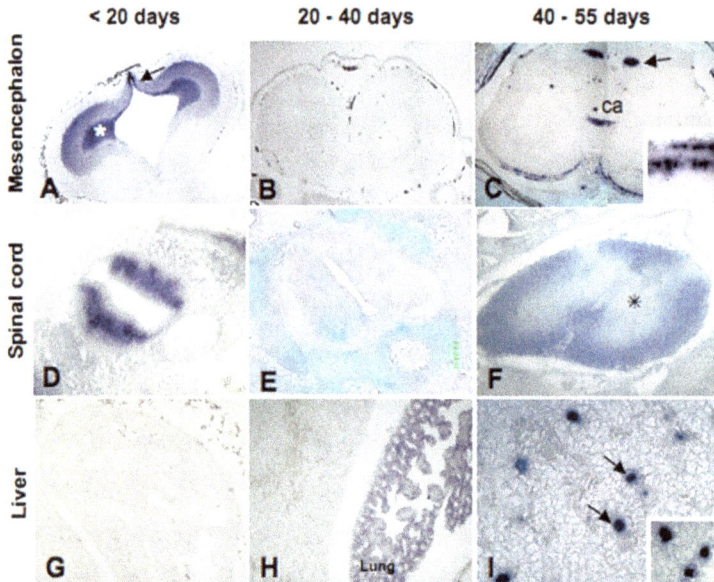

Figure 5. Changes in MT mRNA localisation during *P. sicula* development. Photographs are modified from [31]. (**A–C**) Cross-section of the optic lobes at different stages. (**A**) Mesencephalon at 20 days post deposition (pd): MT mRNA labeling is concentrated in the developing cortex (*); the dorsal ventricular zone is intensely labeled (arrow); (**B**) At 40 days pd, mesencephalon is completely unlabeled; (**C**) In the pre-hatching embryo the transcript is on the cortex (insert), commissura ansulata (ca) and in the basal nuclei (arrow). (**D–F**) Cross-section of spinal cord. (**D**) Neural tube at 5 days pd, with positive cells concentrated in the developing grey matter; (**E**) Unlabeled neural tube; (**F**) Neural tube with unlabeled grey matter (*) and labeled white matter; (**G–I**) Cross-section of liver. Messenger is absent at 10 (**G**) and 40 days pd (**H**); parenchymal lung cells are labeled at 40 days. In the pre-hatching embryos (**I**) Kupffer cells are positive (arrows, insert). Objectives: (**A–F**): 10×; (**G–I**): 20×; (inserts **C,I**): 40×.

Taken together, these results show a complex modulation of *MT* gene expression during lizard development, along with several differences when compared with mammalian embryos, in which the expression of *MT* gene is activated early and abundantly in the visceral organs with detoxifying functions, whereas the gene is activated late, at birth, in the brain [51].

5.2. Metallothionein Expression in P. sicula Embryos Incubated in Cadmium-Contaminated Soil

The absence of MT transcripts in the gut, liver, and kidney of *Podarcis* embryos might be correlated to the peculiarities of their development: indeed, as all terrestrial oviparous vertebrates, they develop and grow using the nutrients inside the eggs, and detoxifying organs are not yet required.

Many studies indicate that MTs protect embryos from intracellular damage caused by harmful substances, particularly heavy metals [52]. Embryo potential exposure and uptake to toxic heavy metals, such as cadmium, varies greatly: in mammals, the placenta is the primary target [53], whereas in oviparous tetrapods, vitellogenesis is considered the primary route of accumulation of toxic metals in eggs [54]. However, it has been demonstrated that metals and other soluble organic contaminants are able to cross the breathable eggshell of many reptiles, affecting embryos' morphology and gene expression [55–57].

So, we decided to evaluate the spatiotemporal changes in MT expression in embryos incubated in Cd contaminated soil (50 mg Cd/Kg soil) [58]. A preliminary histological analysis demonstrated no embryo mortality, but the onset of severe morphological alterations mostly concentrated in the encephalic and optic areas; no alterations were observed in the medulla oblongata, spinal cord and visceral organs throughout development. Interestingly, in situ hybridisation analyses demonstrate a Cd-induced spatiotemporal shift in embryonic *MT* gene expression in some trunk organs but not in the developing brain and retina. In particular, MT transcripts were present in the intestinal mucosa and in the liver sinusoids in 10-day-old embryos (Figure 6), whereas in the control embryos MT mRNA appears only immediately before hatching.

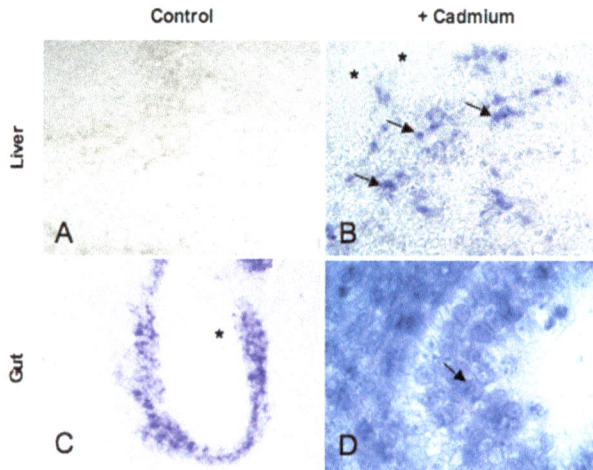

Figure 6. Cd-induction of MT expression in the liver and gut of a *P. sicula* embryo. Photographs are modified from [58]. (**A,C**) Control embryo at 10 days pd. (**A**) Unlabeled liver parenchyma; (**C**) gut mucosa completely unlabeled (*). (**B,D**) Cd-treated embryo at 10 days pd. (**B**) Several sinusoidal cells (arrows) are labeled, liver parenchyma () is unlabeled; (**D**) gut mucosal cells show intensely labeled nuclei (arrow). Objectives: (**A,B**): 20×; (**C**): 10×; (**D**): 40×.

These results support the hypothesis that the MT mRNA is absent in detoxifying organs, since they are not functional during development; however, if solicited by toxic substances, the gene is activated and transcription starts. On the other hand, the morphological alterations observed demonstrate that the cephalic area of Cd-treated embryos is exposed to the toxic effects of cadmium. At this point it is necessary to determine whether the transcripts are translated or, alternatively, whether proteins translated from these transcripts are engaged in functions other than heavy metal detoxification.

6. Conclusions

In adult lizard *Podarcis sicula* the MT protein is present in almost all tissues, as expected considering its fundamental role in modulating intracellular zinc/copper concentrations. Exogenous metals, and cadmium in particular, may trigger a significant upregulation, but the timing and amount of newly synthesised mRNA and/or protein is clearly dose- and tissue- specific. Further analyses are needed to draw conclusions on the relevance of measuring the MT transcripts in some tissues of *P. sicula* to reveal Cd contamination, and to determine if the proteins translated from these transcripts are engaged in functions other than heavy metal detoxification. The particularly complex pattern observed in the tissue distribution and the response to cadmium exposure suggests that MTs might have many different roles, some probably still to be discovered. The elusive MT therefore is far from being known in full, especially in this reptile.

Conflicts of Interest: The authors declare no conflict of interest.

References

1. Margoshes, M.; Vallee, B.L. A cadmium protein from equine kidney cortex. *J. Am. Chem. Soc.* **1957**, *79*, 4813–4814. [CrossRef]
2. Bremner, I. Interactions between metallothionein and trace elements. *Prog. Food Nutr. Sci.* **1987**, *11*, 1–37. [PubMed]
3. Coyle, P.; Philcox, J.C.; Carey, L.C.; Rofe, A.M. Metallothionein: The multipurpose protein. *Cell. Mol. Life Sci.* **2002**, *59*, 627–647. [CrossRef] [PubMed]
4. Palmiter, R.D. The elusive function of metallothioneins. *Proc. Natl. Acad. Sci. USA* **1998**, *95*, 8428–8430. [CrossRef] [PubMed]
5. Hamer, D.H. Metallothionein. *Annu. Rev. Biochem.* **1986**, *55*, 913–951. [CrossRef] [PubMed]
6. Isani, G.; Carpenè, E. Metallothioneins, unconventional proteins from unconventional animals: A long journey from nematodes to mammals. *Biomolecules* **2014**, *4*, 435–457. [CrossRef] [PubMed]
7. Baurand, P.E.; Pedrini-Martha, V.; de Vaufleury, A.; Niederwanger, M.; Capelli, N.; Scheifler, R.; Dallinger, R. Differential expression of metallothionein isoforms in terrestrial snail embryos reflects early life stage adaptation to metal stress. *PLoS ONE* **2015**, *10*, e0116004. [CrossRef] [PubMed]
8. Palacios, O.; Pérez-Rafael, S.; Pagani, A.; Dallinger, R.; Atrian, S.; Capdevila, M. Cognate and noncognate metal ion coordination in metal-specific metallothioneins: The *Helix pomatia* system as a model. *J. Biol. Inorg. Chem.* **2014**, *19*, 923–935. [CrossRef] [PubMed]
9. Höckner, M.; Dallinger, R.; Stürzenbaum, S.R. Nematode and snail metallothioneins. *J. Biol. Inorg. Chem.* **2011**, *16*, 1057–1065. [CrossRef] [PubMed]
10. Palacios, O.; Pagani, A.; Pérez-Rafael, S.; Egg, M.; Höckner, M.; Brandstätter, A.; Capdevila, M.; Atrian, S.; Dallinger, R. Shaping mechanisms of metal specificity in a family of metazoan metallothioneins: Evolutionary differentiation of mollusc metallothioneins. *BMC Biol.* **2011**, *9*, 4. [CrossRef] [PubMed]
11. Sutherland, D.E.; Stillman, M.J. The "magic numbers" of metallothionein. *Metallomics* **2011**, *3*, 444–463. [CrossRef] [PubMed]
12. Hogstrand, C.; Haux, C. Binding and detoxification of heavy metals in lower vertebrates with reference to metallothionein. *Comp. Biochem. Physiol. C* **1991**, *100*, 137–141. [CrossRef]
13. Uchida, Y.; Takio, K.; Titani, K.; Ihara, Y.; Tomonaga, M. The growth inhibitory factor that is deficient in the Alzheimer's disease brain is a 68 amino acid metallothionein-like protein. *Neuron* **1991**, *7*, 337–347. [CrossRef]
14. Quaife, C.J.; Findley, S.D.; Erickson, J.C.; Froelick, G.J.; Kelly, E.J.; Zambrowicz, B.P.; Palmiter, R.D. Induction of a new metallothionein isoform (MT-IV) occurs during differentiation of stratified squamous epithelia. *Biochemistry* **1994**, *33*, 7250–7259. [CrossRef] [PubMed]
15. Moleirinho, A.; Carneiro, J.; Matthiesen, R.; Silva, R.M.; Amorim, A.; Azevedo, L. Gains, losses and changes of function after gene duplication: Study of the metallothionein family. *PLoS ONE* **2011**, *6*, e18487. [CrossRef] [PubMed]
16. Scudiero, R.; Filosa, S.; Trinchella, F. Metallothionein gene evolution in vertebrates: Events of gene duplication and loss during squamates diversification. In *Advances in Medicine and Biology*; Berhardt, L.V., Ed.; Nova Science Publishers: Happauge, NY, USA, 2011; Volume 24, pp. 321–335.
17. Serén, N.; Glaberman, S.; Carretero, M.A.; Chiari, Y. Molecular evolution and functional divergence of the metallothionein gene family in vertebrates. *J. Mol. Evol.* **2014**, *78*, 217–233. [CrossRef] [PubMed]
18. Scudiero, R. Unexpected diversity of metallothionein primary structure in Amphibians: Evolutionary implications for vertebrate metallothioneins. In *Amphibians: Anatomy, Ecological Significance and Conservation Strategies*; Lombardi, M.P., Ed.; Nova Science Publishers: Hauppauge, NY, USA, 2014; pp. 27–38, ISBN 9781633214347.
19. Scudiero, R.; Tussellino, M.; Carotenuto, R. Identification and expression of an atypical isoform of metallothionein in the African clawed frog *Xenopus laevis*. *C. R. Biol.* **2015**, *338*, 314–320. [CrossRef] [PubMed]
20. Trinchella, F.; Riggio, M.; Filosa, S.; Parisi, E.; Scudiero, R. Molecular cloning and sequencing of metallothionein in squamates: New insights into the evolution of the metallothionein genes in vertebrates. *Gene* **2008**, *423*, 48–56. [CrossRef] [PubMed]

21. Trinchella, F.; Esposito, M.G.; Scudiero, R. Metallothionein primary structure in amphibians: Insights from comparative evolutionary analysis in vertebrates. *C. R. Biol.* **2012**, *335*, 480–487. [CrossRef] [PubMed]

22. Riggio, M.; Trinchella, F.; Parisi, E.; Filosa, S.; Scudiero, R. Accumulation of zinc, copper and metallothionein mRNA in lizard ovary proceeds without a concomitant increase in metallothionein content. *Mol. Reprod. Dev.* **2003**, *66*, 347–382. [CrossRef] [PubMed]

23. Trinchella, F.; Riggio, M.; Filosa, S.; Volpe, M.G.; Parisi, E.; Scudiero, R. Cadmium distribution and metallothionein expression in lizard tissues following acute and chronic cadmium intoxication. *Comp. Biochem. Physiol. C Toxicol. Pharmacol.* **2006**, *144*, 272–278. [CrossRef] [PubMed]

24. Scudiero, R.; Filosa, S.; Motta, C.M.; Simoniello, P.; Trinchella, F. Cadmium in the wall lizard *Podarcis Sicula*: Morphological and molecular effects on embryonic and adult tissues. In *Reptiles: Biology, Behavior and Conservation*; Baker, K.J., Ed.; Nova Science Publishers: Happauge, NY, USA, 2011; pp. 147–162, ISBN 978-1-61324-740-2.

25. Verderame, M.; Limatola, E.; Scudiero, R. The terrestrial lizard *Podarcis sicula* as experimental model in emerging pollutants evaluation. In *Ecotoxicology and Genotoxicology: Non-Traditional Terrestrial Models*; Larramendy, M., Ed.; Royal Society of Chemistry (RSC) Publishing: London, UK, 2017; Chapter 12, ISBN 978-1-78262-811-8.

26. Simoniello, P.; Filosa, S.; Riggio, M.; Scudiero, R.; Tammaro, S.; Trinchella, F.; Motta, C.M. Responses to cadmium intoxication in the liver of the wall lizard *Podarcis Sicula*. *Comp. Biochem. Physiol. C Toxicol. Pharmacol.* **2010**, *151*, 194–203. [CrossRef] [PubMed]

27. Simoniello, P.; Filosa, S.; Scudiero, R.; Trinchella, F.; Motta, C.M. Cadmium impairment of reproduction in the female wall lizard *Podarcis sicula*. *Environ. Toxicol.* **2013**, *28*, 553–562. [CrossRef] [PubMed]

28. Carnevali, O.; Mosconi, G.; Angelini, F.; Limatola, E.; Ciarcia, G.; Polzonetti-Magni, A. Plasma vitellogenin and 17β-estradiol levels during the annual reproductive cycle of *Podarcis s. sicula* Raf. *Gen. Comp. Endocrinol.* **1991**, *84*, 337–343. [CrossRef]

29. Borrelli, L.; de Stasio, R.; Motta, C.M.; Parisi, E.; Filosa, S. Seasonal-dependent effect of temperature on the response of adenylate cyclase to FSH stimulation in the oviparous lizard, *Podarcis sicula*. *J. Endocrinol.* **2000**, *167*, 275–280. [CrossRef] [PubMed]

30. Motta, C.M.; Castriota Scanderbeg, M.; Filosa, S.; Andreuccetti, P. Role of pyriform cells during the growth of oocytes in the lizard *Podarcis sicula*. *J. Exp. Zool. A Ecol. Genet. Physiol.* **1995**, *273*, 247–256. [CrossRef]

31. Simoniello, P.; Motta, C.M.; Scudiero, R.; Trinchella, F.; Filosa, S. Spatiotemporal changes in metallothionein gene expression during embryogenesis in the wall lizard *Podarcis sicula*. *J. Exp. Zool.* **2010**, *313A*, 410–420. [CrossRef] [PubMed]

32. De Moor, C.H.; Richter, J.D. Translational control in vertebrate development. *Int. Rev. Cytol.* **2001**, *203*, 567–608. [CrossRef] [PubMed]

33. Pedrini-Martha, V.; Schnegg, R.; Baurand, P.E.; de Vaufleury, A.; Dallinger, R. The physiological role and toxicological significance of the non-metal-selective cadmium/copper-metallothionein isoform differ between embryonic and adult helicid snails. *Comp. Biochem. Physiol. C Toxicol. Pharmacol.* **2017**, *22*. [CrossRef] [PubMed]

34. Waalkes, M.P.; Chernoff, S.B.; Klaassen, C.D. Cadmium-binding proteins of rat testes. Characterization of a low-molecular-mass protein that lacks identity with metallothionein. *Biochem. J* **1984**, *220*, 811–818. [CrossRef] [PubMed]

35. Waalkes, M.P.; Perantoni, A.; Palmer, A.E. Isolation and partial characterization of the low-molecular-mass zinc/cadmium-binding protein from the testes of the patas monkey (*Erythrocebus patas*). Distinction from metallothionein. *Biochem. J.* **1988**, *256*, 131–137. [CrossRef] [PubMed]

36. Nishimura, H.; Nishimura, N.; Tohyama, C. Localization of metallothionein in the genital organs of the male rat. *J. Histochem. Cytochem.* **1990**, *38*, 927–933. [CrossRef] [PubMed]

37. De, S.K.; Enders, G.C.; Andrews, G.K. High levels of metallothionein messenger RNAs in male germ cells of the adult mouse. *Mol. Endocrinol.* **1991**, *5*, 628–636. [CrossRef] [PubMed]

38. Tohyama, C.; Nishimura, N.; Suzuki, J.S.; Karasawa, M.; Nishimura, H. Metallothionein mRNA in the testis and prostate of the rat detected by digoxigenin-labeled riboprobe. *Histochemistry* **1994**, *101*, 341–346. [CrossRef] [PubMed]

39. Suzuki, J.S.; Kodama, N.; Molotkov, A.; Aoki, E.; Tohyama, C. Isolation and identification of metallothionein isoforms (MT-1 and MT-2) in the rat testis. *Biochem. J.* **1998**, *334*, 695–701. [CrossRef] [PubMed]

40. Han, Y.L.; Sheng, Z.; Liu, G.D.; Long, L.L.; Wang, Y.F.; Yang, W.X.; Zhu, J.Q. Cloning, characterization and cadmium inducibility of metallothionein in the testes of the mudskipper *Boleophthalmus pectinirostris*. *Ecotoxicol. Environ. Saf.* **2015**, *119*, 1–8. [CrossRef] [PubMed]

41. Durnam, D.M.; Palmiter, R.D. Transcriptional regulation of the mouse *metallothionein-I* gene by heavy metals. *J. Biol. Chem.* **1981**, *256*, 5712–5716. [PubMed]

42. Rosati, L.; Prisco, M.; Coraggio, F.; Valiante, S.; Scudiero, R.; Laforgia, V.; Andreuccetti, P.; Agnese, M. PACAP and PAC1 receptor in the reproductive cycle of male lizard *Podarcis sicula*. *Gen. Comp. Endocrinol.* **2014**, *205*, 102–108. [CrossRef] [PubMed]

43. Verderame, M.; Angelini, F.; Limatola, E. Spermatogenic waves and expression of AR and ERs in germ cells of *Podarcis sicula*. *Int. J. Zool.* **2014**, *2014*. [CrossRef]

44. Verderame, M.; Limatola, E.; Scudiero, R. Metallothionein expression and synthesis in the testis of the lizard *Podarcis sicula* under natural conditions and following estrogenic exposure. *Eur. J. Histochem.* **2017**, *61*, 90–95. [CrossRef]

45. Vallee, B.L.; Falchuk, K.H. The biochemical basis of zinc physiology. *Physiol. Rev.* **1993**, *73*, 79–118. [PubMed]

46. Sørensen, M.B.; Stoltenberg, M.; Henriksén, K.; Ernst, E.; Danscher, G.; Parvinen, M. Histochemical tracing of zinc ions in the rat testis. *Mol. Hum. Reprod.* **1998**, *4*, 423–428. [CrossRef] [PubMed]

47. Cano-Gauci, D.F.; Sarkar, B. Reversible zinc exchange between metallothionein and the estrogen receptor zinc finger. *FEBS Lett.* **1996**, *386*, 1–4. [CrossRef]

48. Kelly, E.J.; Quaife, C.J.; Froelick, G.J.; Palmiter, R.D. Metallothionein I and II protect against zinc deficiency and zinc toxicity in mice. *J. Nutr.* **1996**, *126*, 1782–1790. [PubMed]

49. Andrews, G.K.; Huet-Hudson, Y.M.; Paria, B.C.; McMaster, M.T.; De, S.K.; Dey, S.K. Metallothionein gene expression and metal regulation during preimplantation mouse embryo development (MT mRNA during early development). *Dev. Biol.* **1991**, *145*, 13–27. [CrossRef]

50. Andersen, R.D.; Piletz, J.E.; Birren, B.W.; Herschman, H.R. Levels of metallothionein messenger RNA in foetal, neonatal and maternal rat liver. *Eur. J. Biochem.* **1983**, *131*, 497–500. [CrossRef] [PubMed]

51. Andrews, G.K.; Lee, D.K.; Ravindra, R.; Lichtlen, P.; Sirito, M.; Sawadogo, M.; Schaffner, W. The transcription factors MTF-1 and USF1 cooperate to regulate mouse metallothionein-I expression in response to the essential metal zinc in visceral endoderm cells during early development. *EMBO J.* **2001**, *20*, 1114–1122. [CrossRef] [PubMed]

52. McAleer, M.F.; Tuan, R.S. Cytotoxicant-induced trophoblast dysfunction and abnormal pregnancy outcomes: Role of zinc and metallothionein. *Birth Defects Res. C Embryo Today* **2004**, *72*, 361–370. [CrossRef] [PubMed]

53. Bush, P.G.; Mayhew, T.M.; Abramovich, D.R.; Aggett, P.J.; Burke, M.D.; Page, K.R. A quantitative study on the effects of maternal smoking on placental morphology and cadmium concentration. *Placenta* **2000**, *21*, 247–256. [CrossRef] [PubMed]

54. Guirlet, E.; Das, K.; Thomé, J.P.; Girondot, M. Maternal transfer of chlorinated contaminants in the leatherback turtles, *Dermochelys coriacea*, nesting in French Guiana. *Chemosphere* **2010**, *79*, 720–726. [CrossRef] [PubMed]

55. Marco, A.; López-Vicente, M.; Pérez-Mellado, V. Arsenic uptake by reptile flexible-shelled eggs from contaminated nest substrates and toxic effect on embryos. *Bull. Environ. Contam. Toxicol.* **2004**, *72*, 983–990. [CrossRef] [PubMed]

56. Marco, A.; López-Vicente, M.; Pérez-Mellado, V. Soil acidification negatively affects embryonic development of flexible-shelled lizard eggs. *Herpetol. J.* **2005**, *15*, 107–111.

57. Trinchella, F.; Cannetiello, M.; Simoniello, P.; Filosa, S.; Scudiero, R. Differential gene expression profiles in embryos of the lizard *Podarcis sicula* under in ovo exposure to cadmium. *Comp. Biochem. Physiol. C Toxicol. Pharmacol.* **2010**, *151*, 33–39. [CrossRef] [PubMed]

58. Simoniello, P.; Motta, C.M.; Scudiero, R.; Trinchella, F.; Filosa, S. Cadmium-induced teratogenicity in lizard embryos: Correlation with metallothionein gene expression. *Comp. Biochem. Physiol. C Toxicol. Pharmacol.* **2011**, *153*, 119–127. [CrossRef] [PubMed]

International Journal of
Molecular Sciences

MDPI

Review

Iron, Oxidative Damage and Ferroptosis in Rhabdomyosarcoma

Alessandro Fanzani * and Maura Poli *

Department of Molecular and Translational Medicine (DMMT), University of Brescia, Viale Europa 11, 25123 Brescia, Italy
* Correspondence: alessandro.fanzani@unibs.it (A.F.); maura.poli@unibs.it (M.P.);
    Tel.: +39-030-371-7303 (A.F. & M.P.); Fax: +39-030-371-7305 (A.F. & M.P.)

Received: 28 June 2017; Accepted: 3 August 2017; Published: 7 August 2017

Abstract: Recent data have indicated a fundamental role of iron in mediating a non-apoptotic and non-necrotic oxidative form of programmed cell death termed ferroptosis that requires abundant cytosolic free labile iron to promote membrane lipid peroxidation. Different scavenger molecules and detoxifying enzymes, such as glutathione (GSH) and glutathione peroxidase 4 (GPX4), have been shown to overwhelm or exacerbate ferroptosis depending on their expression magnitude. Ferroptosis is emerging as a potential weapon against tumor growth since it has been shown to potentiate cell death in some malignancies. However, this mechanism has been poorly studied in Rhabdomyosarcoma (RMS), a myogenic tumor affecting childhood and adolescence. One of the main drivers of RMS genesis is the Retrovirus Associated DNA Sequences/Extracellular signal Regulated Kinases (RAS/ERK)signaling pathway, the deliberate activation of which correlates with tumor aggressiveness and oxidative stress levels. Since recent studies have indicated that treatment with oxidative inducers can significantly halt RMS tumor progression, in this review we covered different aspects, ranging from iron metabolism in carcinogenesis and tumor growth, to mechanisms of iron-mediated cell death, to highlight the potential role of ferroptosis in counteracting RMS growth.

Keywords: iron; ferroptosis; oxidative damage; rhabdomyosarcoma

1. Introduction

Iron is the most abundant heavy metal in mammals (about 3–5 g in human adults), as it is involved in a number of biological processes, ranging from metabolism and oxygen transport to DNA synthesis and antioxidant defense. Many redox enzymes involved in cellular respiration use iron-sulfur (Fe-S) clusters as preferred cofactors, named ferredoxins, such as nicotinamide adenine dinucleotide (NADH) dehydrogenase, hydrogenases, coenzyme Q—cytochrome c reductase, succinate-coenzyme Q reductase and other components of the mitochondrial electron transport chain. In addition, heme iron is used for oxygen transport by hemoglobin and myoglobin and for the detoxification of reactive oxygen species (ROS) by catalase and superoxide dismutase enzymes. Often tumor cells show a marked alteration in metabolism leading to intracellular accumulation of iron, which is strongly utilized for tumor growth and angiogenesis [1]. Accordingly, some anti-tumor strategies using metal chelators have been successfully developed [2]; however, iron deprivation is potentially harmful to non-tumor cells, as it may favor cell death by apoptosis [3–5]. Recently, an iron-dependent type of programmed cell death has been identified, named ferroptosis [6–8]. High intracellular iron concentrations can trigger ferroptosis by enhancing the generation of lipid peroxides, and this can be reverted using iron chelators [9]. Currently, the use of ferroptosis as a weapon against tumors is of increasing interest, as tumor cells traditionally exhibit high endogenous oxidative stress levels due to gene aberrations promoting continuous cell cycles (gain of RAS, and myelocytomatosis viral related oncogene MYC) and resistance to cell senescence (P53 loss) [10]. In the next paragraphs we will describe the complex role of iron in cancer and discuss the available evidence on iron and ferroptosis in rhabdomyosarcoma (RMS), the most frequent soft tissue tumor affecting patients of pediatric and adolescent age.

2. The Pleiotropic Role of Iron in Cancer

Deregulation of iron homeostasis may have a different impact in cancer depending on the stage of tumor progression, as summarized in Figure 1. As detailed below, iron levels influence carcinogenesis, tumor progression and sensitivity to ferroptosis.

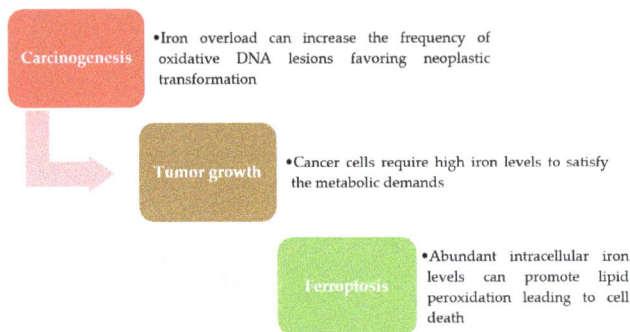

Figure 1. Iron in cancer. Iron can promote carcinogenesis by oxidative stress that increases DNA damage. Following neoplastic transformation, tumors utilize various mechanisms to maintain the high intracellular iron free levels necessary for tumor growth. Over time the iron overload could become deleterious by inducing lipid peroxidation and ferroptosis.

2.1. Iron-Induced Oxidative Stress Plays a Role in Carcinogenesis

Iron overload has been associated with a higher risk of carcinogenesis, as observed in some pathological conditions, including hereditary hemochromatosis, ovarian endometriosis, chronic inflammation induced by viral hepatitis B/C, and exposure to foreign asbestos nanoparticles [11,12]. This is due to the ability of iron to promote DNA damage, as described in a model of renal carcinogenesis [13]. In particular, the most frequent DNA damage is the 8-oxo-7,8-dihydro-2′-deoxyguanosine (8-oxo-dG), resulting from the oxidation of guanine, which potently induces G:C→T:A transversion mutations. Several enzymes involved in DNA repair have been identified, such as the 8-oxoguanine DNA glycosylase 1 (OGG1) [14,15] and a homologue MutT variant first isolated in a mutant strain of *E. coli* [16]. Interestingly, patients with chronic hepatitis C, who have abnormally high levels of 8-oxo-dG and repair enzymes, are protected from the formation of pre-neoplastic lesions and hepatocellular carcinoma by phlebotomy and a low iron diet [17,18]. Other studies further confirmed that phlebotomy twice a year for five years significantly protects against cancer events [19]. Altogether these data indicate that iron-induced oxidative stress can represent a critical factor for carcinogenesis induced by DNA mutagenesis [20–22].

2.2. Iron Addiction Is a Hallmark of Cancer Cells

In humans, iron homeostasis is under the control of mechanisms that coordinate the absorption, export, storage, transport and utilization of iron. The amount of iron circulating in serum and available to tissue may originate from the diet (about 1–2 mg/day), the recycling of hemoglobin by macrophages (about 20 mg/day) and hepatic stores (0.5–1 g) [23]. Iron release from these sources is controlled by hepcidin, a circulating 25 amino acid peptide hormone that reduces systemic iron availability via the binding and degradation of ferroportin (FPN), the only known cellular iron exporter [24]. Dietary iron absorption is mediated by the divalent metal transporter (DMT1) and the duodenal cytochrome b (Dcytb), both iron-regulated [25]. Plasma iron is delivered by transferrin to all tissues presenting the transferrin receptor 1 (TfR1), which mediates its endocytosis [23]. Iron is then reduced and delivered throughout the cytosol to mitochondria for the synthesis of heme groups, Fe/S complexes and iron enzymes, whereas the excess is sequestered and stored by ferritins [26] (Figure 2). The amount of iron bound to ferritins (up to 4500 atoms) can be recycled via a recently identified mechanism mediated by a nuclear receptor coactivator 4 (NCOA4), which targets H-ferritin to lysosomal degradation [27]. As a result of these coordinated events, in non-tumor cells only a minor fraction of free labile iron is present in the cytosol, usually complexed with low molecular weight molecules including glutathione (GSH), citrate, sugars, ascorbate, nucleotides, and also enzymes [23]. On the other hand, an abnormal increase of the intracellular free iron pool is observed in cancer cells, as described in ovarian, breast, lung, prostate, and pancreatic tumors, colorectal hepatoma, gastric and hematological cancers, and melanoma [1]. This effect, commonly referred as to "iron addiction" [1], is the result of the deregulation of different mechanisms. For example, altered MYC expression, which plays a key role in cell transformation, is also responsible for iron metabolism by modulating the activity of the iron responsive protein-2 (IRP2), which in turn orchestrates the expression of different iron proteins [28]. As depicted in Figure 2, cancer cells show increased iron absorption due to high expression of TfR1, a downstream target of MYC oncoprotein [29] and hypoxia inducible factors (HIFs) [30], as observed in breast, renal, and ovarian tumors [31–33]. In addition, the down-regulation of FPN mediated by hepcidin can limit iron export [34], whereas MYC and RAS can promote the release of stored iron by the degradation of H-ferritin [35,36]. Finally, the stroma, endothelial and inflammatory cells composing the tumor niche can release iron to feed the neighboring tumor cells through a concerted upregulation of FPN and down-regulation of ferritin and heme-oxygenase [37].

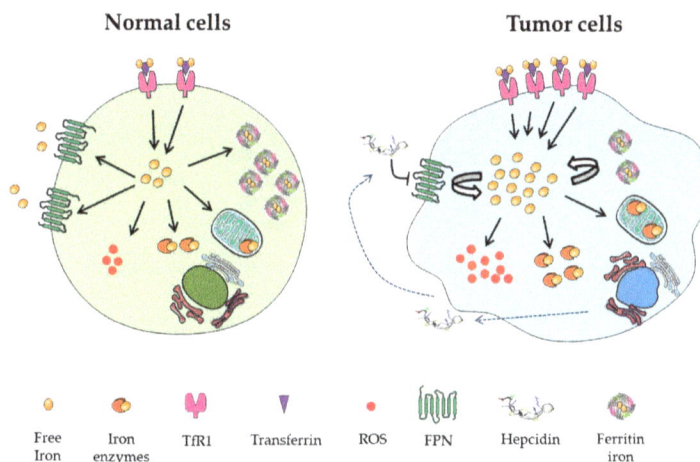

Figure 2. Iron addiction of tumor cells. In normal cells the transferrin receptor 1 (TfR1)-mediated iron absorption is counter balanced by iron efflux via ferroportin (FPN); the free iron pool is used by cytosolic and mitochondrial enzymes and the excess is stored by ferritins to prevent cytotoxicity. As a result, only a minor part of the intracellular iron, present as a free labile pool, can stimulate the formation of Reactive Oxygen Species ROS. In contrast, tumor cells often show higher levels of TfR1, down-regulation of FPN mediated by secreted hepcidin and lower levels of ferritins, which leads to an increased intracellular labile iron pool; despite this it's mostly being utilized for tumor growth by cytosolic and mitochondrial iron enzymes, the exceeding amount can promote increased oxidative stress via ROS accumulation. The figure was adapted using a template on the servier medical art website (available online: www.servier.com) licensed under the creative commons attribution 3.0 unported license (available online: http://creativecommons.org/license/by/3.0/).

2.3. Iron as a Trigger of Ferroptosis in Tumor Cells

Ferroptosis is an iron-dependent form of programmed cell death [6] that differs from canonical apoptosis, necroptosis or autophagy in its morphological features and biochemical pathways [38–40]. As depicted in Figure 3, intracellular iron accumulation yields hydroxyl radicals via the Fenton reaction, therefore promoting the oxidation of polyunsaturated fatty acids (PUFAs, such as linoleic, arachidonic and docosahexaenoic acids). The resulting lipid peroxides and hydroperoxides [41] cause severe structural/functional alterations of cell membranes [8]. Treatments with iron chelators, antioxidant scavengers (like GSH or Vitamin E), and specific inhibitors of lipid peroxidation (like Ferrostatin-1) can prevent ferroptosis activation [6,42]. Moreover, glutathione peroxidase 4 (GPX4), a selenoprotein, protects from ferroptosis [43] as it catalyzes the endogenous neutralization of lipid hydroperoxides (L-OOH) into innocuous lipid alcohols [44,45]. Indeed, the treatment of cells with the GPX4 inhibitor RSL3 (RAS selective lethal 3) rapidly induces ferroptosis [43]. On the other hand, erastin (eradicator of RAS and ST-expressing cells) facilitates ferroptosis preferentially in RAS-positive cancer cells [46,47]. Preliminary studies have shown that erastin inhibits the membrane potential of mitochondria [48], while subsequent studies have elucidated how erastin reduces GPX4 activity [6]. Specifically erastin, similarly to sulfasalazine [6], sorafenib [49,50] and glutamate [6], inhibits cystine absorption from the extracellular space mediated by system Xc^-, a glutamate-cystine antiporter [6,51]; this reduces the intracellular biosynthesis of GSH, leading to subsequent impairment of GPX4 activity. After its recent discovery in 2012, the importance of ferroptosis has attracted much interest as it represents a potential mechanism for controlling tumor growth. To date, different types of tumor have shown sensitivity to ferroptosis inducers [10], including diffuse large B-cell lymphoma, renal cell carcinoma, liver cancer, cervical carcinoma, osteosarcoma, prostate adenocarcinoma, pancreatic carcinoma, and ovarian carcinoma [43,52–54]. In addition, several proteins and pathways have been described as modulators of ferroptosis (Table 1). However, few studies have so far documented the ferroptosis process in sarcomas.

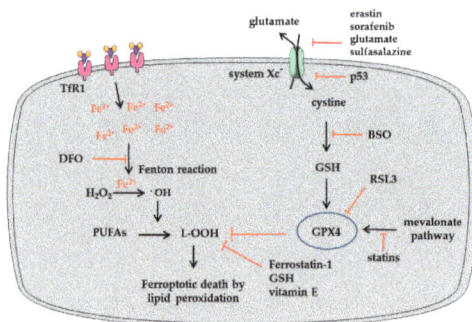

Figure 3. Intracellular levels of iron, glutathione (GSH) and polyunsaturated fatty acids (PUFAs) influence ferroptosis. The abundant intracellular iron through the Fenton reaction can result in higher formation of hydroxyl radicals (•OH), the most reactive ROS (Reactive Oxygen Species) intermediates. These promote conversion of PUFAs into lipid hydroperoxides (L–OOH) that lead to ferroptosis. This process can be exacerbated pharmacologically by the inhibition of glutathione peroxidase 4 (GPX4), the enzyme responsible for L–OOH neutralization, by treatment with RAS selective lethal 3 (RSL3). Alternatively, GPX4 activity may be inhibited by a depletion of GSH via treatment with inhibitors of the system Xc^- responsible for cystine uptake (such as erastin, sorafenib, glutamate and sulfasalazine) or with buthionine-sulfoximine (BSO), an inhibitor of the first reaction of GSH biosynthesis. The system Xc^- is also transcriptionally repressed by p53. In addition, treatment with inhibitors of the mevalonate pathway, such as statins, affect GPX4 synthesis and stimulate ferroptosis. On the other hand, strategies to prevent ferroptosis include treatment with iron chelators such as deferoxamine (DFO) and neutralization of L-OOH by treatment with lipid peroxidation inhibitors (Ferrostatin-1) and antioxidant scavengers (GSH, Vitamin E).

Table 1. Proteins and pathways modulating ferroptosis.

Pro-Ferroptosis	Function	References
ACSL4	Acyl-CoA synthase long-chain 4 increases the fraction of long polyunsaturated ω6 fatty acids in cellular membranes	[55]
CARS	Cysteinyl-tRNA synthetase is an enzyme involved in charging of tRNAs with cysteine for protein translation	[56]
Gln	Glutamine via glutaminolysis is essential for ferroptosis triggered by deprivation of full amino acids or of cystine alone	[57]
HO-1	Heme oxygenase-1 is a heme-degrading enzyme releasing iron	[58]
LOX-5	Lipoxygenase-5 catalyzes the dioxygenation of PUFAs	[8]
NCOA4	Nuclear receptor coactivator 4 promotes H-Ferritin degradation	[27,59]
NOX	NADPH oxidase produces ROS species	[6]
P53	It represses the expression of SLC7A11 encoding a subunit of the system Xc⁻	[60]
SAT1	Spermidine/spermine N-acetyltransferase increases the peroxidation of arachidonic acid	[61]
TfR1	Transferrin receptor 1 is involved in the iron uptake	[57]
Anti-Ferroptosis	Function	References
Ferritin	The main intracellular iron storage protein	[62]
GPX4	Glutathione peroxidase-4 is a selenoenzyme neutralizing lipid hydroperoxides	[43]
HSPA5	Heath shock protein-5 prevents GPX4 degradation	[63]
HSPB1	Heat shock protein β-1 protects from lipid ROS	[52]
IRP2	Iron responsive protein-2 controls the transcription of TfR1, Ferritin and FPN	[62]
MT-1	Metallothionein-1 binds heavy metals	[64,65]
Mevalonate pathway	Pathway controlling the biosynthesis of selenoproteins, such as GPX4	[66]
Mitochondrial Ferritin	Iron-storage protein	[67]
NRF2	Nuclear factor erythroid 2-related factor 2 drives a transcriptional antioxidant program	[68]
System Xc⁻	The antiporter involved in cystine absorption	[60]

3. Ferroptosis and Rhabdomyosarcoma

3.1. Rhabdomyosarcoma Is a Soft Tissue Sarcoma Characterized by Oxidative Stress

Sarcomas are mesenchymal tumors originating from cell precursors committed to form fat, blood vessels, nerves, bones, muscles, deep skin tissues, and cartilage. The latest classification by the World Health Organization [69] divided sarcomas into non-soft tissue sarcomas and soft-tissue sarcomas (STS). The former includes bone sarcomas such as osteosarcoma, Ewing's sarcoma, and chondrosarcoma; while the latter includes a family of more than 50 neoplasms representing about 20% of childhood and adolescence tumors and 1% of all adult cancers. RMS arises from cell progenitors committed to skeletal muscle [70–73] and is the most common STS in patients of pediatric and adolescent age [74,75]. RMS subdivides into four main subtypes depending on histology appearance, tumor location, age of onset, and molecular drivers (Table 2).

Table 2. Histological classification and molecular drivers of rhabdomyosarcom (RMS).

RMS Histotypes	% of All RMS Cases	Location	Age	Prognosis	Dominant Molecular Drivers
Embryonal	60%	Genitourinary tract, head and neck, urinary bladder, prostate, biliary tract, abdomen, pelvis, retroperitoneum	<10	favorable	Activating mutations in PDGFRA, ERBB2, FGFR4, RAS, PIK3CA [76–79] IGF-2 overexpression [80,81] Somatic mutations in p53 [82]
Alveolar	20%	Extremities, head and neck, chest, genital organs, abdomen and anal area	10–20	unfavorable	Chromosomal translocation t(2;13)(q35;q14) [83,84] N-MYC overexpression [85] IGF-2 overexpression [81]
Pleomorphic	10%	Extremities, chest and abdomen	60–80	unfavorable	Complex karyotypes with no recurrent structural alterations
Spindle cell	10%	Paratesticular, head and neck	<10 and >40	favorable (children) unfavorable (adults)	NCOA2 gene rearrangements [86] Mutations in MYOD1 [87]

Abbreviations used are: ERBB2, erb-b2 receptor tyrosine kinase 2; FGFR4, fibroblast growth factor receptor 4; IGF-2, insulin-like growth factor 2; MYOD1, myogenic differentiation 1; NCOA2, Nuclear Receptor Coactivator 2; MYC, myelocytomatosis viral related oncogene; PDGFRA, platelet-derived growth factor receptor A; PIK3CA, phosphatidylinositol-4,5-bisphosphate 3-kinase, catalytic subunit alpha; RAS, retrovirus-associated DNA sequences.

The two predominant histotypes are the embryonal (ERMS) and the alveolar (ARMS) forms that commonly affect children under 10 years or adolescents/young adults, respectively [88]. Currently, chemotherapy, radiotherapy, and surgery are used to treat this aggressive tumor, with a five-year survival rate of higher than 70% in patients with localized disease; however, the overall survival of patients with metastasis remains low [89,90]. Different types of molecular drivers have been identified for each RMS subtype (Table 2) [91]. The most aggressive ARMS is dominated by a chromosomal translocation t(2;13)(q35;q14) that juxtaposes the DNA binding domain of the *PAX3* gene in a frame with the activation domain of the *FOXO1* gene, giving rise to a Pax3-Foxo1 chimeric transcription factor that is found in 70% of ARMS cases and is considered a predictor of poor prognosis [92,93]. ERMS is the most frequent form and is characterized by activating mutations in a number of receptor and transducer molecules, which cause the deliberate activation of the extracellular regulated kinases 1/2 (ERK1/2) and phosphatidylinositol-4,5-bisphosphate 3-kinase (PI3K) signaling pathways [69,71]. Among the oncogenic transducers, RAS is considered a major driver of ERMS etiogenesis [94], as the RAS^{G12V} mutated form is sufficient to convert normal myogenic cell precursors into pre-neoplastic and neoplastic counterparts [82,95]. Accordingly, germline RAS mutations on chromosome 11p15.5 are causative of the Costello syndrome, which predisposes individuals to the formation of embryonic tumors, including ERMS [96]. Sustained RAS activation correlates with ERMS tumor risk and was shown to promote a higher rate of G→T transversions due to high ROS formation [97,98]. As a consequence, RMS tumors were reported to be sensitive to a number of oxidative inducers [97], including auranofin—an inhibitor of thioredoxin reductase [99], cervistatin—a synthetic statin causing mitochondrial impairment [100], and ouabain—a glycoside inhibiting the Na$^+$/K$^+$ ATPase activity (Table 3). According to these findings, the identification of oxidative-stress inducers represents a milestone for the implementation of the therapeutic regimen of RMS.

Table 3. Oxidative stress inducers that have shown efficacy in RMS treatment.

Agents	Targets	Reference
auranofin	Inhibitor of thioredoxin reductase	[97]
buthionine-sulfoximine	Inhibitor of the first step of GSH biosynthesis	[101]
cervistatin	Synthetic statin causing mitochondrial impairment	[97]
NBDHEX	Inhibitor of GSH transferase P1-1	[102]
ouabain	Inhibitor of the Na$^+$/K$^+$ ATPase activity	[97]
sorafenib	Inhibitor of system Xc$^-$	[103]

3.2. Ferroptosis in Rhabdomyosarcoma: State of the Art

Schott et al. found that lentiviral infection of mutated hyperactive RAS forms (NRAS^{G12V}, KRAS^{G12V} and HRAS^{G12V}) in the RMS13 cell line significantly protected against ferroptosis induced by erastin, RSL3 and auranofin, suggesting that activation of the RAS/ERK pathway may confer protection against oxidative stress [104]. However, experiments recently carried out in our laboratories indicated that RMS cell lines with higher basal RAS/ERK activity are preferentially sensitized towards ferroptosis (manuscript under preparation). Thus, more detailed studies must be carried out to verify the relationship between RAS/ERK signaling, oxidative damage and sensitivity to ferroptosis in RMS. High GSH biosynthesis was reported to be necessary for growth, detoxification, and multidrug resistance in RMS. For example, increased GSH levels and GSH-S-transferase (GSTs) activity were observed in high-grade and metastatic STS treated with doxorubicin [105], as well as in ERMS tumors resistant to doxorubicin, topotecan and vincristine [106]. Moreover, high levels of reduced GSH were found in the serum of RMS patients [107]. These findings suggest that lowering GSH levels could affect RMS survival. Recent studies showed that inhibition of the GST isoenzyme, namely GSH transferase P1-1, with 6-(7-nitro-2,1,3-benzoxadiazol-4-ylthio) hexanol (NBDHEX), potentiates cell death in RMS cells treated with chemotherapeutic agents [102]. In the context of ferroptosis, treatment of RMS with sorafenib, an inhibitor of system Xc$^-$ causing the depletion of endogenous GSH, counteracts cell proliferation in vitro and xenograft tumor growth in vivo [103]. However, a Phase 2 trial study did not show consistent effects of sorafenib in RMS tumor cohorts, nonetheless its combination with irinotecan or topotecan is being evaluated in children and young adults with refractory solid tumors [108]. Buthionine-sulfoximine (BSO) is another ferroptosis inducer, which inhibits the first reaction of GSH biosynthesis catalyzed by the γ-glutamylcysteine synthetase (γ-GCSc) [43]. BSO has been shown to be effective in reducing the tumorigenic potential of two rat RMS cell lines in vitro and in vivo [101]. Interestingly, one of the two cell lines used in this study was more resistant to the BSO treatment. Biochemical analysis revealed higher levels of γ-glutamyltranspeptidase (γ-GT) [101], a plasma membrane enzyme of the outer surface responsible for breaking down extracellular GSH to increase cystine absorption [109], ultimately leading to an increase in intracellular GSH levels. Thus, it could be argued that the higher γ-GT levels inhibit ferroptosis by increasing GSH levels. Notably, abnormally high γ-GT enzymatic levels were found in patients with high-grade and metastatic sarcomas [105]. Finally, it has been shown that the treatment of RMS cells with tunicamycin or *N*-glycosidase is sufficient to lower GSH levels and cause cell death, suggesting that protein *N*-glycosylation is required for GSH biosynthesis [110]. In this regard, the folding and auto-catalytic cleavage of γ-GT has been shown to be dependent on *N*-glycosylation [111], suggesting again its potential involvement in ferroptosis resistance.

4. Conclusions

Recent discoveries shed light on a peculiar iron-dependent non-apoptotic form of cell death, namely ferroptosis, the execution of which is dependent on the availability of intracellular free iron, the levels of PUFAs and the levels of enzymes of detoxification from oxidative stress. RMS tumors display hallmarks of oxidative damage, which could predict the susceptibility of RMS to a number of oxidative stress inducers. Despite the important role of iron metabolism and iron proteins in cancer, little work has been done on RMS. Therefore, we discussed the agents involved in ferroptosis activation, believing that a better understanding of this mechanism in RMS may lead to therapeutic improvements.

Acknowledgments: Funding. The work was supported by Fondazione Cariplo and Regione Lombardia project "New Opportunities and ways toward European Research Council ERC" 2014-2256 to Maura Poli We would like to thank Caterina Branca (Arizona State University, Tempe, Arizona, AZ, USA) for text proofreading.

Author Contributions: Alessandro Fanzani and Maura Poli wrote the paper and prepared the figures.

Conflicts of Interest: The authors declare no conflict of interest.

References

1. Torti, S.V.; Torti, F.M. Iron and cancer: More ore to be mined. *Nat. Rev. Cancer* **2013**, *13*, 342–355. [CrossRef] [PubMed]
2. Cazzola, M.; Bergamaschi, G.; Dezza, L.; Arosio, P. Manipulations of cellular iron metabolism for modulating normal and malignant cell proliferation: Achievements and prospects. *Blood* **1990**, *75*, 1903–1919. [PubMed]
3. Ke, J.Y.; Cen, W.J.; Zhou, X.Z.; Li, Y.R.; Kong, W.D.; Jiang, J.W. Iron overload induces apoptosis of murine preosteoblast cells via ros and inhibition of AKT pathway. *Oral Dis.* **2017**. [CrossRef] [PubMed]
4. Tian, Q.; Wu, S.; Dai, Z.; Yang, J.; Zheng, J.; Zheng, Q.; Liu, Y. Iron overload induced death of osteoblasts in vitro: Involvement of the mitochondrial apoptotic pathway. *Peer J.* **2016**, *4*, e2611. [CrossRef] [PubMed]
5. Li, S.W.; Liu, C.M.; Guo, J.; Marcondes, A.M.; Deeg, J.; Li, X.; Guan, F. Iron overload induced by ferric ammonium citrate triggers reactive oxygen species-mediated apoptosis via both extrinsic and intrinsic pathways in human hepatic cells. *Hum. Exp. Toxicol.* **2016**, *35*, 598–607. [CrossRef] [PubMed]
6. Dixon, S.J.; Lemberg, K.M.; Lamprecht, M.R.; Skouta, R.; Zaitsev, E.M.; Gleason, C.E.; Patel, D.N.; Bauer, A.J.; Cantley, A.M.; Yang, W.S.; et al. Ferroptosis: An iron-dependent form of nonapoptotic cell death. *Cell* **2012**, *149*, 1060–1072. [CrossRef] [PubMed]
7. Yang, W.S.; Stockwell, B.R. Ferroptosis: Death by lipid peroxidation. *Trends Cell Biol.* **2016**, *26*, 165–176. [CrossRef] [PubMed]
8. Gaschler, M.M.; Stockwell, B.R. Lipid peroxidation in cell death. *Biochem. Biophys. Res. Commun.* **2017**, *482*, 419–425. [CrossRef] [PubMed]

9. Cao, J.Y.; Dixon, S.J. Mechanisms of ferroptosis. *Cell Mol. Life Sci.* **2016**, *73*, 2195–2209. [CrossRef] [PubMed]

10. Yu, H.; Guo, P.; Xie, X.; Wang, Y.; Chen, G. Ferroptosis, a new form of cell death, and its relationships with tumourous diseases. *J. Cell Mol. Med.* **2017**, *21*, 648–657. [CrossRef] [PubMed]

11. Toyokuni, S. Role of iron in carcinogenesis: Cancer as a ferrotoxic disease. *Cancer Sci.* **2009**, *100*, 9–16. [CrossRef] [PubMed]

12. Toyokuni, S.; Ito, F.; Yamashita, K.; Okazaki, Y.; Akatsuka, S. Iron and thiol redox signaling in cancer: An exquisite balance to escape ferroptosis. *Free Radic. Biol. Med.* **2017**, *108*, 610–626. [CrossRef] [PubMed]

13. Ebina, Y.; Okada, S.; Hamazaki, S.; Ogino, F.; Li, J.L.; Midorikawa, O. Nephrotoxicity and renal cell carcinoma after use of iron- and aluminum-nitrilotriacetate complexes in rats. *J. Natl. Cancer Inst.* **1986**, *76*, 107–113. [PubMed]

14. Van der Kemp, P.A.; Thomas, D.; Barbey, R.; de Oliveira, R.; Boiteux, S. Cloning and expression in escherichia coli of the OGG1 gene of saccharomyces cerevisiae, which codes for a dna glycosylase that excises 7,8-dihydro-8-oxoguanine and 2,6-diamino-4-hydroxy-5-*N*-methylformamidopyrimidine. *Proc. Natl. Acad. Sci. USA* **1996**, *93*, 5197–5202. [CrossRef] [PubMed]

15. Nash, H.M.; Bruner, S.D.; Schärer, O.D.; Kawate, T.; Addona, T.A.; Spooner, E.; Lane, W.S.; Verdine, G.L. Cloning of a yeast 8-oxoguanine DNA glycosylase reveals the existence of a base-excision DNA-repair protein superfamily. *Curr. Biol.* **1996**, *6*, 968–980. [CrossRef]

16. Aburatani, H.; Hippo, Y.; Ishida, T.; Takashima, R.; Matsuba, C.; Kodama, T.; Takao, M.; Yasui, A.; Yamamoto, K.; Asano, M. Cloning and characterization of mammalian 8-hydroxyguanine-specific DNA glycosylase/apurinic, apyrimidinic lyase, a functional mutm homologue. *Cancer Res.* **1997**, *57*, 2151–2156. [PubMed]

17. Kato, J.; Miyanishi, K.; Kobune, M.; Nakamura, T.; Takada, K.; Takimoto, R.; Kawano, Y.; Takahashi, S.; Takahashi, M.; Sato, Y.; et al. Long-term phlebotomy with low-iron diet therapy lowers risk of development of hepatocellular carcinoma from chronic hepatitis C. *J. Gastroenterol.* **2007**, *42*, 830–836. [CrossRef] [PubMed]

18. Kato, J.; Kobune, M.; Nakamura, T.; Kuroiwa, G.; Takada, K.; Takimoto, R.; Sato, Y.; Fujikawa, K.; Takahashi, M.; Takayama, T.; et al. Normalization of elevated hepatic 8-hydroxy-2′-deoxyguanosine levels in chronic hepatitis C patients by phlebotomy and low iron diet. *Cancer Res.* **2001**, *61*, 8697–8702. [PubMed]

19. Zacharski, L.R.; Chow, B.K.; Howes, P.S.; Shamayeva, G.; Baron, J.A.; Dalman, R.L.; Malenka, D.J.; Ozaki, C.K.; Lavori, P.W. Decreased cancer risk after iron reduction in patients with peripheral arterial disease: Results from a randomized trial. *J. Natl. Cancer Inst.* **2008**, *100*, 996–1002. [CrossRef] [PubMed]

20. Kasai, H. Analysis of a form of oxidative DNA damage, 8-hydroxy-2′-deoxyguanosine, as a marker of cellular oxidative stress during carcinogenesis. *Mutat. Res.* **1997**, *387*, 147–163. [CrossRef]

21. Kondo, S.; Toyokuni, S.; Tanaka, T.; Hiai, H.; Onodera, H.; Kasai, H.; Imamura, M. Overexpression of the *HOGG1* gene and high 8-hydroxy-2′-deoxyguanosine (8-OHDG) lyase activity in human colorectal carcinoma: Regulation mechanism of the 8-OHDG level in DNA. *Clin. Cancer Res.* **2000**, *6*, 1394–1400. [PubMed]

22. Okamoto, K.; Toyokuni, S.; Kim, W.J.; Ogawa, O.; Kakehi, Y.; Arao, S.; Hiai, H.; Yoshida, O. Overexpression of human mutt homologue gene messenger RNA in renal-cell carcinoma: Evidence of persistent oxidative stress in cancer. *Int. J. Cancer* **1996**, *65*, 437–441. [CrossRef]

23. Ganz, T.; Nemeth, E. Hepcidin and iron homeostasis. *Biochim. Biophys. Acta* **2012**, *1823*, 1434–1443. [CrossRef] [PubMed]

24. Nemeth, E.; Tuttle, M.S.; Powelson, J.; Vaughn, M.B.; Donovan, A.; Ward, D.M.; Ganz, T.; Kaplan, J. Hepcidin regulates cellular iron efflux by binding to ferroportin and inducing its internalization. *Science* **2004**, *306*, 2090–2093. [CrossRef] [PubMed]

25. Gunshin, H.; Allerson, C.R.; Polycarpou-Schwarz, M.; Rofts, A.; Rogers, J.T.; Kishi, F.; Hentze, M.W.; Rouault, T.A.; Andrews, N.C.; Hediger, M.A. Iron-dependent regulation of the divalent metal ion transporter. *FEBS Lett.* **2001**, *509*, 309–316. [CrossRef]

26. Arosio, P.; Levi, S. Cytosolic and mitochondrial ferritins in the regulation of cellular iron homeostasis and oxidative damage. *Biochim. Biophys. Acta* **2010**, *1800*, 783–792. [CrossRef] [PubMed]

27. Mancias, J.D.; Wang, X.; Gygi, S.P.; Harper, J.W.; Kimmelman, A.C. Quantitative proteomics identifies NCOA4 as the cargo receptor mediating ferritinophagy. *Nature* **2014**, *509*, 105–109. [CrossRef] [PubMed]

28. Maffettone, C.; Chen, G.; Drozdov, I.; Ouzounis, C.; Pantopoulos, K. Tumorigenic properties of iron regulatory protein 2 (IRP2) mediated by its specific 73-amino acids insert. *PLoS ONE* **2010**, *5*, e10163. [CrossRef] [PubMed]

29. O'Donnell, K.A.; Yu, D.; Zeller, K.I.; Kim, J.W.; Racke, F.; Thomas-Tikhonenko, A.; Dang, C.V. Activation of transferrin receptor 1 by c-MYC enhances cellular proliferation and tumorigenesis. *Mol. Cell Biol.* **2006**, *26*, 2373–2386. [CrossRef] [PubMed]

30. Tacchini, L.; Bianchi, L.; Bernelli-Zazzera, A.; Cairo, G. Transferrin receptor induction by hypoxia. Hif-1-mediated transcriptional activation and cell-specific post-transcriptional regulation. *J. Biol. Chem.* **1999**, *274*, 24142–24146. [CrossRef] [PubMed]

31. Habashy, H.O.; Powe, D.G.; Staka, C.M.; Rakha, E.A.; Ball, G.; Green, A.R.; Aleskandarany, M.; Paish, E.C.; Douglas Macmillan, R.; Nicholson, R.I.; et al. Transferrin receptor (CD71) is a marker of poor prognosis in breast cancer and can predict response to tamoxifen. *Breast Cancer Res. Treat.* **2010**, *119*, 283–293. [CrossRef] [PubMed]

32. Jeong, D.E.; Song, H.J.; Lim, S.; Lee, S.J.; Lim, J.E.; Nam, D.H.; Joo, K.M.; Jeong, B.C.; Jeon, S.S.; Choi, H.Y.; et al. Repurposing the anti-malarial drug artesunate as a novel therapeutic agent for metastatic renal cell carcinoma due to its attenuation of tumor growth, metastasis, and angiogenesis. *Oncotarget* **2015**, *6*, 33046–33064. [CrossRef] [PubMed]

33. Basuli, D.; Tesfay, L.; Deng, Z.; Paul, B.; Yamamoto, Y.; Ning, G.; Xian, W.; McKeon, F.; Lynch, M.; Crum, C.P.; et al. Iron addiction: A novel therapeutic target in ovarian cancer. *Oncogene* **2017**, *36*, 4089–4099. [CrossRef] [PubMed]

34. Torti, S.V.; Torti, F.M. Ironing out cancer. *Cancer Res.* **2011**, *71*, 1511–1514. [CrossRef] [PubMed]

35. Wu, K.J.; Polack, A.; Dalla-Favera, R. Coordinated regulation of iron-controlling genes, H-ferritin and Irp2, by c-MYC. *Science* **1999**, *283*, 676–679. [CrossRef] [PubMed]

36. Kakhlon, O.; Gruenbaum, Y.; Cabantchik, Z.I. Repression of ferritin expression modulates cell responsiveness to H-ras-induced growth. *Biochem. Soc. Trans.* **2002**, *30*, 777–780. [CrossRef] [PubMed]

37. Corna, G.; Campana, L.; Pignatti, E.; Castiglioni, A.; Tagliafico, E.; Bosurgi, L.; Campanella, A.; Brunelli, S.; Manfredi, A.A.; Apostoli, P.; et al. Polarization dictates iron handling by inflammatory and alternatively activated macrophages. *Haematologica* **2010**, *95*, 1814–1822. [CrossRef] [PubMed]

38. Latunde-Dada, G.O. Ferroptosis: Role of lipid peroxidation, iron and ferritinophagy. *Biochim. Biophys. Acta* **2017**, *1861*, 1893–1900. [CrossRef] [PubMed]

39. Doll, S.; Conrad, M. Iron and ferroptosis: A still ill-defined liaison. *IUBMB Life* **2017**, *69*, 423–434. [CrossRef] [PubMed]

40. Xie, Y.; Hou, W.; Song, X.; Yu, Y.; Huang, J.; Sun, X.; Kang, R.; Tang, D. Ferroptosis: Process and function. *Cell Death Differ.* **2016**, *23*, 369–379. [CrossRef] [PubMed]

41. Thomas, C.; Mackey, M.M.; Diaz, A.A.; Cox, D.P. Hydroxyl radical is produced via the fenton reaction in submitochondrial particles under oxidative stress: Implications for diseases associated with iron accumulation. *Redox Rep.* **2009**, *14*, 102–108. [CrossRef] [PubMed]

42. Skouta, R.; Dixon, S.J.; Wang, J.; Dunn, D.E.; Orman, M.; Shimada, K.; Rosenberg, P.A.; Lo, D.C.; Weinberg, J.M.; Linkermann, A.; et al. Ferrostatins inhibit oxidative lipid damage and cell death in diverse disease models. *J. Am. Chem. Soc.* **2014**, *136*, 4551–4556. [CrossRef] [PubMed]

43. Yang, W.S.; SriRamaratnam, R.; Welsch, M.E.; Shimada, K.; Skouta, R.; Viswanathan, V.S.; Cheah, J.H.; Clemons, P.A.; Shamji, A.F.; Clish, C.B.; et al. Regulation of ferroptotic cancer cell death by GPX4. *Cell* **2014**, *156*, 317–331. [CrossRef] [PubMed]

44. Seiler, A.; Schneider, M.; Förster, H.; Roth, S.; Wirth, E.K.; Culmsee, C.; Plesnila, N.; Kremmer, E.; Rådmark, O.; Wurst, W.; et al. Glutathione peroxidase 4 senses and translates oxidative stress into 12/15-lipoxygenase dependent- and AIF-mediated cell death. *Cell Metab.* **2008**, *8*, 237–248. [CrossRef] [PubMed]

45. Ran, Q.; Liang, H.; Gu, M.; Qi, W.; Walter, C.A.; Roberts, L.J.; Herman, B.; Richardson, A.; van Remmen, H. Transgenic mice overexpressing glutathione peroxidase 4 are protected against oxidative stress-induced apoptosis. *J. Biol. Chem.* **2004**, *279*, 55137–55146. [CrossRef] [PubMed]

46. Dolma, S.; Lessnick, S.L.; Hahn, W.C.; Stockwell, B.R. Identification of genotype-selective antitumor agents using synthetic lethal chemical screening in engineered human tumor cells. *Cancer Cell* **2003**, *3*, 285–296. [CrossRef]

47. Yang, W.S.; Stockwell, B.R. Synthetic lethal screening identifies compounds activating iron-dependent, nonapoptotic cell death in oncogenic-RAS-harboring cancer cells. *Chem. Biol.* **2008**, *15*, 234–245. [CrossRef] [PubMed]

48. Yagoda, N.; von Rechenberg, M.; Zaganjor, E.; Bauer, A.J.; Yang, W.S.; Fridman, D.J.; Wolpaw, A.J.; Smukste, I.; Peltier, J.M.; Boniface, J.J.; et al. RAS-RAF-MEK-dependent oxidative cell death involving voltage-dependent anion channels. *Nature* **2007**, *447*, 864–868. [CrossRef] [PubMed]

49. Lachaier, E.; Louandre, C.; Godin, C.; Saidak, Z.; Baert, M.; Diouf, M.; Chauffert, B.; Galmiche, A. Sorafenib induces ferroptosis in human cancer cell lines originating from different solid tumors. *Anticancer Res.* **2014**, *34*, 6417–6422. [PubMed]

50. Louandre, C.; Ezzoukhry, Z.; Godin, C.; Barbare, J.C.; Mazière, J.C.; Chauffert, B.; Galmiche, A. Iron-dependent cell death of hepatocellular carcinoma cells exposed to sorafenib. *Int. J. Cancer* **2013**, *133*, 1732–1742. [CrossRef] [PubMed]

51. Dixon, S.J.; Patel, D.N.; Welsch, M.; Skouta, R.; Lee, E.D.; Hayano, M.; Thomas, A.G.; Gleason, C.E.; Tatonetti, N.P.; Slusher, B.S.; et al. Pharmacological inhibition of cystine-glutamate exchange induces endoplasmic reticulum stress and ferroptosis. *Elife* **2014**, *3*, e02523. [CrossRef] [PubMed]

52. Sun, X.; Ou, Z.; Xie, M.; Kang, R.; Fan, Y.; Niu, X.; Wang, H.; Cao, L.; Tang, D. HSPB1 as a novel regulator of ferroptotic cancer cell death. *Oncogene* **2015**, *34*, 5617–5625. [CrossRef] [PubMed]

53. Eling, N.; Reuter, L.; Hazin, J.; Hamacher-Brady, A.; Brady, N.R. Identification of artesunate as a specific activator of ferroptosis in pancreatic cancer cells. *Oncoscience* **2015**, *2*, 517–532. [CrossRef] [PubMed]

54. Greenshields, A.L.; Shepherd, T.G.; Hoskin, D.W. Contribution of reactive oxygen species to ovarian cancer cell growth arrest and killing by the anti-malarial drug artesunate. *Mol. Carcinog.* **2017**, *56*, 75–93. [CrossRef] [PubMed]

55. Doll, S.; Proneth, B.; Tyurina, Y.Y.; Panzilius, E.; Kobayashi, S.; Ingold, I.; Irmler, M.; Beckers, J.; Aichler, M.; Walch, A.; et al. ACSL4 dictates ferroptosis sensitivity by shaping cellular lipid composition. *Nat. Chem. Biol.* **2017**, *13*, 91–98. [CrossRef] [PubMed]

56. Hayano, M.; Yang, W.S.; Corn, C.K.; Pagano, N.C.; Stockwell, B.R. Loss of cysteinyl-tRNA synthetase (cars) induces the transsulfuration pathway and inhibits ferroptosis induced by cystine deprivation. *Cell Death Differ.* **2016**, *23*, 270–278. [CrossRef] [PubMed]

57. Gao, M.; Monian, P.; Quadri, N.; Ramasamy, R.; Jiang, X. Glutaminolysis and transferrin regulate ferroptosis. *Mol. Cell* **2015**, *59*, 298–308. [CrossRef] [PubMed]

58. Kwon, M.Y.; Park, E.; Lee, S.J.; Chung, S.W. Heme oxygenase-1 accelerates erastin-induced ferroptotic cell death. *Oncotarget* **2015**, *6*, 24393–24403. [CrossRef] [PubMed]

59. Gao, M.; Monian, P.; Pan, Q.; Zhang, W.; Xiang, J.; Jiang, X. Ferroptosis is an autophagic cell death process. *Cell Res.* **2016**, *26*, 1021–1032. [CrossRef] [PubMed]

60. Jiang, L.; Kon, N.; Li, T.; Wang, S.J.; Su, T.; Hibshoosh, H.; Baer, R.; Gu, W. Ferroptosis as a p53-mediated activity during tumour suppression. *Nature* **2015**, *520*, 57–62. [CrossRef] [PubMed]

61. Ou, Y.; Wang, S.J.; Li, D.; Chu, B.; Gu, W. Activation of sat1 engages polyamine metabolism with p53-mediated ferroptotic responses. *Proc. Natl. Acad. Sci. USA* **2016**, *113*, E6806–E6812. [CrossRef] [PubMed]

62. Reed, J.C.; Pellecchia, M. Ironing out cell death mechanisms. *Cell* **2012**, *149*, 963–965. [CrossRef] [PubMed]

63. Zhu, S.; Zhang, Q.; Sun, X.; Zeh, H.J.; Lotze, M.T.; Kang, R.; Tang, D. HSPA5 regulates ferroptotic cell death in cancer cells. *Cancer Res.* **2017**, *77*, 2064–2077. [CrossRef] [PubMed]

64. Houessinon, A.; François, C.; Sauzay, C.; Louandre, C.; Mongelard, G.; Godin, C.; Bodeau, S.; Takahashi, S.; Saidak, Z.; Gutierrez, L.; et al. Metallothionein-1 as a biomarker of altered redox metabolism in hepatocellular carcinoma cells exposed to sorafenib. *Mol. Cancer* **2016**, *15*, 38. [CrossRef] [PubMed]

65. Sun, X.; Niu, X.; Chen, R.; He, W.; Chen, D.; Kang, R.; Tang, D. Metallothionein-1g facilitates sorafenib resistance through inhibition of ferroptosis. *Hepatology* **2016**, *64*, 488–500. [CrossRef] [PubMed]

66. Warner, G.J.; Berry, M.J.; Moustafa, M.E.; Carlson, B.A.; Hatfield, D.L.; Faust, J.R. Inhibition of selenoprotein synthesis by selenocysteine tRNA$^{[ser]sec}$ lacking isopentenyladenosine. *J. Biol. Chem.* **2000**, *275*, 28110–28119. [CrossRef] [PubMed]

67. Wang, Y.Q.; Chang, S.Y.; Wu, Q.; Gou, Y.J.; Jia, L.; Cui, Y.M.; Yu, P.; Shi, Z.H.; Wu, W.S.; Gao, G.; et al. The protective role of mitochondrial ferritin on erastin-induced ferroptosis. *Front. Aging Neurosci.* **2016**, *8*, 308. [CrossRef] [PubMed]

68. Sun, X.; Ou, Z.; Chen, R.; Niu, X.; Chen, D.; Kang, R.; Tang, D. Activation of the p62-KEAP1-NRF2 pathway protects against ferroptosis in hepatocellular carcinoma cells. *Hepatology* **2016**, *63*, 173–184. [CrossRef] [PubMed]

69. Fletcher, C.D.M.; Bridge, J.A.; Hogendoorn, P.C.W.; Mertens, F. *Who Classification of Tumours of Soft Tissue and Bone*; IARC: Lyon, France, 2013.

70. Linardic, C.M.; Downie, D.L.; Qualman, S.; Bentley, R.C.; Counter, C.M. Genetic modeling of human rhabdomyosarcoma. *Cancer Res.* **2005**, *65*, 4490–4495. [CrossRef] [PubMed]

71. Hettmer, S.; Liu, J.; Miller, C.M.; Lindsay, M.C.; Sparks, C.A.; Guertin, D.A.; Bronson, R.T.; Langenau, D.M.; Wagers, A.J. Sarcomas induced in discrete subsets of prospectively isolated skeletal muscle cells. *Proc. Natl. Acad. Sci. USA* **2011**, *108*, 20002–20007. [CrossRef] [PubMed]

72. Rubin, B.P.; Nishijo, K.; Chen, H.I.; Yi, X.; Schuetze, D.P.; Pal, R.; Prajapati, S.I.; Abraham, J.; Arenkiel, B.R.; Chen, Q.R.; et al. Evidence for an unanticipated relationship between undifferentiated pleomorphic sarcoma and embryonal rhabdomyosarcoma. *Cancer Cell* **2011**, *19*, 177–191. [CrossRef] [PubMed]

73. Rodriguez, R.; Rubio, R.; Menendez, P. Modeling sarcomagenesis using multipotent mesenchymal stem cells. *Cell Res.* **2012**, *22*, 62–77. [CrossRef] [PubMed]

74. Parham, D.M.; Alaggio, R.; Coffin, C.M. Myogenic tumors in children and adolescents. *Pediatr. Dev. Pathol.* **2012**, *15*, 211–238. [CrossRef] [PubMed]

75. Kashi, V.P.; Hatley, M.E.; Galindo, R.L. Probing for a deeper understanding of rhabdomyosarcoma: Insights from complementary model systems. *Nat. Rev. Cancer* **2015**, *15*, 426–439. [CrossRef] [PubMed]

76. Shern, J.F.; Chen, L.; Chmielecki, J.; Wei, J.S.; Patidar, R.; Rosenberg, M.; Ambrogio, L.; Auclair, D.; Wang, J.; Song, Y.K.; et al. Comprehensive genomic analysis of rhabdomyosarcoma reveals a landscape of alterations affecting a common genetic axis in fusion-positive and fusion-negative tumors. *Cancer Discov.* **2014**, *4*, 216–231. [CrossRef] [PubMed]

77. Taylor, J.G.; Cheuk, A.T.; Tsang, P.S.; Chung, J.Y.; Song, Y.K.; Desai, K.; Yu, Y.; Chen, Q.R.; Shah, K.; Youngblood, V.; et al. Identification of FGFR4-activating mutations in human rhabdomyosarcomas that promote metastasis in xenotransplanted models. *J. Clin. Investig.* **2009**, *119*, 3395–3407. [PubMed]

78. Stratton, M.R.; Fisher, C.; Gusterson, B.A.; Cooper, C.S. Detection of point mutations in N-ras and K-ras genes of human embryonal rhabdomyosarcomas using oligonucleotide probes and the polymerase chain reaction. *Cancer Res.* **1989**, *49*, 6324–6327. [PubMed]

79. Shukla, N.; Ameur, N.; Yilmaz, I.; Nafa, K.; Lau, C.Y.; Marchetti, A.; Borsu, L.; Barr, F.G.; Ladanyi, M. Oncogene mutation profiling of pediatric solid tumors reveals significant subsets of embryonal rhabdomyosarcoma and neuroblastoma with mutated genes in growth signaling pathways. *Clin. Cancer Res.* **2012**, *18*, 748–757. [CrossRef] [PubMed]

80. Scrable, H.; Cavenee, W.; Ghavimi, F.; Lovell, M.; Morgan, K.; Sapienza, C.C.P. A model for embryonal rhabdomyosarcoma tumorigenesis that involves genome imprinting. *Proc. Natl. Acad. Sci. USA* **1989**, *86*, 7480–7484. [CrossRef] [PubMed]

81. Anderson, J.; Gordon, A.; McManus, A.; Shipley, J.; Pritchard-Jones, K. Disruption of imprinted genes at chromosome region 11p15.5 in paediatric rhabdomyosarcoma. *Neoplasia* **1999**, *1*, 340–348. [CrossRef] [PubMed]

82. Taylor, A.C.; Shu, L.; Danks, M.K.; Poquette, C.A.; Shetty, S.; Thayer, M.J.; Houghton, P.J.; Harris, L.C. P53 mutation and MDM2 amplification frequency in pediatric rhabdomyosarcoma tumors and cell lines. *Med. Pediatr. Oncol.* **2000**, *35*, 96–103. [CrossRef]

83. Skapek, S.X.; Anderson, J.; Barr, F.G.; Bridge, J.A.; Gastier-Foster, J.M.; Parham, D.M.; Rudzinski, E.R.; Triche, T.; Hawkins, D.S. PAX-FOXO1 fusion status drives unfavorable outcome for children with rhabdomyosarcoma: A children's oncology group report. *Pediatr. Blood Cancer* **2013**, *60*, 1411–1417. [CrossRef] [PubMed]

84. Missiaglia, E.; Williamson, D.; Chisholm, J.; Wirapati, P.; Pierron, G.; Petel, F.; Concordet, J.P.; Thway, K.; Oberlin, O.; Pritchard-Jones, K.; et al. PAX3/FOXO1 fusion gene status is the key prognostic molecular marker in rhabdomyosarcoma and significantly improves current risk stratification. *J. Clin. Oncol.* **2012**, *30*, 1670–1677. [CrossRef] [PubMed]

85. Mercado, G.E.; Xia, S.J.; Zhang, C.; Ahn, E.H.; Gustafson, D.M.; Laé, M.; Ladanyi, M.; Barr, F.G. Identification of PAX3-FKHR-regulated genes differentially expressed between alveolar and embryonal rhabdomyosarcoma: Focus on mycn as a biologically relevant target. *Genes Chromosomes Cancer* **2008**, *47*, 510–520. [CrossRef] [PubMed]

86. Sumegi, J.; Streblow, R.; Frayer, R.W.; Dal Cin, P.; Rosenberg, A.; Meloni-Ehrig, A.; Bridge, J.A. Recurrent t(2;2) and t(2;8) translocations in rhabdomyosarcoma without the canonical PAX-FOXO1 fuse PAX3 to members of the nuclear receptor transcriptional coactivator family. *Genes Chromosomes Cancer* **2010**, *49*, 224–236. [PubMed]

87. Agaram, N.P.; Chen, C.L.; Zhang, L.; LaQuaglia, M.P.; Wexler, L.; Antonescu, C.R. Recurrent MYOD1 mutations in pediatric and adult sclerosing and spindle cell rhabdomyosarcomas: Evidence for a common pathogenesis. *Genes Chromosomes Cancer* **2014**, *53*, 779–787. [CrossRef] [PubMed]

88. Parham, D.M.; Barr, F.G. Classification of rhabdomyosarcoma and its molecular basis. *Adv. Anat. Pathol.* **2013**, *20*, 387–397. [CrossRef] [PubMed]

89. Ognjanovic, S.; Linabery, A.M.; Charbonneau, B.; Ross, J.A.C.P. Trends in childhood rhabdomyosarcoma incidence and survival in the united states, 1975–2005. *Cancer* **2009**, *115*, 4218–4226. [CrossRef] [PubMed]

90. Hettmer, S.; Li, Z.; Billin, A.N.; Barr, F.G.; Cornelison, D.D.; Ehrlich, A.R.; Guttridge, D.C.; Hayes-Jordan, A.; Helman, L.J.; Houghton, P.J.; et al. Rhabdomyosarcoma: Current challenges and their implications for developing therapies. *Cold Spring Harb. Perspect. Med.* **2014**, *4*, a025650. [CrossRef] [PubMed]

91. Zanola, A.; Rossi, S.; Faggi, F.; Monti, E.; Fanzani, A. Rhabdomyosarcomas: An overview on the experimental animal models. *J. Cell Mol. Med.* **2012**, *16*, 1377–1391. [CrossRef] [PubMed]

92. Shapiro, D.N.; Sublett, J.E.; Li, B.; Downing, J.R.; Naeve, C.W. Fusion of PAX3 to a member of the forkhead family of transcription factors in human alveolar rhabdomyosarcoma. *Cancer Res.* **1993**, *53*, 5108–5112. [PubMed]

93. Galili, N.; Davis, R.J.; Fredericks, W.J.; Mukhopadhyay, S.; Rauscher, F.J.; Emanuel, B.S.; Rovera, G.; Barr, F.G. Fusion of a fork head domain gene to PAX3 in the solid tumour alveolar rhabdomyosarcoma. *Nat. Genet.* **1993**, *5*, 230–235. [CrossRef] [PubMed]

94. Langenau, D.M.; Keefe, M.D.; Storer, N.Y.; Guyon, J.R.; Kutok, J.L.; Le, X.; Goessling, W.; Neuberg, D.S.; Kunkel, L.M.; Zon, L.I.C.P. Effects of ras on the genesis of embryonal rhabdomyosarcoma. *Genes Dev.* **2007**, *21*, 1382–1395. [CrossRef] [PubMed]

95. Olson, E.N.; Spizz, G.; Tainsky, M.A. The oncogenic forms of N-ras or H-ras prevent skeletal myoblast differentiation. *Mol. Cell Biol.* **1987**, *7*, 2104–2111. [CrossRef] [PubMed]

96. Aoki, Y.; Niihori, T.; Kawame, H.; Kurosawa, K.; Ohashi, H.; Tanaka, Y.; Filocamo, M.; Kato, K.; Suzuki, Y.; Kure, S.; et al. Germline mutations in hras proto-oncogene cause costello syndrome. *Nat. Genet.* **2005**, *37*, 1038–1040. [CrossRef] [PubMed]

97. Chen, X.; Stewart, E.; Shelat, A.A.; Qu, C.; Bahrami, A.; Hatley, M.; Wu, G.; Bradley, C.; McEvoy, J.; Pappo, A.; et al. Targeting oxidative stress in embryonal rhabdomyosarcoma. *Cancer Cell* **2013**, *24*, 710–724. [CrossRef] [PubMed]

98. Zhang, M.; Linardic, C.M.; Kirsch, D.G. RAS and ROS in rhabdomyosarcoma. *Cancer Cell* **2013**, *24*, 689–691. [CrossRef] [PubMed]

99. Liu, Y.; Li, Y.; Yu, S.; Zhao, G. Recent advances in the development of thioredoxin reductase inhibitors as anticancer agents. *Curr. Drug Targets* **2012**, *13*, 1432–1444. [CrossRef] [PubMed]

100. Bouitbir, J.; Charles, A.L.; Echaniz-Laguna, A.; Kindo, M.; Daussin, F.; Auwerx, J.; Piquard, F.; Geny, B.; Zoll, J. Opposite effects of statins on mitochondria of cardiac and skeletal muscles: A "mitohormesis" mechanism involving reactive oxygen species and PGC-1. *Eur. Heart J.* **2012**, *33*, 1397–1407. [CrossRef] [PubMed]

101. Castro, B.; Alonso-Varona, A.; del Olmo, M.; Bilbao, P.; Palomares, T. Role of γ-glutamyltranspeptidase on the response of poorly and moderately differentiated rhabdomyosarcoma cell lines to buthionine sulfoximine-induced inhibition of glutathione synthesis. *Anticancer Drugs* **2002**, *13*, 281–291. [CrossRef] [PubMed]

102. Pasello, M.; Manara, M.C.; Michelacci, F.; Fanelli, M.; Hattinger, C.M.; Nicoletti, G.; Landuzzi, L.; Lollini, P.L.; Caccuri, A.; Picci, P.; et al. Targeting glutathione-S transferase enzymes in musculoskeletal sarcomas: A promising therapeutic strategy. *Anal. Cell. Pathol. (AMST)* **2011**, *34*, 131–145. [CrossRef] [PubMed]

103. Maruwge, W.; D'Arcy, P.; Folin, A.; Brnjic, S.; Wejde, J.; Davis, A.; Erlandsson, F.; Bergh, J.; Brodin, B. Sorafenib inhibits tumor growth and vascularization of rhabdomyosarcoma cells by blocking IGF-1R-mediated signaling. *Onco Targets Ther.* **2008**, *1*, 67–78. [CrossRef] [PubMed]
104. Schott, C.; Graab, U.; Cuvelier, N.; Hahn, H.; Fulda, S. Oncogenic RAS mutants confer resistance of RMS13 rhabdomyosarcoma cells to oxidative stress-induced ferroptotic cell death. *Front. Oncol.* **2015**, *5*, 131. [CrossRef] [PubMed]
105. Hochwald, S.N.; Rose, D.M.; Brennan, M.F.; Burt, M.E. Elevation of glutathione and related enzyme activities in high-grade and metastatic extremity soft tissue sarcoma. *Ann. Surg. Oncol.* **1997**, *4*, 303–309. [CrossRef] [PubMed]
106. Seitz, G.; Bonin, M.; Fuchs, J.; Poths, S.; Ruck, P.; Warmann, S.W.; Armeanu-Ebinger, S. Inhibition of glutathione-s-transferase as a treatment strategy for multidrug resistance in childhood rhabdomyosarcoma. *Int. J. Oncol.* **2010**, *36*, 491–500. [CrossRef] [PubMed]
107. Zitka, O.; Skalickova, S.; Gumulec, J.; Masarik, M.; Adam, V.; Hubalek, J.; Trnkova, L.; Kruseova, J.; Eckschlager, T.; Kizek, R. Redox status expressed as GSH: GSSG ratio as a marker for oxidative stress in paediatric tumour patients. *Oncol. Lett.* **2012**, *4*, 1247–1253. [PubMed]
108. Kim, A.; Widemann, B.C.; Krailo, M.; Jayaprakash, N.; Fox, E.; Weigel, B.; Blaney, S.M. Phase 2 trial of sorafenib in children and young adults with refractory solid tumors: A report from the children's oncology group. *Pediatr. Blood Cancer* **2015**, *62*, 1562–1566. [CrossRef] [PubMed]
109. Zhang, H.; Forman, H.J.; Choi, J. γ-glutamyl transpeptidase in glutathione biosynthesis. *Methods Enzymol.* **2005**, *401*, 468–483. [PubMed]
110. Calle, Y.; Palomares, T.; Castro, B.; del Olmo, M.; Alonso-Varona, A. Removal of *N*-glycans from cell surface proteins induces apoptosis by reducing intracellular glutathione levels in the rhabdomyosarcoma cell line s4mh. *Biol. Cell* **2000**, *92*, 639–646. [CrossRef]
111. West, M.B.; Wickham, S.; Quinalty, L.M.; Pavlovicz, R.E.; Li, C.; Hanigan, M.H. Autocatalytic cleavage of human γ-glutamyl transpeptidase is highly dependent on *N*-glycosylation at asparagine 95. *J. Biol. Chem.* **2011**, *286*, 28876–28888. [CrossRef] [PubMed]

International Journal of
Molecular Sciences

MDPI

Review

Drosophila melanogaster Models of Metal-Related Human Diseases and Metal Toxicity

Pablo Calap-Quintana [1], Javier González-Fernández [1,2], Noelia Sebastiá-Ortega [1,3], José Vicente Llorens [1,*] (ORCID) and María Dolores Moltó [1,2,3]

1   Department of Genetics, University of Valencia, Campus of Burjassot, 46100 Valencia, Spain; pablo.calap@uv.es (P.C.-Q.); javier.gonzalez-fernandez@uv.es (J.G.-F.); noelia.sebastia@uv.es (N.S.-O.); dolores.molto@uv.es (M.D.M.)
2   Biomedical Research Institute INCLIVA, 46010 Valencia, Spain
3   Centro de Investigación Biomédica en Red de Salud Mental CIBERSAM, Spain
*   Correspondence: j.vicente.llorens@uv.es; Tel.: +34-963-544-4504

Received: 12 June 2017; Accepted: 30 June 2017; Published: 6 July 2017

Abstract: Iron, copper and zinc are transition metals essential for life because they are required in a multitude of biological processes. Organisms have evolved to acquire metals from nutrition and to maintain adequate levels of each metal to avoid damaging effects associated with its deficiency, excess or misplacement. Interestingly, the main components of metal homeostatic pathways are conserved, with many orthologues of the human metal-related genes having been identified and characterized in *Drosophila melanogaster*. *Drosophila* has gained appreciation as a useful model for studying human diseases, including those caused by mutations in pathways controlling cellular metal homeostasis. Flies have many advantages in the laboratory, such as a short life cycle, easy handling and inexpensive maintenance. Furthermore, they can be raised in a large number. In addition, flies are greatly appreciated because they offer a considerable number of genetic tools to address some of the unresolved questions concerning disease pathology, which in turn could contribute to our understanding of the metal metabolism and homeostasis. This review recapitulates the metabolism of the principal transition metals, namely iron, zinc and copper, in *Drosophila* and the utility of this organism as an experimental model to explore the role of metal dyshomeostasis in different human diseases. Finally, a summary of the contribution of *Drosophila* as a model for testing metal toxicity is provided.

Keywords: *Drosophila*; metal homeostasis; iron; copper; zinc; frataxin; ATP7; dZip99C; neurodegeneration; heavy metal toxicity

1. Introduction

Model organisms have been extensively used to unravel basic and conserved biological processes. The fruit fly, *Drosophila melanogaster* (hereinafter *Drosophila*), is one of the most studied eukaryotic organisms and has made fundamental contributions to different areas of biology. *Drosophila* has also gained appreciation as a useful model organism of human diseases. Comparative genomic studies estimate that up to 75% of the human genes implicated in diseases are conserved in *Drosophila* [1]. The similarity between human and *Drosophila* genomes is not limited only to genetic elements, but also to the relationship between them, with numerous examples of conserved biological mechanisms. The *Drosophila* genome is smaller in size and has a smaller number of genes compared to the human genome, which facilitates genetic studies [2]. Accordingly, many human gene families are composed of paralogues with redundancy of overlapping functions that correspond to a single gene or a smaller gene family in *Drosophila*.

Flies are greatly appreciated in the laboratory because they are undemanding animals, with a short generation interval and inexpensive maintenance. Flies can easily be handled, bred and genetically manipulated in large numbers. Currently, there is a high number of genetic tools available in *Drosophila* that allows researchers to address some of the outstanding questions concerning basic processes underlying human diseases. Furthermore, fly models can be exploited very successfully to discover the genetic modifiers of disease phenotypes by means of modifier screens [3]. This methodology is central in identifying novel genes that function in the same disease processes and to progress in understanding the disease pathology. In turn, this is essential for researching appropriate therapies. Nowadays, different human diseases have been modeled in *Drosophila*, which cover a wide range of physiological alterations, including metabolic dysfunctions and neurodegeneration [4].

Drosophila is also useful for studying potential toxic effects of different compounds, such as metals [5]. As flies have a short biological life cycle, it is possible to address metal toxicity during development and adulthood. Survival, neuronal function and behavior assays are easy to perform in this organism. It is also simple to search for toxicity-mediated mechanisms at the molecular level in *Drosophila* [6,7].

This review focuses first on the conserved components that control metal homeostasis in flies. Following this, it provides instructive examples about what we have learned from *Drosophila* as a model for inherited diseases related to metal dysregulation and as a model for testing metal toxicity.

2. *Drosophila* Genes Involved in Metal Homeostasis

Approximately 25% of proteins require metals to perform their function [8]. Iron (Fe), copper (Cu) and zinc (Zn) are essential transition metals that organisms assimilate from food. Homeostatic control of these metals at systemic and cellular levels is vitally fine-tuned to ensure their availability as cellular nutrients. This fine-tuning also prevents detrimental effects of their deficiency, excess or misplacement. Proteins for import, transport, storage, excretion and regulation of metals are essential components of pathways for metal metabolism and homeostasis. Interestingly, the main players of these pathways are conserved through evolution, with many orthologues of the human metal-related genes having been identified and characterized in *Drosophila*. Here, we provide an overview of the uptake, transport and efflux of Fe, Cu and Zn at the cellular level in flies.

2.1. Iron

Iron (Fe) is required for the survival of almost all organisms, because it plays a crucial role in many biological processes, such as oxygen transport, cellular respiration, gene regulation and DNA biosynthesis. Fe participates in such processes through the heme and iron-sulfur cluster (ISC) prosthetic groups in which Fe shows its ability to exchange electrons with different substrates.

Fe deficit compromises the function of the metabolic pathways in which Fe acts as an important cofactor. A misplaced and increased level of Fe promotes the formation of reactive oxygen species (ROS) through the Fenton–Haber–Weiss reaction [9], including the highly reactive hydroxyl radicals. This results in glutathione consumption, lipid peroxidation and DNA damage, which finally can compromise cell viability.

Drosophila shares many key genes involved in iron metabolism with mammals (Table 1, Figure 1A). *Malvolio* (*Mvl*), the *Drosophila* orthologue of the mammalian *Divalent metal transporter-1* (*DMT1*) [10] is expressed in the fly midgut and has been proposed to function as an importer for dietary iron in a similar way to its mammalian counterpart [11]. Mammals and insects store iron absorbed in a bioavailable form of ferrihydrite inside ferritin, a protein formed of heavy and light chain subunits (coded by the *Fer1HCH* and *Fer2LCH* genes respectively in *Drosophila*). While ferritin is mainly a cytosolic protein in mammals, it is found in the secretory system (endoplasmic reticulum (ER), Golgi apparatus and secretory vesicles) in most insects, including *Drosophila*, and is secreted in the fly hemolymph in large amounts [12,13]. Transferrin 1 is also an iron-binding protein abundant in the hemolymph for which multiple functions have been suggested in insects. Nevertheless, it is still unsolved whether this protein serves as an iron transport carrier between cells in a similar manner to the mammalian transferring [14,15]. Iron absorption and metabolism are post-transcriptional processes regulated by the IRP/IRE system, which is conserved through diverse taxonomic groups [12,16]. When Fe concentration is low, iron regulatory proteins (IRPs) bind to iron-responsive elements (IREs) on the 5′UTR and 3′UTR mRNAs of their target proteins, which controls translation initiation or mRNA stability (reviewed in [17]). Two IRP genes (*Irp-1A* and *Irp-1B*) have been described in *Drosophila* and have both shown aconitase activity, but only Irp-1A functions as an iron regulatory protein [18].

Table 1. Conserved proteins in iron homeostasis.

Human Gene	Primary Metals	Metal-Related Function	*Drosophila* Orthologue	References
Divalent Metal Transporter 1 (DMT1, SLC11A2)	Fe	Divalent metals transport / Iron absorption	*Malvolio (Mvl)*	[10,11]
Ferritin Heavy Chain 1 (FTH1)	Fe	A component of ferritin / Iron storage	*Ferritin 1 heavy chain homologue (Fer1HCH)*	[12,15]
Ferritin Light Chain (FTL)	Fe	A component of ferritin / Iron storage	*Ferritin 2 light chain homologue (Fer2LCH)*	[12,15]
Ferritin Mitochondrial (FTMT)	Fe	Iron storage / Oxidative stress protection	*Ferritin 3 heavy chain homologue (Fer3HCH)*	[19]
Transferrin (TF)	Fe	Serum iron binding transport protein	*Transferrin 1 (Tsf1)*	[13,14]
Aconitase 1 (ACO1)	Fe	Iron sensor	*Iron regulatory protein 1A (Irp-1A)* / *Iron regulatory protein 1B (Irp-1B)*	[16,18]
Mitoferrin 1 (SLC25A37)	Fe	Mitochondrial iron importer	*Mitoferrin (mfrn)*	[19]
Frataxin (FXN)	Fe	Mitochondrial iron chaperone	*Frataxin (fh)*	[20]
Duodenal cytochrome b (DCYTB, CYBRD1)	Fe	Ferric-chelate reductase that reduces Fe^{3+} to Fe^{2+}	*No extended memory (nemy)* / *CG1275*	[21]
Hephaestin (HEPH)	Fe	Ferroxidase activity oxidizing Fe^{2+} to Fe^{3+}	*Multicopper oxidase-1 (Mco1)*	[22]
Ceruloplasmin (CP)	Fe	Ferroxidase activity oxidizing Fe^{2+} to Fe^{3+}	*Multicopper oxidase-3 (MCO3)*	[23]

Gene symbols for human genes are indicated according to the Human Genome Organization Gene Nomenclature Committee (HGNC) (Available online: http://www.genenames.org). An alias is indicated in several cases. *Drosophila* gene symbols are cited in agreement with the Flybase (Available online: http://flybase.org).

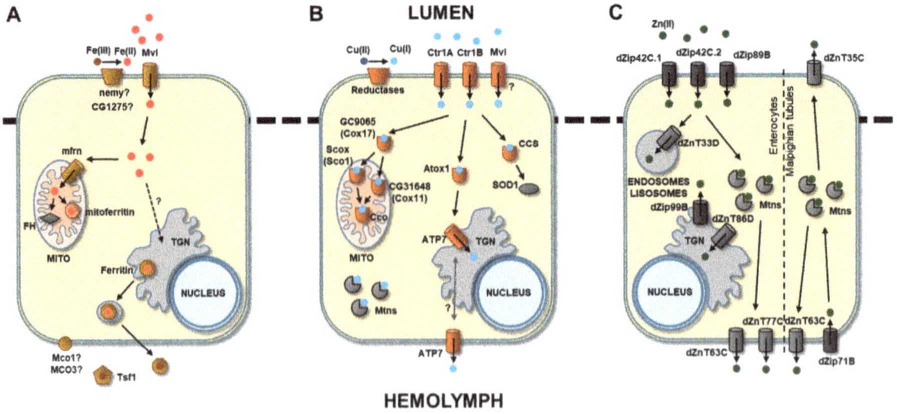

Figure 1. Main pathways of (**A**) iron, (**B**) copper and (**C**) zinc uptake, storage and export in *Drosophila melanogaster*. TGN, trans-Golgi network; MITO, mitochondria. Question marks represent an unknown mechanism. Arrows represent the direction of the metal transport.

Genes involved in mitochondrial Fe metabolism are also conserved between flies and humans. A mitochondrial ferritin has been identified in both organisms (*Fer3HCH* is the encoding gene in *Drosophila*), which has shown an important antioxidant role through the regulation of mitochondrial Fe availability [19,24]. In fact, the expression of human mitochondrial ferritin in frataxin-deficient cells protects mitochondria from oxidative stress, which is a situation that provokes mitochondrial iron overload [25,26]. Similarly, the expression of *Fer3HCH* in frataxin-deficient flies is able to extend the mean lifespan of the adult individuals [27]. Bridwell-Rabb et al. propose that frataxin is an iron chaperone involved in the regulation of ISC biosynthesis [28].

Drosophila also has other orthologues with known iron-related functions in mammals that remain to be fully elucidated in flies. The *Duodenal cytochrome b* (*Dcytb*) is highly expressed in the intestinal epithelia of mammals and has ferric reductase activity to convert the inorganic dietary iron(III) state, Fe^{3+} to the iron(II) state and Fe^{2+} in the absorption process. In *Drosophila*, two candidates for this function have been proposed, which are *no extended memory* (*nemy*) and the gene with the annotation ID of *CG1275* [21]. Fe^{2+} is oxidized to Fe^{3+} for efflux from intestinal enterocytes into the circulatory system. Mammalian hephaestin, a transmembrane copper-dependent ferroxidase, plays this role and mediates iron efflux in cooperation with the basolateral iron transporter, ferroportin 1. Another copper-dependent oxidase enzyme, ceruloplasmin, participates in the iron transport in plasma in association with transferrin, which carries iron in the ferric state. It has been suggested that the *multicopper oxidase-1* (*Mco1*) and the *multicopper oxidase-3* (*MCO3*) may be involved in iron homeostasis in *Drosophila* showing hephaestin-like [22] and ceruloplasmin-like [29] functions, respectively. It has been recently reported that *Mco1* orthologues from diverse insect species, including *Drosophila*, function as ascorbate oxidases [23], and there is accumulating evidence strongly suggesting that ascorbate within mammalian systems can regulate cellular iron uptake and metabolism [30]. Thus, iron homeostasis might also be influenced by the ascorbate redox state in *Drosophila*.

Finally, there are some mammalian genes participating in Fe metabolism with no known orthologues in flies, such as those genes encoding the transferrin receptor, the iron export protein, ferroportin or the iron hormone, hepcidin [31]. It may account for differences in the control of systemic Fe levels between insects and mammals. However, *Drosophila* has become a powerful model for studying human disorders caused by mutations in the main components of cellular Fe homeostasis pathways. Friedreich's ataxia is an example of these human conditions, and we report in this review the contribution of *Drosophila* as an experimental model in the advancement of knowledge in this disease.

2.2. Copper

Copper (Cu) is a vital co-factor for enzymes involved in a wide range of roles, from oxidative stress defense (superoxide dismutase (SOD)) to pigmentation (tyrosinase) or energy production (cytochrome c oxidase). Its usefulness for life comes from its ability to change its oxidation state between the cupric (Cu(II)) and cuprous (Cu(I)) forms.

In mammals, copper is distributed throughout the organism in two steps [32]. First, Cu is absorbed in the apical surface of the enterocytes of the small intestine to be exported to the circulation bound to serum proteins. Following this, Cu is absorbed by the liver, where it is bound to the serum ceruloplasmin and exported again to the circulation for its use by the peripheral tissues. The liver is also responsible for the removal of excess Cu, exporting it to the bile for its elimination through the feces. The main transporter for cellular import of Cu is CTR1 (copper transporter 1), which belongs to a family of small proteins with three transmembrane domains [33]. Before Cu is transported by CTR1, Cu(II) is reduced to Cu(I) by reductases in the plasma membrane. The transporter DMT1, which is involved in iron uptake and reabsorption, could also import Cu, but this possibility requires further research [34,35].

In *Drosophila*, copper is transported in a single step from the midgut to all tissues through the circulation. The multicopper oxidase MCO3 is similar to ceruloplasmin in its ferroxidase activity [29], but it has not been detected in the hemolymph [36,37]. Some authors suggest that another MCO orthologue could still be implicated in Cu homeostasis [38]. As for Cu excretion, it has been observed that in larvae growing in media with a high Cu concentration, the metal accumulates in the Malpighian tubules [39]. These organs are located between the midgut and the hindgut, which regulate fluid and ion balance [40]. Malpighian tubules may absorb excess Cu for its excretion to the hindgut and elimination through the feces [38]. There are three *CTR1* genes in *Drosophila* (Table 2) with some differences in their expression patterns [41]. *Ctr1A* is ubiquitously expressed and is required for survival [42], suggesting that it is the primary transporter for dietary Cu uptake. *Ctr1B* expression in the gut is transcriptionally induced when there is a Cu limitation and repressed in the case of excess [41]. The expression of *Ctr1C* seems to take place mainly in the male germline. In *Drosophila*, *Mvl* (the orthologue of the human *DMT1*) could also contribute to Cu uptake, since a reduction in *Mvl* activity causes an increased sensitivity to the Cu limitation in females and S2 cells [43].

Table 2. Conserved proteins in copper homeostasis.

Human Gene	Primary Metals	Metal-Related Function	*Drosophila* Orthologue	References
Solute carrier family 31 member 1 (SLC31A1, CTR1)	Cu	Copper uptake	Copper transporter 1A (Ctr1A) Copper transporter 1B (Ctr1B) Copper transporter 1C (Ctr1C)	[33,41,42]
Copper chaperone for superoxide dismutase (CCS)	Cu	Chaperone; copper donor to SOD1	Copper chaperone for superoxide dismutase (Ccs)	[44,45]
Cytochrome c oxidase copper chaperone (COX17)	Cu	Chaperone; copper donor to COX11 and SCO1	CG9065	[46]
Cytochrome c oxidase copper chaperone (COX11)	Cu	Chaperone; copper transfer to cytochrome c oxidase	CG31648	[46]
Cytochrome c oxidase assembly protein (SCO1)	Cu	Chaperone; copper transfer to cytochrome c oxidase	Synthesis of cytochrome c oxidase (Scox)	[46,47]
Antioxidant 1 copper chaperone (ATOX1)	Cu	Chaperone; copper donor to ATP7A and ATP7B	Antioxidant 1 copper chaperone (Atox1)	[48]
ATPase copper transporting α (ATP7A)	Cu	Copper delivery to proteins in the secretory pathway; copper efflux	ATP7 (ATP7)	[49–55]
ATPase copper transporting β (ATP7B)	Cu	Copper delivery to proteins in the secretory pathway; copper efflux	ATP7 (ATP7)	[49–56]

Gene symbols for human genes are indicated according to the Human Genome Organization Gene Nomenclature Committee (HGNC) (Available online: http://www.genenames.org). An alias is indicated in several cases. *Drosophila* gene symbols are cited in agreement with the Flybase (Available online: http://flybase.org).

Once copper has entered the cell (Figure 1B), it is transported to its target proteins by Cu-specific chaperones with a high affinity for the metal: CCS, COX17 and ATOX1. The CCS (copper chaperones for superoxide dismutase) provide Cu for SOD1 [44]. COX17 transfers Cu to the proteins, COX11 and SCO1, which are located in the inner mitochondrial membrane and deliver the metal to the cytochrome c oxidase [46]. In mammals, ATOX1 transfers Cu to ATP7A and ATP7B, which are two P-Type ATPase transporters. ATP7A donates Cu to different enzymes in the secretory pathway [33,49,50]. It is also required for Cu transport across the blood brain barrier (BBB) and from intestinal enterocytes into the circulation. ATP7B delivers Cu to ceruloplasmin, and in the case of Cu excess, Cu moves to the biliary canalicular membrane for its excretion to the bile duct and its elimination through the digestive tract [49,50,56]. One important characteristic of both transporters is their capacity for trafficking between the trans-Golgi network and the cell membrane in response to Cu levels.

In *Drosophila*, orthologues of Atox1, CCS, SCO1 and COX17 have been found, and they are thought to perform the same roles as in mammals [45,47,48]. However, in the fly, there is only one Cu transporting P-type ATPase named ATP7, which is expressed in all tissues [51,52]. This transporter is similar to ATP7A, since it is required for transferring Cu to enzymes in the secretory pathway [52–54] and from the gut cells to the circulatory system [55]. Moreover, the similarity between ATP7 and ATP7A has also been observed in functional assays and in the domains that share both proteins [51]. Despite the fact that the efflux activity of ATP7 has been proven [53,55], the translocation of the transporter from the trans-Golgi network to the cell membrane in response to the Cu levels has not been demonstrated yet [38].

Copper levels must be tightly controlled because besides being critical for life, Cu can also be potentially toxic. It can catalyze the formation of hydroxyl radicals through Fenton reactions and alter the structure of proteins [57,58]. As a mechanism of scavenging and/or storage of Cu and other metals, all eukaryotes express a group of proteins known as metallothioneins. These are low weight proteins with a high content of cysteines that have a great metal binding capacity [59–61]. Metallothioneins are expressed at a basal level and are strongly induced in response to heavy metal load in order to cope with their toxicity [59–61]. In mammals, there are four major members of this protein family, namely MT-I, MT-II, MT-III and MT-IV [62]. In *Drosophila*, five different metallothioneins have been described, namely MtnA, MtnB, MtnC, MtnD and MtnE [63,64].

The transcription factor MTF-1 (metal-responsive transcription factor-1) is the main regulator of the response to heavy metal load, including Cu [61,65,66]. In normal conditions, the majority of MTF-1 is located in the cytoplasm. In conditions of heavy metal load, MTF-1 accumulates in the nucleus and binds to the metal response elements (MRE) to induce the expression of its target genes, which include the metallothioneins, both in mammals and *Drosophila* [66,67]. In *Drosophila*, MTF-1 is also responsible for the induction of *Ctr1B* expression under limited Cu [68] and *ATP7* in response to excess Cu [55], while regulation of *CTR1*, *ATP7A* or *ATP7B* by MTF-1 has not been observed in mammals [38].

2.3. Zinc

Zinc (Zn) is an indispensable and ubiquitous trace element present in many tissues in a wide range of organisms and is involved in many general cellular functions. It has multitudinous effects on growth, development, immune system and nervous system function [69]. As a regulatory, catalytic and structural component, Zn is also needed for proteins involved in DNA repair, transcription, translation and protein signaling, which requires tight homeostasis. It modulates the stability of cell membranes by reducing peroxidative damage [70] and provides protection against the disruption of cells [71] by maintaining the structure and function of the membrane barrier [72]. The majority of Zn is bound with high affinity to metalloproteins and subsequently is said to be non-bioavailable [73]. Furthermore, approximately 40% of putative zinc-binding proteins are transcription factors carrying zinc-binding domains, such as zinc finger motifs, zinc cluster and zinc twist. These enable specific gene regulatory processes to be activated [73]. Other studies have reported the ability of Zn to act as a neurotransmitter and a second messenger [74].

The bioavailable Zn concentration is maintained at a nanomolar level through uptake, storage and secretion [75]. Several studies have indicated that Zn transporters are likely to be the main mechanism of Zn movement in cells, although diffusion of Zn ions may occur under physiological conditions [76]. In mammals, there are two Zn transporter families. One is composed of solute-linked carrier genes, namely the SLC39a/Zrt-Irt-like protein family (also known as ZIP proteins), while the other family includes cation diffusion facilitator genes, the SLC30a family (also known as ZnT proteins) [77].

The ZnT family of transporters specifically promotes the transport of Zn from the cytosol to extracellular environment or transfers Zn into organelle compartments within the cell, thereby reducing the concentration of cytosolic zinc. The distribution of ZnTs is widespread, although some are cell-specific [78]. The ZIP family is comprised of genes that are ubiquitously expressed in all organisms. This family promotes the transport of Zn from the extracellular environment into the cytoplasm or from the cellular organelles to the cytoplasm [79]. Thus far, at least fourteen ZIPs and ten ZnTs [77] have been identified in mammals, while ten ZIPs and seven ZnTs orthologues have been identified in *Drosophila* that control the zinc homeostasis (Table 3).

Table 3. Conserved proteins in zinc homeostasis.

Human Gene	Primary Metals	Metal-Related Function	*Drosophila* Orthologue	References
SLC30A1 (hZnT1)	Zn	Exporting cytosolic zinc into the extracellular space	*dZnT63C*	[80]
SLC30A10 (hZnT10)	Mn, Zn	Zinc transporter localized to early/recycling endosomes or Golgi	*dZnT77C*	[81,82]
SLC30A2 (hZnT2)				[83–85]
SLC30A3 (hZnT3)	Zn	Transporting zinc into the lumen of vesicular compartments	*dZnT33D* *dZnT35C*	[86]
SLC30A8 (hZnT8)				[87,88]
SLC30A4 (hZnT4)	Zn	Maintenance of cytosolic zinc; homeostasis by controlling zinc; translocation to the lysosomes	*dZnT41F*	[89]
SLC30A7 (hZnT7)	Zn	Transports zinc into early secretory pathway and contributing to its homeostatic control	*dZnT86D*	[90,91]
SLC30A9 (hZnT9)	Zn	No zinc transport functions; acts as nuclear receptor coactivator	*dZnT49B*	[92]
SLC39A1 (hZIP1)			*dZip42C.1*	[93,94]
SLC39A2 (hZIP2)	Zn	Imports zinc from the extracellular space	*dZip42C.2* *dZip89B*	[95,96]
SLC39A3 (hZIP3)			*dZip88E*	[97,98]
SLC39A5 (hZIP5)	Zn	Zinc importer	*dZip71B*	[99,100]
SLC39A6 (hZIP6)	Zn	Zinc importer that can be a growth factor-elicited signaling molecule	*fear of intimacy (foi)*	[101,102]
SLC39A10 (hZIP10)				[103,104]
SLC39A7 (hZIP7)	Zn	Zinc importer from endoplasmic reticulum and Golgi apparatus; implicated in the glycemic control in skeletal muscle	*Catecholamines* up *(catsup)*	[105,106]
SLC39A9 (hZIP9)	Zn	Zinc importer localized to the Golgi apparatus and the cell surface; plays a crucial role in B-cell receptor	*dZip102B*	[107,108]
SLC39A11 (hZIP11)	Zn	Not well defined	*dZip48C*	[109]
SLC39A13 (hZIP13)	Fe	Mobilizes zinc from the lumen of Golgi apparatus and cytoplasmic vesicles to cytosol and plays a pivotal role in cellular signaling	*dZip99C*	[110]
	Zn			[111]

Gene symbols for human genes are indicated according to the Human Genome Organization Gene Nomenclature Committee (HGNC) (Available online: http://www.genenames.org). An alias is also indicated for them. *Drosophila* gene symbols are cited in agreement with the Flybase (Available online: http://flybase.org).

Catsup (*Catecholamines up*) and foi (*fear of intimacy*) are the best characterized Zn transporters in *Drosophila*. *Catsup* is the orthologue of the human *hZIP7* gene. The encoded protein is localized in the ER and Golgi apparatus, playing a more general role in the maintenance of the function of these organelles. Catsup is present in a punctate pattern in the cell body, axonal and dendritic synapses of neurons in the central brain in *Drosophila*. It has been shown to play a role as a negative regulator of the catecholamine biosynthesis pathway [112], as well as in synaptic transport and release of dopamine [113]. Catsup mutations produce abnormal accumulation of membrane proteins, such as Notch receptor, EGFR and APPL in ER and Golgi apparatus [114].

Foi has been suggested to be the orthologue of the human genes *hZIP6* and *hZIP10*, with the encoded protein located in the basolateral cell membrane. It is implicated in cell migration in the gonad and trachea morphogenesis [115,116], as well as in embryonic glial cell patterning [117]. Foi is also necessary for correct specification and consequent differentiation of mesodermal derivatives, acting as a Zn transporter that regulates Zn cellular homeostasis. Its function can be partially replaced by other Zn transporters of the same family that have a similar subcellular localization [118].

In *Drosophila*, *dZip42C.1*, *dZip42C.2* and *dZip89B* (likely orthologues of *hZIP1*, *hZIP2* and *hZIP3*) are the most important transporters for absorption of Zn from the intestinal lumen. All of them are located in the apical membrane of enterocytes (Figure 1C), and their expression can be affected by dietary Zn levels (*dZip42C.1* and *dZip42C.2*) [119] or can be independent of Zn content acting as a constitutive Zn transporter with a lower affinity for this metal (*dZip89B*) [120].

The Zn release to the hemolymph is mediated through two overlapping Zn exporters, *dZnT63C* and *dZnT77C* (orthologues of *hZnT1* and *hZnt10*) both located in the basolateral membrane of enterocytes. Ubiquitous RNA interference (RNAi) of *dZnT63C* results in developmental arrest at the larval stage, lethality before pupae formation and Zn accumulation under physiological conditions [121]. Knocking down *dZnT77C* also produces a greater sensitivity to Zn deficiency [119], demonstrating that they are key Zn transporters in dietary Zn absorption.

Intracellular Zn homeostasis and transport is poorly understood in *Drosophila*. Even so, *dZnT86D* is expressed in Malpighian tubules, midgut and other organs. The encoded protein is located in the Golgi apparatus [122] and has been shown to be important for the control of Zn levels. Ubiquitous knockdown of *dZnT86D* causes a severe reduction in Zn level in the whole fly, while gut-specific knockdown of this gene produces little effect on Zn metabolism, but local alteration of the Zn level in Golgi apparatus. The larval lethality resulting from ubiquitous silencing of *dZnT86D* is partially rescued by expression of *hZnT7*, which corroborates that it is the orthologue [119].

dZnT33D is another intracellular Zn transporter located in the vesicles, which is the orthologue of *hZnT4*. Knocking down this gene ubiquitously induces larval and embryonic lethality, as well as the expression of MtnB [123]. This indicates that dZnT33D is an important regulator of intracellular Zn homeostasis, although it does not play a role in the dietary Zn absorption.

Zn reabsorption and excretion are also important processes in the homeostasis of this metal. dZnT63C is also localized in the basolateral membrane of cells in Malpighian tubules, which suggests its implication in reabsorption of Zn [121]. *dZnT35C* and *dZip71B* are counterparts in Zn excretion. *dZnT35C* (orthologue of *hZnT3* and *hZnT8*) is located in the apical membrane in the cells and mutant flies that has accumulated Zn, which is in support of a role in Zn excretion [124]. *dZip71B* (orthologue of *hZIP5*) is situated in the basolateral membrane, with gene knockdown producing a hypersensitivity to Zn overload similar to *dZnT35C*, although this does not happen on normal media. In contrast to *dZnT35C*, *dZip71B* knockdown shows decreased Zn content, implying that dZip71B acts upstream of dZnT35C to drive Zn excretion in the body [125]. It is suggested that dZip71B functions in transporting Zn from body into Malpighian tubules and acts together with apically localized dZnT35C to pump Zn out of the body. Meanwhile, dZnT63C is responsible for zinc reabsorption in Malpighian tubules [125]. Finally, dZnT41F is located intracellularly, and its function is not well defined, although the principal expression seems to be in the Malpighian tubules. This suggests dZnT63C may be involved in Zn homeostasis in this organ.

There are other Zn transporters that are predicted to be ZIP and ZnT orthologues in *Drosophila*, and phylogenetic analysis indicated that they are related to mammalian genes. *dZip48C*, *dZip102B* and *dZnT49B* are the orthologues of the human genes *hZIP11*, *hZIP9* and *hZnT9* respectively. Furthermore, *dZip88E* may be an orthologue of *hZIP1* and *hZIP2* or *hZIP3*. Further studies should be carried out to demonstrate their functions.

No studies have so far addressed the question of why there are so many zinc transporters needed to regulate Zn homeostasis. One possible explanation is the vast requirement of Zn within the animal's body compared to other metals, such as Cu or Fe. Additionally, Zn transporters might participate in the homeostasis of different metals, because it has been reported that some of them can co-transport Zn and other metals [126,127]. *dZip99C*, a member of the ZIP family, was generally believed to import zinc, but its main function is as a Fe exporter for the secretory pathway [110]. Mutation in the human orthologue of *dZip99C*, the *hZIP13* gene [110] causes the rare disease spondylocheirodysplasia Ehlers-Danlos syndrome-like [111,128], which is described in this revision.

3. *Drosophila* as a Model of Human Inherited Diseases Related to Metal Homeostasis

Drosophila has become a powerful model for studying human disorders caused by mutations in main components of pathways of cellular metal homeostasis. We find illustrative examples in Friedreich's ataxia, an iron-related disorder; in Menkes and Wilson diseases caused by a deficiency and an excess of copper, respectively; and in spondylocheirodysplasia Ehlers-Danlos syndrome-like due to mutations in a member of the ZIP family of zinc transporters. In addition, *Drosophila* has shown its utility as an experimental model to explore the role of metal dyshomeostasis in different neurodegenerative diseases.

3.1. Friedreich's Ataxia

Friedreich's ataxia (FRDA) is a rare autosomal recessive disease with an estimated prevalence of 2–4/100,000 in populations of European origin [129]. It is a multisystemic condition affecting the central and peripheral nervous systems, the heart and other organs. FRDA is caused by a decrease in frataxin levels due to an intronic GAA triplet repeat expansion within *FXN*, the gene encoding this protein [130]. The GAA repeats can form a variety of unusual DNA structures when the GAA expansion reaches a certain size and have the potential to interfere *FXN* transcription through heterochromatic mechanisms [131].

Many studies in model organisms, such as *Saccharomyces cerevisiae*, *Mus musculus*, *D. melanogaster* and *Caenorhabditis elegans*, have contributed extensively to our current understanding of the frataxin function and the physio-pathological consequences of its deficiency [132]. Frataxin is a mitochondrial nuclear-encoded protein, which has a function that has still not been fully clarified, but it is accepted that frataxin is required for ISC biosynthesis and for cellular iron homeostasis [28]. Its deficit provokes dysfunction of ISC-containing enzymes, such as aconitase and some components of the mitochondrial respiratory chain complexes. Other hallmarks of frataxin deficiency are mitochondrial iron accumulation coupled to cytosolic iron depletion and increased susceptibility to oxidative stress [133]. In FRDA patients, excess of iron was first addressed in the myocardium and later reported in several tissues, such as the spleen, liver and cerebellum [134,135]. In FRDA models, mitochondrial iron accumulation was first described in yeast [136], before being described in mice [137] and *Drosophila* [138]. The role of iron in FRDA pathogenesis has been extensively investigated, specifically examining whether iron accumulation is either a primary event or an end-stage event, and focusing on the mechanism underlying iron-mediated toxicity.

The RNAi methodology has been combined with the GAL4/UAS system to successfully knock down the *Drosophila frataxin homolog* (*fh*) [20]. This experimental strategy has provided tissue-specific and ubiquitous knockdown mutants in which *fh* suppression is dose dependent [139,140]. Such studies have indicated that frataxin function is vital during development and that different tissues show distinct vulnerability to the frataxin deficit. Llorens et al. [140] obtained a knockdown fly line with a systemic three-fold decrease of frataxin to parallel patient's conditions more closely. Flies expressing the 30% residual frataxin have normal embryonic development, but show poor motor coordination and reduced lifespan in adulthood. These phenotypes are more severe under hyperoxic conditions, and aconitase activity is strongly reduced in this scenario. These results support the previous proposal that yeast frataxin can acts as an iron chaperone protecting aconitase against oxidative-mediated inactivation [141]. In addition, they are in favor of an important role of oxidative stress in the progression of FRDA. Most studies have suggested that iron accumulation becomes toxic in FRDA by promoting ROS production through the Fenton reaction [142].

Taking advantage of the valuable genetic screens methodology in *Drosophila*, Soriano et al. investigated whether genetic modification of pathways involved in metal homeostasis could improve the impairment in the motor performance of the FRDA model flies [143]. Knockdown of the iron homeostasis regulators *Irp1-A* and *Irp-1B* and their targets transferrin (*Tsf1* and *Tsf3*) and *Mvl* was sufficient to suppress this phenotype by reducing iron overload in the majority of cases [143]. The function of *Tsf3* has not been characterized yet, and its identification as a suppressor for FRDA phenotypes in flies provides indirect evidence of its involvement in iron metabolism in *Drosophila*. Similarly, genetic suppression of the mitochondrial iron importer mitoferrin [19] was able to counteract physiological and molecular phenotypes of frataxin loss-of-function in flies [27].

A novel ROS independent mechanism that may contribute to neurodegeneration in FRDA has recently been proposed using mutant flies with a chemically-induced missense mutation in *fh* [144]. This severe loss-of-function mutation causes an iron accumulation in adult photoreceptor neurons when it is selectively expressed in this tissue. No increase in ROS was detected, while sphingolipid synthesis was upregulated. This in turn activated the 3-phosphoinositide dependent protein kinase 1 (Pdk1) and myocyte enhancer factor-2 (Mef-2). Interestingly, blocking any step of the iron/sphingolipid/Pdk1/Mef-2 pathway suppresses neurodegeneration in mutant flies. Notably, this pathway is also activated in a mouse model with reduced Fxn levels in nervous systems and in the cardiac tissue of FRDA patients [145].

Overall, *Drosophila* models of FRDA support the proposal that iron plays a major role in FRDA physio-pathology. Although reduction in iron levels by iron chelation has been relatively successful in clinical trials [146,147], genetic or pharmacological intervention through the sphingolipid/Pdk1/Mef-2 pathway and pathways regulating iron homeostasis are new approaches to be explored in preclinical studies.

3.2. Menkes and Wilson Syndromes

Menkes and Wilson syndromes are Mendelian disorders of Cu metabolism caused by deficiency and excess of this metal, respectively [148]. Menkes disease is an X-linked recessive neurodegenerative disorder with reported incidences of 1 in 300,000 in European countries, 1 in 360,000 in Japan and 1 in 50,000–100,000 in Australia [149]. Patients suffer from skeletal defects, growth failure and progressive degeneration of the central nervous system and exists considerable variability in the severity of symptoms. The disease is caused by mutations in the *ATP7A* gene, which are mostly intragenic mutations or partial gene deletions [150]. The encoded protein is critical for intestinal absorption of Cu and its transport across the BBB. Therefore, in Menkes disease, Cu accumulates in the cells of the BBB and choroid plexus, making the metal unable to reach the neuronal tissue [148,151–153]. The decrease in Cu levels in the brain results in deficiencies in the activity of Cu-containing enzymes.

Mutation in the copper transporter *ATP7B* gene causes Wilson disease, an autosomal recessive condition with an estimated wide prevalence of 1–9 in 100,000. Patients suffer from chronic liver disease, cirrhosis and several neurological symptoms due to a failure of copper excretion, which leads to an accumulation of the metal in the liver and brain [154].

As indicated before, *Drosophila* only has one Cu transporting P-type ATPase (named ATP7), which has a function that seems to be more closely related to that of ATP7A [51]. To study the role of ATP7, Norgate and colleagues [52] created a null mutation of the gene using imprecise excision of an inserted P-element. The mutant larvae were extremely lethargic in comparison with the controls, while their mouthparts were smaller and reduced in pigmentation. It was observed in first instar larvae that the upregulation of *Ctr1B* did not take place, even under Cu-limiting conditions. The authors suggested that Cu remained trapped in the gut cells and was not transported to the rest of the organism. As mutant larvae failed to grow and develop to the second instar, it was not possible to conduct any study in adulthood. To solve this problem, Bahadorani et al. [155] obtained conditional mutants in which silencing of *ATP7* by an RNAi construct was restricted to the gut cells. Tissue-specific suppression of *ATP7* also resulted in pre-adult mortality, but about 50% of the affected individuals survived to adulthood. Therefore, in this model, it was possible to obtain some adult survivors to study the effects of the transporter repression in adulthood. These adults showed an average lifespan that was slightly shorter than controls and a reduction in whole-body Cu content. In addition, they were sensitized to oxidative stress, probably due to a decrease in the activity of the Cu/Zn superoxide dismutase [155]. Cu supplementation and *MTF-1* overexpression enhanced pupal survival to the adult stage, and based on this later finding, the authors suggested that induction of *MTF-1* expression could be used as an additional approach for the treatment of Menkes disease. They argued that the metallothioneins induced by MTF-1 could bind to the excess Cu accumulated in the gut, preventing its toxicity.

A more recent work has also shown how *Drosophila* can be a powerful model to study some details of the different gene mutations responsible for a disease. Mutations in *ATB7A* and *ATP7B* are not only responsible for Menkes and Wilson diseases, but can also be involved in other disorders, such as distal motor neuropathy [156] and Alzheimer's disease [157,158]. Mercer et al. [159] introduced different *ATP7A* and *ATP7B* pathogenic mutations into a genomic *ATP7* rescue construct. All mutations abrogated the in vivo function of *ATP7* in *Drosophila*, but the effect produced by each one of them was different. The authors also highlighted that this system can be useful to screen for compounds that would be able to restore the function of the transporters.

3.3. Spondylocheirodysplasia-Ehlers-Danlos Syndrome-Like

Dyshomeostasis of Zn is related to several human disorders, such as diabetes, cancer and neurodegenerative diseases, and is also related to compromised immunity of the host. Moreover, mutations in Zn transporters are associated with hereditary human diseases. Examples of these include the transient neonatal deficiency (*hZnT2* mutations), acrodermatitis enteropathica (*hZIP4* mutations) or the spondylocheirodysplasia form of Ehlers-Danlos syndrome (*hZIP13* mutations).

The Ehlers-Danlos syndrome (EDS) is a heterogeneous group of inherited disorders of connective tissue characterized by articular hypermobility, skin hyperelasticity and tissue fragility. It affects skin, ligaments, joints, bone, blood vessels and internal organs. The natural history and mode of inheritance differ between the six major types [160]. The spondylocheirodysplasia form of Ehlers-Danlos syndrome (SCD-EDS) is a very rare condition with a worldwide prevalence of <1 in 1,000,000. It is an autosomal recessive disease caused by loss-of-function mutations in the *hZIP13* gene, which encodes a protein located in ER/Golgi apparatus [111,128]. The mouse knockout of *mZIP13* shows maturation defects affecting connective tissue development [111]. In these animals, the nuclear translocation of SMAD transcription factors is impaired, while their phosphorylation is unaffected. In addition, Zn levels in serum and isolated primary fibroblasts are not altered. At the same time, the Zn level in Golgi apparatus is upregulated and is downregulated in the nucleus, indicating an accumulation of this metal in ER/Golgi in the SCD-EDS [111]. It is proposed that zinc accumulation in ER/Golgi can compete with iron in the secretory pathway affecting the activity of enzymes in this pathway that use Fe as a cofactor. This is the case of two enzymes, namely lysyl and prolyl hydroxylases, which have a central role in the biosynthesis of collagens [161]. Mutations in the *PLOD1* gene encoding lysyl hydroxylase (LH1) cause Ehlers-Danlos type VI [162], which shares several clinical signs with SCD-EDS. In contrast to a zinc excess in SCD-EDS, a later study using human cell cultures has suggested that this disease may be the result of trapping Zn in vesicles leading to a deficiency of this metal in ER/Golgi rather than an overload and iron failing to compete with zinc in the importing assay [163].

Clarification of the SCD-EDS molecular pathology comes from analyzing the functions of *dZip99C*, the putative fly orthologue of *hZIP13*. Xiao et al. [110] generated transgenic lines of *Drosophila* to study the effects of modulating the expression of *dZip99C* in specific tissues and in the whole organism. Surprisingly, the authors found that in *Drosophila*, *dZip99C* (a presumed zinc importer) is considerably involved in dietary Fe absorption. Ubiquitous reduction of *dZip99C* expression provokes developmental arrest at the pupal stage, which can be rescued by dietary iron supplementation. Furthermore, the addition of the iron-specific chelator worsened the phenotype. Tissue-specific reduction of *dZip99C* expression in the midgut affects Fe content in whole fly, with a reduction of 50%, although the Zn level is unaltered. However, Fe levels are not reduced in the cytosol of enterocytes, which suggests that *dZip99C* is involved in Fe extrusion from the gut to the body. The authors investigated the relation of dZip99C and ferritin on iron assimilation, because *Drosophila* ferritin transport absorbed dietary iron through the secretory pathway for its systemic use [13]. Thus, dZip99C is essential for ferric iron loading in the secretory pathway.

The intracellular location of dZip99C was examined in human intestinal Caco2 cells transfected with a myc-tagged dZip99C construct and in the fly gut. Both cases demonstrated that dZip99C is located in ER/Golgi in a similar way as its human counterpart. By means of radioactive Fe transport assay, it was also shown that dZip99C is able to transport Fe in *Escherichia coli*. The authors demonstrated that the human gene *hZIP13* but not *hZIP7*, another zinc transporter similar to *hZIP13* belonging to the same subfamily of LIV-1 transporters, rescues the developmental defects caused by knocking down *dZip99C* in flies. These results show that *dZip99C and hZIP13* are orthologues. The main function of *dZip99C* and probably for *hZIP13* is transporting iron into the secretory pathways. These results suggest that iron dyshomeostasis instead of zinc is the major cause of SCD-EDS. Mutations in *hZIP13* may compromise iron content into ER/Golgi, affecting the functions of lysyl and prolyl hydroxylases that are essential for the stability of the collagen triple helix. Further research in human tissues and other experimental models are needed to confirm iron dyshomeostasis in SCD-EDS.

3.4. Huntington's Disease

Huntington's disease (HD) is a neurodegenerative disorder characterized by movement alterations, psychiatric disturbances and cognitive dysfunction. It is transmitted in an autosomal dominant manner and its prevalence ranges from 5.96–13.7 cases per 100,000 in North America, northwestern Europe and Australia [164]. HD is caused by a CAG triplet repeat expansion located in exon 1 of the *HTT* gene, which encodes an expanded polyglutamine (polyQ) tract in the huntingtin (Htt) protein [165].

The mutant Htt-exon1 protein, carrying the polyQ tract, forms aggregates in vitro and in vivo [166]. Interestingly, it has been shown that the first 171-amino acid fragment of both normal and mutant proteins interacts with Cu [167] and that the metal favors the aggregation [168]. It is highly relevant that an increase in Cu levels has been reported in the brains of patients, as well as in mouse and rat models of HD [167,169,170].

The important role of Cu in HD has been highlighted using a *Drosophila* model of the disease [171]. Tissue-specific expression of the human Htt exon 1 protein carrying the polyQ expansion has been analyzed in flies. Expression of this mutated protein in the peripheral nervous system of the fly provokes alterations in motor coordination. These defects are ameliorated by reducing Cu import via expression of a *Ctr1B* RNAi. In contrast, these defects are exacerbated when reducing Cu export by an *ATP7* RNAi. Flies expressing the human mutant Htt in the eyes show photoreceptor degeneration, which is rescued by *Ctr1B* RNAi or *ATP7* overexpression and enhanced by *ATP7* RNAi. Analyzing the aggregation of Htt-exon1 polyQ protein in the brains of model flies, it is also observed that aggregation is alleviated by *Ctr1B* RNAi or *ATP7* overexpression and increased by *ATP7* RNAi. To further prove the key role of Cu in HD, Xiao et al. [171] mutated two critical residues potentially implicated in the binding of Cu to Htt exon1-polyQ. By doing so, the toxicity of the mutant protein was drastically reduced, and the subsequent phenotypes seemed to be no longer dependent on Cu. Based on these results, it was shown that HD involves both polyQ toxicity and a Cu-modulating effect, with these two pathways needing consideration in the treatment of the disease [171].

3.5. Parkinson's Disease

Parkinson's disease (PD) is a common neurodegenerative disorder characterized by movement impairments and mood alterations. It is caused by loss of dopaminergic (DAergic) neurons in the substantia nigra pars compacta (SNpc) [172], which is mostly due to a sporadic origin with a contribution of both environmental and genetic factors. There is also a small percentage of familial cases caused by single gene mutations that can be modeled in organisms like *Drosophila*. For instance, mutations in *PARK2* are responsible for a form of juvenile Parkinsonism [173]. *PARK2* codifies for PARKIN, an E3 ubiquitin ligase [174,175]. It has been observed that flies lacking the orthologue gene *parkin* show mitochondrial abnormalities in muscle and germline tissues in addition to increased sensitivity to oxidative stress [176–178]. Saini et al. [179] reported interactions of Fe and Cu with the parkin deficiency in *Drosophila*. Flies carrying a null mutation (*park*$^{25/25}$) of *parkin* [177] show a reduction in lifespan, which is rescued using the Cu chelator bathocuproine disulfonate or the Fe chelator bathophenanthroline sulfonated sodium salt. These flies also show an ROS reduction in their heads after the chelator treatments. The authors argued that the lifespan extension might be due to oxidative stress reduction through the sequestration of redox-active Cu and Fe. They further prove the interaction between parkin and Cu by overexpressing *Ctr1B* in the fly eye using the *GMR* promoter. An increased amount of imported Cu through Ctr1B along with the parkin deficiency was enough to induce a rough eye phenotype. This phenotype was not present in *park*$^{25/+}$ flies overexpressing the Cu transporter. Using the *parkin*-deficient flies, it was also shown that other metals could also influence the severity of PD. *Parkin* mutant flies showed an increase in their survival and eclosion frequency when they were fed on a high-zinc food [180]. These effects could be due to the antioxidant properties that Zn may exert by being able to protect protein sulfhydryl groups, compete against redox-active metals or upregulate the expression of metallothioneins [180,181].

Previous data have already shown that several metals could damage dopaminergic neurons in humans and other model organisms, leading to movement alterations [182–187]. This effect was also further explored in *Drosophila* by treating wild type flies (Canton-S) with iron, copper and manganese [6]. An acute and chronic exposure to high doses of these metals clearly reduced the lifespan of the flies and the number of neurons of dopaminergic neuron clusters in their brains.

One of the main hallmarks of PD is the presence of Lewy bodies, which are mainly composed of protein α-synuclein [188]. The toxicity of α-synuclein in dopaminergic neurons may be related to the formation of protein oligomers. It has been observed that some mutations of the α-synuclein gene (*SNCA*), such as A53T and A30P, provoke a faster aggregation in the mutant protein compared to the wild type one. Moreover, it has been reported that iron binding can increase α-synuclein aggregation [189] and that the mutant proteins may have a higher ferrireductase activity, converting Fe^{3+} to the more reactive Fe^{2+} [190]. This interaction between α-synuclein and Fe has also been recently explored in *Drosophila*. Flies expressing mutant A53T or A30P α-synuclein in dopaminergic neurons show a higher decline in motor function when treated with Fe compared to flies expressing the wild type protein in the same Fe conditions [191]. Similarly, Fe only induces a significantly selective dopaminergic neuron loss in the PPM3 neuron cluster only in the flies expressing the mutant proteins.

3.6. Alzheimer's Disease

Alzheimer's disease (AD) is a progressive neurodegenerative condition, which is the main cause of dementia among aged people [192]. The disorder is characterized by the shrinkage of several regions in the hippocampus and cortex implicated in memory and cognition. Characteristically, there is an accumulation of amyloid plaques rich in the β-amyloid (Aβ) peptide and neurofibrillary tangles with hyperphosphorylated tau in the brain [193]. The toxicity of Aβ aggregates may be due to degradation of the barrier properties of cellular membranes [194,195], binding to cell receptors [196–199] and the generation of oxidative stress, since Aβ aggregates can form complexes with redox active metals, such as Fe and Cu [200,201]. *Drosophila* has also been used to study the toxicity of the Aβ peptide and its interaction with metals. Expressing the 42-amino acid isoform of Aβ (Aβ42) in the neurons of flies resulted in a decrease in lifespan and locomotor performance in comparison with flies expressing the less toxic 40-amino-acid isoform of Aβ (Aβ40) [202]. The Aβ40 isoform does not contain the final two hydrophobic residues that make the Aβ42 isoform more prone to aggregation. Expression of the Aβ42 carrying the Arctic mutation (E22G) [203], which further increases aggregation, produces even more severe phenotypes. By means of microarray analysis and genetic screen, it was shown that oxidative stress plays an important role in the toxicity of Aβ42. In fact, the levels of carbonyl groups in the protein extracts of fly heads increase with the expression of the wild type Aβ42 or the Arctic Aβ42, which indicates the presence of oxidative damage. In the same study, a strong interaction between Fe and the toxicity of the Arctic Aβ42 is pointed out. The expression of the heavy or light subunits of ferritin rescues the phenotypes associated with the Aβ42 toxicity. After this rescue, flies have an increased survival and motor performance, while carbonyl groups' content is reduced. It was suggested that the effect of the ferritin subunits could be due, at least in part, to a reduction in the hydroxyl radical production by sequestering the Fe ions. In a further study, the same group confirmed the relevance of Fe in the Aβ toxicity [204]. Flies treated with the Fe-specific chelator YM-F24 had a reduced toxicity of this peptide, while knocking down the endogenous ferritin worsened it. Moreover, Fe actively affects the aggregation process of Aβ, delaying the formation of well-ordered aggregates and possibly promoting its toxicity by this effect. In a recent study, it was shown that Fe affects the dimerization of Aβ through the interaction with three N-terminal histidines in the peptide [205]. To analyze their position-dependent and position independent effects, histidines were systematically substituted by alanine residues in Aβ. It was found that the susceptibility of the flies to oxidative stress was determined only by the number of histidines.

Most AD clinical forms have sporadic etiology, but there are also some familial forms linked to mutations in the *APP* (amyloid precursor protein), *PSEN1* (presenilin 1) and *PSEN2* (presenilin 2) genes [206]. Presenilins act as catalytic subunits of the γ-secretase multi-protein complex [207], which in turn is involved in the cleavage of the amyloid precursor protein. Presenilins also seem to have a role in Cu and Zn uptake in mammalian systems [208]. This novel role of presenilins has been studied in *Drosophila* by tissue-specific knocking down of the single presenilins orthologue (*PSN*) in the digestive system [209]. It was observed that the *PSN* knockdown flies had an increased Cu tolerance due to a reduced Cu uptake. These flies showed a reduction in total Cu levels and were less sensitive to excess dietary Cu when overexpressing *ATP7*. Taking into account that presenilins are involved in the trafficking and localization of several proteins [210,211], it was suggested that an impaired cellular localization of Ctr1A and Ctr1B could be responsible for reduced Cu uptake.

4. *Drosophila* as Model for Testing Metal Toxicity

Heavy metals are natural components of the Earth's crust. Contamination of heavy metals in the environment is a major global concern, because of toxicity and threat to human life and ecosystems. In humans, exposure to heavy metals is mostly occupational [212]. Metal-containing dusts and fumes are generated along the complete life cycle of metal objects from ore mining through smelting and final product manufacturing to waste management and recycling. These dusts and fumes are found in the workplace atmosphere, sometimes at hazardous concentrations. Many metals are xenobiotics because they used to have a minimal presence (and hence, bioavailability) before emission into the environment from industry. Furthermore, they either are completely useless and toxic for the human organism (e.g., mercury, lead or cadmium) or are essential micronutrients, but toxic when given in excess concentrations (e.g., iron and copper). The toxic effects of heavy metals and their consequences show a broad spectrum [213]. For example, metals, such as arsenic, cadmium, lead and mercury, cause many altered conditions including hypophosphatemia, heart disease, liver damage, cancer, central nervous system injury and sensory disturbances.

Based on the chemical and physical properties of heavy metals, three major molecular mechanisms of toxicity have been distinguished [214]. The first is production of ROS by auto-oxidation and the Fenton reaction, which is characteristic for transition metals, such as iron or copper. The second is blocking of essential functional groups in biomolecules. These reactions have mainly been reported for non-redox-reactive heavy metals, such as cadmium and mercury. The final mechanism involves displacement of essential metal ions from biomolecules, which occurs with different types of heavy metals. As a result of these toxic effects, various cellular metabolic processes are inhibited, which are detrimental to the cell. *Drosophila* has also contributed to our understanding of the underlying mechanisms associated with metal toxicity. Alterations on the cell cycle, circadian rhythms and enzymatic pathways, DNA repair impairment in addition to genotoxicity are several consequences of metal toxicity found in flies. Table 4 summarizes the main sources of metal contamination and human exposition, as well as the clinical symptoms that have been reported in humans and *Drosophila* findings.

Table 4. Main sources, routes of human exposure, symptoms and contributions of research in *Drosophila*.

Metal	Natural and Human Sources	Main Human Exposure	Symptoms	*Drosophila* Findings	Reference
Aluminum	Water treatment agents, aerosol, cosmetics, food additives, beverage cans, cookware, fireworks, explosives, rubber manufacturing	Drinking water, food, inhalation, dermal contact, pharmaceuticals	Mouth ulcers, skin lesions, bone, lung and brain damage, neurodegeneration, loss of memory, problems with balance and loss of coordination	Neurological injury, neurodegeneration, developmental alterations, behavior impairment, lifespan reduction and daily rhythm alterations, increase iron accumulation and ROS production	[215–220]
Arsenic	Arsenic minerals, sedimentary bed rocks, mining, melting, pesticides, fertilizers, drugs, soaps	Drinking contaminated water	Abnormal heart beat, damage in blood vessels, skin lesions, cancer, neurological problems, high rate of mortality	Genotoxicity of methylated metabolites, susceptibility related with genes of the biosynthesis of glutathione, brain injury, developmental alterations	[215,221–226]
Cadmium	Batteries, plastics, pigments, weathering, volcanic eruptions, river transport, fertilizers, pesticides, smelting, mining	Contaminated food and drinking water, inhalation, occupational exposure	Renal dysfunction, bone and lung damage and kidney disease	Changes in transferase enzymatic activity, stress response, cell cycle alterations, interference in DNA repair mechanism	[215,227–230]
Chromium	Burning of petroleum, coil and oil, pigment oxidants, fertilizers, metal planting tanneries, sewage, metallurgy, paper production	Water, occupational exposure	Ulcers, fever, renal failure, liver damage and hemorrhagic diathesis	DNA damage, alterations in pre- and post-replication mechanism implicated in repair DNA, changes in humoral innate immune response	[215,227,231,232]
Lead	Pipes, paints, gasoline, cosmetics, bullets, pesticides, fertilizers, mining, fossil fuel burning	Occupational exposure, food, smoking and water	Arthritis, renal dysfunction, vertigo, hallucinations, birth defects, mental retardation, psychosis, hyperactivity, autism, brain damage	Alterations in presynaptic calcium regulation, identification of QTL associated with behavioral lead-dependent changes, weak mutagenic effect, endocrine disruption	[215,227,233–240]
Manganese	Steel industry, mining, soil erosion, fungicides, fertilizers, dry-cell batteries, fireworks, ceramics, paint, cosmetics	Occupational exposure, water and food	Manganism, tremors, psychosis, fatigue, irritability	Reduced cell viability, induction of ROS, decrease in lifespan and locomotor activity	[6,216,241–246]
Mercury	Agriculture, mining, wastewater discharges, batteries	Contaminated water and marine food	Brain damage, memory problems, depression, hair loss, fatigue, tremors, changes in vision and hearing	Morphometric changes, interference in cellular signaling pathways and enzymatic mechanisms, inhibition of Notch cleavage by γ-secretase	[6,215,243–249]

5. Conclusions

As a model organism, *Drosophila melanogaster* has been intensely studied for over a century. Today, the knowledge related to the biology of the fly and the tools to work with it vastly exceed those available for many other model organisms. The conservation of many cellular and organismal processes between humans and flies has also made *Drosophila* one of the best choices to start studying many issues relevant for human health. The fly contributes greatly to achieving a better understanding about metal uptake, distribution, storage or excretion, as well as in establishing the underlying toxicity mechanisms of metals and their role in the pathophysiology of many human diseases. We have presented here some illustrative examples of how *Drosophila* can be used to model and study rare diseases in which metals are the main players; other highly prevalent neurodegenerative disorders in which the importance of metals is increasingly being considered; or the toxicity of high levels of several metals on their own. Nevertheless, there are still many other metal-related disorders and questions that can be addressed by taking advantage of *Drosophila*, evidencing the great potential of this little organism.

Acknowledgments: This study was supported by a grant from Generalitat Valenciana, Spain (PROMETEOII/2014/067). Pablo Calap-Quintana is a recipient of a fellowship from Generalitat Valenciana, Spain.

Conflicts of Interest: The authors declare no conflicts of interest.

Abbreviations

ACO	Aconitase
AD	Alzheimer's disease
Aβ	β-amyloid peptide
BBB	Blood brain barrier
CCS	Copper chaperones for superoxide dismutase
CP	Ceruloplasmin
CTR1	Copper Transporter-1
Cu	Copper
DAergic	Dopaminergic
DMT1	Divalent metal transporter-1
EDS	Ehlers-Danlos syndrome
ER	Endoplasmic reticulum
Fe	Iron
FRDA	Friedreich's ataxia
FXN	Frataxin
HD	Huntington's disease
Htt	Huntingtin
IRE	Iron-responsive element
IRP	Iron regulatory protein
ISC	Iron-sulfur cluster
LH1	Lysyl hydroxylase
MCO	Multicopper oxidase
Mef-2	Myocyte enhancer factor-2
MRE	Metal response elements
MTF-1	Metal-responsive transcription factor-1
PD	Parkinson's disease
Pdk1	Protein kinase-1
polyQ	Polyglutamine
QTL	Quantitative trait locus
RNAi	RNA interference
ROS	Reactive oxygen species

SCD-EDS   Spondylocheirodysplasia form of Ehlers-Danlos syndrome
SNpc   Substantia nigra pars compacta
SOD   Superoxide dismutase
TF   Transferrin
Zn   Zinc

References

1. Reiter, L.T.; Potocki, L.; Chien, S.; Gribskov, M.; Bier, E. A systematic analysis of human disease-associated gene sequences in *Drosophila melanogaster*. *Genome Res.* **2001**, *11*, 1114–1125. [CrossRef] [PubMed]
2. Adams, M.D.; Celniker, S.E.; Holt, R.A.; Evans, C.A.; Gocayne, J.D.; Amanatides, P.G.; Scherer, S.E.; Li, P.W.; Hoskins, R.A.; Galle, R.F.; et al. The genome sequence of *Drosophila melanogaster*. *Science* **2000**, *287*, 2185–2195. [CrossRef] [PubMed]
3. Bier, E. Drosophila, the golden bug, emerges as a tool for human genetics. *Nat. Rev. Genet.* **2005**, *6*, 9–23. [CrossRef] [PubMed]
4. Pandey, U.B.; Nichols, C.D. Human disease models in *Drosophila melanogaster* and the role of the fly in therapeutic drug discovery. *Pharmacol. Rev.* **2011**, *63*, 411–436. [CrossRef] [PubMed]
5. Hosamani, R. Muralidhara Acute exposure of *Drosophila melanogaster* to paraquat causes oxidative stress and mitochondrial dysfunction. *Arch. Insect Biochem. Physiol.* **2013**, *83*, 25–40. [CrossRef] [PubMed]
6. Bonilla-Ramirez, L.; Jimenez-Del-Rio, M.; Velez-Pardo, C. Acute and chronic metal exposure impairs locomotion activity in *Drosophila melanogaster*: A model to study Parkinsonism. *Biometals* **2011**, *24*, 1045–1057. [CrossRef] [PubMed]
7. Paula, M.T.; Zemolin, A.P.; Vargas, A.P.; Golombieski, R.M.; Loreto, E.L.S.; Saidelles, A.P.; Picoloto, R.S.; Flores, E.M.M.; Pereira, A.B.; Rocha, J.B.T.; et al. Effects of Hg(II) exposure on MAPK phosphorylation and antioxidant system in *D. melanogaster*. *Environ. Toxicol.* **2014**, *29*, 621–630. [CrossRef] [PubMed]
8. Maret, W. Metalloproteomics, metalloproteomes, and the annotation of metalloproteins. *Metallomics* **2010**, *2*, 117–125. [CrossRef] [PubMed]
9. Andrews, N.C.; Schmidt, P.J. Iron homeostasis. *Annu. Rev. Physiol.* **2007**, *69*, 69–85. [CrossRef] [PubMed]
10. Rodrigues, V.; Cheah, P.Y.; Ray, K.; Chia, W. Malvolio, the *Drosophila* homologue of mouse NRAMP-1 (Bcg), is expressed in macrophages and in the nervous system and is required for normal taste behaviour. *EMBO J.* **1995**, *14*, 3007–3020. [PubMed]
11. Folwell, J.L.; Barton, C.H.; Shepherd, D. Immunolocalisation of the *D. melanogaster Nramp* homologue *Malvolio* to gut and Malpighian tubules provides evidence that *Malvolio* and *Nramp2* are orthologous. *J. Exp. Biol.* **2006**, *209*, 1988–1995. [CrossRef] [PubMed]
12. Nichol, H.; Law, J.H.; Winzerling, J.J. Iron metabolism in insects. *Annu. Rev. Entomol.* **2002**, *47*, 535–559. [CrossRef] [PubMed]
13. Tang, X.; Zhou, B. Ferritin is the key to dietary iron absorption and tissue iron detoxification in *Drosophila melanogaster*. *FASEB J.* **2013**, *27*, 288–298. [CrossRef] [PubMed]
14. Dunkov, B.; Georgieva, T. Insect iron binding proteins: Insights from the genomes. *Insect Biochem. Mol. Biol.* **2006**, *36*, 300–309. [CrossRef] [PubMed]
15. Tang, X.; Zhou, B. Iron homeostasis in insects: Insights from *Drosophila* studies. *IUBMB Life* **2013**, *65*, 863–872. [CrossRef] [PubMed]
16. Muckenthaler, M.; Gunkel, N.; Frishman, D.; Cyrklaff, A.; Tomancak, P.; Hentze, M.W. Iron-regulatory protein-1 (IRP-1) is highly conserved in two invertebrate species—Characterization of IRP-1 homologues in *Drosophila melanogaster* and Caenorhabditis elegans. *Eur. J. Biochem.* **1998**, *254*, 230–237. [CrossRef] [PubMed]
17. Hentze, M.W.; Muckenthaler, M.U.; Galy, B.; Camaschella, C. Two to tango: Regulation of Mammalian iron metabolism. *Cell* **2010**, *142*, 24–38. [CrossRef] [PubMed]
18. Lind, M.I.; Missirlis, F.; Melefors, O.; Uhrigshardt, H.; Kirby, K.; Phillips, J.P.; Söderhäll, K.; Rouault, T.A. Of two cytosolic aconitases expressed in *Drosophila*, only one functions as an iron-regulatory protein. *J. Biol. Chem.* **2006**, *281*, 18707–18714. [CrossRef] [PubMed]
19. Metzendorf, C.; Wu, W.; Lind, M.I. Overexpression of *Drosophila* mitoferrin in l(2)mbn cells results in dysregulation of Fer1HCH expression. *Biochem. J.* **2009**, *421*, 463–471. [CrossRef] [PubMed]

20. Cañizares, J.; Blanca, J.M.; Navarro, J.A.; Monrós, E.; Palau, F.; Moltó, M.D. *dfh* is a *Drosophila* homolog of the Friedreich's ataxia disease gene. *Gene* **2000**, *256*, 35–42. [CrossRef]

21. Verelst, W.; Asard, H. A phylogenetic study of cytochrome b561 proteins. *Genome Biol.* **2003**, *4*, R38. [CrossRef] [PubMed]

22. Lang, M.; Braun, C.L.; Kanost, M.R.; Gorman, M.J. Multicopper oxidase-1 is a ferroxidase essential for iron homeostasis in *Drosophila melanogaster*. *Proc. Natl. Acad. Sci. USA* **2012**, *109*, 13337–13342. [CrossRef] [PubMed]

23. Peng, Z.; Dittmer, N.T.; Lang, M.; Brummett, L.M.; Braun, C.L.; Davis, L.C.; Kanost, M.R.; Gorman, M.J. Multicopper oxidase-1 orthologs from diverse insect species have ascorbate oxidase activity. *Insect Biochem. Mol. Biol.* **2015**, *59*, 58–71. [CrossRef] [PubMed]

24. Arosio, P.; Levi, S. Cytosolic and mitochondrial ferritins in the regulation of cellular iron homeostasis and oxidative damage. *Biochim. Biophys. Acta* **2010**, *1800*, 783–792. [CrossRef] [PubMed]

25. Campanella, A.; Isaya, G.; O'Neill, H.A.; Santambrogio, P.; Cozzi, A.; Arosio, P.; Levi, S. The expression of human mitochondrial ferritin rescues respiratory function in frataxin-deficient yeast. *Hum. Mol. Genet.* **2004**, *13*, 2279–2288. [CrossRef] [PubMed]

26. Campanella, A.; Rovelli, E.; Santambrogio, P.; Cozzi, A.; Taroni, F.; Levi, S. Mitochondrial ferritin limits oxidative damage regulating mitochondrial iron availability: Hypothesis for a protective role in Friedreich ataxia. *Hum. Mol. Genet.* **2009**, *18*, 1–11. [CrossRef] [PubMed]

27. Navarro, J.A.; Botella, J.A.; Metzendorf, C.; Lind, M.I.; Schneuwly, S. Mitoferrin modulates iron toxicity in a *Drosophila* model of Friedreich's ataxia. *Free Radic. Biol. Med.* **2015**, *85*, 71–82. [CrossRef] [PubMed]

28. Bridwell-Rabb, J.; Fox, N.G.; Tsai, C.-L.; Winn, A.M.; Barondeau, D.P. Human frataxin activates Fe-S cluster biosynthesis by facilitating sulfur transfer chemistry. *Biochemistry* **2014**, *53*, 4904–4913. [CrossRef] [PubMed]

29. Bettedi, L.; Aslam, M.F.; Szular, J.; Mandilaras, K.; Missirlis, F. Iron depletion in the intestines of Malvolio mutant flies does not occur in the absence of a multicopper oxidase. *J. Exp. Biol.* **2011**, *214*, 971–978. [CrossRef] [PubMed]

30. Lane, D.J.R.; Richardson, D.R. The active role of vitamin C in mammalian iron metabolism: Much more than just enhanced iron absorption! *Free Radic. Biol. Med.* **2014**, *75*, 69–83. [CrossRef] [PubMed]

31. Mandilaras, K.; Pathmanathan, T.; Missirlis, F. Iron absorption in *Drosophila melanogaster*. *Nutrients* **2013**, *5*, 1622–1647. [CrossRef] [PubMed]

32. Van den Berghe, P.V.E.; Klomp, L.W.J. New developments in the regulation of intestinal copper absorption. *Nutr. Rev.* **2009**, *67*, 658–672. [CrossRef] [PubMed]

33. Petris, M.J. The SLC31 (Ctr) copper transporter family. *Pflug. Arch.* **2004**, *447*, 752–755. [CrossRef] [PubMed]

34. Espinoza, A.; Le Blanc, S.; Olivares, M.; Pizarro, F.; Ruz, M.; Arredondo, M. Iron, copper, and zinc transport: Inhibition of divalent metal transporter 1 (DMT1) and human copper transporter 1 (hCTR1) by shRNA. *Biol. Trace Elem. Res.* **2012**, *146*, 281–286. [CrossRef] [PubMed]

35. Illing, A.C.; Shawki, A.; Cunningham, C.L.; Mackenzie, B. Substrate profile and metal-ion selectivity of human divalent metal-ion transporter-1. *J. Biol. Chem.* **2012**, *287*, 30485–30496. [CrossRef] [PubMed]

36. Guedes, S.M.; Vitorino, R.; Tomer, K.; Domingues, M.R.M.; Correia, A.J.F.; Amado, F.; Domingues, P. *Drosophila melanogaster* larval hemolymph protein mapping. *Biochem. Biophys. Res. Commun.* **2003**, *312*, 545–554. [CrossRef] [PubMed]

37. Vierstraete, E.; Cerstiaens, A.; Baggerman, G.; Van den Bergh, G.; De Loof, A.; Schoofs, L. Proteomics in *Drosophila melanogaster*: First 2D database of larval hemolymph proteins. *Biochem. Biophys. Res. Commun.* **2003**, *304*, 831–838. [CrossRef]

38. Southon, A.; Burke, R.; Camakaris, J. What can flies tell us about copper homeostasis? *Metallomics* **2013**, *5*, 1346–1356. [CrossRef] [PubMed]

39. Schofield, R.M.; Postlethwait, J.H.; Lefevre, H.W. MeV-ion microprobe analyses of whole *Drosophila* suggest that zinc and copper accumulation is regulated storage not deposit excretion. *J. Exp. Biol.* **1997**, *200*, 3235–3243. [PubMed]

40. O'Donnell, M.J.; Maddrell, S.H. Fluid reabsorption and ion transport by the lower Malpighian tubules of adult female *Drosophila*. *J. Exp. Biol.* **1995**, *198*, 1647–1653. [PubMed]

41. Zhou, H.; Cadigan, K.M.; Thiele, D.J. A copper-regulated transporter required for copper acquisition, pigmentation, and specific stages of development in *Drosophila melanogaster*. *J. Biol. Chem.* **2003**, *278*, 48210–48218. [CrossRef] [PubMed]

42. Turski, M.L.; Thiele, D.J. *Drosophila* Ctr1A functions as a copper transporter essential for development. *J. Biol. Chem.* **2007**, *282*, 24017–24026. [CrossRef] [PubMed]

43. Southon, A.; Farlow, A.; Norgate, M.; Burke, R.; Camakaris, J. Malvolio is a copper transporter in *Drosophila melanogaster*. *J. Exp. Biol.* **2008**, *211*, 709–716. [CrossRef] [PubMed]

44. Schmidt, P.J.; Kunst, C.; Culotta, V.C. Copper activation of superoxide dismutase 1 (SOD1) in vivo. Role for protein-protein interactions with the copper chaperone for SOD1. *J. Biol. Chem.* **2000**, *275*, 33771–33776. [CrossRef] [PubMed]

45. Kirby, K.; Jensen, L.T.; Binnington, J.; Hilliker, A.J.; Ulloa, J.; Culotta, V.C.; Phillips, J.P. Instability of superoxide dismutase 1 of *Drosophila* in mutants deficient for its cognate copper chaperone. *J. Biol. Chem.* **2008**, *283*, 35393–35401. [CrossRef] [PubMed]

46. Horng, Y.-C.; Cobine, P.A.; Maxfield, A.B.; Carr, H.S.; Winge, D.R. Specific copper transfer from the Cox17 metallochaperone to both Sco1 and Cox11 in the assembly of yeast cytochrome C oxidase. *J. Biol. Chem.* **2004**, *279*, 35334–353340. [CrossRef] [PubMed]

47. Porcelli, D.; Oliva, M.; Duchi, S.; Latorre, D.; Cavaliere, V.; Barsanti, P.; Villani, G.; Gargiulo, G.; Caggese, C. Genetic, functional and evolutionary characterization of scox, the *Drosophila melanogaster* ortholog of the human SCO1 gene. *Mitochondrion* **2010**, *10*, 433–448. [CrossRef] [PubMed]

48. Hua, H.; Günther, V.; Georgiev, O.; Schaffner, W. Distorted copper homeostasis with decreased sensitivity to cisplatin upon chaperone Atox1 deletion in *Drosophila*. *Biometals* **2011**, *24*, 445–453. [CrossRef] [PubMed]

49. Prohaska, J.R.; Gybina, A.A. Intracellular copper transport in mammals. *J. Nutr.* **2004**, *134*, 1003–1006. [PubMed]

50. Harris, E.D. Basic and clinical aspects of copper. *Crit. Rev. Clin. Lab. Sci.* **2003**, *40*, 547–586. [CrossRef] [PubMed]

51. Southon, A.; Palstra, N.; Veldhuis, N.; Gaeth, A.; Robin, C.; Burke, R.; Camakaris, J. Conservation of copper-transporting P(IB)-type ATPase function. *Biometals* **2010**, *23*, 681–694. [CrossRef] [PubMed]

52. Norgate, M.; Lee, E.; Southon, A.; Farlow, A.; Batterham, P.; Camakaris, J.; Burke, R. Essential roles in development and pigmentation for the *Drosophila* copper transporter DmATP7. *Mol. Biol. Cell* **2006**, *17*, 475–484. [CrossRef] [PubMed]

53. Binks, T.; Lye, J.C.; Camakaris, J.; Burke, R. Tissue-specific interplay between copper uptake and efflux in *Drosophila*. *J. Biol. Inorg. Chem.* **2010**, *15*, 621–628. [CrossRef] [PubMed]

54. Sellami, A.; Wegener, C.; Veenstra, J.A. Functional significance of the copper transporter ATP7 in peptidergic neurons and endocrine cells in *Drosophila melanogaster*. *FEBS Lett.* **2012**, *586*, 3633–3638. [CrossRef] [PubMed]

55. Burke, R.; Commons, E.; Camakaris, J. Expression and localisation of the essential copper transporter DmATP7 in *Drosophila* neuronal and intestinal tissues. *Int. J. Biochem. Cell. Biol.* **2008**, *40*, 1850–1860. [CrossRef] [PubMed]

56. Wijmenga, C.; Klomp, L.W.J. Molecular regulation of copper excretion in the liver. *Proc. Nutr. Soc.* **2004**, *63*, 31–39. [CrossRef] [PubMed]

57. Valko, M.; Morris, H.; Cronin, M.T.D. Metals, toxicity and oxidative stress. *Curr. Med. Chem.* **2005**, *12*, 1161–1208. [CrossRef] [PubMed]

58. Uriu-Adams, J.Y.; Keen, C.L. Copper, oxidative stress, and human health. *Mol. Asp. Med.* **2005**, *26*, 268–298. [CrossRef] [PubMed]

59. Bonneton, F.; Théodore, L.; Silar, P.; Maroni, G.; Wegnez, M. Response of *Drosophila* metallothionein promoters to metallic, heat shock and oxidative stresses. *FEBS Lett.* **1996**, *380*, 33–38. [CrossRef]

60. Andrews, G.K. Regulation of metallothionein gene expression by oxidative stress and metal ions. *Biochem. Pharmacol.* **2000**, *59*, 95–104. [CrossRef]

61. Egli, D.; Selvaraj, A.; Yepiskoposyan, H.; Zhang, B.; Hafen, E.; Georgiev, O.; Schaffner, W. Knockout of "metal-responsive transcription factor" MTF-1 in *Drosophila* by homologous recombination reveals its central role in heavy metal homeostasis. *EMBO J.* **2003**, *22*, 100–108. [CrossRef] [PubMed]

62. Palmiter, R.D. The elusive function of metallothioneins. *Proc. Natl. Acad. Sci. USA* **1998**, *95*, 8428–8430. [CrossRef] [PubMed]

63. Egli, D.; Domènech, J.; Selvaraj, A.; Balamurugan, K.; Hua, H.; Capdevila, M.; Georgiev, O.; Schaffner, W.; Atrian, S. The four members of the *Drosophila* metallothionein family exhibit distinct yet overlapping roles in heavy metal homeostasis and detoxification. *Genes Cells* **2006**, *11*, 647–658. [CrossRef] [PubMed]

64. Atanesyan, L.; Günther, V.; Celniker, S.E.; Georgiev, O.; Schaffner, W. Characterization of MtnE, the fifth metallothionein member in *Drosophila*. *J. Biol. Inorg. Chem.* **2011**, *16*, 1047–1056. [CrossRef] [PubMed]
65. Lichtlen, P.; Schaffner, W. Putting its fingers on stressful situations: The heavy metal-regulatory transcription factor MTF-1. *Bioessays* **2001**, *23*, 1010–1017. [CrossRef] [PubMed]
66. Zhang, B.; Egli, D.; Georgiev, O.; Schaffner, W. The *Drosophila* homolog of mammalian zinc finger factor MTF-1 activates transcription in response to heavy metals. *Mol. Cell. Biol.* **2001**, *21*, 4505–4514. [CrossRef] [PubMed]
67. Balamurugan, K.; Egli, D.; Selvaraj, A.; Zhang, B.; Georgiev, O.; Schaffner, W. Metal-responsive transcription factor (MTF-1) and heavy metal stress response in *Drosophila* and mammalian cells: A functional comparison. *Biol. Chem.* **2004**, *385*, 597–603. [CrossRef] [PubMed]
68. Selvaraj, A.; Balamurugan, K.; Yepiskoposyan, H.; Zhou, H.; Egli, D.; Georgiev, O.; Thiele, D.J.; Schaffner, W. Metal-responsive transcription factor (MTF-1) handles both extremes, copper load and copper starvation, by activating different genes. *Genes Dev.* **2005**, *19*, 891–896. [CrossRef] [PubMed]
69. MacDonald, R.S. The role of zinc in growth and cell proliferation. *J. Nutr.* **2000**, *130*, 1500S–1508S. [PubMed]
70. Srivastava, R.C.; Farookh, A.; Ahmad, N.; Misra, M.; Hasan, S.K.; Husain, M.M. Reduction of cis-platinum induced nephrotoxicity by zinc histidine complex: The possible implication of nitric oxide. *Biochem. Mol. Biol. Int.* **1995**, *36*, 855–862. [PubMed]
71. Hennig, B.; Wang, Y.; Ramasamy, S.; McClain, C.J. Zinc protects against tumor necrosis factor-induced disruption of porcine endothelial cell monolayer integrity. *J. Nutr.* **1993**, *123*, 1003–1009. [PubMed]
72. Rodriguez, P.; Darmon, N.; Chappuis, P.; Candalh, C.; Blaton, M.A.; Bouchaud, C.; Heyman, M. Intestinal paracellular permeability during malnutrition in guinea pigs: Effect of high dietary zinc. *Gut* **1996**, *39*, 416–422. [CrossRef] [PubMed]
73. Vallee, B.L.; Falchuk, K.H. The biochemical basis of zinc physiology. *Physiol. Rev.* **1993**, *73*, 79–118. [PubMed]
74. Kitamura, H.; Morikawa, H.; Kamon, H.; Iguchi, M.; Hojyo, S.; Fukada, T.; Yamashita, S.; Kaisho, T.; Akira, S.; Murakami, M.; et al. Toll-like receptor-mediated regulation of zinc homeostasis influences dendritic cell function. *Nat. Immunol.* **2006**, *7*, 971–977. [CrossRef] [PubMed]
75. Bozym, R.A.; Chimienti, F.; Giblin, L.J.; Gross, G.W.; Korichneva, I.; Li, Y.; Libert, S.; Maret, W.; Parviz, M.; Frederickson, C.J.; et al. Free zinc ions outside a narrow concentration range are toxic to a variety of cells in vitro. *Exp. Biol. Med.* **2010**, *235*, 741–750. [CrossRef] [PubMed]
76. Reyes, J.G. Zinc transport in mammalian cells. *Am. J. Physiol.* **1996**, *270*, C401–C410. [PubMed]
77. Kambe, T.; Yamaguchi-Iwai, Y.; Sasaki, R.; Nagao, M. Overview of mammalian zinc transporters. *Cell. Mol. Life Sci.* **2004**, *61*, 49–68. [CrossRef] [PubMed]
78. Redenti, S.; Chappell, R.L. Localization of zinc transporter-3 (ZnT-3) in mouse retina. *Vis. Res.* **2004**, *44*, 3317–3321. [CrossRef] [PubMed]
79. Liuzzi, J.P.; Cousins, R.J. Mammalian zinc transporters. *Annu. Rev. Nutr.* **2004**, *24*, 151–172. [CrossRef] [PubMed]
80. Palmiter, R.D.; Findley, S.D. Cloning and functional characterization of a mammalian zinc transporter that confers resistance to zinc. *EMBO J.* **1995**, *14*, 639–649. [PubMed]
81. Bosomworth, H.J.; Thornton, J.K.; Coneyworth, L.J.; Ford, D.; Valentine, R.A. Efflux function, tissue-specific expression and intracellular trafficking of the Zn transporter ZnT10 indicate roles in adult Zn homeostasis. *Metallomics* **2012**, *4*, 771–779. [CrossRef] [PubMed]
82. Leyva-Illades, D.; Chen, P.; Zogzas, C.E.; Hutchens, S.; Mercado, J.M.; Swaim, C.D.; Morrisett, R.A.; Bowman, A.B.; Aschner, M.; Mukhopadhyay, S. SLC30A10 is a cell surface-localized manganese efflux transporter, and parkinsonism-causing mutations block its intracellular trafficking and efflux activity. *J. Neurosci.* **2014**, *34*, 14079–14095. [CrossRef] [PubMed]
83. Falcón-Pérez, J.M.; Dell'Angelica, E.C. Zinc transporter 2 (SLC30A2) can suppress the vesicular zinc defect of adaptor protein 3-depleted fibroblasts by promoting zinc accumulation in lysosomes. *Exp. Cell. Res.* **2007**, *313*, 1473–1483. [CrossRef] [PubMed]
84. Itsumura, N.; Inamo, Y.; Okazaki, F.; Teranishi, F.; Narita, H.; Kambe, T.; Kodama, H. Compound heterozygous mutations in SLC30A2/ZnT2 results in low milk zinc concentrations: A novel mechanism for zinc deficiency in a breast-fed infant. *PLoS ONE* **2013**, *8*, e64045. [CrossRef] [PubMed]
85. Palmiter, R.D.; Cole, T.B.; Findley, S.D. ZnT-2, a mammalian protein that confers resistance to zinc by facilitating vesicular sequestration. *EMBO J.* **1996**, *15*, 1784–1791. [PubMed]

86. Cole, T.B.; Wenzel, H.J.; Kafer, K.E.; Schwartzkroin, P.A.; Palmiter, R.D. Elimination of zinc from synaptic vesicles in the intact mouse brain by disruption of the ZnT3 gene. *Proc. Natl. Acad. Sci. USA* **1999**, *96*, 1716–1721. [CrossRef] [PubMed]

87. Chimienti, F.; Devergnas, S.; Favier, A.; Seve, M. Identification and cloning of a β-cell-specific zinc transporter, ZnT-8, localized into insulin secretory granules. *Diabetes* **2004**, *53*, 2330–2337. [CrossRef] [PubMed]

88. Chimienti, F.; Devergnas, S.; Pattou, F.; Schuit, F.; Garcia-Cuenca, R.; Vandewalle, B.; Kerr-Conte, J.; van Lommel, L.; Grunwald, D.; Favier, A.; et al. In vivo expression and functional characterization of the zinc transporter ZnT8 in glucose-induced insulin secretion. *J. Cell Sci.* **2006**, *119*, 4199–4206. [CrossRef] [PubMed]

89. Kukic, I.; Lee, J.K.; Coblentz, J.; Kelleher, S.L.; Kiselyov, K. Zinc-dependent lysosomal enlargement in TRPML1-deficient cells involves MTF-1 transcription factor and ZnT4 (Slc30a4) transporter. *Biochem. J.* **2013**, *451*, 155–163. [CrossRef] [PubMed]

90. Fukunaka, A.; Kurokawa, Y.; Teranishi, F.; Sekler, I.; Oda, K.; Ackland, M.L.; Faundez, V.; Hiromura, M.; Masuda, S.; Nagao, M.; et al. Tissue nonspecific alkaline phosphatase is activated via a two-step mechanism by zinc transport complexes in the early secretory pathway. *J. Biol. Chem.* **2011**, *286*, 16363–16373. [CrossRef] [PubMed]

91. Ishihara, K.; Yamazaki, T.; Ishida, Y.; Suzuki, T.; Oda, K.; Nagao, M.; Yamaguchi-Iwai, Y.; Kambe, T. Zinc transport complexes contribute to the homeostatic maintenance of secretory pathway function in vertebrate cells. *J. Biol. Chem.* **2006**, *281*, 17743–17750. [CrossRef] [PubMed]

92. Chen, Y.-H.; Kim, J.H.; Stallcup, M.R. GAC63, a GRIP1-dependent nuclear receptor coactivator. *Mol. Cell. Biol.* **2005**, *25*, 5965–5972. [CrossRef] [PubMed]

93. Dufner-Beattie, J.; Langmade, S.J.; Wang, F.; Eide, D.; Andrews, G.K. Structure, function, and regulation of a subfamily of mouse zinc transporter genes. *J. Biol. Chem.* **2003**, *278*, 50142–50150. [CrossRef] [PubMed]

94. Higashi, Y.; Segawa, S.; Matsuo, T.; Nakamura, S.; Kikkawa, Y.; Nishida, K.; Nagasawa, K. Microglial zinc uptake via zinc transporters induces ATP release and the activation of microglia. *Glia* **2011**, *59*, 1933–1945. [CrossRef] [PubMed]

95. Gaither, L.A.; Eide, D.J. Functional expression of the human hZIP2 zinc transporter. *J. Biol. Chem.* **2000**, *275*, 5560–5564. [CrossRef] [PubMed]

96. Inoue, Y.; Hasegawa, S.; Ban, S.; Yamada, T.; Date, Y.; Mizutani, H.; Nakata, S.; Tanaka, M.; Hirashima, N. ZIP2 protein, a zinc transporter, is associated with keratinocyte differentiation. *J. Biol. Chem.* **2014**, *289*, 21451–21462. [CrossRef] [PubMed]

97. Kelleher, S.L.; Lönnerdal, B. Zn transporter levels and localization change throughout lactation in rat mammary gland and are regulated by Zn in mammary cells. *J. Nutr.* **2003**, *133*, 3378–3385. [PubMed]

98. Kelleher, S.L.; Lopez, V.; Lönnerdal, B.; Dufner-Beattie, J.; Andrews, G.K. Zip3 (Slc39a3) functions in zinc reuptake from the alveolar lumen in lactating mammary gland. *Am. J. Physiol. Regul. Integr. Comp. Physiol.* **2009**, *297*, R194–R201. [CrossRef] [PubMed]

99. Dufner-Beattie, J.; Kuo, Y.-M.; Gitschier, J.; Andrews, G.K. The adaptive response to dietary zinc in mice involves the differential cellular localization and zinc regulation of the zinc transporters ZIP4 and ZIP5. *J. Biol. Chem.* **2004**, *279*, 49082–49090. [CrossRef] [PubMed]

100. Geiser, J.; De Lisle, R.C.; Andrews, G.K. The zinc transporter Zip5 (Slc39a5) regulates intestinal zinc excretion and protects the pancreas against zinc toxicity. *PLoS ONE* **2013**, *8*, e82149. [CrossRef] [PubMed]

101. Kong, B.Y.; Duncan, F.E.; Que, E.L.; Kim, A.M.; O'Halloran, T.V.; Woodruff, T.K. Maternally-derived zinc transporters ZIP6 and ZIP10 drive the mammalian oocyte-to-egg transition. *Mol. Hum. Reprod.* **2014**, *20*, 1077–1089. [CrossRef] [PubMed]

102. Shen, R.; Xie, F.; Shen, H.; Liu, Q.; Zheng, T.; Kou, X.; Wang, D.; Yang, J. Negative correlation of LIV-1 and E-cadherin expression in hepatocellular carcinoma cells. *PLoS ONE* **2013**, *8*, e56542. [CrossRef] [PubMed]

103. Lichten, L.A.; Ryu, M.-S.; Guo, L.; Embury, J.; Cousins, R.J. MTF-1-mediated repression of the zinc transporter Zip10 is alleviated by zinc restriction. *PLoS ONE* **2011**, *6*, e21526. [CrossRef] [PubMed]

104. Hojyo, S.; Miyai, T.; Fujishiro, H.; Kawamura, M.; Yasuda, T.; Hijikata, A.; Bin, B.-H.; Irié, T.; Tanaka, J.; Atsumi, T.; et al. Zinc transporter SLC39A10/ZIP10 controls humoral immunity by modulating B-cell receptor signal strength. *Proc. Natl. Acad. Sci. USA* **2014**, *111*, 11786–11791. [CrossRef] [PubMed]

105. Huang, L.; Kirschke, C.P.; Zhang, Y.; Yu, Y.Y. The ZIP7 gene (Slc39a7) encodes a zinc transporter involved in zinc homeostasis of the Golgi apparatus. *J. Biol. Chem.* **2005**, *280*, 15456–15463. [CrossRef] [PubMed]

106. Myers, S.A.; Nield, A.; Chew, G.-S.; Myers, M.A. The zinc transporter, Slc39a7 (Zip7) is implicated in glycaemic control in skeletal muscle cells. *PLoS ONE* **2013**, *8*, e79316. [CrossRef] [PubMed]

107. Matsuura, W.; Yamazaki, T.; Yamaguchi-Iwai, Y.; Masuda, S.; Nagao, M.; Andrews, G.K.; Kambe, T. SLC39A9 (ZIP9) regulates zinc homeostasis in the secretory pathway: Characterization of the ZIP subfamily I protein in vertebrate cells. *Biosci. Biotechnol. Biochem.* **2009**, *73*, 1142–1148. [CrossRef] [PubMed]

108. Taniguchi, M.; Fukunaka, A.; Hagihara, M.; Watanabe, K.; Kamino, S.; Kambe, T.; Enomoto, S.; Hiromura, M. Essential role of the zinc transporter ZIP9/SLC39A9 in regulating the activations of Akt and Erk in B-cell receptor signaling pathway in DT40 cells. *PLoS ONE* **2013**, *8*, e58022. [CrossRef] [PubMed]

109. Gaither, L.A.; Eide, D.J. Eukaryotic zinc transporters and their regulation. *Biometals* **2001**, *14*, 251–270. [CrossRef] [PubMed]

110. Xiao, G.; Wan, Z.; Fan, Q.; Tang, X.; Zhou, B. The metal transporter ZIP13 supplies iron into the secretory pathway in *Drosophila melanogaster*. *Elife* **2014**, *3*, e03191. [CrossRef] [PubMed]

111. Fukada, T.; Civic, N.; Furuichi, T.; Shimoda, S.; Mishima, K.; Higashiyama, H.; Idaira, Y.; Asada, Y.; Kitamura, H.; Yamasaki, S.; et al. The zinc transporter SLC39A13/ZIP13 is required for connective tissue development; its involvement in BMP/TGF-β signaling pathways. *PLoS ONE* **2008**, *3*, e3642. [CrossRef]

112. Stathakis, D.G.; Burton, D.Y.; McIvor, W.E.; Krishnakumar, S.; Wright, T.R.; O'Donnell, J.M. The catecholamines up (Catsup) protein of *Drosophila melanogaster* functions as a negative regulator of tyrosine hydroxylase activity. *Genetics* **1999**, *153*, 361–382. [PubMed]

113. Wang, Z.; Ferdousy, F.; Lawal, H.; Huang, Z.; Daigle, J.G.; Izevbaye, I.; Doherty, O.; Thomas, J.; Stathakis, D.G.; O'Donnell, J.M. Catecholamines up integrates dopamine synthesis and synaptic trafficking. *J. Neurochem.* **2011**, *119*, 1294–1305. [CrossRef] [PubMed]

114. Groth, C.; Sasamura, T.; Khanna, M.R.; Whitley, M.; Fortini, M.E. Protein trafficking abnormalities in *Drosophila* tissues with impaired activity of the ZIP7 zinc transporter Catsup. *Development* **2013**, *140*, 3018–3027. [CrossRef] [PubMed]

115. Mathews, W.R.; Ong, D.; Milutinovich, A.B.; Van Doren, M. Zinc transport activity of Fear of Intimacy is essential for proper gonad morphogenesis and DE-cadherin expression. *Development* **2006**, *133*, 1143–1153. [CrossRef] [PubMed]

116. Van Doren, M.; Mathews, W.R.; Samuels, M.; Moore, L.A.; Broihier, H.T.; Lehmann, R. Fear of intimacy encodes a novel transmembrane protein required for gonad morphogenesis in *Drosophila*. *Development* **2003**, *130*, 2355–2364. [CrossRef] [PubMed]

117. Pielage, J.; Kippert, A.; Zhu, M.; Klämbt, C. The *Drosophila* transmembrane protein Fear-of-intimacy controls glial cell migration. *Dev. Biol.* **2004**, *275*, 245–257. [CrossRef] [PubMed]

118. Carrasco-Rando, M.; Atienza-Manuel, A.; Martín, P.; Burke, R.; Ruiz-Gómez, M. Fear-of-intimacy-mediated zinc transport controls the function of zinc-finger transcription factors involved in myogenesis. *Development* **2016**, *143*, 1948–1957. [CrossRef] [PubMed]

119. Qin, Q.; Wang, X.; Zhou, B. Functional studies of *Drosophila* zinc transporters reveal the mechanism for dietary zinc absorption and regulation. *BMC Biol.* **2013**, *11*, 101. [CrossRef] [PubMed]

120. Richards, C.D.; Burke, R. Local and systemic effects of targeted zinc redistribution in *Drosophila* neuronal and gastrointestinal tissues. *Biometals* **2015**, *28*, 967–974. [CrossRef] [PubMed]

121. Wang, X.; Wu, Y.; Zhou, B. Dietary zinc absorption is mediated by ZnT1 in *Drosophila melanogaster*. *FASEB J.* **2009**, *23*, 2650–2661. [CrossRef] [PubMed]

122. Dechen, K.; Richards, C.D.; Lye, J.C.; Hwang, J.E.C.; Burke, R. Compartmentalized zinc deficiency and toxicities caused by ZnT and Zip gene over expression result in specific phenotypes in *Drosophila*. *Int. J. Biochem. Cell. Biol.* **2015**, *60*, 23–33. [CrossRef] [PubMed]

123. Lye, J.C.; Richards, C.D.; Dechen, K.; Paterson, D.; de Jonge, M.D.; Howard, D.L.; Warr, C.G.; Burke, R. Systematic functional characterization of putative zinc transport genes and identification of zinc toxicosis phenotypes in *Drosophila melanogaster*. *J. Exp. Biol.* **2012**, *215*, 3254–3265. [CrossRef] [PubMed]

124. Yepiskoposyan, H.; Egli, D.; Fergestad, T.; Selvaraj, A.; Treiber, C.; Multhaup, G.; Georgiev, O.; Schaffner, W. Transcriptome response to heavy metal stress in *Drosophila* reveals a new zinc transporter that confers resistance to zinc. *Nucleic Acids Res.* **2006**, *34*, 4866–4877. [CrossRef] [PubMed]

125. Yin, S.; Qin, Q.; Zhou, B. Functional studies of *Drosophila* zinc transporters reveal the mechanism for zinc excretion in Malpighian tubules. *BMC Biol.* **2017**, *15*, 12. [CrossRef] [PubMed]

126. Liuzzi, J.P.; Aydemir, F.; Nam, H.; Knutson, M.D.; Cousins, R.J. Zip14 (Slc39a14) mediates non-transferrin-bound iron uptake into cells. *Proc. Natl. Acad. Sci. USA* **2006**, *103*, 13612–13617. [CrossRef] [PubMed]

127. Wang, C.-Y.; Jenkitkasemwong, S.; Duarte, S.; Sparkman, B.K.; Shawki, A.; Mackenzie, B.; Knutson, M.D. ZIP8 is an iron and zinc transporter whose cell-surface expression is up-regulated by cellular iron loading. *J. Biol. Chem.* **2012**, *287*, 34032–34043. [CrossRef] [PubMed]

128. Giunta, C.; Elçioglu, N.H.; Albrecht, B.; Eich, G.; Chambaz, C.; Janecke, A.R.; Yeowell, H.; Weis, M.; Eyre, D.R.; Kraenzlin, M.; et al. Spondylocheiro dysplastic form of the Ehlers-Danlos syndrome-An autosomal-recessive entity caused by mutations in the zinc transporter gene SLC39A13. *Am. J. Hum. Genet.* **2008**, *82*, 1290–1305. [CrossRef] [PubMed]

129. Palau, F.; Espinós, C. Autosomal recessive cerebellar ataxias. *Orphanet J. Rare Dis.* **2006**, *1*, 47. [CrossRef] [PubMed]

130. Campuzano, V.; Montermini, L.; Moltò, M.D.; Pianese, L.; Cossée, M.; Cavalcanti, F.; Monros, E.; Rodius, F.; Duclos, F.; Monticelli, A.; et al. Friedreich's ataxia: Autosomal recessive disease caused by an intronic GAA triplet repeat expansion. *Science* **1996**, *271*, 1423–1427. [CrossRef] [PubMed]

131. Yandim, C.; Natisvili, T.; Festenstein, R. Gene regulation and epigenetics in Friedreich's ataxia. *J. Neurochem.* **2013**, *126*, 21–42. [CrossRef] [PubMed]

132. Puccio, H. Multicellular models of Friedreich ataxia. *J. Neurol.* **2009**, *256*, 18–24. [CrossRef] [PubMed]

133. Schmucker, S.; Puccio, H. Understanding the molecular mechanisms of Friedreich's ataxia to develop therapeutic approaches. *Hum. Mol. Genet.* **2010**, *19*, R103–R110. [CrossRef] [PubMed]

134. Bradley, J.L.; Blake, J.C.; Chamberlain, S.; Thomas, P.K.; Cooper, J.M.; Schapira, A.H. Clinical, biochemical and molecular genetic correlations in Friedreich's ataxia. *Hum. Mol. Genet.* **2000**, *9*, 275–282. [CrossRef] [PubMed]

135. Koeppen, A.H.; Mazurkiewicz, J.E. Friedreich ataxia: Neuropathology revised. *J. Neuropathol. Exp. Neurol.* **2013**, *72*, 78–90. [CrossRef] [PubMed]

136. Foury, F.; Cazzalini, O. Deletion of the yeast homologue of the human gene associated with Friedreich's ataxia elicits iron accumulation in mitochondria. *FEBS Lett.* **1997**, *411*, 373–377. [CrossRef]

137. Puccio, H.; Simon, D.; Cossée, M.; Criqui-Filipe, P.; Tiziano, F.; Melki, J.; Hindelang, C.; Matyas, R.; Rustin, P.; Koenig, M. Mouse models for Friedreich ataxia exhibit cardiomyopathy, sensory nerve defect and Fe-S enzyme deficiency followed by intramitochondrial iron deposits. *Nat. Genet.* **2001**, *27*, 181–186. [CrossRef] [PubMed]

138. Soriano, S.; Llorens, J.V.; Blanco-Sobero, L.; Gutiérrez, L.; Calap-Quintana, P.; Morales, M.P.; Moltó, M.D.; Martínez-Sebastián, M.J. Deferiprone and idebenone rescue frataxin depletion phenotypes in a *Drosophila* model of Friedreich's ataxia. *Gene* **2013**, *521*, 274–281. [CrossRef] [PubMed]

139. Anderson, P.R.; Kirby, K.; Hilliker, A.J.; Phillips, J.P. RNAi-mediated suppression of the mitochondrial iron chaperone, frataxin, in *Drosophila. Hum. Mol. Genet.* **2005**, *14*, 3397–3405. [CrossRef] [PubMed]

140. Llorens, J.V.; Navarro, J.A.; Martínez-Sebastián, M.J.; Baylies, M.K.; Schneuwly, S.; Botella, J.A.; Moltó, M.D. Causative role of oxidative stress in a *Drosophila* model of Friedreich ataxia. *FASEB J.* **2007**, *21*, 333–344. [CrossRef] [PubMed]

141. Bulteau, A.-L.; O'Neill, H.A.; Kennedy, M.C.; Ikeda-Saito, M.; Isaya, G.; Szweda, L.I. Frataxin acts as an iron chaperone protein to modulate mitochondrial aconitase activity. *Science* **2004**, *305*, 242–245. [CrossRef] [PubMed]

142. Chiang, S.; Kovacevic, Z.; Sahni, S.; Lane, D.J.R.; Merlot, A.M.; Kalinowski, D.S.; Huang, M.L.-H.; Richardson, D.R. Frataxin and the molecular mechanism of mitochondrial iron-loading in Friedreich's ataxia. *Clin. Sci.* **2016**, *130*, 853–870. [CrossRef] [PubMed]

143. Soriano, S.; Calap-Quintana, P.; Llorens, J.V.; Al-Ramahi, I.; Gutiérrez, L.; Martínez-Sebastián, M.J.; Botas, J.; Moltó, M.D. Metal Homeostasis Regulators Suppress FRDA Phenotypes in a *Drosophila* Model of the Disease. *PLoS ONE* **2016**, *11*, e0159209. [CrossRef] [PubMed]

144. Chen, K.; Lin, G.; Haelterman, N.A.; Ho, T.S.-Y.; Li, T.; Li, Z.; Duraine, L.; Graham, B.H.; Jaiswal, M.; Yamamoto, S.; et al. Loss of Frataxin induces iron toxicity, sphingolipid synthesis, and Pdk1/Mef2 activation, leading to neurodegeneration. *Elife* **2016**, *5*, e16043. [CrossRef] [PubMed]

145. Chen, K.; Ho, T.S.-Y.; Lin, G.; Tan, K.L.; Rasband, M.N.; Bellen, H.J. Loss of Frataxin activates the iron/sphingolipid/PDK1/Mef2 pathway in mammals. *Elife* **2016**, *5*, e20732. [CrossRef] [PubMed]

146. Boddaert, N.; Le Quan Sang, K.H.; Rötig, A.; Leroy-Willig, A.; Gallet, S.; Brunelle, F.; Sidi, D.; Thalabard, J.-C.; Munnich, A.; Cabantchik, Z.I. Selective iron chelation in Friedreich ataxia: Biologic and clinical implications. *Blood* **2007**, *110*, 401–408. [CrossRef] [PubMed]

147. Pandolfo, M.; Arpa, J.; Delatycki, M.B.; Le Quan Sang, K.H.; Mariotti, C.; Munnich, A.; Sanz-Gallego, I.; Tai, G.; Tarnopolsky, M.A.; Taroni, F.; et al. Deferiprone in Friedreich ataxia: A 6-month randomized controlled trial. *Ann. Neurol.* **2014**, *76*, 509–521. [CrossRef] [PubMed]

148. Kodama, H.; Fujisawa, C.; Bhadhprasit, W. Inherited copper transport disorders: Biochemical mechanisms, diagnosis, and treatment. *Curr. Drug Metab.* **2012**, *13*, 237–250. [CrossRef] [PubMed]

149. Tümer, Z.; Møller, L.B. Menkes disease. *Eur. J. Hum. Genet.* **2010**, *18*, 511–518. [CrossRef] [PubMed]

150. Tümer, Z. An overview and update of ATP7A mutations leading to Menkes disease and occipital horn syndrome. *Hum. Mutat.* **2013**, *34*, 417–429. [CrossRef] [PubMed]

151. Kodama, H.; Sato, E.; Gu, Y.-H.; Shiga, K.; Fujisawa, C.; Kozuma, T. Effect of copper and diethyldithiocarbamate combination therapy on the macular mouse, an animal model of Menkes disease. *J. Inherit. Metab. Dis.* **2005**, *28*, 971–978. [CrossRef] [PubMed]

152. Donsante, A.; Johnson, P.; Jansen, L.A.; Kaler, S.G. Somatic mosaicism in Menkes disease suggests choroid plexus-mediated copper transport to the developing brain. *Am. J. Med. Genet. A* **2010**, *152A*, 2529–2534. [CrossRef] [PubMed]

153. Donsante, A.; Yi, L.; Zerfas, P.M.; Brinster, L.R.; Sullivan, P.; Goldstein, D.S.; Prohaska, J.; Centeno, J.A.; Rushing, E.; Kaler, S.G. ATP7A gene addition to the choroid plexus results in long-term rescue of the lethal copper transport defect in a Menkes disease mouse model. *Mol. Ther.* **2011**, *19*, 2114–2123. [CrossRef] [PubMed]

154. Rodriguez-Castro, K.I.; Hevia-Urrutia, F.J.; Sturniolo, G.C. Wilson's disease: A review of what we have learned. *World J. Hepatol.* **2015**, *7*, 2859–2870. [CrossRef] [PubMed]

155. Bahadorani, S.; Bahadorani, P.; Marcon, E.; Walker, D.W.; Hilliker, A.J. A *Drosophila* model of Menkes disease reveals a role for DmATP7 in copper absorption and neurodevelopment. *Dis. Model. Mech.* **2010**, *3*, 84–91. [CrossRef] [PubMed]

156. Kennerson, M.L.; Nicholson, G.A.; Kaler, S.G.; Kowalski, B.; Mercer, J.F.B.; Tang, J.; Llanos, R.M.; Chu, S.; Takata, R.I.; Speck-Martins, C.E.; et al. Missense mutations in the copper transporter gene ATP7A cause X-linked distal hereditary motor neuropathy. *Am. J. Hum. Genet.* **2010**, *86*, 343–352. [CrossRef] [PubMed]

157. Bucossi, S.; Polimanti, R.; Mariani, S.; Ventriglia, M.; Bonvicini, C.; Migliore, S.; Manfellotto, D.; Salustri, C.; Vernieri, F.; Rossini, P.M.; et al. Association of K832R and R952K SNPs of Wilson's disease gene with Alzheimer's disease. *J. Alzheimers Dis.* **2012**, *29*, 913–919. [PubMed]

158. Squitti, R.; Polimanti, R.; Bucossi, S.; Ventriglia, M.; Mariani, S.; Manfellotto, D.; Vernieri, F.; Cassetta, E.; Ursini, F.; Rossini, P.M. Linkage disequilibrium and haplotype analysis of the ATP7B gene in Alzheimer's disease. *Rejuvenation Res.* **2013**, *16*, 3–10. [CrossRef] [PubMed]

159. Mercer, S.W.; Wang, J.; Burke, R. In Vivo Modeling of the Pathogenic Effect of Copper Transporter Mutations That Cause Menkes and Wilson Diseases, Motor Neuropathy, and Susceptibility to Alzheimer's Disease. *J. Biol. Chem.* **2017**, *292*, 4113–4122. [CrossRef] [PubMed]

160. Beighton, P. The Ehlers-Danlos syndromes. In *McKusick's Heritable Disorders of Connective Tissue*, 5th ed.; Beighton, P., Ed.; Mosyb: Maryland Heights, MO, USA, 1993; pp. 189–251.

161. Myllyharju, J. Prolyl 4-hydroxylases, key enzymes in the synthesis of collagens and regulation of the response to hypoxia, and their roles as treatment targets. *Ann. Med.* **2008**, *40*, 402–417. [CrossRef] [PubMed]

162. Yeowell, H.N.; Walker, L.C. Mutations in the lysyl hydroxylase 1 gene that result in enzyme deficiency and the clinical phenotype of Ehlers-Danlos syndrome type VI. *Mol. Genet. Metab.* **2000**, *71*, 212–224. [CrossRef] [PubMed]

163. Jeong, J.; Walker, J.M.; Wang, F.; Park, J.G.; Palmer, A.E.; Giunta, C.; Rohrbach, M.; Steinmann, B.; Eide, D.J. Promotion of vesicular zinc efflux by ZIP13 and its implications for spondylocheiro dysplastic Ehlers-Danlos syndrome. *Proc. Natl. Acad. Sci. USA* **2012**, *109*, E3530–E3538. [CrossRef] [PubMed]

164. Baig, S.S.; Strong, M.; Quarrell, O.W. The global prevalence of Huntington's disease: A systematic review and discussion. *Neurodegener. Dis. Manag.* **2016**, *6*, 331–343. [CrossRef] [PubMed]

165. MacDonald, M.; Ambrose, C.; Duyao, M.; Myers, R.; Lin, C. A novel gene containing a trinucleotide repeat that is expanded and unstable on Huntington's disease chromosomes. The Huntington's Disease Collaborative Research Group. *Cell* **1993**, *72*, 971–983. [CrossRef]

166. Scherzinger, E.; Lurz, R.; Turmaine, M.; Mangiarini, L.; Hollenbach, B.; Hasenbank, R.; Bates, G.P.; Davies, S.W.; Lehrach, H.; Wanker, E.E. Huntingtin-encoded polyglutamine expansions form amyloid-like protein aggregates in vitro and in vivo. *Cell* **1997**, *90*, 549–558. [CrossRef]

167. Fox, J.H.; Kama, J.A.; Lieberman, G.; Chopra, R.; Dorsey, K.; Chopra, V.; Volitakis, I.; Cherny, R.A.; Bush, A.I.; Hersch, S. Mechanisms of copper ion mediated Huntington's disease progression. *PLoS ONE* **2007**, *2*, e334. [CrossRef] [PubMed]

168. Hands, S.L.; Mason, R.; Sajjad, M.U.; Giorgini, F.; Wyttenbach, A. Metallothioneins and copper metabolism are candidate therapeutic targets in Huntington's disease. *Biochem. Soc. Trans.* **2010**, *38*, 552–558. [CrossRef] [PubMed]

169. Dexter, D.T.; Carayon, A.; Javoy-Agid, F.; Agid, Y.; Wells, F.R.; Daniel, S.E.; Lees, A.J.; Jenner, P.; Marsden, C.D. Alterations in the levels of iron, ferritin and other trace metals in Parkinson's disease and other neurodegenerative diseases affecting the basal ganglia. *Brain* **1991**, *114 Pt 4*, 1953–1975. [CrossRef] [PubMed]

170. Pérez, P.; Flores, A.; Santamaría, A.; Ríos, C.; Galván-Arzate, S. Changes in transition metal contents in rat brain regions after in vivo quinolinate intrastriatal administration. *Arch. Med. Res.* **1996**, *27*, 449–452. [PubMed]

171. Xiao, G.; Fan, Q.; Wang, X.; Zhou, B. Huntington disease arises from a combinatory toxicity of polyglutamine and copper binding. *Proc. Natl. Acad. Sci. USA* **2013**, *110*, 14995–15000. [CrossRef] [PubMed]

172. Schneider, S.A.; Obeso, J.A. Clinical and pathological features of Parkinson's disease. *Curr. Top. Behav. Neurosci.* **2015**, *22*, 205–220. [PubMed]

173. Kitada, T.; Asakawa, S.; Hattori, N.; Matsumine, H.; Yamamura, Y.; Minoshima, S.; Yokochi, M.; Mizuno, Y.; Shimizu, N. Mutations in the parkin gene cause autosomal recessive juvenile parkinsonism. *Nature* **1998**, *392*, 605–608. [PubMed]

174. Shimura, H.; Hattori, N.; Kubo, S.I.; Mizuno, Y.; Asakawa, S.; Minoshima, S.; Shimizu, N.; Iwai, K.; Chiba, T.; Tanaka, K.; et al. Familial Parkinson disease gene product, parkin, is a ubiquitin-protein ligase. *Nat. Genet.* **2000**, *25*, 302–305. [PubMed]

175. Zhang, Y.; Gao, J.; Chung, K.K.; Huang, H.; Dawson, V.L.; Dawson, T.M. Parkin functions as an E2-dependent ubiquitin-protein ligase and promotes the degradation of the synaptic vesicle-associated protein, CDCrel-1. *Proc. Natl. Acad. Sci. USA* **2000**, *97*, 13354–13359. [CrossRef] [PubMed]

176. Greene, J.C.; Whitworth, A.J.; Andrews, L.A.; Parker, T.J.; Pallanck, L.J. Genetic and genomic studies of *Drosophila* parkin mutants implicate oxidative stress and innate immune responses in pathogenesis. *Hum. Mol. Genet.* **2005**, *14*, 799–811. [CrossRef] [PubMed]

177. Greene, J.C.; Whitworth, A.J.; Kuo, I.; Andrews, L.A.; Feany, M.B.; Pallanck, L.J. Mitochondrial pathology and apoptotic muscle degeneration in *Drosophila* parkin mutants. *Proc. Natl. Acad. Sci. USA* **2003**, *100*, 4078–4083. [CrossRef] [PubMed]

178. Pesah, Y.; Pham, T.; Burgess, H.; Middlebrooks, B.; Verstreken, P.; Zhou, Y.; Harding, M.; Bellen, H.; Mardon, G. *Drosophila* parkin mutants have decreased mass and cell size and increased sensitivity to oxygen radical stress. *Development* **2004**, *131*, 2183–2194. [CrossRef] [PubMed]

179. Saini, N.; Oelhafen, S.; Hua, H.; Georgiev, O.; Schaffner, W.; Büeler, H. Extended lifespan of *Drosophila* parkin mutants through sequestration of redox-active metals and enhancement of anti-oxidative pathways. *Neurobiol. Dis.* **2010**, *40*, 82–92. [CrossRef] [PubMed]

180. Saini, N.; Schaffner, W. Zinc supplement greatly improves the condition of parkin mutant *Drosophila*. *Biol. Chem.* **2010**, *391*, 513–518. [CrossRef] [PubMed]

181. Bray, T.M.; Bettger, W.J. The physiological role of zinc as an antioxidant. *Free Radic. Biol. Med.* **1990**, *8*, 281–291. [CrossRef]

182. Südmeyer, M.; Saleh, A.; Wojtecki, L.; Cohnen, M.; Gross, J.; Ploner, M.; Hefter, H.; Timmermann, L.; Schnitzler, A. Wilson's disease tremor is associated with magnetic resonance imaging lesions in basal ganglia structures. *Mov. Disord.* **2006**, *21*, 2134–2139. [CrossRef] [PubMed]

183. Perl, D.P.; Olanow, C.W. The neuropathology of manganese-induced Parkinsonism. *J. Neuropathol. Exp. Neurol.* **2007**, *66*, 675–682. [CrossRef] [PubMed]

184. Sengstock, G.J.; Olanow, C.W.; Dunn, A.J.; Barone, S.; Arendash, G.W. Progressive changes in striatal dopaminergic markers, nigral volume, and rotational behavior following iron infusion into the rat substantia nigra. *Exp. Neurol.* **1994**, *130*, 82–94. [CrossRef] [PubMed]

185. Morello, M.; Zatta, P.; Zambenedetti, P.; Martorana, A.; D'Angelo, V.; Melchiorri, G.; Bernardi, G.; Sancesario, G. Manganese intoxication decreases the expression of manganoproteins in the rat basal ganglia: An immunohistochemical study. *Brain Res. Bull.* **2007**, *74*, 406–415. [CrossRef] [PubMed]

186. Cass, W.A.; Grondin, R.; Andersen, A.H.; Zhang, Z.; Hardy, P.A.; Hussey-Andersen, L.K.; Rayens, W.S.; Gerhardt, G.A.; Gash, D.M. Iron accumulation in the striatum predicts aging-related decline in motor function in rhesus monkeys. *Neurobiol. Aging* **2007**, *28*, 258–271. [CrossRef] [PubMed]

187. Burton, N.C.; Guilarte, T.R. Manganese neurotoxicity: Lessons learned from longitudinal studies in nonhuman primates. *Environ. Health Perspect.* **2009**, *117*, 325–332. [CrossRef] [PubMed]

188. Stefanis, L. α-Synuclein in Parkinson's Disease. *Cold Spring Harb. Perspect. Med.* **2012**, *2*, a009399. [CrossRef] [PubMed]

189. Ostrerova-Golts, N.; Petrucelli, L.; Hardy, J.; Lee, J.M.; Farer, M.; Wolozin, B. The A53T α-synuclein mutation increases iron-dependent aggregation and toxicity. *J. Neurosci.* **2000**, *20*, 6048–6054. [PubMed]

190. Davies, P.; Moualla, D.; Brown, D.R. A-synuclein is a cellular ferrireductase. *PLoS ONE* **2011**, *6*, e15814. [CrossRef]

191. Zhu, Z.-J.; Wu, K.-C.; Yung, W.-H.; Qian, Z.-M.; Ke, Y. Differential interaction between iron and mutant α-synuclein causes distinctive Parkinsonian phenotypes in *Drosophila*. *Biochim. Biophys. Acta* **2016**, *1862*, 518–525. [CrossRef] [PubMed]

192. Alzheimer's Association. 2016 Alzheimer's disease facts and figures. *Alzheimer's Dement.* **2016**, *12*, 459–509.

193. Overk, C.R.; Masliah, E. Pathogenesis of synaptic degeneration in Alzheimer's disease and Lewy body disease. *Biochem. Pharmacol.* **2014**, *88*, 508–516. [CrossRef] [PubMed]

194. Demuro, A.; Mina, E.; Kayed, R.; Milton, S.C.; Parker, I.; Glabe, C.G. Calcium dysregulation and membrane disruption as a ubiquitous neurotoxic mechanism of soluble amyloid oligomers. *J. Biol. Chem.* **2005**, *280*, 17294–17300. [CrossRef] [PubMed]

195. Yoshiike, Y.; Kayed, R.; Milton, S.C.; Takashima, A.; Glabe, C.G. Pore-forming proteins share structural and functional homology with amyloid oligomers. *Neuromol. Med.* **2007**, *9*, 270–275. [CrossRef]

196. Barry, A.E.; Klyubin, I.; Mc Donald, J.M.; Mably, A.J.; Farrell, M.A.; Scott, M.; Walsh, D.M.; Rowan, M.J. Alzheimer's disease brain-derived amyloid-β-mediated inhibition of LTP in vivo is prevented by immunotargeting cellular prion protein. *J. Neurosci.* **2011**, *31*, 7259–7263. [CrossRef] [PubMed]

197. Kudo, W.; Lee, H.-P.; Zou, W.-Q.; Wang, X.; Perry, G.; Zhu, X.; Smith, M.A.; Petersen, R.B.; Lee, H. Cellular prion protein is essential for oligomeric amyloid-β-induced neuronal cell death. *Hum. Mol. Genet.* **2012**, *21*, 1138–1344. [CrossRef] [PubMed]

198. Laurén, J.; Gimbel, D.A.; Nygaard, H.B.; Gilbert, J.W.; Strittmatter, S.M. Cellular prion protein mediates impairment of synaptic plasticity by amyloid-β oligomers. *Nature* **2009**, *457*, 1128–1132. [CrossRef] [PubMed]

199. Younan, N.D.; Sarell, C.J.; Davies, P.; Brown, D.R.; Viles, J.H. The cellular prion protein traps Alzheimer's Aβ in an oligomeric form and disassembles amyloid fibers. *FASEB J.* **2013**, *27*, 1847–1858. [CrossRef] [PubMed]

200. Jiang, D.; Men, L.; Wang, J.; Zhang, Y.; Chickenyen, S.; Wang, Y.; Zhou, F. Redox reactions of copper complexes formed with different β-amyloid peptides and their neuropathological [correction of neuropathalogical] relevance. *Biochemistry* **2007**, *46*, 9270–9282. [CrossRef] [PubMed]

201. Everett, J.; Céspedes, E.; Shelford, L.R.; Exley, C.; Collingwood, J.F.; Dobson, J.; van der Laan, G.; Jenkins, C.A.; Arenholz, E.; Telling, N.D. Ferrous iron formation following the co-aggregation of ferric iron and the Alzheimer's disease peptide β-amyloid (1–42). *J. R. Soc. Interface* **2014**, *11*, 20140165. [CrossRef] [PubMed]

202. Rival, T.; Page, R.M.; Chandraratna, D.S.; Sendall, T.J.; Ryder, E.; Liu, B.; Lewis, H.; Rosahl, T.; Hider, R.; Camargo, L.M.; et al. Fenton chemistry and oxidative stress mediate the toxicity of the β-amyloid peptide in a *Drosophila* model of Alzheimer's disease. *Eur. J. Neurosci.* **2009**, *29*, 1335–1347. [CrossRef] [PubMed]

203. Nilsberth, C.; Westlind-Danielsson, A.; Eckman, C.B.; Condron, M.M.; Axelman, K.; Forsell, C.; Stenh, C.; Luthman, J.; Teplow, D.B.; Younkin, S.G.; et al. The "Arctic" APP mutation (E693G) causes Alzheimer's disease by enhanced Aβ protofibril formation. *Nat. Neurosci.* **2001**, *4*, 887–893. [CrossRef] [PubMed]

204. Liu, B.; Moloney, A.; Meehan, S.; Morris, K.; Thomas, S.E.; Serpell, L.C.; Hider, R.; Marciniak, S.J.; Lomas, D.A.; Crowther, D.C. Iron promotes the toxicity of amyloid β peptide by impeding its ordered aggregation. *J. Biol. Chem.* **2011**, *286*, 4248–4256. [CrossRef] [PubMed]

205. Ott, S.; Dziadulewicz, N.; Crowther, D.C. Iron is a specific cofactor for distinct oxidation-and aggregation-dependent Aβ toxicity mechanisms in a *Drosophila* model. *Dis. Model. Mech.* **2015**, *8*, 657–667. [CrossRef] [PubMed]

206. Karch, C.M.; Cruchaga, C.; Goate, A.M. Alzheimer's disease genetics: From the bench to the clinic. *Neuron* **2014**, *83*, 11–26. [CrossRef] [PubMed]
207. Vetrivel, K.S.; Zhang, Y.; Xu, H.; Thinakaran, G. Pathological and physiological functions of presenilins. *Mol. Neurodegener.* **2006**, *1*, 4. [CrossRef] [PubMed]
208. Greenough, M.A.; Volitakis, I.; Li, Q.-X.; Laughton, K.; Evin, G.; Ho, M.; Dalziel, A.H.; Camakaris, J.; Bush, A.I. Presenilins promote the cellular uptake of copper and zinc and maintain copper chaperone of SOD1-dependent copper/zinc superoxide dismutase activity. *J. Biol. Chem.* **2011**, *286*, 9776–9786. [CrossRef] [PubMed]
209. Southon, A.; Greenough, M.A.; Ganio, G.; Bush, A.I.; Burke, R.; Camakaris, J. Presenilin promotes dietary copper uptake. *PLoS ONE* **2013**, *8*, e62811. [CrossRef] [PubMed]
210. Zhang, M.; Haapasalo, A.; Kim, D.Y.; Ingano, L.A.M.; Pettingell, W.H.; Kovacs, D.M. Presenilin/γ-secretase activity regulates protein clearance from the endocytic recycling compartment. *FASEB J.* **2006**, *20*, 1176–1178. [CrossRef] [PubMed]
211. Wang, R.; Tang, P.; Wang, P.; Boissy, R.E.; Zheng, H. Regulation of tyrosinase trafficking and processing by presenilins: Partial loss of function by familial Alzheimer's disease mutation. *Proc. Natl. Acad. Sci. USA* **2006**, *103*, 353–358. [CrossRef] [PubMed]
212. He, Z.L.; Yang, X.E.; Stoffella, P.J. Trace elements in agroecosystems and impacts on the environment. *J. Trace Elem. Med. Biol.* **2005**, *19*, 125–140. [CrossRef] [PubMed]
213. Tchounwou, P.B.; Yedjou, C.G.; Patlolla, A.K.; Sutton, D.J. *Heavy Metal Toxicity and the Environment*; Springer: Basel, Switzerland, 2012; pp. 133–164.
214. Tamás, M.J.; Sharma, S.K.; Ibstedt, S.; Jacobson, T.; Christen, P. Heavy metals and metalloids as a cause for protein misfolding and aggregation. *Biomolecules* **2014**, *4*, 252–267. [CrossRef] [PubMed]
215. Jaishankar, M.; Tseten, T.; Anbalagan, N.; Mathew, B.B.; Beeregowda, K.N. Toxicity, mechanism and health effects of some heavy metals. *Interdiscip. Toxicol.* **2014**, *7*, 60–72. [CrossRef] [PubMed]
216. Martinez-Finley, E.J.; Chakraborty, S.; Fretham, S.J.B.; Aschner, M. Cellular transport and homeostasis of essential and nonessential metals. *Metallomics* **2012**, *4*, 593–605. [CrossRef] [PubMed]
217. Kijak, E.; Rosato, E.; Knapczyk, K.; Pyza, E. *Drosophila melanogaster* as a model system of aluminum toxicity and aging. *Insect Sci.* **2014**, *21*, 189–202. [CrossRef] [PubMed]
218. Wu, Z.; Du, Y.; Xue, H.; Wu, Y.; Zhou, B. Aluminum induces neurodegeneration and its toxicity arises from increased iron accumulation and reactive oxygen species (ROS) production. *Neurobiol. Aging* **2012**, *33*, 199.e1–199.e12. [CrossRef] [PubMed]
219. Krewski, D.; Yokel, R.A.; Nieboer, E.; Borchelt, D.; Cohen, J.; Harry, J.; Kacew, S.; Lindsay, J.; Mahfouz, A.M.; Rondeau, V. Human health risk assessment for aluminium, aluminium oxide, and aluminium hydroxide. *J. Toxicol. Environ. Health. B Crit. Rev.* **2007**, *10*, 1–269. [CrossRef] [PubMed]
220. Huang, N.; Yan, Y.; Xu, Y.; Jin, Y.; Lei, J.; Zou, X.; Ran, D.; Zhang, H.; Luan, S.; Gu, H. Alumina nanoparticles alter rhythmic activities of local interneurons in the antennal lobe of *Drosophila*. *Nanotoxicology* **2013**, *7*, 212–220. [CrossRef] [PubMed]
221. Smith, A.H.; Lingas, E.O.; Rahman, M. Contamination of drinking-water by arsenic in Bangladesh: A public health emergency. *Bull. World Health Organ.* **2000**, *78*, 1093–1103. [PubMed]
222. Rizki, M.; Kossatz, E.; Velázquez, A.; Creus, A.; Farina, M.; Fortaner, S.; Sabbioni, E.; Marcos, R. Metabolism of arsenic in *Drosophila melanogaster* and the genotoxicity of dimethylarsinic acid in the *Drosophila* wing spot test. *Environ. Mol. Mutagen.* **2006**, *47*, 162–168. [CrossRef] [PubMed]
223. Ortiz, J.G.M.; Opoka, R.; Kane, D.; Cartwright, I.L. Investigating arsenic susceptibility from a genetic perspective in *Drosophila* reveals a key role for glutathione synthetase. *Toxicol. Sci.* **2009**, *107*, 416–426. [CrossRef] [PubMed]
224. Muñiz Ortiz, J.G.; Shang, J.; Catron, B.; Landero, J.; Caruso, J.A.; Cartwright, I.L. A transgenic *Drosophila* model for arsenic methylation suggests a metabolic rationale for differential dose-dependent toxicity endpoints. *Toxicol. Sci.* **2011**, *121*, 303–311. [CrossRef] [PubMed]
225. Niehoff, A.-C.; Schulz, J.; Soltwisch, J.; Meyer, S.; Kettling, H.; Sperling, M.; Jeibmann, A.; Dreisewerd, K.; Francesconi, K.A.; Schwerdtle, T.; et al. Imaging by Elemental and Molecular Mass Spectrometry Reveals the Uptake of an Arsenolipid in the Brain of *Drosophila melanogaster*. *Anal. Chem.* **2016**, *88*, 5258–5263. [CrossRef] [PubMed]

226. Meyer, S.; Schulz, J.; Jeibmann, A.; Taleshi, M.S.; Ebert, F.; Francesconi, K.A.; Schwerdtle, T. Arsenic-containing hydrocarbons are toxic in the in vivo model *Drosophila melanogaster*. *Metallomics* **2014**, *6*, 2010–2014. [CrossRef] [PubMed]

227. Griswold, W.; Martin, S. Human Health Effects of Heavy Metals. *Environ. Sci. Technol.* **2009**, *15*, 1–6.

228. Guan, D.; Mo, F.; Han, Y.; Gu, W.; Zhang, M. Digital gene expression profiling (DGE) of cadmium-treated *Drosophila melanogaster*. *Environ. Toxicol. Pharmacol.* **2015**, *39*, 300–306. [CrossRef]

229. Bernard, A. Cadmium & its adverse effects on human health. *Indian J. Med. Res.* **2008**, *128*, 557–564. [PubMed]

230. Rizki, M.; Kossatz, E.; Creus, A.; Marcos, R. Genotoxicity modulation by cadmium treatment: Studies in the *Drosophila* wing spot test. *Environ. Mol. Mutagen.* **2004**, *43*, 196–203. [CrossRef] [PubMed]

231. Pragya, P.; Shukla, A.K.; Murthy, R.C.; Abdin, M.Z.; Kar Chowdhuri, D. Characterization of the effect of Cr(VI) on humoral innate immunity using *Drosophila melanogaster*. *Environ. Toxicol.* **2015**, *30*, 1285–1296. [CrossRef] [PubMed]

232. Mishra, M.; Sharma, A.; Negi, M.P.S.; Dwivedi, U.N.; Chowdhuri, D.K. Tracing the tracks of genotoxicity by trivalent and hexavalent chromium in *Drosophila melanogaster*. *Mutat. Res.* **2011**, *722*, 44–51. [CrossRef] [PubMed]

233. Brochin, R.; Leone, S.; Phillips, D.; Shepard, N.; Zisa, D.; Angerio, A. The Cellular Effect of Lead Poisoning and Its Clinical Picture. *Georg. Undergrad. J. Health Sci.* **2008**, *5*, 1–8.

234. Hirsch, H.V.B.; Lnenicka, G.; Possidente, D.; Possidente, B.; Garfinkel, M.D.; Wang, L.; Lu, X.; Ruden, D.M. *Drosophila melanogaster* as a model for lead neurotoxicology and toxicogenomics research. *Front. Genet.* **2012**, *3*, 68. [CrossRef] [PubMed]

235. He, T.; Hirsch, H.V.B.; Ruden, D.M.; Lnenicka, G.A. Chronic lead exposure alters presynaptic calcium regulation and synaptic facilitation in *Drosophila* larvae. *Neurotoxicology* **2009**, *30*, 777–784. [CrossRef] [PubMed]

236. Hirsch, H.V.B.; Possidente, D.; Averill, S.; Despain, T.P.; Buytkins, J.; Thomas, V.; Goebel, W.P.; Shipp-Hilts, A.; Wilson, D.; Hollocher, K.; et al. Variations at a quantitative trait locus (QTL) affect development of behavior in lead-exposed *Drosophila melanogaster*. *Neurotoxicology* **2009**, *30*, 305–311. [CrossRef]

237. Ruden, D.M.; Chen, L.; Possidente, D.; Possidente, B.; Rasouli, P.; Wang, L.; Lu, X.; Garfinkel, M.D.; Hirsch, H.V.B.; Page, G.P. Genetical toxicogenomics in *Drosophila* identifies master-modulatory loci that are regulated by developmental exposure to lead. *Neurotoxicology* **2009**, *30*, 898–914. [CrossRef] [PubMed]

238. Carmona, E.R.; Creus, A.; Marcos, R. Genotoxicity testing of two lead-compounds in somatic cells of *Drosophila melanogaster*. *Mutat. Res.* **2011**, *724*, 35–40. [CrossRef] [PubMed]

239. Hirsch, H.V.B.; Possidente, D.; Possidente, B. Pb^{2+}: An endocrine disruptor in *Drosophila*? *Physiol. Behav.* **2010**, *99*, 254–259. [CrossRef] [PubMed]

240. Peterson, E.K.; Wilson, D.T.; Possidente, B.; McDaniel, P.; Morley, E.J.; Possidente, D.; Hollocher, K.T.; Ruden, D.M.; Hirsch, H.V.B. Accumulation, elimination, sequestration, and genetic variation of lead (Pb^{2+}) loads within and between generations of *Drosophila melanogaster*. *Chemosphere* **2017**, *181*, 368–375. [CrossRef] [PubMed]

241. Horning, K.J.; Caito, S.W.; Tipps, K.G.; Bowman, A.B.; Aschner, M. Manganese Is Essential for Neuronal Health. *Annu. Rev. Nutr.* **2015**, *35*, 71–108. [CrossRef]

242. Herrero Hernandez, E.; Discalzi, G.; Valentini, C.; Venturi, F.; Chiò, A.; Carmellino, C.; Rossi, L.; Sacchetti, A.; Pira, E. Follow-up of patients affected by manganese-induced Parkinsonism after treatment with CaNa2EDTA. *Neurotoxicology* **2006**, *27*, 333–339. [CrossRef] [PubMed]

243. Ternes, A.P.L.; Zemolin, A.P.; da Cruz, L.C.; da Silva, G.F.; Saidelles, A.P.F.; de Paula, M.T.; Wagner, C.; Golombieski, R.M.; Flores, É.M.M.; Picoloto, R.S.; et al. *Drosophila melanogaster*—An embryonic model for studying behavioral and biochemical effects of manganese exposure. *EXCLI J.* **2014**, *13*, 1239–1253. [PubMed]

244. Adedara, I.A.; Abolaji, A.O.; Rocha, J.B.T.; Farombi, E.O. Diphenyl Diselenide Protects Against Mortality, Locomotor Deficits and Oxidative Stress in *Drosophila melanogaster* Model of Manganese-Induced Neurotoxicity. *Neurochem. Res.* **2016**, *41*, 1430–1438. [CrossRef] [PubMed]

245. Bonilla, E.; Contreras, R.; Medina-Leendertz, S.; Mora, M.; Villalobos, V.; Bravo, Y. Minocycline increases the life span and motor activity and decreases lipid peroxidation in manganese treated *Drosophila melanogaster*. *Toxicology* **2012**, *294*, 50–53. [CrossRef] [PubMed]

246. Bianchini, M.C.; Gularte, C.O.A.; Escoto, D.F.; Pereira, G.; Gayer, M.C.; Roehrs, R.; Soares, F.A.A.; Puntel, R.L. Peumus boldus (Boldo) Aqueous Extract Present Better Protective Effect than Boldine Against Manganese-Induced Toxicity in D. melanogaster. *Neurochem. Res.* **2016**, *41*, 2699–2707. [CrossRef] [PubMed]

247. Chen, C.-W.; Chen, C.-F.; Dong, C.-D. Distribution and Accumulation of Mercury in Sediments of Kaohsiung River Mouth, Taiwan. *APCBEE Procedia* **2012**, *1*, 153–158. [CrossRef]

248. Trasande, L.; Landrigan, P.J.; Schechter, C. Public health and economic consequences of methyl mercury toxicity to the developing brain. *Environ. Health Perspect.* **2005**, *113*, 590–596. [CrossRef] [PubMed]

249. Morais, S.; Costa, F.G.; Lourdes Pereir, M. De Heavy Metals and Human Health. In *Environmental Health—Emerging Issues and Practice*; InTech: Rijeka, Croatia, 2012.

International Journal of
Molecular Sciences

MDPI

Review

Roles of Copper-Binding Proteins in Breast Cancer

Stéphanie Blockhuys and Pernilla Wittung-Stafshede *

Department Biology and Biological Engineering, Chalmers University of Technology,
412 96 Gothenburg, Sweden; steblo@chalmers.se
* Correspondence: pernilla.wittung@chalmers.se; Tel.: +46-766072283

Academic Editor: Reinhard Dallinger
Received: 20 March 2017; Accepted: 18 April 2017; Published: 20 April 2017

Abstract: Copper ions are needed in several steps of cancer progression. However, the underlying mechanisms, and involved copper-binding proteins, are mainly elusive. Since most copper ions in the body (in and outside cells) are protein-bound, it is important to investigate what copper-binding proteins participate and, for these, how they are loaded with copper by copper transport proteins. Mechanistic information for how some copper-binding proteins, such as extracellular lysyl oxidase (LOX), play roles in cancer have been elucidated but there is still much to learn from a biophysical molecular viewpoint. Here we provide a summary of copper-binding proteins and discuss ones reported to have roles in cancer. We specifically focus on how copper-binding proteins such as mediator of cell motility 1 (MEMO1), LOX, LOX-like proteins, and secreted protein acidic and rich in cysteine (SPARC) modulate breast cancer from molecular and clinical aspects. Because of the importance of copper for invasion/migration processes, which are key components of cancer metastasis, further insights into the actions of copper-binding proteins may provide new targets to combat cancer.

Keywords: copper-binding protein; copper transport; lysyl oxidase; SPARC; MEMO1; ATOX1; cancer; breast cancer; metastasis

1. Copper Proteins in Biology

Copper (Cu) is an indispensable metal ion that plays a crucial role in the biochemistry of every living organism. In the body, Cu can exist in oxidized (Cu^{2+}) and reduced (Cu^{2+}) redox states. The unique electronic structure of Cu makes it useful as cofactor in redox-reactions of enzymes that perform biological functions required for normal growth and development. The recommended daily intake of Cu in healthy adults is 0.9 mg/day. Uptake of Cu in the human body depends on many factors and nutritional components. Cu is absorbed in the small intestine by amino acid transporters mainly for Met, His, and Cys amino acids. Cu is then transported primarily by serum albumin to the liver for subsequent delivery to enzymes and different parts of the body.

The redox activity that allows Cu to contribute functionalities to proteins also provides a risk for toxicity [1]. To minimize toxic effects of free Cu and to regulate the distribution of Cu in time and space, organisms have evolved elaborate protein-based systems for uptake, cellular transport, protein loading, and storing of Cu. During the process of cellular Cu uptake, oxidized Cu is first reduced and then it enters the cell through the copper transporter CTR1 [2]. In the cytoplasm, Cu is shuttled to targets by at least three pathways [3]: in the secretory path, the cytoplasmic Cu chaperone ATOX1 delivers Cu to the Cu-transporting P_{1B}-type ATPases ATP7A and ATP7B (Wilson and Menkes disease proteins) in the trans-Golgi network. After transfer to ATP7A/B, fueled by ATP hydrolysis, Cu is channeled to the lumen and loaded onto target Cu-dependent enzymes, e.g., ceruloplasmin (CP) and lysyl oxidase (LOX) [4]. CP is the major Cu transporting protein in blood plasma: it functions as a ferroxidase, oxidizing iron, and it also delivers Cu to other cells [5,6]. In fact, CP contains more than

75% of total plasma Cu, and albumin and transcuprein are known to carry the remaining plasma Cu. LOX is also secreted, and is a Cu-dependent amino oxidase enzyme that crosslinks extracellular matrix (ECM) proteins such as collagen, elastin, and other ECM proteins. In addition to the secretory pathway for intracellular Cu transport, two other pathways for Cu transport in the cytoplasm are known. In one, the Cu chaperone for superoxide dismutase 1 (CCS) delivers Cu to cytoplasmic Cu/Zn superoxide dismutase 1 (SOD1); in the other, Cu is directed to the mitochondria by COX17 and some additional proteins (i.e., SCOs, COX11), for incorporation in cytochrome c oxidase (COX1 and COX2).

We recently established the human Cu proteome (i.e., the collection of all identified human Cu-binding proteins) [7]. We revealed a total of 54 proteins and, of these proteins, 12 are classified as Cu transporters (CTR1 and CTR2 for import, CCS and ATOX1 for cytoplasmic transport, and COX11, COX17, SCO1, and SCO2 that provide Cu to the mitochondria). Also classified as transport proteins are COMMD1 (copper metabolism domain containing 1), an enigmatic protein that may be involved in the regulation of exocytosis of Cu-loaded vesicles [8,9], and cutC copper transporter (CUTC) that, in addition to Cu transport, has been proposed to be an enzyme [10]. About half of the identified Cu-binding proteins are enzymes and these proteins are found in all but Golgi intracellular compartments, extracellularly, and in the plasma membrane. The remaining proteins in the Cu proteome have either non-enzymatic or unknown functions. Notably, none but ATOX1 (see below) of the identified Cu-binding proteins are transcription factors.

2. Copper in Cancer

Because Cu is important for the function of many enzymes [1,11,12], it is reasonable that Cu is required for characteristic phenomena involved in cancer progression such as proliferative immortality, angiogenesis, and metastasis [13,14]. In fact, cancer tissue and cancer patients' serum have been found to contain increased Cu but levels of other metals (e.g., iron, zinc) are often lower than normal [13,15,16]. Cancer is a multifactorial collection of diseases that involves uncontrolled growth of cells, followed by cancer cell invasion, dissemination, and secondary tumor formation at local and distant sites; these processes are often connected with an immune response. For a tumor to grow larger than a few mm, angiogenesis is needed meaning that new blood vessels must form. Cu can induce a pro-angiogenic response [17] by direct binding to angiogenic factors as well as by promoting the expression of such factors [18,19]. Cu can also activate metabolic and proliferative enzymes that enhance the ability of cancerous cells to metastasize, as will be described below [13].

Because of the recognized importance of Cu in many of the cancer hallmarks, there have been attempts to develop general Cu-chelating compounds as anticancer therapies [13,20,21]. Reducing systemic Cu levels decreased the activity of COX1 and reduced ATP levels, which leads to lessened oxidative phosphorylation and thereby reduces growth of proliferating cancer cells [22]. Moreover, Cu deprivation was found to inhibit the so-called epithelial-to-mesenchymal transition (EMT) of cells, which is a process in which cells loose cell polarity and cell-cell adhesion and instead gain migratory and invasive properties [23]. However, it is not clear if this type of cell-culture results are transferrable to human cancer patients. Also, as essentially all Cu in the body is protein-bound, drugs targeted to Cu-transport and Cu-dependent proteins important for cancer-promoting processes may constitute more efficient future drug targets. In accordance with this, it was recently shown that small molecules that inhibited cellular Cu transport, by specific targeting of CCS and ATOX1, acted as a selective approach to reducing proliferation of cancer cells but not normal cells [24].

3. Copper Proteins Involved in Diverse Aspects of Cancer

For a global view of Cu protein expression changes in various cancers, we recently analyzed the RNA transcript level changes of the Cu proteome (i.e., the identified 54 Cu-binding proteins) in different cancer tissues using information from the Cancer Genome Atlas, or TCGA, database [7]. Many of the proteins in the human Cu proteome exhibit either up- or downregulation in the different cancers [7]. To give some examples, LOX, LOX-like proteins 1 and 2, secreted protein acidic and rich in cysteine (SPARC), and ENOX2 are upregulated in more than 6 out of 18 cancers analyzed. With respect to ATOX1, it is upregulated in breast, colorectal, uterus, and liver tumors [7]. Several studies using sources other than TCGA have also assessed transcript levels in different cancers, but these will not be discussed here. To understand if transcript expression changes (from our analysis and reported by others) are related to cancer-promoting events, molecular studies of defined biological systems are needed. Indeed, some of the enzymes in the Cu proteome have been reported to play mechanistic roles in cancer, as summarized below.

As mentioned earlier, LOX is secreted by cancer cells to create pre-metastatic niches by stimulating collagen cross-linking and fibronectin synthesis [25,26]. Adaptation to hypoxia is a driving force for tumor progression whereby hypoxia stimulates the hypoxia-inducible factor 1α (HIF-1α)-mediated LOX expression [27,28]. Several studies have investigated the mechanistic role of extracellular LOX in cancer cell invasion and metastasis [29,30]. LOX-mediated ECM cross-linking and stiffening induce integrin-mediated focal adhesion formation and PI3K signaling, promoting tumor growth and invasion [31,32]. In addition to extracellular modifications, LOX also appears to regulate cancer progression within cells such as via modulation of actin polymerization that promotes migratory phenotypes [33]. Notably, recent data have indicated LOX and LOX-like proteins as histone-modifying enzymes, which regulate gene expression [34].

MAP2K1 (mitogen-activated protein kinase kinase 1), also called MEK1, is an intra-cellular Cu-dependent kinase involved in the mitogen-activated protein kinase (MAPK) signaling pathway, and thereby is related to cancer growth [35], invasion, and metastasis [36,37]. Another intracellular Cu-dependent protein is mediator of cell motility 1 (MEMO1), a redox enzyme that has been reported to facilitate tumor cell migration and metastasis through several molecular mechanisms. MEMO1 promotes cell migration by stimulating the dynamics of the cell cytoskeleton via activation of cofilin [38] and interaction with the RhoA-mDia1 signaling complex [39] by sustaining reactive oxygen species production upon nitric oxide stimulation in cell protrusions [40], as well as by upregulation of the EMT regulator SNAIL1 through interaction with the insulin-like receptor substrate 1 (IRS1) and activation of the PI3K/Akt signaling pathway [41]. Using a yeast two-hybrid screen, nuclear LOX was found to interact with, among others, MEMO1 [42], suggesting a link between these two Cu-binding proteins.

SPARC is classified as a secreted Cu-binding glycoprotein with counter-adhesive properties and functions extracellularly to promote invasion. SPARC plays a role in tumor invasion and metastasis via modulation of cell–cell and cell–matrix interactions [43,44]. It was shown that the Cu-binding domain of SPARC mediated cell survival via interactions with integrin β1 and activation of integrin-linked kinase [45]. The role of SPARC seems to depend on the cancer type: it is associated with highly aggressive tumor phenotypes in some cancer [46], but in others it appeared as a tumor suppressor [47]. Taken together, SPARC expression appears to correlate with invasion and progression of gliomas and melanomas but in many epithelial cancers, hyper-methylation of the SPARC promoter reduces the amount of SPARC produced by the tumor cells. A clearer description of the molecular mechanisms of SPARC action is needed to understand its divergent effects on human cancers. Another of the Cu-binding proteins, COMMD1, is interesting as it is often found downregulated in invasive human cancers. It was shown that COMMD1 inhibits both NFκB and HIF-1 mediated gene expression (that both promote tumor growth, survival, and invasion) and, in the case of HIF-1, by direct interaction that disrupted HIF dimerization [48]. Whereas LOX obtains Cu from the secretory pathway, it is not known how SPARC, MEMO1, COMMD1, and MEK1 are loaded with Cu.

Most of the above discussed proteins are classified as Cu-dependent enzymes. However, unprecedented findings have suggested that the Cu chaperone ATOX1 has new activities connected to cancer [35]. ATOX1 was found to act as a Cu-dependent transcription factor [49] that stimulates expression of the proliferation protein cyclin D1 [50] and the extracellular antioxidant protein SOD3 [51,52]. ATOX1 also acts as a Cu-dependent transcription factor for NADPH oxidase promoting inflammatory neovascularization [53] and, ATOX1 may potentially regulate malignant angiogenesis as it is necessary for platelet-derived growth factor-induced Cu-dependent cell migration [54].

Also another cytoplasmic Cu chaperone, CCS, appears to have additional functions in the cell. CCS was shown to enter the nucleus and regulate the HIF-1 transcriptional complex in a Cu-dependent manner. This resulted in expression of the vascular endothelial growth factor (VEGF) which in turn promoted tumor growth [55].

Our research group have also visualized ATOX1 in the nuclei of mammalian cells, but we found no binding to the DNA promotor sequence proposed for ATOX1 upon using in-vitro biophysical experiments [56]. Therefore, we searched for protein partners that may mediate a transcriptional response together with ATOX1. From a large yeast two-hybrid screen, using ATOX1 as bait, we identified several new human protein partners [57] and several of the confident hits are proteins with functions related to signaling and cancer (e.g., CPEB4 (cytoplasmic polyadenylation element binding protein 4) [58], DNMT1 (DNA methyl transferase 1), and PPM1A (protein phosphatase, Mg^{2+}/Mn^{2+} dependent 1A) [59–62]).

4. Copper-Binding Proteins in Breast Cancer: A Clinical Perspective

Breast cancer is the most common cancer among women and can be categorized into four molecular subtypes—i.e., luminal A, luminal B, HER2, and triple negative/basal-like—depending on specific biomarkers. Breast cancer tissue and serum have been reported to contain higher Cu levels than control samples and, moreover, the Cu levels were highest in the most advanced breast cancers [63]. Mapping of gene expression signatures in breast cancer cell lines has noted ATP7A as well as LOX as upregulated genes [64]. Also ATP7B was found upregulated in breast cancer but not in normal adjacent tissue [65]. In the latter study, it was suggested that high ATP7B protein levels will increase cell resistance to platinum-based anticancer drugs. Several in vitro studies from our lab have demonstrated that cisplatin, and other platinum compounds, can bind to Cu transport proteins such as ATOX1 and ATP7B [66–68]. In accord, it was shown that targeting of cysteine-containing Cu-transporting proteins (e.g., ATP7A/B, ATOX1) with ammonium tetrathiomolybdate enhanced sensitivity of breast cancer cells to cisplatin [69].

Our analysis of the TCGA data for breast cancer [7] revealed upregulation of 26% of the Cu proteome proteins: F5, ATP7B, SLC31A1, SCO2, HEPHL1, CUTA, ATOX1, COX17, TYRP1, MT3, LOX-like proteins 1 and 2, SPARC, and MOXD1. With respect to molecular mechanistic studies of Cu proteins' roles in breast cancer, there is such information reported for ten proteins (described in detail in Table S1) and below we discuss findings for selected proteins of the ten (MEMO1, SPARC, LOX, and some LOX-like proteins), followed by a separate section about ATOX1.

In approximately 85% of patients with advanced breast cancer, metastasis affects the bone which results in osteolytic lesions and renders the cancer largely untreatable. It was recently demonstrated that LOX secretion was specifically associated with metastasis to the bone in patients with estrogen-receptor negative breast cancer [29]. LOX enzymatic activity promoted TWIST transcription, thereby mediating EMT of cancer cells [70]. Moreover, in breast cancer cells, LOX was found intracellularly, both in the nucleus and cytoplasm, and hydrogen peroxide generated as a side-product upon LOX activity appeared to facilitate cell adhesion and migration through activation of the FAK/Src signaling pathway [71].

LOXL2, LOXL3, and LOXL4 all belong to the LOX-like (LOXL) family of proteins. Like LOX, LOXL2 was shown to promote invasive/metastatic phenotypes in breast cancer cells which was explained by altered LOXL2 protein processing and localization [72]. It was found that LOXL2 and LOXL3 interacted with, and stabilized, SNAIL1, which induced EMT and promoted invasion through repression of, among others, E-cadherin expression [73]. In addition to SNAIL1 effects, LOXL2 influences expression of SPARC and the extracellular proteins TIMP1 and MMP9 [74,75]; in fact, LOXL2 was noted as prognostic factor in breast cancer [74]. LOXL2 activity in basal-like carcinoma cells were found to affect tight junction and cell polarity complexes by a mechanism which involves downregulation of involved genes and, LOXL2 is required for cell invasion, tumor growth, and lung metastasis of basal-like breast carcinoma cells [76].

As already mentioned in the previous section, MEMO1 facilitates tumor cell migration, metastasis, and pathways related to EMT, and these results were obtained using breast cancer cells [39–41]. Because many breast cancer types are hormone-dependent it is important to probe cross-interactions between growth factor- and steroid hormone-mediated signaling pathways in order to find suitable drug targets. In this respect, MEMO1 was reported to act at the intersection between growth factor (heregulin and IGF1) and estrogen signaling in breast cancer cells. Specifically, MEMO1 was found to control estrogen receptor α (ERα) sub-cellular localization, phosphorylation, and function downstream of ErbB2/ER or IGFIR/ER thereby activating MAPK and PI3K signaling pathways that promoted breast cancer cell migration and/or proliferation [77,78].

SPARC stimulates breast cancer growth and metastasis in in-vivo models [79]. Statistical analysis of tissue microarray data showed that upregulation of SPARC occurs in basal-like breast tumors [80]. Moreover, SPARC levels were found to be inversely correlated with the content of the estrogen receptor. This result suggests that SPARC levels are associated with more aggressive breast cancer tumors [81,82].

5. ATOX1 in Breast Cancer

Mapping of gene expression signatures in breast cancer cell lines has noted upregulation of ATOX1 via proteomics [83]. Using immunohistochemistry, we investigated ATOX1 in 67 breast cancer sections in tissue microarrays (TMAs). In agreement with the TCGA data (mentioned above, [7]), we found ATOX1 levels to be increased in cancer as compared to normal breast tissue. Scoring of the 67 breast cancer samples revealed that the highest ATOX1 intensities were found in samples of all cancer molecular subtypes but the HER2 subtype [7]. When we turned to cell line studies, we stumbled on a putative functional role for ATOX1 in breast cancer cells. Using an aggressive breast cancer cell line, we made the discovery that ATOX1 accumulates at lamellipodia borders of migrating breast cancer cells and ATOX1 silencing resulted in migration defects as evidenced from reduced wound closure. Thus, ATOX1 may have an unknown role in breast cancer cell migration [84], that parallels the reported role for ATOX1 in endothelial cell wound healing [85].

Interestingly, CPEB4 has been reported to play a promoting role in breast cancer by modulating mRNA transcript translation in the cytoplasm [58]. As mentioned above, our yeast two-hybrid screen identified an interaction between CPEB4 and ATOX1 [57] which calls for further investigation of putative synergistic cancer-promoting effects between these proteins.

6. Summary and Outlook

Clearly, Cu-dependent processes are of importance for breast (and other) cancer development. Thus it becomes important to elucidate the molecular mechanisms and pathways for how involved Cu-dependent proteins are loaded with Cu—i.e., how the flow of Cu via Cu transport proteins are directed to Cu-dependent proteins in breast cancer cells. Based on the available data for Cu-binding proteins in breast cancer, speculations can be made that, of course, should be tested experimentally in the future in controlled cell lines [86]. In Figure 1, we have compiled the known paths involving the Cu-binding proteins LOX, MEMO1, and SPARC in breast cancer, using a migrating cancer cell as the model.

One interesting possibility is that ATOX1 delivers Cu to MEMO1 at the lamellipodia edges such that MEMO1 becomes activated and, in turn, can activate cofilin (direct interaction between MEMO1 and cofilin has been reported [38]) resulting in actin dynamics modulation and thereby promotion of cell migration. We further speculate that this intra-cellular scenario may be coupled to integrin-mediated ECM-induced signaling, e.g., from SPARC, which can stimulate small GTPases that can play roles in cofilin/actin function. We imagine that a wealth of new molecular, mechanistic knowledge in the coming years will allow for the development of new breast cancer drugs directed towards selected Cu-binding proteins.

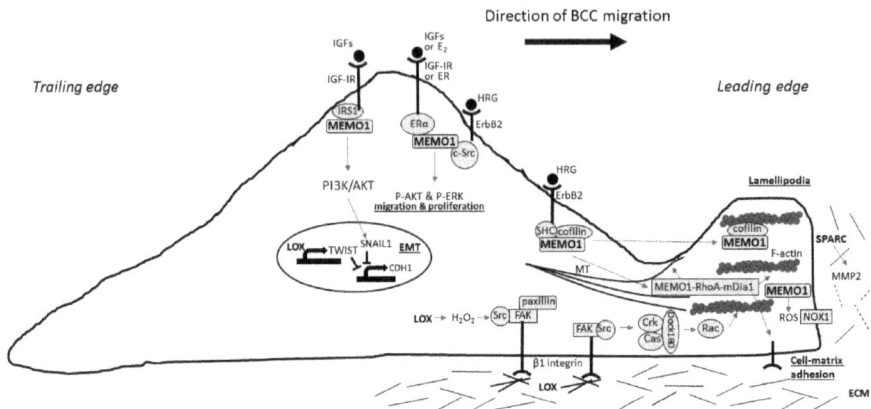

Figure 1. Model of a migrating breast cancer cell (BCC) with reported molecular signaling pathways for LOX, SPARC, and MEMO1. (IGF, insulin growth factor; IGF-IR, insulin growth factor 1 receptor; E2, estrogen; ER, estrogen receptor; HRG, heregulin; IRS1, insulin receptor substrate 1; ERα, estrogen receptor α; PI3K, phosphoinositide 3-kinase; EMT, epithelial-mesenchymal transition; MT, microtubuli; ROS, reactive oxygen species; NOX1, NADPH oxidase 1; ECM, extracellular matrix; MMP2, matrix metalloproteinase 2; FAK, focal adhesion kinase); SHC, Src homology 2 domain containing. The arrows indicate the direction of the molecular signaling pathways.

Supplementary Materials: Supplementary materials can be found at www.mdpi.com/1422-0067/18/4/871/s1.

Acknowledgments: The Swedish Natural Research Council, the Knut and Alice Wallenberg Foundation, and Chalmers Foundation provided financial support.

Conflicts of Interest: The authors declare no conflict of interest.

References

1. Grubman, A.; White, A.R. Copper as a key regulator of cell signalling pathways. *Expert Rev. Mol. Med.* **2014**, *16*, e11. [CrossRef] [PubMed]
2. Ohrvik, H.; Thiele, D.J. How copper traverses cellular membranes through the mammalian copper transporter 1, CTR1. *Ann. N. Y. Acad. Sci.* **2014**, *1314*, 32–41. [CrossRef] [PubMed]
3. Puig, S.; Thiele, D.J. Molecular mechanisms of copper uptake and distribution. *Curr. Opin. Chem. Biol.* **2002**, *6*, 171–180. [CrossRef]
4. Koch, K.A.; Pena, M.M.; Thiele, D.J. Copper-binding motifs in catalysis, transport, detoxification and signaling. *Chem. Biol.* **1997**, *4*, 549–560. [CrossRef]
5. Campbell, C.H.; Brown, R.; Linder, M.C. Circulating ceruloplasmin is an important source of copper for normal and malignant animal cells. *Biochim. Biophys. Acta* **1981**, *678*, 27–38. [CrossRef]
6. Ramos, D.; Mar, D.; Ishida, M.; Vargas, R.; Gaite, M.; Montgomery, A.; Linder, M.C. Mechanism of copper uptake from blood plasma ceruloplasmin by mammalian cells. *PLoS ONE* **2016**, *11*, e0149516. [CrossRef] [PubMed]
7. Blockhuys, S.; Celauro, E.; Hildesjo, C.; Feizi, A.; Stal, O.; Fierro-Gonzalez, J.C.; Wittung-Stafshede, P. Defining the human copper proteome and analysis of its expression variation in cancers. *Met. Integr. Biomet. Sci.* **2017**, *9*, 112–123. [CrossRef] [PubMed]
8. Wang, Y.; Hodgkinson, V.; Zhu, S.; Weisman, G.A.; Petris, M.J. Advances in the understanding of mammalian copper transporters. *Adv. Nutr.* **2011**, *2*, 129–137. [CrossRef] [PubMed]
9. Phillips-Krawczak, C.A.; Singla, A.; Starokadomskyy, P.; Deng, Z.; Osborne, D.G.; Li, H.; Dick, C.J.; Gomez, T.S.; Koenecke, M.; Zhang, J.S.; et al. COMMD1 is linked to the WASH complex and regulates endosomal trafficking of the copper transporter ATP7A. *Mol. Biol. Cell* **2015**, *26*, 91–103. [CrossRef] [PubMed]
10. Li, Y.; Du, J.; Zhang, P.; Ding, J. Crystal structure of human copper homeostasis protein CUTC reveals a potential copper-binding site. *J. Struct. Biol.* **2010**, *169*, 399–405. [CrossRef] [PubMed]

11. Matson Dzebo, M.; Arioz, C.; Wittung-Stafshede, P. Extended functional repertoire for human copper chaperones. *Biomol. Concepts* **2016**, *7*, 29–39. [CrossRef] [PubMed]

12. Turski, M.L.; Thiele, D.J. New roles for copper metabolism in cell proliferation, signaling, and disease. *J. Biol. Chem.* **2009**, *284*, 717–721. [CrossRef] [PubMed]

13. Denoyer, D.; Masaldan, S.; La Fontaine, S.; Cater, M.A. Targeting copper in cancer therapy: "Copper that cancer". *Met. Integr. Biometal Sci.* **2015**, *7*, 1459–1476. [CrossRef] [PubMed]

14. Hanahan, D.; Weinberg, R.A. Hallmarks of cancer: The next generation. *Cell* **2011**, *144*, 646–674. [CrossRef] [PubMed]

15. Byrne, C.; Divekar, S.D.; Storchan, G.B.; Parodi, D.A.; Martin, M.B. Metals and breast cancer. *J. Mammary Gland Biol. Neoplas.* **2013**, *18*, 63–73. [CrossRef] [PubMed]

16. Gupte, A.; Mumper, R.J. Elevated copper and oxidative stress in cancer cells as a target for cancer treatment. *Cancer Treat. Rev.* **2009**, *35*, 32–46. [CrossRef] [PubMed]

17. Rigiracciolo, D.C.; Scarpelli, A.; Lappano, R.; Pisano, A.; Santolla, M.F.; de Marco, P.; Cirillo, F.; Cappello, A.R.; Dolce, V.; Belfiore, A.; et al. Copper activates HIF-1α/GPER/VEGF signalling in cancer cells. *Oncotarget* **2015**, *6*, 34158–34177. [PubMed]

18. Pan, Q.; Kleer, C.G.; van Golen, K.L.; Irani, J.; Bottema, K.M.; Bias, C.; de Carvalho, M.; Mesri, E.A.; Robins, D.M.; Dick, R.D.; et al. Copper deficiency induced by tetrathiomolybdate suppresses tumor growth and angiogenesis. *Cancer Res.* **2002**, *62*, 4854–4859. [PubMed]

19. Kenneth, N.S.; Hucks, G.E., Jr.; Kocab, A.J.; McCollom, A.L.; Duckett, C.S. Copper is a potent inhibitor of both the canonical and non-canonical NFκB pathways. *Cell Cycle* **2014**, *13*, 1006–1014. [CrossRef] [PubMed]

20. Alvarez, H.M.; Xue, Y.; Robinson, C.D.; Canalizo-Hernandez, M.A.; Marvin, R.G.; Kelly, R.A.; Mondragon, A.; Penner-Hahn, J.E.; O'Halloran, T.V. Tetrathiomolybdate inhibits copper trafficking proteins through metal cluster formation. *Science* **2010**, *327*, 331–334. [CrossRef] [PubMed]

21. Brewer, G.J. The use of copper-lowering therapy with tetrathiomolybdate in medicine. *Expert Opin. Investig. Drugs* **2009**, *18*, 89–97. [CrossRef] [PubMed]

22. Ishida, S.; Andreux, P.; Poitry-Yamate, C.; Auwerx, J.; Hanahan, D. Bioavailable copper modulates oxidative phosphorylation and growth of tumors. *Proc. Natl. Acad. Sci. USA* **2013**, *110*, 19507–19512. [CrossRef] [PubMed]

23. Li, S.; Zhang, J.; Yang, H.; Wu, C.; Dang, X.; Liu, Y. Copper depletion inhibits CoCl$_2$-induced aggressive phenotype of MCF-7 cells via downregulation of HIF-1 and inhibition of SNAIL/TWIST-mediated epithelial-mesenchymal transition. *Sci. Rep.* **2015**, *5*, 12410. [CrossRef] [PubMed]

24. Wang, J.; Luo, C.; Shan, C.; You, Q.; Lu, J.; Elf, S.; Zhou, Y.; Wen, Y.; Vinkenborg, J.L.; Fan, J.; et al. Inhibition of human copper trafficking by a small molecule significantly attenuates cancer cell proliferation. *Nat. Chem.* **2015**, *7*, 968–979. [CrossRef] [PubMed]

25. Siddikuzzaman; Grace, V.M.; Guruvayoorappan, C. Lysyl oxidase: A potential target for cancer therapy. *Inflammopharmacology* **2011**, *19*, 117–129. [CrossRef] [PubMed]

26. Xiao, Q.; Ge, G. Lysyl oxidase, extracellular matrix remodeling and cancer metastasis. *Cancer Microenviron.* **2012**, *5*, 261–273. [CrossRef] [PubMed]

27. Erler, J.T.; Bennewith, K.L.; Cox, T.R.; Lang, G.; Bird, D.; Koong, A.; Le, Q.T.; Giaccia, A.J. Hypoxia-induced lysyl oxidase is a critical mediator of bone marrow cell recruitment to form the premetastatic niche. *Cancer Cell* **2009**, *15*, 35–44. [CrossRef] [PubMed]

28. Pez, F.; Dayan, F.; Durivault, J.; Kaniewski, B.; Aimond, G.; Le Provost, G.S.; Deux, B.; Clezardin, P.; Sommer, P.; Pouyssegur, J.; et al. The HIF-1-inducible lysyl oxidase activates HIF-1 via the Akt pathway in a positive regulation loop and synergizes with HIF-1 in promoting tumor cell growth. *Cancer Res.* **2011**, *71*, 1647–1657. [CrossRef] [PubMed]

29. Cox, T.R.; Rumney, R.M.; Schoof, E.M.; Perryman, L.; Hoye, A.M.; Agrawal, A.; Bird, D.; Latif, N.A.; Forrest, H.; Evans, H.R.; et al. The hypoxic cancer secretome induces pre-metastatic bone lesions through lysyl oxidase. *Nature* **2015**, *522*, 106–110. [CrossRef] [PubMed]

30. Cox, T.R.; Gartland, A.; Erler, J.T. Lysyl oxidase, a targetable secreted molecule involved in cancer metastasis. *Cancer Res.* **2016**, *76*, 188–192. [CrossRef] [PubMed]

31. Levental, K.R.; Yu, H.; Kass, L.; Lakins, J.N.; Egeblad, M.; Erler, J.T.; Fong, S.F.; Csiszar, K.; Giaccia, A.; Weninger, W.; et al. Matrix crosslinking forces tumor progression by enhancing integrin signaling. *Cell* **2009**, *139*, 891–906. [CrossRef] [PubMed]

32. Erler, J.T.; Bennewith, K.L.; Nicolau, M.; Dornhofer, N.; Kong, C.; Le, Q.T.; Chi, J.T.; Jeffrey, S.S.; Giaccia, A.J. Lysyl oxidase is essential for hypoxia-induced metastasis. *Nature* **2006**, *440*, 1222–1226. [CrossRef] [PubMed]

33. Payne, S.L.; Hendrix, M.J.; Kirschmann, D.A. Lysyl oxidase regulates actin filament formation through the p130(Cas)/Crk/Dock180 signaling complex. *J. Cell. Biochem.* **2006**, *98*, 827–837. [CrossRef] [PubMed]

34. Iturbide, A.; Garcia de Herreros, A.; Peiro, S. A new role for LOX and LOXL2 proteins in transcription regulation. *FEBS J.* **2015**, *282*, 1768–1773. [CrossRef] [PubMed]

35. Brady, D.C.; Crowe, M.S.; Turski, M.L.; Hobbs, G.A.; Yao, X.; Chaikuad, A.; Knapp, S.; Xiao, K.; Campbell, S.L.; Thiele, D.J.; et al. Copper is required for oncogenic braf signalling and tumorigenesis. *Nature* **2014**, *509*, 492–496. [CrossRef] [PubMed]

36. Lemieux, E.; Bergeron, S.; Durand, V.; Asselin, C.; Saucier, C.; Rivard, N. Constitutively active MEK1 is sufficient to induce epithelial-to-mesenchymal transition in intestinal epithelial cells and to promote tumor invasion and metastasis. *Int. J. Cancer* **2009**, *125*, 1575–1586. [CrossRef] [PubMed]

37. Turski, M.L.; Brady, D.C.; Kim, H.J.; Kim, B.E.; Nose, Y.; Counter, C.M.; Winge, D.R.; Thiele, D.J. A novel role for copper in RAS/mitogen-activated protein kinase signaling. *Mol. Cell. Biol.* **2012**, *32*, 1284–1295. [CrossRef] [PubMed]

38. Meira, M.; Masson, R.; Stagljar, I.; Lienhard, S.; Maurer, F.; Boulay, A.; Hynes, N.E. MEMO is a cofilin-interacting protein that influences PLCγ1 and cofilin activities, and is essential for maintaining directionality during ErbB2-induced tumor-cell migration. *J. Cell Sci.* **2009**, *122*, 787–797. [CrossRef] [PubMed]

39. Zaoui, K.; Honore, S.; Isnardon, D.; Braguer, D.; Badache, A. MEMO-Rhoa-mDia1 signaling controls microtubules, the actin network, and adhesion site formation in migrating cells. *J. Cell Biol.* **2008**, *183*, 401–408. [CrossRef] [PubMed]

40. MacDonald, G.; Nalvarte, I.; Smirnova, T.; Vecchi, M.; Aceto, N.; Dolemeyer, A.; Frei, A.; Lienhard, S.; Wyckoff, J.; Hess, D.; et al. MEMO is a copper-dependent redox protein with an essential role in migration and metastasis. *Sci. Signal.* **2014**, *7*, ra56. [CrossRef] [PubMed]

41. Sorokin, A.V.; Chen, J. MEMO1, a new IRS1-interacting protein, induces epithelial-mesenchymal transition in mammary epithelial cells. *Oncogene* **2013**, *32*, 3130–3138. [CrossRef] [PubMed]

42. Okkelman, I.A.; Sukaeva, A.Z.; Kirukhina, E.V.; Korneenko, T.V.; Pestov, N.B. Nuclear translocation of lysyl oxidase is promoted by interaction with transcription repressor p66β. *Cell Tissue Res.* **2014**, *358*, 481–489. [CrossRef] [PubMed]

43. Arnold, S.A.; Brekken, R.A. SPARC: A matricellular regulator of tumorigenesis. *J. Cell Commun. Signal.* **2009**, *3*, 255–273. [CrossRef] [PubMed]

44. Nagaraju, G.P.; Dontula, R.; El-Rayes, B.F.; Lakka, S.S. Molecular mechanisms underlying the divergent roles of SPARC in human carcinogenesis. *Carcinogenesis* **2014**, *35*, 967–973. [CrossRef] [PubMed]

45. Weaver, M.S.; Workman, G.; Sage, E.H. The copper binding domain of SPARC mediates cell survival in vitro via interaction with integrin β1 and activation of integrin-linked kinase. *J. Biol. Chem.* **2008**, *283*, 22826–22837. [CrossRef] [PubMed]

46. Morrissey, M.A.; Jayadev, R.; Miley, G.R.; Blebea, C.A.; Chi, Q.; Ihara, S.; Sherwood, D.R. SPARC promotes cell invasion in vivo by decreasing type IV collagen levels in the basement membrane. *PLoS Genet.* **2016**, *12*, e1005905. [CrossRef] [PubMed]

47. Bhoopathi, P.; Gondi, C.S.; Gujrati, M.; Dinh, D.H.; Lakka, S.S. SPARC mediates src-induced disruption of actin cytoskeleton via inactivation of small GTPases Rho-Rac-Cdc42. *Cell Signal.* **2011**, *23*, 1978–1987. [CrossRef] [PubMed]

48. Van de Sluis, B.; Mao, X.; Zhai, Y.; Groot, A.J.; Vermeulen, J.F.; van der Wall, E.; van Diest, P.J.; Hofker, M.H.; Wijmenga, C.; Klomp, L.W.; et al. COMMD1 disrupts HIF-1α/β dimerization and inhibits human tumor cell invasion. *J. Clin. Investig.* **2010**, *120*, 2119–2130. [CrossRef] [PubMed]

49. Itoh, S.; Kim, H.W.; Nakagawa, O.; Ozumi, K.; Lessner, S.M.; Aoki, H.; Akram, K.; McKinney, R.D.; Ushio-Fukai, M.; Fukai, T. Novel role of antioxidant-1 (ATOX1) as a copper-dependent transcription factor involved in cell proliferation. *J. Biol. Chem.* **2008**, *283*, 9157–9167. [CrossRef] [PubMed]

50. Klein, E.A.; Assoian, R.K. Transcriptional regulation of the cyclin D1 gene at a glance. *J. Cell Sci.* **2008**, *121*, 3853–3857. [CrossRef] [PubMed]

51. Itoh, S.; Ozumi, K.; Kim, H.W.; Nakagawa, O.; McKinney, R.D.; Folz, R.J.; Zelko, I.N.; Ushio-Fukai, M.; Fukai, T. Novel mechanism for regulation of extracellular sod transcription and activity by copper: Role of antioxidant-1. *Free Radic. Biol. Med.* **2009**, *46*, 95–104. [CrossRef] [PubMed]

52. Ozumi, K.; Sudhahar, V.; Kim, H.W.; Chen, G.F.; Kohno, T.; Finney, L.; Vogt, S.; McKinney, R.D.; Ushio-Fukai, M.; Fukai, T. Role of copper transport protein antioxidant 1 in angiotensin II-induced hypertension: A key regulator of extracellular superoxide dismutase. *Hypertension* **2012**, *60*, 476–486. [CrossRef] [PubMed]

53. Chen, G.F.; Sudhahar, V.; Youn, S.W.; Das, A.; Cho, J.; Kamiya, T.; Urao, N.; McKinney, R.D.; Surenkhuu, B.; Hamakubo, T.; et al. Copper transport protein antioxidant-1 promotes inflammatory neovascularization via chaperone and transcription factor function. *Sci. Rep.* **2015**, *5*, 14780. [CrossRef] [PubMed]

54. Kohno, T.; Urao, N.; Ashino, T.; Sudhahar, V.; McKinney, R.D.; Hamakubo, T.; Iwanari, H.; Ushio-Fukai, M.; Fukai, T. Novel role of copper transport protein antioxidant-1 in neointimal formation after vascular injury. *Arterioscler. Thromb. Vasc. Biol.* **2013**, *33*, 805–813. [CrossRef] [PubMed]

55. Qiu, L.; Ding, X.; Zhang, Z.; Kang, Y.J. Copper is required for cobalt-induced transcriptional activity of hypoxia-inducible factor-1. *J. Pharmacol. Exp. Ther.* **2012**, *342*, 561–567. [CrossRef] [PubMed]

56. Kahra, D.; Mondol, T.; Niemiec, M.S.; Wittung-Stafshede, P. Human copper chaperone ATOX1 translocates to the nucleus but does not bind DNA in vitro. *Protein Pept. Lett.* **2015**, *22*, 532–538. [CrossRef] [PubMed]

57. Ohrvik, H.; Wittung-Stafshede, P. Identification of new potential interaction partners for human cytoplasmic copper chaperone ATOX1: Roles in gene regulation? *Int. J. Mol. Sci.* **2015**, *16*, 16728–16739. [CrossRef] [PubMed]

58. Sun, H.T.; Wen, X.; Han, T.; Liu, Z.H.; Li, S.B.; Wang, J.G.; Liu, X.P. Expression of CPEB4 in invasive ductal breast carcinoma and its prognostic significance. *OncoTargets Ther.* **2015**, *8*, 3499–3506.

59. Ortiz-Zapater, E.; Pineda, D.; Martinez-Bosch, N.; Fernandez-Miranda, G.; Iglesias, M.; Alameda, F.; Moreno, M.; Eliscovich, C.; Eyras, E.; Real, F.X.; et al. Key contribution of CPEB4-mediated translational control to cancer progression. *Nat. Med.* **2012**, *18*, 83–90. [CrossRef] [PubMed]

60. Callebaut, I.; Courvalin, J.C.; Mornon, J.P. The BAH (bromo-adjacent homology) domain: A link between DNA methylation, replication and transcriptional regulation. *FEBS Lett.* **1999**, *446*, 189–193. [CrossRef]

61. Das, A.K.; Helps, N.R.; Cohen, P.T.; Barford, D. Crystal structure of the protein serine/threonine phosphatase 2C at 2.0 A resolution. *EMBO J.* **1996**, *15*, 6798–6809. [PubMed]

62. Lin, X.; Duan, X.; Liang, Y.Y.; Su, Y.; Wrighton, K.H.; Long, J.; Hu, M.; Davis, C.M.; Wang, J.; Brunicardi, F.C.; et al. PPM1A functions as a Smad phosphatase to terminate tgfβ signaling. *Cell* **2006**, *125*, 915–928. [CrossRef] [PubMed]

63. Kuo, H.W.; Chen, S.F.; Wu, C.C.; Chen, D.R.; Lee, J.H. Serum and tissue trace elements in patients with breast cancer in taiwan. *Biol. Trace Elem. Res.* **2002**, *89*, 1–11. [CrossRef]

64. Nagaraja, G.M.; Othman, M.; Fox, B.P.; Alsaber, R.; Pellegrino, C.M.; Zeng, Y.; Khanna, R.; Tamburini, P.; Swaroop, A.; Kandpal, R.P. Gene expression signatures and biomarkers of noninvasive and invasive breast cancer cells: Comprehensive profiles by representational difference analysis, microarrays and proteomics. *Oncogene* **2006**, *25*, 2328–2338. [CrossRef] [PubMed]

65. Kanzaki, A.; Toi, M.; Neamati, N.; Miyashita, H.; Oubu, M.; Nakayama, K.; Bando, H.; Ogawa, K.; Mutoh, M.; Mori, S.; et al. Copper-transporting P-type adenosine triphosphatase (ATP7B) is expressed in human breast carcinoma. *Jpn. J. Cancer Res.* **2002**, *93*, 70–77. [CrossRef] [PubMed]

66. Palm-Espling, M.E.; Wittung-Stafshede, P. Reaction of platinum anticancer drugs and drug derivatives with a copper transporting protein, ATOX1. *Biochem. Pharmacol.* **2012**, *83*, 874–881. [CrossRef] [PubMed]

67. Palm-Espling, M.E.; Andersson, C.D.; Bjorn, E.; Linusson, A.; Wittung-Stafshede, P. Determinants for simultaneous binding of copper and platinum to human chaperone ATOX1: Hitchhiking not hijacking. *PLoS ONE* **2013**, *8*, e70473. [CrossRef] [PubMed]

68. Palm, M.E.; Weise, C.F.; Lundin, C.; Wingsle, G.; Nygren, Y.; Bjorn, E.; Naredi, P.; Wolf-Watz, M.; Wittung-Stafshede, P. Cisplatin binds human copper chaperone ATOX1 and promotes unfolding in vitro. *Proc. Natl. Acad. Sci. USA* **2011**, *108*, 6951–6956. [CrossRef] [PubMed]

69. Chisholm, C.L.; Wang, H.; Wong, A.H.; Vazquez-Ortiz, G.; Chen, W.; Xu, X.; Deng, C.X. Ammonium tetrathiomolybdate treatment targets the copper transporter ATP7A and enhances sensitivity of breast cancer to cisplatin. *Oncotarget* **2016**, *7*, 84439–84452. [CrossRef] [PubMed]

70. El-Haibi, C.P.; Bell, G.W.; Zhang, J.; Collmann, A.Y.; Wood, D.; Scherber, C.M.; Csizmadia, E.; Mariani, O.; Zhu, C.; Campagne, A.; et al. Critical role for lysyl oxidase in mesenchymal stem cell-driven breast cancer malignancy. *Proc. Natl. Acad. Sci. USA* **2012**, *109*, 17460–17465. [CrossRef] [PubMed]

71. Payne, S.L.; Fogelgren, B.; Hess, A.R.; Seftor, E.A.; Wiley, E.L.; Fong, S.F.; Csiszar, K.; Hendrix, M.J.; Kirschmann, D.A. Lysyl oxidase regulates breast cancer cell migration and adhesion through a hydrogen peroxide-mediated mechanism. *Cancer Res.* **2005**, *65*, 11429–11436. [CrossRef] [PubMed]

72. Hollosi, P.; Yakushiji, J.K.; Fong, K.S.; Csiszar, K.; Fong, S.F. Lysyl oxidase-like 2 promotes migration in noninvasive breast cancer cells but not in normal breast epithelial cells. *Int. J. Cancer* **2009**, *125*, 318–327. [CrossRef] [PubMed]

73. Moon, H.J.; Finney, J.; Xu, L.; Moore, D.; Welch, D.R.; Mure, M. MCF-7 cells expressing nuclear associated lysyl oxidase-like 2 (LOXL2) exhibit an epithelial-to-mesenchymal transition (EMT) phenotype and are highly invasive in vitro. *J. Biol. Chem.* **2013**, *288*, 30000–30008. [CrossRef] [PubMed]

74. Ahn, S.G.; Dong, S.M.; Oshima, A.; Kim, W.H.; Lee, H.M.; Lee, S.A.; Kwon, S.H.; Lee, J.H.; Lee, J.M.; Jeong, J.; et al. LOXL2 expression is associated with invasiveness and negatively influences survival in breast cancer patients. *Breast Cancer Res. Treat.* **2013**, *141*, 89–99. [CrossRef] [PubMed]

75. Barker, H.E.; Chang, J.; Cox, T.R.; Lang, G.; Bird, D.; Nicolau, M.; Evans, H.R.; Gartland, A.; Erler, J.T. LOXL2-mediated matrix remodeling in metastasis and mammary gland involution. *Cancer Res.* **2011**, *71*, 1561–1572. [CrossRef] [PubMed]

76. Moreno-Bueno, G.; Salvador, F.; Martin, A.; Floristan, A.; Cuevas, E.P.; Santos, V.; Montes, A.; Morales, S.; Castilla, M.A.; Rojo-Sebastian, A.; et al. Lysyl oxidase-like 2 (LOXL2), a new regulator of cell polarity required for metastatic dissemination of basal-like breast carcinomas. *EMBO Mol. Med.* **2011**, *3*, 528–544. [CrossRef] [PubMed]

77. Frei, A.; MacDonald, G.; Lund, I.; Gustafsson, J.A.; Hynes, N.E.; Nalvarte, I. MEMO interacts with c-src to control estrogen receptor α sub-cellular localization. *Oncotarget* **2016**, *7*, 56170–56182. [CrossRef] [PubMed]

78. Jiang, K.; Yang, Z.; Cheng, L.; Wang, S.; Ning, K.; Zhou, L.; Lin, J.; Zhong, H.; Wang, L.; Li, Y.; et al. Mediator of ErbB2-driven cell motility (MEMO) promotes extranuclear estrogen receptor signaling involving the growth factor receptors IGF1R and ErbB2. *J. Biol. Chem.* **2013**, *288*, 24590–24599. [CrossRef] [PubMed]

79. Guttlein, L.N.; Benedetti, L.G.; Fresno, C.; Spallanzani, R.G.; Mansilla, S.F.; Rotondaro, C.; Raffo Iraolagoitia, X.L.; Salvatierra, E.; Bravo, A.I.; Fernandez, E.A.; et al. Predictive outcomes for HER2-enriched cancer using growth and metastasis signatures driven by SPARC. *Mol. Cancer Res.* **2017**, *15*, 304–316. [CrossRef] [PubMed]

80. Sarrio, D.; Rodriguez-Pinilla, S.M.; Hardisson, D.; Cano, A.; Moreno-Bueno, G.; Palacios, J. Epithelial-mesenchymal transition in breast cancer relates to the basal-like phenotype. *Cancer Res.* **2008**, *68*, 989–997. [CrossRef] [PubMed]

81. Watkins, G.; Douglas-Jones, A.; Bryce, R.; Mansel, R.E.; Jiang, W.G. Increased levels of SPARC (osteonectin) in human breast cancer tissues and its association with clinical outcomes. *Prostaglandins Leukot Essent Fat. Acids* **2005**, *72*, 267–272. [CrossRef] [PubMed]

82. Graham, J.D.; Balleine, R.L.; Milliken, J.S.; Bilous, A.M.; Clarke, C.L. Expression of osteonectin mRNA in human breast tumours is inversely correlated with oestrogen receptor content. *Eur. J. Cancer* **1997**, *33*, 1654–1660. [CrossRef]

83. Choong, L.Y.; Lim, S.; Chong, P.K.; Wong, C.Y.; Shah, N.; Lim, Y.P. Proteome-wide profiling of the MCF10AT breast cancer progression model. *PLoS ONE* **2010**, *5*, e11030. [CrossRef] [PubMed]

84. Blockhuys, S.; Wittung-Stafshede, P. Copper chaperone ATOX1 plays role in breast cancer cell migration. *Biochem. Biophys. Res. Commun.* **2017**, *483*, 301–304. [CrossRef] [PubMed]

85. Das, A.; Sudhahar, V.; Chen, G.F.; Kim, H.W.; Youn, S.W.; Finney, L.; Vogt, S.; Yang, J.; Kweon, J.; Surenkhuu, B.; et al. Endothelial antioxidant-1: A key mediator of copper-dependent wound healing in vivo. *Sci. Rep.* **2016**, *6*, 33783. [CrossRef] [PubMed]

86. Neve, R.M.; Chin, K.; Fridlyand, J.; Yeh, J.; Baehner, F.L.; Fevr, T.; Clark, L.; Bayani, N.; Coppe, J.P.; Tong, F.; et al. A collection of breast cancer cell lines for the study of functionally distinct cancer subtypes. *Cancer Cell* **2006**, *10*, 515–527. [CrossRef] [PubMed]

MDPI AG
St. Alban-Anlage 66
4052 Basel
Switzerland
Tel. +41 61 683 77 34
Fax +41 61 302 89 18
www.mdpi.com

IJMS Editorial Office
E-mail: ijms@mdpi.com
www.mdpi.com/journal/ijms

www.ingramcontent.com/pod-product-compliance
Lightning Source LLC
Chambersburg PA
CBHW051710210326
41597CB00032B/5430